A STUDENT'S TEXT-BOOK

OF

ZOOLOGY.

BY

ADAM SEDGWICK, M.A., F.R.S.,

FELLOW AND TUTOR OF TRINITY COLLEGE, CAMBRIDGE;
AND READER OF ANIMAL MORPHOLOGY IN THE UNIVERSITY.

VOL. I.

LONDON:
SWAN SONNENSCHEIN AND CO., LTD.,
NEW YORK: THE MACMILLAN CO.

PREFACE.

In preparing the present work I have been actuated mainly by the desire to place before English students of Zoology a treatise in which the subject was dealt with on the lines followed with so much advantage by Claus and his predecessors in their works on Zoology. My original intention was to bring out a new edition of Claus' *Lehrbuch*, revised and brought up to date, and a trace of this intention may be seen in a few pages of the present volume. But this plan was, for various reasons, soon given up, and the present treatise is, with the exception of about twenty pages, an entirely new work.

For a successful study of Zoology it is necessary that the student should begin by making a thorough examination of individual animals, of their structure, of the functions of their parts, of their relation to the external world and to each other. This method of study by types, which was largely introduced into this country by Huxley, and is admirably exemplified by that author's book on the *Crayfish*, is absolutely necessary as a preliminary to any thorough study of Zoology. By pursuing it the student acquires, if the animals are properly selected, a knowledge of the principal forms of animal life, and a basis from which more extended studies can be made. It is to assist these more extended studies that the present work is designed. At the same time it is hoped that the book will be of value to all interested in Natural History, whether professedly students of Zoology or not, if in no other way than as a handy book of reference in which, by means of the index, information may be gained of the general nature and habits of a large number of animals, and of the more important and striking of the phenomena of animal life. To assist in giving the book utility in this direction, I have endeavoured, in the index, to refer the reader to the page on which technical terms are used for the first time and explained. At one time I thought of adding to each chapter a detailed account of some easily accessible species belonging to the group with which it dealt, but on reflection it appeared to me that such accounts were not required: for we possess them in an excellent form in many useful and well-known books, which are accessible to everyone. Moreover, to have done so would have either unduly increased the size of the book or rendered necessary the omission of much interesting matter concerning the infinite variation of animal structure and habits not found in works easily accessible to students.

Small print has been used for those parts of the work which deal with disputed matters, or with subjects of a more recondite character. It has also been employed for the accounts of the families and genera which will be used mainly for reference. By this means I have been able to give far more information than would otherwise have been possible. To further the same object I have used, in the small print dealing with families and genera, not only abbreviations, but also what has been called by a friendly critic the style of the note-book. I very carefully considered my critic's objection to this "note-book style," but I decided that so long as I did not become unintelligible, the employment of it was justified by the object in view. Moreover, I have been careful to limit its use to those parts of the work which were apart from the main narrative, and would be almost entirely used for purposes of reference. Most of the abbreviations are explained in the index, and it is hoped that no inconvenience will arise on account of their use. Authors' names for genera are given throughout. It has been pointed out to me, too late however for alteration, that the customary abbreviations for those names are not always used, and that my abbreviations have varied even on the same page. I am afraid that this charge is true. I can only hope that my carelessness in this respect will not cause my readers serious annoyance.

To make the book more complete as a work of reference, I have endeavoured to mention and give an account of as many families as possible. For the same reason I have named a large number of genera without giving any account of them. It may appear to some absurd to name without describing so many genera. My object in doing it has been to make the book more useful in enabling students to track as many unknown names as possible to their place in the system.

I must ask the indulgence of my readers towards the many imperfections of this work. It is impossible to have a specialist's knowledge of every group, and in a book of this size, dealing with an enormous number of facts and names, it is beyond human capacity to avoid mistakes. Every care has been taken, and it is hoped that they will be found to be neither numerous nor important. The errors would undoubtedly have been much more numerous had it not been for the kind assistance given me by my friends. To Mr. J. J. Lister I am especially indebted. He has looked through all the proof sheets, and has brought to bear a critical power and discrimination which have been invaluable. I cannot be sufficiently grateful for his assistance. Mr. Lister has also contributed the account of Coral Reefs, and of the reproduction of the Foraminifera. My thanks are also owing to Mr. Heape and Mr. Graham Kerr, who looked through the greater part of the proofs, and to Mr. Shipley, Dr. Benham, Professor Haddon, and Dr. Harmer, who looked over the proofs of portions of the book, and gave me the benefit of their special knowledge. I have thus often been saved from errors into which I should otherwise have fallen. My principal sources of information are acknowledged in the foot-notes; but I must not omit to make mention here of works from which I have obtained special help; these are Bütschli's *Protozoa* in Bronn's *Thierreich*, Chun's introductory account of the *Coelenterata* also in Bronn,

Wasiliewski's *Sporozoa*, Pelseneer's *Mollusca*, and Benham's *Polychaeta* in the Cambridge Natural History.

Of the illustrations about fifty are new, of the remainder the majority are from Claus' *Lehrbuch;* but some, which I have been permitted to make use of by the courtesy of the author and publishers, are from Bronn's *Thierreich*, Perrier's *Zoologie*, Korschelt and Heider's *Embryology*, and Lang's *Text-book of Comparative Anatomy*.

In the classification the principal departures from precedent concern the group of *Amphineura*, which has been given up, and the *Gephyrea*, which has been broken up into four independent phyla. The reasons for these innovations are given in the body of the work.

The work will be issued in two volumes. The present volume deals with the whole of the animal kingdom except the *Arthropoda*, the *Echinodermata*, and the *Chordata*. The treatment of these will be included in the second volume, in the production of which I have been fortunate enough to gain the co-operation of Mr. Lister. The second volume is in preparation, and will, we hope, appear without any great delay. It will, if possible, contain a part dealing generally with the facts and principles of Zoology, but it may be necessary, from considerations of size, to reserve this for a third volume.

TABLE OF CONTENTS.

CHAPTER IV.

CHAPTER IV. COELENTERATA—*contd.*

CHAPTER V.

CHAPTER VI.

CHAPTER VII.

CHAPTER VIII.

CHAPTER IX.

CHAPTER X.

CHAPTER X. MOLLUSCA—*contd.*

CHAPTER XI.

CHAPTER I.

PROTOZOA.*

Animals in which there is one nucleus, or, if more than one nucleus, in which the nuclei are disposed apparently irregularly and without relation to the functional tissues of the animal. Conjugating cells of the form of ova and spermatozoa are never formed.

Structurally the Protozoa are so simple that the reproduction of the species is effected either by division of the body into two or more parts, or by a separation off of a small portion, which so nearly resembles the parent in structure that the phenomenon of embryonic development is almost, if not completely, absent from the life-history.

The body is always composed of a contractile granular substance, filled with vacuoles; it may also contain a *pulsating vacuole*, and present the phenomenon of granule currents. The *pulsating vacuole* consists of a space without differentiated walls filled with a clear fluid. This space apparently diminishes and disappears through the contraction of the surrounding plasma, and then reappears.

There are, however, differentiations, both in the interior of the body and in its external boundary, on which a classification may be founded. In the simplest cases, the entire body consists of a small lump of protoplasm (or sarcode, as it was at first called), the contractility of which is confined by no firm external membrane. This lump of protoplasm is sometimes semi-fluid, and protrudes and retracts processes. It is sometimes of tougher consistence in parts, and protrudes thread-like rays (*Rhizopoda*). Nutrition takes place through the intussusception of extraneous bodies, which can be surrounded and enclosed by the protoplasmic substance at any portion whatsoever of the periphery of the body. In other cases the body which sends out slender processes (*pseudopodia*) secretes silicious or calcareous needles, lattice-work shells, or shells perforated by holes, to shelter and protect the body (*Foraminifera*, *Radiolaria*).

* O. Bütschli, "Protozoa" in Bronn's *Thierreich*, 1880-2.

In the *Infusoria* the body is bounded by an external membrane, and is capable of quick and varied locomotion by means of the movements of the cilia, hairs, bristles, etc., which it possesses. The nourishing matter may be solid, in which case it is taken in through a mouth, and the remainder after digestion cast out through an anal aperture (*holozoic* nutrition), or it may be in a fluid form as a putrescent solution, in which case the name *Infusoria* is well applied, and be taken in by simple osmosis through the walls of the body (*saprophytic* nutrition).

The conjugation, or fusion, of two or more individuals has been observed at some period or another of the life-history in most of the groups of Protozoa; and in most groups the power of withdrawing the pseudopodia or cilia and of forming stout membranes round the body (encystment) to protect the organism against adverse external influences is generally present.

The Protozoa fall into three main classes. Of these the first (*Gymnomyxa*) possess the power of thrusting out processes of their body as pseudopodia; the second (*Infusoria*) are, for the most part, without pseudopodia, but bear cilia, or flagella; while the third (*Sporozoa*) possess neither pseudopodia nor cilia, and are parasitic in habit.

Table showing the classification of the *Protozoa.*

CLASS I. GYMNOMYXA (SARCODINA).

- Sub-class 1. RHIZOPODA.
  - Order 1. Amœboidea.
  - ,, 2. Testacea.
- Sub-class 2. MYCETOZOA.
- ,, 3. HELIOZOA.
- ,, 4. RADIOLARIA.

CLASS II. INFUSORIA.

- Sub-class 1. MASTIGOPHORA.
  - Order 1. Flagellata.
    - Sub-order 1. *Monadina.*
    - ,, 2. *Euglenoidea.*
    - ,, 3. *Heteromastigoda.*
    - ,, 4. *Isomastigoda.*
    - ,, 5. *Phytomastigoda.*
  - Order 2. Choanoflagellata.
  - ,, 3. Dinoflagellata.
  - ,, 4. Silicoflagellata.
  - ,, 5. Cystoflagellata.

CLASS II. INFUSORIA—*Continued.*

- Sub-class 2. CILIATA.
  - Order 1. Gymnostomata.
  - ,, 2. Trichostomata.
    - Sub-order 1. *Aspirotricha.*
    - ,, 2. *Spirotricha.*
      - *a.* Heterotricha.
      - *b.* Oligotricha.
      - *c.* Hypotricha.
      - *d.* Peritricha.
- Sub-class 3. ACINETARIA.

CLASS III. SPOROZOA.

- Order 1. Gregarinida.
  - Sub-order 1. *Polycystidea.*
  - ,, 2. *Monocystidea.*
- Order 2. Coccidiidea.
- ,, 3. Haemosporidia.
  - Sub-order 1. *Drepanidiidea.*
  - ,, 2. *Acystosporidea.*
- Order 4. Myxosporidia.
- ,, 5. Sarcosporidia.

Class 1. GYMNOMYXA (SARCODINA).*

Protozoa possessing the power of thrusting out pseudopodia. An investing membrane is absent, or, if present, is incomplete, and leaves a considerable portion of the protoplasm exposed. A calcareous shell, or silicious skeleton, is very usually secreted.

The body-substance, which is richly granulated, and may contain pigment, contracts slowly, and sends out at the same time the processes called *pseudopodia;* and these serve not only as a means of movement, but also for the reception of nourishment. The pseudopodia may be broad, lobed, or finger-like processes (Fig. 2), by means of which a quick and flowing motion can be imparted to the body mass; or they may be filiform radiating processes (Fig. 1); or, lastly, they may anastomose with one another, and form networks. A tougher, clear homogeneous external layer (*exoplasm*) is usually to be distinguished as the peripheral boundary from a more fluid and more granular internal mass (*endoplasm*). During motion the former is projected in processes into which the granules of the latter stream more or less quickly. In the stiffer pseudopodia streams of granules are observable, slow but regular, passing from the base to the extremity and *vice versâ.* The explanation of

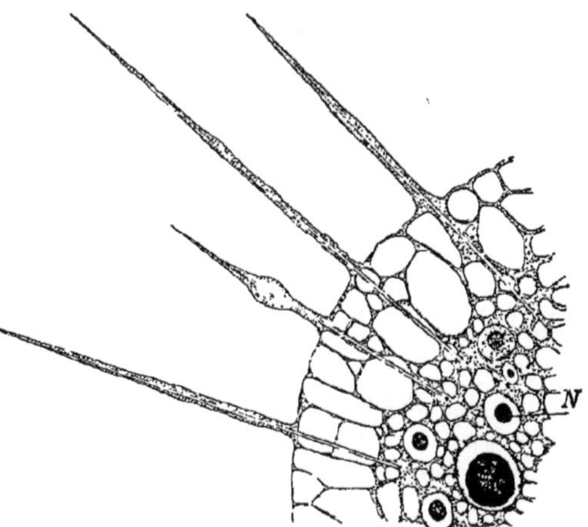

FIG. 1.—Optical section through portion of the body of *Actinosphaerium Eichhornii* (after Hertwig and Lesser). *N* nuclei in the endoplasm, from which the vacuolated ectoplasm is clearly distinguishable. In the centre of the pseudopodia the axial thread is visible.

* Dujardin, "Observations sur les Rhizopodes." *Comptes rendus*, 1835. Ehrenberg, "Uber noch jetzt zahlreich lebende Thierarten der Kreidebildung und den Organismus der Polythalamien." *Abhandlung der Akad. zu Berlin*, 1839. Max Sigm. Schultze, "Uber den Organismus der Polythalamien." Leipzig, 1854. Joh. Müller, "Uber die Thalassicolen, Polycystinen und Acanthometren." 1858. E. Haeckel, "Die Radiolarien." Eine Monographie. Berlin, 1862.

these movements is to be sought in the contractility of the surrounding portions of protoplasm (Fig. 1).

A pulsating space, the contractile vacuole, is not unfrequently to be found in the protoplasm, *e.g.*, *Difflugia*, *Actinophrys*, *Arcella* (Fig. 2). Nuclei are usually to be made out, but there are forms in which no trace of a nucleus has yet been found. In such cases either our methods of observation are faulty, or the protoplasm of the nucleus is not yet differentiated as a separate structure (the *Monera* of E. Haeckel), or we have to do with a transient, non-nucleated stage in the life-history.

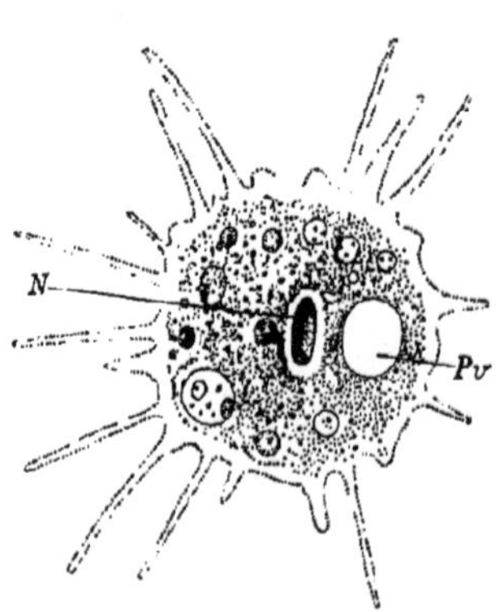

FIG. 2.—*Amœba (Dactylosphæra) polypodia* (after Fr. E. Schulze). *N* nucleus. *Pv* pulsating vacuole.

The protoplasm usually secretes silicious or calcareous structures, either as fine spicula and hollow spines, which are directed from the centre to the periphery in regular order and number, or as lattice-work chambers (*Radiolaria*), which often bear points and spines, or finally as single and many-chambered shells with, in some cases, finely perforated walls (*Foraminifera*) and one larger opening. Through this last (Fig. 4), as well as through the countless pores of the small shells (Fig. 3), the slender threads of sarcode pass out to the exterior as pseudopodia, changing without intermission in form, size, and number, and often joining themselves together in delicate networks. (Figs. 3, 4.)

The pseudopodia, by their slow contraction, afford a means of locomotion, while they also serve for the taking up of nourishment by surrounding and transporting into the interior of the body small vegetable organisms as *Bacillaria*. Among the shell-bearing forms, the reception and digestion of food takes place outside the shell in the peripheral threads and networks of sarcode; for each spot on the surface can for the time being assume the functions of mouth, and also of anus, by rejecting the undigested remnants. While the power of emitting pseudopodia is characteristic of the form in which the *Gymnomyxa* are usually met with, it is usual to find a stage in their life-history in which locomotion is effected by means of flagella. To such flagellated forms the term *mastigopod* has been applied, as opposed to *myxopod*, a word indicating a form with the power of emitting pseudopodia.

Reproduction is commonly effected by simple division, but in some cases encystment occurs, and the protoplasm breaks up into a number of minute portions called *spores*. The spores may be either naked or provided with a wall, forming the spore-case.

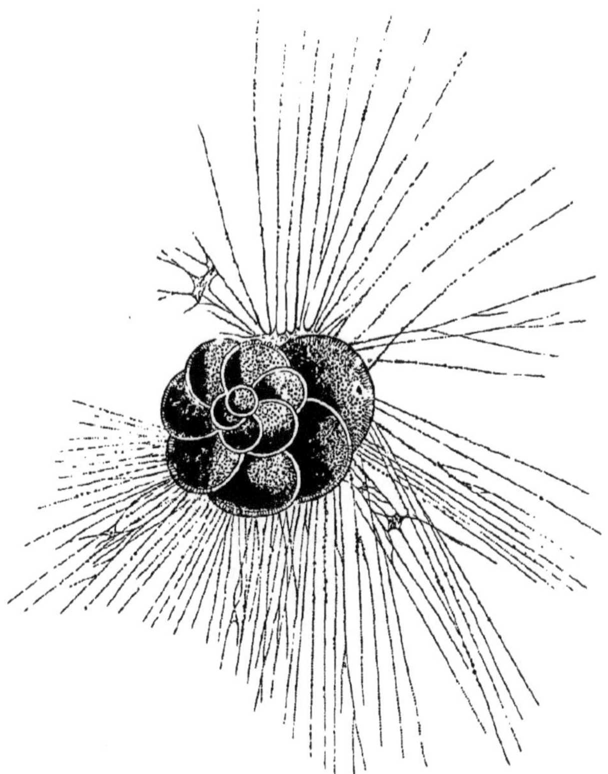

FIG. 3.—*Rotalia veneta* (after M. Schultze), with a Diatom taken in the network of pseudopodia.

The marine shell-bearing forms contribute by the accumulation of their shells to the formation of the sea sand, and even to the deposition of thick strata. An innumerable quantity of fossil forms from various and very ancient formations are known.

Sub-class I. RHIZOPODA.*

Gymnomyxa, either naked or with a shell, the shell generally calcareous and often pierced with fine pores for the exit of the pseudopodia.

Only in rare cases is the shell substance of a silicious nature; in all other forms it is membranous, with or without adhering sand particles, or consists of a calcareous deposit in a basis of organic

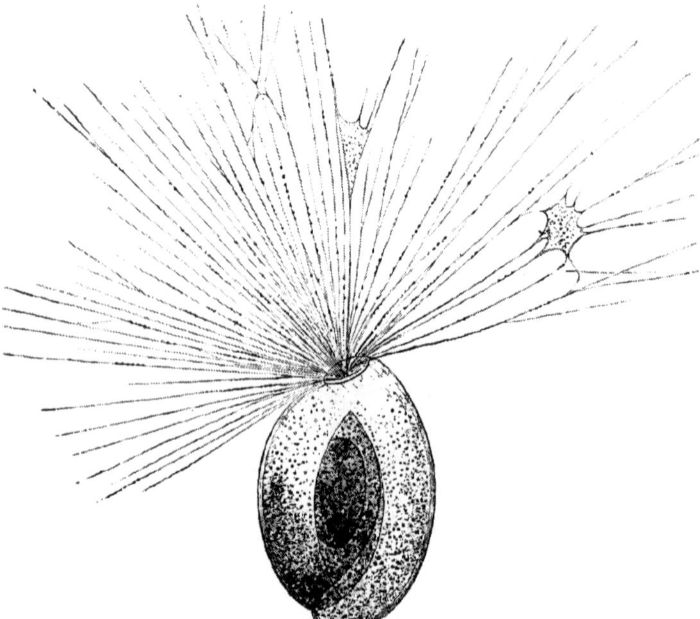

Fig. 4.—*Miliola tenera*, with network of pseudopodia (after M. Schultze).

matter. The shell is either a simple chamber, usually provided with a large opening, or is many-chambered, that is, is composed of numerous chambers arranged upon one another in a definite order

* Besides D'Orbigny, Max Schultze, l. c., compare W. C. Williamson, "On the recent Foraminifera of Great Britain," London, 1858. Carpenter, "Introduction to the Study of the Foraminifera," London, 1862. Reuss, "Entwurf einer system. Zusammenstellung der Foraminiferen," Wien, 1861. H. B. Brady, "Report on the Foraminifera," *Challenger Reports*, 1884. J. J. Lister, "Contributions to the Life-History of the Foraminifera," *Phil. Trans.*, 186, 1895.

according to definite laws. The spaces of these chambers communicate by means of narrow passages and large openings in the partition walls. In like manner those portions of the protoplasm which are enclosed in the individual chambers are in direct communication with one another by means of processes which pass through the passages and openings in the septa, and connect one portion with another. In the perforated forms there may be an additional complication owing to the formation of a secondary shell by the protoplasm. This secondary shell is placed outside the primary, and is traversed by canals containing the protoplasm. The multinucleate condition is very generally found, and probably constitutes a phase in the life-history. For instance, in *Gromia* and *Difflugia* specimens have been found with a large number of nuclei, though the usual condition appears to be with one nucleus. The quality of the body-substance, the mode of movement and nourishment, agree closely with those which have been depicted as characteristic of the class. Our knowledge of the mode of **reproduction** is imperfect. Amongst the forms without a shell, fission has been observed. In *Haliphysema* one of the nuclei with a small portion of protoplasm is said to become marked off from the rest and to form a small body not unlike an ovum imbedded in the protoplasm. This may be called *internal* budding.

In several of the families of our order (*Miliolidæ*, *Lagenidæ*, *Rotalidæ*, and *Nummulinidæ*) we meet with the phenomenon of dimorphism, that is to say, the species is made of two distinct kinds of individuals, which appear to alternate with one another in the life-history. (Fig. 5.) The one of these forms—the *microspheric*—is distinguished by the small size of the initial chamber and by the possession of a large number of small nuclei; while the other—the *megalospheric*—has a large initial chamber and one large nucleus. In the microspheric form, reproduction is effected by the simple emergence of the protoplasm from the shell and its division into a number of spherical bodies; each of these secretes a shell which constitutes the first chamber of a megalospheric form. In the megalospheric form, on the other hand, the protoplasm in the shell breaks up into a number of small masses, each of which issues as a flagellated zoospore. The further history of these is not known, but they probably give rise to the microspheric form.

In *Polystomella crispa* L., one of the *Nummulinidæ*, whose life-history is better known than that of other Foraminifera, the initial chamber, or *microsphere*, occupying the centre of the shells of the microspheric form, has a diameter of about 10 μ. It is succeeded by a series of chambers of gradually increasing size, which are added one after another in a spiral manner. Numbers of small nuclei which increase in number by simple division, are scattered apparently irregularly through the protoplasm. The existence of the microspheric form, as an individual, is terminated in the following manner. The animal becomes firmly attached to some object by its pseudopodia, and the protoplasm emerges from the shell. After involved streaming movements which continue for some hours, the protoplasm divides simultaneously into a number of spherical masses, having generally a diameter of about 70 μ., which, after secreting a shell, rapidly draw apart from one another, and each sets up an independent existence. These form the initial chambers of the *megalospheric* form of the species. The

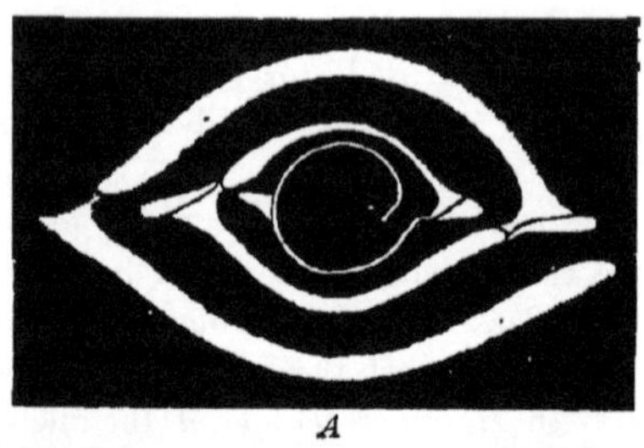

A

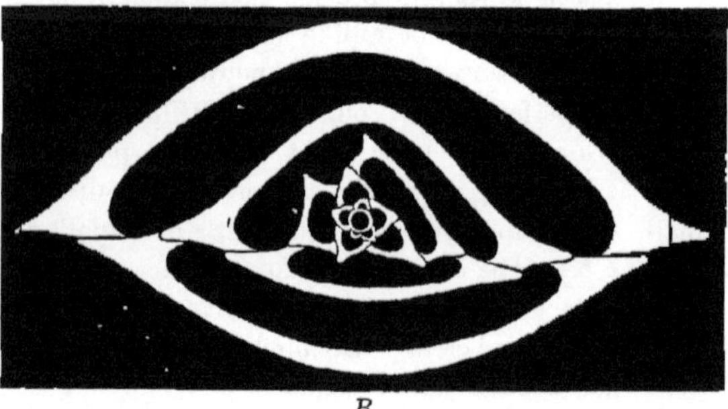

B

FIG. 5.—*Biloculina depressa* d'Orb. Sections of the shell: *A*, of the megalospheric form (megalosphere 200–400 μ.) longitudinal. *B*, of the microspheric form (microsphere 20 μ.) transverse. The two terminal chambers are omitted in *B* (from Lister, after Schlumberger).

whole of the protoplasm of the parent is thus divided up to form the brood of young, its empty shell falling to the bottom. The second chamber of the megalospheric form of *Polystomella* has a peculiar shape, and the other chambers added in succession build up the spiral shell. In this form a single large nucleus is present, whose size increases with the growth of the protoplasm. Prior to the reproduction of the megalospheric form the large nucleus disappears, and in place of it, though the way in which they are produced has not been followed, great numbers of minute nuclei are found scattered through the protoplasm. After a preliminary division of these nuclei by karyokinesis, the protoplasm breaks up into flagellated zoospores having a diameter of 5 μ. The intervening stages between the zoospore, produced by the megalospheric form, and the microsphere in which the microspheric form takes its origin, have not been followed; but there is some, though at present inconclusive evidence, in favour of the supposition that the microsphere results from the conjugation of two zoospores. The relative sizes of the microsphere and zoospore—10 μ. and 5 μ., agrees fairly well with this view. One of the earliest observations relating to the dimorphism of the Foraminifera was of the well-established fact that the microspheric form is much less abundant than the megalospheric, and this admits of easy explanation on the supposition that the union of two separate organisms is required for its production. Finally, Schaudinn's observation of the conjugation of the zoospores of *Hyalopus*, which is, however, a form not known to be dimorphic, supports the hypothesis.

From the foregoing account of the life-history of *Polystomella* it appears that the dimorphism of the Foraminifera is due to the occurrence of alternating or recurring generations, and there is some, though at present inconclusive evidence in support of the view that the megalospheric generation arises asexually, and the microspheric generation as the result of conjugation. In the genus *Orbitolites* (*Miliolidæ*), while the megalospheric form has been found to be produced from a microspheric form as in *Polystomella*, it has also been seen to arise from a megalospheric parent. In this case, then, it must be supposed that the generations do not regularly alternate, but that the megalospheric form may be repeated before the brood of zoospores is produced.

Besides the difference in the size of the initial chambers, the shells of the two forms present in some cases marked differences in the mode of growth. Thus in the genus *Biloculina** (*Miliolidæ*), while the mode of growth of the megalospheric form is, as shown by Schlumberger, on the biloculine plan from the first (Fig. 5), that of the microspheric form is at first on the quinqueloculine plan, and it is not until many chambers have been formed that the biloculine plan, characteristic of the genus, is assumed.

The application of the term dimorphism to the phenomenon above described is in accordance with its general use in zoology and botany. It has, however, been used, together with the terms *trimorphism* and *polymorphism*, in another and quite different sense, namely to indicate the occurrence of two (three, or more) different modes of growth in the building up of the shell of a single individual. Thus, in the shell of the genus *Peneroplis*, the chambers

* The mode of growth of the shells of the genera *Biloculina*, *Triloculina*, and *Quinqueloculina* is a modification of the spiral. The chambers are elongated and increase in size as they succeed one another, each occupying half a turn of the spiral. The result is that the mouths of the chambers are directed successively in opposite directions, and a *long axis* of the shell is thus established.
In the genus *Biloculina* (Fig. 5) the chambers lie in the same plane, and each overlaps its predecessors at the sides. Hence only two chambers are exposed in the outer contour of the shell. In the other genera the chambers are narrower and do not lie in the same plane, the median plane of each being directed at a definite angle, which is constant for the genus, to that of its predecessor. The result is that in one case three, and in the other five chambers are exposed in the contour of the shell, and the triloculine and quinqueloculine forms of shell are respectively produced. The genera *Triloculina* and *Quinqueloculina* (d'O.) were by Williamson included in the genus *Miliolina*.

are at first arranged spirally, but in the later stages of growth in a rectilinear series. Such forms are called dimorphous.

Conjugation has been observed to take place in a few cases (*Arcella*, *Difflugia*, *Euglypha*, etc.), but details concerning it are unknown. In the Polythalamous forms, it is possible, as hinted above, that conjugation takes place between the free-swimming zoospores.

The Foraminifera present four main varieties of shell: (1), the chitinous, *e.g.*, *Gromia*, imperforate; (2), the porcellaneous, *e.g.*, *Miliola*, imperforate, and characterised by their opaque white colour and abundant organic basis; (3), the hyaline, *e.g.*, *Globigerina*, perforate, and with but little organic basis; and (4), the arenaceous, perforate, as in *Psammosphæra*, but generally imperforate. The last are formed of small foreign particles united by a cementing substance. It is the fact that perforate and imperforate arenaceous forms are found within the limits of the same family, which has rendered necessary the abandonment of the old division of the order into *Perforata* and *Imperforata*. Specimens of *Biloculina ringens* living in the red clay at 3000 fathoms, a depth at which calcareous organisms are generally absent, were found by Brady to have a shell composed of silica.

FIG. 6.—Nummulitic Limestone, with horizontal section of *N. distans* (after Zittell).

Analysis of the calcareous shells shows that the mineral constituents consist of carbonate of lime and carbonate of magnesia, the latter varying from five to ten per cent. There is besides a trace (generally under 0·5 per cent.) of silica.

In spite of their small size, the shells of our simple organisms may lay claim to no small consequence, since they not only accumulate in enormous quantity in the sea sand (M. Schultze calculated their number for an ounce of sea sand from Molo di Gaëta at about one and a half millions), but are also found as fossils in different formations (the cretaceous and tertiary), and have yielded an essential material to the construction of rocks. Silicious nodules of *Polythalamia* are even found in Silurian deposits. The most remarkable, on account of their considerable size, are the *Nummulites* (Fig. 6) in the thick formation of the so-called Nummulite limestone (Pyrenees and elsewhere). A coarse chalk of the

Paris basin, which makes an excellent building stone, contains the *Triloculina trigonula* (*Miliolite chalk*).

The greater number of *Foraminifera* are marine, and move by creeping on the bottom of the sea, but some of them are pelagic and *Globigerina* and *Orbulina* live at the surface. Some of the genera are found in the brackish water of estuaries, and they may even extend into fresh water. The bottom of the sea at very considerable depths is also covered with a rich abundance of forms, especially with *Globigerina*, the remains of the shells of which give rise to an enduring deposit.

Order I. **Amœboidea.**

Naked Amœba-like Rhizopoda, with lobose or reticulate pseudopodia, usually with nucleus and contractile vacuole.

Amœba, nucleated forms, with contractile vacuole and lobose pseudopodia. ***Protamœba*** Haeckel, small forms, with lobose pseudopodia; nucleus and contractile vacuole not observed. ***Hyalodiscus*** Hertwig and Lesser, disc-shaped, nucleated forms, without pseudopodia; locomotion by a flowing movement without change of form. ***Protomyxa*** Haeckel, ***Myxodictyum*** Haeckel, ***Protogenes*** Haeckel, are forms with anastomosing pseudopodia, and without observed nucleus. ***Pelomyxa*** Greef, large amœba-like forms, containing many nuclei, refractile bodies, and cylindrical crystals. Other genera are ***Gloidium*** Sorokin, ***Chætoproteus*** Stein, ***Plakopus*** F. E. Schulze, ***Dactylosphæra*** Hertw. and Lesser, ***Podostoma*** Clap. and Lachm., ***Amphizonella*** Greef, ***Gymnophrys*** Cienk. ***Bathybius*** Huxley, found in the deep sea mud of the Atlantic, if it is indeed a living organism (and not simply a deposit of gypsum). ***Coccoliths***, ***Coccospheres***, and ***Rhabdospheres*** are found in the gelatinous mud of the bottom of the Atlantic, and were supposed by Huxley* to be portions of the body of Bathybius. Similar bodies have been found in the chalk. They are in reality small marine vegetable organisms, with calcareous walls, and are oceanic in habit. They form with Diatoms and pelagic *Oscillatoriæ*, a large part of the vegetable food of marine animals.†

Order II. **Testacea (Foraminifera).**

Principally marine Rhizopods, with a shell which is either membranous, calcareous, or rarely silicious, and may be single-chambered (*Monothalamia*) or many-chambered (*Polythalamia*), usually with more than one nucleus.

In this order the pseudopodia are generally slender and anastomosing, but in the fresh water *Arcella* and *Difflugia* (Fig. 8) they are lobed. The protoplasm has the power of exuding through the openings in the shell and of forming a layer which may be much vacuolated on the outside of the shell (Fig. 10). In such cases the shell may be regarded as an internal structure, and appropriately

* *Q.J.M.S.*, vol. 8. † Vide *Nature*, 1897, p. 510.

compared with the internal capsule of the *Radiolaria*. The shell also often possesses long, delicate, spiny processes, which are broken off unless the animal be very carefully handled (Fig. 10).

The *Foraminifera* are mainly bottom organisms, but a few genera of the hyaline forms are pelagic. The pelagic genera are important because of their extraordinary abundance. Whether these pelagic forms have the power of supporting life on the surface of the bottom-ooze is unknown.

The definition of species and genera in this order is rendered difficult by the large number of intermediate varieties, which, indeed, often constitute a complete series. Carpenter* on this subject writes as follows: "The ordinary notion of species as assemblages of individuals marked out from each other by definite

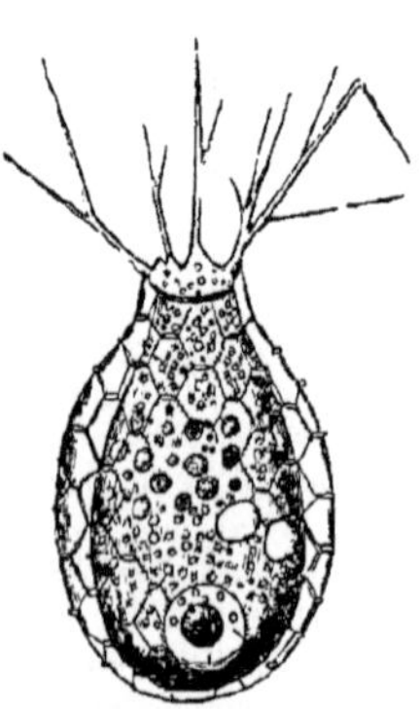

FIG. 7.—*Englypha globosa* (after Hertwig and Lesser).

FIG. 8.—*Difflugia oblonga* (after Stein).

characters that have been genetically transmitted from original prototypes similarly distinguished, is quite inapplicable to this group; since, even if the limits of such assemblages were extended so as to include what would elsewhere be accounted genera, they would still be found so intimately connected by gradational links, that definite lines of demarcation could not be drawn between them."

Fam. 1. **Arcellina.** Shell watch-glass shape, membranous; with more than one nucleus; with contractile vacuoles; gas vacuoles with a hydrostatic function may be secreted by the protoplasm. Fresh water. *Cochliopodium* Hertw. and Lesser, *Pyxidicula* Ehrbg., *Arcella* Ehrbg., *Hyalosphenia* Stein, *Quadrula* F. E. Schulze, *Difflugia* Leclerc, shell incrusted by foreign particles, usually pyriform in shape (fig. 8).

* W. R. Carpenter, "Introduction to the Study of the Foraminifera," preface, p. x.

Fam. 2. **Euglyphina**. Shell chitinous or silicious, composed of hexagonal or roundish plates; pseudopodia pointed, branched. Fresh water. *Euglypha* Duj. (Fig. 7), *Trinema* Duj.

Fam. 3. **Gromidæ**. Test chitinous, smooth or incrusted with foreign bodies, imperforate, with a pseudopodial aperture at one or both extremities; pseudopodia long, branching, reticulated. Fresh water and marine. *Gromia* Duj., *Lieberkühnia* C. and L., *Mikrogromia* R. Hertwig, *Diaphoropodon* Archer, *Shepherdella* Siddal.

Fam. 4. **Miliolidæ**. Test imperforate, mono- or polythalamic; normally calcareous and porcellaneous, sometimes incrusted with sand; under starved conditions (*e.g.* in brackish water) becoming chitinous or chitino-arenaceous; at abyssal depths occasionally consisting of a thin homogeneous, imperforate silicious film. Marine. *Squamulina* Schultze, *Nubecularia* Defrance, *Biloculina* d'Orb., *Fabularia* Defrance, *Spiroloculina* d'Orb., *Miliolina* Williamson, *Hauerina* d'Orb., *Vertebralina* d'Orb., *Orbiculina* Lamarck, *Orbitolites* Lamarck.

Fam. 5. **Astrorhizidæ**. Test invariably composite, usually of large size and monothalamic; often branched or radiate, sometimes segmented by constriction of the walls, but seldom or never truly septate; polythalamic forms never symmetrical.

Sub-fam. 1. *Astrorhizinæ*. Walls thick, composed of loose sand or mud very slightly cemented. *Astrorhiza* Sandahl, *Pelosina* Brady, *Syringammina* Brady.

Sub-fam. 2. *Pilulinæ*. Test monothalamic; walls thick, composed chiefly of felted sponge-spicules and fine sand, without calcareous or other cement. *Pilulina* Carpenter.

Sub-fam. 3. *Saccammininæ*. Chambers nearly spherical; walls thin, composed of firmly cemented sand grains. *Saccammina* M. Sars, *Psammosphæra* Sch., *Sorosphæra* Brady.

Sub-fam. 4. *Rhabdammininæ*. Test composed of firmly cemented sand grains, often with sponge-spicules intermixed; tubular, straight, radiate, branched, or irregular; free or adherent; with one or more apertures; rarely segmented. *Rhabdammina* M. Sars, *Rhizammina* Brady, *Sagenella* Brady, *Botellina* Carp., *Haliphysema* Bowerbank.

Fam. 6. **Lituolidæ**. Test arenaceous, usually regular in contour; septation of the polythalamic forms often imperfect, chambers frequently labyrinthic. Comprises sandy isomorphs of the simple porcellaneous and hyaline types (*Cornuspira*, *Miliolina*, *Lagena*, *Globigerina*, *Rotalia*, etc.), together with some adherent species. *Lituola* Lamarck, *Thurammina* Brady, *Ammodiscus* Reuss, *Trochammina* Parker and Jones, *Webbina* d'Orb., *Cyclammina* Brady.

Fam. 7. **Textularidæ**. Test of the larger species arenaceous, either with or without a perforate calcareous basis; smaller forms hyaline and conspicuously perforated. Chambers arranged in two or more alternating series, or spiral, or confused; often dimorphous. *Textularia* Defrance, *Cuneolina* d'Orb., *Verneuilina* d'Orb., *Tritaxia* Reuss, *Pavonina* d'Orb., *Valvulina* d'Orb., *Clavulina* d'Orb., *Bulimina* d'Orb., *Virgulina* d'Orb., *Bolivina* d'Orb., *Pleurostomella* Reuss.

Fam. 8. **Chilostomellidæ**. Test calcareous, finely perforate, polythalamous. Segments following each other from the same end of the long axis, or alternately at the two ends, or in cycles of three; more or less embracing. Aperture, a curved slit at the end or margin of the final segment. *Ellipsoidina* Seguenza, *Chilostomella* Reuss, *Allomorphina* Reuss.

Fam. 9. **Lagenidæ.** Test calcareous, very finely perforated; either monothalamous, or consisting of a number of chambers joined in a straight, curved, spiral, alternating or (rarely) branching series. Aperture simple or radiate, terminal. No interseptal skeleton nor canal system. *Lagena* Walker and Boys, monothalamous; *Nodosaria* Lamarck, chambers in linear series; *Lingulina* d'Orb., *Frondicularia* Defrance, *Rhabdogonium* Reuss, *Marginulina* d'Orb., *Vaginulina* d'Orb., *Rimulina* d'Orb., *Cristellaria* Lam., chambers in spiral series; *Amphicoryne* Schlumberger, *Lingulinopsis* Reuss, *Flabellina* d'Orb., *Polymorphina* d'Orb., *Dimorphina* d'Orb., *Uvigerina* d'Orb.

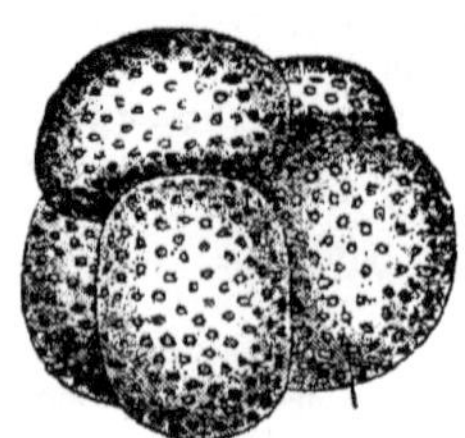

Fig. 9.—*Planorbulina* (*Acervulina*) *globosa* (after M. Schultze).

Fam. 10. **Globigerinidæ.** Test free, calcareous perforate; chambers few, inflated, arranged spirally; aperture single or multiple, conspicuous. No supplementary skeleton nor canal system. All the larger species pelagic in habit. *Globigerina* d'Orb. when taken in a spoon shows very long fine calcareous spines projecting from the shell (Fig. 10). These are not found in forms taken in the tow-net or in ooze. Dimorphism has not been observed in this genus. *Orbulina* d'Orb., spherical shell, often double, containing Globigerina-like shell; it is formed by the external protoplasm as a kind of large last chamber. *Hastigerina* Wy. Thomson, *Pullenia* P. and J., *Sphæroidina* d'Orb., *Candeina* d'Orb.

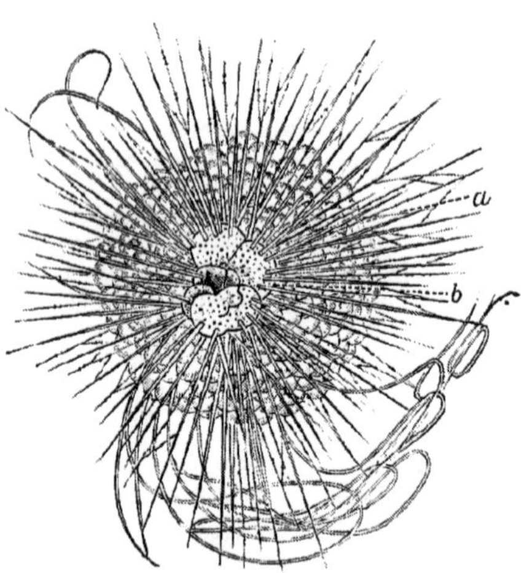

Fig. 10.—*Hastigerina* (*Globigerina*) *Murrayi* W. Thomson. *a* bubbly vacuolated protoplasm, enclosing *b*, the perforated Globigerina-like shell. (From the peripheral protoplasm project not only fine pseudopodia, but hollow calcareous spines, which are set on the shell (from Murray).

Fam. 11. **Rotalidæ.** Test calcareous perforate; free or adherent. Typically spiral and "Rotaliform," *i.e.*, coiled in such a manner that the whole of the segments are visible on the superior surface, those of the last convolution only on the inferior or apertural side, sometimes one face being more convex, sometimes the other. Aberrant forms evolute, outspread, acervuline, or irregular. Some of the higher modifications with double chamber walls, supplemental skeleton, and a system of canals. *Spirillina* Ehrbg., *Cymbalopora* Hagenow, *Discorbina* P. and J., *Planorbulina* d'O. (Fig. 9), *Truncatulina* d'O., *Anomalina* P. and J., *Carpenteria* Gray, *Pulvinulina* P. and J., *Rotalia* Lam., *Calcarina* d'O., *Tinoporus* Carpenter, *Gypsina* Carter, *Thalamopora* Roemer, *Polytrema* Risso, *Patellina*.

Fam. 12. **Nummulinidæ.** Test calcareous and finely tubulated; typically free, polythalamous, and symmetrically spiral. The higher modifications all possessing a supplemental skeleton and a canal system of greater or less complexity. *Fusulina* Fischer and *Schwagerina* Möller extinct, palæozoic, *Nonionina* d'O., *Polystomella* Lam., *Archædiscus* Brady, carboniferous, *Amphistegina* d'O., *Operculina* d'O., *Heterostegina* d'O., *Nummulites* Lam. mostly extinct (Carboniferous and Eocene), *Cycloclypeus* Carp., *Orbitoides* d'O., *Eozoon* (?) Dawson.

Testamœbiformia (Carter. An. and Mag. 1880). Adherent testaceous Rhizopoda with the general form of *Amœba*. Systematic position doubtful. *Holocladina* test calcareous and branched with pustuliferous surface, *Cystcodictyina* test calcareous, sessile, unbranched, uniformly punctate surface, *Ceratcstina* test chitinous and polythalamous.

SUB-CLASS II. MYCETOZOA.*

Gymnomyxa of large size, formed by the fusion of many small amœba-like forms. Fruit-like cysts or sporophore, and coated spores are always formed.

The *Mycetozoa* were formerly regarded as Plants, and under the name *Myxomycetes* placed amongst the Fungi. After the discovery that the spores do not produce a mycelium, but hatch out as swarm cells, de Bary introduced the name *Mycetozoa*. They have the form of large masses of a granular protoplasm bounded by a clear protoplasm—the ectoplasm—without granules. These masses, which are called *plasmodia* and may attain a surface extension of several square inches, infest the damp surface and interstices of vegetable substances, such as dead leaves, rotten wood, or even sound wood. They spread out into a network on their substratum, over which they advance with a creeping movement due partly to the formation of pseudopodia at the advancing margin and partly to the flow of the dark granular endoplasm. The flow of this latter is a rhythmic one, being reversed at nearly regular intervals. The endoplasm contains numerous nuclei and vacuoles, and sometimes granules of calcium carbonate. The vacuoles often contract and expel their contents which is either watery or of refuse matter. Occasionally the streaming ceases, the plasmodium breaks up into masses containing ten to twenty nuclei, a superficial membrane is formed round each of these masses, and they enter a resting state: this is the stage of the **Sclerotium.** The sclerotium can be made to reassume the active condition by the addition of moisture.

* De Bary, "Die Mycetozoen." 1864. A. Lister, "A Monograph of the Mycetozoa." London, 1894. W. Zopf, "Die Pilzthiere oder Schleimpilze." *Encyklopædie der Naturwissenschaften*, 1885.

The **reproduction** is always effected by spore-formation. The spores are either contained within fruit-like cysts—the *Sporangia* —(*Endosporeæ*), which are either simple (spore fruits) (Fig. 11) or combined into an *æthalium* (fruit-cake) of a cushion-like shape consisting of numerous convoluted sporangia (Fig. 12); or they are not contained in a cyst, but are produced upon the surface of upgrowths of the plasmodium—the *sporophore*—(*Exosporeæ*). In the latter case the spores divide, after issuing as amœboid organisms, by three successive bipartitions into eight cells, which soon obtain flagella and separate; they are comparable to the just-hatched spores of the *Endosporeæ*. The spores are always enclosed in a coat of a cellulose-like material and possess a single nucleus. They are contained as a rule in the meshes of a network of supporting fibres—the *capillitium*—which is formed within the

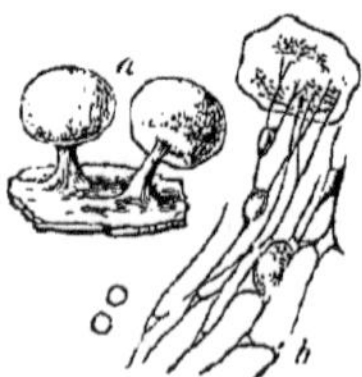

FIG. 11.—*Physarum nutans* (from Lister), *a*, two sporangia magnified nine times; *b*, capillitium threads, with lime-knots attached to a fragment of the sporangium-wall. × 110.

FIG. 12.—*Fuligo septica*, *a*, Aethalium, one third natural size; *b*, capillitium threads with lime-knots and two spores. × 120.

sporangial cyst by the spore-protoplasm. The division into spores is in the *Endosporeæ* preceded by a single division of the nuclei of the sporeplasm by karyokinesis.* On the *germination* of the spore the spore coat bursts and the contents issues as an amœboid organism which soon protrudes one flagellum. (Fig. 13.) The swarm-cells so formed swim by their flagellum, ingest solid food by their pseudopodia (at the non-flagellate end) and undergo frequent bipartition. They may also withdraw the flagellum and encyst (*microcysts*), but this is only temporary: they emerge and re-assume the swarm-cell form. After a time the flagellum is withdrawn and they creep about in an amœboid manner, and ultimately several of them fuse together to form the multinucleated plasmodium. (Fig. 14.)

* The nuclei of the plasmodium sometimes multiply simultaneously by karyokinesis, though it is highly probable that simple division occurs as well.

Order 1. **Exosporeæ.** Spores produced on the sporophores and not enclosed in a cyst. *Ceratiomyxa* Schroeter, plasmodium in rotten wood fruiting on the outside.

Order 2. **Endosporeæ.** Spores produced in a sporangium. *Badhamia* Berkeley, *Physarum* Persoon, *Fuligo* Haller, sporangia combined into an æthalium. *F. septica* Gmelin, flowers of tan. *Cienkowskia* Rostafinski, *Physarella* Peck, *Craterium* Trentepohl, *Leocarpus* Link, *Chondrioderma* Rost., *Trichamphora* Junghuhn, *Diachæa* Fries, *Didymium* Schrader, *Lepidoderma* de Bary, *Stemonitis* Gleditsch, *Comatricha* Preuss, *Enerthenema* Bowman, *Lamproderma* Rost., *Clastoderma* Blytt, *Amaurochæte* Rost., *Brefeldia* Rost., *Lindbladia* Fries, *Cribraria* Pers., *Dictydium* Schrad., *Licea* Schrad., *Orcadella* Wingate, *Tubulina* Pers., *Siphoptychium* Rost., *Alwisia* Berkeley and Broome, *Dictydiæthalium* Rost., *Enteridium* Ehrenb., *Reticularia* Bull, *Trichia* Haller, *Oligonema* Rost., *Cornuvia* Rost., *Arcyria* Hill, *Lachnobolus* Fries, *Perichæna* Fries, *Margarita* Lister, *Dianema* Rex, *Prototrichia* Rost., *Lycogala* Micheli.

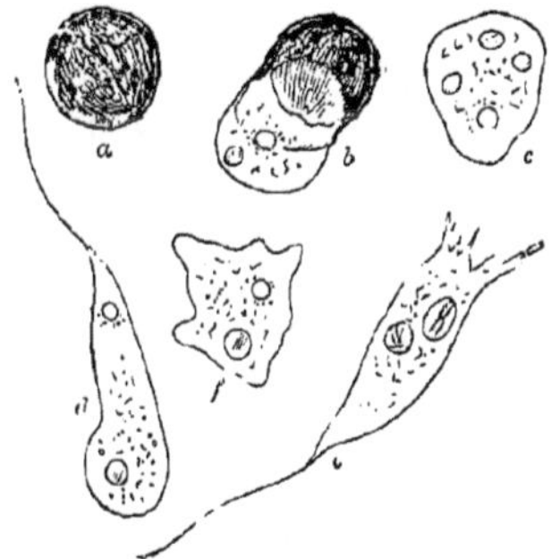

FIG. 13.—*Didymium difforme* (after Lister), *a*, spore; *b*, swarm-cell escaping from the spore-case; *c*, newly hatched swarm-cell with nucleus and three vacuoles; *d*, flagellated swarm-cell; *e*, swarm-cell with two vacuoles containing bacteria, and produced at the posterior end into pseudopodia; *f*, amœboid swarm-cell. × 720.

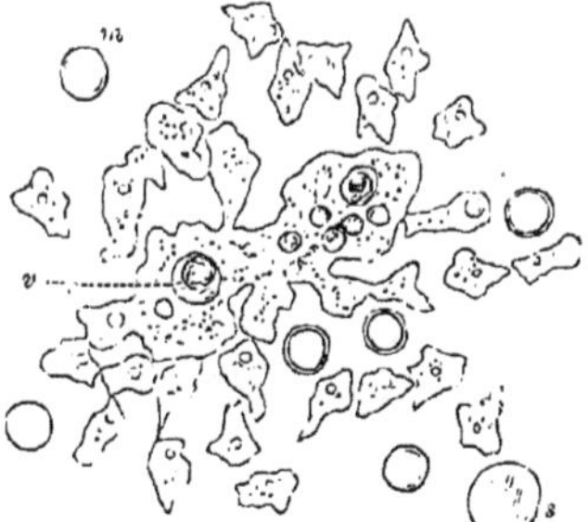

FIG. 14.—*Didymium difforme* (after Lister), young plasmodium with attendant amœboid swarm-cells, some of which have turned into microcysts (*m*); one microcyst being digested in a vacuole (*v*). An empty spore-shell at *s*. × 470.

In the neighbourhood of *Mycetozoa* may be placed provisionally the peculiar marine form *Labyrinthula*, described by Cienkowsky (*Arch. f. M. Anat.*, III.), from the harbour of Odessa. This animal consists of aggregations of roundish to spindle-shaped cells placed in a finely granular substance. From this mass, hyaline or finely fibrous processes are given off. These processes branch and anastomose so as to form a labyrinthic network along which the cells glide. *Chlamydomyxa* Archer (*Q.J.M.S.*, XIX.), seems to be a fresh-water organism of the same nature.

The *Sorophora* which are classed by some authors with the Mycetozoa, appear to be more nearly allied to *Labyrinthula*. In the vegetative phase they live on the dung of various animals, and are formed by the coming together of numbers of amœbulæ produced from spores. The amœbulæ, however, retain their distinctness, and do not fuse to form a homogeneous plasmodium as in the Mycetozoa, nor are there streaming movements throughout the mass. The mode of increase of the amœbulæ in this stage has not been followed. In the spore formation of *Dictyostelium* the mass rises up into club-shaped prominences, in the axis of which a septate stalk is formed, and the amœbulæ become gradually aggregated at the summit of the stalk, where they encyst. The amœbulæ which escape from the spores divide by fission, but they do not pass through a flagellate stage.

Genera. *Copromyxa* Zopf., *Cynthulina* Cienk., *Dictyostelium* Brefeld, *Acrasis* Van Tieghem, *Polyspondylium* Brefeld.

SUB-CLASS III. HELIOZOA.*

For the most part fresh-water Gymnomyxa with stiff radiating pseudopodia, and one or more nuclei; usually with contractile vacuole. A radial silicious skeleton sometimes present.

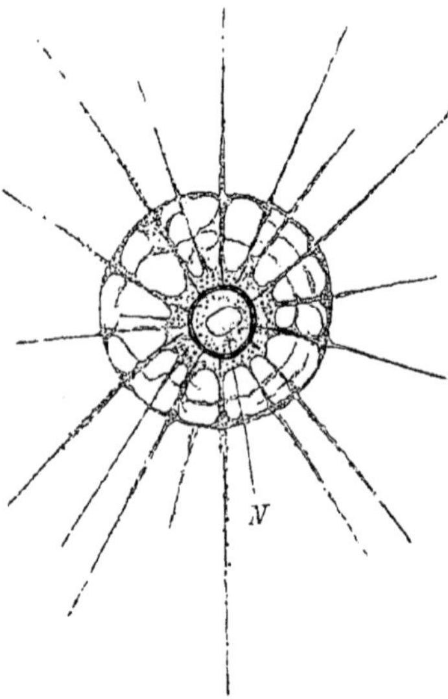

FIG. 15.—Young *Actinosphærium*, still with a single nucleus (after F. E. Schulze). *N* nucleus.

The characteristic pseudopodia give the name to the group. (Fig. 15.)

When a skeleton is secreted, it consists either of radially arranged silicious spines (*Acanthocystis*) or of latticed silicious shells (*Clathrulina*), and so closely resembles the skeleton of the Radiolaria that the Heliozoa have been actually described as *fresh-water Radiolaria.*

They differ from the Radiolaria in the absence of the complicated differentiations of the protoplasm, particularly of the central capsule. One or more nuclei may be present in the central mass. An important distinguishing mark is afforded by the presence of the pulsating vacuoles, which have not been observed in any marine Radiolarian.

Reproduction is effected by fission. Encystment and subsequent

* L. Cienkowski, "Ueber *Clathrulina.*" *Archiv. für mikrosk. Anatomie,* Tom III., 1867. R. Greeff, "Ueber Radiolarien und radiolarienähnliche Rhizopoden des süssen Wassers." *Ibid.* Tom V. & XI. R. Hertwig und Lesser, "Ueber Rhizopoden und denselben nahe stehende Organismen." *Ibid.* Suppl. Tom X., 1874. Also Archer and F. E. Schulze, etc.

spore-formation sometimes takes place. This has been observed in *Actinosphærium*, in which form the spores acquire a silicious coat. In some genera, *e.g.*, *Clathrulina* the spores are hatched as flagellate* forms.

Conjugation of two or more individuals has been observed. According to Brauer the nuclei of *Actinosphærium* fuse with one another, so that their number may be much reduced. Whether this nuclear fusion takes place only in forms which have resulted from conjugation, is not known.

Schaudinn has recently published a preliminary account of the encystment and conjugation of *Actinophrys sol*,† which is of special interest in view of a similar process described by Wolters‡ in the Gregarines (*Monocystis*), and the division of nuclei preceding conjugation in the *Ciliata*. He finds that after two individuals have come together and formed a cyst wall, the nucleus of each divides by karyokinesis into two daughter nuclei, of which one is extruded from the protoplasm, while the other unites with its fellow to form the conjugation nucleus. The subsequent division of this nucleus, together with the protoplasm, into two or four gives rise to the resting cysts, from which after some days the young *Actinophrys* escapes. The analogy of this process with the formation of the polar bodies in the maturing ova of higher animals is striking.

Order 1. **Aphrothoraca.** Heliozoa without a skeleton (sometimes temporarily invested by a gelatinous membrane). *Nuclearia* Cienk. body-form changeable, multinucleate; (*Monobia*, *Vampyrella*, *Myxastrum* sometimes placed here). *Actinophrys* Ehrb. (the sun animalcule) pseudopodia with axial fibre which can be traced to the single central nucleus; may form colonies by incomplete fission. *Actinosphærium* Stein, large, multinucleate; *Actinolophus* F. E. Sch.

Order 2. **Chlamydophora.** Heliozoa with gelatinous envelope. *Heterophrys* Archer; *Sphærastrum* Greef.

Order 3. **Chalarothoraca.** Heliozoa with skeleton of loosely arranged isolated silicious spicules; *Pompholyxophrys* Archer; *Raphidiophrys* Archer; *Pinacocystis* H. and L.; *Pinaciophora* Greef; *Acanthocystis* Carter.

Order 4. **Desmothoraca.** Heliozoa with a stalked or unstalked shell perforated by numerous pores; *Clathrulina* Cienk.; *Orbulinella* Entz.

* Cienkowski, *Arch. f. Mic. Anat.*, 1867, p. 311.

† Fr. Schaudinn. "Uber die Copulation von Actinophrys sol." *Sitzungsber. d. K. pr. Akad. d. Wiss. zu Berlin*, 1896, V.

‡ "Die conjugation und sporenbildung bei Gregarinen." *Arch. f. Mikr. Anat.* XXXVII. p. 99.

SUB-CLASS IV. RADIOLARIA.*

Marine Gymnomyxa with radiating pseudopodia, central capsule, and usually a skeleton of silica or acanthin.

The body contains a membranous porous capsule (the *central capsule*), in which is contained a slimy protoplasm with vacuoles and granules (*intracapsular sarcode*), fat and oil globules, and albuminous bodies, and more rarely crystals and concretions. The intracapsular mass contains also a single large nucleus or several small nuclei. The

FIG. 16.—*Thalassicolla pelagica*, with central capsule and single large nucleus, also numerous alveoli in the protoplasm (after E. Haeckel).

extracapsular sarcode which communicates with the intracapsular through the pores in the capsule, and which emits on all sides simple or anastomosing pseudopodia, contains numerous yellow cells, sometimes pigment masses; and in some cases it is much vacuolated like the external protoplasm of some pelagic Foraminifera (*Thalassicolla pelagica*. Fig. 16). The yellow cells are Algæ living symbiotically with

* Joh. Müller, "Ueber die Thalassicollen, Polycystinen und Acanthometren." *Abh. der Berl. Akad.* 1858. E. Haeckel, "*Die Radiolarien.*" Eine Monographie, Berlin, 1862. R. Hertwig, "*Der Organismus der Radiolarien*," Jena, 1879. E. Haeckel, "Report on the Radiolaria." *Challenger Reports*, 1887.

the Radiolarian. They have been named *Zooxanthella nutricola*, and contain chlorophyl, a nucleus and a cellulose wall. The central capsule, which may be either conical or spherical, is either perforated by fine pores over its whole circumference (*Peripylaria*), or the pores are limited to a definite part of its surface (*Monopylaria*), or there are only a few, usually three, large pores (*Tripylaria*).

Many Radiolaria form colonies, and are composed of numerous individuals. In such colonies the extracapsular protoplasm is united with that of neighbouring individuals, so that the whole colony may be described as consisting of a common mass of vacuolated protoplasm containing in itself, not as in the monozoic Radiolaria a single central capsule, but a number of capsules.

The whole animal—in the solitary as well as in the colonial forms—is embedded in a structureless jelly, called the *Calymma*. The extracapsular protoplasm may be described as consisting of the following parts: (1) the *Sarcomatrix*, the layer which surrounds the central capsule; (2) the *Sarcodictyum*, the layer which bounds the outer surface of the calymma; (3) the *Sarcoplegma*, or anastomosing threads which traverse the calymma and connect the sarcodictyum and sarcomatrix. From the extracalymmar sarcodictyum proceed the pseudopodia.

Only a few species remain naked and without firm deposits; as a rule, the soft body possesses a skeleton, which is composed either of silica or of an organic substance called Acanthin, and either lies entirely outside the central capsule (*Ectolithia*) or is partially within it (*Entolithia*). In the most simple cases the skeleton consists of small, simple, or toothed silicious needles (spicula) united together, which sometimes give rise to a fine sponge-work round the periphery of the protoplasm, *e.g.*, *Physematium*. In a higher grade we find stronger hollow silicious spicules, which radiate from the middle point of the body to the periphery in regular number and order, *e.g.*, *Acanthometra*. (Fig. 17.) A fine peripheral framework of spicules may be added to these. In other cases simple or compound lattice-works, and perforated

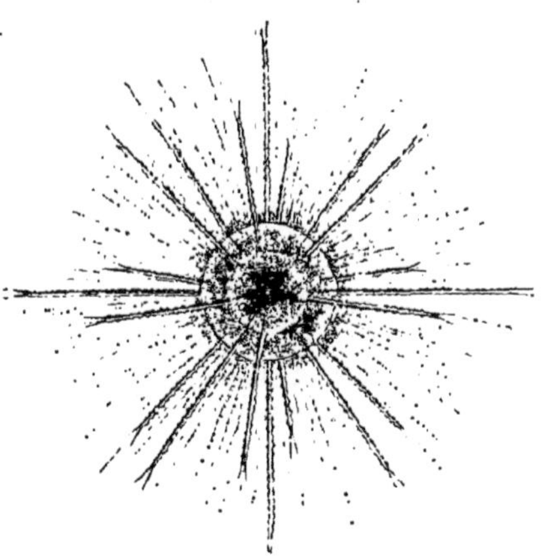

FIG. 17.—*Acanthometra Mülleri* (after E. Haeckel).

shells of various external form (like helmets, bird-cages, shells, etc.) are found, and on the periphery of these, spicules and needles, and even external concentric shells of similar shape may be formed, *e.g.*, *Polycystina*. (Figs. 18 and 19.)

Up to the present time but little has been made out about the reproduction of these animals. Besides fission of the central capsule (*Polycyttaria*), the formation of spores has been observed. These are formed from the contents of the central capsule, and, after the bursting of the latter, become free-swimming mastigopods. The spore-formation is of two kinds: in the one it results in the development of a mastigopod containing a crystal—the *crystalligerous* swarmer; in the other (dimorphous) two kinds of swarmers are formed—the macrospores and microspores, being distinguished from one another by their size. The further history of the spores is not known. Conjugation has not been observed in the Radiolaria; but it has been suggested that the macrospores and microspores may turn out to be conjugating cells. Radiolaria are inhabitants of the sea, and swim at the surface, but some live on the bottom, even at great depths.

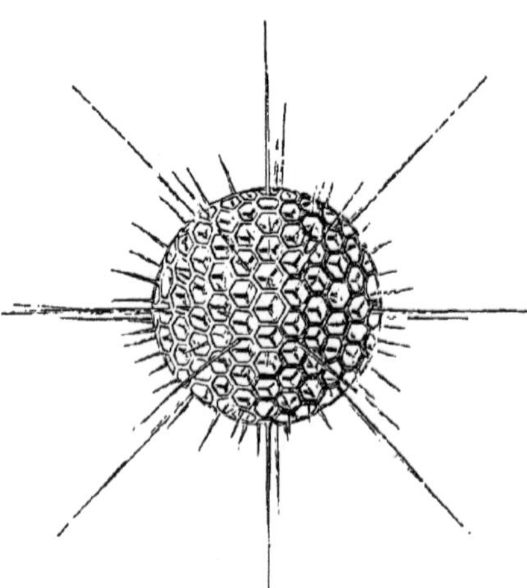

FIG. 18.—*Heliosphæra echinoides* (after E. Haeckel).

Fossil remains of Radiolaria have been made known in great numbers by Ehrenberg, *e.g.*, from the chalky marl and polishing slate found at certain parts of the coast of the Mediterranean (Caltanisetta in Sicily, Zante and Ægina in Greece), and in particular from the rocks of Barbados and Nikobar, where the Radiolaria have given rise to widely extended rock formations. Samples of sand also from very considerable depths have shown themselves rich in Radiolarian shells.

Order 1. **PERIPYLARIA.** Central capsule uniformly perforated by numerous fine pores, with or without silicious skeleton.

Fam. 1. **Colloidea** without skeleton.—Solitary are *Thalassolampe* H., *Thalasso-*

pila H., *Thalassicolla* Huxl., *Thalassophysa* H.; colonial (*Polycyttaria*) is *Collozoum* H.

Fam. 2. **Beloidea**, skeleton of loose silicious needles.—Solitary are *Thalassosphæra* H., *Thalassoplancta* H., *Physematium* Meyen; colonial (*Polycyttaria*) are *Belonozoum* H., *Sphærozoum* Meyen.

Fam. 3. **Sphæroidea**, with one to numerous concentric spherical shells. Colonial (*Polycyttaria*) are *Collosphæra* Müller one-shelled, *Clathrosphæra* H. two concentric shells; solitary are *Stigmosphæra* H. 1-shelled, *Carposphæra* H. 2-shelled, *Thecosphæra* H. 3-shelled, *Cromyosphæra* H. 4-shelled, *Xiphosphæra* H. 1-shelled with two radial spines, *Stylosphæra* H. 2-shelled with two radial spines, *Staurosphæra* H. 1-shelled with four radial spines. *Hexastylus* H. one-shelled with six spines, *Haliomma* H. 2-shelled and numerous spines, *Heliosphæra* H. (Fig. 18) with numerous radial spines of two sizes and a fenestrated shell.

Fam. 4. **Prunoidea** with ellipsoidal to cylindrical latticed shells and central capsule. *Ellipsis* H., *Druppula* H., *Spongurus* H., *Artiscus* H., *Cyphinus* H., *Panartus* H., *Zygartus* H.

Fam. 5. **Discoidea**, shell and central capsule discoidal or lenticular. *Cenodiscus* H., *Phacodiscus* H., *Coccodiscus* H., *Porodiscus* H., *Polydiscus* H., *Spongodiscus* H.

Fam. 6. **Larcoidea.**

Order 2. **ACANTHARIA.** Skeleton of acanthin, in the form of spines radiating from the central point; central capsule uniformly perforated (Peripylaria type).

Fam. 7. **Actinelida**, with a variable number of usually irregularly arranged spines. *Astrolophus* H., *Litholophus* H., *Chiastolus* H.

Fam. 8. **Acanthonida**, with 20 spines arranged according to Müller's law. *Acanthometra* J. Müller (Fig. 17), *Astrolonche* H., *Quadrilonche* H., *Amphilonche* H.

Fam. 9. **Sphærophracta**, with 20 equal quadrangular spines, and a complete fenestrated spherical shell. *Sphærocapsa* H., *Dorataspis* H., *Phractaspis* H., *Phractopelta* H.

Fam. 10. **Prunophracta**, with ellipsoidal, lenticular, or double-coned shell, and 20 spines of different size arranged according to Müller's law. *Belonaspis* H., *Hexalaspis* H., *Diploconus* H.

Order 3. **MONOPYLARIA.** Skeleton silicious, rarely without skeleton; central capsule monaxonic to bilateral, with simple wall and single polar perforated area; extracapsular plasma without pigment.

Fam. 11. **Nassoidea**, without skeleton. *Nassela* H.

Fam. 12. **Plectoidea.** Skeleton of 3 or more spines radiating from one point (placed beneath the basal pole of the c.c.) or from a central rod; a complete latticed shell is never formed. *Plagoniscus* H., *Plagonium* H., *Plectanium* H.

Fam. 13. **Stephoidea.** Skeleton of one to several fused rings, which may be connected by a loose network. *Lithocircus* H., *Zygocircus* H., *Cortina* H., *Stephanium* H., *Semantis* H., *Coronidium* H., *Tympanidium* H.

Fam. 14. **Spyroidea.** Fam. 15. **Botryoidea.**

Fam. 16. **Cyrtoidea.** *Eucyrtidium* H. (Fig. 19), helmet-shaped latticed shell outside central capsule, *Eucecryphalus* H.

Order 4. **TRIPYLARIA (PHÆODARIA).** Central capsule with double membrane, at one pole with a spout-like main opening on a striped field, and frequently an accessory opening on each side of the main axis of the opposite pole; sometimes several central capsules in one individual; always with extra-capsular pigment mass (*Phæodium* H.), which covers the region of the main opening. Skeleton either purely silicious, or weakly silicated with much organic substance, always extracapsular, rarely absent.

Fam. 17. **Phæocystina.** Partly without skeleton, partly with loose skeletal structures; central capsule in the centre of the spherical body. *Phæodinia* H. *Aulacantha* H., *Aulactinium* H.

Fam. 18. **Phæosphæria.** Fam. 19. **Phæogromia.** Fam. 20. **Phæoconchia.**

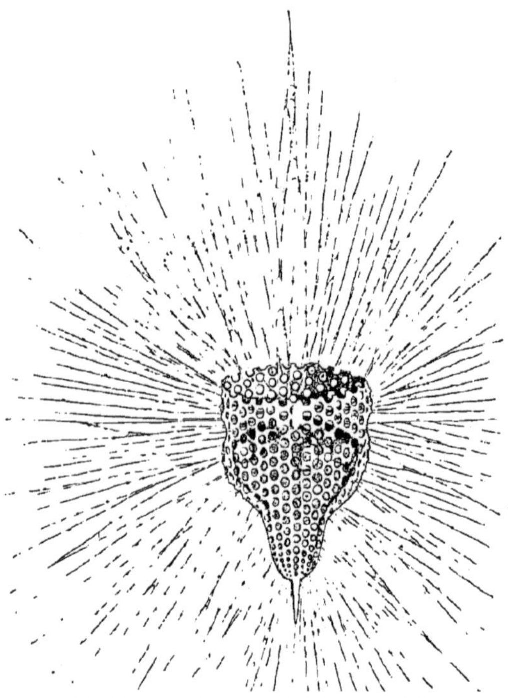

FIG. 19.—*Eucyrtidium cranoides* (after E. Haeckel).

Class II. INFUSORIA. *

Protozoa with a definite form, provided with an external membrane, and bearing either flagella or cilia. Mouth and anus usually, contractile vacuole and one or more nuclei always present.

* Ehrenberg, *Die Infusionsthierchen als vollkommene Organismen*, 1838. Balbiani, "Études sur la Reproduction des Protozoaires." *Journ. de la Phys.*, Tom. III. Balbiani, "Recherches sur les phénomènes sexuels des Infusoires." *Journ. de la Phys.*, Tom. IV. Claparède und Lachmann, *Études sur les Infusoires et les Rhizopodes*, 2 vol. Génève, 1858-1861. E. Haeckel, "Zur Morphologie

Infusoria were discovered towards the end of the 17th century in a vessel of stagnant water by A. von Leeuwenhoek, who made use of a magnifying glass for the examination of small organisms. The name Infusoria, which was at first used to denote all animalculæ which appear in infusions and are only visible with the aid of a microscope, was first brought into use by Ledermüller and Wrisberg in the last century. Later on the Danish naturalist O. Fr. Müller made valuable additions to our knowledge of Infusoria. He observed their conjugation and their reproduction by fission and gemmation, and wrote the first systematic work on the subject. O. Fr. Müller included a much larger number of forms than we do nowadays, for he placed among the Infusoria all invertebrate water animalculæ without jointed organs of locomotion and of microscopical size.

The knowledge of Infusoria received a new impulse from the comprehensive researches of Ehrenberg. The principal work of this investigator, "Die Infusionsthierchen als vollkommene Organismen," discovered a kingdom of organisms hardly thought of. These were observed and portrayed under the highest microscopic powers. Many of Ehrenberg's drawings may even yet be taken as patterns, and are hardly surpassed by later representations, but the significance of the facts observed has been essentially corrected by more recent investigations. Ehrenberg also conceded too great an extent to the group of Infusoria, including not only the lowest plants such as *Diatomaceæ*, *Desmidiaceæ*, under the name of *Polygastrica anentera*, but also the much more highly organized *Rotifera*. As he chose the organization of the last-named for the basis of his explanations, he was led into numerous errors. Ehrenberg ascribed to the Infusoria mouth and anus, stomach and intestines, testis and ovary, kidneys, sense-organs, and a vascular system, without being able to give reliable proofs of the nature of these organs. There very soon came a reaction in the way of regarding the Infusorian structure; for the discoverer of the *Rhizopoda*, Dujardin, as well as von Siebold and Kölliker (the latter taking into consideration the so-called *nucleus* and *nucleolus*), referred the Infusorian body to the simple cell. In the subsequent works of Stein, Claparède, Lachmann, and Balbiani numerous differentiations

der Infusorien." *Jen. Zeitschrift*, Tom. VII., 1873. O. Bütschli, *Studien über die ersten Entwickelungsvorgänge des Eizelle, die Zelltheilung und die Conjugation des Infusorien*, Frankfurt, 1876. Saville Kent, *A Manual of the Infusoria*, London, 1880-2. Maupas, "Sur la multiplication des Infusoires Ciliés." *Arch. d. Zool. Exp.* (2), 6, and "La Rajeunissement Karyogamique chez les Ciliés, *ibid.* (2), 7.

were certainly shown to exist, which, however, can all be referred to differentiation of the body of the cell. This view is supported by the more recent work of Bütschli and Maupas, who have shown that in their reproduction these animals resemble other *Protozoa:* that is to say, that the whole body participates in the reproductive fission, that the parent disappears in the offspring, and

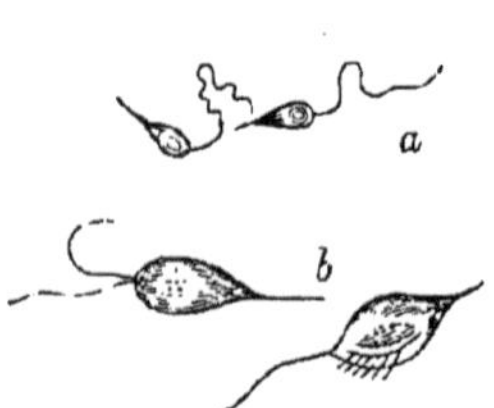

Fig. 21.—*a*, *Cercomonas intestinalis*; *b*, *Trichomonas vaginalis* (after Leuckart).

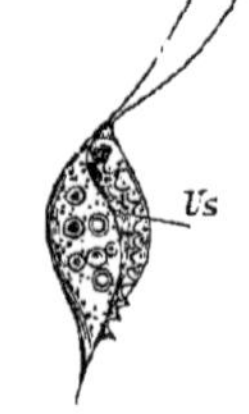

Fig. 22.—*Trichomonas batrachorum* (after Stein). *Us* undulating membrane.

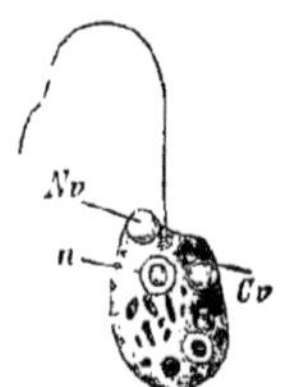

Fig. 23.—*Oikomonas termo* (after Bütschli). *n* nucleus; *Cv* contractile vacuole; *Nv* vacuole which takes up the food (oral vacuole).

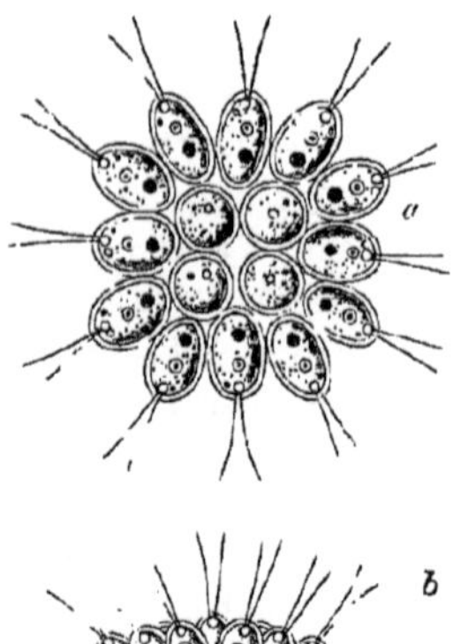

Fig. 24.—*Gonium pectorale* (after Stein). The colony *a* from above, *b* from the side.

that special conjugating cells of the nature of ova and spermatozoa are not formed. Maupas especially, by following the history of the individual resulting from conjugation, has definitely established the fundamental distinction between conjugation and reproduction, and has thrown a flood of light upon the meaning of the whole phenomenon of conjugation.

The outer boundary of the body is usually formed by a cuticle—a delicate transparent membrane, the surface of which is beset with vibratile and moving appendages of various kinds. In the smallest *Infusoria*—the *Mastigophora*, we find only one or two long whip-like cilia; while the more highly differentiated *Ciliata* are usually richly provided with cilia. Finally, in the *Acinetaria* the young forms have cilia, and the adults a number of delicate tentacle-like processes, which either end in suctorial discs or are pointed.

Sub-class I. MASTIGOPHORA.

Infusoria generally of small size provided with flagella.

This sub-class includes forms which live in putrefying infusions, parasitic forms, and forms which live freely. They all have contractile vacuoles, and some of them have an opening at the base of the flagellum for the reception of solid substance (holozoic nutrition). Encystment and spore formation are very commonly found; and it has been shown by Dallinger and Drysdale* that the spores are capable of resisting a temperature above boiling point. Dallinger has also shown, in the case of some of the infusion forms, that it is possible by very gradually raising the temperature in which the animals are living during a number of successive generations to produce a race for which the optimum temperature is considerably above the normal killing temperature for the species.

Conjugation is known to occur very generally, and in some cases the conjugating individuals (or gametes)† are especially differentiated. This is notably the case in *Volvox*, in which the gametes are of two kinds, and recall the spermatozoa and ova of the higher animals. Many members of our sub-class are difficult to distinguish from the swarm-spores of certain *Rhizopoda* and of the *Mycetozoa*, and even from the zoospores of unicellular *Algæ*. It is necessary, therefore, to point out that in the *Mastigophora* the flagellated stage covers the main, if not the entire, period of the life of the organism.

The nucleus is almost invariably single.

Order 1.—Flagellata.‡

Mastigophora with flagella, without collar or cilia.

This order includes holozoic, holophytic, and saprophytic forms. Many of them are parasitic and many live in infusions. It is not infrequent to find an amœboid condition of the body combined with the possession of the flagellum (*Mastigamœba*), or to find these two conditions alternating in the life-history. Many of them form colonies, and an outer cuticular skeleton in the form of a cup or investing membrane may be present; and in some forms a gelatinous layer is secreted. In the holozoic forms food may be taken up by

* "Researches on the Life-history of the Monads." *Monthly Mic. Journal*, 10-13.

† An organism which conjugates, whether specially differentiated or not, is called a *gamete*, and the product of the conjugation is a *zygote*.

‡ G. Klebs, "Flagellatenstudien." *Z. f. w. Z.*, 55, 1892.

means of pseudopodia in an amœboid fashion, or there is a definite spot at the base of the main flagellum where the food enters: this spot is either marked by a mouth-vacuole (Fig. 23 *Nr*) into which the food slips, or by the presence of a mouth-opening with or without a pharyngeal continuation. The expulsion of undigested remains of food appears to be localised and often to take place by the bursting of a vacuole; but the position of the temporary anus, which seems to be variable in the different forms, has only been determined in a few cases. Contractile vacuoles close to the body-surface seem to be always present, and in the *Euglenina* they appear to open into a receptacle which is in communication with the hind end of the pharynx. **Chromatophores** of the same character and function as those of plants are present in the holophytic forms, and vary in colour from a light green to a brown (Chlorophyll and Diatomin). They contain amylum bodies, which consist of a central mass of a highly stainable plasma—the **pyrenoid**—and of an outer zone of amylum. The pyrenoids may increase by division. Amylum bodies are also found in colourless saprophytic forms. Paramylum, a substance more nearly allied to cellulose, is sometimes present in the protoplasm. Chromatophores may be present or absent in closely allied forms, and even in the same form at different times: their presence is of no systematic importance.

The **nucleus** is always single except in *Trepomonas*, which sometimes has two. **Stigmata** as red pigment spots are often present in the protoplasm, usually at the base of the flagellum.

Reproduction takes place by fission, which is usually, if not always, longitudinal, in both the active and resting state, and sometimes by continued fission (spore formation) in the resting state. In the first case the fission may be into two, or by successive binary fissions into four, eight, sixteen, or even thirty-two before the young separate. When the fission is into two, the flagella become doubled in number before the body divides. The manner in which this doubling occurs is disputed: very likely a new set of flagella and of the other organs of the body is formed before the division occurs. When the first binary fission is succeeded by others, the successive fissions take place within the cuticle, while the animal continues to move by the two original flagella which remain attached to one of the products of fission.

Finally reproduction may take place by continued fission (spore formation), during the resting state (*Bodo*, *Tetramita*). This has been observed to follow conjugation.

Conjugation* is of very common occurrence. In certain small *Bodonineæ* it takes place between several individuals in the amœboid state, so as to give rise to a plasmodium, which, as in the *Mycetozoa*, encysts and divides into spores. In *Cercomonas* the gametes become amœboid and conjugate in pairs. Sometimes the gametes, or one of them, differ in size or history from the ordinary forms. In the *Chlamydomonadina* small forms (*microgametes* or *microgonidia*) often arise by fission and conjugate with similar microgametes or with the ordinary form, or with forms which have been produced from the ordinary form by a smaller amount of fission. In *Polytoma* gametes are produced by tetrapartite fission, and are all alike.

The differentiation of the gametes is carried furthest in some of the *Volvocina*. In *Volvox* both gametes are specially differentiated individuals of the colony. The megagamete or macrogonidium is an ovum-like cell without flagella, while the microgamete or microgonidium is a small flagellated organism produced by fission from the ordinary form. In some species these two kinds of gametes are not found in the same individuals, so that there are unisexual colonies. The zygote resulting from their conjugation secretes a thick membrane, and remains in the resting state for some time. On the death of the mother it falls to the bottom, and when conditions are favourable it develops into a new colony.

Conjugation is, so far as is known, followed by encystment, the zygote remaining for some time quiescent. While in this condition it is capable of resisting drought, which indeed seems to be favourable to it. When placed in suitable conditions the contents of the cyst divide into two or four parts, which issue as young forms to begin a new cycle of life. The breaking up of the zygote into minute spores is only described for a few forms (Dallinger, *Bodo*, *Tetramitu*).

The *Flagellata* are allied to the *Gymnomyxa* by such forms as the *Monadina*, in which the amœboid condition may occur as an important phase in the life-history, and by those of the *Gymnomyxa* in which the young leave the spore-case in the flagellate form (mastigopod). In other words the *Flagellata* and *Gymnomyxa* agree with one another in including forms which pass through both the myxopod and mastigopod condition in the life-history, the difference between these forms consisting mainly in the relative duration of the two states. *Paramœba*, a form recently described by Schaudinn (*Sitzb. K. Preuss. Akad. W.*,

* Klebs throws doubt upon the occurrence of conjugation in any true Flagellate. He admits that it takes place in the *Volvocina*, which he removes from the *Flagellata*, and in certain other forms such as *Ciliophrys* and *Protomonas*, all of which he refers to the *Heliozoa*; but more recently Dill (*Jahrb. wiss. Bot.*, 28, 1895) has not only stated that transverse fission occurs, but also that gametes are formed in *Chlamydomonas*, which is a member of the **Flagellata.**

1896), seems to be a form exactly intermediate between the two groups. On the other hand the *Flagellata* show distinct affinities to the lower plants by such forms as the *Volvocina*, which Klebs replaces amongst the *Algæ*, and by the *Chrysomonadina* and *Cryptomonadina*, which he unites into a special group—the *Chromomonadina*—of the *Flagellata*.

Sub-order 1. **MONADINA.**

Small to very small forms of simple structure, naked and often more or less amœboid, sometimes with tests; usually colourless, rarely with chromatophores; with one anterior large flagellum, to which may be added one or two small flagella; special mouth-opening sometimes absent, sometimes present, but never continued into a well-developed pharynx.

Fam. 1. **Rhizomastigina.** Simple, mouthless forms with one or two flagella, and in some cases the power of thrusting out pseudopodia; in other cases the amœboid condition may be assumed with or without retraction of the flagella. *Mastigamœba* F. E. Sch., fresh-water; *Ciliophrys* Cienk., fresh-water, heliozoon-like; *Dimorpha* Gruber; *Actinomonas* Kent; *Trypanosoma* Gruby, parasitic forms from blood of *Amphibia*, *Pisces* and *Chelonia*, with undulating membrane.

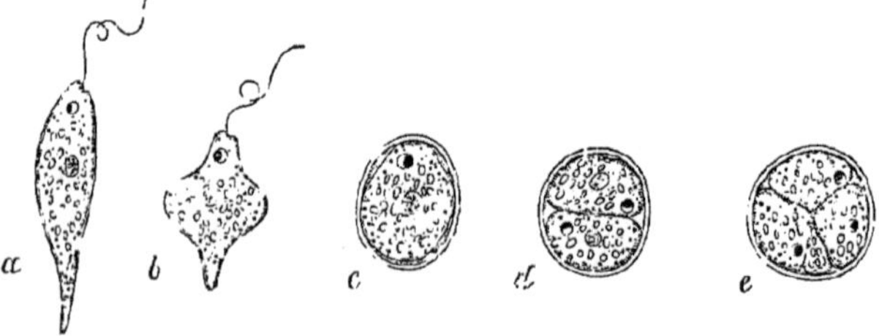

FIG. 25.—*Euglena viridis.* *a* and *b*, free-swimming in different stages of contraction; *c*, *d*, *e*, encysted and in process of division.

Fam. 2. **Cercomonadina.** Form oval to elongated, often amœboid; one large, forwardly-directed flagellum with mouth, as a vacuole for taking up food, at its base. *Cercomonas* Dujardin (Fig. 21), fresh-water and infusions, hind end continued into pseudopod-like fibre; *Herpetomonas* Kent, parasitic, gut of *Musca*; and *Trilobus*, blood of *Mus*; *Oikomonas* Kent (Fig. 23), fresh-water, infusions and marine; *Ancyromonas* Kent.

Fam. 3. **Codonœcina.** Monad with attached gelatinous or membranous cup. *Codonœca* James Clark, salt and fresh-water; *Platytheca* Stein.

Fam. 4. **Bikœcina.** Monads with cup; hind end fastened to base of cup with contractile stalk; cup usually fastened by stalk; some colonial. *Bicosœca* Clark, fresh-water and marine; *Posteriodendron* Stein, fresh-water.

Fam. 5. **Heteromonadina.** Small, colourless monads with an anterior large flagellum, one or two contiguous smaller accessory flagella; often colonial and then stalked. *Monas* Ehrb.; *Dendromonas* Stein, fresh-water; *Cephalothamnium* Stein, attached to Cyclops; *Anthophysa* Bory d. Vinc., fresh-water; *Dinobryon* Ehrb.; *Uroglena* Ehrb.

Sub-order 2. **EUGLENOIDEA.**

Monoflagellate forms; body contractile or stiff; mouth and well-developed pharynx at base of flagellum; contractile vacuole near pharynx, often with

Fam. 6. **Cœlomonadina.** Euglenoids coloured with numerous chlorophyll bodies, or one to two larger plate-like chromatophores; usually no true pharynx. *Cœlomonas* Stein, ectoplasm with chlorophyll grains; *Gonyostomum* Diesing, like preceding, with trichocysts in ectoplasm; *Microglena* Ehrb.; *Chromulina* Cienk.; *Cryptoglena* Ehrb.

Fam. 7. **Euglenina.** Elongated; hind end usually pointed; spirally striped cuticle; reservoir with usually several contractile vacuoles and simple stigma close behind pharynx; chromatophores, usually green, almost always present. *Euglena* Ehrb. (Fig. 25); *Colacium* Ehrb.; *Eutreptia* Perty; *Ascoglena* Stein; *Trachelomonas* Ehrb.

Fam. 8. **Chloropeltina.** Like preceding, but with thicker cuticle. *Lepocinclis* Perty; *Phacus* Nitzsch.

Fam. 9. **Menoidina.** Like *Euglenina*, but without chlorophyll and stigma; saprophytic. *Astasiopsis* Bütschli; *Menoidium* Perty; *Rhabdomonas* Fresenius.

Fam. 10. **Peranemina.**

Fam. 11. **Petalomonadina.**

Fam. 12. **Astasiina,** with small or large second flagellum. *Astasia* Ehrb.; *Heteronema* Duj.; *Sphenomonas* Stein.

Sub-order 3. HETEROMASTIGODA.

With two flagella of different character and size, the one directed forward and the other (sometimes two) trailed behind; with at least a mouth-spot, which in the larger forms becomes a mouth with pharynx; colourless, holozoic; sometimes amœboid.

Fam. 13. **Bodonina.** *Bodo* Ehrb. (*Heteromita*), the hooked monad and the springing monad. *Phyllomitus* Stein; *Colponema* Stein; *Dallingeria* Kent.

Fam. 14. **Anisonemina.**

Sub-order 4. ISOMASTIGODA.

With two, four, rarely five equal flagella at the anterior end; rarely with mouth opening and pharynx.

Fam. 15. **Amphimonadina.**

Fam. 16. **Spongomonadina.** Biflagellate; colonial, the individuals living in a granular jelly, or at the end of branched gelatinous tubes. *Spongomonas* Stein; *Cladomonas* Stein; *Rhipidodendron* Stein; *Diplomita* Kent.

Fam. 20. **Tetramitina.** Naked and sometimes amœboid *Isomastigoda*, with finely pointed hind end; anterior end with four equal flagella, one of which may be larger and directed backwards; the latter may have the form of an undulating membrane; distinct mouth only rarely discernible; holozoic. *Collodictyon* Carter; *Tetramitus* Perty; *Monocercomonas* Grassi; *Trichomonas* Donné (Figs. 21 *b* and 22); *Trichomastix* Blochmann.

Fam. 21. **Polymastigina.** Somewhat oval, with broader or pointed hind end, which is continued into two flagella. At the anterior end of the body are two or three flagella on each side. Holozoic, and perhaps in part saprophytic. *Hexamitus* Duj.; *Megastoma* Grassi.

Fam. 22. **Trepomonadina.** The two anterior flagella arise far apart from one another at the sides of the body. *Trepomonas* Duj.

Fam. 23. **Cryptomonadina.** Coloured or colourless; usually laterally compressed, without true cuticle, with two long anterior flagella; anterior end obliquely truncated, and pitted inwards slightly on one side; the pit may lead into a pharynx. *Cyathomonas* Fromentel, possibly having affinities with the *Dinoflagellata*; *Chilomonas* Ehrb.; *Cryptomonas* Ehrb.; *Oxyrrhis* Duj.

The following three families of the *Isomastigoda* are separated as a sub-order named Phytomastigoda on account of their plant-like features.

Sub-order 5. **PHYTOMASTIGODA.**

Holophytic, vegetable-like *Isomastigoda* with chlorophyll, without mouth.

Fam. 17. **Chrysomonadina.** Solitary or colonial; rarely with test; with two, rarely one, brown to greenish-brown chromatophore; usually with eye-spot at the base of flagella; colonies free-swimming, with spherical grouping of individuals. *Stylochrysalis* Stein; *Chrysopyxis* Stein; *Nephroselmis* Stein; *Synura* Ehrb., colonial, with cuticle often growing out into fine spines; *Syncrypta* Ehrb.

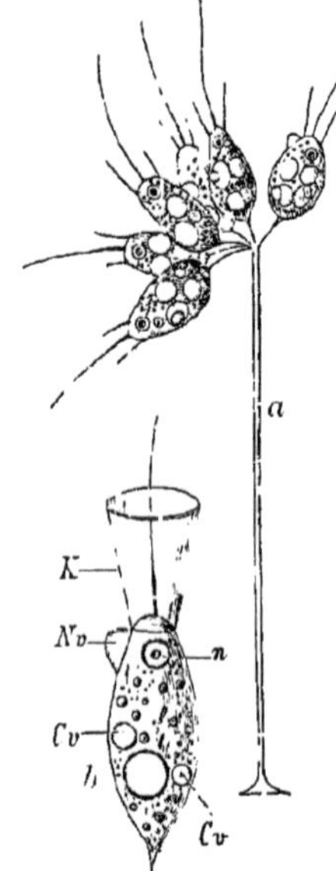

Fig. 23.—*Codosiga botrytis* (after Bütschli). *a*, colony, *b*, one individual. *K* collar; *n* nucleus; *Cv* contractile vacuole; *Nv* oral vacuole.

Fam. 18. **Chlamydomonadina.** Almost always green on account of considerable, usually single, chromatophore; usually delicate membranous shell without large opening; one to two contractile vacuoles at base of flagella; usually one eye-spot (stigma). Reproduce by continued division within the shell-membrane; usually forming macro- and microgonidia; mostly solitary. *Hymenomonas* Stein: *Chlorangium* Stein; *Chlorogonium* Ehrb.; *Polytoma* Ehrb., saprophytic, without chromatophores, with amylum bodies: *Chlamydomonas* Ehrb.; *Hæmatococcus* Agardh.; *Spondylomorum* Ehrb., colonial; *Coccomonas* Stein; *Phacotus* Perty.

Fam. 19. **Volvocina.** Biflagellate colonial *Phytomastigoda*, intermediate in structure between *Chlamydomonas* and *Hæmatococcus*. Reproduction by continued division of all, or of certain, individuals of the colony (*parthenogonidia*) to form daughter colonies. In some, probably in all, conjugation occurs between definite individuals of the colony, with or without differentiation of the colonies and gametes into male and female. The result of conjugation is a resting zygote, which develops later into one or into several new colonies. *Gonium* O. F. Müller (Fig. 24), colonies of four or sixteen individuals united to a quadrangular plate-like group; reproduction by simultaneous division of all the individuals to form daughter colonies; *Stephanosphæra* Cohn; *Pandorina* Bory de Vincent: *Eudorina* Ehrb.; *Volvox* L., spherical colonies with numerous individuals, which are placed at equal distances within the common thick colonial membrane, and lie in special membranes which stand off from the cells and are compressed against one another into the form of hexagons.

Order 2. CHOANOFLAGELLATA.

Mastigophora with a collar-like process of protoplasm round the base of the single flagellum.

Solitary and colonial forms are included in this group. The

individuals may be naked, or may secrete a cup, or may be embedded in jelly (*Proterospongia*). Their nutrition is holozoic. The food-particles, brought by the currents set up by the flagellum, adhere to the outer side of the collar, down which they move until they are swallowed by a kind of vacuole-like elevation of the body protoplasm at the base of the collar (*oral vacuole*). Ejection of fæcal matter takes place in the collar area, though there is no distinct anal spot. The collar is protoplasmic and retractile.

Fam. 1. **Phalansterina.** Colonial, each individual in a granular gelatinous tube. Colonies either a lamellar expansion or a dichotomously branching stock; collars narrow, conical, and of constant shape. *Phalansterium* Cienk.

Fam. 2. **Craspedomonadina.** Solitary or colonial, collars considerable, conical and of changeable form. Individuals naked or with incomplete cup, or in gelatinous mass. *Monosiga* Kent; *Codosiga* J. Clark (Fig. 26); *Codonocladium* Stein; *Hirmidium* Perty; *Proterospongia* Kent, colonial, the individuals are embedded in jelly and readily assume the amœboid condition; *Salpingœca* J. Clark, and *Polyœca* Kent, with thin-walled cups.

Order 3. Dinoflagellata.*

Bilateral, asymmetrical Mastigophora with a ventral groove and two flagella. A membrane consisting of cellulose is generally present.

The *Dinoflagellata* are most nearly allied to the *Cryptomonadina.* Like these they possess two flagella, which arise from a groove. The flagella are always distinguishable from one another: the one as the longitudinal flagellum directed forwards (*Adinida*) or backwards (*Dinifera*), and the other as the transverse flagellum because it has a transverse circular course round the base of the first (Fig. 27). The transverse flagellum moves by very short waves, and was until lately taken to be a transverse ring of fine cilia (hence *Cilioflagellata*). A membrane or shell consisting of cellulose and often prolonged into processes (Fig. 28), is nearly always present. They closely resemble the *Flagellata* in their internal structure. Except in one genus (*Polykrikos*), there is never more than a single nucleus. Chromatophores of a green to brown colour are

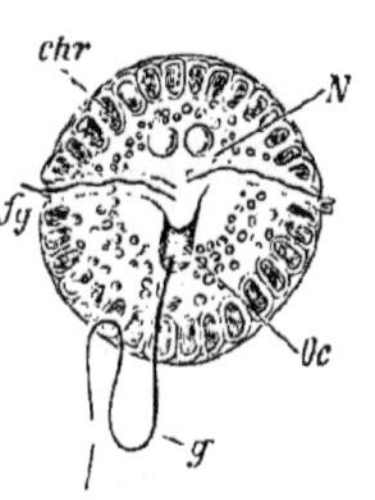

Fig. 27.—*Glenodinium cinctum* (after Bütschli) ventral view. *g* longitudinal flagellum; *fg* transverse flagellum; *N* nucleus; *Oc* stigma (eye-spot); *chr* chromatophores.

* R. S. Bergh, "Der Organismus der Cilioflagellaten." *Morph. Jahrb.*, 7, 1881. Fr. Schütt, "Die Peridineen d. Plankton-Expedition," Th. 1. *Ergeb. Plankton-Exp.*, 4, 1895.

generally present, and contain chlorophyll and diatomin*; but there are colourless forms. Amylum, fat, red pigment, and stigmata may also be present. There is always a longitudinal groove upon what we call the ventral surface, and in the *Dinifera* there is a second groove —the transverse groove—encircling the body; the latter generally has a slightly spiral course (so that the two ventral ends of it are not quite at the same level, Figs. 27, 28) and in one genus makes two complete turns. The two flagella arise, as a rule, close together, where the two grooves cross one another. The transverse flagellum lies wrapped round the body in the circular groove. The flagella project through a hole in the cuticle, which in some genera at least extends as a slit along the left side of the ventral groove, being enlarged posteriorly where the longitudinal flagellum projects. Reproduction takes place by transverse fission. Conjugation has been observed in a few forms, and in a few cases individuals have been observed to unite in chains (? a form of conjugation). In *Polykrikos*, which has four nuclei, there are eight transverse furrows, each with a flagellum. The presence of a mouth is doubtful. Fresh-water and marine.

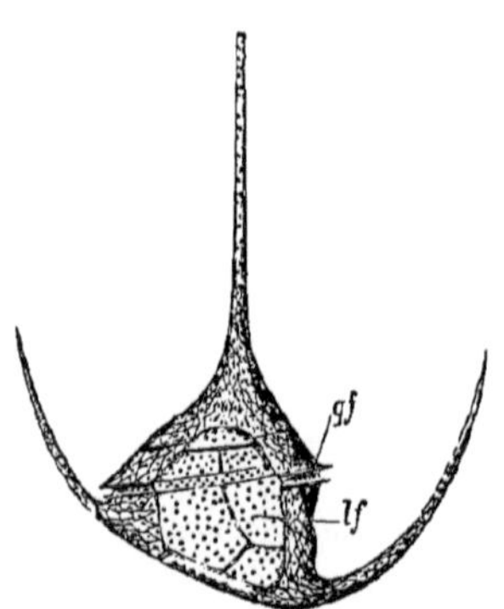

FIG. 28.—Shell of *Ceratium tripos* (after Stein). *lf* longitudinal furrow; *gf* transverse furrow.

Sub-order 1. **ADINIDA.**

Longish bilaterally-symmetrical forms with inclination to asymmetry; the two flagella arise at the anterior pole, and the transverse furrow is not developed; with bivalved porous membrane; two vacuoles near one another at the anterior end; chromatophores.

Fam. 1. **Prorocentrina**, with characters of sub-order. *Exuviaella* Cienk.: *Prorocentrum* Ehrb.

Sub-order 2. **DINIFERA.**

With a more or less distinct transverse furrow containing a flagellum. Longitudinal flagellum directed backwards.

Fam. 2. **Peridinida**, with the transverse groove at or near the centre of the body. *Podolampas* Stein; *Peridinium* Ehrb.; *Goniodoma* Stein; *Ceratium* Schrank (Fig. 28); *Pyrophacus* Stein; *Glenodinium* Ehrb. (Fig. 27); *Gymnodinium* Stein, without cuticle; *Ceratocorys* Stein.

* By some naturalists there is supposed to be affinity between the *Dinoflagellata* and the silicious Algæ, the *Diatomaceæ*.

Fam. 3. **Dinophysida.** *Phalacroma* Stein; *Dinophysis* Ehrb.; *Amphisolenia* Stein; *Ornithocercus* Stein; *Histioncis* Stein.

Fam. 4. **Polydinida,** with several transverse grooves and flagella; naked. *Polykrikos* Bütschli.

Order 4. SILICOFLAGELLATA.*

Marine mastigophora with one flagellum and a latticed silicious cap on one side of the body.

The forms included in this order were formerly regarded by some naturalists as *Radiolaria*, on account of their silicious latticed-skeleton, and by others as parts of the skeleton of *Radiolaria* (large species of *Phæodaria*). Their soft parts were imperfectly known, owing to the fact that they are small, exceedingly sensitive to adverse influences, and consequently very difficult to get under observation in the living state. The body lies within the skeleton, and contains some small brownish-yellow spherical bodies. There is a nucleus in the centre of the protoplasm, bounded by a membrane, and containing a nucleolus; it has been called the *central body*. There is one long vibratile flagellum. The body is without a bounding membrane, and there are no pseudopodia. The reproductive processes are unknown. The skeleton consists of hollow silicious rods, and has the form of two circles united by rods; it invests the body in a cap-like manner.

Fam. **Dictyochidæ,** with the characters of the order. *Mesocena* Ehrb.; *Dictyocha* Ehrb.; *Distephanus* H.; *Cannopilus* H.

Order 5. CYSTOFLAGELLATA (RHYNCHOFLAGELLATA).

Mastigophora of large size with a single nucleus, reticular protoplasm, and a stout membrane.

Noctiluca is a nearly spherical form of large size (1 mm. in diameter), and has on its ventral side a groove, called the peristome, at the base of which is the elongated slit-like mouth. From the anterior end of the peristome projects a large transversely striated flagellum, sometimes called the tentacle, and a little further back on the right side are two organs, the *tooth* and the *lip*. The flagellum is a contractile organ and varies in form: it moves slowly from side to side. The tooth is a protoplasmic projection, and is probably actively movable. From the lip there projects forward a smaller and vibratile flagellum. The greater part of the protoplasm with the nucleus is aggregated on the ventral side at the base of

* A. Borgert. "Ueber die Dictyochiden," etc. *Z. f. w. Z.*, 51, 1891, p. 629.

the peristomial groove, and from it there radiate in all directions to the periphery branching strands of protoplasm. At the periphery there is a thin layer of reticulated phosphorescent protoplasm immediately underlying the stout cuticle which bounds the body. The nutrition is holozoic, and food vacuoles are formed. *Leptodiscus* has the form of a flattened disc, concave on one side and convex on the other. On the convex side is the mouth somewhat eccentrically placed, and from the same side there is a tubular depression from which projects a flagellum. The protoplasm is much vacuolated as in *Noctiluca*, and the principal part of it is aggregated at the centre of the concave or aboral side of the disc.

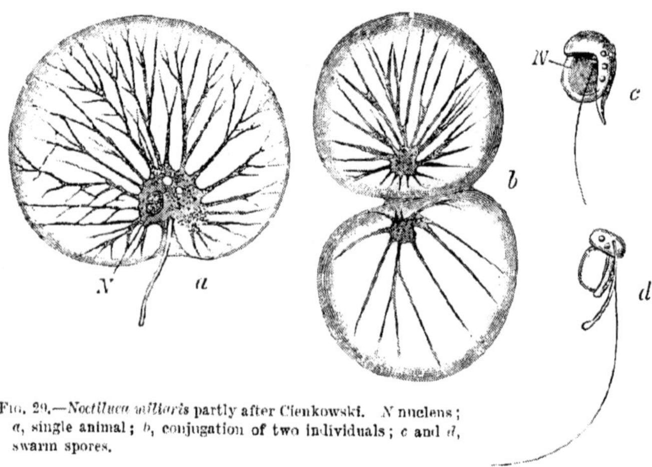

FIG. 29.—*Noctiluca miliaris* partly after Cienkowski. *N* nucleus; *a*, single animal; *b*, conjugation of two individuals; *c* and *d*, swarm spores.

Noctiluca occasionally draws in its flagella and loses its peristome and enters the resting state, but has not been observed to encyst. It reproduces in two ways: by binary fission and by the formation of small swarmers. If by fission, the division of the nucleus and central plasma is soon followed by that of the whole body. Before the latter is completed the development of the peristome and its organs in the new individuals begins. Individuals with two central masses and two nuclei are occasionally found. The formation of spores takes place during the resting state. The central plasma projects slightly on the surface, and divides by a superficial constriction into two prominences. The nucleus karyokinetically participates in this and future divisions. These two again divide into four, and so on, until a large number of small prominences are formed; these eventually become free, and constitute the spores. The latter are provided with a flagellum, a pointed process—the so-called spine—and a nucleus. Their later history has not been followed, and it is not known whether they conjugate. The process of division by which these spores arise is incomplete, and resembles the cleavage of a meroblastic egg. While it is taking place a part of the protoplasm of the rest of the body seems to pass into the dividing disc, but the superficial layer beneath the cuticle always remains. The fate of the maternal body after the separation of the spores is not known.

Conjugation takes place between two individuals, and results in complete fusion; but the fate of the zygote has not been traced.

Noctiluca owes its name to its phosphorescent power. The light is emitted from numerous points in the surface protoplasm, and principally when the animal is disturbed. It sometimes appears in enormous numbers on the sea surface. It is cosmopolitan, and is principally confined to the coasts, but it has been taken in the open ocean.

There are two genera—both marine. *Noctiluca* Suriray, with very slight power of change of form; cosmopolitan. *Leptodiscus* R. Hertwig, body contractile and movements energetic; Mediterranean.

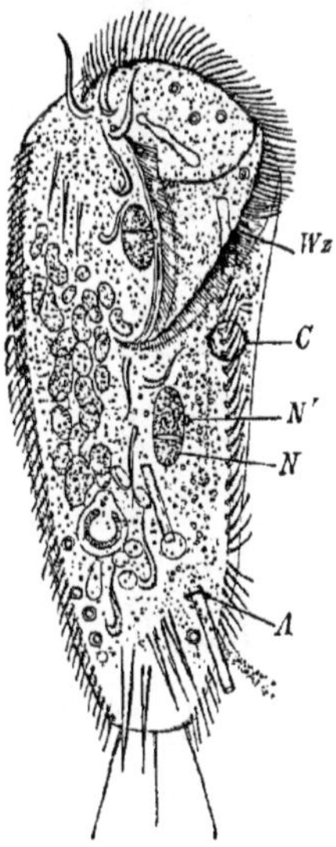

SUB-CLASS II. CILIATA.*

Infusoria provided with cilia; mouth and anus, nucleus and paranucleus are generally present.

This group contains the most highly organised of the Protozoa. The locomotive appendages are cilia, which are modified in diverse ways, as will be explained below; and immovable hairs and stiff bristles, which may even have the form of bent hooks and be employed in locomotion and attachment, are often present. The power of forming

FIG. 30.—*Stylonychia mytilus* (after Stein), seen from ventral side. *Wz* adoral zone of cilia; *C* contractile vacuole; *N* nucleus; *N'* paranucleus; *A* anus.

* O. Bütschli, *loc. cit.* W. Saville Kent, *A Manual of the Infusoria*. London, 1880-82. E. Maupas, "Sur la multiplication des Infusoires Ciliés." *Arch. Zool exp. et gén.* (2), Vol. VI.; and "La Rajeunissement Karyogamique chez les Ciliés. *Ibid.*, Vol. VII.

pseudopodia is almost always absent, and the body is generally bounded by a thin pellicle or cuticle. Certain fixed *Ciliata*, as *Stentor* (Fig. 31) and *Cothurnia*, secrete external coverings or shells into which they can be retracted. Nourishment is in a few cases taken in by endosmosis through the whole surface of the body; but as a general rule there is an oral aperture usually near the anterior pole of the body, through which solid food is introduced. A second aperture, which acts as anus, and which can be seen as a slit during the exit of the excreta, is often present at a definite part of the body.

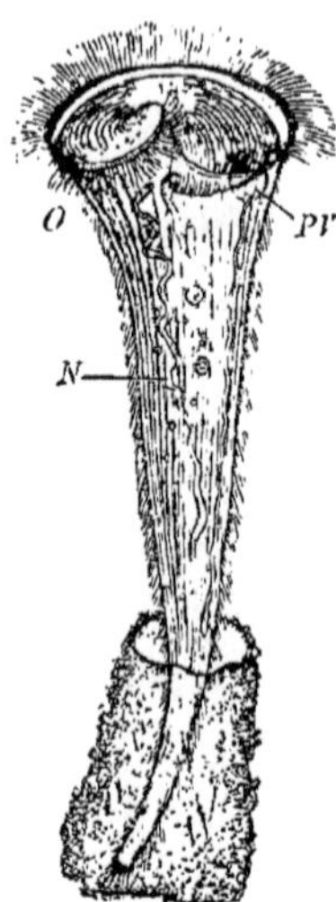

FIG. 31.—*Stentor rœselii* Ehrb., after Stein. *O* oral aperture, with gullet; *PV* contractile vacuole; *N* nucleus.

The pellicle seems to be part of the living tissue, and frequently has an alveolated structure, in which case it forms the *alveolar layer*. Beneath it there is generally a layer of cortical plasma—the ectoplasm—which has a firmer consistence than the more fluid endoplasm. The **pharynx**, or **œsophagus**, projects into the endoplasm as a tubular prolongation of the ectoplasm and pellicle. Through this the food-stuff passes into the endoplasm, in which it gives rise to food-balls. The latter undergo a slow rotating movement round the body in the endoplasm, during which the food is digested; and finally the solid useless remainder is ejected through the anal aperture. A digestive canal, bounded by distinct walls, exists no more than do the numerous stomachs which Ehrenberg, who was deceived by the food vacuoles, ascribed to his *Infusoria polygastrica*. In some cases there is a tube leading in from the anus towards the œsophagus (*Nyctotherus*), but it ends in the digestive endoplasm, and does not join the œsophagus. In some cases (the *Enchelina* and *Chlamydodonta*, Fig. 32), the œsophagus is surrounded by a layer of stiff rods forming the so-called **rod-apparatus** of the œsophagus.

The firmer, more viscid ectoplasm is to be regarded pre-eminently as the motor and sensory layer of the body. In it we may find fibrillæ resembling muscular fibres (*Stentor*, stalk of *Vorticella*). These fibrillæ are differentiations of the alveolar layer and are sometimes varicose. When they shift into the cortical layer (ectoplasm), they form the so-called *myophane* layer. Small rod-shaped bodies—the

trichocysts—are sometimes present in the cortical plasma. It appears that on suitable stimulation they have the power of everting needle-shaped structures which probably have a stunning action upon other organisms. In one form, *Epistylis umbellaria*, definite *nematocysts*, like those of the *Cœlenterata*, are present in the ectoplasm.

The **contractile vacuoles** are fixed in position and contained in the ectoplasm, generally near the surface of the body. When more than one is present they increase in number with the size of the individual. They open outward through pores in the pellicle, or in some cases (Vorticellines) into a reservoir, or there may be a long excretory canal as in *Lembadion*. The contractile vacuoles are re-formed by the fusion of formative vacuoles which appear in the neighbouring plasma during the diastole of the preceding vacuole, or there may be canals leading even from distant parts of the body, which collect the fluid for the formation of the new vacuole (*Paramecidæ*). The walls of these canals are, in some cases at any rate, contractile. These vacuoles and canal-systems are probably excretory in function, collecting waste matters in solution from the body generally and discharging it externally. They have been compared, probably with some justice, to the excretory organs of *Platyhelminthes* and *Rotifera*.

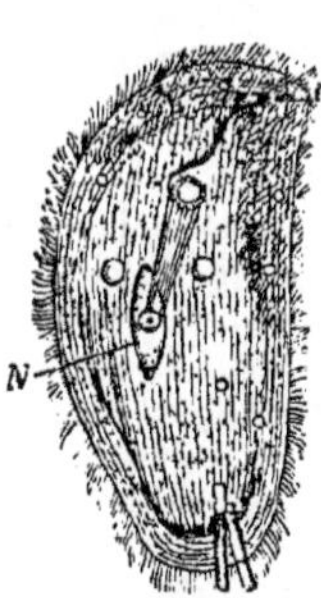

FIG. 32.—*Chilodon cucullus*, after Stein; with rod-apparatus round the œsophagus. *N* nucleus; excreta are passing out per anum.

The vibratile appendages of the body are of four kinds, and they all seem to be processes of the cuticle or alveolar layer.

(1) The fine cilia which serve for swimming.

(2) The **Cirri**: stouter vibratile processes which taper towards their free ends. They are placed on the ventral surface of the body, and serve for locomotion in a pediform manner (hence called *legs* or *styles*), or for attachment.

(3) **Membranellæ**: short flattened cilia which, when ending in a point, are not clearly distinguishable from cirri; they form the *adoral zone* of the peristome of the *Spirotricha*, in which they create the whirlpools by which the food is brought to the mouth.

(4) **Undulating membranes** placed in the neighbourhood of the mouth, and assisting in the prehension of food.

The fibre which can often be made out running from the base of the cilium, apparently to the myoplane layer, is continuous with the cuticle as well as with the cilium. In a few forms (*Actinobolus*) retractile tentacular processes are present.

In the simplest forms the mouth is at the anterior end of the body, the ciliary covering is uniform, and the cilia are nearly all the same size (*Enchelina*). An advance upon this is presented by those forms in which the mouth is on the ventral surface, and in which there is a triangular area—the *peristome*—leading back from the front end of the body to the mouth (*Paramœcium*). There may be a special row of strong cilia, cirri, or membranellæ—the *adoral row*, or zone—running along the left side of the peristome (*Hypotricha*, Fig. 30). The peristome, when ventral, is very commonly asymmetrical. In some cases the adoral row has the form of a spiral round the mouth region, and the peristome may be placed at the front end of the body (*Stentor*, *Vorticella*). In the *Hypotricha* the following regions may be distinguished on the ventral surface of the body: a pre-oral and a post-oral region. The pre-oral region consists of the peristome on the left side, and the *frontal area* on the right (Fig. 41.)

FIG. 33.—*Opalina ranarum* (after W. Engelmann).

The majority of the *Ciliata* are multinuclear, and the nuclei are almost, if not quite, always differentiated into two kinds. One of these—the *macronucleus*—is larger than the other, the *micronucleus*. The macronucleus, which lies in the endoplasm, is generally single, and presents considerable variation of form; it may be spheroidal, band-shaped, moniliform, or even branched. In some cases the macronucleus is represented by a great number of small nuclei. It is probable that these small nuclei are all parts of one nucleus, and connected with one another by fine fibres; so that the condition is really that of a much-branched nucleus. Fragmentation of the macronucleus into fine pieces has been observed, and is probably a stage in its final disappearance; but the real significance of the phenomenon is unknown. The micronucleus is often multiple; it varies in position, form, and number, in different species. It is always smaller than the macronucleus, and usually lies close to the latter. It appears only to be absent in certain *Opalinæ* and in some of the so-called multinuclear *Ciliata* just referred to. In the normal reproduction of the animal it appears that, while the micronucleus divides by karyokinesis, the macronucleus divides directly.

The most usual method of **reproduction** is by fission. When the forms reproduced remain connected together, a colony is formed, *e.g.*, the colonies of *Epistylis* and *Carchesium*. Fission usually takes place by a transverse division (at right angles to the long axis), and the products may be equal or unequal. In cases of inequality we have transitions to budding, which is really only a modified fission. Less frequently (*Vorticella*) the fission takes place through the long axis (Fig. 36), and still more rarely in a diagonal direction. The onset of fission does not appear to depend upon the size of the individual, for it may take place in large or in small specimens. The nuclei always participate in the division, but do not always lead the way; for in many cases undoubted new formations appear in the protoplasm (rudiments of new ciliary structures, of a mouth and contractile vacuole), before any changes are observed in the macro- and micronuclei. As to the other organs they do not participate in the division (unless they extend the whole length of the body), but one of the fission products develops them afresh, generally before the fission is completed. In the *Hypotricha* the ciliary structures of both the products of fission appear to be new formations.

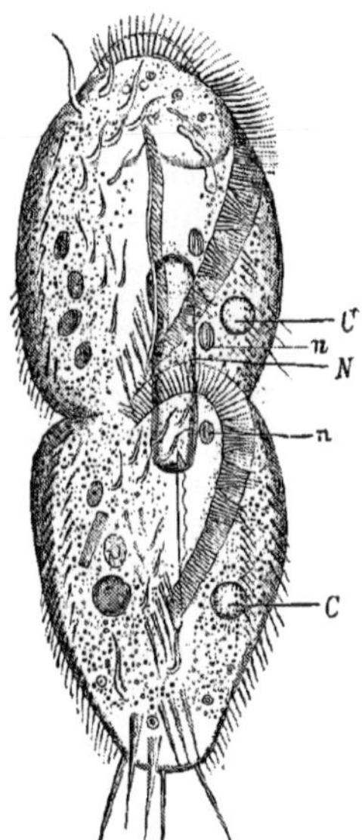

FIG. 34.—*Stylonychia mytilus* in division, from Stein. *C* contractile vacuole; *N* macronucleus; *n* micronuclei.

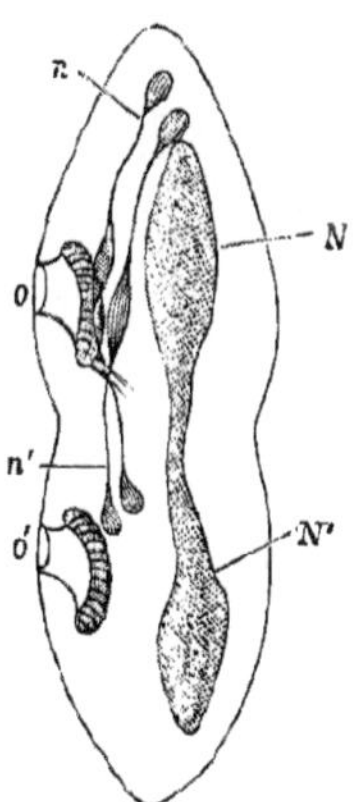

FIG. 35.—*Paramæcium aurelia* in division, after R. Hertwig. *N* macronucleus; *n* micronucleus; *o.* mouth of the anterior portion; *N'*, *n'*, *o'* the same of the hinder portion.

In *Opalina* the nucleus divides many times, so that a multinucleate condition is produced (Fig. 33). The gradual division of the whole animal by a series of binary fissions into a number of small pieces which encyst, takes place subsequently. From the cyst a small uninucleate form eventually emerges.

Reproduction is often preceded by *encystment*, which appears to be of great importance for the preservation of the body from desiccation.

The animal retracts its cilia, contracts its body to a globular mass, and then secretes a transparent cyst which hardens and protects it and enables it to survive in damp air. In water the contents of the cyst divide into a number of parts, which attain freedom by the bursting of the cyst, each one becoming a young animal.

The rapidity of fission depends upon the temperature and upon the food, and seems to be fairly constant for each species. Thus an individual of *Stylonychia pustulata*, if well supplied with food, will divide once in twenty-four hours in a temperature of 5° to 10° C., and once in twelve hours if the temperature be from 10° to 15° C. Although the rapidity of fission is less in senile individuals, there does not appear to be any special increase in it after conjugation.

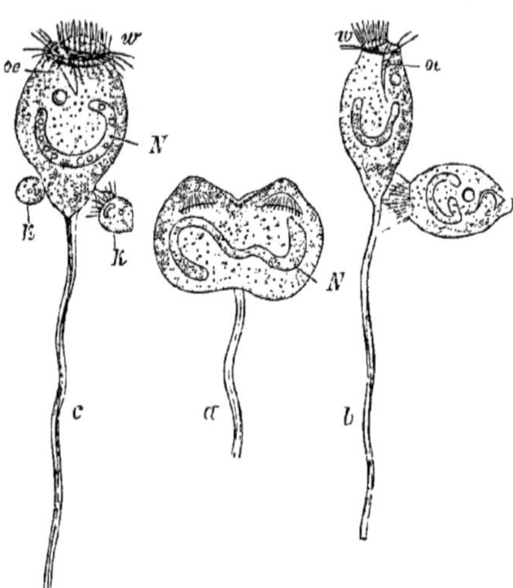

FIG. 36.—*Vorticella microstoma*, after Stein. *a*, in process of fission; *N* nucleus, the mouth apparatus in each portion is formed afresh; *b*, fission is completed, one product of it is set free after the formation of a posterior circlet of cilia; *w* adoral zone of cilia; *oe* œsophagus; *c*, Vorticella in process of bud-like conjugation; *k* the bud-like individuals (microgametes) attached.

The **Conjugation** of the *Ciliata* is generally of a temporary nature. Two individuals (rarely more) apply themselves together and acquire protoplasmic continuity. After a few hours they separate and lead their ordinary life. In the Vorticellines the conjugation is complete and permanent. As a general rule conjugation takes place between two ordinary individuals of the species, but in some cases the conjugating individuals, *i.e.*, the gametes, are specially produced by division from the ordinary individuals. Thus *Leucophrys patula* divides rapidly several times, and produces dwarfed forms uniformly ciliated and incapable of taking nourishment. These small forms are the gametes. In the *Vorticellinæ* the gametes differ; one of the conjugating individuals is the ordinary fixed form, and is called the *megagamete*, while the other is a small free-swimming form, produced by two or three successive fissions (or in some species by budding) of a fixed form into four or eight. These are the *microgametes;* they become free-swimming and conjugate with a fixed form (Fig. 36). In *Zoothamnium* the megagamete is also modified, being larger than the ordinary individual.

The results of conjugation may be considered under two heads—the physiological and the morphological. The physiological results of, and the conditions favourable to conjugation, have been mainly elucidated by Maupas. He found that it actually effected, as had long been suspected, a rejuvenescence of the conjugating individuals; that without it senile degeneration, followed by death, ensued, and that to be effective it must take place at a particular stage in the life-history. His course of procedure was as follows: he isolated an *exconjugate, i.e.*, an individual which had just been released from conjugation, and he kept it and the products of its bipartition under continuous observation.

He found that after a certain number of bipartitions the vigour and size of the descendants or products of his initial exconjugate diminished, and that their nuclear and ciliary apparatus became imperfect; and that these changes, to which he applies the term senile, eventually led to the death of the whole stock. He further found that it was not possible to induce conjugation during the period of the earlier bipartitions; this is the period of *immaturity;* but that after a certain number of bipartitions *puberty* is attained and conjugation can be induced; this is the period of *eugamy.* This stage merges into the last, the period of senescence, during which the individuals become gradually reduced in size, the ciliary apparatus more and more imperfect, the micronuclei absorbed, and conjugation ineffective. The period of senescence ends in death. In *Stylonychia pustulata* puberty occurred at the 130th bipartition, and senile degeneration began at about the 170th, and death took place at about the 316th. Conjugation in the period of eugamy conferred the power of again beginning the cycle; but the conjugation of the period of senescence had no result in retarding the degenerative changes or in averting death. Maupas further asserts that fertile conjugation between descendants of the same exconjugate cannot be effected in the period of eugamy; but this statement requires confirmation. These facts explain why it is that when conjugating individuals are met with they generally occur in large numbers (epidemics of conjugation), for the inclination to conjugation comes only at a particular stage in the life of the stock; and the divisions being rapid, there will always be a considerable number of descendants of the same exconjugate in juxtaposition.

Maupas also found that exhaustion of the food supply, which usually determines encystment, causes conjugation among individuals of mixed origin and of suitable age.

The morphological changes, our knowledge of which we owe mainly to Balbiani, Bütschli, and particularly Maupas, appear to consist of the breaking-up and disappearance of the macronucleus, the particles of which—in some cases, at any rate—are excreted

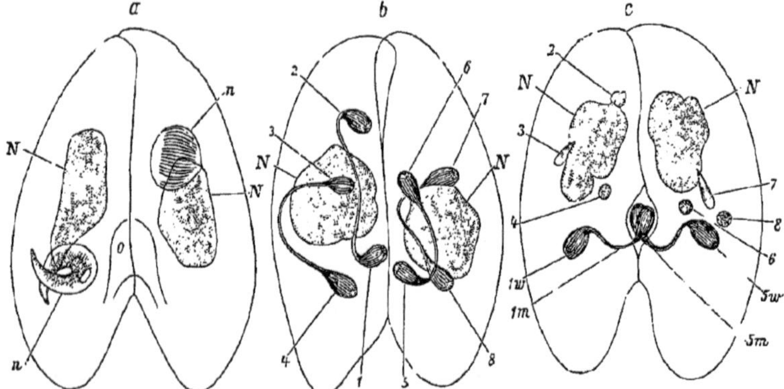

Fig. 37.—Conjugation of *Paramæcium caudatum* after R. Hertwig. *N* macronucleus; *n* micronucleus; *o* mouth. In *a* the micronucleus is forming a spindle; *b* illustrates the second division of the micronucleus, resulting in the formation of four micronuclei in each animal; of these, those marked 2, 3, 4, 6, 7, 8 will atrophy, while 1 and 5 will persist and divide; *c*, the micronuclei 2, 3, 4, 6, 7, 8 are disappearing, while 1 and 5 are dividing into two, one of which, *1m* and *5m*, migrate respectively through the protoplasmic connection into the other conjugating individual, while the other, *1w* and *5w*, remains in its own individual.

through the anus; and in the increase by successive bipartitions of the micronucleus. Of the micronuclei so produced, all abort except two, one of which is migratory and the other stationary; the migratory nucleus passes into the other gamete, and fuses with its stationary nucleus to form the *zygote-nucleus.* The two gametes, each with a zygote-nucleus, now separate, and the zygote-nucleus gives rise by successive bipartitions to the nuclear apparatus of the exconjugate. The details differ in different species. As an example of the process we may take the simple case of *Colpidium colpoda*, which possesses one macro- and one micronucleus. Soon after the onset of conjugation, which lasts some hours, the micronucleus of each gamete undergoes two successive bipartitions, and so gives rise to four micronuclei; of these three abort, and one divides again to form the two *pronuclei*, as we may call them. One of these migrates into the other gamete, and fuses with its stationary pronucleus to form the zygote-nucleus. This undergoes two bipartitions, during which the gametes separate and the old macronucleus disappears. The exconjugates now possess four nuclei, all derived from the zygote-nucleus. Of these, two become micronuclei and two macronuclei of the first bipartition, which takes place at from four to nine days after the separation. So that the exconjugate has for some time double the ordinary nuclear apparatus, and the first bipartition takes place quite independently of the nuclear division. For some days after conjugation the exconjugates take no food, and appear to be without a mouth and gullet. They acquire the latter and begin to feed about twenty-four hours before the first bipartition.

The mode of life of the *Ciliata* is very various. Most of them lead an independent life; some are carnivorous and others herbivorous. The former are very rapacious, and may take up even *Rotifera*. Some, as *Amphileptus*, select fixed Infusorians, as *Carchesium* and *Epistylis*, for their prey, and swallow them down as far as the origin of their stalk. They then, while fixed on the stalk, secrete a capsule, and divide into two or more individuals. Some, as the mouthless *Opalina* and many *Bursaridæ*, are parasitic in the intestines and bladder of Vertebrata. To these belong the *Balantidium coli*, from the large intestine of man. (Fig. 40.)

Order 1. Gymnostomata.

The mouth is usually closed except during inception of food, and is without undulating membranes. The pharynx when distinctly developed is without ciliary structures, but is usually provided with a rod-apparatus or a modification of one. The food is always swallowed, never taken in by a whirlpool. Ciliation usually holotrichous, but often more or less reduced.

Fam. 1. **Enchelina.** The mouth is terminal or sub-terminal, is usually round, rarely slit-like. Anus usually terminal. Conjugation terminal.

Sub-fam. 1. **Holophryina.** Without tentacular structures. *Holophrya* Ehrb., m. and f.w.*; *Urotricha* Clap. and L. with caudal bristle, f.w.; *Enchelys* Hill, m. and f.w.; *Spathidium* Duj., f.w.; *Chænia* Quennerstedt, m.; *Prorodon* Ehrb.; *Dinophrya* Bütschli, f.w.; *Lacrymaria* Ehrb.

Sub-fam. 2. **Actinobolina.** With retractile tentacular organs and cilia. *Actinobolus* Stein, f.w.

* m., f.w., and infus. are abbreviations for marine, fresh-water, and infusions.

Sub-fam. 3. **Colepina.** Mouth terminal, surrounded by a circle of cirri. Body sometimes surrounded by armour composed of pieces of the pellicle arranged in an annular manner. *Plagiopogon* Stein, f.w.; *Coleps* Nitzsch. f.w.; *Tiarina* Berg, m.; *Stephanopogon* Entz, m.

Sub-fam. 4. **Cyclodinina.** Cilia confined to one or several rings. Mouth on a papilliform projection. *Didinium* Stein, f.w.; *Mesodinium* Stein, a large oral papilla with one or more cirri arising from its base, m. and f.w.

Sub-fam. 5. **Prorotrichina.** Bluntly truncated oral end; ciliation confined to front end, or there are in addition incomplete rings of cilia. *Butschlia* Schuberg, rumen of Ruminants.

Fam. 2. **Trachelina.** Mouth is either a long slit which extends from the front end along the ventral surface backwards, or its hinder part is alone developed as a short slit-like or roundish opening. The oral end of the body usually tapers in a proboscis-like manner. Pharynx short or absent.

Sub-fam. 1. **Amphileptinæ.** Mouth on the convex ventral edge of the dorsalwards-bent proboscis, sometimes as long slit, sometimes as round opening. *Amphileptus* Ehrb., m. and f.w.; *Lionotus* Wrzesniowski, mouth along whole ventral edge of long proboscis, f.w. and m.; *Loxophyllum* Duj., f.w. and m.; *Trachelius* Schrank, mouth as round opening at base of proboscis, endoplasm vacuolated, f.w.; *Dileptus* Duj., f.w. and m.

Sub-fam. 2. **Loxodina.** Proboscis bent ventrally, and mouth on its concave edge. Ciliation confined to the right side. *Loxodes* Ehrb., large size, endoplasm vacuolated, on dorsal side a row of excretion vacuoles, each with a dark body, f.w.

Fam. 3. **Chlamydodonta.** Body never elongated; mouth always at a distance from the front end. Pharynx always with well-developed rod-apparatus, or a smooth, sometimes peculiarly-formed œsophageal tube.

Sub-fam. 1. **Nassulina.** Ciliation complete; *Nassula* Ehrb., m. and f.w.

Sub-fam. 2. **Chilodontina.** Ciliation confined to, or stronger on the ventral surface than on the back. *Orthodon* Gruber, m.; *Chilodon* Ehrb. (Fig. 32), f.w. and infus., m.; *Chlamydodon* Ehrb., m.; *Opisthodon* Stein, mouth far back, f.w.; *Phascolodon* Stein; *Scaphidiodon* Stein, m.

Sub-fam. 3. **Erviliina.** Ciliation confined to ventral surface, or a small field of it. Caudal end with a well-developed movable style usually arising a little ventrally of hind end. *Aegyria* Clap. and L., m.; *Onychodactylus* Entz, m.; *Trochilia* Stein, f.w. and m.; *Dysteria* Huxley, f.w. and m.

Order 2. Trichostomata.

Mouth as a rule always open, rarely closed when not in use; pharynx always tubular and open; edges of mouth provided with undulating membranes which are continued into the pharynx, or the latter is provided with cilia. Food rarely swallowed, usually brought by a whirlpool or by special ciliary structures.

Sub-order 1. ASPIROTRICHA.

The mouth in the most primitive forms extends as a slit from the front end along the ventral surface; but is usually removed from the front end as a reniform or crescent-shaped opening. Pharynx, when present, without rod-apparatus. At the edges of the mouth or in the pharynx are one or two undulating membranes.

Fam. 1. **Chilifera.** Mouth in the anterior half of the body, or close behind the middle. Pharynx either scarcely developed or short. The undulating membranes stand either at the edges of the mouth or in the pharynx. A so-called peristomial field leading to the mouth absent or little developed.

Leucophrys Ehrb., f.w.; *Glaucoma* Ehrb., f.w. and infus.; *Dallasia* Stokes, f.w.; *Frontonia* Clap. and L., f.w. and m.; *Ophryoglena* Ehrb., f.w.; *Colpidium* Stein, f.w., infus., m.; *Chasmatostoma* Engelmann, f.w.; *Uronema* Duj., f.w., m., par. on skin of star-fishes; *Urozona* Schewiakoff, f.w.; *Loxocephalus* Kent, f.w. and infus.; *Colpoda* Müller, f.w. and infus. (hay).

Fam. 2. **Microthoracina.** Asymmetrical; mouth in hinder part of body, placed somewhat laterally at the anterior end of a peristomial furrow, which begins behind; ciliation sometimes complete, sometimes confined to ventral surface, always sparse. *Cinetochilum* Perty, f.w., m.; *Microthorax* Engelmann, f.w.; *Ptychostomum* Stein, paras. in intestine of Oligochætes; *Ancistrum* Maupas, in mantle cavity of *Mytilus*, *Venus*, and probably *Ostrea*; *Drepanomonas* Fresenius, f.w.

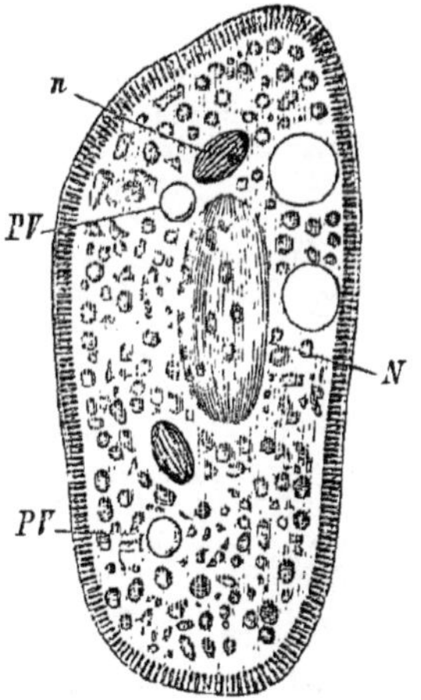

FIG. 38.—*Paramæcium bursaria* about one hour after conjugation (after Bütschli). *n* micronucleus; *N* macronucleus; *PV* contractile vacuole.

Fam. 3. **Paramæcina.** Mouth sometimes in the anterior, sometimes in the hinder half of the body, with considerable triangular shallow peristomial pit passing to it from the left side of the body. Pharynx tubular, with long undulating membranes or row of cilia. Ciliation close and uniform. *Paramæcium* Hill (Fig. 37–39), f.w. and m.

Fam. 4. **Urocentrina.** Mouth in middle of ventral surface with long tubular pharynx, like that of *Paramæcina*. Ciliation reduced to two broad annular zones, one in front and one behind. *Urocentrum* Nitzsch, f.w. and m.

Fam. 5. **Pleuronemina.** Ciliation complete and usually considerable; mouth at the end of a peristome placed at varying distances from the front end on the ventral surface. The left edge of the peristome with considerable undulating membrane; the right edge with a weaker membrane, or with a row of closely-placed cilia. Pharynx slightly developed or absent. *Lembadion* Perty, f.w.; *Pleuronema* Duj., f.w. and m.; *Calyptotricha* Phillips; *Lembus* Cohn, f.w., infus., m.

Fam. 6. **Isotrichina.** Pellicle thick; ciliation total and close. Mouth posterior. Parasitic in rumen of Ruminants. *Isotricha* Stein; *Dasytricha* Schuberg.

Fam. 7. **Opalinina.** Ciliation complete and almost always uniform; mouth and pharynx absent. *Anoplophrya* intest. of Oligochætes, Polychætes, Clepsine, Paludina: *Hoplitophrya* Stein, with two hook-like structures, Oligochætes and Planarians; *Discophrya* Stein, with anterior sucker, Planarians and Anura; *Opalinopsis* Foettinger, macronucleus in young forms elongated and twisted, later an irregular, even branched and anastomosing mass breaking up into pieces; venous appendages of *Sepia*, *Octopus*, or in liver of *Sepiola* and *Octopus*:

Opalina Park. and Val. (Fig. 33), in young state a roundish nucleus. Micronucleus not observed; no contractile vacuole; rectum of *Anura*.

Sub-order 2. SPIROTRICHA.

Always with distinct adoral zone, which usually consists of membranellæ, and has a more or less spiral course round a peristomial area. The latter is distinguished by other peculiarities from the rest of the body-surface.

Section 1. Heterotricha.

With well-developed adoral zone or spiral, and a complete ciliary covering (except *Gyrocorys*).

Fam. 1. **Plagiotomina.** Peristome as a narrow furrow, which usually begins close to the front end and passes along the ventral surface to the mouth, which is either at the middle of the body or at the hind end. The adoral zone stretches from the mouth along the left side of the peristome, and is usually straight. Pharynx tubular. *Conchopthirus* Stein, ma.n.* single or multiple, f.w., ectopar. in slime of different land and f.w. molluscs, in gastral cavity of Actiniæ; *Plagiotoma* Duj., paras.; *Nyctotherus* Leidy, with anal tube, paras. in intestine of Anura, Insects, and Myriapods; *Blepharisma* Perty, f.w.; *Metopus* Clap. and L., f.w. and m.; *Spirostomum* Ehrb., ma.n. single, mi. n. numerous, f.w. and m.

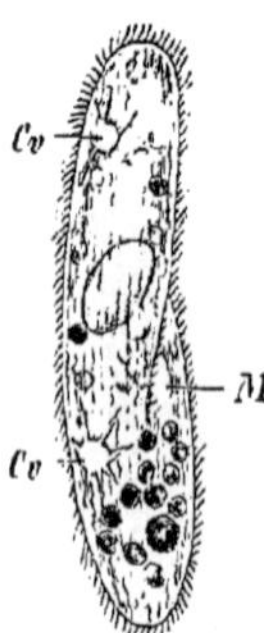

FIG. 39.—*Paramæcium aurelia*, after Ehrenberg. *M* mouth; *Cv* contractile vacuoles with canals.

Fam. 2. **Bursarina.** Peristome as a more or less triangular (apex oralwards) field, and not a furrow, as in *Plagiotomina*. Pharynx absent or but slightly developed. The adoral row extends along the left peristome-edge only, or crosses over in front as far as the right anterior angle of the peristome. *Balantidium* Cl. and L., paras. rectum of man, pig, Amphibia, body cavity of Polychætes; *Balantidiopsis* Bütschli, intestine of Rana; *Condylostoma* Duj., mi. n. numerous, f.w. and m.; *Bursaria*, O. F. Müller, large size, ma. n. long, mi. n. numerous, f.w.

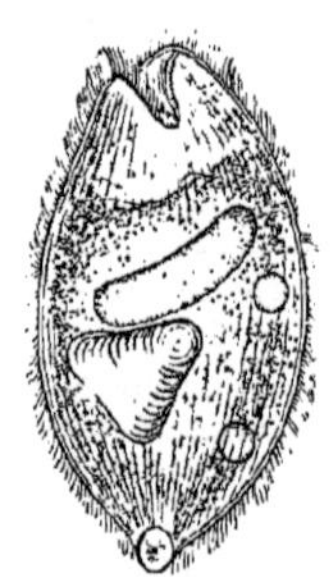

FIG. 40.—*Balantidium coli*, with two contractile vacuoles after Stein. Near the nucleus lies a starch grain which has been eaten; a ball of excrement is passing out per anum at the hind end.

Fam. 3. **Stentorina.** Body elongated. Peristome short and at front end. Its two edges sometimes prolonged into wings. The adoral spiral passes either across the front end of the peristome to the right corner of the same, or completely surrounds the peristomial area. Undulating membrane absent. The peristomial surface is ciliated and spirally striped. Pharynx tubular. *Climacostomum* Stein, f.w.; *Stentor* Oken (Fig. 31), ma.n. long. mi.n. numerous, sometimes fixed and with gelatinous tube, f.w.; *Folliculina* Lam. with peristomial wings, usually inhabits chitinous tubes, f.w. and m.

Fam. 4. **Gyrocoryna.** Bell-shaped, anterior end rounded, posterior end as caudal appendage projecting from the bell; a ventral furrow with cilia, a row of

* ma.n., mi.n. are abbreviations of macro- and micronucleus respectively; f.w. and m. similarly standing for fresh-water and marine.

cilia at edge of bell leading to mouth at base of appendage. *Cœnomorpha* Perty, f.w. and m.

Section 2. **Oligotricha.**

Never elongated, usually spherical or conical. Peristomial field at front end and at right angles to the long axis. Adoral row nearly or completely a closed circle. Ciliation of body partly well developed, partly much reduced.

Fam. 1. **Lieberkühnina,** possibly a young form of Stentor.

Fam. 2. **Halterina.** Peristomial surface without cilia. Body with few scattered or no cilia; sometimes with scattered immovable setæ. No shell. *Strombidium* Cl. and L., f.w. and m.; *Halteria* Duj., f.w.

Fam. 3. **Tintinnoina.** Provided with a tubular shell, to the base of which the body is fastened by a stalk. Adoral row as a circle of large membranellæ, inside which is a row of fine cilia (paroral). *Tintinnidium* Kent, f.w. and m.; *Tintinnus* Schrank, m.; *Tintinnopsis* Stein, m.; *Codonella* Häck., f.w. and m.; *Dictyocysta* Ehrb., m.

Fam. 4. **Ophryoscolecina,** with thick pellicle, hinder end often with spine-like processes, deep funnel-shaped peristomial region. Anus terminal, usually with anal tube. Parasitic in rumen of Ruminants. *Entodinium* Stein; *Diplodinium* Schuberg; *Ophryoscolex* Stein.

Section 3. **Hypotricha.**

Body dorso-ventrally flattened, ventral surface usually flat, dorsal convex; peristomial field usually triangular and in same plane as rest of ventral surface. Dorsal surface without cilia, but with stiff bristles. The ventral cilia uniform or in various ways reduced and differentiated. Pharynx little developed or absent.

Fam. 1. **Peritromina.** Peristome but little marked off from frontal area. Ciliation of ventral surface close and uniform without differentiation of stronger cilia or cirri. *Peritromus* Stein, m. and f.w.

Fam. 2. **Oxytrichina.** Peristome distinctly marked off from frontal area. Ventral ciliation in the most primitive forms uniform in oblique longitudinal rows; but some stronger cilia are almost always present on the frontal area (frontal cirri) and at the hind end (anal cirri). Usually the ventral cilia are cirri. A right and left row are distinguished as *marginal cirri* from the imperfect median rows which are called *ventral cirri* (Fig. 41). These rows of cirri must not be confused with the adoral row of membranellæ (*a.*) on the left side of the peristome. There is an undulating membrane on the right side of the peristomial area (the preoral membrane, Fig. 41 *m.*), and in many forms a row of cilia on the right side of the adoral row (the *paroral* row, Fig. 41 *b*); both these structures are, like the adoral row, continued into the pharynx. *Trichogaster* Sterki, f.w.; *Urostyla* Ehrb., f.w. and m.; *Kerona* Ehrb., commensal on Hydra; *Epiclintes* Stein, m.; *Stichotricha* Perty, f.w. and m.; *Strongylidium* Sterki, f.w.; *Holosticha* Wrzesn., m.; *Amphisia* Sterki, f.w. and m.; *Uroleptus* Ehrb., f.w. and m.; *Onychodromus* Stein (Fig. 41), f.w.; *Pleurotricha* Stein, f.w.; *Gastrostyla* Engelm., f.w.; *Gonostomum* Sterki, f.w. and m.; *Urosoma* Kowalewsky, f.w.; *Oxytricha* Ehrb., f.w. and m.; *Stylonychia* Ehrb. (Fig. 30), f.w. and m.; *Actinotricha* Cohn, m.; *Balladina* Kow., f.w.; *Psilotricha* Stein, f.w.

Fam. 3. **Euplotina.** Ciliation much reduced; the anal cirri are always present, but the marginal frontal and ventral cirri may be absent; encuirassed. *Euplotes* Ehrb., f.w. and m.; *Diophrys* Duj., m.; *Uronychia* Stein, m.; *Aspidisca* Ehrb. (Fig. 42), f.w. and m.

Section 4. **Peritricha.**

Cilia confined to an adoral spiral and a posterior circlet which is not always present. The adoral spiral with the peristomial area is placed at the front end of the body (except in one family). The mouth, anus, and reservoir of the contractile vacuole all open into a depression of the peristome called the vestibule. The peristome and adoral cilia are generally surrounded by a projecting lip-like ridge.

Fam. 1. **Spirochonina.** With a peculiar peristomial funnel rolled into a spiral at the front end. Reproduction by budding near the peristome. Attached by an adhesive disc at the hind end. Total conjugation between small animals with undeveloped peristome. No posterior circlet of cilia. Parasitic on the legs of *Gammarus*, *Limnoria*, *Nebalia*. *Spirochona* Stein.

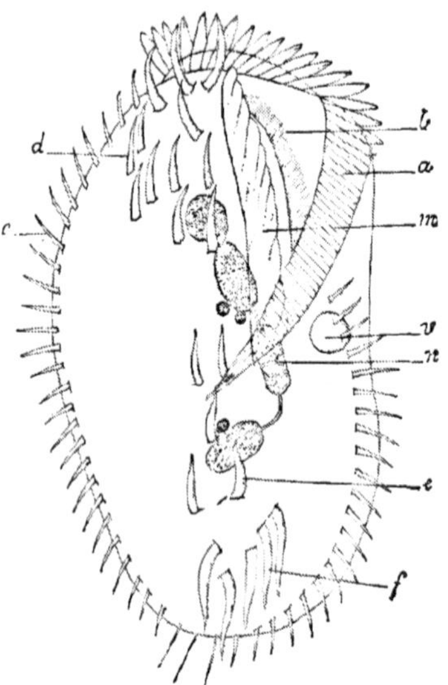

Fig. 41.—Ciliary apparatus of a Hypotrich *Onychodromus grandis* (from Perrier after Maupas). *a* adoral membranellæ; *b* paroral cilia; *c* marginal cirri; *d* frontal cirri; *e* ventral cirri; *f* anal cirri; *m* preoral undulating membrane; *n* nucleus; *v* contractile vacuole.

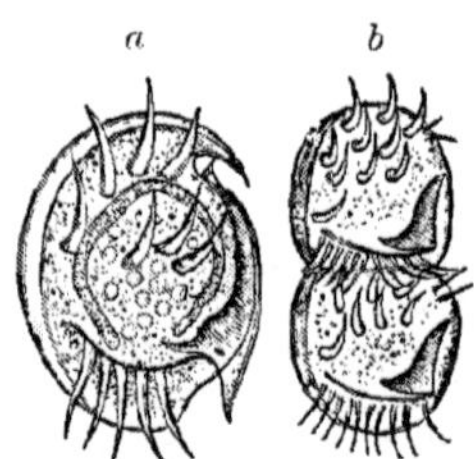

Fig. 42.—*a*, *Aspidisca lyncaster*, and *b*, *Aspidisca polystyla* during fission (after Stein).

Fam. 2. **Lichnophorina.** Peristome and adoral spiral ventral. Hind end as a sucker for attachment, and surrounded by the posterior circlet. Ectoparasitic on the skin of Medusæ, Opisthobranchs, worms, and Asteroids. *Lichnophora* Clap.

Fam. 3. **Urceolarina.** Free-swimming, with posterior circlet of cilia, which encloses an adhesive disc; without peristomial lip. *Trichodina* Ehrb., ectoparasitic on the skin of fresh-water and marine animals, *e.g.*, Hydra, Planarians, etc.—also endoparasitic in bladder of fishes and Amphibia. *Cyclochæta* Jackson, on the surface of sponges, gills of *Scorpæna* and *Trigla*, etc.; *Trichodinopsis* Clap. and L., in gut and mantle cavity of *Cyclostoma*.

Fam. 4. **Vorticellina.** For the most part attached; without permanent posterior circlet; with peristomial lip, which can be closed over the peristome in a sphincter-like fashion; a large undulating membrane continued into the vestibule. In addition to the adoral zone there is an inner circle of cilia corresponding to the paroral row of other types; it extends into the vestibule.

Scyphidia Lachmann, ectopar.; *Gerda* Clap. and L., f.w.; *Astylozoon* Engelmann, f.w.; *Vorticella* L., with long contractile fibre for attachment, f.w. and m.; *Carchesium* Ehrb., colonial, f.w.; *Zoothamnium* Ehrb., colonial, f.w. and m.; *Glossatella* Bütschli, attached, but without stalk; *Epistylis* Ehrb., colonial, and *Rhabdostyla* Kent, solitary, stalk without contractile fibre; *Opercularia* Goldf.; *Ophrydium* Bory; *Cothurnia* Ehrb.; *Vaginicola* Lam.; *Lagenophrys* Stein, with lorica.

Forms of uncertain position:

Multicilia Cienk., covered by long flagella-like cilia, m. and f.w.; *Grassia* Fisch, covered with long cilia, parasitic in stomach of frog and in blood of *Hyla viridis*; *Magosphæra* Haeckel, free-swimming ciliated forms occurring in spherical colonies.

SUB-CLASS III. ACINETARIA.

Infusoria with knobbed tentacle-like processes which serve as sucking tubes. Ciliated in the young state.

These animals are always sedentary in habit, and either free or attached; when the latter they may be sessile or stalked. They prey upon the living tissues of other organisms by means of their tentacles. The latter are processes of the cortical protoplasm, and are of two main kinds, connecting which there are intermediate forms: (1) the so-called *prehensile* tentacles which taper distally, although they do not end in a sharp point, and (2) the *suctorial*, which are cylindrical in shape and rounded at the end, which may even be swollen into a distinct knob.

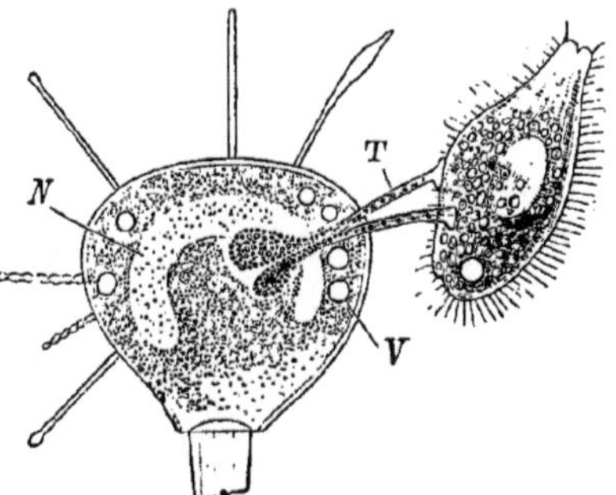

FIG. 43.—*Acineta ferrumequinum* Ehrb., sucking the body of a small infusorian (*Enchelys*), after Lachmann. *T* suctorial tentacle; *V* vacuole; *N* nucleus.

The tentacles of both kinds appear to contain a canal which opens distally to the exterior, and leads at the other end into the central protoplasm of the body. The fluid or semi-fluid contents of their prey pass down these canals in a current, the cause of which is not quite understood. Maupas has suggested that the transparent ectoplasm of the Acinetan first passes in an invisible current by the tentacle into the body of the prey, there absorbs the protoplasm, and then returns with its burden to its own body in a current which can be traced by the granule contents of the protoplasm. All tentacles are retractile, and the prehensile do not seem to differ materially in function from the suctorial.* In their retraction they often become marked by a peculiar spiral wrinkling of their surfaces—possibly due to torsion.

Mouth (other than the tentacular pores) and anus are not present. Contractile vacuoles are present and vary in number. The macronucleus may be elongated or branched, attaining a great extension and complexity in such a form as *Dendrosoma*, which looks like a colony of Acinetans attached to a creeping stolon. A micronucleus is certainly present in some forms, and probably in all. The body has a pellicle which is in some cases thickened to form a shell or theca. The stalk of attachment, when present, is not contractile.

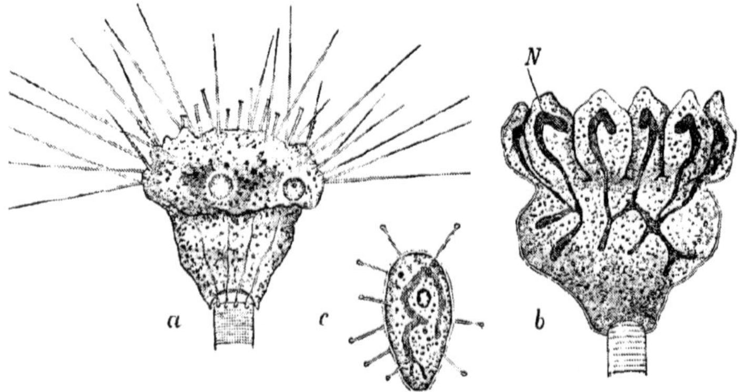

FIG. 44.—*Ephelota* (*Podophrya*) *gemmipara*, after R. Hertwig. *a*, with extended tentacles (both prehensile and suctorial) and two contractile vacuoles; *b*, the same, with ripe buds into which processes of the branched nucleus *N* enter; *c*, free young form.

Reproduction takes place in one of three ways:

1. Equal transverse fission; as a result of this the organism divides into two equal pieces, one of which—the distal—retracts its tentacles, acquires cilia, and swims away, while the other, or basal piece, keeps the old attachment and condition. Such reproduction, in which the products are more or less equal in size, is found in *Hypocoma*, *Sphærophrya*, *Podophrya*, *Urnula*, etc.

2. Simple to multiple budding; this is characteristic of the genus *Ephelota* (Fig. 44). One or more buds, each containing a process of

* The prehensile tentacles when present first seize the prey, bring it within reach of the knobbed tentacles; but it is by no means clear that they do not share in the suction act.

the nucleus, are formed, and eventually nipped off as free-swimming ciliated forms.

3. Internal budding is the most common method of increase. It takes place in *Tokophrya quadripartita* as follows: a funnel-shaped invagination of the apical surface is formed, the opening of which narrows to a pore. The bud arises from the base of the pit or brood-pouch so formed, as a projection containing a part of the nucleus. Eventually the bud becomes constricted off so as to lie freely in the brood-pouch, acquires some cilia, and escapes through the opening as the free-swimming young form. In *Dendrocometes* the bud is evaginated through the pore before separation, so that it forms a projection of the body of the mother. The young are always ciliated, and the ciliation may be holotrichous, hypotrichous, or peritrichous. It occasionally happens (*Podophrya fixa*, etc.) that a whole individual retracts its tentacles, acquires cilia, and breaks away as a free-swimming swarmer.

Conjugation of the temporary kind has been observed, and is probably a general phenomenon in the group. The changes of the macro- and micronucleus accompanying it seem to be of the same nature as in the *Ciliata*.

Encystment is of common occurrence, but its relation to other vital phenomena is not known. The cyst wall often possesses annular thickenings.

Fam. 1. **Hypocomina**. Freely movable, not attached; with permanently ciliated ventral surface and one suctorial tentacle. *Hypocoma* Gruber, m., ectopar. on *Zoothamnium*.

Fam. 2. **Urnulina**. With one or two (rarely more) tentacles not distinctly knobbed. *Rhyncheta* Zenker, freely motile, without theca, on ventral side of Cyclops; *Urnula* Clap. and L., attached and with theca, on stalk of Epistylis.

Fam. 3. **Metacinetina**. With stalked funnel-shaped theca, the walls of which are perforated by slits for the exit of the knobbed tentacles. *Metacineta* Bütschli, f.w.

Fam. 4. **Podophryina**. Tentacles numerous and usually considerable, on the whole surface or only apical, either all distinctly knobbed, or some of them without knobs serving as prehensile tentacles. *Spærophrya* Clap. and L., without stalk, endoparasitic in *Ciliata; Endosphæra* Engelm., *Podophrya* Ehrb., stalked, with knobbed similar tentacles from all parts of body; *Ephelota* Wright (Fig. 44), tentacles both knobbed and pointed, chiefly from free end, m., on Hydroids, Polyzoa, and Crustacea; *Podocyathus* Kent, like the last, but with theca.

Fam. 5. **Acinetina**. Stalked, or with stalked or unstalked theca with simple opening. Tentacles numerous, all alike, and usually distinctly knobbed. *Tokophrya* Bütschli; *Acineta* Ehrb. (Fig. 43), stalk continued as theca; *Solenophrya* Clap. and L., theca sessile.

Fam. 6. **Dendrosomina**. Without stalk or theca. Tentacles numerous, all alike and knobbed, arranged in tufts, which may be numerous and placed at the ends of branch-like lobes. *Trichophrya* Clap. and L., *Dendrosoma* Ehrb., large animal resembling a colony, with branched macronucleus extending throughout the body.

Fam. 7. **Dendrocometina.** Sessile, with numerous knobbed tentacles on branched arms or over the whole free surface. Attached by the whole basal surface or by a part of it. *Dendrocometes* Stein, gills of *Gammarus pulex; Stylocometes* Stein, on gill-plates of *Asellus aquaticus.*

Fam. 8. **Ophryodendrina.** With short or long stalk, tentacles rarely distinctly knobbed, and borne by one to several proboscis-like processes of the body. *Ophryodendron* Clap. and L.

Class III. SPOROZOA.*

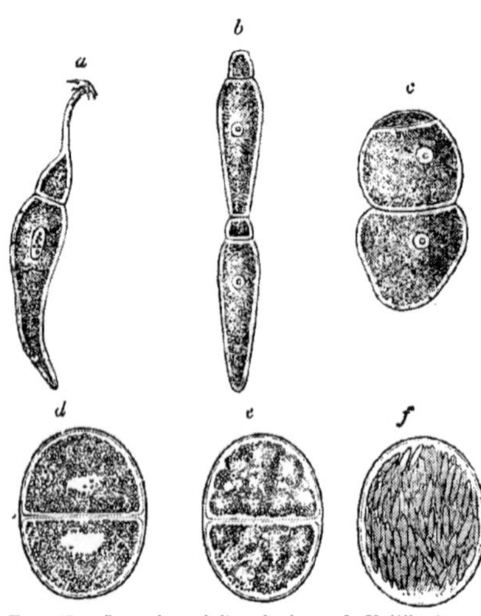

FIG. 45.—*Gregarines* (after Stein and Kölliker). *a*, *Stylorhynchus oligacanthus* from the intestine of *Calopteryx; b*, *Gregarina* (Clepsidrina) *polymorpha*, from the intestine of the meal-beetle, two forms in "association"; *c*, two forms of the same in conjugation; *d*, encystment completed; *e*, sporulation; *f*, cyst with completely formed spores (pseudonavicellæ).

Parasitic Protozoa which reproduce by spore-formation. Nutriment is taken up in the liquid state. In most, probably in all, the young stages at least are intracellular in habit.

The *Sporozoa* are found in all the great groups of animals except the *Protozoa* and *Cœlenterata.* In the young state at least they live embedded in the protoplasm of their host, into which they make their way when hatched out from the minute spores. They are therefore described as intracellular parasites. In some forms they remain within the cell throughout life, but more often they outgrow the cell and come to lie free in the tissues or spaces of the body. They live entirely on the nutritive juices of their host, and their power of movement is limited. Some have little or no power of

* Balbiani, *Lecons sur les Sporozoaires*, Paris, 1884. Bütschli, "Sporozoa," in Bronn's *Klassen u. Ord. d. Thierreiches*, 1880-82. L. Pfeiffer, *Die Protozoen als Krankheitserreger*, ed. 2, Jena, 1891, and *Nachträge*, Jena, 1895. Idem, *Die Zellerkrankungen etc. durch Sporozoen*, Jena, 1893. V. Wasielewski, *Sporozoenkunde*, Jena, 1896.

changing their form, while others are amœboid. They vary much in size, some not exceeding ten micromillimetres in length, while others may attain a length of sixteen mm. The protoplasm usually exhibits a differentiation into ectoplasm and endoplasm, and the endoplasm is often highly granular—a feature which becomes more marked with the age of the individual.

Conjugation occurs in some groups (*Gregarinida*, *Drepanidiidia*, etc.), but has not been observed in all.

Reproduction is effected by the division of the protoplasm into spores, which may be coated or naked. In most cases encystment precedes sporulation, but in some of them the spores are produced gradually during the ordinary life of the individual. In the *Gregarinida* and *Coccidiidea* the spore-protoplasm divides to form the young forms. In others the whole spore becomes the young form. There is often a little residual protoplasm—generally non-nucleated—left over after the sporulation. The young which issue from the spores are either falciform or amœboid.

In *Gregarinida* and *Coccidiidea* the sporulation usually takes place after the cyst has left the host (*exogeny*), but in the other forms it is effected within the host (*endogeny*).

A process which may or may not be analogous to the formation of polar bodies of the metazoan ovum (or speaking more generally, to the reduction-divisions of the progametes) has been observed by Wolters in Gregarines (*Arch. Mic. Anat.* Bd. 37) and by Labbé (*loc. cit.*) in *Coccidia*. The nucleus divides; one half remains in the animal, while the other passes to the surface and disappears. This phenomenon precedes the fusion of the nuclei in the conjugating gregarine, and sporulation in the *Coccidia*.

In *Gregarinida* and *Coccidiidea* the cysts pass out with the fæces and enter another host in its food. In the other orders the method of transference from host to host is not certainly known, but probably in some cases the spores are not able to leave the host until its death, after which they may enter a new host in the food; while in other cases it is possible that the infection is transmitted by blood-sucking insects, or through the lungs in dust.

It is possible that there may be—in the exogenous forms at least—some other mode of reproduction besides that of spore-formation. Very little is known on this head; but unless there is some other reproductive process, it is difficult to see how in exogenous forms, such as *Coccidium oviforme* of the rabbit, the enormous number of individuals which characterise acute coccidiosis is produced. It is also probable that in some cases, *e.g.*, the forms which live in blood, there is an intermediate host, or that the spores have the power of developing and living outside the body.

As a general rule they do not inflict serious damage on their hosts; but in some cases they are very injurious, and may cause death. This is sometimes due to the destruction of large tracts of cells or of great numbers of blood corpuscles. Whether in such cases the injury is due to any other cause than merely eating out the cell, such as the production of an injurious substance as the result of their vital activity, is not known. In some endogenous forms, *e.g.*, *Myxosporidia*, in which the spores cannot escape from the host, extensive tumours may be formed.

In the classification of the class adopted in this work, the blood-parasites have been united in the order *Hæmosporidia*, and the *Coccidiidea* have been separated from the *Gregarinida*. It is very probable that these three orders, which are more closely allied to each other than to the other two orders, should be united in one group.

Order 1. Gregarinida.*

Parasitic Protozoa which are embedded during the whole or a part of their lives in the protoplasm of their hosts. They are without mouth or anus, and they usually reproduce by coated spores. Cilia and pseudopodia are absent.

The *Gregarinida* live as parasites in the alimentary canal, and in the tissues of most animals. They are not found in Protozoa, Cœlenterates, or Vertebrata. In the young state they lie entirely within the protoplasm of a cell of their host, usually an epithelial cell of the intestine; but as they grow older and increase in size they project from the cell, to which they remain attached for a time. Eventually they become free, and lie in the cavity of the intestine or other organ of their host. The body is generally elongated in a vermiform manner, and consists of a granular semi-fluid endoplasm containing a nucleus, a thin external layer of clear ectoplasm, and a thick external cuticle. Hooks for attachment, and hair-like processes may be present as modifications of the cuticle. The structure of the body may be complicated by the presence of a partition wall dividing the endoplasm into an anterior portion called the *protomerite*, and a posterior—the *deutomerite*. The partition consists of a prolongation of the ectoplasm, and the

* Aimé Schneider, "Contributions à l'histoire des Gregarines des Invertebrés de Paris et de Roscoff." *Arch. de Zool. expér. et gén.*, tom. iv., 1875; and in various succeeding volumes of the same Journal. O. Bütschli, "Protozoa," in Bronn's *Klassen und Ordnungen d. Thierreichs*, 1880-2.

nucleus lies in the deutomerite. In such forms there is generally a small third division of the body in front of the protomerite called the *epimerite.* It contains a small quantity of endoplasm, and is the part of the body by which attachment to the cell-host is effected. When the animal loses this attachment and lies free in the alimentary canal the epimerite disappears.

The epimerite varies much in form, and bears appendages of very different kinds (hooks, spikes, filaments, etc.). It is usually found only in septate forms (with a proto- and deutomerite), but occasionally it is present on the front end of a monocystid (non-septate) form.

Occasionally after detachment from their cell-host they penetrate the gut-wall, and form cysts which project into the body-cavity — these are known as the *body-cavity forms;* but as a rule they lie freely in the alimentary canal.

Free Gregarines—particularly the Polycystid (septate) forms—have a peculiar habit of adhering to one another, end to end, in rows (Fig. 45). These are the so-called *associations* of Gregarines. As many as a dozen may be so joined together. The anterior individual of an association is called the *primite*, the rest the *satellites.* The phenomenon has nothing to do with conjugation.

Nourishment is effected by endosmosis through the body-walls, and motion is confined to a slow gliding forward of the body. The body, however, has sometimes the power of change of form, though such change is not of an amœboid character. It consists rather of a contraction of the ectoplasm which causes movements of the endoplasm, and is not unlike the streaming movements seen in the *Mycetozoa.*

In many forms there are longitudinal fibrillar thickenings of the cuticle, and occasionally a special superficial layer of the ectoplasm immediately beneath the cuticle is distinguished as the *sarcocyte.* The sarcocyte, which is sometimes found only on the anterior part of the body, may contain a layer of transversely-placed fibres, which are very possibly contractile in function, and form the part of the ectoplasm called the *myocyte.*

Reproduction is effected by the division of the body while in an encysted condition into spores, and seems generally, if not universally, to be preceded by conjugation of two or more individuals. The procedure is as follows: two (rarely more) individuals apply themselves together in various ways, and gradually flatten out against one another so as to form a rounded syzygium. In some cases, after remaining in this condition for some little time, it appears that they separate and encyst separately. As a rule, however, they remain together, and a cyst-wall is secreted round the syzygium; but even in this case the fusion of the conjugating individuals does not appear always to be complete, for the line between them can frequently be seen until the spore-formation has begun. In *Diplocystis* and *Gamocystis* conjugation occurs very early, so that it is rare to find any but very small forms not united with another in conjugation. This precocious conjugation may take place some time before encystment. The internal processes accompanying conjugation have not been thoroughly made out, but it appears that the nuclei of the conjugating pair meet in the bridge of protoplasm which connects the two bodies, and there fuse to form a zygote-nucleus. This divides and gives rise to a number of nuclei, which travel to the surface and are budded off with a certain portion of the protoplasm as the small rounded spores. The protoplasm of the syzygium is more or less used up in this process, but there generally appears to be a certain amount (residual protoplasm) left over, unchanged, in the cyst. The uninucleated spores acquire a chitin-like coat, which has often a spindle-shaped, oval, or oblong form, and in this condition are called *pseudonavicellæ*.* The contents of the pseudonavicellæ divides longitudinally into a number, generally six to eight, of sickle-shaped structures called the *falciform bodies*, and into a small amount of residual protoplasm, which disappears. The falciform bodies on the bursting of the cyst and the solution of the spore-case become young Gregarines. It thus appears that the young Gregarines arise by the division of the spore-protoplasm. When first liberated from the case, the young often perform active serpentining movements and make their way into a cell of their host (generally a new host, but in endogenous sporulation it may be the same host, see below), in which they increase in size, until they grow too large for the cell and project into the neighbouring space. Eventually they become detached and lie freely in the organ they infest—alimentary canal, testis, kidney, or whatever it may be. Some of the forms are only known in the full-grown free condition, and we cannot therefore be certain that they are intracellular in habit in the young stage. The spores escape by the bursting of the cyst, or in some cases through sporoducts which are formed in the protoplasm of the syzygium and everted through perforations in the cyst-wall.

As a general rule conjugation precedes the formation of the cyst, but encystment of solitary forms and subsequent sporulation has been

* In *Porospora gigantea* of the lobster the spores are naked.

observed. It is possible, however, that such solitary forms have escaped from conjugation.* The division of the cyst-contents into spores in Polycystids usually takes place outside the body of the host after evacuation with the fæces (*exogenous* sporulation); but in the Monocystids and *body-cavity* forms, sporulation is endogenous, *i.e.*, it takes place in the body of the host. In the latter case the spores may either be evacuated and produce their germs in a new host, or the germs may be hatched out in and reinfest the same host.

Endogenous cysts found in organs which do not communicate with the exterior can only set free their spores for a new host after the death of their host. The body-cavity cysts are found only in insects, and mainly in females, which often die after laying their eggs.

Sub-order 1. **MONOCYSTIDEA.**

Without differentiation of the body into protomerite and deutomerite, etc. Generally of considerable size; when full grown lying free in the body-spaces of their host. The protoplasm after encystment breaks up into spores, each of which becomes coated and divides into several falciform bodies. Conjugation appears to be general. *Monocystis* Stein, elongated, one end with cuticular hairs, sporulation incomplete; body-cavity, gut, and testis of earthworm; *Gamocystis* Stein, cyst with gelatinous coat and sporoducts, intestine of Blatta lapponica and Ephemerid-larvæ; *Conorhynchus* Greef, almost always in syzygial condition, gut of Echiurus; *Gonospora* Schn., like *Monocystis*, Annelids, *Urospora* Schn., like *Monocystis*, gut of Nemertines, body-cavity of Sipunculus, testis of Tubifex.

Sub-order 2. **POLYCYSTIDEA (SEPTATA).**

With differentiation of the body into protomerite, deutomerite, and sometimes epimerite. Other characters as in *Monocystidea*. In alimentary canal of Arthropoda. *Dufouria* Schn., Colymbetes larva; *Bothriopsis* Schn., various Dytiscidæ; *Porospora* Schn., Homarus; *Stenocephalus* Schn., Julus; *Hyalospora* Schn., Thysanura; *Euspora* Schn., Melolontha larva; *Clepsidrina* (Gregarina) Hammersch. (Fig. 45), various insects; *Pileocephalus* Schn.; *Echinocephalus* Schn., Lithobius; *Stylorhynchus* Schn. (Fig. 45), Opatrum, Asida, Blaps; *Actinocephalus* Schn., Coleoptera, Locusta, Sciara larva.

Order 2. Coccidiidea.†

Minute Gregarine-like forms which mainly infest epithelial cells, and do not outgrow their cell-host. They produce falciform young.

This is a provisional division to include certain small oval parasites with a single nucleus found in the cells of various animals. They have so far been found in Vertebrates, Arthropods, and Molluscs. They bring about the destruction of their cell-host, and those which

* According to Léger there is a Gregarine in the body-cavity of Glycera which produces spores without encystment.

† A. Labbé, "Recherches sur les Coccidies," *Arch. Zool. Exp.* (3), 4, 1896.

infest the epithelium of the alimentary canal may cause the death of their host. They encyst within the cell which they inhabit, and their protoplasm gives rise to coated spores. Conjugation is unknown, and there are no cuticular structures until the cyst-wall is formed.

The cyst-wall may be thin, in which case sporulation takes place in the same host, and the germs (falciform bodies) are set free to reinfest the same host (endogenous sporulation), or it is thick and double-contoured, in which case the cyst, on the breaking down of its cell-host, falls into the alimentary canal and is ejected with the fæces. In the latter case the sporulation (exogenous) takes place outside the host, and the germs are set free after the spores have entered a new host.

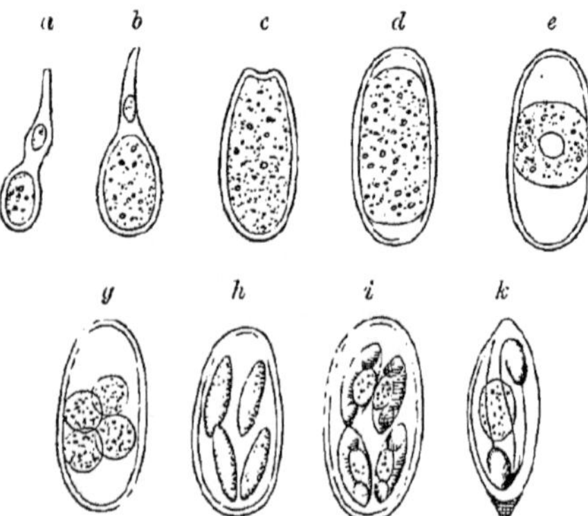

Fig. 46.—*Coccidium oviforme* from the liver of the rabbit (from Wasielewski, after Balbiani). *a* and *b*, young coccidia in epithelial cells of the bile duct, the cell-nucleus lies in the upper process of the cell host; *c*, encysted form; *d* and *e*, contraction of the protoplasm to a sphere; *g*, *h*, *i*, spore-formation; *k*, ripe spore with two falciform young and a residual body.

In some *Coccidiidea* only one spore* is produced; in such cases the whole animal becomes the spore, and its cyst-wall the spore case. In the rest the contents of the cyst divides into two or more spores (leaving as usual a residue) which acquire cuticular coats. In all cases the contents of the spore gives rise to the falciform bodies (one or more, leaving a small unused residue) which are the *germs* or *young*.

Tribe 1. **Monosporea.** The whole contents of the cyst forms only one spore, and the spore has no spore-case distinct from the cyst-wall. *Orthospora* A. Schn., spore with four falciform bodies, intest. epithelium of *Triton*; *Eimeria* Schn., numerous falc. bodies, intest. epithelium of mouse, frog, fishes, myriapods, etc.

* The *archespore* of Labbé seems to be merely the spore before the formation of the spore-case. The *sporozoites* are the bodies which proceed from the spore; they are called in the text falciform bodies, and are the young forms. In Labbé's nomenclature a spore which produces one falciform body is called *monozoic*; that which produces two falciform bodies is called *dizoic*, and so on to *polyzoic*.

Tribe 2. **Oligosporea.** Cyst produces but few spores. *Cyclospora* Schn., intest. epithelium of *Glomeris*, and the cat. *Coccidium* Leuck., in each of the four spores only two falciform bodies; in cells of the intestine and liver of mammals, birds, reptiles, amphibia, and fishes; *C. oviforme* Leuck. (Fig. 46), epithelium of bile-duct and intestine of rabbit.*

Tribe 3. **Polysporea.** With many spores. *Klossia* Schn., about sixty spores each with four to six falc. bodies; kidney Helix, Succinea, and Neretina; *Minchinia* Labbé, spores with two long filaments, in liver and connective tissue of Chiton; *Barroussia* Schn., each spore with one falc. body; *Adelea* Schn., in intestinal epithelium of *Lithobius forficatus*.

Order 3. Hæmosporidia.

Minute intracellular parasites which mainly infest the corpuscles of the blood. The spores are naked and do not sub-divide.

The *Hæmosporidia*, like the *Coccidiidea*, never become so large as to outgrow their cell-host; but they differ from them in possessing naked spores. Moreover the spores do not sub-divide, but become directly transformed into the young forms.† It is not known how they leave their host or how they enter new hosts.

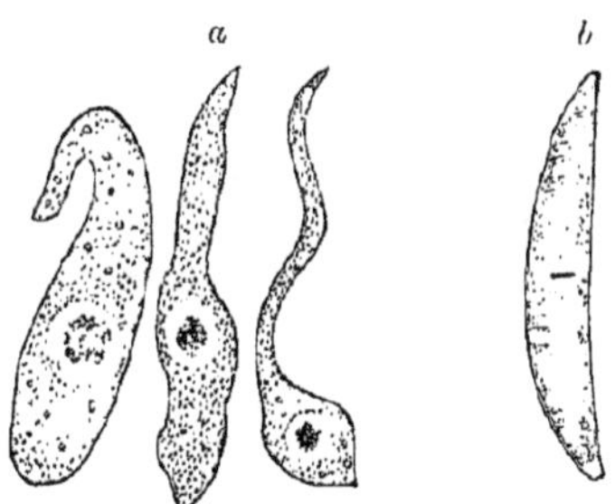

Fig. 47.—*Danilewskya lacazei* from the lizard. *a*, three free forms with gregarine-like appearance, fresh; *b*, after treatment with gold-chloride and hæmatoxylin, showing myocyte-fibrillæ.

Sub-order 1. DREPANIDIIDIA.

The *Drepanidiidia* infest the blood-corpuscles and are found in Amphibia, Reptilia, and Aves. They have not yet been found in fishes and mammals, or in invertebrates. They were discovered by Ray Lankester‡ in the frog.

They are small uninucleated gregarine-like creatures which infest the red blood-corpuscles (rarely the white blood-corpuscles and cells of spleen, liver, and kidney) of their host. They cause the destruction of their cell-host, from which they pass to live for a time freely in the blood. They then enter another blood-corpuscle, secrete a cyst-wall, and break up into elongated oval

* This forms an exogenous cyst, and it is a question how the rabbit becomes infested with the innumerable coccidia which are present in acute coccidiosis. It has been suggested that these coccidia reproduce in two ways—the one by exogenous sporulation as described in the text, and the other endogenously in the cell-host by the direct breaking up of the coccidium into numerous falciform bodies without the formation of spores, or by its simple division.

† This statement perhaps requires qualification, for in some of the Acystosporidia—the so-called two-spored forms—the nucleus of the form about to sporulate divides into two, and gives rise to two centres of spore-formation. (See below.)

‡ *Vide Quart. J. Mic. Sci.*, xi., 1871, and *ibid.*, vol. xxii., 1882.

bodies which are naked and do not further sub-divide. These constitute the young form, and may be compared to the spores of the other forms. They are set free in the blood on the breaking down of the cell-host, and soon enter another corpuscle, where they acquire their full size. They then leave the corpuscle and enter upon the stage described above, during which they live freely in the blood. They have been observed to conjugate completely in pairs in this free stage. In the free state they move either by serpentine bending of the body or by wave-like contractions of the body substance. It is not known how they are transported to a new host.

Drepanidium Lankester, Rana, Aves; *Karyolysus* Labbé, Lacerta; *Danilewskya* Labbé, Lacerta, Cistudo, Rana.

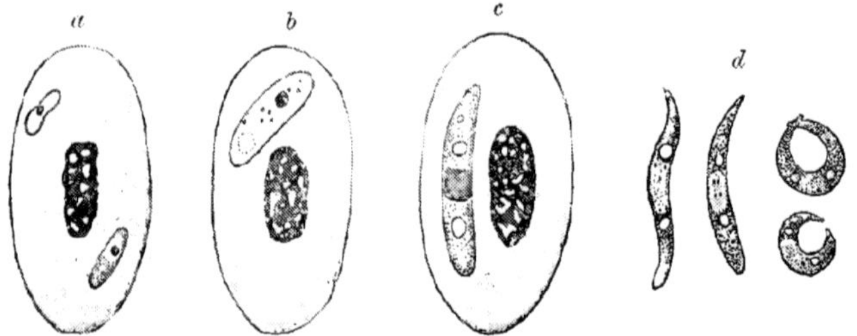

FIG. 48.—*a*–*c*, red blood-corpuscles of the frog infected with *Drepanidium princeps*; *d*, free forms in movement (from Wasielewski after Labbé).

Sub-order 2. ACYSTOSPORIDIA.*

These are amœboid cell-parasites, found only in vertebrata (except fishes and reptiles). They do not form a cyst-wall. They mainly infest red blood-corpuscles, but are also sometimes found in the kidney, liver, and intestinal epithelium. They cause hypertrophy of the corpuscle and a diminution of its hæmoglobin.

The *Acystosporidia* are specially interesting, because they include forms which are associated with some important diseases in certain of the larger mammals, *e.g.*, malaria in man and Texas-fever in cattle. They are minute amœboid organisms, with a nucleus and pigment-grains (Fig. 49). They never leave their cell-host, and their movements, which are always amœboid, take place within the cell. They readily undergo degenerative changes in preparations of the blood of infected animals. These changes consist in the disruption of their corpuscular host and in the thrusting out of vibratile processes, which break off from the body (Fig. 50): this is the so-called *Polymitus*-form. Finally the body itself breaks up.

They reproduce in their cell-host by the formation of minute spores without encystment. They first assume a rounded form; the nucleus then breaks up (so-called one-spored forms) into a number of minute fragments, around which a portion of the protoplasm becomes segregated (Fig. 51); these small nucleated fragments constitute the naked spores of the animals. In some cases the nucleus divides (so-called two-spored forms) into two before breaking up, so

* A. Labbé, "Parasites endoglobulaires." *Arch. Zool. expér.* (3), 2. Laveran, *L'Hematozoaire du paludisme*, Paris, 1891. J. Mannaberg, *Die Malariaparasiten*, Vienna, 1893.

that two groups of spores are formed. The spores are generally amœboid, but in *Karyophagus* they are sickle-shaped. They pass into the blood, and then enter other blood-corpuscles.

How these organisms are carried from host to host is not known; but it has been suggested that they may be taken into the lungs in dust, and be carried by parasitic insects and ticks. It appears certain that the organism associated with Texas-fever is carried by a tick, *Boophilus bovis*.*

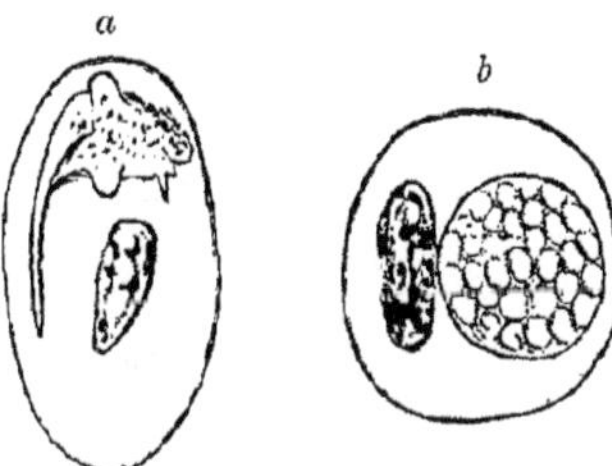

FIG. 49.—*Cytamœba bacterifera*, from the blood of *Rana esculenta* (from Wasielewski, after Labbé). *a*, amœboid form with long movable pseudopodia; *b*, rounded form with numerous spores.

Fam. 1. **Acystidæ.** Epithelial parasites which form falciform germs. *Karyophagus* Steinhaus, amphibia.

Fam. 2. **Hæmamœbidæ.** Mainly in blood-corpuscles; form amœboid germs. *Halteridium* Labbé, with two spores, birds, health unaffected; *Proteosoma* Labbé, with one spore, birds, produces fever and may cause death; *Hæmamœba* Grassi, with one spore, man, occurs in two forms, the one amœboid (variety *tertiana*), and the other semilunar and immovable (variety *quaterna*). *H. laverani* Labbé (Figs. 52, 53), discovered in 1880 by Laveran in the blood of malaria patients, causes destruction of the red corpuscles, period of development of germs 48-72 hours. Golgi showed the connection between the attacks of fever and the development of this parasite; *Dactylosoma* Labbé and *Cytamœba* Labbé, in *Rana esculenta*; *Apiosoma bigeminum* Smith, associated with Texas-fever in

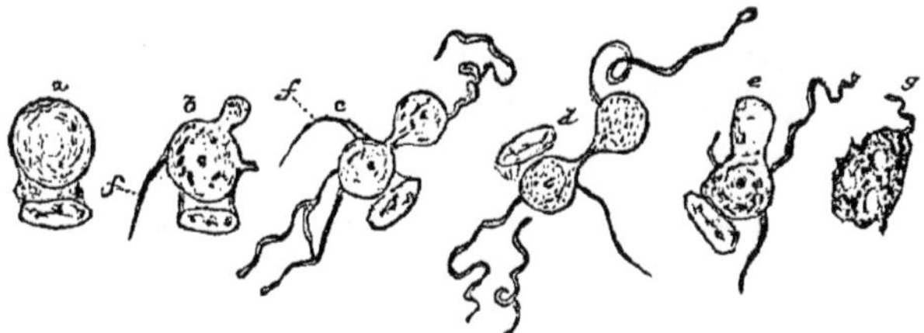

FIG. 50.—Successive stages of degeneration (Polymitus-form) of *Halteridium danilewskyi*, from the blood of the lark; *b*, *c* stages with vibratile flagella; in *d* these are being cast off (after Labbé).

cattle, infection carried by ticks, amœboid organisms in the red blood-corpuscles, high fever, anæmia, bloody urine, the number of red blood-corpuscles is diminished in one week to one-sixth.

Babesia bovis Babes, in blood of the ox, causing hæmoglobinurea; *Amœbosporidium polyphagum* Bonome, associated with Icterus-hæmaturia of the sheep, are probably allied here.

* Th. Smith, Centralbl. Bakt. Parasitk. 13, 1893, p. 511.

Order 4. Myxosporidia.*

Multinucleated, amœboid sporozoa parasitic in the cells, tissues, or spaces of animals. The spores possess polar capsules and fibres. The contents of the spores do not divide, but issue as amœboid young.

These are the so-called fish-psorosperms of J. Müller. They are parasitic in worms, arthropods, polyzoa, and vertebrates. They

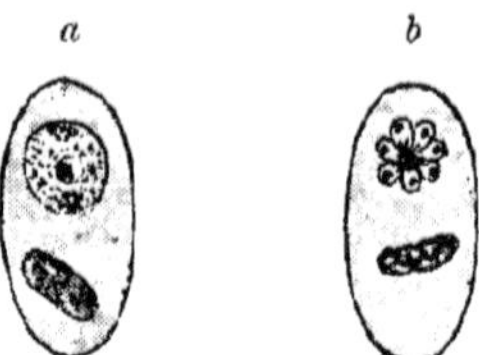

Fig. 51.—*Proteosoma grassii* Labbé, from the blood of a finch. Sporulation of the one-spored parasite. *a*, showing the parasite in the corpuscle; *b*, showing the parasite broken up into spores (after Labbé).

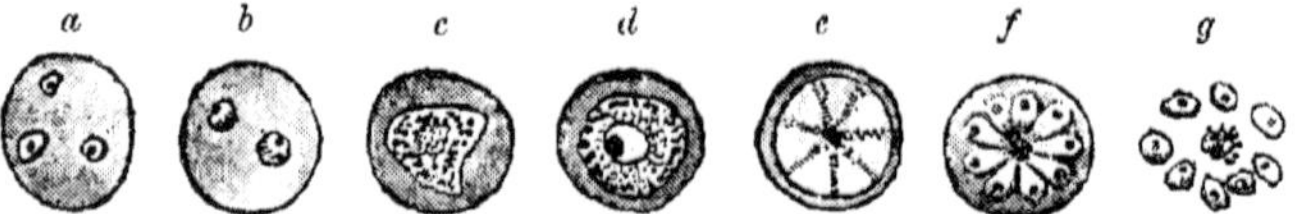

Fig. 52.—*Hæmamœba laverani*, variety quaterna, from the blood of a man with malaria (after Labbé). *a*, newly-infected blood-corpuscles; *b*, *c*, *d*, successive stages in the growth of the parasite in the corpuscle; *e*, beginning of sporulation; *f*, rosette-shaped group of spores round a central residual body; *g*, spores (young forms) set free in the blood by the breaking up of the corpuscle.

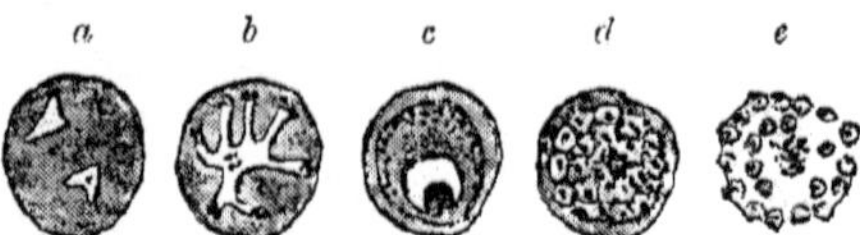

Fig. 53.—*Hæmamœba laverani*, variety tertiana, from the blood of a man with malaria. *a*, young parasites in a corpuscle; *b*, amœboid form; *c*, rounded form; *d*, sporulation; *e*, free spores (after Labbé).

have attained some notoriety from the fact that the organism (*Glugea bombycis*), which causes the *Pebrin*-disease of silk-worms, belongs to the order.

They include two principal varieties, viz., those which lead a free life, creeping about on the surface of the gall-bladder, urinary bladder, and kidney tubes (Fig. 54), and those which are embedded

* R. R. Gurley, "The Myxosporidia, or Psorosperms of fishes and the epidemics produced by them." *Bull. U.S. Fish Commission*, Part 18, 1894. P. Thélohan, "Recherches sur les Myxosporidies." *Bull. Sci. de la France et Belgique*, 26, 1895.

in the tissues. The tissue-forms occur either in *cysts*, which are often visible to the naked eye, or in an *infiltrated form*, and they may infest almost any tissue (bone, cartilage excepted).

It is not quite clear to what extent they are cell-parasites. Many of them are certainly intracellular in the young state, and it is possible that all may be so; but the youngest stages have not been studied in all forms. Sporulation in the intracellular forms may begin in the quite young forms, even before they have outgrown their cell-host. The *Myxosporidia* differ from other Sporozoa in their tendency to cause tumours in their host. They may cause serious diseases: the silk-worm disease has been referred to, and they have been known to cause serious mortality among fishes.

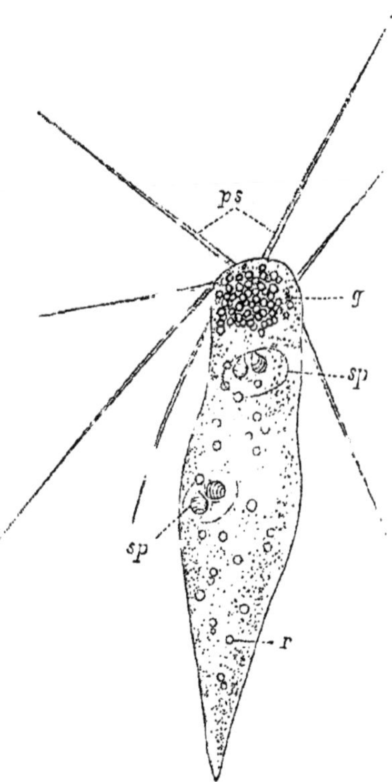

FIG. 54.—*Leptotheca agilis* as a type of free-living Myxosporidia, from the gall-bladder of *Trygon vulgaris*. *ps* pseudopodia; *g* fat-drops; *r* refractile granules; *sp* spores (after Thélohan).

Those which lead a free life are amœboid and vary much in shape (Fig. 54). The shape of the cysts and of the infiltration-forms depends upon the physical condition of the tissues.

The *Myxosporidia*, though amœboid, do not take up solid food, but resemble other *Sporozoa* in absorbing the nutritive juices of their hosts. The body shows a division into ecto- and endoplasm. The nuclei are numerous and lie in the endoplasm, which contains granules and fat-drops.

Reproduction takes place by spore-formation. They differ, however, from other *Sporozoa* in the fact that the whole body does not break up into spores at one time, but the spores are formed gradually in the endoplasm while the parent still continues to grow and to move. When a spore is about to be formed, a small spherical mass of endoplasm containing one nucleus is marked off from the rest by a delicate membrane; it thus constitutes an ovum-like body lying in the endoplasm. It is called a *primitive sphere* (Fig. 55); its nucleus gradually divides, karyokinetically, into ten nuclei, and it then itself divides into two parts, each of which contains three of the ten nuclei of the primitive sphere; the four remaining nuclei together with a portion of protoplasm forms a small residual body, which soon disappears (*cf.* the residual bodies of other *Sporozoa*). The two trinucleated bodies thus formed are the *sporoblasts;* they are enclosed in the membrane of the original primitive sphere, which soon thickens into a resistent spore-case. Each sporoblast divides into three cells (Fig. 55*e*), of which one gives rise to one spore, and the other two to the two polar-capsules (Fig. 55*f*). The polar-capsules are formed in and at the expense of the protoplasm of the polar-capsule cells, which wholly disappear in the process. The polar-capsules are ovoid bodies containing a long spirally-coiled thread (Fig. 56), which is everted with considerable force when the spore is acted upon by the digestive

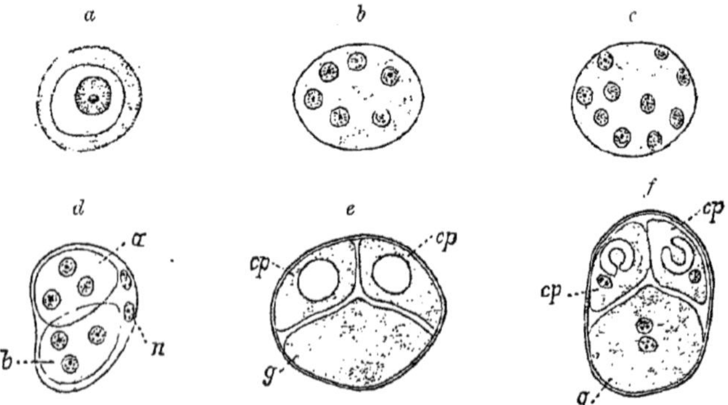

FIG. 55.—Spore-formation of *Myxobolus* (after Thélohan). *a*, primitive sphere with nucleus, a little endoplasm of the parent is shown; *b*, stage with six nuclei, and *c*, with ten nuclei. *d*, division of primitive sphere into two sporoblasts *a* and *b*, each with three nuclei, and a residual body with four nuclei *n*; *e*, division of one of the sporoblasts into two smaller capsule-forming cells *cp*, each with a vacuole, and a larger cell (the spore) *g*. *f*, formation of polar-capsules.

juices of the animal which swallows it. It is probably everted with such force that it pierces the wall of the alimentary canal, and thus effects the attachment of the spore to its new host (Fig. 56). The spore-case bursts in the course of twenty-four hours after this attachment, and the contained germ makes its way as an amœboid form through the intestinal wall and migrates to the tissue in which it is to live.

In the case just described, which is that of the genus *Myxobolus*, each primitive sphere gives rise to two spores; but in some cases only one spore results, and in others three or more are formed. In the tissue-forms a considerable number of spores proceed from each primitive sphere.

The species are distributed by the spores, which are carried to the exterior, or when this is impossible, as in the case of the tissue-forms, are set free on the death of their host.

Fam. 1. **Myxididæ.** Spores variously formed with two polar-capsules; includes the forms least degenerated by parasitic life. Principally in gall-bladder and kidney tubes of fishes and amphibia. *Leptotheca* Thélohan (Fig. 54), gall-bladder or kidney of various fishes and amphibia; *Ceratomyxa* Thél., gall-bladder of fishes; *Sphærospora* Thél., kidney-tubes of fishes, *S. elegans* Thél., kidney-tubes and ovary of stickleback; *Myxidium* Bütschli, *M. lieberkühni* Bütschli, urinary bladder of pike; *Sphæromyxa* Thél.; *Myxosoma* Thél.

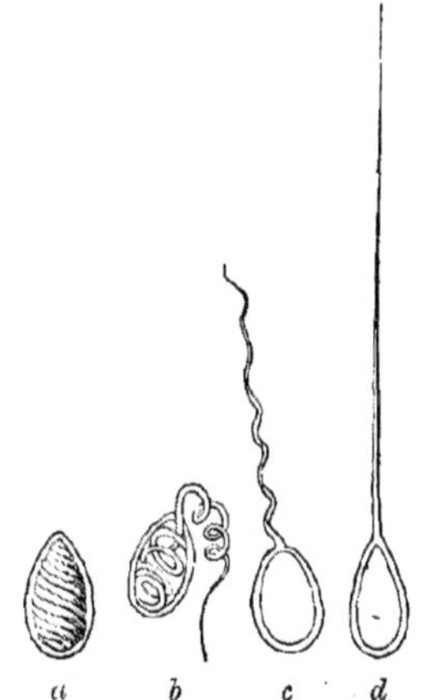

FIG. 56.—Polar-capsules of *Myxobolus ellipsoides*. *a*, polar-capsule with spirally-coiled thread; *b*, *c*, *d*, eversion of the fibre (after Balbiani).

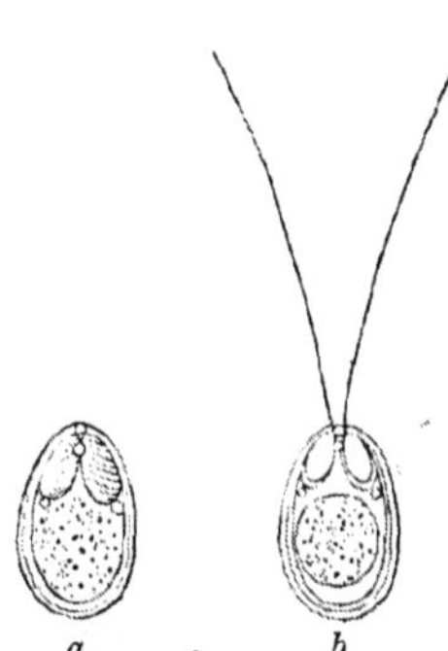

FIG. 57.—Spores of *Myxobolus ellipsoides* (after Balbiani), showing polar-capsules in *b* the threads are everted.

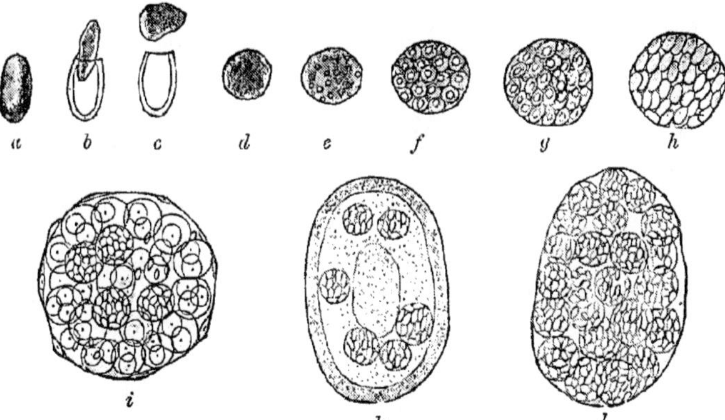

FIG. 58.—*Glugea bombycis* (after Balbiani). *a*, ripe spore; *b*, *c*, hatching o. amœboid young; *d*, *e*, growth stages; *f*, *g*, *h*, sporulation; *i*, testis-follicle of silk-worm caterpillar, strongly infested with *Glugea*; *k*, *l*, two infected stomach-epithelial cells of the caterpillar of *Attacus (Saturnia) pernyi*; *k*, beginning of infection; *l*, the cell is completely filled with spores.

Fam. 2. **Chloromyxidæ.** Spores with four polar-capsules. *Chloromyxum* Mingazzini: *C. incisum* Gurley, gall-bladder *Raja batis.*

Fam. 3. **Myxobolidæ.** Almost all tissue-parasites, principally gills, spleen, etc., of fishes; one or two polar-capsules. *Myxobolus* Bütschli (Figs. 55–57): *M. piriformis* Thél., gills, spleen, kidney of the tench (*Tinca vulgaris*); *M. dispar* Thél., gills of Carp (*Cyprinus rutilus*); *Henneguya* Thél.; *H. psorospermica* Thél., gills, eye-muscles, ovary of pike; *H. media* Thél., kidney and ovary of stickleback.

Fam. 4. **Glugeidae.** With very small oviform spores, having at the broad end a non-colourable vacuole, at the narrow end a polar-capsule usually invisible. *Glugea* Thél., mainly tissue-parasites; *Gl. bombycis* Thél. (*Microsporidium bombycis* Balb.), in all tissues of *Bombyx mori*, is the cause of the Pebrine disease of silk-worms (Fig. 58), which between the years 1854–67 caused the loss of one milliard francs to the French silk-worm industry; combated by microscopical examination of eggs and rejection of those infected (Pasteur and Balbiani); *Gl. bryozoides* Korotneff, sexual organs and body-cavity of *Alcyonella fungosa*; *Pleistophora* Gurley; *Thelohania* Henneguy, muscles of Palæmon, Crangon, Astacus; *Th. contejeani* Hen., muscles of *Astacus fluviatilis.*

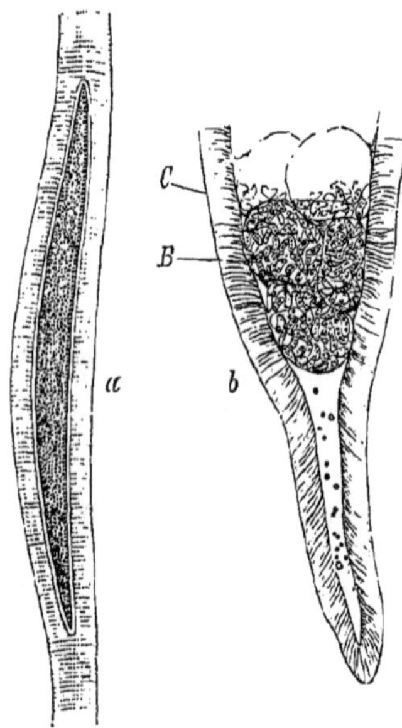

FIG. 59.—Rainey's Corpuscles from the flesh of a pig. *a*, an animal inside a muscle-fibre; *b*, posterior end of same strongly magnified; *C* cuticle; *B* spores.

Order 5. SARCOSPORIDIA.*

Cylindrical intracellular parasites infesting the striped muscular fibres of certain vertebrates.

These are the so-called Mischer's tubes, the contents of which are known as Rainey's Corpuscles (Fig. 59). They contain a number of more or less spherical bodies, which divide up into still smaller bodies—the germs. These latter become sickle-shaped and appear to constitute the young. Very little is known about this group. We are ignorant of the manner in which the transference from host to host is effected, and of the young stages of infection.

* Bertram, "Beiträge z. Kennt. d. Sarcosporidien." *Zool. Jahrb. Abth. f. Anat.*, 5. Ai. Schneider, "Ophryocystis bütschlii." *Arch. Zool. Exp.* (2), 2, 1884.

Sarcocystis Lankester, in the muscles of pig, sheep, gecko. *Sarcosporidia* have been described in man, cat, dog, mouse, rat, hare, rabbit, ox, deer, horse, etc.

The *Amœbosporidia* and *Serosporidia* may be taken here. The *Amœbosporidia* are multinucleated amœboid forms, which conjugate, encyst, and produce one coated spore. The spore divides into eight falciform young; they may also increase by dividing directly; they live in the Malpighian tubules of some beetles. *Ophryocystis* A. Schn. in Blaps, and Akis. The *Serosporidia* are long oval parasites which infest the body-cavity of some crustacea. They increase by direct division, and by encysting without conjugation and subsequent breaking up into numerous amœboid young. *Serosporidium* L. Pfeiffer, in Cypris, Daphnia, and Gammarus.

CHAPTER II.

THE METAZOA.

All animals above the *Protozoa* have been classed together as *Metazoa*, and possess the following characters in common:—

There is always more than one nucleus, and the nuclei are for the most part arranged with a definite relation to the functional tissues.

Conjugation always takes place, but the structure is so complex that conjugation between the ordinary individuals of the species is impossible. Consequently special individuals—the gametes—are produced for the purpose of conjugating. These individuals, which have a very similar form throughout the group, are simple in structure and unicellular in character; there are always two kinds of them in every species, called respectively *ova* and *spermatozoa.* They arise by a process of unequal fission from their parent, and may both be produced by one individual or by different individuals. When they are both produced by the same individual, that individual is said to be *hermaphrodite.* When they are produced by different individuals, that parent which produces the ova is called the *female*, while that which produces the spermatozoa is called the *male;* and the individuals are said to be unisexual and the species *dioecious.* The conjugating individuals, or gametes, produced by the male never have the power of assuming the ordinary form of the species, and though they have, as a rule, the power of independent locomotion, soon die unless placed in the most favourable circumstances. The gametes produced by the female, on the other hand, while they are without the power of locomotion and have a rather greater power of independent life, are in rare cases capable of becoming more complex in structure, and of assuming the form of the adult. To females which produce such ova the term *parthenogenetic* is applied. In the vast majority of cases, however, the ovum has not the power of changing its form and of developing into the ordinary form of the species unless it first conjugates with the spermatozoon. The zygote so produced is *uninuclear*, and has the property of developing into the ordinary form of the species. This method of reproduction, in which a new individual arises from the combination of two independent individualities in the zygote, is called the *sexual method of reproduction*, as opposed to the *asexual method*, in which a *multinucleated* mass is separated off from the parent with the power of assuming more or less directly the ordinary form of the species. The asexual method, though common in the vegetable kingdom, is comparatively rarely found amongst animals (*Coelenterata*, *Polyzoa*, *Tunicata*, *Annelida*, etc.).

It thus appears that the *Metazoa* may be defined as—

Animals in which the ordinary (so-called adult) form of the species has always more than one nucleus, and in which the nuclei are for the most part arranged regularly and with a definite relation to the functional tissues of the animal (so-called cellular arrangement). Special conjugating individuals of the form of ova and spermatozoa are always formed.

CHAPTER III.

PORIFERA.*

THE Porifera present a great variety of external form. They may be cup-shaped, saucer-shaped, tubular, rod-shapëd, foliaceous, trumpet-shaped, fan-shaped, mushroom-shaped, lobed, digitate, branched, or irregular, etc. (Figs. 60–62.) As a general rule, the form is extremely variable even in the same species, and is therefore of little use in identification. They are almost, if not quite, always attached to foreign objects; this may be effected by a broad basal surface, or they may be stalked. In some cases they are rooted in sand or in mud by basal processes or by special rooting spicules. With the exception of the fresh-water *Spongillidæ*, they are marine, and are found at all depths. One family—the *Clionidæ*—bore into shells and stones.

As the name implies, the surface of the body presents a large number of pores, which are minute in size and inhalent in function. These pores lead into a system of channels which, after permeating almost the whole body, open to the exterior by one or more—but always a few—larger exhalent openings called *oscula*. This system of spaces connecting the inhalent pores with the exhalent oscula is the canal system. Through it there passes—maintained, as we shall see, by ciliary action—a continual stream of water, which enters by the inhalent pores and passes out through the oscula. The sponge is covered by an epithelial layer which we may call *ectoderm;* the canal system is lined by an epithelium, which as we shall see is usually partly ectoderm and partly *endoderm:* but the main mass of the body is formed of a soft tissue which we shall call *mesoderm.* The mesoderm consists of a gelatinous basis (though no gelatine has been detected in it), containing a protoplasmic network holding

* For principal literature see classes and orders. R. Hanitsch, "Revision of the generic nomenclature and classification in Bowerbank's 'British Spongiadæ.'" *Proceedings and Transactions of the Liverpool Biol. Soc.*, vol. 8, 1893, p. 173.

nuclei, and presenting differentiations of various kinds—so-called muscle cells, amœboid cells, generative cells, scleroblasts (spicule-forming cells). The mesodermal network is continuous both with the ectoderm and with the endoderm; indeed, these layers may fairly be regarded as superficial bounding expansions of the meso-dermal mass. Skeletal structures, the main function of which is

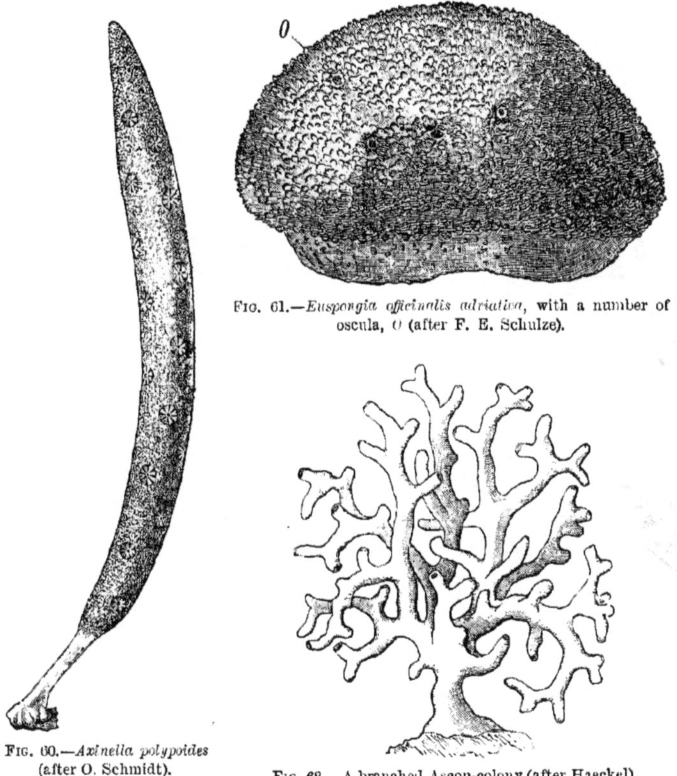

FIG. 61.—*Euspongia officinalis adriatica*, with a number of oscula, O (after F. E. Schulze).

FIG. 60.—*Axinella polypoides* (after O. Schmidt).

FIG. 62.—A branched Ascon-colony (after Haeckel).

to support the sponge-body, are contained in the mesoderm. These may be calcareous or silicious, in which case we get the so-called sponge-spicules; or they may consist of a horny material called spongin (common bath sponge, Fig. 63); or finally, spongin fibres and silicious spicules may co-exist. The generative cells are budded off from the mesodermal network, and are eventually dehisced into some part of the canal system, whence they are carried to the exterior.

It is probable that all the protoplasmic tissues of the sponge are contractile, *i.e.*, both the epithelial layers and the mesodermal network; but special structures in the course of the inhalent and exhalent parts of the canal system have been described as muscular sphincters. Ectodermal cells carrying hair-like sensory projections have been described; these, like other epithelial structures, are connected below with the mesodermal network, parts of which have therefore been interpreted as nerve-fibres and nerve-cells; but

FIG. 63.—Piece of network of horny fibres from *Euspongia equina*.

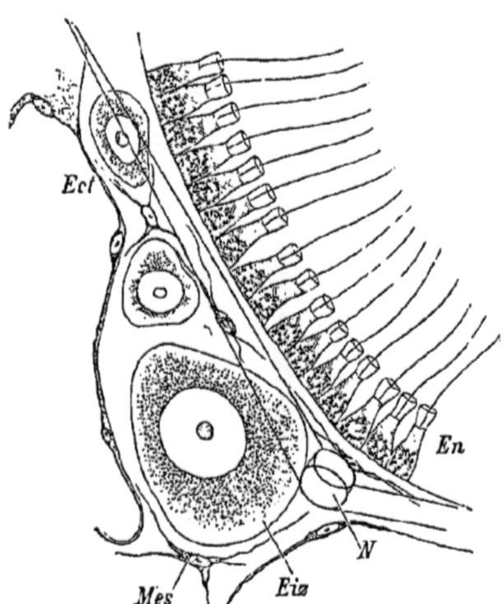

FIG. 64.—Section through a calcareous sponge (*Sycon raphanus*), after F. E. Sch. *Ect* ectoderm; *En* endoderm of a flagellated chamber; *Mes* mesoderm; *N* calcareous spicule in the mesoderm; *Eiz* ovum.

there is no reason why one part of the network should be considered as more especially adapted for nervous conduction and reflection than another. The ectodermal epithelium consists of flat cells; the endoderm is partly formed of flat cells and partly of somewhat cylindrical cells, each with a flagellum and collar. These are the **choanocytes** (Fig. 64). They are perhaps the most characteristic constituents of the sponge-body; the collar is a membranous prolongation of the cell at its free end round the base of the flagellum; and the whole cell resembles an individual of the *Choanoflagellata*. It contains one or more contractile vacuoles, and its base is prolonged into processes which join similar processes of neighbouring cells and the mesodermal network. The collared cells are, as a rule, confined to special parts of the canal system called the *flagellated* or *ciliated chambers*. Their main function is, no doubt, to cause the current of water which is continually flowing through the sponge. Ciliated epithelium is not found in the *Porifera*, though in some sponges (*Oscarella lobularis, Plakina monolopha*, etc.) the ectoderm cells carry flagella.

We may now proceed to describe in greater detail the various parts of the sponge-body.

The simplest form of sponge—we do not say the most primitive, though it may be so—is presented by the Ascon type of the order *Calcarea*.

The *Ascon-person*, which is characteristic of the genus *Leucosolenia*,* consists of a cup- or vase-shaped animal attached by one end, and presenting at the other an opening—the osculum. The walls are thin, and consist of ectoderm outside, flagellated endoderm inside, lining the cavity of the cup, and thin mesoderm, containing triradiate calcareous spicules, between the two. They are further pierced by numerous pores, the *prosopyles*.

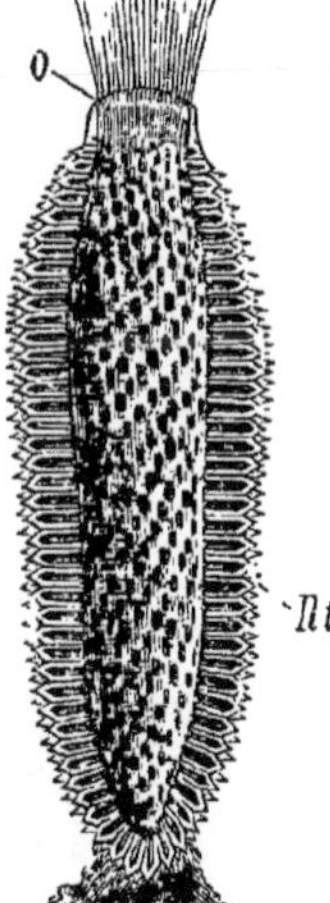

Fig. 65.—Longitudinal section through *Sycon raphanus*, slightly magnified. *O* osculum with collar of spicules; *Rt* radial tubes which open into the central cavity.

In the *Sycon-person* (Fig. 65), which is characteristic of the *Heterocœla*, there is a tube or cup open by the osculum at one end and attached at the other. This tube is lined by flat cells, and gives off all around and throughout its length numerous short diverticula lined by flagellated cells. These are the radial flagellated chambers: they possess, in addition to the one main opening into the cavity of the central tube, which we may call the gastric cavity, numerous minute pores—the *prosopyles*—through which water passes from the exterior into the flagellated chambers. These radial tubes, in short, resemble an Ascon in structure, except for the absence of an osculum at their free end. In the simplest *Heterocœla* the radial chambers stand out freely from the central tube, and do not touch at any

* This genus comprises *Ascetta primordialis* and all Haeckel's Ascons. The *Olynthus* is a hypothetical animal imagined by Haeckel, and closely approaching the Ascon-person in form.

point (*Sycetta*). But in the more complex forms the walls of adjacent chambers fuse more or less completely where they touch, and the spaces between them are broken up and called the *inhalent canals* or *intercanals*. These intercanals open by the prosopyles into the chambers, and outwards on the surface of the sponge (*Sycon*). In still more complex forms the outer ends of the chambers and the openings of the intercanals are covered by a membrane called the *dermal membrane* or *cortex*. This cortex is perforated by numerous pores—the *dermal pores*—(to be distinguished from the prosopyles) which lead into the intercanals

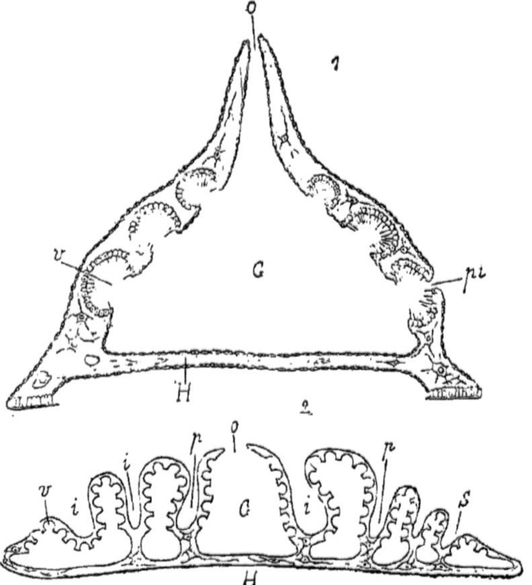

FIG. 66.—*1*, Diagrammatic section of a *Rhagon*. *H* hypophare; *O* osculum; *G* gastral cavity; *v* flagellated chambers; *pi* prosopyle (from Perrier, after Sollas).
2, Diagram of a simple form of the Eurypylous type of canal-system. The spongophare is folded. *H* hypophare; *G* gastral cavity; *o* osculum; *i* incurrent sinuses; *p* dermal pores.

(*Grantia*, *Ute*, *Sycyssa*, *Heteropegma*, *Amphoriscus*, etc.). In this type the chambers may branch, and the dermal cortex may attain a considerable thickness, as also may the wall of the central tube (gastral cortex). A further complication is effected by the retreat of the flagellated cells towards the distal (outer) ends of the chambers, their place being taken by pavement epithelium. We thus get *exhalent canals* coming off from the gastral cavity and lined by flat cells, together with a reduction in length of the flagellated chambers (*Leucilla uter*, *Leucandra aspera*, etc.). Finally the chambers are small and spherical and irregularly scattered through the sponge-wall; the inhalent and exhalent canals being much developed. This is the *Leucon* stage of Haeckel and is found in *Leucandra*, *Leucilla*, etc. In the non-calcareous sponges the canal-system is generally on a somewhat different plan. The simplest form is the so-called **Rhagon** type (Fig. 66) found in the embryos of certain forms (*Plakina*, *Reniera*). The Rhagon has the form of a flattened pyramid, attached by its broad base and opening by the osculum at its apex. The sac is lined by flat cells, but possesses on its upper wall a number of small flagellated chambers into which the prosopyles open. The lower basal wall of the Rhagon, which is

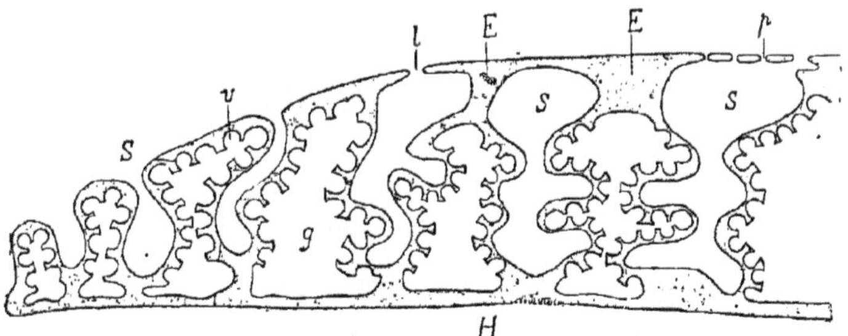

FIG. 67.—Diagram showing the relations of the ectosome. *H* hypophare; *E* ectosome; *l* fold of the ectosome roofing over the incurrent sinus; *p* a sieve-plate of dermal ostia or inhalent pores; *g* excurrent sinuses; *v* flagellated chambers; *S* incurrent sinuses (from Perrier).

without flagellated chambers, is called the *hypophare*, and the upper wall, with the chambers, the *spongophare*. The openings of the chambers into the gastric cavity of the Rhagon are called the *apopyles*. The Rhagon condition is not found in any adult sponge, but the nearest approach to it is presented by *Plakina monolopha* and *Oscarella lobularis*, in which the spongophare is folded, so as to give rise to *incurrent sinuses* or canals similar to the intercanals of the *Calcarea* (Fig. 66, *2*).

As a result of this same folding the chambers open, not into the central or gastral cavity, but into diverticula of it. These diverticula are the *excurrent sinuses*. We thus get a modification of the Rhagon canal-system called the **Eurypylous** type. As a general rule in the Eurypylous type, there is concrescence between the folds of the spongophare, and the openings of the incurrent sinuses are roofed over by a membrane (Fig. 67, *l*) which is exactly comparable to the dermal cortex of the *Calcarea*, and is pierced by inhalent pores leading into the incurrent sinuses. (Figs. 67, 68.) This membrane is called the *ectosome*, as opposed to the rest of the sponge, which contains the chambers and is called the *choanosome;* and the incurrent sinuses beneath it constitute the subdermal cavities, which, as is obvious, correspond to the intercanals of the Calcarea. By further folding of the spongophare and suppression of chambers on the main excurrent sinuses an increase in complexity is obtained.

In the Eurypylous type the chambers open directly into the excurrent sinuses; but in many sponges they are removed from the surface of the latter, and the apopyle of each of them is prolonged into a canaliculus—the *aphodus*—lined by a prolongation of the epithelium of the excurrent sinus into which it opens. Further,

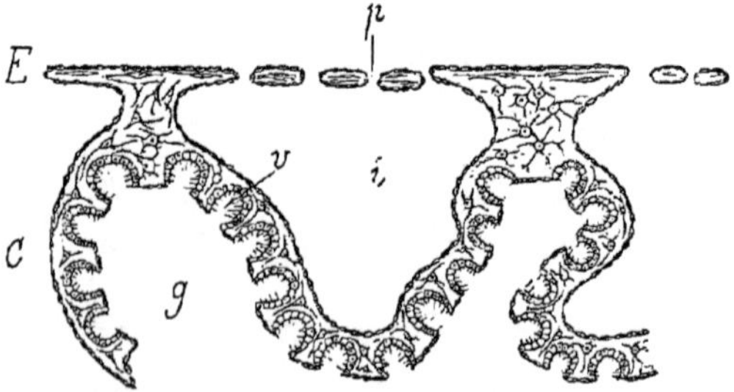

FIG. 68.—Diagram of a section of the outer part of *Tetilla pedifera* (after Sollas, from Perrier). *C* choanosome; *E* ectosome; *i* incurrent sinus; *v* flagellated chamber; *g* excurrent sinus.

there is only one prosopyle to each chamber. This is the **Aphodal** modification of the Rhagon type (*Stelletidae*, *Geodinidae*, etc.).

Finally there is the **Diplodal** chamber system (Fig. 69), in which there is a canaliculus leading to the prosopyle of each chamber—the *prosodus*—as well as an *aphodus* leading from it (*Chondrosina*, *Corticium candelabrum*, etc.).

In Eurypylous sponges the ectosome never attains any special development; but in other types it becomes greatly thickened and histologically differentiated, and is called the **cortex**. In such cases the tubes leading through the cortex from the sieve-pores to the subcortical (subdermal) cavities are called **chones**, and are provided at some part of their course with a muscular sphincter, the *velum*. The inhalent pores may be diffuse or collected into pore-areas forming the so-called sieve-plates.

Histology. There is but little to add to the statements in the general account. The most remarkable feature is the flagellated cells or choanocytes. They are larger in the *Calcarea* than in most other sponges. They present the feature of projecting into the chambers without contact with their neighbours.

It was supposed that in life their collars were everted so as to touch the collars of neighbouring cells and form a continuous thin membrane lining the chamber; but this has been shown to be a post-mortem phenomenon.

The mesoderm contains many tissue elements which are all connected together, and are embedded in a kind of gelatinous basis.

To revert to the language of the cell theory, those cells of this tissue, to which no special function can be assigned—such as reproduction, contraction, skeletal formation—vary somewhat in their arrangement, and so give rise to different forms of mesoderm; we may mention *Collenchyme* with scattered stellate cells; *Sarcenchyme* gelatinous basis reduced, and granular cells closely packed; *Cystenchyme* with vesicular, vacuolate cells close together or separated by a matrix; *Chondrenchyme* like hyaline cartilage.

There are also pigment cells, fusiform connective tissue cells, muscle cells as granular fusiform cells round the openings of the water-canals.

Spicules originate in one cell, or more than one cell may be associated in their production (*Lithistida*). It has been recently stated that three cells are concerned in the formation of the triradiate spicules of the *Calcarea*, and that they are derived from the ectoderm (Minchin).

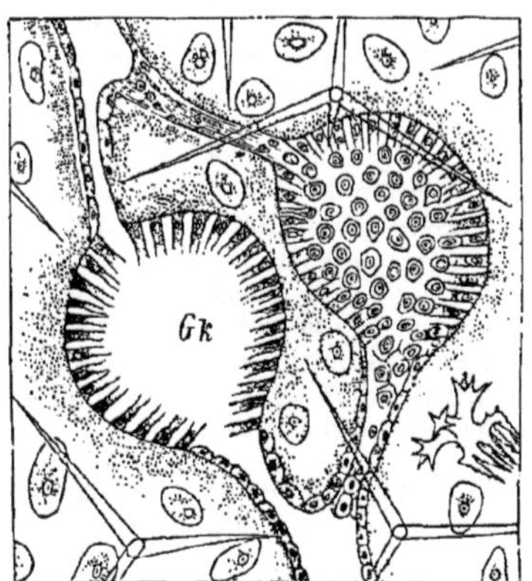

FIG. 69.—Section of *Corticium candelabrum* (after F. E. Sch.). *Gk* flagellated chamber, with *prosodus* leading to it and *aphodus* from it.

The ectoderm is composed of flattened cells, though the cell limits are not always discernible. In some sponges at least the cells of the ectoderm seem to be capable of assuming, when contracted, a mushroom shape.

The ingestion* of food and foreign bodies by sponges is effected by the collared cells, and probably also by the flat cells of the inhalent and exhalent canals, and of the surface ectoderm of the body. From these the solid bodies pass into the subjacent mesoderm cells, this transit being probably effected by protoplasmic flow along the strands connecting the surface epithelia with the subjacent tissues, in very much the same manner as food passes along the pseudopodial network

* *Vide* E. Metschnikoff, "Spongiologische Studien." *Z. f. w. Z.*, Bd. xxxii. p. 372. A. Dendy, "Studies on the Comparative Anatomy of Sponges." *Q. J. M. S.*, vol. xxxv. p. 216.

into the body of a Foraminiferan. But although the choanocytes may have this ingestive capacity, their main function is probably to cause currents through the canal system.

The question of the ingestion of solid food by sponges has been much disputed. There can be no doubt that solid bodies are introduced into the mesoderm cells, for apart from the fact that foreign bodies such as sand-grains, etc. are found in the mesoderm cells and spongin fibres of many sponges, Metschnikoff and other observers have found carmine particles in carmine-fed sponges. It is not, however, certain that the flat epithelia co-operate with the choanocytes in this introduction. Metschnikoff (*loc. cit.*, p. 372) states that *Halisarca*, when overfed with carmine, loses its canals and becomes a mass of amœboid cells containing swallowed matter and surrounded by a common envelope of ectoderm. The same fact has been observed by Lieberkühn* in *Spongilla* in winter. From these observations, and others by Haeckel and Carter, it appears that under certain nutritive conditions the choanocytes may lose their flagella and collars (according to Topsent and others these structures are retractile like the pseudopods of an Amœba) and become amœboid, and the whole sponge may revert to the condition of the larva of *Aplysina* (F. E. Schulze, *Z. f. w. Z.* xxx. Pl. 24, Fig. 30), of a protoplasmic network with nuclei at the nodes surrounded by a cortical layer of ectoderm.

The collared cells are thus inconstant, and appear to be merely parenchyma cells specially modified and capable, under certain nutritive conditions, of passing back to their original form. When they vanish the canal system also goes, and the sponge becomes solid so far as the latter is concerned. Inasmuch as the parenchyma cells and the ectoderm cells are all connected by their processes (except in the cases in which they break away and become amœboid), it is clear that the sponge in this condition, and in the case of Schulze's larva already referred to, is but little more than a multinucleated Protozoon, differing from the latter in the greater development of the vacuoles of the central portions, and in the presence of a distinct cortical layer of nuclei.

Skeleton. Skeletal structures are found in almost all sponges (absent in certain *Hexaceratina* and *Carnosa*), and are of considerable importance in the classification. They may consist of calcareous spicules, of silicious spicules, or of spongin. Spongin is a horny substance, resembling silk in chemical composition. It is usually found in the form of fibres connecting together the silicious spicules (many *Monaxonida*), or forming the entire skeleton (*Ceratosa*). In a few cases it is present as separate horny spicules (*Darwinella*). In the *Ceratosa*, and probably in other sponges, it is secreted in concentric layers by a number of mesodermal cells, called spongoblasts, which are found coating the fibres (Fig. 70). In some cases the fibres enclose foreign bodies, such as sand grains. In the *Monaxonida* the amount of spongin present is, roughly speaking, inversely proportional to the number of spicules, and there are all variations between

* "Beiträge zur Entw. d. Spongillen." *Müller's Arch.*, 1856.

forms with no spongin and forms with no spicules in the spongin fibres (*Ceratosa*). Which is the primitive condition—if either is—it is impossible to say. The *Ceratosa* are an artificial order, its families being related to different families of the *Halichondrina*. It may, indeed, be looked upon as an assemblage of halichondrine forms, in which the reduction of silica and development of spongin have reached their extreme limits. The calcareous and silicious spicules are secreted in the protoplasm of the mesodermal network; *Scleroblast* is the term applied to the special uninucleated part of the network in which they arise. They consist of an organic axis which is generally continuous through the points of the spicules with the adjacent organic structures, and of an organic sheath which is presumably a remnant of the parent scleroblast. The silicious spicules are formed of opal (colloid silica).

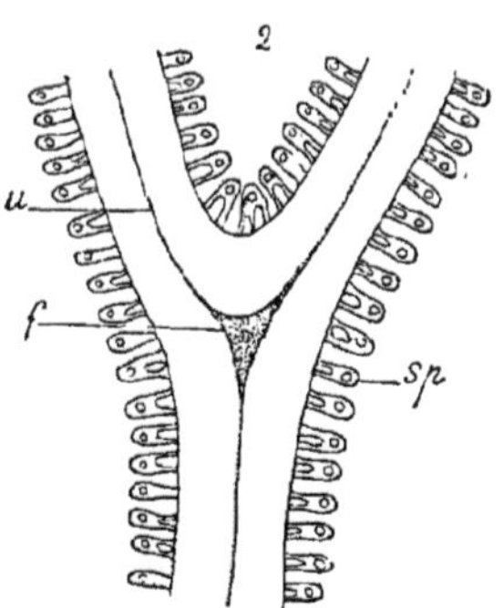

FIG. 70.—Spongin-fibre of *Euspongia irregularis* (from Perrier). *u*, *f* axial medulla of the fibre; *sp* spongoblasts.

The spicules are of two kinds—the large spicules or *megascleres* (essential spicules of Bowerbank, skeletal spicules of Carter), and the small spicules or *microscleres* (auxiliary spicules of Bowerbank, flesh spicules of Carter). The megascleres (Figs. 71, 72, and 78) are embedded in spongin fibres which may be either reticulate or radiating in arrangement; or if there is no spongin, they are held together by strong connective tissue bands. In addition there is usually a number of megascleres scattered irregularly through the tissues. In the *Lithistida* the spicules which are called *desmas* are articulated together so as to form a network. The microscleres (Fig. 73), which are not really sharply distinguishable from the megascleres—for the two pass into one another—do not, as a rule, take part in forming the supporting skeleton; they are embedded in the mesoderm and sometimes project into the canals. The spicules near the surface of the body are often differently arranged to those of the main skeleton. It is well known that some sponges shed an immense number of spicules; and it appears that in many cases the spicules are continually being moved towards the surface, where they are cast off, and replaced by spicules formed in the central parts of the sponge.

The spicules of sponges in the diversity, symmetry, and intricacy of their form, in the perfection and finish of their architecture, constitute some of the most astonishing objects in natural history. In view of them it is impossible to regard sponges as low in the scale of evolution: such finish and such perfection of structure can only have been reached as the result of a long process of evolutionary changes. While it is pretty clear that the main function of the skeletal structures is the support and protection of the sponge body, it is by no means easy to give explanations of the diversity and complexity of form which they present. The form of the megascleres is probably connected with the form of the canal system, with which they are in relation (F. E. Schulze); but the form and even the existence of the microscleres defies any reasonable explanation. By some spongologists the small spicules are regarded as functionless, and as having on that account a greater value for classificatory purposes. However this may be, all classes of skeletal structures are utilised in classification, and although no single character is by itself of much use for this purpose, the form of the microscleres is perhaps as important as the form of the megascleres.

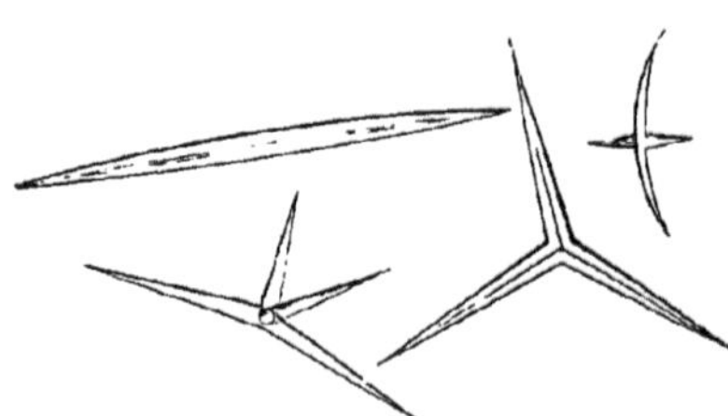

FIG. 71.—Calcareous spicules (triods) of *Sycon*.

It is therefore necessary to consider in some detail the forms of sponge spicules.

MEGASCLERES.

There are five principal kinds of megascleres: (1) *Monaxons*—rod-like megascleres; (2) *Tetraxons*—megascleres with four axes, proceeding from a central point, with four rays; (3) *Triaxons*—megascleres with three axes crossing at right angles, with six rays; (4) *Polyaxons*—megascleres with several axes; (5) *Spheres*—megascleres in which growth is concentric about the origin.

Each of these classes contains many varieties, the most important of which must be dealt with.

I. **Monaxons,** megascleres in which growth is directed from a single origin in one or both directions along a single axis. The ray or rays of a monaxon are called *actines*.

1. **Rhabdus,** growth proceeds in both directions along the axis—hence *diactine*. The principal varieties of Rhabdus are as follows:—

With similar ends:

a. *Oxea*—needle-shaped, pointed at both ends (Fig. 72, *7*).

b. *Tornote*—abruptly pointed at each end (*8*).

c. *Strongyle*—rounded at each end (*9*).
d. *Tylote*—knob-like thickening at each end (*10*).
With dissimilar ends:
e. *Strongyle-oxea*—oxeate externally.
f. *Tylotoxea* ,, ,, (Fig. 72, *12*).
g. *Oxystrongyle* ,, internally.
h. *Oxytylote* ,, ,, (Fig. 72, *11*).
i. *Oxyclad*—externally ends in two or more secondary actines or **cladi.**
k. *Strongyloclad*—ectactine cladose.
l. *Tyloclad* ,, ,,

The two latter have generally three cladi, in which case we have the characteristic spicule of the *Tetractinellida*, the **Triæne.**

The **Triæne** consists of the *rhabdome*, or shaft, and the *cladome*, which consists of the three cladi, a straight line joining the ends of the two cladi is the *chord.* The *sagitta* is a perpendicular from the origin of the cladome to the chord.

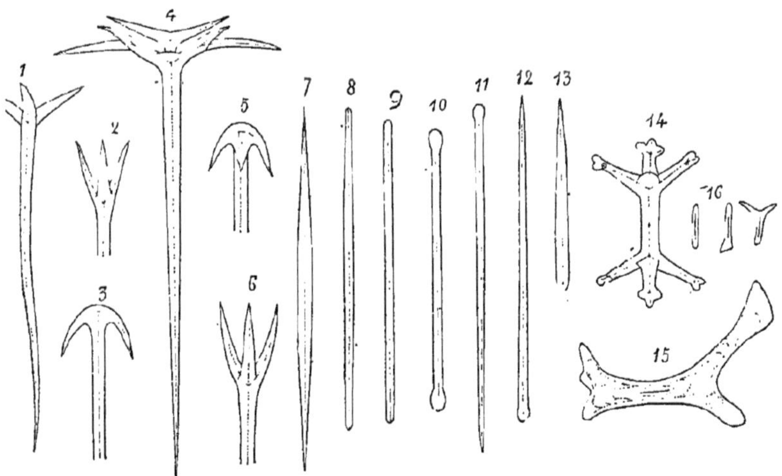

FIG. 72.—Megascleres of the *Monactinellida* and of the *Tetractinellida*. *1*, plagiotriæne; *2*, *6*, protriæne; *3*, *5*, anatriæne; *4*, dichotriæne; *7*, oxea; *8*, tornote; *9*, strongyle; *10*, tylote; *11*, oxytylote; *12*, tylotoxea; *13*, style; *14*, *15*, *16*, desmas (after Sollas, from Perrier).

Varieties of Triæne:
a. *Anatriæne* (anchor)—cladi directed backwards (Fig. 72, *3*, *5*).
b. *Protriæne* (Fig. 72, *2*, *6*), } cladi directed forwards at different angles with the rhabdome.
c. *Plagiotriæne* (Fig. 72, *1*), }
d. *Orthotriæne* }
e. *Dichotriæne*—cladi dichotomous (Fig. 72, *4*).
f. *Trichotriæne*—cladi trifurcate.
g. *Phyllotriæne*—with lamellar cladi, found only in *Lithistida.*
h. *Discotriæne*—cladome is a disc in which the separate cladi are not distinguishable; only in *Lithistida.*
i. *Amphitriæne*—both ends of rhabdome end in a triæne.
k. *Centrotriæne*—the cladi arise from the centre of the rhabdome.

2. **Stylus**, growth proceeds along the axis in one direction only—hence *monactine* :

a. *Style*—strongylote at origin (Fig. 72, *13*).

b. *Tylostyle*—tylote at origin.

Desmas (Fig. 72, *14*, *15*, *16*) are megascleres which form the skeletal network of the *Lithistida*. They are usually formed by the deposition of successive layers of silica upon an ordinary spicule—the *crepis*—the axial rod of which is arrested in development. When the crepis is a monaxon we get a *rhabdocrepid desma*, and when a tretraxon the desma is called *tetracrepid*.

II. **Tetraxons** are megascleres with four actines inclined at an angle of about 110° to one another. Growth proceeds in one direction only along each of the four axes. There are two kinds: (a) the *calthrops*, when all four actines are present; and (b) the *triod* (Fig. 71), when one actine is suppressed, the remaining three lying in one plane. The triod is characteristic of the calcareous sponges.

III. The **Triaxons** (Fig. 78) are megascleres with six actines, consisting typically of a system of three equal axes intersecting at right angles. They are characteristic of the *Hexactinellida*.

IV. **Polyaxons** are megascleres with several axes proceeding from a centre.

V. **Spheres** are megascleres in which growth is concentric round a central point.

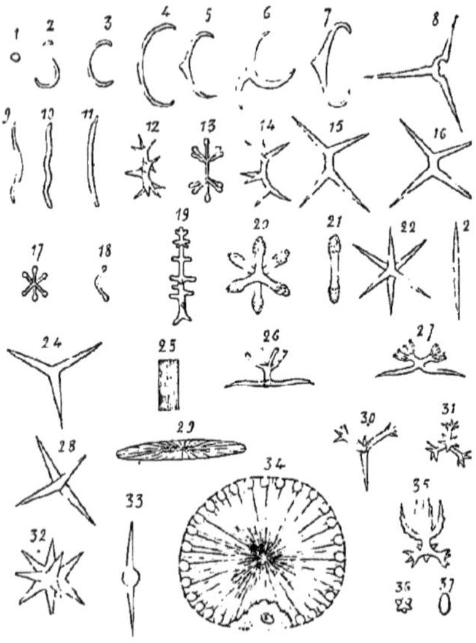

FIG. 73.—Microscleres (from Perrier, after Sollas). *1*, globule; *2*, *3*, sigma spires; *4*, *5*, *7*, sigmas; *6*, microtriod; *8*, transition between sigma and microtriod; *9*, toxa; *10*, spirula; *11*, microstrongyle; *12*, spiraster; *13*, amphiaster; *14*, metaster; *15*, *16*, plesiaster; *17*, chiaster; *18*, spiraster; *19*, samidaster; *20*, anthaster; *21*, microxea; *22*, oxyaster; *23*, microxea; *24*, microtriod; *25*, orthodragma; *26*, monolophous microcalthrops; *27*, dilophous microcalthrops; *28*, simple microcalthrops; *29*, elongated sterraster; *30*, trilophous microcalthrops; *31*, tetralophous microcalthrops; *32*, spheraster; *33*, centrotylote; *34*, sterraster; *35*, candelabrum; *36*, pycnaster; *37*, microstrongyle.

MICROSCLERES.

Microscleres (auxiliary spicules of Bowerbank, flesh spicules of Carter) are of two chief kinds (Fig. 73).

1. **Spires** are microscleres with a spiral twist. There are many varieties, of which we may mention :

a. *Sigmaspire*—a C- or S-shaped spicule according to the aspect (*2*, *3*).

b. *Sigma*—a C-shaped spicule, not spirally twisted (*5*, *6*, *7*).

c. *Toxa*—a bow-shaped spicule, not spirally twisted (*9*).

d. *Chela*—a more or less curved shaft, bearing at each end a variable number of recurved processes.

2. **Asters** are multiactinate microscleres. There are two chief kinds; (A) *asters* or *euasters*, in which the actines proceed from a centre, and (B) *streptasters*, in which the actines proceed from an axis which is usually spiral.

A. **Euasters** are in many varieties; but of these we need only note the *sterraster*, in which the actines are numerous and soldered together by subsequently deposited silica, which extends almost as far as their extremities (*29*). Other varieties are the *chiaster* (*17*), the *pycnaster* (*36*), the *oxyaster* (*22*), and the *spheraster* (*32*).

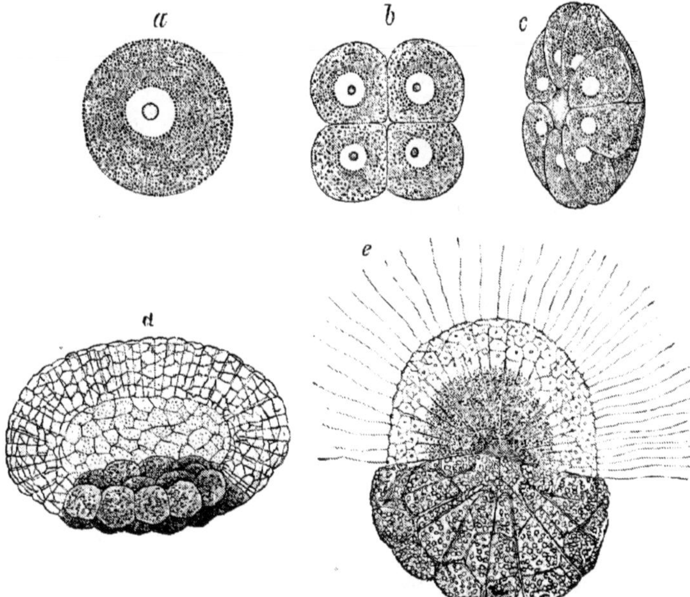

FIG. 74.—Development of *Sycon raphanus* (after F. E. Sch.). *a*, ripe ovum; *b*, stage with four segments; *c*, stage with sixteen segments; *d*, blastosphere with large dark granular cells at the open pole; *e*, free-swimming larva, one half of the body (endodermal) being formed of long ciliated cells, the other (ectodermal) of large granular cells.

B. The **Streptasters** are also various. There are the *spiraster* (*12*), the *metaster* (*14*), the *plesiaster* (*15, 16*), the *sanidaster* (*19*), and the *amphiaster* (*13*). In the *amphiaster* the actines form a whorl at each end of the axis, which is straight.

C. **Reduced asters**, in which the actines are few and variable. Thus we get *microrhabds*, *microcalthrops* (*26–31*), *microtriods* (*24*). There are several varieties of the microcalthrops depending on the branching of some of the actines. Thus there is the *monolophous microcalthrops* (*26*) with one cladose (branched) actine, the *dilophous* (*27*), *trilophous* (*30*), and *tetralophous* (*31*) *microcalthrops* with two, three, and four cladose actines. The *candelabrum* (*35*) is a tetralophous microcalthrops, in which one actine differs from the three others, which are similar to one another. The *bacillus* of Carter is a microstrongyle.

Rhaphides are long hair-like microscleres not in sheaves.

Dragmas are microscleres, several of which are secreted in a single cell or scleroblast. They lie in sheaves.

Trichodal is a term applied to any hair-like spicule.

Sexual reproduction was first demonstrated with certainty by Lieberkühn for *Spongilla*, but more recently has been shown to exist throughout the group. In most cases the ova and spermatozoa seem to reach maturity at different times in the same sponge. The spermatozoa have a small head, and lie in small spaces lined with cells. The ova, like the mother-cells of the spermatozoa, are modified cells of the mesoderm. They are naked amœboid cells, and are fertilised and undergo their first development in the mesoderm. They leave the sponge by passing into the canal-system as ciliated larvæ, which after a brief free-swimming life attach themselves and develop into a young sponge.

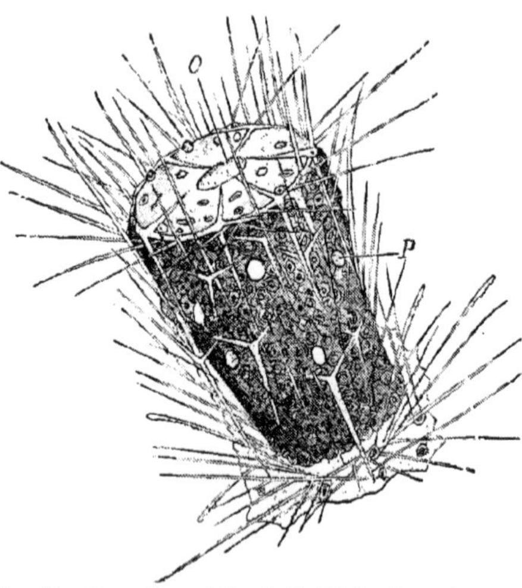

FIG. 75.—Young *Sycon* (after F. E. Sch.). *O* osculum, or exhalent aperture; *P* pores (inhalent) of the wall.

An invaginate gastrula is sometimes formed (*Sycon*, *Oscarella*, Fig. 76), in which case it is observed that the Sponge-larva attaches itself by the blastopore surface, and develops the osculum as a new formation. In other cases a solid morula is formed. The recent important researches of Delage* have shown that the mode of development first revealed by the researches of F. E. Schulze in the *Calcarea* is found throughout the group. According to the results of these observers it appears that the locomotive cells of the larva, which are the first differentiated tissues (Fig. 74*e*), become internal and transformed into the flagellated cells of the endoderm, while the other cells, more or less indifferent in the larva, become transformed into the ectoderm and mesoderm of the adult.

Asexual reproduction is found throughout the Porifera. The presence of

* "Embryogenie des Éponges etc." *Arch. Zool. expér. et gén.* Tom. x., 1892.

more than one osculum is often regarded as a case of incomplete budding. More unequivocal cases are, however, furnished by the external budding found in *Thenea*, *Tethya*, *Lophocalyx*, *Polymastia*, *Oscarella*, etc., and the internal budding, known as the formation of *gemmules*, found in fresh-water and in some marine sponges (Topsent). Gemmules are masses of parenchyma cells containing yolk grains and surrounded by a shell composed of a thick cuticular layer, to which silicious structures are often added. The shell possesses an aperture or micropyle, and the whole structure is to be regarded as a portion of the mesoderm cut off from the rest of the sponge. In *Spongilla* the shell contains the characteristic spicules known as amphidiscs (Fig. 78, 9).

That the power of asexual increase and repair of lost parts is probably a widespread phenomenon in the group is indicated by the fact that sponges can be propagated by artificial fission. It has been attempted, with a certain amount of success, to utilise this property for the purpose of increasing the number of

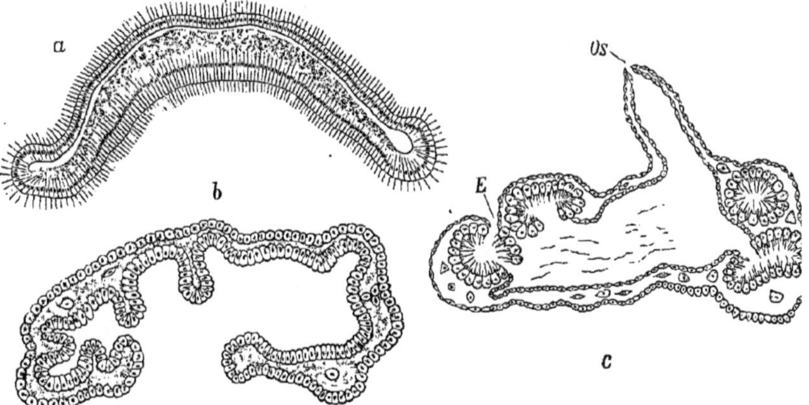

FIG. 76.—Sections through three stages of the development of *Halisarca* (*Oscarella*) *lobularis* (after C. Heider). *a*, gastrula after its fixation; *b*, formation of mesoderm; *c*, development of the osculum (*Os.*), and of the flagellated chambers; *E*, inhalent pore.

marketable sponges in the Mediterranean and in other sponge-growing seas. The marketable sponges belong to the species *Euspongia officinalis*, the Turkey or Levant sponge; *Hippospongia equina*, the horse sponge; and *Euspongia zimocca*, the zimocca sponge. They occur all along the Mediterranean coast to a depth of 200 fathoms, and in many other parts of the world. The sand found in new sponges is an adulteration to increase the weight.

Sponges are found all over the world and at all depths of the ocean. One family only (*Spongillidæ*) is found in fresh-water. The *Tetractinellida* are found in deep and shallow water; but when in deep water, generally near land. The *Monaxonida*, which comprise by far the greatest number of living sponges, also cling to the land; while the *Hexactinellida*, though found in the middle of the great oceans, are more numerous near the land. The *Hexactinellida* are mainly deep-water forms, the characteristic depth being between 200 and 1000 fathoms. For the *Tetractinellida* the characteristic depth is 50 to 200 fathoms, though they are also found in shallow water. The *Calcarea* and *Ceratosa* are mainly shallow-water forms (to a depth of 200 fathoms). The *Monaxonida*, though characteristically inhabitants of shallow water, are found in considerable numbers at all depths. The deep-sea *Monaxonida* are distinguished by their symmetrical and definite shapes.

All spongologists agree as to the immense difficulty of classifying sponges. Not only are the boundaries of the great groups difficult to lay down, but the limits of species, and even of genera, often defy definition. Almost all characters are highly variable, and the number of intermediate forms and of collateral affinities is immense. On the whole we may distinguish three main types, which we shall exalt to the dignity of classes—not because they deserve that rank, for they do not; but because the term fits in more conveniently with the terminology generally used for the group. The three classes are as follows:—

Calcarea. With calcareous spicules and large choanocytes.

Order 1. *Calcarea.*

Triaxonia. With triaxonic (sex-radiate) spicules and large flagellated chambers.

Order 2. *Hexactinellida.*
,, 3. *Hexaceratina.*

Demospongiæ. Without triaxonic spicules; with small choanocytes and ciliated chambers; skeleton of silicious spicules or spongin, or both combined.

Order 4. *Tetractinellida.*
,, 5. *Carnosa.*
,, 6. *Monaxonida.*
,, 7. *Ceratina.*

While these groups stand out fairly sharply, their division into sub-groups, or orders, as we must call them, is fraught with some difficulty. This is particularly the case with the *Demospongiæ*, which include the great majority of living sponges.

The sub-divisions of the *Demospongiæ* must be regarded as entirely artificial, and only to be established for the convenience of the student. The four orders into which we have divided the class are by no means sharply marked off from one another, nor do they form a single series, but rather three or four series running parallel to one another, and connected together at several points. In fact, if we had to set forth in a pictorial manner the true affinities of the families of this sub-class, we should be obliged to use a network or, better still, a spongework, rather than a tree or a line, because so many of the families present affinities of apparently equal importance in more than one direction. And this applies with equal force to genera and species. If this view of a reticular arrangement rather than that of a tree arrangement were generally held, we feel assured that a good many of the difficulties of sponge classification would be disposed of. It is the genealogical tree idea which makes the difficulty.

Fossil sponges are found in various formations, *e.g.*, in the chalk, and these remains appear to differ considerably from those now living. The *Hexactinellida*, however, agree so fully with the ancient forms that they might be direct descendants of them. Finally many of the principal groups extend back into the palæozoic age, in which the *Lithistida* and *Hexactinellida* especially are found in the most ancient Silurian strata.

Class I. CALCAREA.*

Sponges with a calcareous skeleton and large choanocytes.

Order 1. Calcarea.

With the characters of the class.

Sub-order 1. HOMOCŒLA.

Calcarea without flagellated chambers; the internal surface entirely lined by collared cells.

Fam. 1. **Asconidæ.** Gastric cavity a simple sac. *Leucosolenia* Bow.

Poléjaeff and Dendy both unite all *Homocœla* in one genus, for "the spicules of the *Calcarea*, being very variable in every direction, could not serve as a basis for the distinction of genera." Haeckel's genera therefore go. Dendy distinguishes three sections of the genus: (1) Olynthus types which do not form colonies, or, if they do form colonies, in which the individuality of the members of the colony (Ascon-persons) is always recognisable (Fig. 77); (2) colonies the members of which anastomose and form a network; (3) colonies consisting of a central Ascon tube from which other tubes are radially budded off.

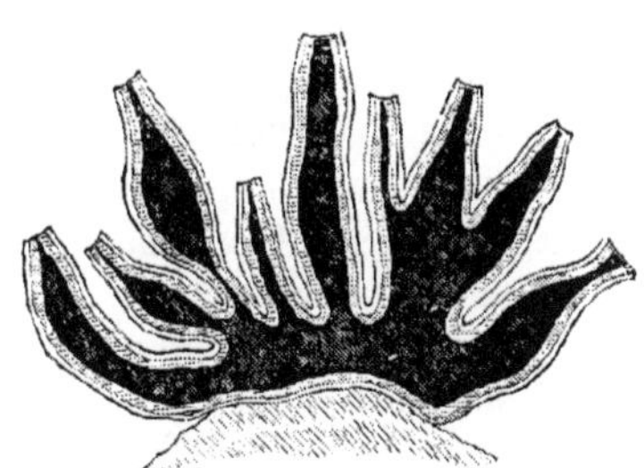

Fig. 77.—Section through an Ascon-colony, diagrammatic (after Haeckel).

* A. Dendy, "A Monograph of the Victorian Sponges," part 1. *Transactions of the Royal Society of Victoria*, vol. 3, part 1. A. Dendy, "Observations on the Structure and Classification of the Calcarea Heterocœla." *Q. J. M. S.*, vol. 35, 1893, p. 159. N. Poléjaeff, "Report on the Calcarea." *Challenger Reports*, vol. 8, 1883.

Sub-order 2. **HETEROCŒLA.**

Calcareous sponges in which the collared cells are confined to more or less well-defined flagellated chambers.

Fam. 1. **Leucascidæ.** Flagellated chambers branched, opening into exhalent canals which converge towards the oscula; their outer ends covered over by a dermal poriferous membrane. Skeleton of irregularly scattered radiate spicules. *Leucascus* Dendy.

Fam. 2. **Sycettidæ.** The blind outer ends of the flagellated chambers projecting freely on the surface and not covered by a dermal cortex. Chamber skeleton articulate. *Sycetta* H.; *Sycon* Risso; *Sycantha* Lendf.

Fam. 3. **Grantiidæ.** With poriferous dermal cortex covering the chamber layer; without subdermal sagittal triradiates. Chamber skeleton varies from regularly articulate to irregularly scattered. *Grantia* Fleming; *Ute* O.S.; *Utella* Dendy; *Anamixilla* Pol.; *Sycyssa* H.; *Leucandra* H.; *Lelapia* Gray; *Leucyssa* H.

Fam. 4. **Heteropidæ.** With poriferous dermal cortex; with subdermal sagittal triradiates. An articulate chamber skeleton present or absent. *Grantessa* Lendf.; *Heteropia* Carter; *Vosmæropsis* Dendy.

Fam. 5. **Amphoriscidæ.** With poriferous dermal cortex; conspicuous subdermal quadriradiate spicules with inwardly directed apical rays are present. *Heteropegma* Pol.; *Amphoriscus* H.; *Syculmis* H.; *Leucilla* H.

Class II. TRIAXONIA.

With triaxonic (sex-radiate) spicules and large flagellated chambers.

Order 1. HEXACTINELLIDA.*

Sponges with very loose soft parts, and with silicious spicules which are either isolated or united into a connected framework, and belong or are reducible to the triaxial system. Canal system simple, with syconate chambers.

In the *Hexactinellida* the ectoderm is a thin layer containing nuclei, but without discernible cell outlines. The lining cells of the flagellated chambers project into the cavity of the chamber, and stand some distance apart from one another. Their bases are connected by basal strands which are apparently processes of the cells themselves. No flagella or collars have so far been detected, but there can be but little doubt that they exist as in other sponges.

The spicules contain a central canal filled with a soft granular substance which is continuous with adjacent structures through openings at the end of the rays. These are closed when the spicule has ceased to grow. The spicules (Fig. 78) typically consist of a system of three equal axes intersecting at right angles, and variations in this type are due to the unequal development or branching of

* F. E. Schulze, "Report on the Hexactinellida." *Challenger Reports*, vol. 21, 1887.

the rays, or to the suppression of some of them. The variations in the form of spicule may be classed in six main groups—hexacts (Fig. 78, *2*), pentacts, tetracts, triacts (*7*), diacts, and monacts. The spicules when united are generally bound together by a fine laminated silicious substance.

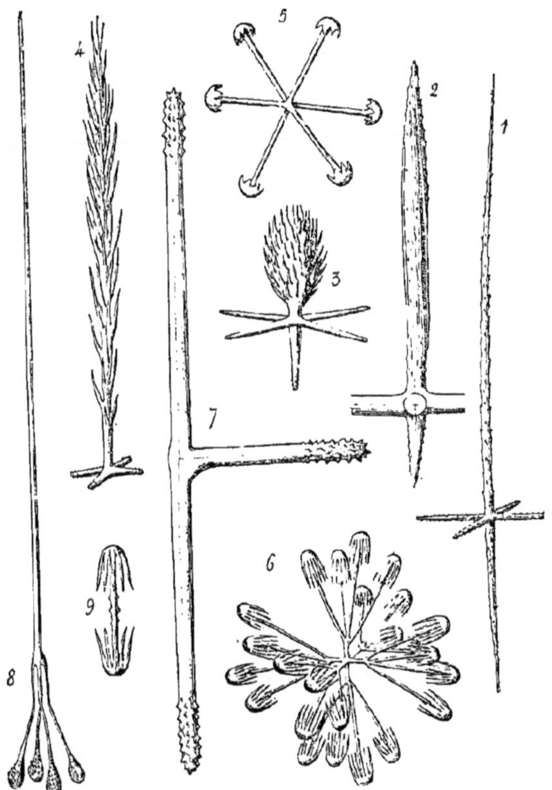

FIG. 78.—Spicules of *Hexactinellida*. *1*, an auto-gastral pinulus of *Sympagella nux*; *2*, hexact of *Holascus fibulatus* to which are applied small diacts; *3*, pinulus of *Caulophacus latus*; *4*, pentact pinulus of *Hyalonema lusitanicum*; *5*, discohexact of *Tægeria pulchra*; *6*, discohexacter of *Dictyocalyx gracilis*; *7*, triact of *Hyalonema gracilis*; *8*, scopula of *Eurete semperi*; *9*, amphidisc of *Hyalonema Sieboldii*. (From Perrier after Schulze.)

Prostalia are the spicules which occur over the outer surface of the sponge. They are of three kinds: *basalia* or rooting spicules, *pleuralia* at the sides, and *marginalia* round the osculum.

Dermalia are spicules in relation with the bounding membrane of the sponge, and are of two kinds: *autodermalia* in the dermal membrane, and *hypodermalia* beneath the dermal membrane.

Gastralia are spicules on the gastral membrane or membrane lining the central chamber of the sponge.

Parenchymalia are spicules in the parenchyma.

Dictyonalia are the parenchymalia which become fused to form the continuous skeletal framework of the *Dictyonina*.

Synapticula are bridges of silica connecting neighbouring spicules.

Oxyhexacts, hexacts with axes running to a point.

Discohexacts, hexacts with axes enlarged at the extremity.

Hexasters, hexacts with axes branching into rays at their extremity.

Pinulus, a pentact or hexact in which one ray bears oblique lateral teeth or prickles (Fig. 78, *1*).

Amphidiscs, a diact at each end of which a convex expansion occurs, which bears six or more backwardly bent marginal teeth (Fig. 78, *9*).

Uncinate, a straight rod, pointed at both ends and beset all over with barbs pointing in the same direction.

Clavula, a rod which bears at one end a club-shaped or transverse discoidal expansion.

Scopula, a rod which bears at one end a number of rays.

Sub-order 1. **LYSSACINA.**

Hexactinellida in which the spicules either remain altogether isolated, or are in part subsequently and irregularly united by siliceous matter or transverse synapticula.

Tribe 1. **HEXASTEROPHORA.** Hexasters always present in the parenchyma. The ciliated chambers are sharply separated from one another, and thimble-shaped.

Fam. 1. **Euplectellidæ.** Dermal skeleton contains sword-shaped oxyhexacts with long proximal ray. *Euplectella* Owen; *Regadrella* O. S.; *Holascus* F. E. S.; *Malacosaccus* F. E. S.; *Taegeria* F. E. S.; *Walteria* F. E. S.; *Habrodictyum* W. Th.; *Eudictyum* Marshall; *Dictyocalyx* F. E. S.; *Rhabdodictyum* O. S.; *Rhabdopectella* O. S.; *Hertwigia* O. S.; *Hyalostylus* F. E. S.

Fam. 2. **Asconematidæ.** The dermal and gastral skeletons contain pentact or hexact pinuli. The hypodermalia and hypogastralia are pentacts with parenchymal discohexasters. *Asconema* S. Kent; *Aulascus* F. E. S.; *Sympagella* O. S.; *Polyrhabdus* F. E. S.; *Balanites* F. E. S.; *Caulophacus* F. E. S.; *Trachycaulus* F. E. S.

Fam. 3. **Rossellidæ.** Dermalia always without a distal radial ray. *Lanuginella* O. S.; *Polylophus* F. E. S.; *Rossella* Carter; *Acanthascus* F. E. S; *Bathydorus* F. E. S.; *Rhabdocalyptus* F. E. S.; *Crateromorpha* Gray; *Aulochone* F. E. S.; *Caulocalyx* F. E. S.; *Aulocalyx* F. E. S.; *Euryplegma* F. E. S.

Tribe 2. **AMPHIDISCOPHORA.** Amphidiscs always present in limiting membranes. Parenchyma without hexasters. Anchoring basalia always present. Chambers not thimble-shaped nor sharply marked off from one another, but forming irregular diverticula of the membrana reticularis.

Fam. 1. **Hyalonematidæ.** Both dermal and gastral membranes contain numerous pentact pinuli. *Hyalonema* Gray; *Pheronema* Leidy; *Poliopogon* Wy. Th.; *Semperella* Gray.

Sub-order 2. DICTYONINA.

Hexactinellida in which the large parenchymal hexacts are from the first more or less regularly united as dictyonalia in a firmly-connected framework.

Tribe 1. **UNCINATARIA.** With uncinates.

A. *With clavulæ.*

Fam. 1. **Farreidæ.** *Farrea* Bow.

B. *With radially disposed scopulæ.*

Fam. 2. **Euretidæ.** *Eurete* Carter; *Periphragella* Marshall; *Lefroyella* Wy. Th.

Fam. 3. **Melittionidæ.** *Aphrocallistes* Gray.

Fam. 4. **Coscinoporidæ.** *Chonelasma* F. E. S.

Fam. 5. **Tetrodictyidæ.** *Hexactinella* Carter; *Cyrtaulon* F. E. S.; *Fieldingia* S. Kent; *Sclerothamnus* W. Marshall.

Tribe 2. **INERMIA.** Without uncinates or scopulæ.

Fam. 1. **Mæandrospongidæ.** *Dactylocalyx* Stutchbury; *Margaritella* O. S.; *Scleroplegma* O. S.; *Myliusia* Gray; *Aulocystis* F. E. S.

Order 2. Hexaceratina.

Sponges with large and saccular ciliated chambers, with simple canals. Skeleton composed of soft horny fibres, sometimes accompanied by horny spicules. The skeleton may be absent.

Fam. 1. **Darwinellidæ.** Skeleton comprises both horny fibres and horny spicules. *Darwinella* Fr. Müller.

Fam. 2. **Aplysillidæ.** Skeleton is composed of horny fibres only. *Ianthella* Gray; *Aplysilla* F. E. S.; *Dendrilla* Lendf.

Fam. 3. **Halisarcidæ.** Skeletal structures absent; chambers syconate. *Halisarca* Duj.; *Bajulus* Lend.

Class III. DEMOSPONGIÆ.

Without triaxonic spicules, with small choanocytes and small ciliated chambers. Skeleton of silicious spicules, or of spongin, or of both combined.

Order 1. Tetractinellida.*

Skeleton characterised by triæne or tetraxon megascleres, or lithistid desmas.

The genus *Placospongia*, which is without the skeletal characters, is included here because of the presence of sterrasters which are not found outside the group.

Sub-order 1. CHORISTIDA.

Tetractinellida in which lithistid desmas are absent, and the megascleres are never articulated to form a coherent skeleton.

Tribe 1. **SIGMATOPHORA.** Microsclere when present a sigmaspire.

* W. J. Sollas, "Report on the Tetractinellida." *Challenger Reports*, vol. 25, 1888. E. Topsent, "Étude Monographique des Spongiaires de France. I. Tetractinellida." *Arch. d. Zool. expér. et gén.* (3) T. 2, 1894.

Fam. 1. **Tetillidæ.** The characteristic megasclere a protriæne. *Tetilla* O. Schm.; *Chrotella* Sollas; *Cinachyra* Soll.; *Craniella* O. Schm.

Fam. 2. **Samidæ.** Characteristic megasclere an amphitriæne. *Samus* Gray.

Tribe 2. **ASTROPHORA.** One or more of the microscleres is an aster.

Fam. 1. **Theneidæ.** Microscleres are spirasters or amphiasters, and oxyasters or microxeas. Without cortex. *Thenea* Gray; *Characella* Soll.; *Pœcillastra* Soll.; *Sphinctrella* O. Schm.; *Triptolemus* Soll.; *Stuba* Soll.; *Nethea* Soll.; *Placinastrella* F. E. Sch.

Fam. 2. **Pachastrellidæ.** The chief megascleres are calthrops; triænes absent. Microscleres may be spirasters, spherasters, or microrabds. *Pachastrella* O. Schm.; *Dercitus* Gray; *Calthropella* Soll.

Fam. 3. **Stellettidæ.** Euasters always present, but never spirasters or sterrasters. With triænes; without calthrops. Chamber-system aphodal, and choanosomal mesoderm sarcenchymatous. Megascleres generally arranged on the radiate type. *Myriastra* Soll.; *Pilochrota* Soll.; *Astrella* Soll.; *Anthastra* Soll.; *Stelletta* O. Schm.; *Dragmastra* Soll.; *Aurora* Soll.; *Ancorina* O. Schm.; *Tribrachium* Weltner; *Tethyopsis* Stewart; *Disyringa* Soll.; *Stryphnus* Soll.; *Ecionema* Bow.; *Papyrula* O. Schm.; *Psammastra* Soll.; *Algol* Soll.

Fam. 4. **Geodiidæ.** The characteristic microsclere is a sterraster. With triæne megascleres. The sterrasters in the cortex are united together by fusiform fibrillated cells. *Erylus* Gray; *Caminus* O. Schm.; *Pachymatisma* Bow.; *Cydonium* Fleming; *Geodia* Lamarck; *Synops* Vos.; *Isops* Soll.

Fam. 5. **Placospongidæ.** The characteristic microsclere is a sterraster. The only megascleres are tylostyles; triænes absent. Often placed with the *Suberitidæ* (Monaxonid) on account of tylostyles. *Placospongia* Gray.

Tribe 3. **MEGASCLEROPHORA.** Without microscleres.

Fam. **Tethyopsillidæ.** *Proteleia* R. and D.; *Tethyopsilla* Lendf.

Sub-order 2. LITHISTIDA.

Tetractinellida provided with a consistent skeleton by the zygosis of modified spicules or desmas.

Tribe 1. **HOPLOPHORA.** With special ectosomal spicules, and usually some form of microsclere.

A. *Ectosomal spicules as triænes. Aphodal.*

Fam. 1. **Tetracladidæ.** Desma tetracrepid. *Theonella* Gray; *Discodermia* Bocage; *Racodiscula* Zittel; *Kaliapsis* Bow *Neosiphonia* Soll.; *Rimella* O. Schm.; *Collinella* O. Schm.; *Sulcastrella* O. Schm.

Fam. 2. **Corallistidæ.** Desma monocrepid (*i.e.*, with crepis as monaxon) and tuberculated. Aphodal. *Corallistes* O. Schm.; *Macandrewia* Gray; *Dædalopelta* Soll.; *Heterophymia* Pomel; *Callipelta* Soll.

Fam. 3. **Pleromidæ.** Desma monocrepid and smooth. Aphodal. *Pleroma* Soll.; *Lyidium* O. Schm.

B. *Ectosomal spicules as microstrongyles or modified microstrongyles (discs). Desmas monocrepid.*

Fam. 4. **Neopeltidæ.** Ectosomal spicules as monocrepid discs. *Neopelta* O. Schm.

Fam. 5. **Scleritodermidæ.** Ectosomal spicules as microstrongyles and the other microscleres as sigmaspires. *Scleritoderma* O. Schm.; *Aciculites* O. Schm.

Fam. 6. **Cladopeltidæ.** Ectosomal spicules as a monocrepid desma highly branched in a plane parallel to the surface. Microscleres absent. *Siphonidium* O. Schm.

Tribe 2. **ANOPLIA.** Special ectosomal spicules and microscleres absent.

Fam. 1. **Azoricidæ.** Desmas monocrepid. *Azorica* Carter; *Tretolophus* Soll.; *Gastrophanella* O. S.; *Setidium* O. S.; *Poritella* O. S.; *Amphibleptula* O. S.; *Tremaulidium* O. S.; *Leiodermatium* O. S.; *Sympyla* Soll.

Fam. 2. **Anomocladidæ.** Desma acrepid, consisting of a variable number of smooth cylindrical cladi radiating from a thickened centrum. Zygosis between the expanded ends of the cladi of one desma and the centrum of another. *Vetulina* O. S.

Order 2. Carnosa.*

Sponges without megascleres; microscleres belonging to the tetraxial type present or absent.

Tribe 1. **MICROSCLEROPTERA.** Microscleres are variously modified tetractinose asters, candelabras, or minute triænes.

Fam. 1. **Plakinidæ.** With tetractinose, triactinose, diactinose asters, and sometimes mono-, di-, or trilophous candelabra. Chamber-system eurypylous or aphodal; mesoderm chiefly collenchymatous. The sponge is divided into a hypomere and a spongomere. *Plakina* F. E. S.; *Placortis* F. E. S.

Fam. 2. **Corticidæ.** Tetractinose asters and candelabras. Aphodal or diplodal. In part sarcenchymatous, in part collenchymatous. *Corticium* O. Schm.; *Calcabrina* Soll.; *Corticella* Soll.; *Rhacella* Soll.

Fam. 3. **Thrombidæ.** Trichotriænes, and sometimes a peculiar form of amphiaster. Ectosome thin and not sharply marked off from choanosome. Diplodal. Mesoderm densely collenchymatous with numerous large granular cells in addition to collencytes. *Thrombus* Soll.

Fam. 4. **Astropeplidæ.** With microxeas and asters; the microxeas are arranged tangentially to the walls of the canal-system, forming a loose felt; eurypylous; ectosome not a cortex. *Astropeplus* Soll.

Tribe 2. **OLIGOSILICINA.** With asters only.

Fam. **Chondrillidæ.** *Chondrilla* O. S.

Tribe 3. **MYXOSPONGIDA.** Without spicules.

Fam. **Gumminidæ.** *Chondrosia* O. S., diplodal rhagon chambers; *Oscarella* Vos., chambers eurypylous and rhagose.

Order 3. Monaxonida.†

Silicious skeleton with uniaxial megascleres.

The *Monaxonida* are related to the *Tetractinellida* through the *Astropeplidæ* on the one hand, and *Plakina* on the other. In fact it is difficult to see why these forms are placed in separate orders.

Sub-order 1. HALICHONDRINA.

Typically non-corticate; skeleton usually reticulate; megascleres usually either oxea or styli.

Fam. 1. **Homorraphidæ.** Megascleres all diactinal, either oxea or strongyla; no microsclera.

* E. Topsent, *Arch. Zool. expér.* (3), 3, 1895.

† Ridley and Dendy, "Report on the Monaxonida." *Challenger Reports*, vol. 20, 1887. E. Topsent, "Exposé des Principes actuels de la Classification des Spongiaires." *Revue Biologique du Nord de la France*, t. 4. *Id.*, "La classification des Halichondrina." *Mém. Soc. Zool. France*, 1894.

Sub-fam. 1. **Renierinæ.** Spicules may be enveloped by a small proportion of spongin, but are never completely enveloped in it. *Halichondria* Fleming, littoral; *Petrosia* Vos.; *Reniera* Nardo; *Calyx* Vos.

Sub-fam. 2. **Chalininæ.** Spongin plentiful; spicules enveloped and united by it. *Pachychalina* Schm.; *Chalina* Grant; *Siphonochalina* O. Schm.; *Cacochalyna* O. S.; *Chalinorrhaphis* Lend.; *Hoplochalina* Lend.

Fam. 2. **Heterorraphidæ.** Megascleres of various forms; microscleres commonly present, but never chelæ.

Sub-fam. 1. **Phlœodictyinæ.** Sponge massive, with tubular processes (fistulæ) projecting from it. With a well-marked external rind. Megasclera oxea, passing into strongyla in some species. *Rhizochalina* Schmidt; *Oceanapia* Norman.

Sub-fam. 2. **Gelliinæ.** Megascleres all diactinal, oxea, or strongyla. Microsclera as sigmata or toxa; no rind or fistulæ. *Gellius* Gray; *Gelliodes* Ridley; *Toxochalina* Ridley.

Sub-fam. 3. **Tedaniinæ.** Megascleres of two forms—monactinal (styli) forming the main skeleton, diactinal (tylota or tornota) dermal. Microsclera as rhaphides. *Tedania* Gray; *Trachytedania* Ridley.

Sub-fam. 4. **Desmacellinæ.** Megascleres all monactinal, stylote to tylostylote. Microscleres sigmata or toxa, or both. *Desmacella* Schm.: *Biemma* Gray.

Sub-fam. 5. **Hamacanthinæ.** Megascleres oxea or styli. Microscleres large diancistra, and sometimes others. *Vomerula* Schm.; *Hamacantha* Gray.

Fam. 3. **Desmacidonidæ.** Megascleres of various forms, usually monactinal. Microscleres always present and always including chelæ.

Sub-fam. 1. **Esperellinæ.** Skeleton fibre not echinated by laterally projecting spicules. *Esperella* Vos.; *Esperiopsis* Carter; *Cladorhiza* M. Sars, megascleres long, often projecting radially, like spines, deep-sea; *Axoniderma* R. and D.; *Chondrocladia* W. Thomson, deep-sea; *Meliiderma* R. and D.; *Desmacidon* Bow.; *Artemisina* Vos.; *Phelloderma* R. and D.; *Sideroderma* R. and D.; *Iophon* Gray; *Amphilectus* Vos.; *Dendoryx* Gray; *Forcepia* Carter; *Yvesia* Topsent; *Melonanchora* Carter; *Damiria* Keller, etc.

Sub-fam. 2. **Ectyoninæ.** Skeleton fibre echinated by laterally projecting spicules. *Myxilla* Schmidt; *Pytheas* Tops.; *Clathria* Schm.; *Rhaphidophlus* Ehlers; *Stylostichon* Tops.; *Microciona* Bow.; *Hymeraphia* Bow.; *Ploeamia* O.S.; *Plumohalichondria* Carter; *Acarnus* Gray; *Echinoclathria* Carter; *Agelas* Duch. and Mich.; *Echinodictyum* Ridley.

Fam. 4. **Axinellidæ.** Skeleton typically non-reticulate, consisting of ascending axes of fibres from which arise subsidiary fibres radiating to the cortex. Megasclera chiefly styli, to which oxea or strongyla may be added. Microsclera rarely present, never chelate. *Hymeniacidon* Bow.; *Phakellia* Bow.; *Ciocalypta* Bow.; *Acanthella* Schm., *Axinella* Schm., generally branched; *Raspailia* Nardo, branched, whip-like; *Dendropsis* R. and D.; *Thrinacophora* Ridley; *Dictyocylindrus* Bow.

Fam. 5. **Spongillidæ.** Fresh-water sponges. Asexual reproduction by gemmules which are often surrounded by a special kind of spicules called amphidiscs. *Spongilla* Lamarck; *Ephydatia* Lamouroux; *Tubella* Carter; *Parmula* Carter; *Heteromeyenia* Potts; *Lubomirskia* Dybowski; *Lessepsia* Keller; *Uruguaya* Carter; *Potamolepis* Marshall.

Sub-order 2. **SPINTHAROPHORA.**

Usually corticate. Megascleres, as a rule, collected in fibres radially arranged from the base to the surface. Microscleres, when present, as some form of aster, never a sigma, sigma spire, or chela.

Tribe 1. **ACICULINÆ.** With diactinal megascleres.

Fam. 1. **Epallacidæ.** Megascleres as oxeas, and microscleres as some form of asters. *Epallax* Soll.; *Scolopes* Soll.; *Dorypleres* Soll.; *Amphius* Soll.; *Asteropus* Soll.; *Coppatias* Soll.

Fam. 2. **Stylocordylidæ.** Sponge differentiated into a head and stalk. Skeleton in head radiately arranged with a cortical layer of smaller spicules set radiately to the surface. Spicules in stalk set longitudinally in a dense axis. Oxeas only. *Stylocordyla* W. Th.

Fam. 3. **Tethyidæ.** Megascleres are strongyloxeas radially arranged. Microscleres are spherasters, and sometimes other forms of Euasters. *Tethya* Lam.; *Columnitis* O. S.; *Xenospongia* Gray; *Magog* Soll.; *Sollasella* Lend.

Tribe 2. **CLAVULINÆ.** With monactinal megascleres.

Fam. 1. **Spirastrellidæ.** Non-boring sponges with numerous microscleres (asters, spirasters, or discasters), typically forming a more or less continuous dermal crust. *Hymedesmia* Bow.; *Spirastrella* O. S.; *Latrunculia* Bocage; *Podospongia* Bocage, etc.

Fam. 2. **Suberitidæ.** Without microscleres of the aster type.

Sub-fam. 1. **Suberitinæ.** Skeleton not radially arranged. *Suberites* Nardo; *Weberella* Vos.; *Poterion* Schlegel, etc.

Sub-fam. 2. **Polymastinæ.** Skeleton radially arranged. *Polymastia* Bow.; *Quasillina* Norman; *Tentorium* Vos.; *Ridleia* Dendy; *Trichostemma* M. Sars, free-living, deep-sea, symmetrical forms with fringe of hair-like spicules for attachment in the mud; *Tethyspira* Tops., etc.

Fam. 3. **Clionidæ.** Boring sponges, generally with microscleres of the aster form. *Cliona* Grant; *Thoosa* Carter; *Alectona* Carter.

Order 4. CERATINA.*

Sponges with a supporting skeleton formed of spongin fibres which are without spicules; ciliated chambers saccular or piriform; microscleres present or absent.

This is undoubtedly an artificial order, and contains forms which are more closely allied to various families of the *Monaxonida* than to each other. It includes the so-called horny sponges.

Fam. 1. **Aulenidæ.** Vestibular spaces complicated; ciliated chambers very small; skeleton reticulate, formed of more or less areniferous horny fibres which do not contain proper spicules. Proper echinating spicules may, however, be attached to the superficial fibres. Allied to the *Desmacidonidæ*. *Aulena* Lendf.; *Hyattella* Lendf.

Fam. 2. **Spongidae.** With small piriform or spherical ciliated chambers; without proper spicules; not clathriform. Allied to the *Homorrhaphidæ*.

Sub-fam. 1. **Eusponginae.** Skeletal network close-meshed, fibres solid, generally with foreign bodies in the main fibres. *Chalinopsilla* Lendf.;

* R. von Lendenfeld, *A Monograph of the Horny Sponges.* Royal Society, London, 1889.

Phyllospongia Ehlers; *Leiosella* Lendf.; *Euspongia* Brown; *E. officinalis* L., the fine Turkey or Levant sponge; *Hippospongia* F. E. S.; *H. equina* O. S., the horse sponge or common bath sponge. *Coscinoderma* Carter; *Heteronema* Keller.

Sub-fam. 2. **Aplysinae.** Skeletal network loose, the axis of the fibres is occupied by a kind of pith. *Aplysina* Nardo; *Luffaria* Pol.; *Thorectandra* Lendf.; *Thorecta* Lendf.; *Aplysinopsis* Lendf.

Sub-fam. 3. **Druinellinae.** Ciliated chambers with long special efferent and afferent canals; fibres thick with irregular lobose surfaces. *Druinella* Lendf.

Sub-fam. 4. **Halminae.** Skeleton as a network of slender fibres containing at the nodes large sand-grains; or with a skeleton of loose sand-grains and dendritically-branched areniferous fibres. *Oligoceras* F. E. S.; *Dysideopsis* Lendf.; *Halme* Lendf.

Sub-fam. 5. **Stelosponginae.** Skeletal network wide-meshed, composed of solid fibres more or less fasciculated. *Stelospongia* Schmidt; *Hircinia* Nardo.

Fam. 3. **Spongelidae.** With large saccular ciliated chambers without efferent canals, a clear ground substance, and a skeleton of solid fibres without proper spicules, but containing sand-grains. Sometimes the fibres are replaced entirely by large scattered sand-grains. More closely allied to the *Heterorraphidae* than to any family of *Ceratina*.

Sub-fam. 1. **Phoriosponginae.** With microscleres. *Phoriospongia* Marshall; *Sigmatella* Lendf.

Sub-fam. 2. **Spongelinae.** Without microscleres. *Spongelia* Nardo; *Psammopemma* Marshall; *Haustia* Lendf.; *Psammoplysilla* Keller.

The deep-sea *Ceratosa* of Haeckel (*Challenger Reports*, vol. xxxii., 1889), belong to this order. He distributes the forms into four families:

Ammoconidae, without spongin skeleton; **Psamminidae**, without spongin skeleton; the **Stannomidae**, with spongin skeleton; and the **Spongelidae**. Most of them were taken at a depth of from 2000 to 3000 fathoms.

CHAPTER IV.

COELENTERATA.*

Radially symmetrical animals with only one cavity in the body—the gastrovascular space—which serves alike for digestion and circulation. The generative cells are always either ectodermal or endodermal.

The *Coelenterata*, which include polyps, corals, sea-anemones, jelly-fishes, etc., are multinucleate animals, in which the greater number of the nuclei are arranged in regular layers at the body surfaces, and constitute, with the protoplasmic layer which contains them, the external and internal epithelia, commonly called the *ectoderm* and *endoderm* respectively. Between these two layers the protoplasm is reduced to a sparse reticulum, without or with only a few nuclei, the spaces between the strands of the reticulum being filled by a gelatinous matter, the *jelly*.† When this interposed gelatinous layer is thin and inconspicuous, as in most polyps, it is called the *supporting lamella*, or *structureless lamella;* when it is thick and bulky, as in some parts of the jelly-fishes, it is called simply the *jelly*. Further, these layers are always differentiated to a greater or less extent into functional tissues—contractile tissues and nervous tissues; and nematocysts are always present as differentiations of the ectoderm or endoderm, or of both layers; the generative cells are always products of one of these layers; and finally, the ectoderm or the protoplasm of the jelly very commonly secretes a skeletal tissue, which may be either cuticular, horny, or calcareous. On the other hand, the internal surface of the body is not differentiated into organs of circulation, of digestion, or of coelom distinct from each other. The vegetative processes are performed by the internal surface of the enteric cavity, or gastro-vascular space as we shall call it, of which the central part functions as stomach and intestine, the peripheral as vascular system.

* R. Leuckart, *Zoologische Untersuchungen*, I., Giessen, 1853. C. Chun, Coelenterata in Bronn's *Klassen u. Ordnungen*, Bd. 2, Abth. 2.

† Sometimes called the mesoglaea—an unsuitable term, because it suggests an ectoglaea and entoglaea, which do not exist.

R. Leuckart was the first to recognise the importance of these characters, and made use of them to separate the *Polyps* and the *Medusae* from the *Echinoderms*, thus resolving Cuvier's type of *Radiata* into the types of *Coelenterata* and *Echinodermata*.

The *Coelenterata* are divided into two main sub-phyla, the *Cnidaria* and the *Ctenophora*, distinguished from one another by a number of characters, of which, perhaps, the most comprehensive is the presence or absence of nematocysts. The entire structure of the body is generally speaking disposed in radial symmetry, although amongst the *Cnidaria* transitions towards bilateral symmetry are sometimes apparent.

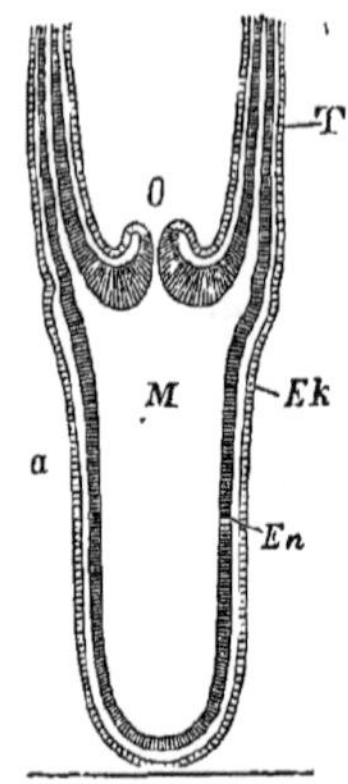

FIG. 79.—Diagrammatic longitudinal section of a hydroid polyp. *O* mouth; *M* enteric space or coelenteron; *Ek* ectoderm; *En* endoderm; *T* tentacle.

Three distinct types of body-form are met with amongst the *Coelenterata*, viz., that of the *Polyp*; of the *Medusa*; and of the *Ctenophore*.

The Polyp type. The Polyp has the form of a cylinder or sac (Fig. 79), of which the posterior or lower end is fixed and the opposite end is free and pierced by an opening—the *mouth*—placed on a flat or conical prominence—the *oral cone* or *hypostome*, and leading into the cavity of the body, or enteric space (*coelenteron*). Around the mouth are placed a number of regularly or irregularly arranged contractile processes—the tentacles, which always contain endoderm, either solid or traversed by a prolongation of the enteric space. The tentacles may be reduced to knob-like warts or be absent altogether (siphonozooids of the *Stylasteridae*, etc.). In rare cases (*Arachnactis*, *Minyas*) the polyp is free-swimming.

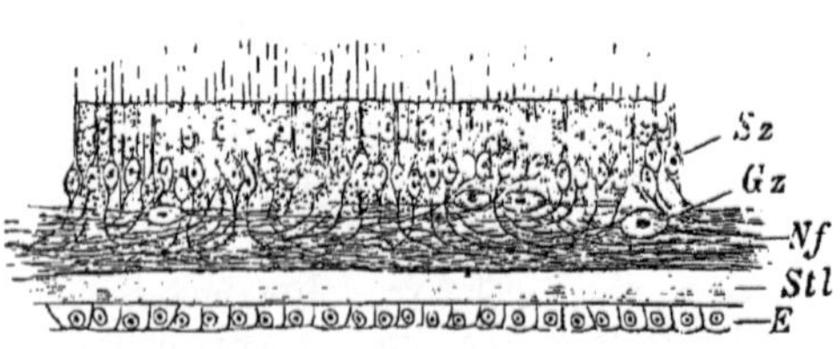

FIG. 80.—Longitudinal section through the nerve ring of *Charybdea*. *Sz* sense cells in the ectoderm; *Gz* ganglion cells; *Nf* nerve-fibres; *Stl* supporting lamella; *E* endoderm cells.

The ectoderm, which is the part of the polyp in closest relation with the outer world, possesses, to use the ordinary parlance of histology, *sense-cells* (Fig. 80) provided with sensory hairs; *nerve* or *ganglion-cells* with branching processes; *epithelio-muscle* cells, with contractile processes (Fig. 81) arranged along the long axis of the body; and *cnidoblasts*, which form the thread-cells, or nematocysts, and carry sensory processes, the *cnidocils* or triggers, projecting on the surface. The thread-cells (Fig. 82) are small capsules consisting of a highly refractile cuticular material, and containing some fluid and a spirally-coiled thread. Under certain mechanical conditions, *e.g.*, under slight pressure produced by contact with a foreign body, these capsules suddenly protrude the thread, which either fastens on to the causes of the disturbance, or pierces it, carrying into it a part of the fluid contents of the capsule. In many parts of the body, and especially on the tentacles, which serve for the capture of prey, these microscopic weapons are present in great numbers, and are often grouped in a peculiar arrangement to form batteries of thread-cells.

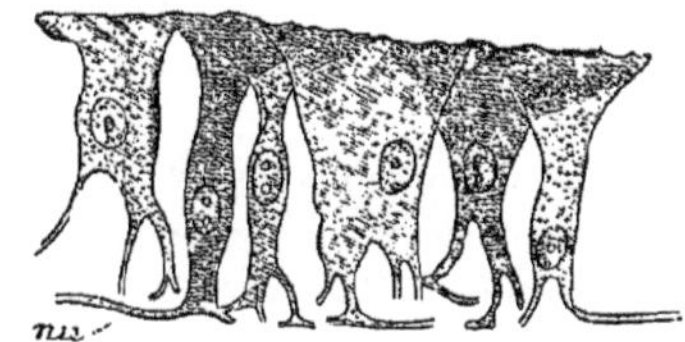

FIG. 81.—Ectoderm cells of *Hydra* with contractile processes (*nu*). From Chun, after Kleinenberg.

The endoderm cells are principally concerned with the processes of digestion and secretion. They often bear cilia for the movement of the contents of the gastrovascular space, and their deeper ends are sometimes prolonged into contractile processes which are transversely arranged. The endoderm of the tentacles is sometimes solid, and modified for a skeletal function by the development of vacuoles and cuticular structures into a form resembling the vegetable parenchyma of plants, or the tissue of the vertebrate notochord (Fig. 83). The food is digested in the Protozoan manner, being surrounded by protoplasmic processes of the endoderm. The digestion is therefore intracellular, and the indigestible remains are cast out into the enteron and ejected through the mouth by the help of the cilia of the endoderm cells.

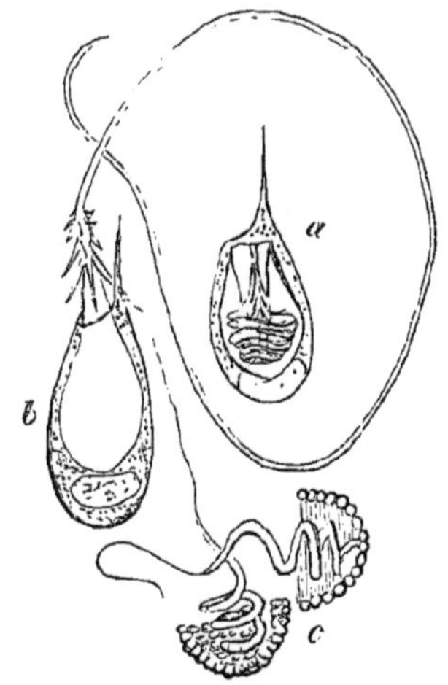

FIG. 82.—*a*, *b* Nematocysts and cnidoblasts of *Cordilophora*, with the cnidocil of the cell (cnidoblast); *c* adhesive cells of a ctenophore (from Lang).

As a rule there is no localised organ for the excretion of the nitrogenous waste, but in a few cases deposits of guanin or urea crystals may be seen in some of the endoderm cells. Some polyps possess pores at the apices of the tentacles or through the body wall, which may serve for excretion (Fig. 84).

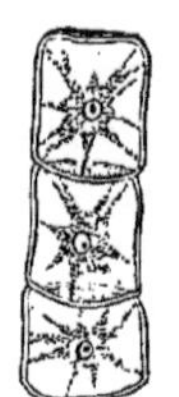

FIG. 83.—Axial cells from the tentacles of *Campanularia.*

Asexual reproduction by budding or fission is very commonly found. If the individuals so produced remain united they give rise to the colonies which are so widely distributed amongst the *Cnidaria*, and which by the continued multiplication of their members may attain a considerable size.

Sexual reproduction is always met with. The sexual cells arise in the endoderm or ectoderm. They sometimes remain at their place of origin (*Hydra*, *Anthozoa*), but often they pass through the supporting lamella and wander to another place where they ripen (most *Hydromedusae*).

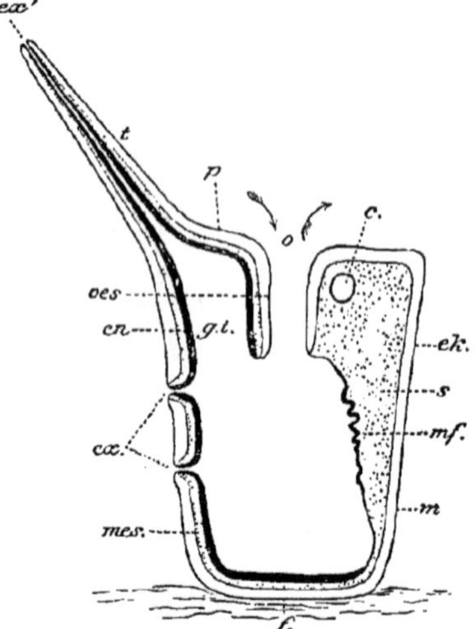

FIG 84.—Diagrammatic longitudinal section through an Anthozoan polyp, passing through an enteric pouch on the left side and through a mesentery on the right. *c* perforation in mesentery; *ek* ectoderm; *en* endoderm (black); *ex* excretion-pores (cinclides); *ex'* excretion pore at the end of a tentacle; *f* pedal disc; *g.t.* enteric pouch; *m* wall of body; *mes* supporting lamella; *mf* mesenterial filament; *o* mouth; *oes* oesophageal tube; *p* edge of peristome; *s* supporting lamella of a mesentery; *t* tentacle (from Chun).

There are two kinds of polyps, characterised by the structure of the enteric cavity —the *Hydroid* polyp and the *Anthozoan* or *coral* polyp. As types of the former we may mention *Hydra* with its single row of hollow tentacles, and *Tubularia* with its double row of solid tentacles. In such polyps the gastrovascular space is simple, and there is no oesophageal tube.

The **Anthozoan Polyps** are usually of a larger size than the Hydroids, and possess a more complicated gastrovascular cavity (Fig. 84). In the first place the mouth leads into a tube which projects into the gastrovascular cavity, and opens into the latter at its lower end. This tube is lined with ectoderm and is called the oesophageal tube, or stomodaeum (Fig. 84 *oes*). In the second place there projects inwards from the side walls of the body a number of vertical partitions formed of folds of the endoderm and a prolongation of the supporting lamella. These are the *mesenteries*. Internally they are attached to the oesophageal tube in the region of that structure, while below it they end in free edges which are somewhat folded and thickened and are called the *mesenterial ridges* and *filaments*. The mesenteries break up the enteric

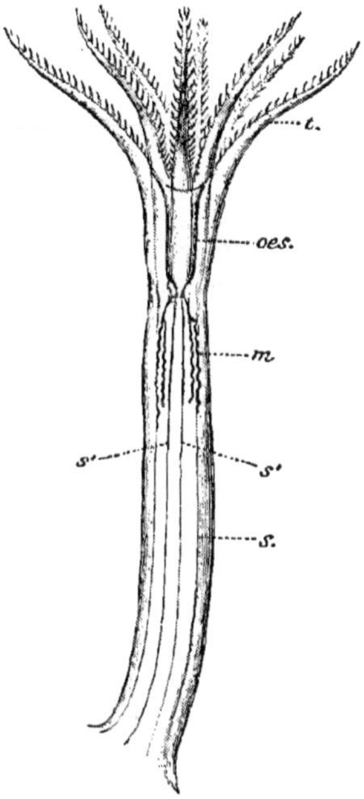

FIG. 85.—Alcyonarian polyp (*Veretillum cynomorium*), from Chun. *t* pinnate tentacles; *oes* oesophageal tube; *m* mesenterial filaments; *s* mesenteries; *s'*, *s'* two adjacent mesenteries with feebly developed filaments.

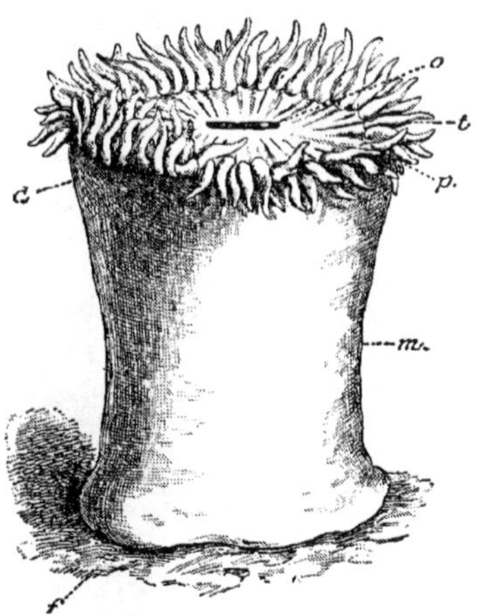

FIG. 86.—Diagram of a hexactinian polyp (from Chun, after Andres). *f* pedal disc; *m* wall of body; *p* peristome; *d* edge of peristome; *t* tentacles; *o* mouth.

cavity into a number of circumferential pouches which, in the lower part of the polyp, open into the central part of the enteric cavity or stomach, but in the upper part of the polyp where the mesenteries are attached to the oesophagus form separate chambers (Figs. 88 and 89). Each of these communicates above with the cavity of a tentacle (Fig. 84). The gastral pouches then are the peripheral parts of the enteric cavity (coelenteron) between the mesenteries. An opening may be present in each mesentery just below the oral disc putting the adjacent chambers in communication (Fig. 84, *c*). Between these mesenteries, which join the oesophageal tube and are called *primary mesenteries*, there may be intercalated mesenteries which, as a rule, do not reach the oesophagus and are called *accessory mesenteries*. There are secondary, tertiary, etc., accessory mesenteries. All the mesenteries decrease in breadth towards the base of the polyp (Fig. 89).

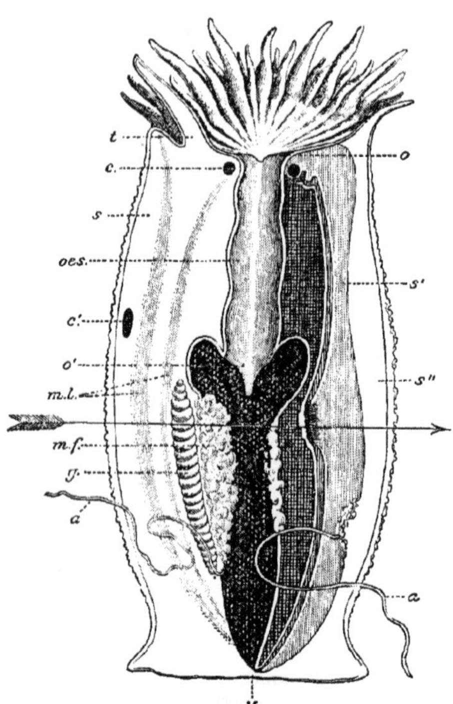

FIG. 87.—Longitudinal section through a Hexactinian (*Phellia limicola*), from Chun, after Andres. *a* acontia; *c*, *c'* septal ostia; *g* gonad; *m.f* mesenterial filament; *m.l* longitudinal muscles; *o* mouth; *o'* internal opening of oesophageal tube; *oes* oesophageal tube: *s* primary mesentery; *s'* secondary mesentery; *s''* tertiary mesentery; *t* tentacle.

The mouth opening is rarely round, but usually has the form of a slit, at the two ends of which (or sometimes at one end) there is a groove lined with long cilia. These grooves are the oesophageal grooves or *gonidial grooves* (*siphonoglyphes*). They remain open when the walls of the rest of the oesophagus are applied to one another (Figs. 88 and 89).

The tentacles are hollow, and may be smooth (*Zoantharia*, Fig. 86) or pinnate (*Alcyonaria*, Fig. 85), and are placed in one or in several rows. Pores may be present at the tips of the hollow tentacles, and on the side walls of the body, in which case they are called **cinclides** (Fig. 84). In the genus *Cerianthus* and its allies there is a large aboral pore. The **mesenterial thickenings** or filaments are specially characteristic of the *Anthozoa*. They consist of thickenings of the endoderm containing gland-cells and thread-cells.

In some sea-anemones contractile fibres—the **acontia**—arise from the edges of the lower ends of the mesenteries: they are closely set with thread-cells and can be protruded from the lateral pores (*cinclides*) in the contraction of the polyp and serve as weapons of defence (Fig. 87).

The muscular system is much more complicated than in Hydroids. The muscles are both ectodermal and endodermal. The ectodermal muscles of the body-wall (longitudinal) are generally feebly developed, while those of the peristomial disc (radial) and of the tentacles (longitudinal) are powerfully developed. The endodermal circular muscles of the pedal disc, the side body-wall, and of the oesophageal

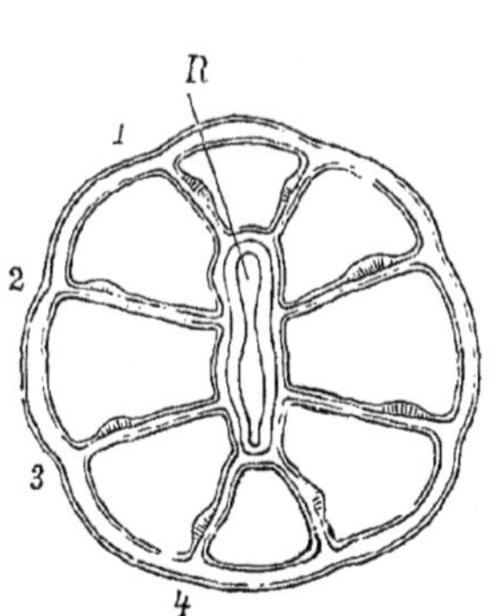

FIG. 88.—Transverse section through an Alcyonarian (after R. Hertwig). *R* gonidial groove; 1, 2, 3, 4 the four pairs of mesenteries with their muscles.

FIG. 89.—Section through an Actinian (*Adamsia*), after R. Hertwig. *Hf* the unpaired (dorsal and ventral) chambers; *R*, *R* gonidial grooves.

tube are well developed. To this system there is added a characteristic and well-developed set of endodermal muscles on the mesenteries. Each septum is provided on one face with transverse (radial) fibres, and on the other with longitudinal (Fig. 90). The lowest section of the transverse muscles are often independent of the rest, and pass from the side body-wall to the pedal disc (*m.t*). The longitudinal muscles are well developed and cause a projection on the face of the mesentery (Figs. 88, 89). While a nucleus is generally associated with each of the ectodermal muscular fibres, the endodermal muscles are in close connection with the base of a cylindrical epithelial cell.

In correspondence with the powerful formation of the musculature, the nervous system reaches a considerable development. It has the form of a diffuse plexus of much-branched glanglion cells, which are contained in both ectoderm and endoderm between the lower ends of the epithelial cells, and are especially developed in the peristome, tentacles, and oesophagus.

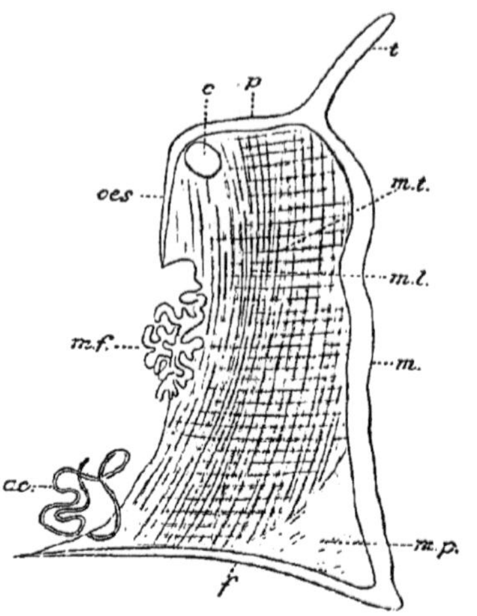

Fig. 90.—Primary mesentery of a Hexactinian (*Sagartia parasitica*), and the parts of the body to which it is attached. *ac* Acontia; *c* Septal ostium; *f* pedal disc; *m.l* longitudinal muscular fibres; *m.t* transverse muscular fibres; *m.p* parietal muscles; *p* peristome, *t* tentacle.

The *Anthozoa* are almost always dioecious, rarely hermaphrodite (*Cerianthus*). Ova and spermatozoa arise from the endodermal cells of the mesenteries, and lie in follicles in the jelly of the same structures. They cause swellings on the faces of the mesenteries, a short distance from their free ends (Fig. 91). Asexual reproduction by budding and fission is very generally present, and often leads to the formation of colonies. The Anthozoan polyps are much inclined to the formation of skeletal structures, which consist of slimy (*Cerianthus*), horny, or calcareous substances.

The symmetry of the Anthozoan polyps is almost always radial. The *Octactinia* (*Alcyonaria*) indicate their 8-radiate structure by their eight feathered tentacles,

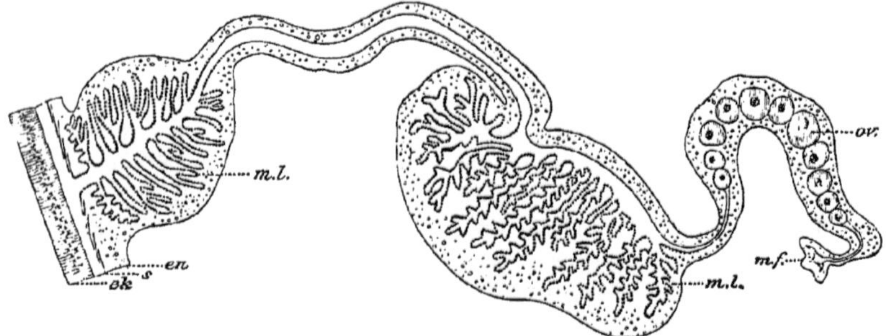

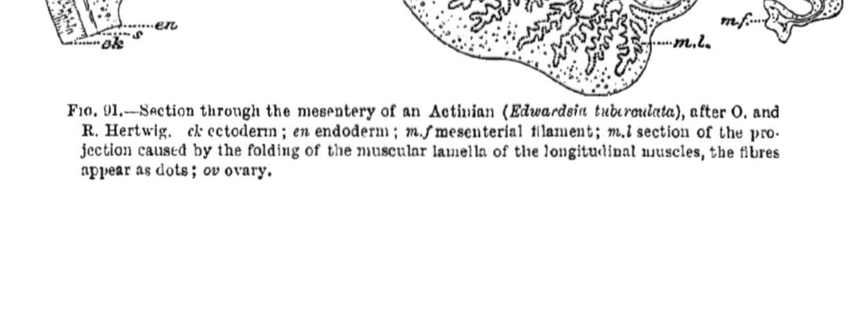

Fig. 91.—Section through the mesentery of an Actinian (*Edwardsia tuberculata*), after O. and R. Hertwig. *ek* ectoderm; *en* endoderm; *m.f* mesenterial filament; *m.l* section of the projection caused by the folding of the muscular lamella of the longitudinal muscles, the fibres appear as dots; *ov* ovary.

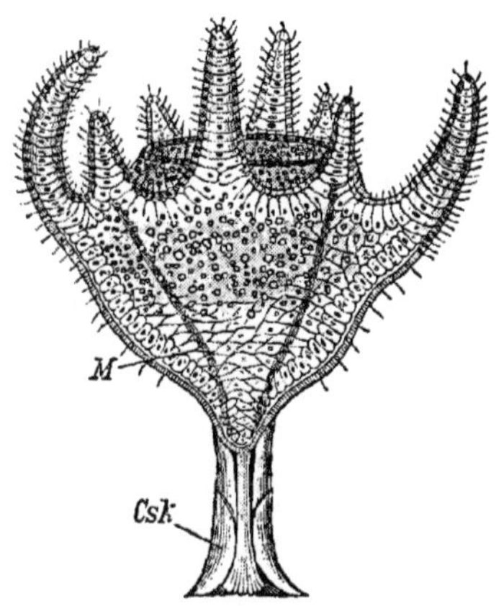

FIG. 92.—Eight-armed Scyphistoma-polyp with wide mouth. *M* longitudinal muscles of the gastral ridges; *Csk* chitinous tube.

and the horny corals (*Antipatharia*) their 6-radiate symmetry by their six tentacles; multiradiate—usually in a multiple of six—are the *Actiniaria* and stone-corals. Nevertheless, hardly a single anthozoan-polyp can be found in which all the organs are strictly arranged according to one and the same number. The fact that the mouth-opening is not usually round, but slit-like, indicates an inclination to a biradiate or bilateral symmetry, and there are transitional forms between a strictly radiate and a bilateral structure. The tentacles rarely show a tendency to the bilateral arrangement, but the mesenteries, as shown by the arrangement of the longitudinal muscular bands, are generally grouped in a bilateral manner (Figs. 88 and 89).

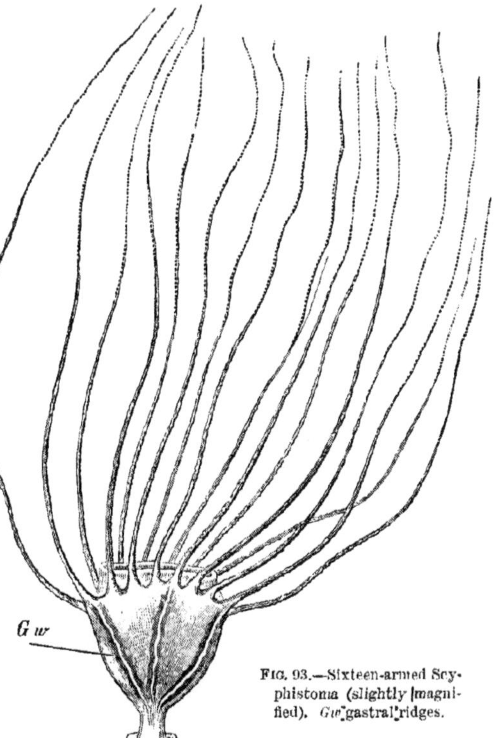

FIG. 93.—Sixteen-armed Scyphistoma (slightly magnified). *Gw* gastral ridges.

The **Scypho-polyp** or **Scyphistoma** is a transitional form between the Hydro- and Anthozoan-polyp. It resembles the hydroid in the absence of the oesophageal tube* and mesenteries, but it recalls the Anthozoan-polyp in the fact that the structureless lamella is developed into a gelatinous layer, and in the presence of gastral ridges (*taenioles*) or folds of endoderm resembling rudimentary mesenteries, into which the jelly is continued. It is somewhat cup-shaped (Fig. 92), and is attached by the aboral end which is elongated and narrow, and often secretes a chitinous tube for fixation. There are eight or sixteen tentacles (Fig. 93) round the mouth supported by a central axis of stiff endoderm cells. There are four gastral ridges or taenioles (Fig. 94), each of which is accompanied by a longitudinal muscle derived from the endoderm.

The Medusa is a free-swimming animal, and consists of a flattened disc or arched bell of gelatinous consistence, from the under or *sub-umbrella* surface of which hangs a central stalk—the *manubrium*—bearing at its free end the mouth. The greater part of the umbrella consists of the jelly or enlarged structureless lamella; this is often traversed by protoplasmic strands which may contain nuclei. In the normal position the medusa swims—by the contraction of the bell—with its convex or *ex-umbrella* surface upwards. The manubrium is frequently prolonged in the region of the mouth into lobes and tentacle-like structures, while the edge of the umbrella is beset with a variable number of true tentacles.

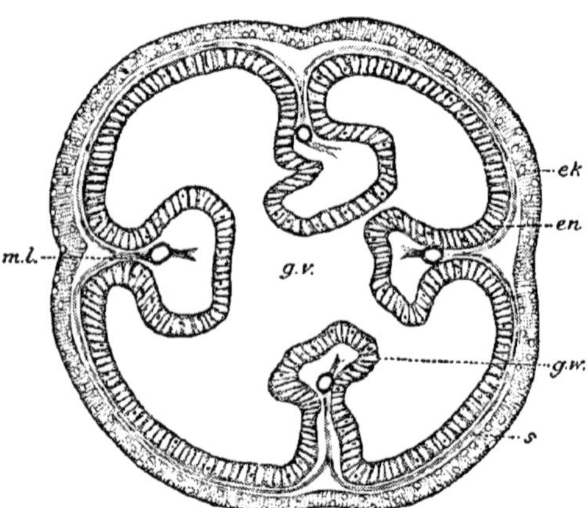

FIG. 94.—Transverse section through the middle part of a Scyphistoma. *ek* ectoderm; *en* endoderm; *s* jelly (structureless lamella); *ml* longitudinal muscle; *gw* gastral ridge; *gv* enteron (from Chun, after Claus).

In a few cases the medusae are attached by the ex-umbrella surface; in the *Lucernaridae* (Fig. 134) by a styliform prolongation of the aboral pole; in some *Rhizostomas* by a sucker-like plate of the ex-umbrella. Some of them can creep

* Götte asserts that there is an oesophageal depression of ectoderm in the Scyphopolyp.

by means of small suckers on their tentacles (*Clavatella prolifera*, *Pectanthus asteroides*), or can adhere by the suctorial action of their mouth openings (*Pelagia*), and even in rare cases are able to lead a parasitic life (*Mnestra parasitica* on a pelagic snail *Phyllirhoë bucephala*).

The contraction of the bell is effected by the circular muscles of the sub-umbrella surface. The ejection of water caused by this contraction drives the medusa along. The **velum** is a muscular membrane at the edge of the bell. It consists of a fold of ectoderm, and assists in the movement of the medusa by elongating the umbrella cavity and narrowing its aperture. When the velum is absent the margin of the bell is lobed (*Acraspedote Medusae*, as opposed to *Craspedote Medusae* in which a velum is present).

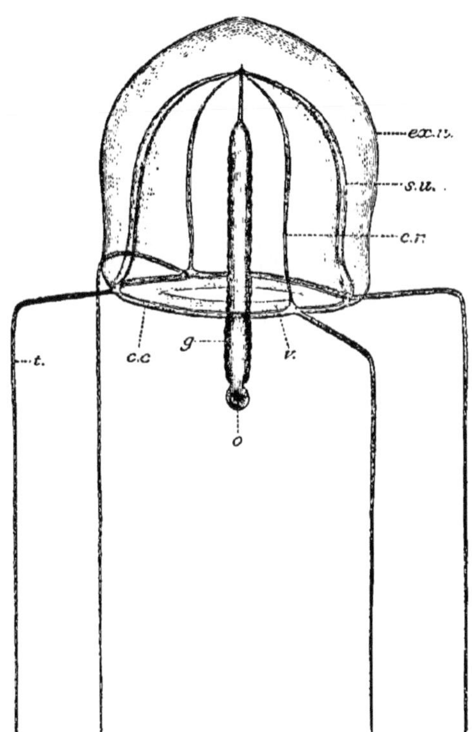

FIG. 95.—*Sarsia mirabilis* (from Chun, after Agassiz). A craspedote ocellate medusa budded from *Coryne mirabilis*. *c.r* radial canal; *c c* circular canal; *ex.u* ex-umbrella; *g* manubrium containing the elongated stomach and surrounded by the gonads (manubrial gonads); *o* mouth; *t* tentacles; *v* velum; *s.u* sub-umbrella.

The marginal tentacles are rarely absent (*Rhizostoma*); occasionally there is only one (*Steenstrupia*) or two (*Aeginopsis*, *Gemmellaria*); more frequently there are four or some multiple of four. Occasionally there are six or a multiple of six (*Carmarina*, Fig. 101); or the tentacles may be numerous, in which case they are either uniformly distributed round the edge of the bell (*Tiaropsis*, *Aurelia*, Fig. 100), or grouped in bundles (four bundles in *Bougainvillea*, eight in *Lucernaridae*, Fig. 134, and in *Cyanea*). Occasionally the tentacles are removed from the edge of the umbrella, and inserted either on the ex-umbrella (*Narcomedusae*, Fig. 96), or on the sub-umbrella (*Cyaneidae*). The tentacles are

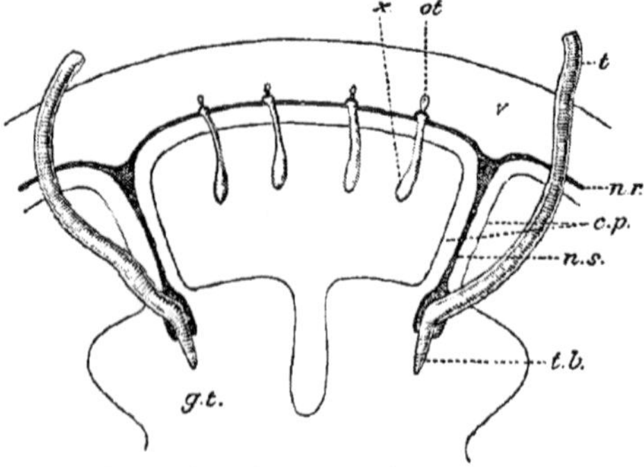

FIG. 96.—Portion of the edge of the umbrella of a Narcomedusan (*Cunina lativentris*) from Chun, after O. and R. Hertwig. *v* velum; *t* tentacles with stiff endodermal axes; *t.b* roots of tentacles; *n.r* nerve ring; *n.s* radial nerve which passes to the base of the tentacles; *ot* marginal bodies; *x* otoporpa; *g.t* gastral pouches; *c.p*, *c.p* the two peronial vessels; they form originally a part of the circular canal, which in the shifting of the tentacles dorsalwards on to the ex-umbrella is festooned and opens into the gastral pouches (after Hertwig, from Chun).

usually hollow and unbranched, more rarely they have a solid endodermal axis (*Narcomedusae*, Fig. 96, *Tesseridae*, *Ephyridae*), or are dichotomously branched (*Cladonemidae*). Besides the larger main tentacles there are often smaller intermediate tentacles at the umbrella edge (*Tessera*, Fig. 133). Occasionally the accessory tentacles are confined to the young stages and drop off in later life, or they may become transformed into the marginal bodies.

In the **gastrovascular apparatus** a central stomach for digestion, and a carrying or circulating system of peripheral canals and pouches

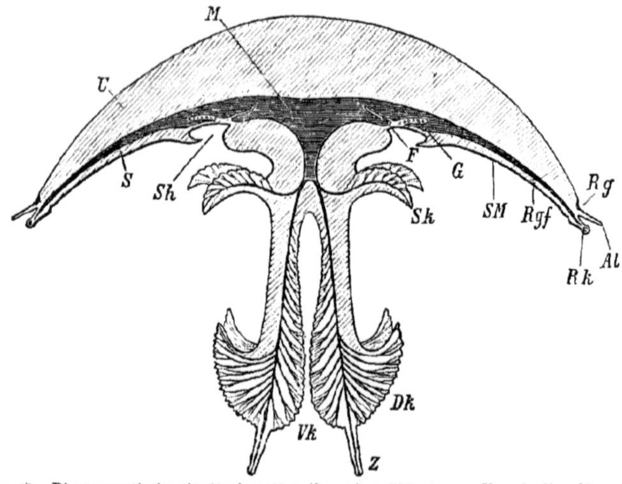

FIG. 97.—Diagrammatic longitudinal section through a *Rhizostoma*. *U* umbrella; *M* gastric cavity; *S* sub-umbrella; *G* gonad; *Sh* sub-genital pit; *F* gastral filament; *SM* muscular system of the sub-umbrella; *Rgf* radial canal; *Rk* marginal body; *Rg* olfactory pit; *Al* ocular lobe; *Sk* shoulder tufts, *Dk* dorsal tufts, *Vk* ventral tufts of the eight arms; *Z* terminal parts of the arms.

can always be distinguished. The mouth leads directly (Fig. 101, *Carmarina*), or by a tube into the stomach, and its lips are often drawn out into four grooved processes—the oral arms (Fig. 100). The edges of the grooves are often frilled, and carry small tentacular filaments.

In *Rhizostoma* the four arms bifurcate (Fig. 97), giving rise to eight, the frilled edges of which fuse and bring about the closure of the mouth. The fusion is not, however, complete, but numerous small openings are left—the suctorial *mouthlets*, in which digestion of the food takes place. From these openings arise small vessels, which gradually uniting with each other form a system of canals passing up the oral arms to open into the stomach (Fig. 97).

The stomach is generally flattened, rarely elongated (*Lucernaria*, *Periphylla*, *Tessera*) in its main axis. Occasionally it is divided by a constriction into two sections, an aboral basal section and a central stomach (*Tessera*, *Pericolpa*, *Periphylla* (Fig. 99), the stalk tube of *Lucernaria*). Gastral filaments (*phacellae*) are highly characteristic of the acraspedote medusae, and recall the same structures in the *Anthozoa*, as do the gastral ridges or taenioles, which to the number of four are placed interradially in the basal and central stomach in the *Lucernaridae* and *Tesseridae* (Figs. 98, 99).

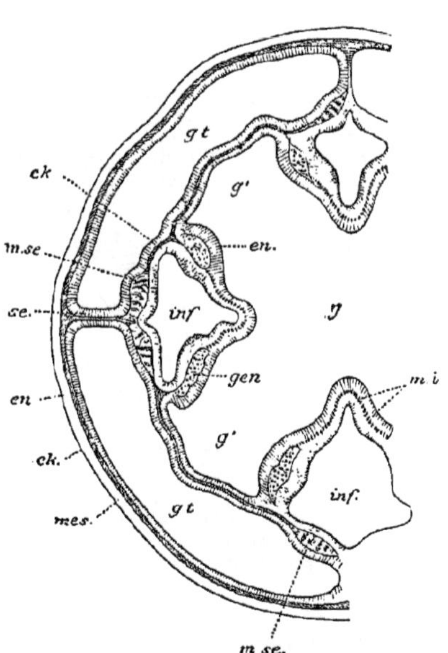

FIG. 98.—Section through the peripheral part of the umbrella of a young Lucernarian (*Craterolophus thetys*). *ek* ectoderm; *mes* jelly; *en* endoderm; *g* central stomach; *g'* radial diverticula of the stomach between the funnels (subgenital pits); *g.t* radial pouches; *se* septa; *inf* funnels lined by ectoderm; *m.se* septal muscles; *m.i* longitudinal muscles of the funnels; *gen* gonads (after Claus).

The peripheral part of the gastrovascular system may appear in the young form as a continuous dorso-ventrally flattened space extending almost to the edge of the umbrella. Later the dorsal and ventral walls of this space come together along certain radially directed lines and fuse with one another, so that the originally continuous space is broken up into a number of pouches or vessels, all leading outwards from the central stomach. At the places where the concrescence of the two walls occurs the endoderm epithelium is retained in the form of a layer of cells called the **vascular** or **endoderm lamella** (Fig. 107).

In many medusae these concrescent places appear as four small *septal unions* (*Cathammata*, Fig. 98 *se*). They delimitate four wide radial pouches of the stomach, which however communicate with one another peripherally beyond the unions by a kind of circular canal.

In the *Lucernaridae* and *Charybdeidae* the septal unions are extended into elongated partitions (Fig. 98, *se*), the *septa*, which delimitate the four radial pouches in nearly their whole extent. But the septa never quite reach the umbrella edge, so that there the four canals are connected by a narrow circular canal. The gastric pouches open into the stomach by the *gastral ostia* (Fig. 99 *g.o*).

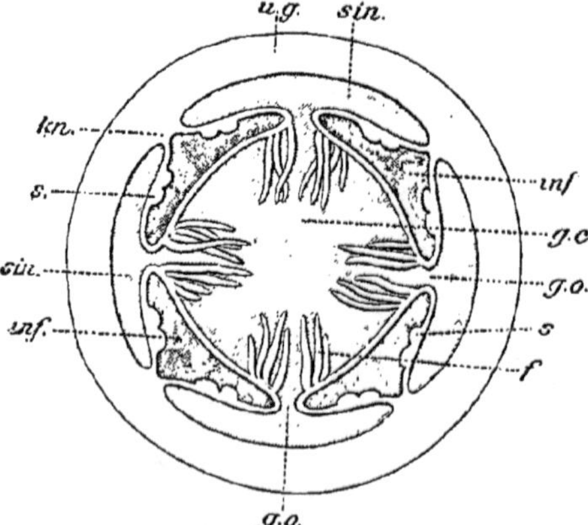

FIG. 99.—Section through the central stomach of an acraspedote medusa (*Periphylla mirabilis*). *u.g* umbrella-jelly; *inf* the four funnels; *g.c* basal stomach; *f* gastral filaments; *g.o* gastral ostia; *sin* circular sinus; *kn* septal unions; *s* gonads.

When the four places of fusion of the dorsal and ventral stomach walls are of considerable extent, the gastral pouches have the form of four narrow radial vessels: this is characteristic of the tetra-radiate craspedote medusae. A circular canal connecting them peripherally is however retained.

In many acraspedote medusae there are, beyond and independent of the four septal unions, sixteen lines of adhesion between the dorsal and ventral walls of the gastric pouches. In some cases the four septal unions are not formed, and only the peripheral lines of adhesion appear: in this way the more or less complicated form of the peripheral gastrovascular apparatus of the *Semostomae* and *Rhizostomae* is introduced. Sometimes the adhesion-lines are narrow and extend up to the edge of the disc (*Pelagidae, Cyaneidae*), sometimes they are broad and leave a circular canal at the periphery. In the first case broad, blindly-ending gastric pouches are formed; in the latter the pouches are reduced to narrow vessels which frequently branch peripherally, or even anastomose (*Aurelia*, Fig. 100, *Rhizostoma*).

A similar branching of the radial vessels occurs also in the *Craspedota*. The endoderm lamella may become secondarily excavated, and thereby new radial vessels may arise, vessels dichotomously branching in a peripheral direction, or centripetal vessels arising at the umbrella edge and ending blindly (*Carmarina*, Fig. 101).

Excretion pores may be present at the umbrella-edge in both craspedote and acraspedote medusae. In the *Craspedota* (most frequently in the *Leptomedusae*) they are placed on warts on the circular canal and open on the sub-umbrella surface. The endoderm cells near the openings are of a glandular nature, and contain concretions which are emptied through the pore. In the *Acraspeda* there are eight excretion pores at the distal end of the eight adradial canals.

The ectoderm which is composed of a flat epithelium on the dorsal surface, and of a muscular epithelium on the sub-umbrella surface and velum, is thickened and stiffened along certain lines and contains a large number of

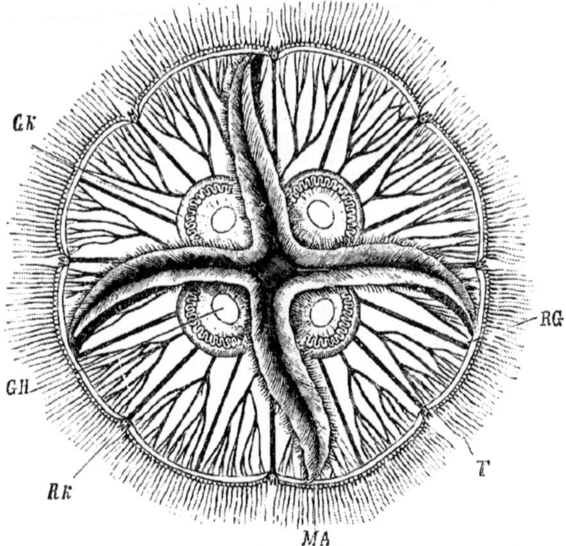

Fig. 100.—*Aurelia aurita* from the oral surface. *MA* the four oral tentacles with the mouth in the centre; *GK* gonads; *GH* aperture of sub-genital pit; *RK* marginal body; *RG* radial vessel; *T* marginal tentacles.

nematocysts. An annular thickening of this kind is found round the umbrella margin in some forms (Fig. 101), and similar thickenings ascend for a short distance on the ex-umbrella surface from the roots of the tentacles—the **peroniums** (Fig. 101, *pe*)—and from the roots of the auditory tentacles—the **otoporpas** (Fig. 96, *o*). The peroniums and otoporpas are continuations of the first-mentioned circular thickening.

The gastrovascular apparatus is lined by endoderm, which in places carries cilia for the movement of the contained matter. The digestion seems to be, partly at least, intracellular. The endoderm is without muscular structures and nematocysts, except on the gastral filaments, which are actively moveable and contain both. Possibly the four longitudinal muscular bands, found beneath the four taenioles and septa in some acraspedote medusae, are of endodermal origin.

The **nervous system** has the form of a plexus of multipolar ganglion cells placed between the muscle-fibres and the ectoderm cells. In addition to this peripheral plexus there are special

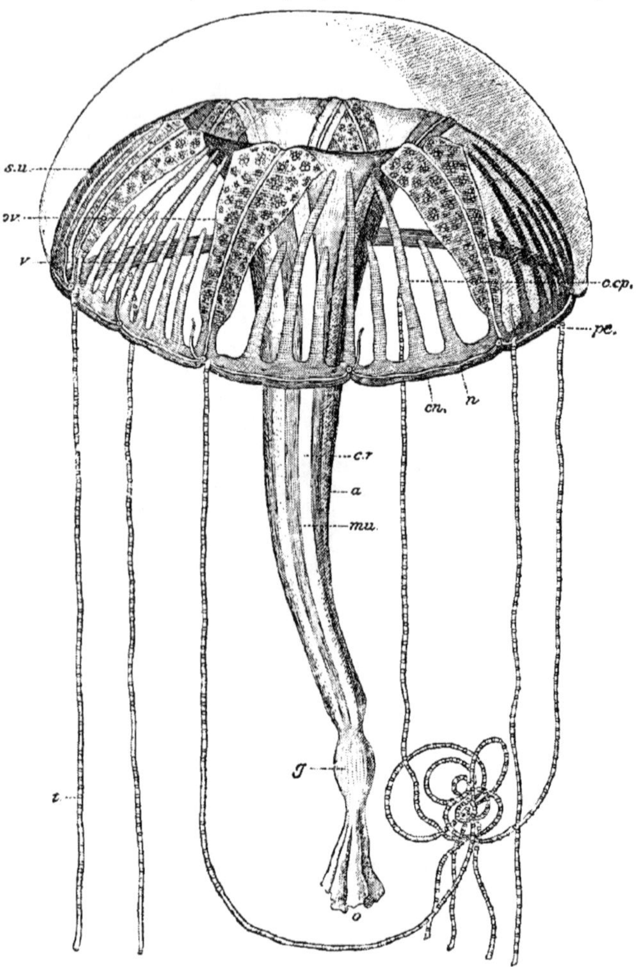

Fig. 101.—*Carmarina (Geryonia) hastata* (from Chun, after Haeckel)—a sex-radiate Trachomedusan. *s.u* sub-umbrella; *a* manubrium; *v* velum; *n* nerve-ring; *rv* circular vessel; *cn* thickened tract of ectoderm at edge of bell; *t* marginal tentacle; *mu* interradial longitudinal muscular bands of the manubrium; *o* mouth; *g* stomach; *cr* one of the six radial canals; *ov* ovaries on the sub-umbrella-wall of the radial vessels; *c.cp* centripetal, blindly ending radial vessel; *pe* peronium. The tentaculocysts are shown near the base of aud between the tentacles.

aggregations of nervous tissue in certain regions. In the craspedote medusae there are two annular nervous tracts in the ectoderm at the base of the velum, one on each side of the supporting lamella (Fig. 102 *n′*, *n″*). They consist of ganglion-cells, nerve-fibres, some of which perforate the supporting lamella to put the two rings in communication, and of sensory epithelial cells bearing stiff projecting sensory hairs, and ending internally in fibres which pass into the nerve-rings.

In the *Acraspeda* there is a special aggregation (Fig. 103 *F*) of the nervous tissue, of the same structure as the nerve-rings of the *Craspedota*, round the base of each of the sense tentacles (marginal bodies, or *rhopalia*); and in the *Charybdeidae* there is a sub-umbrella marginal nerve with a zigzag course connecting together these rudimentary ganglia, which in the other forms are apparently independent of one another. In the *Acraspedota*, pits—the so-called *olfactory pits*—lined by a sensory epithelium are found on the edge of the umbrella above the marginal bodies (Fig. 103 *R*).

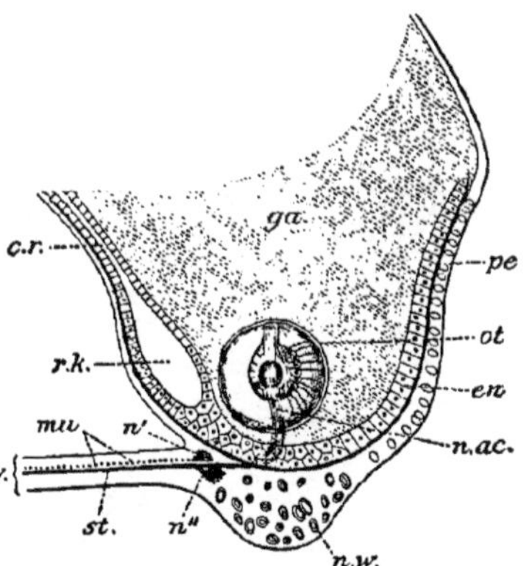

Fig. 102.—Diagrammatic transverse section through the edge of the umbrella of *Carmarina hastata* (from Chun). *v* velum; *st* supporting lamella of velum; *mu* circular muscles of velum in section; *n′* lower, *n″* upper nerve ring; *n.w* thickening of the ectoderm of the edge of the umbrella; *c.r* radial vessel; *r.k* circular vessel; *en* solid cord of endoderm beneath the peronium *pe*; *ot* tentaculocyst; *n.ac* auditory nerve; *ga* umbrella jelly.

At the edge of the umbrella are always placed special sense organs—the **marginal bodies.** They are of two kinds—those which by their structure indicate an auditory function, and those which indicate a visual purpose. In the craspedote medusae they are mutually exclusive, so that we find *ocellate* medusae with sense organs of the visual type and *vesiculate* medusae with sense organs of the auditory type. In the *Acraspeda* the marginal bodies may carry organs of both types.

The sense organs of the auditory type always contain concretions—the *otoliths*—of an organic or inorganic material, which concretions are contained inside cells; and two types of auditory organ are distinguished according as these concrements are contained in cells of the ectoderm (Fig. 104) or in cells of the endoderm (Fig. 105). In the *Craspedota* we find auditory organs of both types. In the *Leptomedusae* they consist of ectodermal pits, which may be closed into vesicles, on the under side of the base of the velum; some of the ectoderm cells of the pit or vesicle contain otoliths, while others bear sensory hairs (Fig. 104). In the other type, found in the *Trachomedusae* and *Narcomedusae*, the auditory organs have the form of short reduced tentacles of the edge of the disc, containing a solid endodermal axis and surrounded at their base by cells bearing long auditory sense hairs. The otoliths are contained in the endoderm cells of the tentacles. These sense tentacles may project freely on the surface (*Aeginidae*, Fig. 96), or they may be sunk in pits, or the pits may be closed and they may lie in vesicles embedded in the jelly of the umbrella (Fig. 102), in which case they may be fitly termed *tentaculocysts*. In both types the hair-bearing sensory cells are prolonged into fine fibres which join the nerve rings (Fig. 105).

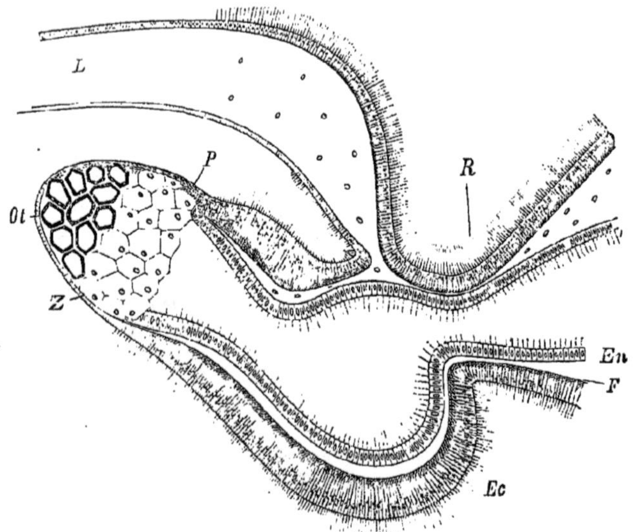

FIG. 103.—Section through the olfactory pit (*R*), the sense tentacle (marginal body), and its nerve centre of *Aurelia aurita*. *L* lobe of umbrella covering the sense tentacle; *P* eye-spot; *Ot* otoliths in the endoderm of the sense tentacle; *Z* endoderm cells after solution of the otoliths; *En* endoderm; *Ec* ectoderm with the underlying tissue of the nerve centre *F*.

The *eye-spots* of the *Craspedota* (ocellate medusae or *Anthomedusae*) consist of a pigmented patch of ectoderm cells just dorsal to the insertion of the velum; a lens-like cuticular thickening over them is sometimes developed (*Lizzia*). The ectoderm patches contain two kinds of cells, namely peripheral cells with pigment surrounding some colourless cells, the bases of which are prolonged into nerve fibres.

The marginal bodies of the *Acraspeda* or the **rhopalia** are absent in the *Tesseridae*, and in most *Lucernaridae*. When they are present they have the form of short marginal tentacles and are distinguished from those of the *Craspedota* by the fact that they contain a hollow prolongation of the gastrovascular system, and are covered on the dorsal side by a hood-like prolongation of the umbrella (hence *Steganophthalmata*, Fig. 103). In some *Lucernaridae* the marginal bodies are present as the so-called marginal anchors (*Haliclystus*, Fig. 134). There are eight of them, four being called radial in position, and the other four interradial; they are placed respectively between the eight marginal lobes. They end in a small swelling with thread cells; their middle portion is surrounded by a kind of collar formed of sticky glandular ectoderm cells, and on the sub-umbrella side of their basal part is an eye-spot.

In other *Acraspeda* the endoderm at the end of the sense tentacle is thickened and loaded with concretions, and the ectoderm round its base is columnar and provided with cilia and sense hairs (Fig. 103). The fibrillar continuations of these cells form a nervous network, with ganglion cells, round the base of the tentacle (Fig. 103, *F*).

Visual organs are sometimes present in this sensory epithelium. They are most complicated in *Charybdea*, where there are two large terminal eyes with a retina, vitreous humour, lens, and cornea.

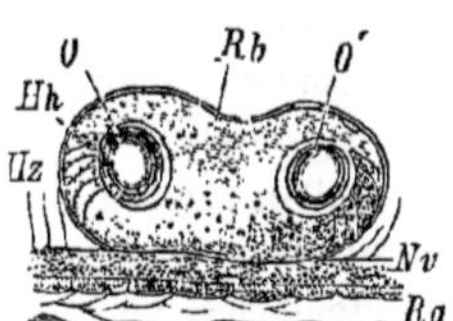

Fig. 104.—Sense organ on the nerve-ring and circular vessel of *Octorchis* (after the Hertwigs). *Rb* Sense organ; *O*, *O* two otoliths; *Hh* auditory hairs; *Hz* auditory cells; *Nv* upper nerve-ring; *Rg* circular vessel (type of *Vesiculata*).

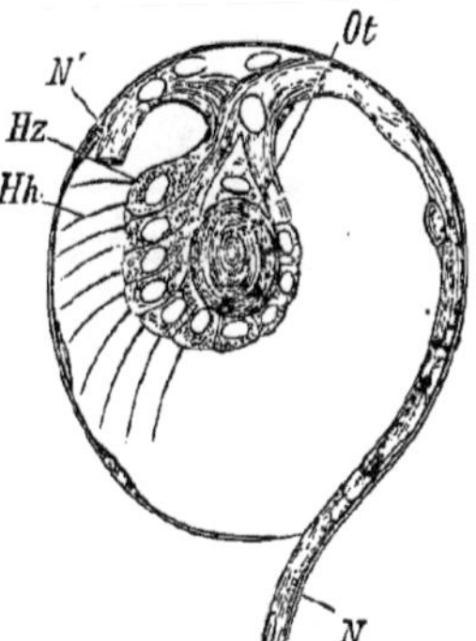

Fig. 105.—Tentaculocyst of *Geryonia* (after O. and R. Hertwig). *N*, *N'* the auditory nerves; *Ot* otolith; *Hz* auditory cells; *Hh* auditory hairs (type of the *Trachomedusae*).

Reproductive organs. The medusae are almost always dioecious, only in rare cases are they hermaphrodite (*Chrysaora*).

The sexual cells of the *Craspedota* appear always to arise in ectoderm, and those of the *Acraspeda* in endoderm. In the *Anthomedusae* they are developed in the ectoderm of the manubrium (Fig. 95) in four radial streaks, each of which later splits into two. In the other *Craspedota* (*Leptomedusae*) they are placed in the ectoderm of the sub-umbrella beneath the radial canals (Fig. 106). It appears, however, that in many medusae (*Obelia*) the generative cells arise in the ectoderm of the manubrium, and then later migrate to the ectoderm below the radial canals, where they occasionally wander into the endoderm and ripen partly in the ectoderm and partly in the endoderm.

In the *Acraspeda* the sexual cells arise in the endodermal walls of the central stomach or gastral pouches. They are interradially* placed, and have the form of four horseshoe-shaped glands, which may secondarily divide into eight adradial halves. In all cases the sexual cells are dehisced into the enteron and pass out by the mouth.

In connection with the genital glands there are very generally present ectodermal invaginations in the form of pits of the sub-umbrella surface. These pits are for respiratory purposes, and constitute the so-called *sub-genital pits* (*Discomedusae*, Fig. 100), and *funnels* (*Lucernaridae* and *Periphyllidae*, Fig. 98). They occur interradially beneath the genital glands.

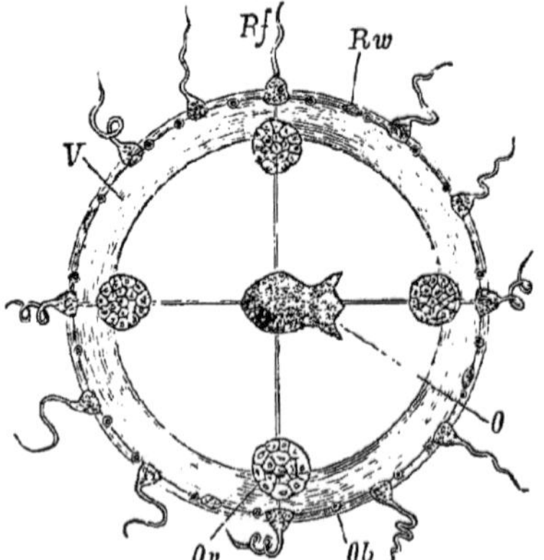

FIG. 106.—*Phialidium variabile* represented from the under side of the umbrella. *V* velum; *O* mouth; *Ov* ovary; *Ob* auditory vesicle; *Rf* marginal tentacles.

Asexual reproduction by budding is of very general occurrence in the craspedote medusae. The buds are formed, usually in groups, rarely singly, on the wall of the manubrium, more rarely at the base of the tentacles, on the circular canal, or on other places. In the *Narcomedusae* the budding is confined to larval life, when a bud-nurse is formed which attaches itself parasitically to other Medusae.

As hinted often in the above description the medusae fall into two well-marked divisions, viz., into the *Craspedota* and *Acraspedota* (*Acraspeda*) or *Acalephae*. Since the celebrated discovery of the alternation of generations in the medusae by Sars we have known that the former are budded off by Hydroid-polyps, the latter by the Scypho-polyp or *Scyphistoma*.

The *Craspedote Medusae* are distinguished by the following characters: a smooth umbrella edge; a muscular velum, without endoderm; a double nerve-ring; gastral filaments and taenioles are

* In describing a Scyphomedusa it is usual to speak of eight primary radii. These are the radii of the marginal bodies. These eight primary radii are further distinguished into four *perradii* and four *interradii*. The oral arms lie in the four perradii, and the gastral ridges and gonads in the four interradii. The *adradii* are the eight radii between the eight primary radii.

absent; sexual cells arise in the ectoderm of the manubrium, or along the radial canals; they are rarely developed directly from the egg.

The *Acraspeda*, on the other hand, have a lobed umbrella edge and no double nerve ring; the marginal sense bodies are always tentacular in form, and contain a prolongation of the gastrovascular system; gastral filaments are always developed on the stomach wall, and taenioles are generally present; the sexual cells are endodermal, and the development is either direct or by fission from a Scypho-polyp.

Connection between the Polyp and the Medusa. The medusa is very generally produced by budding from a polyp, although in certain orders it does rise directly from the egg. For a long time it was considered a remarkable circumstance, hardly admitting of a satisfactory explanation, that organisms which differed so widely as polyps and medusae—they had indeed been systematically separated in different classes—should only form different stages in the life-history of a single cycle of development. The theory of "Alternation of Generations" contained only a description of the matter, and offered no explanation. The discovery of the mode of origin of the medusa as a bud on the body of the polyp first clearly demonstrated the direct relation of the two forms; for it proved that the medusa is a flattened disc-shaped polyp with a shallow, but wide gastric cavity, the peripheral part of which has, by the fusion (Fig. 107, *c*) of its upper and lower walls along four, six, or more radiating areas, become divided into the vascular pouches, or as they are often called, radial canals. The differences consist mainly in the presence of the oral cone or hypostome of the polyp as a well-marked external appendage, the manubrium; in the enlargement of the oral disc to form the muscular sub-umbrella surface (Fig. 107, *c*, *S*); in the collapse of the coelenteron (leaving the endoderm lamella) above referred to; and in the greater development of the nervous and muscular systems in connection with the free-swimming habit of life. To these may be added the thickening of the structureless lamella between the dorsal endoderm and ectoderm, to form the jelly-like tissue which is so characteristic of medusae.

As stated above the medusa is generally produced by budding from a polyp-colony, but however produced it always develops the generative cells. For this reason it is frequently called a **gonophore**. This name, though not employed when the polyp stage is unimportant (*Acalephae*) or absent (some *Acalephae*, *Trachomedusae*, etc.), is applied to all medusae or medusa-like organisms which are produced by budding from a polyp. Typically a gonophore becomes detached from its origin, and as a free-swimming medusa distributes the generative products far and wide; but in some colonies the gonophore does not become detached. In such cases it may acquire the full development of a medusa, but more often its development is not complete, and a degenerate medusa—or as it is often called, a gonophore with concealed medusiform structure—is produced. These degenerate medusae (*hedrioblasts, adelocodonic gonophores, medusoids,* as they are variously called) vary considerably in structure from the stage of an almost perfect medusa, to a stage in which they consist of little more than a bud containing the generative cells (sporosacs).

The **Ctenophore.** The description of the fundamental form of the Ctenophora will be more conveniently dealt with under the description of the class (see below, p. 197).

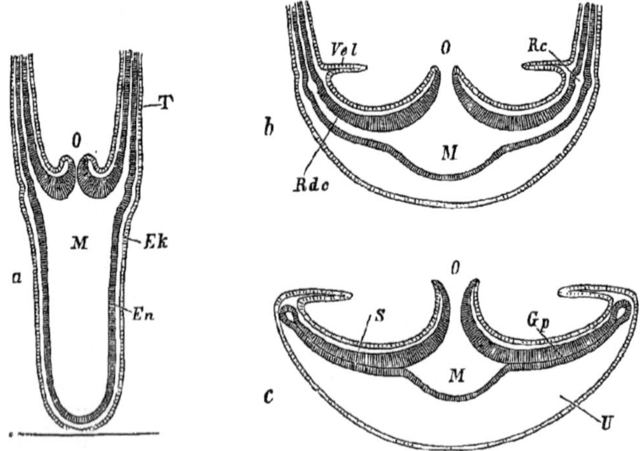

FIG. 107.—Diagrammatic longitudinal sections of *a*, a hydroid polyp, and *b*, *c*, a medusa. Section *b* passes through two radial canals, section *c* between two radial canals. *O* mouth; *T* tentacle; *M* enteron; *Ek* ectoderm; *En* endoderm; *Rdc* radial canals; *Rc* circular canal; *vel* velum; *Gp* vascular or endoderm lamella; *S* sub-umbrella; *U* umbrella.

Asexual reproduction by budding and fission is very widely spread. It is found in all groups with the exception of the Ctenophora, and it frequently leads to the formation of colonies, the component members of which are in bodily continuity. In most cases both layers of the body wall participate in this mode of reproduction, but it has recently been asserted that the ectoderm alone participates in the gemmation of the buds of Hydra; and in the so-called *sporogony*, which is found in some *Narcomedusae*, a single ovum-like cell of the body has the power of reproducing the whole organism. The latter should perhaps be regarded as a case of parthenogenesis which otherwise has not been observed in the *Coelenterata*. With this phenomenon of asexual reproduction must be connected the power of regenerating lost parts—a property which is probably widely diffused in the *Coelenterata*, and reaches, perhaps, its highest expression in *Hydra*. The smallest pieces into which the body of *Hydra* can be cut are able to reproduce the whole polyp.

In the asexual reproduction budding and fission may achieve results varying from the complete formation of a new individual which separates from the parent, to the mere repetition of a portion of the growing organism, which remains attached to the parent. As an instance of the one extreme we may mention *Hydra;* of the other, the *Siphonophora;* and between these two extremes we have the colonial *Hydromedusae* and *Anthozoa.*

In *Hydra* the whole organism is reproduced; in the Hydromedusan colonies the whole parent except the rooting portion or *hydrorhiza;* in the Maeandrine Corals the reproduction is confined to the oral disc and stomodaeum, and perhaps a few tentacles and mesenteries; finally, in the *Siphonophora* the manubrium of the parent Medusa (or polyp, according to the view taken) may produce almost any portion of an entire organism from an umbrella to a tentacle.

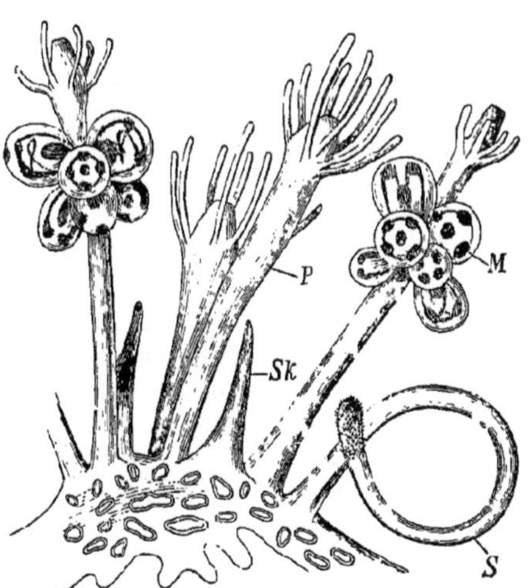

FIG. 108.—*Podocoryne carnea* (after C. Grobben). *P* polyp; *M* medusa bud on the blastostyle; *Sk* skeleton polyp; *S* spiral zooid.

Connected with this asexual reproduction are two important phenomena which must shortly be referred to here. The one is *polymorphism*, and the other *alternation of generations.* In the colonial *Coelenterata* the individuals associated in the colony are rarely all alike in structure. A certain number have one form, others another form, and with this difference in form is often associated a difference in the part which they respectively play in the life of the colony. As an instance we may take the Tubularian colony called *Podocoryne* (Fig. 108). Here from a creeping coenosark or hydrorhiza arise several individuals, some of which are ordinary polyps, with a mouth and tentacles; others are polyps lacking the mouth and tentacles, and acting as defensive organs for the colony (the dactylozooids); others again, while lacking the mouth and possessing only vestiges of tentacles, have the property of budding off the gonophores or sexual members of the colony—these are the blastostyles, and the sexual individuals they produce are imperfectly developed medusae, which characterize this particular species. In some species the medusa acquires a complete development, and breaking away from the colony swims in the sea and distributes its products far and wide.

The phenomenon of alternation of generations depends upon the fact that in many *Hydromedusae* and *Acalephae* the fertilized ovum gives rise to an organism (the polyp) which can produce buds, but not ova and spermatozoa. Some of the buds become Medusae or medusa-like individuals in which the sexual elements are formed. This kind of alternation of generations in which a sexually reproducing generation succeeds one or more generations of asexual forms is called *metagenesis.*

The ovum is generally fertilized outside the parent, but in some *Anthozoa* the fertilization and the early stages of development take place in the enteric cavity. In many forms, particularly amongst the Medusae, all the individuals of the same species discharge their eggs at the same time of day, but this time may alter slightly according to the time of year. It is remarkable that in the most nearly allied species the eggs are discharged at the most different times. It has been suggested that this is an arrangement to prevent the crossing of closely allied species.

Almost all Coelenterates are marine. *Hydra, Microhydra, Cordylophora* amongst polyp forms; and *Limnocodium,* found in the Victoria Regia tank of the Royal Botanical Society, London, and *Limnocnida* from Lake Tanganyika, amongst Medusae, are the most striking of the few fresh-water members of the group. The polyps and polyp colonies are for the most part attached to foreign bodies and lead sedentary lives, while the *Medusae, Siphonophora,* and *Ctenophora* are free-swimming pelagic organisms, striking for their beauty and for the extreme fragility and transparency of their tissues. Many of them are phosphorescent. The greater number are littoral or pelagic in habit, but a few forms, often characterized by important peculiarities of structure, are from the deep sea.

A few are parasitic, *e.g.*, *Polypodium hydriforme* parasitic in the young stage in the ova of the Sterlet (*Acipenser ruthenus*), (Ussow, Ann. and Mag. N. Hist. (5), 18, 1886); and *Mnestra*, a medusa parasitic on *Phyllirhoe*.

The few fossil forms known belong for the most part to the corals, the hard tissues of which have left traces in the rocks of most periods. There are traces of Medusa remains in the Solenhofen slates, and the Graptolites of the Cambrian formation may possibly have been hydroid colonies, resembling the *Sertularidae*.

The following table shows the classification of the *Coelenterata* adopted in this work:—

Sub-phylum I. CNIDARIA.

Class I. HYDROMEDUSAE (CRASPEDOTA).

- Order 1. Hydrida.
- ,, 2. Hydrocorallinae.
- ,, 3. Tubulariae. Anthomedusae.
- ,, 4. Campanulariae. Leptomedusae.
- ,, 5. Trachomedusae.
- ,, 6. Narcomedusae.
- ,, 7. Siphonophora.

Class II. ACALEPHAE (ACRASPEDOTA).

- Order 1. Scyphomedusae (Tetrameralia).

Class II. ACALEPHAE—*Continued.*

- Order 2. Ephyroniae (Octomeralia).

Class III. ANTHOZOA.

- Order 1. Rugosa (Tetracoralla).
- ,, 2. Alcyonaria (Octactinia).
- ,, 3. Zoantharia (Hexactinia).

Sub-phylum II. CTENOPHORA.

Class CTENOPHORA.

- Order 1. Tentaculata.
- ,, 2. Nuda.

Sub-phylum I. CNIDARIA.

Coelenterata with thread-cells.

It is on the whole convenient to use the characteristic thread-cells to divide the *Coelenterata* into two sub-phyla—the *Cnidaria* and *Ctenophora*—although it must not be forgotten that these structures have been observed in one Ctenophore, and are found outside the limits of the Coelenterata altogether (*Platyhelminthes*, *Mollusca*, *Protozoa*).

Class I. HYDROMEDUSAE (CRASPEDOTA).*

Cnidaria in which the medusa has a velum and the polyp is without gastral ridges or filaments.

This class includes polyps; colonies of polyps, which produce

* The Hydromedusae coincides with the old group *Hydrozoa*, excluding the *Acalephae*. It appears convenient to place the latter in a separate class to mark their intermediate position between the *Hydromedusae* and *Actinozoa*.

medusae or modified medusae by budding; and medusae which arise directly from the egg. The polyps are generally attached permanently to foreign bodies, and are of small size; but in the *Siphonophora* the whole colony is free-swimming. The medusae are also of small size and possess a velum. When skeletal structures are present they consist, as a rule, of a more or less horny secretion of the ectoderm (*perisark*). In one order, however, the secretion is calcareous (*Hydrocorallinae*).

The colonies very often present, in a well-marked manner, the phenomenon of polymorphism, and in describing the various

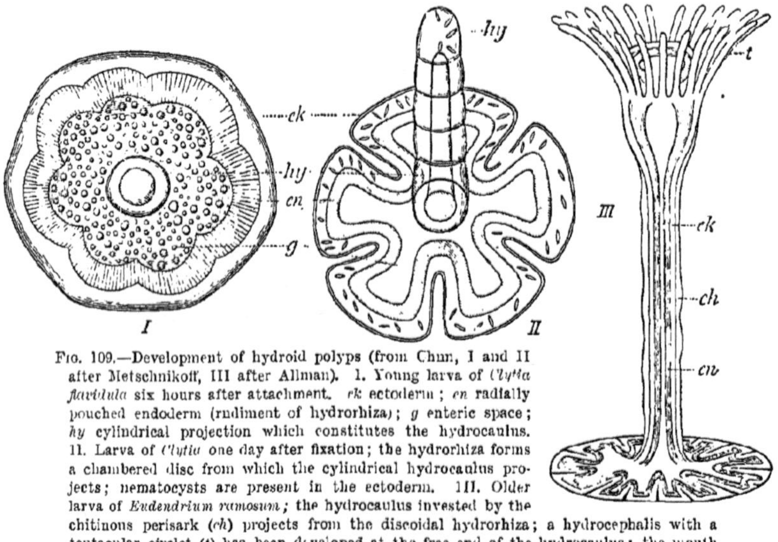

Fig. 109.—Development of hydroid polyps (from Chun, I and II after Metschnikoff, III after Allman). I. Young larva of *Clytia flavidula* six hours after attachment. *ek* ectoderm; *en* radially pouched endoderm (rudiment of hydrorhiza); *g* enteric space; *hy* cylindrical projection which constitutes the hydrocaulus. II. Larva of *Clytia* one day after fixation; the hydrorhiza forms a chambered disc from which the cylindrical hydrocaulus projects; nematocysts are present in the ectoderm. III. Older larva of *Eudendrium ramosum*; the hydrocaulus invested by the chitinous perisark (*ch*) projects from the discoidal hydrorhiza; a hydrocephalis with a tentacular circlet (*t*) has been developed at the free end of the hydrocaulus; the mouth is not yet formed.

modifications which may be present it will be convenient to explain some of the more common terms which are used in the cumbrous and complicated nomenclature of the group.

The fertilized ovum very generally gives rise to an oviform free-swimming larva—the **planula** (Dalyell*), consisting of an outer layer of ciliated ectoderm and an inner hollow mass of endoderm. After a short time it loses its cilia and secretes a thin cuticular covering—the perisark—and becomes attached by one end (Fig. 109). The free end elongates and develops a terminal mouth and

* Sir J. Graham Dalyell, *Rare and Remarkable Animals of Scotland*, London, 1847.

tentacles, while the attached end often spreads itself into a discoidal root (Fig. 109), the *hydrorhiza.** This is the first polyp. The free upstanding portion is correctly termed the *hydranth*, though the word polyp is sometimes loosely applied to it. The hydranth elongates and begins to bud, and two parts become distinguishable in it—a terminal part called the polyp-head or *hydrocephalis* with the mouth and tentacles, and a lower part with the buds, the stem or *coenosark*. The buds lengthen, remain attached to the parent, and themselves become differentiated into hydrocephalis and coenosark. The whole colony of hydranths thus formed is called the *hydrosoma*. The term *hydrophyton* seems to be applied to the coenosark plus hydrorhiza, while *hydrocaulus*† appears to be synonymous with coenosark. In some forms the perisark does not extend on to the polyp-head but stops short at its base; in others it is continued round the polyp-head as a cup—the *hydrotheca* (Fig. 114), which, however, stands off from the polyp, being only connected with it by pseudopodial prolongations of the ectoderm. The polyp-head can shrink into the hydrotheca for protection. Some of the buds formed by the colony develop

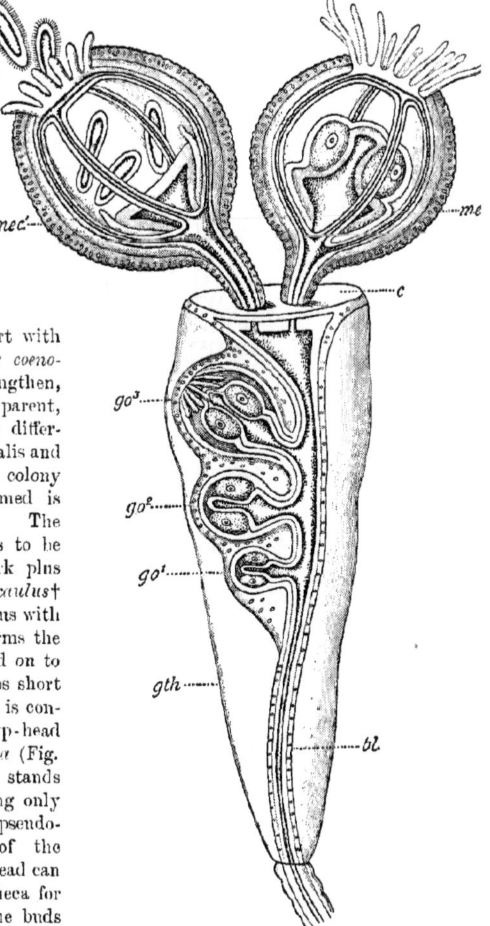

Fig. 110.—Gonangium with hedrioblasts of *Gonothyraea loveni* (from Chun, after Allman). *bl* blastostyle; *gth* gonangium (gonotheca); *c* operculum of gonangium; go^{1-3} budding gonophores (hedrioblasts); mec^1 hedrioblast with planulae *pl*; mec^2 hedrioblast with ova in manubrium.

* The hydrorhiza has sometimes the form of a branching and anastomosing tube, from which the hydranths arise by budding.

† The coenosark or hydrocaulus is said to be fascicled or *polysiphonic* when it is composed of several adherent tubes, monosiphonic when consisting of a single tube.

into medusae which become free-swimming, or *medusoids* which are imperfectly developed medusae and do not become free. These medusae and medusoids are called the *gonophores* because the gonads are contained in them. A *phanerocodonic* gonophore is a free-swimming medusa, and is sometimes termed a *planoblast*; an *adelocodonic* gonophore is a medusoid, and is sometimes called a *hedrioblast*. Sometimes the gonophores are budded only from special hydranths, which are then modified by the absence or diminution of size of the tentacles and mouth. Such a proliferous hydranth is a *blastostyle*. In the forms with hydrothecae, the hydrotheca of the blastostyle forms a chitinous capsule enclosing the blastostyle and gonophores; it is called the *gonangium* or *gonotheca* (Figs. 110, 114). Colonies with gonangia are called *calyptoblastic;* those without *gymnoblastic*. The word *zooid* or *person* is sometimes applied to any individual of a colony, whether hydranth, gonophore, or blastostyle, etc. *Trophosome* means the entire assemblage of zooids which are concerned with the nutrition of the colony, *i.e.*, all the hydranths; *gonosome* the entire assemblage of zooids which are concerned with the sexual reproduction of the colony, *i.e.*, all the gonophores. A *dactylozooid* or *spiral zooid* is a hydranth without or with reduced mouth and tentacles, specially developed in some genera (*Hydractinia* and *Podocoryne*, Fig. 108) for defensive purposes. A *nematophore* (*macho-polyp, guard-polyp, sarcostyle*) is a special stalked projection of the coenosark found in the *Plumularidae* (Fig. 111). It consists of ectoderm, which contains thread-cells and is highly amoeboid, and of a solid axis of endoderm; the whole is enclosed in a theca (*sarcotheca*) of perisark. Nematophores are probably nutritive, catching food as an Amoeba does, by the pseudopodia of their ectoderm.

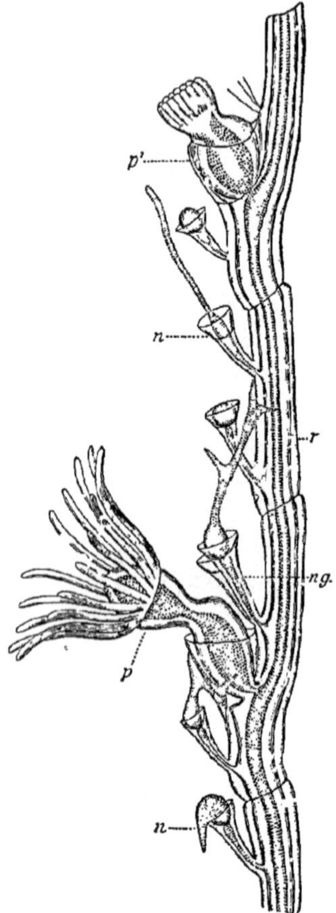

FIG. 111.—*Antennularia antennina*. Branch (hydrocladium) of colony with hydranths and nematophores (from Chun, after Allman). *p* extended, *p'* retracted hydranth; *n*, *ng* nematophores; *r* coenosark (hydrocladium).

Hydrocladium is a special term applied to the hydrotheca-bearing branches (*ramuli*) of the coenosark of *Plumularidae*. *Phylactocarps* are specially modified *hydrocladia* bearing gonangia as well as nematophores and sometimes hydrothecae; they often develop protective branches—the *costae*—which form the walls of an open basket-work and enclose the gonangia. When the costae are well developed and the hydrothecae are suppressed, the phylactocarp is called a *corbula* (*Aglaophenia*).

The generative cells, though nearly always ripening in and discharging from the ectoderm of gonophores, make their first appearance in a great variety of places: in the ectoderm of the free Medusa, in the endoderm or ectoderm of the fixed gonophore, in the ectoderm of the coenosark, near or remote from the budding gonophore, or even in the endoderm. But wherever they arise they eventually migrate—passing, if necessary, through the structureless lamella—to the gonophore, in the ectoderm of which they are generally, though not always, contained.

The Hydromedusae feed chiefly on animal substances, and Hydra possesses chlorophyll bodies in its endoderm. Whether these are chlorophyll bodies like those of plants or symbiotic *Algae*, like those of *Radiolaria*, seems uncertain. The free-swimming *Medusae* and *Siphonophora* are phosphorescent. With a few exceptions they are marine organisms.

Order 1. Hydrida.

Solitary polyps without medusoid buds. Both generative products are developed in the ectoderm of the polyp.

In this order colonies are not formed, and there are no medusoid individuals. The generative cells are produced by the ectoderm of the polyp itself. There is only one genus, *Hydra* L., the fresh-water polyp.

The genera *Protohydra* Greeff, and *Microhydra* Potts, are possibly allied here. *Protohydra* is marine, and reproduces by transverse fission. *Microhydra* is fresh-water, and invested by a coat of mud. Both genera are without tentacles, and the sexual reproduction is unknown.

Order 2. Hydrocorallinae.*

Colonial Hydromedusae consisting of a meshwork of coenosarcal canals, the ectoderm of which secretes a hard calcareous matter filling up the spaces of the meshwork. Polyps of two forms, gastrozooids and dactylozooids. Gonophores of the form of rudimentary medusae are generally present.

For a long time the position of the *Hydrocorallinae* was uncertain. L. Agassis first suggested that *Millepora* was not an Anthozoon but a Hydroid, but it was not until Moseley had worked out the anatomy

* H. N. Moseley, "Report on Corals," *Challenger Reports*, vol. 2. S. J. Hickson, "The Medusae of Millepora Murrayi, etc," *Q.J.M.S.*, vol. 32. 1891.

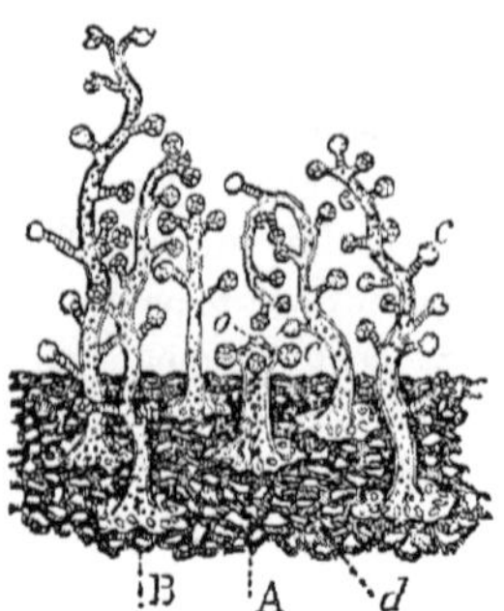

FIG. 112.—Group of polyps of *Millepora nodosa*. *A* gastrozooid surrounded by five dactylozooids *B*; *o* mouth; *c* tentacles; *d* tubes (after Moseley).

of the soft parts of this genus and of the *Stylasteridae*, that the Hydromedusan affinities of these animals were clearly proved, and that the order *Hydrocorallinae* was definitely established.

The colonies have the form of encrusting or arborescent masses consisting of a network of coenosarcal canals (Fig. 113) permeating a hard calcareous support called the *coenosteum*. The canals consist of ectoderm and endoderm, and the coenosteum is calcareous matter secreted by the ectoderm and comparable to the perisark of other Hydroid colonies. From the surface of the colony which is covered by a layer of ectoderm the polyps project. They are of two kinds—*gastrozooids* with a mouth and tentacles, and more numerous *dactylozooids* (Fig. 112) with or without tentacles, but always without a mouth. They are lodged in pits or calycles excavated in the surface of the coenosteum, and lined by a continuation of the surface ectoderm. The relation of the surface ectoderm to the ectoderm of the polyps and of the canals is not known. The polyps of the two kinds are either scattered irregularly over the surface of the colony, or gathered into groups more or less regular, in each of which a centrally placed gastrozooid is surrounded by a

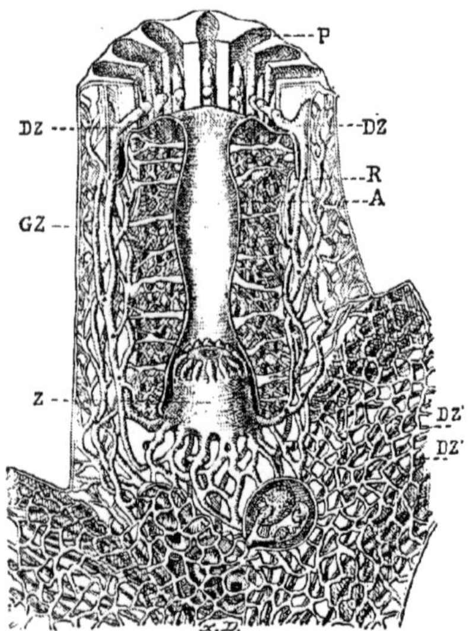

FIG. 113.—Vertical section through a group of retracted zooids of *Allopora profunda* (after Moseley). *DZ* dactylozooids; *P* calycles of dactylozooids separated by pseudosepta; *Z* gastrozooid giving off canals which join the coenosarcal network; *GZ* calycle of gastrozooid; *DZ'* dactylozooids of an adjacent group; *G* gonozooid (gonophore).

ring of dactylozooids. The cavities of the zooids communicate with the coenosarcal meshwork by large canal offsets. Gonophores having the form of rudimentary medusae are developed on the coenosarcal tubes, and are often lodged in special ampullae of the coenosteum. The tentacles generally possess knobbed extremities armed with thread cells. From coral reefs and warm seas.

Fam. 1. **Milleporidae.** Coenosteum arborescent or encrusting, composed of a thin superficial living layer, lying upon dead layers of former growth. Pores without styles, but divided by tabulae marking the successive layers of growth. Dactylozooids with knobbed tentacles. *Millepora* L.

Fam. 2. **Stylasteridae.** Coenosteum arborescent, with a strong tendency to assume a fan-like form, and to the development of the pores on one face only or on the lateral margins of its branches. In some genera a superficial layer only of the coenosteum is living, in others nearly the entire mass retains its vitality. Pores with tabulae in two genera only. Gastropores provided with a conical calcareous projection—the *style*—at their bases. Pseudosepta, arising from the partial confluence of the dactylopores and gastropores, sometimes present. Dactylozooids without tentacles. Colonies dioecious. Found in all seas in shallow and deep waters.

Sporadopora Moseley; *Pliobothrus* Pourtalès; *Errina* Gray; *Distichopora* Lamarck; *Labiopora* Moseley; *Spinipora* Moseley; *Allopora* Ehrenberg; *Stylaster* Gray; *Stenohelia* S. Kent; *Conopora* Moseley; *Astylus* Moseley; *Cryptohelia* M. Edw. and Haime.

Order 3. Tubulariae* (Gymnoblastea).

Without hydrothecae and gonangia. Polyps, when more than one, forming permanent colonies. Generative individuals, when set free, are Anthomedusae.

The *Tubulariae* are almost all colonial (the *Corymorphinae*, *Myriothela*, etc., are solitary), and they all produce medusoid gonophores by budding. The gonophores are either set free as Medusae, or only become partially developed as medusoids, with rudiments of the medusan organs, *e.g.*, manubrium, gastrovascular canals, and marginal tentacles. The *Medusae* have ocelli, and their gonads are in the manubrium.

Section 1. **Tubularinae.**

Colonial forms with a perisark destitute of investing layer of coenosark.

Fam. 1. **Clavidae.** Polyps with scattered filiform tentacles. *Clava* Gmelin; *Cordylophora* Allm., fresh and brackish water; *Tubiclava* Allm.; *Merona* Norman; *Rhizogeton* Ag.; *Clavula* Wright; *Dendroclava* Weissman; *Campaniclava* Allm.; *Corydendrium* v. Ben.

* G. J. Allman, *A Monograph of the Gymnoblastic or Tubularian Hydroids*, Ray Society, 1871. G. J. Allman, "Report on the Hydroidea," Pts. 1 and 2 *Challenger Reports*, 1883 and 1888. T. Hincks, *A Monograph of the British Hydroid Zoophytes*, London, 1868. E. Haeckel, *Monographie der Medusen*, Jena, 1879.

Fam. 2. **Corynidae.** Polyps with scattered, more or less spirally disposed, capitate tentacles. *Coryne* Gärtner; *Actinogonium* Allm.; *Syncoryne* Ehrb.; *Gymnocoryne* Hincks; *Gemmaria* McCrady.

Fam. 3. **Bougainvillidae.** Hypostome not abruptly differentiated (conical). Tentacles filiform in a single circle round the base of the hypostome. *Bougainvillia* Lesson; *Perigonimus* Sars; *Bimeria* S. Wright; *Dicoryne* Allm.; *Stylactis* Allm.; *Atractylis* S. Wright; *Diplura* Green; *Hydranthea* Hincks; *Cionistes* S. Wright; *Heterocordyle* Allm.; *Wrightia* Allm.; *Garveia* S. Wright.

Fam. 4. **Eudendridae.** Hypostome abruptly differentiated from the body (everted). Tentacles filiform in a single row. *Eudendrium* Ehrb.

Fam. 5. **Pennaridae.** Polyps with filiform and capitate tentacles. *Pennaria* Goldfuss; *Haloeordyle* Allm.; *Stauridium* Duj.; *Vorticlava* Alder; *Heterostephanus* Allm.; *Acharadria* S. Wright; *Acaulis* Stimpson; *Cladonema* Duj.

Fam. 6. **Cladocorynidae.** Polyps with simple and ramified capitate tentacles. *Cladocoryne* Rotch.

Fam. 7. **Clavatellidae.** With simple capitate tentacles in a single row. *Clavatella* Hincks.

Fam. 8. **Tubularidae.** Polyp with a proximal and distal row of simple filiform tentacles. *Tubularia* L.; *Hybocodon* Ag.; *Ectopleura* Ag.

Fam. 9. **Myriothelidae.** Polyp solitary. Tentacles scattered, capitate. *Myriothela* Sars.

Section 2. **Hydractininae.**

Colonial. Perisark invested by a superficial covering of naked coenosark: with spiral zooids. Very frequently found coating the gastropod shell of a Hermit Crab.

Fam. 10. **Hydractinidae.** With sessile gonophores. *Hydractinia* v. Ben.

Fam. 11. **Podocorynidae.** Gonophores as free medusae. *Podocoryne* Sars (Fig. 108); *Corynopsis* Allm.

Section 3. **Corymorphinae.**

Polyps solitary, without perisark.

Fam. 12. **Corymorphidae.** Gonophores as free medusae. *Corymorpha* Sars; *Halatractus* Allm.; *Amalthaea* O. Schmidt.

Fam. 13. **Monocaulidae.** Gonophores as fixed sporosacs. *Monocaulus* Allm.

Section 4. **Hydrolarinae.**

Polyps unsymmetrical, with the tentacles, one or two in number, springing from one side of the body.

Fam. 14. **Hydrolaridae.** *Lar* Gosse.

The **Medusae of this order** are arranged by Haeckel as follows:—

ANTHOMEDUSAE.

Craspedota without otocysts, with ocelli at the base of the tentacles, and with manubrial gonads: radial canals usually 4, rarely 6 or 8; budded from Polyps of the Tubulariae.

Fam. 1. **Codonidae.** Mouth-opening simple; gonads not radially divided; 4 narrow radial canals; unbranched tentacles. The polyps of most *Sarsiadae* belong to the genus *Syncoryne*, of *Ectopleura* to *Tubularia*, of *Euphysidae* to *Corymorpha*, of *Globiceps* to *Pennaria*.

Sub-fam. 1. **Sarsiadae.** With 4 radial tentacles. *Codonium* H.*; *Sarsia* Lesson; *Syndictyon* A. Ag.; *Ectopleura* L. Ag.; *Dipurena* McCrady; *Bathycodon* H.

Sub-fam. 2. **Dinemidae.** With 2 radial tentacles. *Dicodonium* H.; *Dinema* v. Ben.

Sub-fam. 3. **Euphysidae.** With 3 rudimentary tentacles, and one strongly developed. *Steenstrupia* Forbes; *Euphysa* Forbes; *Hybocodon* L. Ag.; *Amphicodon* H.

Sub-fam. 4. **Amalthaeidae.** All four tentacles rudimentary. *Amalthaea* O. Schmidt; *Globiceps* Ayres.

Fam. 2. **Tiaridae.** With 4 frilled buccal lobes; with 4 manubrial gonads, which may be split into 8; with 4 wide radial canals; and unbranched tentacles. Ontogeny only known in two species. The polyp of *Turris neglecta* is *Clavula Gossei*, that of *Corynetes Agassizii* is *Halocharis spiralis.*

Sub-fam. 1. **Protiaridae.** With 4 perradial tentacles. *Protiara* H.; *Modeeria* Forbes; *Corynetes* McCrady.

Sub-fam. 2. **Amphinemidae.** With 2 opposite radial tentacles. *Amphinema* H.; *Codonorchis* H.; *Stomotoca* L. Ag.

Sub-fam. 3. **Pandaeidae.** With numerous tentacles. *Pandaea* Lesson; *Conis* Brandt; *Tiara* Lesson; *Turris* Lesson; *Catablema* H.; *Turritopsis* McCrady; *Callitiara* H.

Fam. 3. **Margelidae.** With 4 or more simple or branched oral tentacles, with 4 or 8 separate manubrial gonads; with simple unbranched tentacles, which may be uniformly distributed or grouped in 4 or 8 bundles. Development known in a few species; polyp usually belongs to *Bougainvillea* (*Lizusa, Margelis, Hippocrene, Rathkea*), polyp of *Dysomorpha* to *Podocoryne*, of *Cytaeandra* to *Rhizocline*. The polyp of *Lizzia* (*blondina?*) is said by Allman to be a Campanularian, *Laomedia* (*Leptoscyphus*) *tenuis.*

Sub-fam. 1. **Cytaeidae.** With unbranched oral, and uniformly distributed marginal tentacles. *Cytaeis* Eschsch; *Cubogaster* H.; *Dysmorphosa* Phillipi; *Cytaeandra* H.

Sub-fam. 2. **Lizusidae.** With unbranched oral, and 4 or 8 bundles of marginal tentacles. *Lizusa* H.; *Lizzia* Forbes; *Lizzella* H.

Sub-fam. 3. **Thamnostomidae.** With 4 branched oral, and uniformly distributed marginal tentacles. *Thamnitis* H., *Thamnostylus* H.; *Thamnostoma* H.; *Limnorea* Péron.

Sub-fam. 4. **Hippocrenidae.** Oral tentacles branched, marginal tentacles in 4 or 8 groups. *Margelis* Steenstrup; *Hippocrene* Mertens; *Nemopsis* L. Ag.; *Margellium* H.; *Rathkea* Brandt.

Fam. 4. **Cladonemidae.** With dichotomously branched or feathered tentacles, with 4 to 8 narrow, simple or bifurcated radial canals, with 4 or 8 separate manubrial gonads. Oral arms 4, numerous, or absent. Polyp form known in 3 genera, viz., *Gemmaria* is the medusa of *Gemellaria, Eleutheria* of *Clavatella, Cladonema* of *Stauridium.*

Sub-fam. 1. **Pteronemidae.** Radial canals simple. *Pteronema* H.; *Zanclea* Gegenb.; *Gemmaria* McCrady; *Eleutheria* Quatref.

Sub-fam. 2. **Dendronemidae.** Radial canals bifurcated. *Ctenaria* H. resembles the Ctenophora, ex-umbrella with 8 adradial ribs of thread cells; *Cladonema* Duj.; *Dendronema* H.

* Throughout the Cnidaria H. stands for Haeckel.

Order 4. CAMPANULARIAE* (CALYPTOBLASTEA).

With hydrothecae and gonangia. Colonial. Generative individuals, when set free, are Leptomedusae.

The *Campanulariae* are colonial *Hydromedusae*, and they all produce gonophores by budding. The zooids are provided with hydrothecae (Fig. 114), which, in the case of the blastostyles, form gonangia. The gonophores are either set free as Medusae, or are only partially developed as Medusoids (hedrioblastic), with rudiments of the medusan organs. The Medusae generally have marginal auditory organs of the vesiculate type, and their gonads lie beneath the radial canals (Fig. 106).

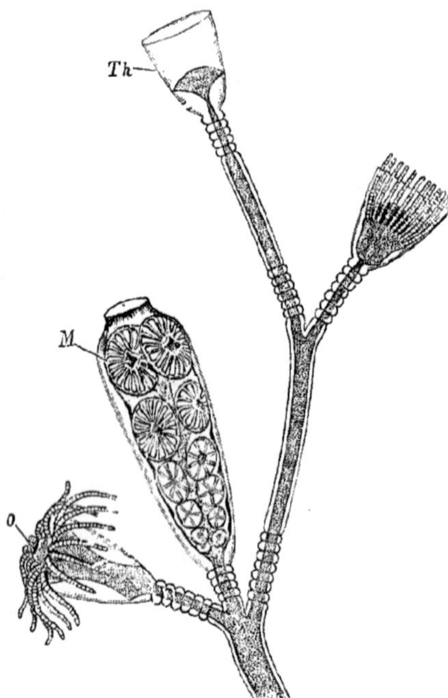

FIG. 114.—Branch of an *Obelia* colony (*O. gelatinosa*). *O* mouth of a nutritive polyp with extended tentacles; *M* medusa-buds on a blastostyle in a gonatheca; *Th* hydrotheca.

Section 1. **Campanularinae.**

Hydrothecae at least in the proximal part of the colony never adnate by their sides to the hydrocaulus.

Fam. 1. **Haleciidae.** Hydrothecae reduced to shallow saucer-shaped pedunculate appendages (hydrophores). Hydranths with conical hypostome. Gonophores hedrioblastic. *Halecium* Oken; *Diplocyathus* Allm.; *Ophioides* Hincks.

Fam. 2. **Campanularidae.** Hydrothecae borne by peduncles, campanulate or tubular. Hydrocaulus not enveloped by peripheral tubes. Gonophores free-swimming or sessile. *Campanularia* Lamarck; *Obelia* Péron and Lesueur; *Thyroscyphus* Allm.; *Hypanthea* Allm.; *Calamphora* Allm.; *Hebella* Allm.; *Halisiphonia* Allm.; *Coppinia* Hassall; *Calycella* Hincks; *Clytia* Lamouroux; *Campanopsis* Claus; *Melicertaria* Haeckel; *Lovénella* Hincks; *Cuspidella* Hincks; *Thaumantias* Esch.; *Campanulina* v. Ben.; *Leptoscyphus* Allm.

* T. Hincks, *A Monograph of the British Hydroid Zoophytes*, London, 1868. G. J. Allman, "Report on the Hydroidea," Pts. 1 and 2, *Challenger Reports*, 1883 and 1888. E. Haeckel, *Monographie der Medusen*, Jena, 1879.

Fam. 3. **Perisiphonidae.** Hydrocaulus enveloped by peripheral tubes; the hydrothecae are never adnate and are carried by the axial tube only. *Lafoëa* Lamouroux; *Lictorella* Allm.; *Cryptolaria* Busk; *Perisiphonia* Allm.

Section 2. **Sertularinae.**

Hydrothecae developed from more than one side of the hydrocaulus, to which they are all adnate for a greater or less extent by their sides.

Fam. 4. **Grammaridae.** Hydrocaulus consisting of an axial tube which carries the hydrothecae and is surrounded by and inseparably coalesced with peripheral tubes without hydrothecae. Hydrothecae adnate to axial tube. *Grammaria* Stimpson.

Fam. 5. **Sertularidae.** Hydrothecae in two or more series, adnate to hydrocaulus. Gonophores sessile. *Sertularia* L.; *Diphasia* Agassiz; *Thuiaria* Fleming; *Desmoscyphus* Allm.; *Hypopyxis* Allm.; *Staurotheca* Allm.; *Dictyocladium* Allm.; *Synthecium* Allm.; *Thecocladium* Allm.; *Hydrallmania* Hincks.

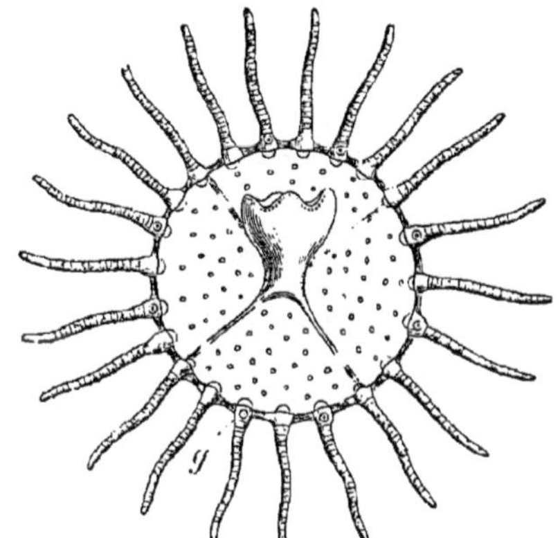

Fig. 115.—Free Medusa of *Obelia gelatinosa* as yet without gonads; *g* otocysts.

Section 3. **Idiinae.**

Hydrothecae adnate to hydrocaulus. Coenosark divided into segments which form two longitudinal series of intercommunicating chambers, each of which communicates with the gastral cavity of a hydranth.

Fam. 6. **Idiidae.** *Idia* Lamouroux.

Section 4. **Plumularinae.**

Hydrothecae developed from one side only of the hydrocaulus, to which they are adnate by their sides. Sarcostyles (nematophores) are always present.

Fam. 7. **Plumularidae.** *Antennularia* Lamarck; *Sciurella* Allm.; *Acanthella* Allm.; *Plumularia* Lamk.; *Schizotricha* Allm.; *Polyplumaria* Sars; *Heteroplon* Allm.; *Halicornaria* Busk; *Azygoplon* Allm.; *Streptocaulus* Allm.; *Diplocheilus* Allm.; *Lytocarpus* Kirchenpauer; *Acanthocladium* Allm.; *Cladocarpus* Allm.; *Aglaophenia* Lamouroux.

The **Medusae of this order** are arranged by Haeckel as follows:—

LEPTOMEDUSAE.

Craspedota partly with, partly without, otocysts; ocelli present or absent; gonads on radial canals; budded from polyps of the Campanulariae.

Fam. 1. **Thaumantidae.** Radial canals simple, unbranched, without otocysts. Polyp form known only in *Laodice* (*Thaumantaria*) *calcarata* as *Thaumantias inconspicua*, and in *Melicertum campanula* as *Melicertum campanula*.

Sub-fam. 1. **Laodicidae.** With 4 radial canals and 4 gonads. *Tetranema* H.; *Dissonema* H.; *Octonema* H.; *Thaumantias* Eschsch.; *Staurostoma* H.; *Laodice* Lesson.

Sub-fam. 2. **Melicertidae.** With 8 radial canals and 8 gonads. *Melicertella* H.; *Melicertissa* H.; *Melicertum* A. Ag.; *Melicertidium* H.

Sub-fam. 3. **Orchistomidae.** With numerous radial canals. *Orchistoma* H.

Fam. 2. **Cannotidae.** Without marginal vesicles, with 4 or 6 radial canals which are branched, bifurcated, or pinnate. Development unknown.

Sub-fam. 1. **Polyorchidae.** With 4 or 8 pinnate radial canals, the side branches ending blindly. *Staurodiscus* H.; *Gonynema* A. Ag.; *Ptychogena* A. Ag.; *Staurophora* Brandt; *Polyorchis* A. Ag.

Sub-fam. 2. **Berenicidae.** With 4 or 6 branched radial canals, branches and main canals open into circular canal. *Cannota* H.; *Dyscannota* H.; *Berenice* Péron and Lesueur; *Dipleurosoma* Axel Boeck.

Sub-fam. 3. **Williadae.** With 4 or 6 bifurcated radial canals, the branches only open into circular canals. *Dicranocanna* H.; *Toxorchis* H.; *Willetta* H.; *Willia* Forbes; *Proboscidactyla* Brandt; *Cladocanna* H.

Fam. 3. **Eucopidae.** With marginal vesicles and 4 simple unbranched radial canals, in whose course 4 or 8 gonads lie. The polyps when known belong to the genera *Campanularia*, *Obelaria*, *Clytia*, *Campanulina*, etc.

Sub-fam. 1. **Obelidae.** 8 adradial marginal vesicles, stomach without stalk. *Eucopium* H.; *Saphenella* H.; *Eucope* Gegenbaur; *Obelia* Péron and Les. (Fig. 115); *Tiaropsis* L. Ag.; *Euchilota* McCrady.

Sub-fam. 2. **Phialidae.** Numerous marginal vesicles, stomach without stalk. *Phialium* H.; *Phialis* H.; *Mitrocomium* H.; *Epenthesis* McCrady; *Mitrocomella* H.; *Phialidium* Leuckart; *Mitrocoma* H.

Sub-fam. 3. **Eutimidae.** 8 adradial marginal vesicles, a distinct, often long, stomach-stalk. *Eutimium* H.; *Eutima* McCrady; *Saphenia* Eschsch.; *Eutimeta* H.; *Eutimalphes* H.; *Octorchidium* H.; *Octorchis* H.; *Octorchandra* H.

Sub-fam. 4. **Irenidae.** Numerous marginal vesicles, a distinct stomach-stalk. *Irenium* H.; *Irene* Esch.; *Tima* Esch.

Fam. 4. **Aequoridae.** With marginal vesicles and with numerous (often over 100) simple or branched radial canals. Polyp of *Polycanna* (*Zygodactyla*) *vitrina* only known as a very small *Campanaria*.

Sub-fam. 1. **Octocannidae.** With 8 simple radial canals. *Octocanna* H.

Sub-fam. 2. **Zygocannidae.** With 8 or more radial canals bifurcated at their base. *Zygocanna* H.; *Zygocannota* H.; *Zygocannula* H.; *Halopsis* A. Ag.

Sub-fam. 3. **Polycannidae.** With numerous simple radial canals which arise separately from the stomach (12 or more to over 100). *Aequorea* Péron and Les.; *Rhegmatodes* A. Ag.; *Stomobrachium* Brandt; *Staurobrachium* H.; *Mesonema* Esch.; *Polycanna* H.

Limnocodium,* a fresh-water medusa from the Victoria Regia tanks of the Royal Botanic Society, London, and of unknown habitat, is probably allied here.

* G. H. Fowler, *Quart. J. Mic. Sci.*, vol. 30, 1889, p. 507.

Order 5. Trachomedusae.*

Hydromedusae without hydrosome (polyp stage); with marginal sense-tentacles in pits or vesicles, with endodermal otoliths. Ocelli usually absent. Gonads radial. Radial canals 4, 6, or 8, often with centripetal canals. With thread-cell thickening of ectoderm round the edge of the umbrella.

The medusae of this order (Fig. 101) develop directly from the egg, and no polyps are known.

Fam. 1. **Petasidae.** With 4 radial canals and 4 gonads, stomach without stalk, with sense-tentacles sometimes free, sometimes in vesicles.

Sub-fam. 1. **Petachnidae.** Without centripetal canals. *Petasus* H., *Dipetasus* H.; *Petasata* H.; *Petachnum* H.; *Aglauropsis* F. Müller; *Gossea* L. Ag.

Sub-fam. 2. **Olindiadae.** With blind centripetal canals. *Olindias* F. Müller.

Fam. 2. **Trachynemidae.** 8 radial canals and 8 gonads, without stomach-stalk, sense-tentacles rarely free, usually in vesicles.

Sub-fam. 1. **Marmanemidae.** Tentacles without suckers, mesogonions† absent. *Trachynema* Gegenbaur; *Marmanema* H.; *Rhopalonema* Gegenbaur.

Sub-fam. 2. **Pectyllidae.** Tentacles with suckers, with 8 radial mesogonions.† *Pectyllis* H.; *Pectis* H.; *Pectanthis* H.

Fam. 3. **Aglauridae.** With 8 radial canals, stomach with stalk, sense tentacles free.

Sub-fam. 1. **Aglanthidae.** With 8 radial gonads (sometimes on the stomach-stalk, sometimes on the sub-umbrella). *Aglantha* H.; *Aglaura* Péron and Les.; *Aglisera* H.

Sub-fam. 2. **Persidae.** 4 or only 2 opposite gonads. *Stauraglaura* H.; *Persa* McCrady.

Fam. 4. **Geryonidae.** With 4 or 6 radial canals and flattened gonads; with long stomach-stalk; with 8 or 12 marginal umbrella-clasps or peroniums; and with 8 or 12 closed tentaculocysts (sense tentacles enclosed in vesicles), which are embedded in the jelly on the axial side of the peroniums (Fig. 102).

The tentacles are in three different groups which appear at three different periods of the development. (1) The primary tentacles are transitory larval organs, and filled with solid endoderm. They are *perradial* and 4 or 6 in number. They pass on to the ex-umbrella, and remain connected with the edge by a peronium. (2) The secondary tentacles are *interradial* and also solid, and 4 or 6 in number. They pass on to the ex-umbrella, and remain connected with the edge by peroniums. They may fall off or persist. (3) The tertiary tentacles develop last and persist. They arise beneath the perradial primary tentacles just to one side of the peroniums of these latter. They are hollow and long, and their cavity opens into the circular canal. There is a solid

* E. Haeckel, *Monographie der Medusen.* Jena, 1879.

† *Mesogonions* are thin vertical radial folds of the sub-umbrella, which underlie the radial canals and divide the gonads into two separate halves. They incompletely divide the umbrella cavity into spaces recalling the funnel-cavities of the *Periphyllidae.*

cartilaginous strip of endoderm beneath each of the 8 or 12 peroniums, close to which are the tentaculocysts.

Sub-fam. 1. **Liriopidae.** 4 gonads and 4 radial canals; 8 tentaculocysts (4 primary perradial and 4 secondary interradial). Permanent tentacles, 4 or 8. *Lirivantha* H.; *Liriope* Lesson; *Glossoconus* H.; *Glossocodon* H.

Sub-fam. 2. **Carmarinidae.** 6 gonads in the course of the 6 radial canals, 12 tentaculocysts (6 primary interradial, and 6 primary perradial). Tentacles 6 or 12. *Geryones* H.; *Geryonia* Pér. and Les., without centripetal canals; *Carmaris* H.; *Carmarina* H. (Fig. 101), with centripetal canals.

Order 6. Narcomedusae.*

Craspedota with free auditory tentacles. Tentacles inserted dorsally on the ex-umbrella, and connected with its edge by peroniums. Radial canals when present in the form of flat radial gastric pouches.

So far as is known the Narcomedusae are without the hydroid phase. There is a thickened ectodermal ring at the umbrella edge which is prolonged on to the ex-umbrella to the insertion of the tentacles as the peroniums (Fig. 96). The peripheral part of the umbrella is lobed. The gonads are primitively in the ventral or lateral wall of the stomach, whence they are often spread out on the radial gastral pouches. The circular canal is either obliterated, or else in festoons (Fig. 96), following the edge of the lobes to open into the gastral pouches. The radial structures (tentacles, lobes, and pouches) vary in number—they may be rarely 4, usually 8 or more to 32. *Otoporpae* or peronial streaks of ectoderm passing from the auditory tentacles may be present (Fig. 96).

Fam. 1. **Cunanthidae.** With wide, pouch-like radial canals, which are connected with the circular canal by double peronial canals (festoon canals, Fig. 96). With otoporpae. *Cunantha* H.; *Cunarcha* H.; *Cunoctantha* H.; *Cunina* Esch.; *Cunissa* H.

Fam. 2. **Peganthidae.** Without radial canals and gastral pouches, but with a festoon canal, with otoporpae. *Polycolpa* H.; *Polyxenia* Esch.; *Pegasia* Pér. and Les.; *Pegantha* H.

Fam. 3. **Aeginidae.** With a circular canal which communicates with the stomach by double peronial canals; with internemal† gastral pouches; without otoporpae. *Aegina* Esch.; *Aeginella* H.; *Aegineta* Gegenbaur; *Aeginopsis* Brandt; *Aeginura* H.; *Aeginodiscus* H.; *Aeginodorus* H.; *Aeginorhodus* H.

Fam. 4. **Solmaridae.** Without circular canal and peronial canals; sometimes without radial canals, sometimes with modified radial canals (pernemal or inter-

* E. Haeckel, *Monographie d. Medusen*. Jena, 1879.

† *Internemal* gastral pouches are really interradial pouches projecting from a radial (*pernemal*) pouch (suppressed in the *Aeginidae*) into the lobes of the peripheral part of the umbrella. Each original radial pouch gives off two of these internemal pouches, one into one lobe and the other into the adjacent lobe. The two internemal pouches of one radial gastric pouch are therefore separated by the double peronial canal, or festoon-like loop of the circular canal, which runs into the central stomach radially.

nemal gastral pouches); without otoporpae. *Solmissus* H.; *Solmundus* H.; *Solmundella* H.; *Solmoneta* H.; *Solmaris* H.

Limnocnida,* a fresh-water Medusa from Lake Tanganyika, is probably allied here.

Order 7. SIPHONOPHORA.†

Free-swimming polymorphic colonies of Hydromedusae produced by budding from an original, probably medusoid, individual. Gonads in gonophores which, as a rule, are not set free.

The colonies of the *Siphonophora* are characterised by the extreme specialization of the individuals composing them. So great indeed is this specialization that some zoologists (Eschscholtz, Huxley, Metscknikoff) have held the view that their component parts are really organs of a single medusoid individual, which is distinguished from an ordinary medusa by the fact that its various parts—manubrium, tentacles, umbrella—have multiplied independently of one another, and have become differentiated and in part dislocated from their primitive positions; in short, that a siphonophore, in possessing in a marked degree the power of vegetative increase of its parts, resembles a plant more than an animal.

This multiplication of the parts of an organism, often independently of one another, is not however by any means exclusively a vegetable characteristic. It must have happened largely in the animal kingdom, and have been a potent factor in determining the forms of animal life.

Another view, and the one more generally held, is that they are free-swimming polymorphic colonies of highly specialized polyps, with the power of producing medusae (Vogt, Leuckart, Gegenbaur, Claus, Chun).

According to it, all the parts of a siphonophore are either modified polyps or medusae, and the primitive zooid of the colony is of the polyp type. Just as the first theory errs too much in denying the colonial origin of our group, so the second theory probably goes too far in affirming it. It is probable that the truth lies between the two views. We hold, with Haeckel and Balfour, that the colonial theory is the true one, but that the primitive zooid of the colony was probably a medusa which has produced other medusae by budding, and that the parts of these medusae possess the power of becoming discrete and removed from the bud to which they belong, and of becoming in some cases secondarily multiplied. So

* R. T. Günther, *Quart. J. Mic. Sci.*, vol. 36, p. 284.

† E. Haeckel, "Report on the Siphonophorae," *Challenger Reports*, vol. 28, 1888. C. Chun, "Die Canarischen Siphonophoren," I. and II., *Abhandlungen d. Senckenbergischen naturf. Gesellsch.* 1891-2.

that many organs of the colony—which on the old colonial theory are modified polyps—are on this view nothing more than parts of medusiform individuals which have shifted their attachment, and are therefore really organs. For instance, the structures called *palpons* (hydrocysts, dactylozooids) are to be looked upon as mouthless manubria of medusoids, the umbrellas of which have become modified as bracts, or are entirely degenerate. The *siphons* (trumpet-shaped polyps, nutritive polyps) are the manubria of medusoids, of which the umbrella is a bract, or a nectocalyx or degenerate. The tentacle, on the other hand, is to be looked upon as the only surviving marginal tentacle of the medusoid of the siphon, which has shifted so as to be attached to the base of its manubrium. This theory then agrees with the second theory in asserting the colonial nature of the Siphonophora, but admits that there has been that vegetative

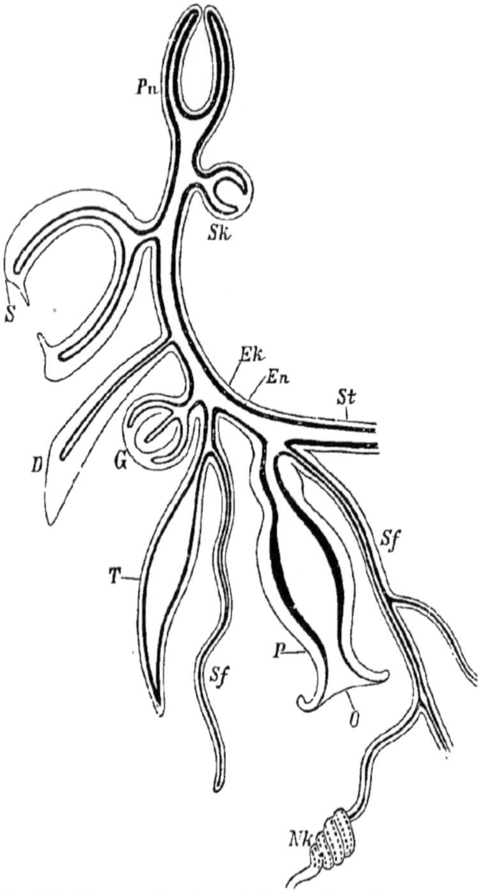

FIG. 116.—Diagram of a colony of *Siphonanthae*. *St* coenosome or stem; *Ek* ectoderm; *En* endoderm; *Pn* pneumatophore; *Sk* budding nectocalyx; *S* nectocalyx; *T* palpon (hydrocyst, dactylozooid); *Sf* tentacle and palpacle; *P* siphon (polyp); *O* mouth of siphon; *Nk* battery of nematocysts; *D* hydrophyllium; *G* gonophore.

repetition and specialization of certain organs which is demanded by the first view.

The diagram (Fig. 116) shows nearly all the possible parts found in colonies of *Siphonanthae*, the largest of the two sub-orders of the *Siphonophora*. We may briefly enumerate these and consider their relation to the colony on this "*medusome*" theory of Haeckel. The stem (*St*) or trunk is the coenosark or **coenosome** of the colony; it is the elongated manubrium of the original larval medusoid, and produces by budding all the parts of the colony. Two parts may be distinguished in it—an upper part, the **nectosome,** to which the swimming organs (nectocalyces and pneumatophores) are attached, and a lower part, the **siphosome,** bearing the nutritive and reproductive organs (siphons, palpons, gonophores). All parts are budded off from the same surface of the stem (the so-called ventral median line), their apparent radial disposition in some forms being due to a spiral twisting of the stem.*

The **swimming organs.** The **nectocalyx** (*S*) is a medusa with canal system and velum but without a manubrium. The **pneumatophore** (*Pn*) is more difficult of interpretation; it may either be regarded as a medusa, in which the umbrella cavity is the air-chamber or **pneumatocyst,** or it may be simply regarded—and this is Haeckel's view—as a part of the ex-umbrella region of the original medusoid larva, the ectoderm of which has become invaginated upon the contained enteric system to form the pneumatocyst (Haeckel distinguishes the ectodermal invagination as the **pneumatosac,** the secreted chitinous lining as the **pneumatocyst).** The space round the pneumatocyst lined by endoderm is the **pericystic space.**

The **siphons** (*P*) may be regarded as polyps, or as the manubria of medusoids. The **palpons** (tasters, hydrocysts, dactylozooids) are mouthless manubria. The **tentacles** (*Sf*) are organs of the siphons (see above). The **palpacles** (*Sf*) are similar organs of the palpons found in one order. The **hydrophyllia** (bracts) are the umbrellas of medusae which are cleft on one side, or which have simply lost their umbrella cavity and of which the manubria are either degenerate or slightly shifted as siphons and palpons. In many forms bracts have undergone a large secondary increase.

Gonophores (*G*) are either budded from the stem or from processes of the latter called **gonostyles,** which may, or may not be, mouthless polyps.

* The twisting when present takes place in opposite directions in the nectosome and siphosome.

The zooids are generally attached to the stem in groups called **cormidia** (Fig. 119). The points of attachment of the cormidia are called nodes, the part of the stem between being internodes; in such cases the cormidia are said to be *ordinate*. Sometimes this regular grouping does not occur, and the various zooids bud off separately from the stem; the cormidia are then said to be *irregular* or *dissolved*.

The nectocalyces by their contractions move the colony through the water; they have a deeply concave muscular sub-umbrella surface. The pneumatophore is a hydrostatic apparatus, and, in those forms which have a long spiral stem, serves to keep the body in an upright condition. The gaseous contents is secreted by some of the cells lining the pneumatocyst and can, in some cases, be expelled freely by contraction of the walls of the pneumatophore through one or more openings—the **stigmata.**

The enteric or gastrovascular system is continuous throughout the colony. The gastral zooids are without oral tentacles, but possess a tentacle arising at their base. This tentacle can be extended to a considerable length and be retracted into a spiral coil. It rarely has a simple form, but, as a rule, it bears a number of unbranched lateral twigs—the **tentilla,** which are also very contractile. These tentacles are invariably beset with a great number of nematocysts, which in many places are closely packed and have a regular arrangement. These aggregations of thread-cells are especially found upon the tentilla, where they give rise to large brightly-coloured swellings, the **cnidosacs** or batteries.

The **gonophores** have a velum, a complete gastrovascular system, and a manubrium; but the mouth is nearly always absent. The generative cells are ectodermal, and arise in the manubrium; they are without radial divisions (as in the *Codonidae* of the *Anthomedusae*). The colonies are generally hermaphrodite, but the gonophores are male and female. The sexual medusoids frequently become separate from the colony when ripe, but are only rarely liberated as small medusae (*Velellidae*), which produce the generative cells during their free life.

The hydrophyllia are leaf-shaped, and composed of a stiff gelatinous substance; they are protective in function. All the appendages are developed as buds formed of ectoderm and endoderm, and containing an endoderm-lined cavity which communicates with the cavity of the stem.

The *Siphonophora* are extremely beautiful transparent, marine organisms, with here and there spots of colour (the hepatic cells of the siphon—the apex of the pneumatophore, the cnidosacs of the tentacles, etc.). They are mostly pelagic in habit, but some come from the deep sea.

The ova are large, generally without vitelline membrane, and undergo a complete and regular segmentation. A free-swimming, solid planula is formed.

There are two main sub-orders*—the *Disconanthae*, in which the primary form is an 8-radiate medusa, the *Disconula*, which produces buds on the ventral side of its umbrella; and the *Siphonanthae*, in which the promorph is a bilateral medusa which produces buds on the ventral side of the base of its manubrium.

Sub-order 1. **Disconanthae.**

The body (coenosome) formed by the umbrella of the original octoradial medusa, which includes a polythalamous pneumatocyst; the buds arise in concentric rings from the sub-umbrella. Larva octoradial **(Disconula).**

This sub-class includes one order.

Section 1. Disconectae. Velellidae.

Siphonophora with a permanent primary umbrella, without nectocalyces and bracts.

The *Disconectae* are medusae with a large manubrium (Fig. 117, *ms*), hollow marginal tentacles, and radial canals opening into a circular canal. There is no velum, and the ex-umbrella surface is pitted inwards in the centre to form an ectodermal sac, the *pneumatocyst* (*lk*). From the under side there hangs downwards a number of accessory manubria (*gm*) which bud the gonophores and are called *gonostyles* (*gm*). The gonostyles open into the radial canals at their basal ends, while distally they may be closed or open. Beneath the pneumatocyst there is a large cellular mass, the **centradenia** (*cd*), or so-called liver. The pneumatocyst opens on the upper surface by

* Chun, who objects to Haeckel's separation of the *Disconanthae* from the rest of the class, arranges the Siphonophora as follows:

Order 1. **CALYCOPHORIDAE.** *With nectocalyces without pneumatophore.*

Order 2. **PHYSOPHORIDAE.** *With pneumatophore.*

Sub-order 1. **Haplophysae.** *Physophoridae with unchambered pneumatocyst, which is partly lined by gas-secreting ectoderm and is without tracheae.*

Tribe 1. *Physonectae.* With pneumatophore and nectocalyces.

Tribe 2. *Pneumatophoridae* (*Physalidae*). Without nectocalyces.

Sub-order 2. **Tracheophysae** (Haeckel's *Disconectae*). With chambered, chitin-lined pneumatocyst, which gives off tracheae to the polyps. Gonophores set free as medusae (*Chrysomitra*).

small apertures—the *stigmata*—and communicates with the centradenia and adjacent tissues by a number of fine tubes—the *tracheae*—which project from its base. The tracheae are lined by a prolongation of the chitinous lining of the pneumatocyst.

The centradenia is separated from the dorsal endoderm of the stomach by a gelatinous plate—the *gastrobasal plate* (*sp*)—which is pierced by the gastral ostia (primitively 8) for the passage dorsalwards of the radial canals (*rk*), which arise from the fundus of the stomach. It is a composite organ, partly consisting of a dense network of endodermal gastral canals, and partly of a parenchyma of ectoderm cells with many cnidoblasts. The function of the former is probably partly digestive and partly renal, of the latter gas-producing. In the simpler *Disconectae*, *i.e.*, in the *Discalidae* the centradenia is composed solely of a compact mass of ectoderm cells and cnidoblasts without canals. The only canals of the centradenia of such forms are eight simple radial canals which arise from the eight ostia of the stomach and run on its upper face—between it and the pneumatosac—uniting in its centre to form a typical octoradial liver star. They are to be looked upon as ascending branches of the eight primary radial canals of the subumbrella. The canals which perforate the centradenia of the more complex forms are branches of these (Fig. 117).

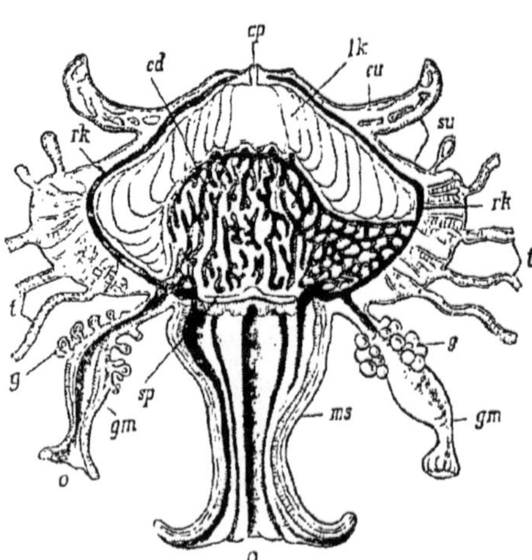

FIG. 117.—*Porpalia prunella* (after Haeckel from Lang). *cd* centradenia (central gland); *lk* pneumatocyst; *cp* central stigma of pneumatocyst; *rk* radial canal; *sp* gastrobasal plate; *eu* ex-umbrella; *su* sub-umbrella; *t* tentacles; *g* gonophores; *o* mouth; *ms* chief siphon; *gm* accessory siphons, in this case gonostyles.

The canal system is primitively octoradiate; it consists of the radial canals above mentioned, which branch as they pass outwards to be united at the margin of the umbrella by a circular canal (not marked in the figure). This system gives off, in addition to the centradenial canals, a set of canals—the *pallial canals*—on the ex-umbrella or upper surface of the pneumatocyst, which unite over the centre of the pneumatophore (round *cp*). This pallial system is the result of the invagination of ectoderm to form the pneumatophore.

The constitution and function of the centradenia, tracheae, and pneumatocyst of the *Disconectae* is disputed. Haeckel's account has been followed in the text. He regards the supposed ectodermal portion of the centradenia or liver as gas-secreting in function, and corresponding to the gas-secreting portion of the pneumatocyst of the *Siphonanthae* (see below), the tracheae being for the purpose of carrying the gas, so secreted, into the pneumatocyst, which is entirely hydrostatic. Chun, on the other hand, holds that in the *Disconanthae*, which habitually float on the surface, the pneumatocyst has no gas-gland, and that the cell mass of the centradenia, which Haeckel calls ectoderm, is endoderm with a rich development of thread cells; further, that the tracheae often end in places where this tissue is absent. He considers that these tubes are really *tracheae for the conveyance of oxygen to the thick glandular endoderm, and that the pneumatocyst in this group is a breathing organ.* In confirmation of this view he states that the living *Velella* does periodically contract its body as though it were expelling air from the air-sac. The elastic chitinous lining receives its explanation also on this view, as it would by its elasticity tend, in regaining its original form, to suck air in through the stigmata.

The gas-secreting ectoderm of the pneumatocyst is present in the young forms, which apparently live below the surface, and probably in the deep sea forms; and no doubt the function of the air-sac is, in these cases, purely hydrostatic, as in other Siphonophora in which there is a gas-gland and no tracheae.

Chun C., *Bericht üb. eine nach d. Canarischen Inseln ausgef. Reise.* Sitzb. Acad. Wiss. Berlin, 1888.

The gonophores which are produced on the gonostyles are small 4-radiate medusae, which do not produce sexual cells until after detachment, when they are known as *Chrysomitra.*

It is probable that the young of the *Disconectae* pass through a larval stage resembling in structure *Discalia.* This would be the so-called *Disconula*, a form actually met with in *Discalia* and presenting an 8-radiate medusiform structure with eight radial canals, eight marginal tentacles, and a dorsal 8-radiate invagination of ectoderm—the pneumatocyst.

Fam. 1. **Discalidae.** From the deep sea. Ex-umbrella without crest, gonostyles without mouth, pneumatocyst divided into a central chamber surrounded by 8 radial chambers, to which may be added a still more circumferential arrangement of 5 to 10 concentric ring-chambers. These chambers communicate with each other by the apertures called *pneumothyrae*, and some of them with the exterior by stigmata. The tentacles have terminal cnidospheres.† *Discalia* H.; *Disconalia* H.

Fam. 2. **Porpitidae.** Circular umbrella without crest; pneumatocyst divided into an octoradiate central part and numerous concentric rings. The gonostyles have mouths. Pneumothyrae* are present. With many stalked cnidospheres† on the tentacles.

* Pneumothyrae are communications between the concentric chambers of the pneumatocyst. † Cnidospheres are spherical knobs composed of cnidoblasts.

Sub-fam. 1. **Porpalidae.** Umbrella highly vaulted. Pneumatocyst campanulate, with a radially lobate margin. *Porpalia* H. (Fig. 117); *Porpema* H.

Sub-fam. 2. **Porpitellidae.** Umbrella flat, slightly vaulted. Pneumatocyst discoidal, without prominent radial marginal lobes. *Porpitella* H.; *Porpita* Lamarck.

Fam. 3. **Velellidae.** With an elliptical, often nearly quadrangular, umbrella including a polythalamous pneumatocyst of the same form, composed of numerous concentric rings, and usually bearing in its diagonal a vertical crest. Marginal tentacles simple, without cnidospheres; gonostyles with mouths. A chitinous prolongation of the pneumatocyst-lining into the crest is generally present. The 8-radiate character of the canal system exists but is much hidden and has become in part bilateral. The pneumatocyst consists of a central chamber sometimes markedly 8-radiate and of many concentric elliptical rings, with stigmata and pneumothyrae. *Rataria* Esch., pneumatocyst without crest. *Velella* Lamarck; *Armenista* H.

Fig. 118.—*Diphyes acuminata* magnified about 8 times. *Sb* somatocyst.

Sub-order 2. Siphonanthae.

Stem (coenosome) formed by the manubrium of the original bilateral medusa. The buds arise in the ventral line of this manubrium. Larva bilateral **(Siphonula).**

Section 1. Calyconectae. Calycophoridae.

Siphonanthae with one or more nectocalyces, without pneumatocyst and palpons. Cormidia ordinate.

A typical member of this group such as *Diphyes* (Fig. 118) consists of a long contractile hollow stem bearing at its upper end two opposed nectocalyces or swimming bells without manubria, but with four radial canals, a circular canal, and velum; and at regular intervals along its course groups of individuals called *cormidia* (Figs. 119, 120).

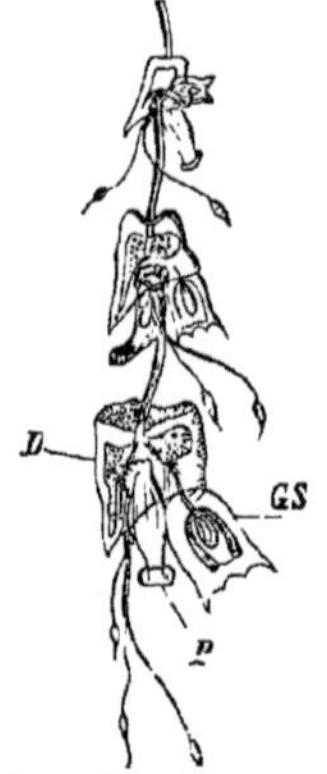

Fig. 119.—Three cormidia attached to the stem (coenosome) of a Diphyid (after Leuckart). *D* hydrophyllium; *GS* gonophore; *P* polyp (siphon), with tentacle. The cormidia separate from the stem to form eudoxids.

At the point where the stem joins the nectocalyx (Fig. 121), or between the nectocalyces if two are present (Fig. 118), there is a deep groove or pit in the jelly called the *hydroecium*, into which the contractile stem with its cormidia can be retracted. In the jelly of the uppermost nectocalyx is a space lined by large vacuolated cells, and sometimes containing an oil-drop. This is a dilatation of the upper end of the central canal of the stem and is called the *somatocyst*.

Each cormidium consists of two medusoid individuals—the one of these is a sterile and the other a fertile medusoid. The sterile medusoid consists of a bract or *hydrophyllium*, a siphon (trumpet-shaped polyp), and a tentacle, while the fertile one is a gonophore. The hydrophyllium is the bell of the sterile medusoid; it possesses rudiments of the radial canals which radiate from an apical dilatation — the *phyllocyst* (Fig. 120) —which corresponds to the somatocyst of the nectocalyx and is connected with the central canal of the stem. The siphon is the manubrium of the sterile medusoid, which is displaced from its umbrella and has a trumpet-shaped mouth at its free end. The

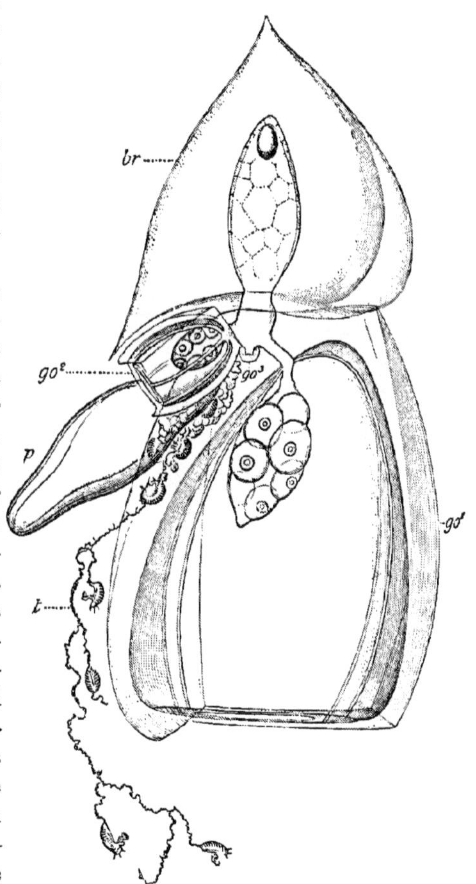

FIG. 120.—*Eudoxia Eschscholtzii*—a female eudoxid of *Muggiaea Kochii* (after Chun), with helmet-shaped hydrophyllium, *br* containing phyllocyst, siphon *p*, tentacle *t*, and three gonophores of different ages, go^1, go^2, go^3, with eggs in the manubrium.

tentacle is the single marginal tentacle of the medusoid which has shifted on to the base of the manubrium. In some forms the gonophores become sexually mature while still attached to the stem, but in the greater number the cormidia are detached before maturity and become free-swimming. Such free-swimming groups are called *Eudoxia* (Fig. 120), and are distinguished as monogastric forms (there being only one mouth and stomach) from the polygastric colonies from which they arise, and when found free are classified separately from the polygastric forms, just as the medusae of the *Anthomedusae* are classified separately from the polyp colonies. The nutritive canals of all the parts of a cormidium unite in the bract (Fig. 120), from which point a bracteal canal passes to join the canal of the stem. The phyllocyst (Fig. 120), which corresponds to the somatocyst (Fig. 122), arises from the same point. The tentacle is tubular and is beset with a series of lateral *tentilla*, also tubular. Each tentillum is composed of three parts —(1) a thin pedicle or proximal part, (2) a dilated middle part the *cnidosac*, and (3) a slender terminal filament. The swelling of the cnidosac is due to a rich development of nematocysts of various kinds, forming the battery.

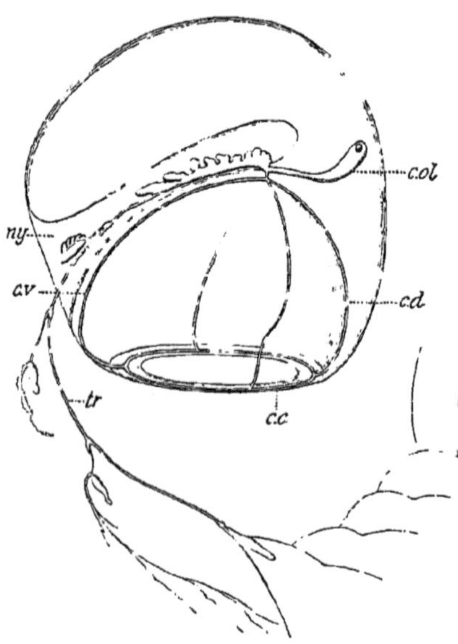

FIG. 121.—*Sphaeronectes gracilis* (from Chun), seen from the side. *ny* hydroecium: *c.ol* somatocyst; *c.v* ventral radial canal, *c.d* dorsal radial canal, *c.c* circular canal of nectocalyx; *tr* coenosome with cormidia.

The gonophore has a 4-radiate canal system and a velum, but is without tentacles or mouth (Fig. 120). The sexual cells originate in the ectoderm of its manubrium. It forms the swimming organ of the cormidium. In some forms it becomes detached, and then a secondary gonophore is formed. In some species (of *Abyla*) a cluster of small gonophores is developed in a single cormidium, in which case a special nectocalyx is developed as a swimming organ. The gonophores are of separate sexes, but the same stem is usually hermaphrodite, bearing male and female cormidia.

The variations in structure of the order depend principally upon the number of the nectocalyces. In the *Monophyidae* there is one nectocalyx, in the *Diphyidae* two, and in the *Polyphyidae* several pairs of nectocalyces.

The first formed mouthless siphon is supposed by Haeckel to elongate and form the stem of the future colony. The oldest cormidium is that which is placed furthest from the nectocalyces (Fig. 122). Chun states that the first nectocalyx (cap-shaped) is retained only in *Monophyes* and *Sphaeronectes* (Fig. 121); in the other *Calyconectae* it is thrown off and replaced by a differently shaped (pyramidal) secondary nectocalyx (Fig. 122), to which more secondary nectocalyces are added later. All *Calyconectae* pass therefore through a monophyid stage.

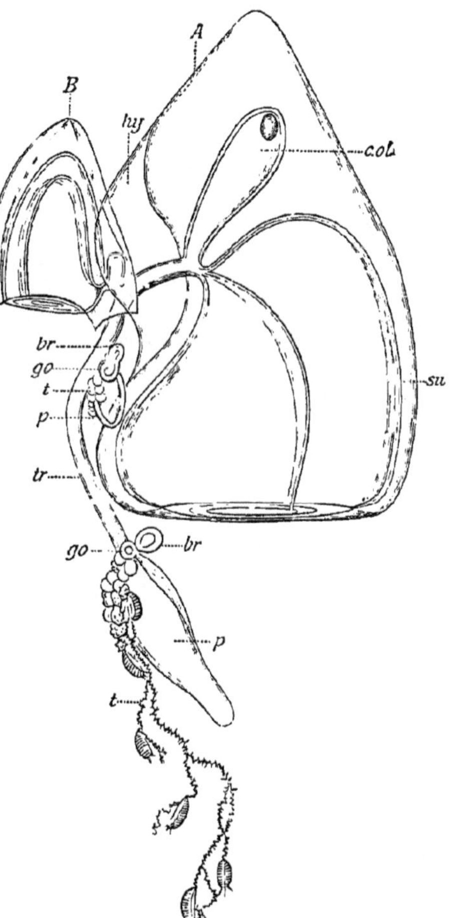

FIG. 122.—Young colony of *Muggiaea kochii* with the primary cap-shaped nectocalyx (*A*) which is soon cast off, and the secondary pyramidal nectocalyx (*B*). *c.ol* somatocyst with oil-drop; *hy* hydroecium; *tr* coenosome (stem) with two cormidia, the upper one being the younger; *br* bud of hydrophyllium; *go* bud of gonophore; *p* siphon; *t* tentacle, *su* sub-umbrella.

As stated above the buds of the cormidia are always formed at the upper end of the stem, so that the oldest cormidium is the lowest (Fig. 122). The law as to the formation of the nectocalyces is not clear in the *Calyconectae*, but in the *Physonectae* the pneumatophore is at the top of the stem, the youngest nectocalyx bud next to it, while the oldest nectocalyx is at the lowest end of the siphosome, *i.e.*, next the youngest cormidium bud.

The development of the egg leads to the formation of a variety of the *Siphonula* larva (Fig. 123). It is a medusoid composed of a nectocalyx (*A*)—the umbrella, of a cylindrical mouthless process—the manubrium, and of a tentacle. The mouthless process is attached to the ex-umbrella surface of the nectocalyx (to what is called in Siphonophoran parlance its ventral side), and is regarded by Haeckel as the original siphon. It is supposed to have protruded through a fissure in the ventral wall of the nectocalyx. This dislocation* of the siphon (if it really exists) from its proper position in the nectocalyx is an example of a widespread phenomenon in the Siphonophora, which accounts for a good many of the peculiar features of the group.

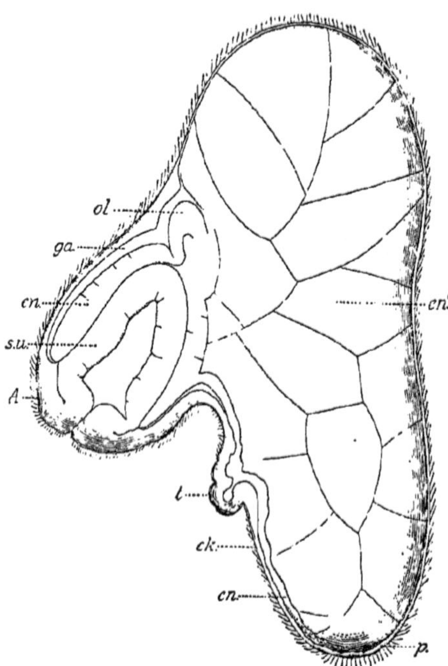

FIG. 123.—Developing *Siphonula* larva of *Muggiaea Kochii* with buds on the ventral surface. *A*, rudiment of primary nectocalyx, with somatocyst *ol*, and commencing jelly *ga* between ectoderm *ek* and endoderm *en*; *su* sub-umbrella; *t* budding tentacle; *p* siphon; *en'* yolky endoderm cells which are absorbed later (after Chun).

The alternation of generations in this order is between the polygastric colony and the monogastric cormidium, produced by budding from the former and when detached known as a eudoxid.† The gonophore of the eudoxid after shedding its genital cells is cast off and a new gonophore is formed.

Fam. 1. **Eudoxidae.** Monogastric, cormidium composed of two medusoids, a sterile and a

* This attachment of the primary siphon to the ex-umbrella surface of the primary nectocalyx is a serious difficulty to the medusome theory. The difficulty may be got over by supposing that the primary siphon is the manubrium of an umbrella which has disappeared, and that the primary nectocalyx is the first bud from the persistent manubrium.

† In some forms the primary gonophore loses its sexual manubrium, and is developed into a special nectocalyx, and a secondary gonophore is formed. In this case the cormidium is composed of three medusoids, and is called an **Ersaeid.**

fertile; without special nectocalyx. *Diplophysa* Gegenbaur; *Eudoxella* H.; *Cucubalus* Q. and G.; *Cucullus* Q. and G.; *Cuboides* Q. and G.; *Amphiroa* Blainville; *Sphenoides* Huxley; *Aglaisma* Esch.

Fam. 2. **Ersaeidae.** Monogastric, cormidium composed of three medusoids, a sterile, a fertile, and a special nectocalyx. *Ersaea* Esch.; *Lilaea* H.

Fam. 3. **Monophyidae.** *Calyconectae Polygastricae* with a single nectophore at the apex of the long tubular stem. Cormidia eudoxiform, separated by equal free internodes; each siphon with a bract. *Monophyes** Claus; *Sphaeronectes** Huxley (Fig. 121); *Mitrophyes* H.; *Cymbonectes*† H.; *Muggiaea*† Busch. (Fig. 122); *Cymba* Esch.; *Doramasia*† Chun; *Halopyramis*† Chun.

Fam. 4. **Diphyidae.**‡ Polygastric *Calyconectae* with two nectocalyces at the apex of the long tubular trunk. Cormidia eudoxiform separated by free equal internodes; each siphon with a bract. *Praya* Blainville; *Galeolaria* (confounded with *Epibulia* a Cystonect) Lesueur; *Diphyes* Cuvier (Fig. 118); *Diphyopsis* H.; *Abyla* Q. and G.; *Bassia* Q. and G.; *Calpe* Q. and G.

Fam. 5. **Stephanophyidae.** Polygastric, with several apical nectocalyces and a special nectocalyx on each cormidium. With small palpons with long tentacles on the internodes. Cormidia not set free as ersaeids. *Stephanophyes* Chun.

Fam. 6. **Desmophyidae.** Polygastric *Calyconectae* with four or more nectocalyces, opposite, in pairs. Cormidia eudoxiform or ersaeiform, separated by equal internodes; each siphon with a bract. *Desmalia* H.; *Desmophyes* H.; umbrella edge of special nectocalyx with 8 ocelli and 8 short tentacles.

Fam. 7. **Polyphyidae.** Polygastric *Calyconectae* with four or

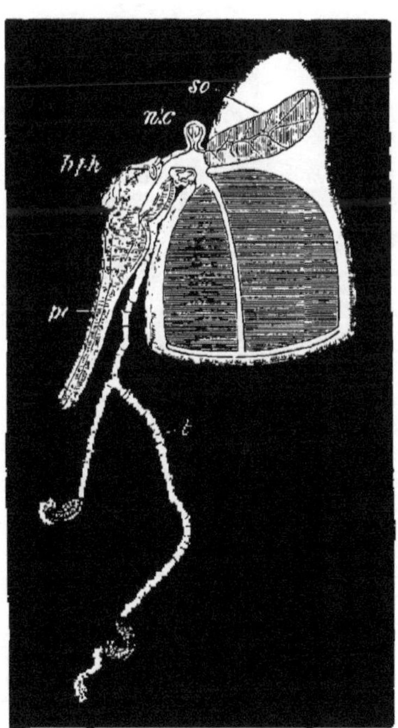

FIG. 124.—An advanced *Siphonula* larva of *Epibulia aurantiaca* with one large nectocalyx (after Metschnikoff, from Balfour). *So* somatocyst; *nc* second imperfectly developed nectocalyx; *hph* hydrophyllium; *po* siphon; *t* tentacle.

* According to Chun the primary nectocalyx of the larva persists in these genera, and there are no secondary or replacement nectocalyces.

† Chun states that in these genera the primary cap-like nectocalyx is thrown off and replaced by a pyramidal secondary nectocalyx.

‡ In this family the primary nectocalyx is replaced by two secondary bells, which are themselves replaced by a succession of similar bells formed from similar buds.

more nectocalyces opposite in pairs. Cormidia without bracts. Gonophores reach maturity while attached to the stem; no free eudoxids or ersaeids. *Hippopodius* Q. and G.; *Polyphyes* H.; *Vogtia* Kölliker.

Section 2. PHYSONECTAE. PHYSOPHORA.

Siphonanthae with a pneumatocyst and several nectocalyces (or bracts instead), and palpons. Cormidia ordinate or irregular.

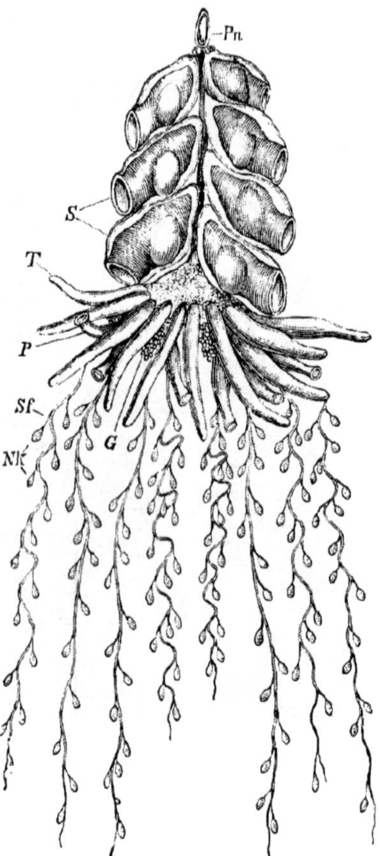

FIG. 125.—*Physophora hydrostatica.* *Pn* pneumatophore; *S* nectocalyces arranged in a double row on the stem; *T* palpon; *P* siphon with tentacles *Sf*; *Nk* groups of nematocysts (cnidosacs); *G* clusters of gonophores.

The *Physonectae* include monogastric and polygastric forms.

The stem carrying the cormidia is either short, sometimes spread out in the form of a sac (Fig. 125), or elongated and spirally twisted (Fig. 126).

The small, often brightly coloured apical pneumatophore is without a terminal opening of the *pneumatocyst*, though sometimes an opening near its base may be made out. The endodermal space of the pneumatophore itself is usually divided by a number of radial septa into pouches, while the invaginated pneumatocyst is divided into two communicating parts—an upper part with a chitinous lining, and a lower part with a thick glandular lining. The latter is called the *air funnel.* It is the gas gland and its lining secretes the gas of the pneumatocyst. The nectocalyces (except when replaced by paddling bracts with rudimentary nectocalyces at their ends) are usually numerous. They have four radial canals, a circular canal, a velum, and sometimes ocelli. The cormidia are rarely dissolved: *i.e.*, the parts are rarely scattered along the stem, but generally ordinate, *i.e.*, in groups (Fig. 126). Each cormidium consists of one siphon (sometimes two or four) with tentacle; several palpons each with a tentacle called a *palpacle;* several bracts which may even in the forms with ordinate cormidia occur in the internodes of the stem (Fig. 126); two gonostyles, one bearing male gonophores and the other female, and very often a **cyston.** A cyston is a structure like a palpon but with a terminal opening: it acts as an anus to the colony, expelling fluid and crystalline excretions through its aperture.

The batteries of the tentilla of the tentacles are enveloped in an involucrum or fold of ectoderm arising at their proximal end. The female gonophore produces only one egg.

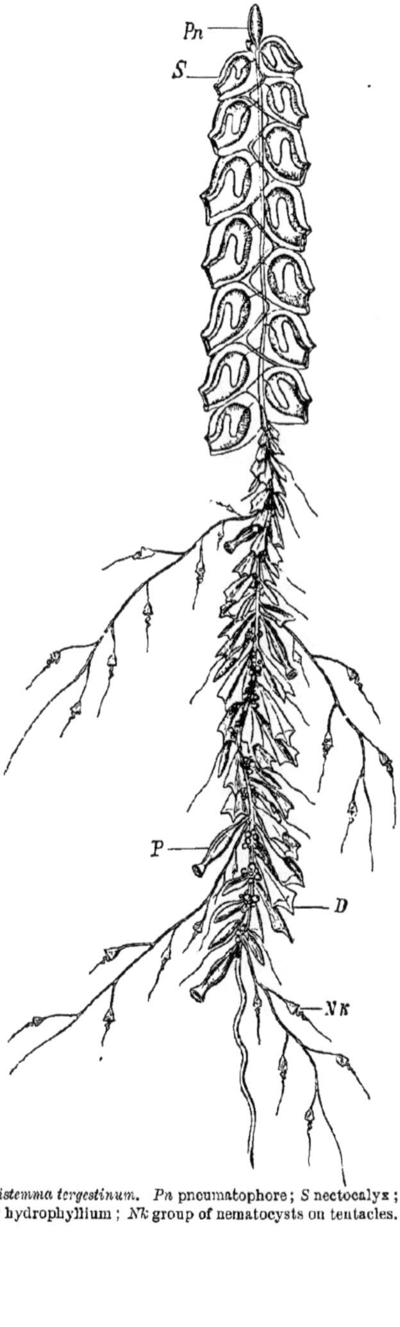

Pn
S
P
D
NK

FIG. 126.—*Halistemma tergestinum.* *Pn* pneumatophore; *S* nectocalyx; *P* siphon; *D* hydrophyllium; *Nk* group of nematocysts on tentacles.

The planula develops at one pole a pneumatocyst as a thickening and involution of ectoderm (like the entocodon of a medusa bud), and at the other a siphon (Fig. 127 *d*). The pneumatocyst is supposed by Chun to be homologous with the primary nectocalyx of the *Calyconectae*. In some forms the upper part of the body gives rise to a cap-shaped hydrophyllium as well as to a pneumatophore (Fig. 127). The crown of hydrophyllia which is sometimes formed persists only in *Athorybia*, where nectocalyces are not formed. In *Agalmopsis* and *Physophora* the primary hydrophyllia of the larva (Fig. 128) fall off as the stem becomes larger, and are replaced by nectocalyces.

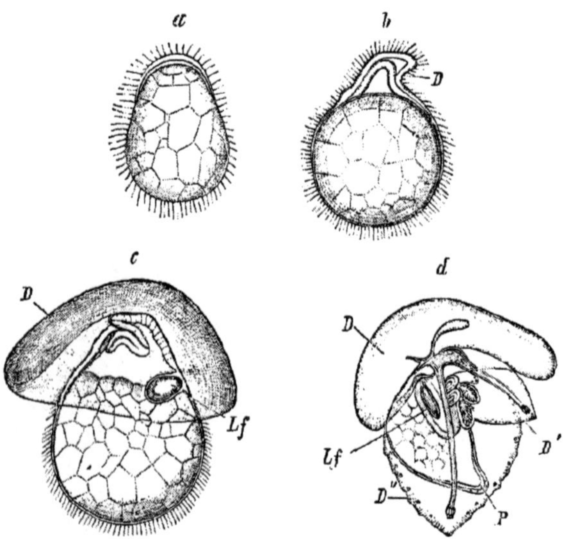

FIG. 127.—Development of *Agalmopsis Sarsii* (after Metschnikoff). *a*, planula; *b*, stage with developing hydrophyllium *D*; *c*, stage with cap-shaped hydrophyllium (*D*) and developing pneumatocyst *Lf*; *d*, stage with three hydrophyllia (*D*, *D'*, *D''*), siphon *P* and tentacle; *Lf* pneumatocyst.

Fam. 1. **Circalidae.** Monogastric *Physonectae* with a corona of nectocalyces, without bracts. *Circalia* H.

Fam. 2. **Athoridae.** Monogastric *Physonectae* with a corona of bracts, without nectocalyces. *Athoria* H.; *Athoralia* H.

Fam. 3. **Apolemidae.** Polygastric *Physonectae* with a long tubular stem bearing numerous siphons, palpons, and bracts; each siphon with unbranched tentacle. Nectocalyces biserial; either two opposite nectocalyces, or two alternate series of opposite nectocalyces. Pneumatophore without radial pouches. Nectocalyces with tentacles arising from the stem.

Sub-fam. 1. **Dicymbidae.** Two opposite nectocalyces only. Cormidia monogastric; each with a single cyston. *Dicymba* H.

Sub-fam. 2. **Apolemopsidae.** Two opposite rows of nectocalyces. Cormidia polygastric; each with several cystons. *Apolemia* Esch.; *Apolemopsis* Brandt.

Fam. 4. **Agalmidae.** Polygastric *Physonectae* with a long tubular stem, bearing numerous siphons, palpons, and bracts. Nectocalyces numerous, biserial. Pneumatophore with radial pouches. All the genera except four, viz., *Stephanomia* Pér. and Les., *Crystallodes* H., *Anthemodes* H., *Cuneolaria* Eysenhardt, have dissolved cormidia. *Phyllophysa* L. Ag.; *Agalma* Esch.: *Halistemma* Huxley (Fig. 126); *Cupulita* Q. and G.; *Agalmopsis* Sars (Fig. 128); *Lychnagalma* H.

Fam. 5. **Forskalidae.** Polygastric, with a long tubular stem bearing numerous siphons, palpons, bracts; each siphon with a branched tentacle. Nectocalyces numerous; multiserial, strobiliform in several spiral rows. Pneumatophore with radial pouches. The largest and most splendid of all *Physonectae.* Cormidia dissolved in all except *Strobalia* H. *Forskaliopsis* H. has palpacles among the nematocalyces. *Forskalia* Kölliker; *Bathyphysa* Studer.

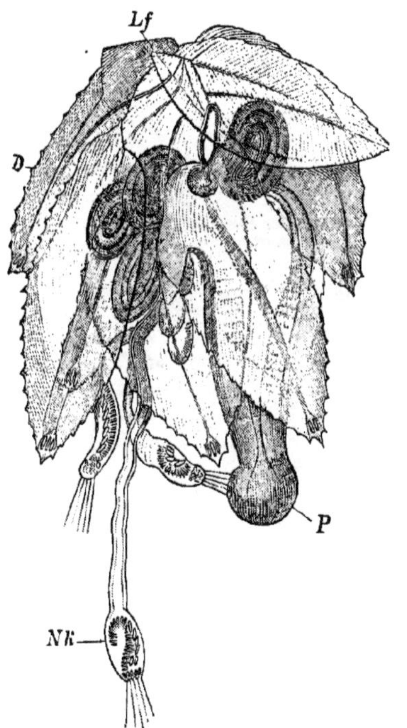

Fig. 128.—Small larval colony of *Agalmopsis* after the type of *Athorybia.* *Lf* pneumatophore; *D* the crown of hydrophyllia; *Nk* groups of nematocysts; *P* siphon.

Fam. 6. **Nectalidae.** Polygastric, with a short vesicular stem, bearing numerous siphons, palpons, and bracts; tentacles branched. 2 or 4 rows of nectocalyces. Pneumatophore with radial pouches. *Nectalia* H.; *Sphyrophysa* L. Ag.

Fam. 7. **Discolabidae.** Like the preceding but with a corona of palpons instead of bracts; without bracts. *Physophora* Forskal, biserial nectocalyces (Fig. 125); *Discolabe* Esch., quadriserial nectocalyces; *Stephanospira* Gegenbaur, multiserial nectocalyces.

Fam. 8. **Anthophysidae.** Polygastric, with short vesicular stem bearing numerous siphons and palpons; each siphon with a branched tentacle. Nectocalyces replaced by corona of bracts (as in Fig. 128). Pneumatophore with radial pouches. *Rhodophysa* Blainville; *Melophysa* H.; *Athorybia* Esch.; *Anthophysa* Mertens; *Ploeophysa* Fewkes.

Section 3. Auronectae.

Siphonanthae with a large pneumatophore, a corona of nectocalyces, a peculiar aurophore, and a network of canals in the jelly of the thickened trunk. Siphosome spheroidal, ovate, or turnip-shaped.

Deep sea forms. The **aurophore** (Fig. 129) is an appendage of the pneumatocyst, and contains a central tube putting the cavity of the pneumatocyst in communication with the exterior. It is placed on one side of the pneunatophore, and its central tube (*pistillum*) is surrounded by a number of radial chambers, which are separated by septa and communicate with the pericystic (endodermal) space of the pneumatophore. Very possibly the aurophore is a gas-secreting gland.

Fam. 1. **Stephalidae.** *Stephalia* H.; *Stephonalia* H.
Fam. 2. **Rhodalidae.** *Auralia* H.; *Rhodalia* H.

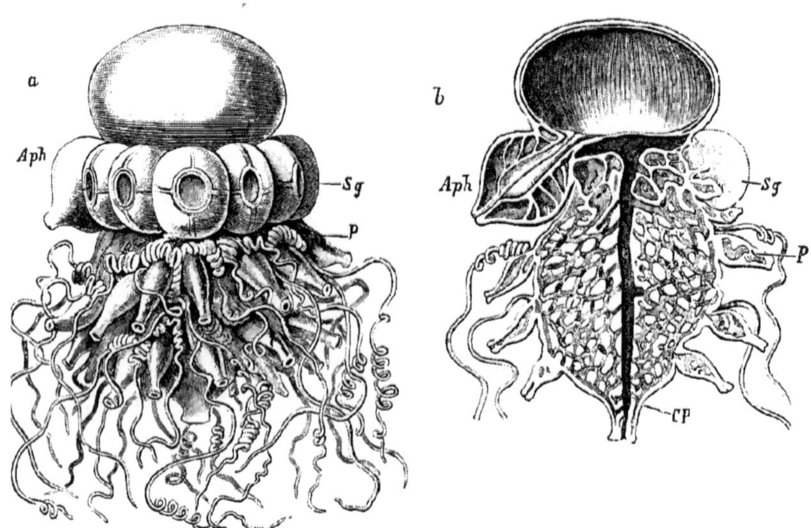

FIG. 120.—*Stephalia corona* (after Haeckel). *a*, side view; *b*, section. *Aph* Aurophore; *Sg* corona of nectocalyces; *P* siphons with their tentacles; *CP* the large central siphon, the enteron of which forms the central tube of the siphosome (coenosome or stem).

Section 4. CYSTONECTAE. PHYSALIDAE.

Siphonanthae with a large apical pneumatophore without nectocalyces and without bracts. Pneumatocyst with an apical stigma.

This order includes *Physalia*, the well-known Portuguese Man of War, which we may take as type.

Physalia possesses a large pneumatophore lying nearly horizontally and bearing posteriorly and ventrally the numerous siphons, palpons, and branched gonostyles. The stigma is at the front end of the pneumatophore, and leads into a large pneumatocyst. The pericystic cavity is simple and not divided. The air-secreting cells or *pneu-*

madenia are confined to the under part of the pneumatocyst. The float bears on its dorsal side a crest formed by a fold of the trunk, *i.e.*, of the part of the body which projects behind the sac (as well as ventral to it), and which carries the cormidia. The pneumatocyst extends into the crest, and becomes divided up by a number of transverse septa into air-chambers. The cormidia are very numerous; they appear to be dissolved in the old individuals, but in the younger stages they are ordinate. When they can be made out they may be seen to consist of palpons, siphons, and branched gonostyles arising from a common stem. The tentacles arise from the palpons. The gonostyles are hermaphrodite, and the female gonophores break away and develop their ova as free-swimming *Anthomedusae*.

The youngest larva of the *Physalidae* is known as a *Cystonula*. It has a float and one siphon with tentacle hanging below it. Later it elongates horizontally and produces on the ventral side, anterior to the first siphon (*i.e.*, nearer the morphological apex), the cormidia. The primary siphon or cormidium persists at the hinder end of the float, *i.e.*, at the end opposite to the stigma, and is in some forms always marked off from the numerous secondary cormidia.

There are monogastric and polygastric forms in the order. The pneumatocyst has generally a gas-secreting thick epithelium or **pneumadenia** in its basal part. This may be partly constricted off as a hypocystic air-funnel. The epithelium of the pneumadenia in many forms sends out branching villi of its ectoderm which project into the pericystic space, and are covered towards the latter by its ciliated endodermal lining. These *hypocystic villi* are composed of large cells and are solid. They are probably mechanical in function helping to support the air-vessel.

In many forms the gonostyles bear palpons (gono-palpons); and in the *Physalidae* the palpons have a tentacle. The tentacles are often branched, but without a cnidosac (battery). The sting of *Physalia* is particularly poisonous.

Fam. 1. **Cystalidae.** Monogastric, with one large siphon bearing a tentacle and surrounded by a corona of siphons. Pneumatophore without radial septa or hypocystic villi. *Cystalia* H.

Fam. 2. **Rhizophysidae.** Polygastric, with a long stem bearing in its ventral median line numerous monogastric cormidia with single palpon and tentacle. Pneumatophore large with radial pericystic pouches, but with hypocystic villi.

Sub-fam. 1. **Cannophysidae.** Cormidia ordinate. Gonostyles attached to the stem at the base of the siphons. *Aurophysa* H.; *Cannophysa* H.

Sub-fam. 2. **Linophysidae.** Cormidia dissolved. Gonostyles attached to stem between the siphons. *Linophysa* H.; *Nectophysa* H.; *Pneumophysa* H.; *Rhizophysa* Pér. and Les.

Fam. 3. **Salacidae.** Polygastric with long stem bearing in its ventral median line numerous polygastric cormidia. Pneumatophore large, without radial pericystic pouches, but with hypocystic villi. *Salacia* H.

Fam. 4. **Epibulidae.** Polygastric with a short inflated spirally convoluted stem. Cormidia ordinate in a spiral ring protected by a corona of palpons. Pneumatophore without pericystic radial pouches, but with hypocystic villi. *Epibulia* Esch.; *Angela* Lesson.

Fam. 5. **Physalidae.** Polygastric, with a short inflated stem horizontally expanded along the ventral side of the large horizontal pneumatophore. Cormidia in a multiple series along the ventral side of the trunk, usually dissolved. Pneumatophore large, with a chambered dorsal crest, without radial septa or hypocystic villi. *Alophota* Brandt; *Arethusa* H.; *Physalia* Lamarck; *Caravella* H.

Class II. ACALEPHAE.* ACRASPEDA.

Medusae of considerable size with gastral filaments (phacellae); with endodermal gonads; with lobed umbrella-edge; without true velum.

The medusae of this class are distinguished from those of the

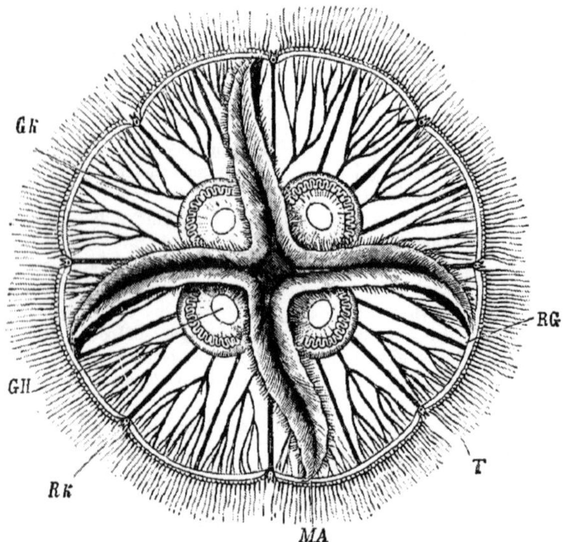

Fig. 130.—*Aurelia aurita*, from the oral surface. *MA* the four oral tentacles with the mouth in the centre; *Gk* generative organs; *GH* aperture of subgenital pit; *Rk* sense organ (marginal body); *RG* radial vessel; *T*, tentacle at edge of the disc.

* E. Haeckel, *Monographie der Medusen*, Jena, 1879. L. Agassiz, *Contributions to the Natural History of the United States, Acalephae*, vol. 3, 1860, vol. 4, 1862. C. Claus, "Studien über Polypen u. Quallen der Adria," *Denks. d. k. Akad. Wiss. Wien*, 1877. *Id.*, *Unters. üb. d. Organisation u. Entw. Acalephen*, Prag, 1883. A. Götte, *Üb. d. Entwickelung v. Aurelia aurita u. Cotylorhiza tuberculata*, 1887. C. Claus, "Üb. die Entwick. des Scyphostoma, etc," I. and II. *Arb. a. d. Zool. Inst. Wien*, 9 and 10, 1891 and 1893.

Hydromedusae by their larger size and the greater thickness and stiffness of their umbrella, the gelatinous tissue of which contains a quantity of strong fibrillae, and a network of elastic fibres.

Another characteristic of the group is derived from the structure of the edge of the umbrella. This is divided by a regular number of indentations usually into eight groups of lobes between which the sense organs are contained in special pits (Fig. 130).

The marginal lobes of the *Acalephae*, like the continuous velum of the *Hydromedusae*, appear to be secondary formations at the edge of the disc. In the young stage known as *Ephyra* (Fig. 131), which is common to most of the *Ephyroninae*, they are present as eight pairs of relatively long tongue-like processes, and grow out from the disc-like segments of the *Strobila* as marginal processes. An undivided marginal membrane (the *velarium*), differing from the velum of the *Craspedota* in containing prolongations of the canals of the gastrovascular system, is present in the *Charybdeidae* alone.

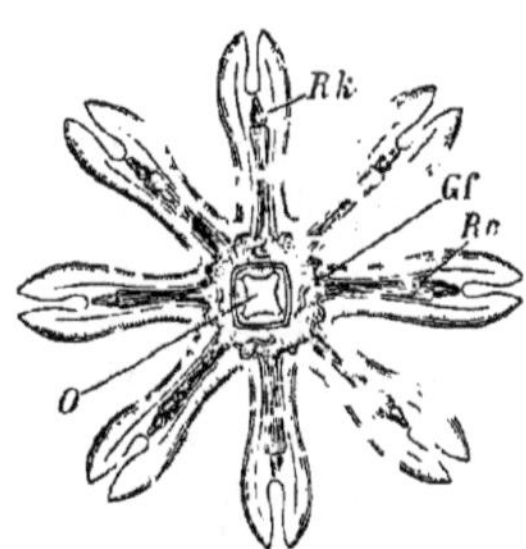

FIG. 131.—An *Ephyra* seen from the oral side. *Rk* marginal body; *Gf* gastral filament; *Rt* radial pouch of enteron; *O* mouth.

The *Acalephae* differ from the *Hydromedusae* in possessing, as a rule, large oral tentacles at the free end of the wide manubrium. These may be regarded as being derived from an unequal growth of the edges of the mouth. They grow as four arm-like processes of the manubrium from the angles of the mouth, and are placed perradially (see p. 118, note), *i.e.*, they alternate with the genital organs and gastric filaments. In some cases the arms become forked at an early period, and four pairs of arms are formed, the lobed tufted edges of which may again divide and sub-divide into many branches. In this case, the margins of the mouth and the opposed surfaces of each pair of arms fuse in early life as described above, p. 111 (*Rhizostomidae*, Fig. 97).

Further, there is to be observed in the *Acalephae* two types of structure, which we may term the type of the *Scyphistoma* and the type of the *Ephyra*, respectively. To the former type, which is called the **Scyphomedusae**, belong the sessile forms (*Lucernaridae*, Fig. 134), and swimming forms which perhaps possess a direct development (*Tesseridae*, Fig. 133, *Charybdeidae*, Fig. 136). The edge of the umbrella is only incompletely divided into lobes; marginal bodies are absent in the *Lucernaridae* and *Tesseridae*, their place being taken by tentacles or marginal anchors, which are homologous with marginal bodies. The stomach is surrounded by four wide gastric pouches separated from one another by septa (Fig. 99). The sexual organs are placed in the sub-umbrella wall of the gastric pouches. The umbrella is usually much arched and frequently prolonged into a stalk.

The second type, which is called the **Ephyroninae**, is a modification of that of the *Ephyra* itself. The *Ephyra* (Fig. 131) possesses eight marginal lobes, which are forked distally and carry in the clefts of the forks eight rhopalia. Each marginal lobe has a radial prolongation of the stomach The *Ephyra* therefore shows while it is still on the *Strobila*, a predominance of the 8-radiate structure in opposition to the 4-radiate build of the Scyphistoma type. At the same time it should be noted that the more centrally placed organs (buccal arms, gonads, gastral filaments) are 4-radiate. Between the eight lobes of the Ephyra there grow out later eight additional lobes (sometimes more), also bifurcated distally and carrying tentacles in place of the marginal bodies.

The *Ephyroninae* then are distinguished from *Scyphomedusae* by the lobed structure of the umbrella edge, by the presence of eight or more rhopalia, and by the division of the peripheral part of the coelenteron into eight or more radial vessels, which are seldom widened in a pouch-like manner. The gonads are interradial and placed in the ventral wall of the central stomach. The umbrella is flattened, usually discoidal. The *Periphyllidae* (Fig. 135) and *Pericolpidae* are intermediate between these two groups in that the central part of the umbrella presents the characters of the *Scyphomedusae*, while the peripheral parts recall the structure of the *Ephyra*. They possess four knot-like septa which bound the four gastric pouches, and at the edge of the umbrella there are sixteen places of adhesion between the dorsal and ventral endoderm. Moreover, the *Periphyllidae* have four taenioles (gastral ridges); their gonads are in the circular sinus; the umbrella is bell-shaped or flat, and marked on its dorsal surface by an annular constriction which indicates the junction between the ventral Scyphistoma-like and the distal Ephyra-like parts.

The *Scyphistoma*, which may be regarded as the promorph of the *Scyphomedusae*, is the larval form of the *Ephyroninae*.

The gastric filaments which are worm-like and movable, and are not found in *Hydromedusae*, afford a distinctive mark. They correspond to the mesenteric filament of the *Anthozoa*, and lend the same aid to digestion by the secretion of their glandular endodermal covering. They are always attached to the sub-umbrella wall of the stomach, and fall in the four interradii, *i.e.*, the radii of the generative organs which alternate with the radii of the angles of the mouth (radii of the first order). They usually follow the inner edge of the generative frill in a simple or curved line.

The nervous system and rhopalia have already been described (p. 115).

The four generative organs of the Acalephae can be easily distinguished in consequence of their size and their bright colouring. In some cases, at any rate in the *Discomedusae*, they protrude as folded bands into special cavities in the umbrella, the so-called sub-genital pits (hence the term *Phanerocarpae* Esch.). In all cases these bands lie on the lower (sub-umbrella) wall of the digestive cavity, from which they originate as leaf-like prominences. Their upper surface is covered with gastric epithelium; their under, which is turned towards the sub-umbrella, with germinal epithelium, the elements of which in the process of development, pass into the gelatinous substance of the band. The subgenital pits have already been described (p. 118): they may be completely absent (*Ephyra*, *Nausithoë*): their lining consists of sub-umbrella ectoderm and is quite distinct from the generative epithelium, which is of endodermal origin. The mature generative products are dehisced into the gastric cavity, and pass out through the mouth; but in many cases the ova undergo their embryonic development either in the ovary (*Chrysaora*) or in the oral tentacles (*Aurelia*). Separate sexes are the rule. Male and female individuals, however, apart from the colour of their generative organs, have only slight sexual differences; as, for instance, the form and length of the tentacles (*Aurelia*). *Chrysaora* is hermaphrodite.

In the *Ephyroninae* the development is generally accompanied by an alternation of generations; the asexual generations being represented by the *Scyphistoma* (*Hydra tuba*) and *Strobila*; but in exceptional cases it is direct (*Pelagia*). In all cases a complete segmentation leads to the formation of a ciliated larva (Fig. 132 *a*), the so-called *planula*, which attaches itself by the pole which is directed forwards in swimming. This pole is, however, opposite to the gastrula mouth, which in the meantime becomes closed, while round the mouth, which is formed as a perforation at the free end, the tentacles appear (Fig. 132 *b*, *c*). As in the embryo *Actinia*, two opposite tentacles first make their appearance; not, however, simultaneously, the one appearing after the other, so that the young larva about to develop into the Scyphistoma presents a bilaterally symmetrical structure. Subsequently the second pair appears in a plane at right angles to the plane of the first tentacles. These four tentacles mark the radii of the first order. Then alternating with these, but in a less regular succession, the third and fourth pairs appear; and soon after, in the plane of these latter, four longitudinal folds (Fig. 132 *d*) of the gastric cavity are developed (radii of the second order or of the gastric filaments and genital organs; often called the *interradii*, see above, p. 118, note).

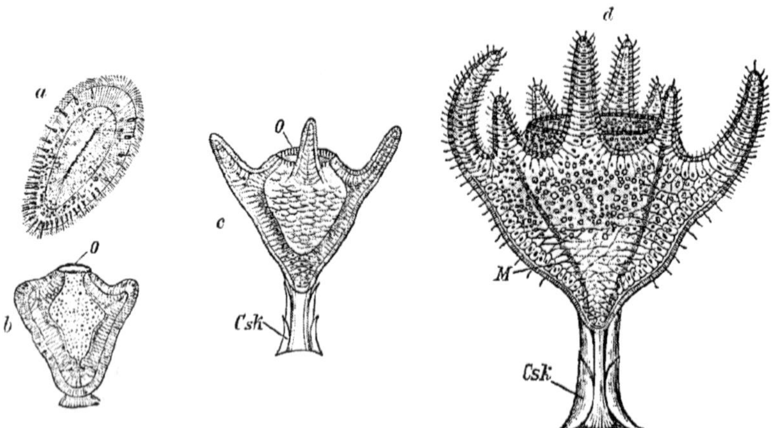

FIG. 132 *a–h*.—Larval development of Chrysaora; *a*, planula with narrow enteron; *b*, the same after attachment and formation of mouth *O* and commencing tentacles; *c*, young scyphistoma with 4 tentacles; *Csk* periderm; *d*, eight-armed scyphistoma with wide mouth; *M* longitudinal muscles of gastral ridges.

The eight-armed *Scyphistoma* soon produces eight fresh tentacles (Fig. 132 *e*), which succeed one another in irregular succession, and alternate with the tentacles already present. Their position determines the intermediate radii, or *adradii*, of the future young *Discophor* or *Ephyra*. After the formation of the circle of tentacles and the secretion of a clear basal periderm (*Chrysaora*), the *Scyphistoma* is capable of reproduction by gemmation and fission. At first the *Scyphistoma* appears to multiply only by budding; the second mode of reproduction, the process of *strobilization*, begins later. This consists essentially in the fission and division of the anterior half of the body into a number of segments, thus changing the *Scyphistoma* into a *Strobila* (Fig. 132 *f*). The separation of the segments progresses continuously from the anterior end to the base of the *Strobila*, so that after the disappearance of the tentacles (Fig. 132 *g*), first the terminal segment, then the second, and so forth, attain independent existence (Fig. 132 *h*). Each segment becomes an *Ephyra* (Fig. 131), developing eight pairs of elongated marginal lobes, with a marginal

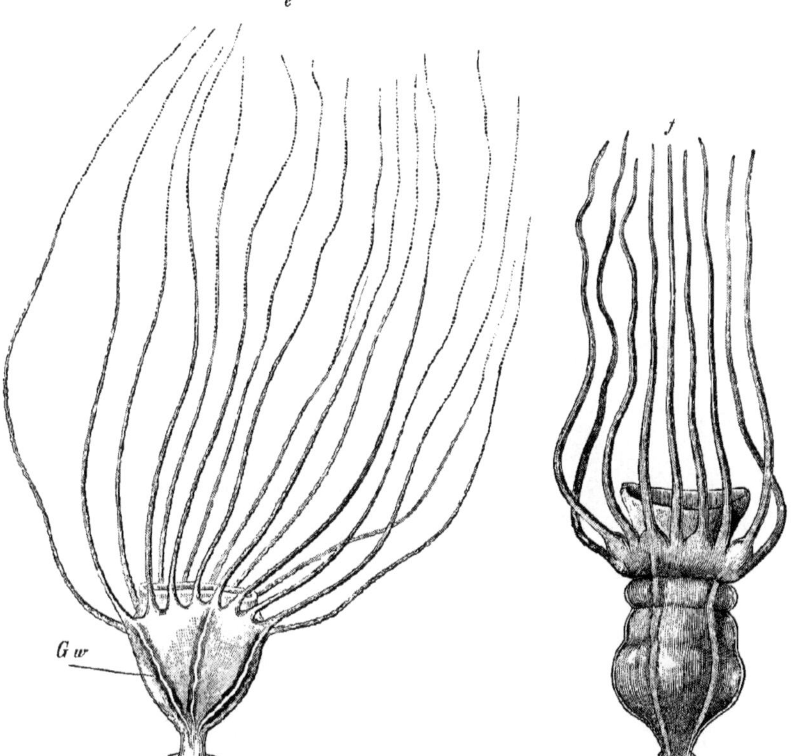

FIG. 132.—*e*, scyphistoma with sixteen arms (slightly magnified); *Gw* gastral ridges; *f*, commencing strobilisation.

body in the notch which separates the two lobes of the same pair. It is these marginal lobes which give to the edge of the umbrella of the *Ephyra* its characteristic appearance. The young *Ephyra* gradually acquires the special peculiarities of form and organization of the sexually mature animal.

The number of nematocysts accumulated on the upper surface of the disc and on the tentacles of many *Medusae* enables them to cause a perceptible stinging sensation on contact. Many, *e.g. Pelagia*, are phosphorescent. According to Panceri, this phenomenon originates in the fat-like contents of certain epithelial cells on the surface. The Acalephae may attain a large size, the bell in some *Rhizostomae* and *Cyaneidae* reaching a diameter of from two to six feet.

In spite of the delicacy of their tissues, certain large *Medusae* have left impressions in the lithographic slate of Sohlenhofen (*Medusites circularis*, etc.).

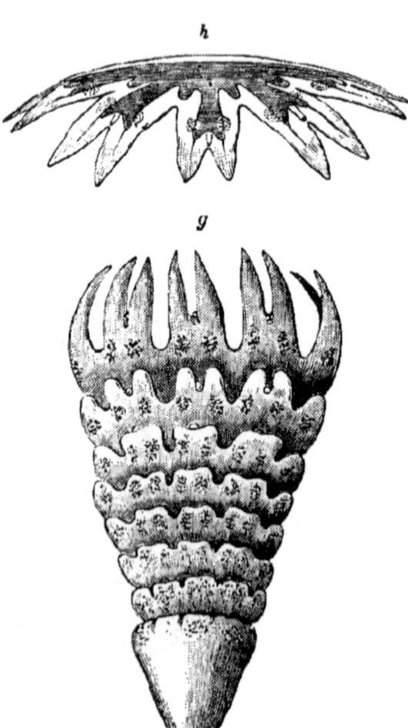

Fig. 132.—*g*, fully formed strobila with separating ephyrae; *h*, free ephyra (1·5 to 2 mm. in diameter).

Order 1.

Scyphomedusae.

Tetrameralia. Tesseronae.

Acalephae with or without four rhopalia (sense-tentacles); stomach with four gastral pouches separated by short or long septal-unions. Gonads in sub-umbrella wall of the gastral pouches. Umbrella highly vaulted.

The *Scyphomedusae* are best considered in their relation to the *Scyphistoma*. They may be looked upon as Scyphistomas deprived of their tentacles, which indeed are only transitory structures, and elongated so as to assume the form of a cup, and changed in several particulars which are characteristic of the medusa stage. The four septa (Fig. 98) arise by the fusion of the four gastric folds with the wide oral disc, which becomes drawn in and concave like a sub-umbrella. These four septa separate the same number of gastrovascular pouches; while the margin of the cup may be drawn out into eight arm-like processes from which groups of short, knobbed tentacles arise (Fig. 134).

The genital organs extend on the oral wall of the umbrella into the arms as eight band-shaped, plicated ridges (Fig. 134, *I.*). They run along in pairs at the lower part of each septum in the gastric cavity (Fig. 98). The ovum, according to Fol, undergoes a complete segmentation, which results in a single-layered blastosphere. This becomes an oval, two-layered larva, which becomes ciliated, swims freely about, and finally attaches itself. The further development probably takes place directly without alternation of generations.

The *Scyphomedusae* are without exception marine animals, and are remarkable for their great reproductive power. According to A. Meyer, if the stalk of *Lucernaria* be cut off, the cup reproduces a new one, and injured individuals, and even excised pieces, can become perfect animals.

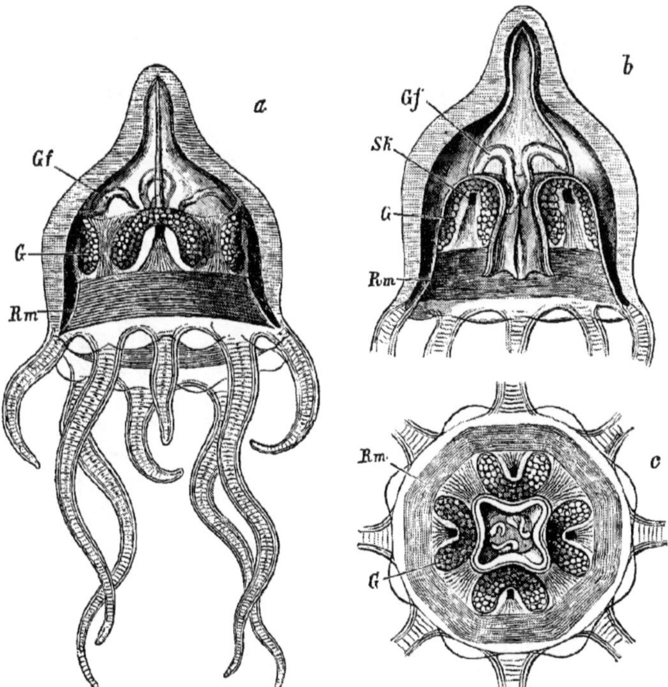

FIG. 133.—*Tessera princeps* (after Haeckel). *a*, external view magnified about 20 times; *b*, longitudinal section through two perradii; *c*, view of sub-umbrella. *Gf* gastral filaments; *G* gonads; *Rm* circular muscle; *Sk* septal unions.

Sub-order 1. STAUROMEDUSAE.

Without rhopalia, but in their place 8 simple principal tentacles (Fig. 133) or marginal anchors (Fig. 134). Stomach with 4 wide perradial pouches connected peripherally by a ring-canal. Gonads as 4 interradial horse-shoe shaped thickenings (Fig. 133), or 4 pairs of adradial ridges in the ventral walls of the gastric pouches (Fig. 134).

Fam. 1. **Tesseridae.** Umbrella edge without lobes; 8 principal tentacles (4 perradial and 4 interradial); marginal anchors absent. Apex of the ex-umbrella surface prolonged into a hollow process or stalk for attachment. *Tessera* H. (Fig. 133) and *Tesserantha* H., with hollow, ex-umbrella process not used for attachment; *Depastrella* H., *Depastrum* H., with ex-umbrella stalk for attachment.

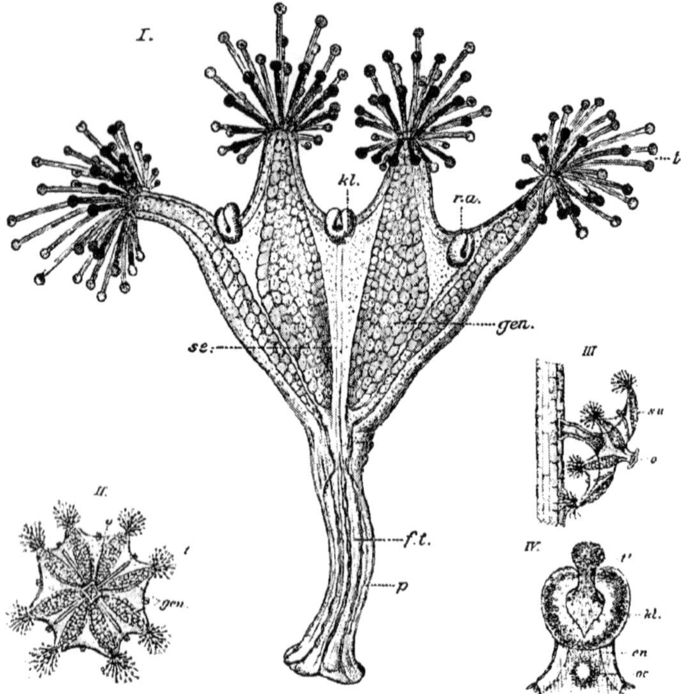

Fig. 134.—*Haliclystus auricula* (from Chun, after Clark). I. From the side. II. From the oral face. III. From the side with evaginated umbrella and protruded mouth. IV. Marginal anchor from the axial side. *p* stalk; *su* sub-umbrella; *t* one of the eight tufts of knobbed tentacles on the eight hollow triangular marginal lobes; *ra* one of the eight marginal anchors; *t'* anchor tentacle; *kl* collar of adhesive glandular ectoderm; *oc* eye-spot; *en* vessel of marginal anchor; *o* mouth; *se* interradial septal ridge passing into the taenioles (*ft*) of the stalk; *gen* one of the eight adradial gonads on the sub-umbrella wall of the four radial gastric pouches, representing four interradial horse-shoe gonads connected at the oral end of the septal ridge.

Fam. 2. **Lucernaridae.** With 8 adradial umbrella lobes, and tufts of short knobbed tentacles at end of each lobe. With 8 principal tentacles (4 per- and 4 interradial) as marginal anchors (Fig. 134), or absent. An ex-umbrella stalk for attachment. *Haliclystus* Clark, *Halicyathus* Clark, with marginal anchors; *Lucernaria* O. F. Müller, *Craterolophus* Clark, without marginal anchors.

Sub-order 2. PEROMEDUSAE.

With 4 interradial rhopalia (with otoliths and eyes); with 4 perradial tentacles, or with 12 tentacles (4 per- and 8 adradial, Fig. 135); with 8 or 16 marginal lobes. The 4 radial gastral pouches are separated from one another by very short septal-unions or septa, so that the stomach may be said to be surrounded by a wide circular sinus, communicating with it by 4 ostia (Fig. 99). The circular sinus gives off towards the periphery of the umbrella 8 or 16 flat pockets, each of which gives off two lobe-pockets, and between these one pocket to a tentacle or sense-body. 4 horse-shoe shaped gonads in the ventral wall of the circular sinus.

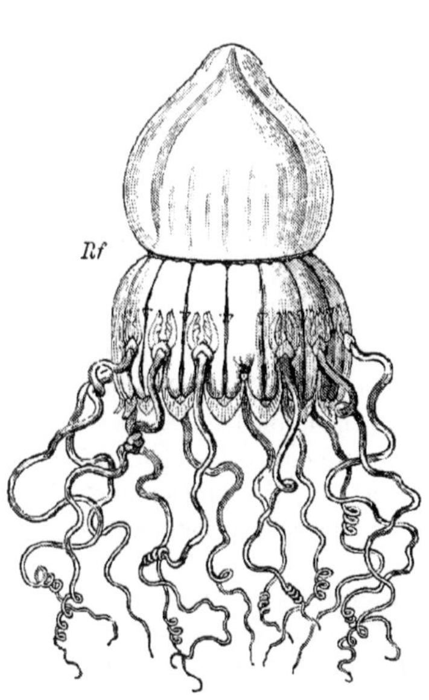

Fig. 135.—*Periphylla hyacinthina* (after Haeckel). *Rf* annular groove dividing the umbrella into a proximal conical part, and a ventral lobed region.

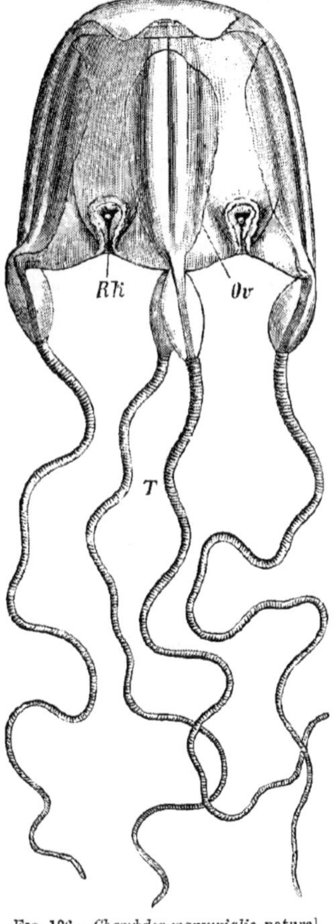

Fig. 136.—*Charybdea marsupialis*, natural size. *T* tentacles; *Rk* marginal bodies; *Ov* gonads.

Fam. 3. **Pericolpidae.** With 4 perradial tentacles, 4 interradial rhopalia, and 8 adradial marginal lobes. *Pericolpa* H.; *Pericrypta* H.

Fam. 4. **Periphyllidae.** 12 tentacles, 4 rhopalia, 16 marginal lobes. *Peripalma* H.; *Periphylla* H. (Fig. 135).

Sub-order 3. **CUBOMEDUSAE** (*Marsupialida*).

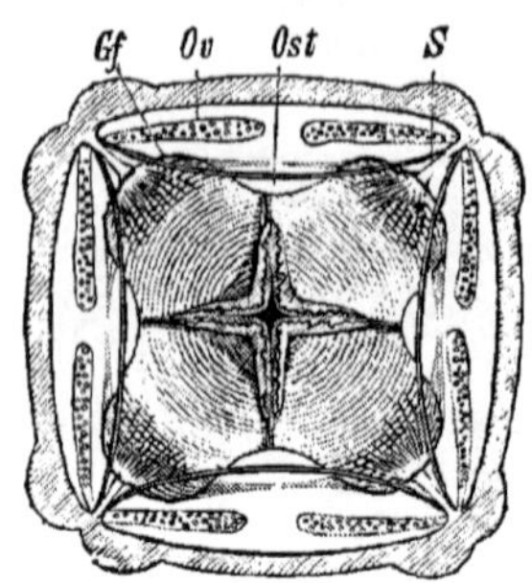

FIG. 137.—The apical half of a *Charybdea* divided transversely, seen from the sub-umbrella side. The four oral arms are visible. *Ov* ovaries on the four septa; *Ost* ostia of the gastric pouches; *Gf* gastric filaments; *S* septa.

With quadrangular umbrella (Fig. 136), 4 perradial rhopalia (with otoliths and eyes), 4 interradial marginal tentacles, 4 perradial gastric pouches, separated by long and narrow interradial septa. Gonads as 4 pairs of broad plates fastened by one edge to the radial septa (Fig. 137), and projecting into the pouches. With a smooth-edged *velarium* containing prolongations of the gastrovascular system. With a nerve ring on the sub-umbrella side of the edge of the bell, having a zig-zag course.

Fam. 5. **Charybdeidae.** *Procharagma* H.; *Procharybdis* H.; *Charybdea* Pér. and Les.; *Tamoya* F. Müller.

Fam. 6. **Chirodropidae.** *Chiropsalmus* L. Ag.; *Chirodropus* H.

Order 2. EPHYRONINAE. OCTOMERALIA. DISCOMEDUSAE.

Acalephae with 8 or more rhopalia (sense tentacles) (4 per- and 4 interradial, and often several accessory). Stomach with 8, 16, 32, or more radial pockets or canals. Gonads sub-gastral (in ventral wall of central stomach). Umbrella flat, generally discoidal. Larval form Ephyra.

The *Ephyroninae* can at once be distinguished from the *Scyphomedusae* by the discoidal lobed umbrella, and usually by the large size of the oral tentacles. The lobes of the umbrella, however much they may differ in detail, can always be reduced to the eight pairs of lobes of the *Ephyra* (Fig. 131), which, as the promorph of the *Ephyroninae*, presents most clearly the 8-rayed symmetry characteristic of the group.

The gonads have the form of horse-shoe shaped frills (Fig. 130), and project into the widely open subgenital pits. The germinal epithelium, which is always embedded in the gelatinous substance of the umbrella, is covered with an endodermal layer. Development takes place by alternation of generations. In rare cases (*Pelagia*) the development is simplified, and the larva passes directly into the *Ephyra*, missing out the attached *Scyphistoma* and *Strobila* stage.

The gastrovascular system may be pouch-like or canalicular. In *Aurelia* (Fig. 130), in which it is canalicular, the eight primary radial canals (*i.e.*, the four perradial and four interradial) are branched, while the eight secondary radial canals (adradial) are unbranched. The parts of the stomach from which the eight adradial and the four interradial canals arise are pouched outwards.

Sub-order 1. CANNOSTOMAE.

With simple quadrangular manubrium without oral arms; with short solid marginal tentacles.

Fam. 7. **Ephyridae.** Usually 16 wide gastric pouches, rarely 32–64, without terminal branches; usually 8 rhopalia, rarely 16–32; usually 16, rarely 32–64 marginal lobes. This family may be described as consisting of sexually mature Ephyrae. *Ephyra* Pér. and Les.; *Palephyra* H.; *Zonephyra* H.; *Nausicaa* H.; *Nausithoë* Kölliker; *Nauphanta* H.; *Atolla* H.; *Collaspis* H.

Fam. 8. **Linergidae.** With wide radial gastric pouches, and branched, blind lobe-canals; without circular canal. *Linantha* H.; *Linerges* H.; *Liniscus* H.; *Linuche* Esch.

Sub-order 2. SEMOSTOMAE.

With 4 large perradial oral arms, and with long hollow tentacles.

Fam. 9. **Pelagidae.** *Semostomae* with 16 simple wide gastric pouches, without branched distal canals, without circular canal. *Pelagia* Pér. and Les., with 8 adradial tentacles, 8 rhopalia, and 16 marginal lobes; *Chrysaora* Pér. and Les., with 24 tentacles (and 8 rhopalia), and 32 marginal lobes; *Dactylometra* L. Ag.

Fam. 10. **Cyaneidae.** *Semostomae* with 16 or 32 wide gastral pouches, and branched, blind lobe-canals, without circular canal. 16–32 or more marginal lobes; 8 or 16 rhopalia (4 per-, 4 inter-, and 8 adradial); 8 or more long hollow tentacles. *Procyanea* H.; *Medora* Couthouy; *Stenoptycha* L. Ag.; *Desmonema* L. Ag.; *Cyanea* Pér. and Les.; *Patera* Lesson; *Melusina* H.

Fam. 11. **Flosculidae.** *Semostomae* with 16 or more simple unbranched narrow radial canals and with a circular canal. 8 rhopalia; 8–24 or more long hollow tentacles. *Floscula* H.; *Floresea* H.

Fam. 12. **Ulmaridae.** *Semostomae* with 16 or more narrow radial canals, which branch and often anastomose, and are connected by a circular canal. 8 or 16 rhopalia; 8–24 or more hollow tentacles. *Ulmaris* H.; *Umbrosa* H.; *Undosa* H.; *Sthenonia* Esch.; *Phacellophora* Brandt; *Aurelia* Pér. and Les.; *Aurosa* H.

Sub-order 3. RHIZOSTOMAE.*

With 8 large adradial, root-like, simple or branched, oral arms, with numerous suctorial mouths, without central mouth opening, and without marginal tentacles.

Fam. 13. **Toreumidae.** *Rhizostomae* with 4 separated subgenital pits, and with ventral suctorial frills on the 8 oral arms (no dorsal frills). 8, 12, or 16 rhopalia; 8–16 or more narrow radial canals, branched and anastomosing. *Archirhiza* H.; *Toreuma* H.; *Polyclonia* L. Ag.; *Cassiopea* Pér. and Les.; *Cephea* Pér. and Les.; *Polyrhiza* L. Ag.

Fam. 14. **Pilemidae.** *Rhizostomae* with 4 separated subgenital pits, and with dorsal as well as ventral sucking frills on the 8 oral arms. 8 rhopalia; 8–16 or more branched and anastomosing radial canals, with circular canal.

* The forms commonly called by the generic name *Rhizostoma* belong to the genus *Pilema.* The term Rhizostomae is kept for the sub-order.

Toxoclytus L. Ag.; *Lychnorhiza* H.; *Phyllorhiza* L. Ag.; *Eupilema* H.; *Pilema* H.; *Rhopilema* H.; *Brachiolophus* H.; *Stomolophus* L. Ag.

Fam. 15. **Versuridae.** *Rhizostomae* with a single central subgenital porticus (*i.e.*, subgenital pits united), with ventral suctorial oral frills only. 8 rhopalia; 8-16 or more narrow, branched, anastomosing, radial canals. *Haplorhiza* H.; *Cannorhiza* H.; *Versura* H.; *Crossostoma* L. Ag.; *Cotylorhiza* L. Ag.; *Stylorhiza* H.

Fam. 16. **Crambessidae.** *Rhizostomae* with a single central subgenital porticus, oral arms with dorsal and ventral frills. 8 rhopalia; 8–16 or more anastomosing radial canals; usually a circular canal. *Crambessa* H.; *Mastigias* L. Ag.; *Eucrambessa* H.; *Thysanostoma* L. Ag.; *Himantostoma* L. Ag.; *Leptobrachia* Brandt; *Leonura* H.

Class III. ACTINOZOA* (ANTHOZOA).

Polyps colonial or solitary, with oesophageal tube, mesenteric folds, and endodermal gonads. A medusoid sexual generation is unknown.

FIG. 138.—Branch of a polyparium of *Corallium rubrum* (after Lacaze Duthiers). *P* polyp.

The polyp of the *Actinozoa* has already been described (p. 102). It differs from that of the *Hydromedusae* in being larger, in having a greater muscular development, a better developed structureless lamella or jelly which often contains muscular and skeletal elements. The development of this jelly, which has a tough, dense character, is, in the colonial forms, greater in the lower parts of the polyps than in the upper, the result of which is the formation of the branched or massive **coenenchyme** (Fig. 138), from the surface of which the free ends of the polyps project. A calcareous skeleton is very generally present, but its form and method of formation vary in the different groups.

The mesenteries and tentacles vary much in number. In the *Alcyonaria* there are always eight; in the *Zoantharia*, in which there are primary and secondary mesenteries, the number is sometimes six

* Ehrenberg, "Beiträge zur physiologischen Kenntniss der Korallenthiere im Allgemeinen u. besonders des rothen Meeres, etc," *Abhand. d. Berliner Akad.*, 1832. Ch. Darwin, *The Structure and Distribution of Coral Reefs*, London, 1842. J. D. Dana, *United States Expl. Expedition, Zoophytes*, Philadelphia, 1846. M. Edwards and J. Haime, *Histoire Naturelle des Coraillaires*, 3 vols., Paris, 1857-60. Lacaze Duthiers, *Histoire Naturelle du Corail*, Paris, 1864.

or some multiple of six; but it may be different: indeed, the greatest variety is found in this character in the order *Zoantharia*.

The gonads are produced on the mesenteries (Fig. 91), and the embryos sometimes undergo the early stages of development within the parent.

Asexual reproduction by budding and fission is of great importance. Buds can be formed in various positions, even at the oral end, in which case a strobila-like form appears. In *Blastotrochus* the buds appear at right angles to the axis of the parent (Fig. 139).

In *Gonactinia prolifera* the polyp divides transversely, a new set of tentacles arising on the lower half (Fig. 139*a*). In some cases a portion of the basal expansion is separated off by contraction of the body, and develops into a new polyp. This is called *laceration*.

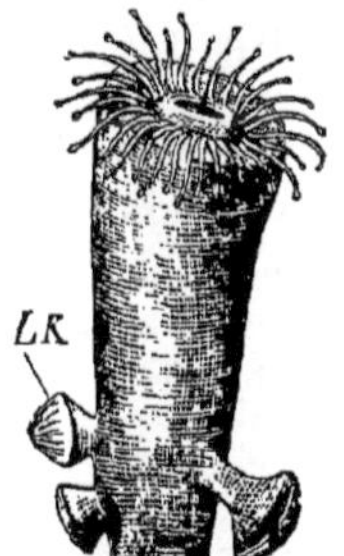

FIG. 139.—*Blastotrochus nutrix* (after C. Semper). *LK* lateral bud.

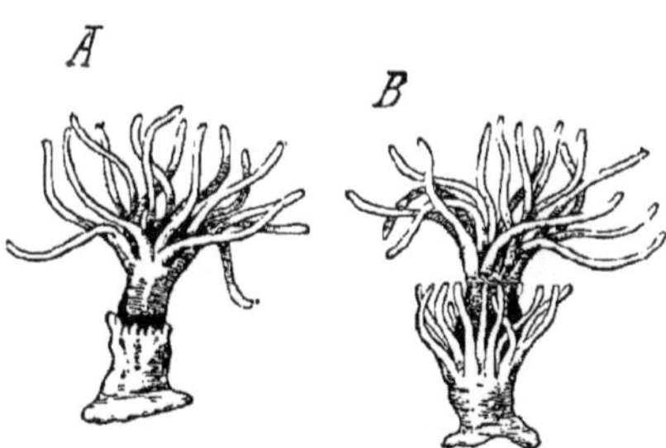

FIG. 139*a*.—Two stages of transverse fission of *Gonactinia prolifera*, Sars (after Blochman and Hilger).

If the individuals so produced remain connected with one another, a polyp-colony is formed, which may attain very various forms and great size. As a rule the individuals are embedded in a common body mass, the *coenenchyme*,* and their gastric cavities communicate more or less directly, so that the juices acquired by the individual polyps penetrate through the whole stock. This stock affords us an excellent example of an animal community built up out of similar members. The formation of the generative products alone is sometimes confined to special polyps, which, however, discharge all other functions of polyp life.

The skeletal formations of the polyps are specially noteworthy. In almost every case, with the exception of *Actinia*, there is a deposit of solid calcareous matter, and according to the density of this deposit, there is produced a leathery, chalky, or even stony framework.

* This word is used in a different sense in the *Madreporaria*, which see.

If the skeleton has the form of isolated needles or toothed rods (Fig. 140) of calcareous substance deposited in the jelly of the coenenchyma (or polyp), the polyp-stock has a fleshy, leathery nature (*Alcyonaria*); but if, on the contrary, the calcareous structures are fused or cemented together, a solid, more or less firm calcareous skeleton is developed (*Corallium*, *Tubipora*). Finally, the skeleton may be of a stony character and secreted by the ectoderm of the lower part of the polyp (*Madreporaria*).

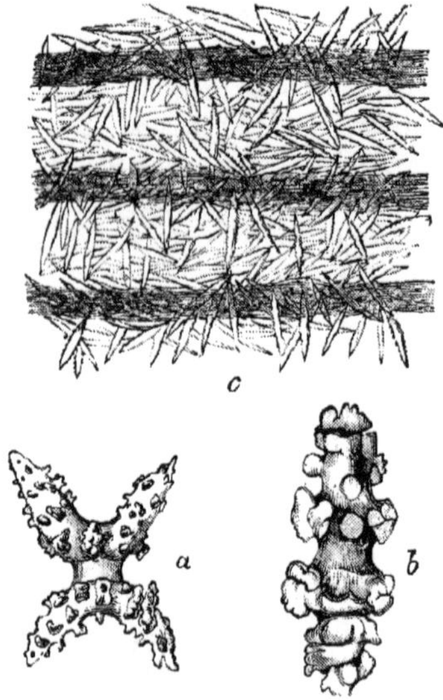

Fig. 140.—Calcareous bodies (*Sclerodermites*) of *Alcyonaria* (after Kölliker). *a*, of *Plexaurella*; *b*, of *Gorgonia*; *c*, of *Alcyonium*.

The important diversities of form in the polyp-stocks are not only occasioned by the differences of structure of the skeleton of the polyp, but are also the resultant of varying methods of growth by gemmation and imperfect fission. According to the method, numerous modifications of branched stocks are distinguished, *e.g.*, *Madrepores* (Fig. 141), *Oculinidae* (Fig. 142), and the lamellar and massive stocks as *Astraea* (Fig. 143) and the *Maeandrinidae* (Fig. 144).

Fig. 141.—*Madrepora verrucosa* (after Ed. H.).

The *Anthozoa* are all inhabitants of the sea, and live mostly in the warmer zones, but certain types of the fleshy *Octactinia* and *Actinia* are distributed in all latitudes. Some genera of the Madreporaria are found in the deep sea, where they may form accumulations of considerable extent, but the polyps which take the

Fig. 142.—Branch of *Oculina speciosa* (after Ed. H.).

principal share in the formation of coral reefs live near the surface, being rarely found alive at a greater depth than 40 fathoms. They are confined to a zone extending about 28 degrees on either side of the equator, and only here and there extend beyond these bounds. Their calcareous skeletons, together with those of mille-

FIG. 143.—*Astraea* (*Goniastraea*) *pectinata* Ehrbg. (after Klunzinger).

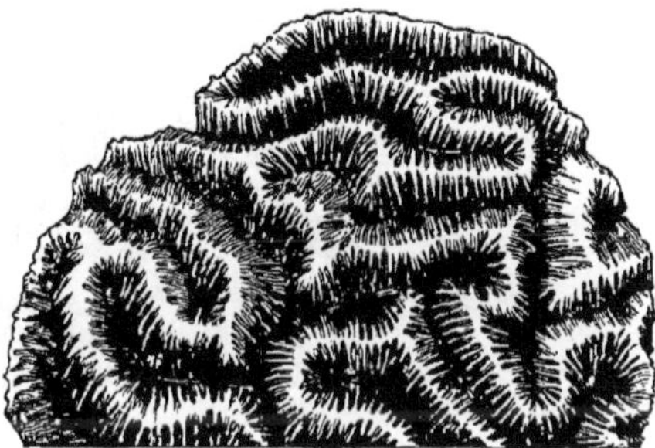

FIG. 144.—*Maeandrina* (*Coeloria*) *arabica* Klz. (after Klunzinger).

pores, the shells of molluscs, echinoderms, annelids, foraminifera, cemented together into a compact rock by encrusting organisms and by deposited lime, build up in the course of time masses of colossal extent.

Coral reefs are generally met with in one of three forms, *fringing reefs*, *barrier reefs*, and *atolls*.

A **fringing or shore reef** is a platform of rock skirting the shore and ending, seawards, in an abrupt edge from which there is a steep slope down to the sea bottom. To a depth of 20 or 30 fathoms the slope is covered with growing coral, the upper part being bathed in the surf of the breaking rollers. The upper surface of the outer edge of the reef, which is generally uncovered at low water, is higher than the part nearer the shore, and, in exposed reefs, is largely formed by encrusting calcareous algae (*Nullipores*), which thrive in the freshly aerated water.

In reefs in protected situations some corals, especially the *Madrepores*, grow on the edge of the reef where they are laid bare at low tide. The reef is frequently broken here and there by channels which have a depth of a few fathoms and run in towards the land in branching and tortuous courses, often expanding into irregular pools floored by coral sand. These protected pools and channels, whose vertical sides are formed by a luxuriant growth of hard corals, millepores, and the leathery colonies of *Alcyonaria*, are tenanted by an abundant reef fauna often displaying the most brilliant and varied colours. Between the channels the surface of the reef is formed of beds of corals which have nearly reached the surface, or of tracts of dead coral-rock, or sand often consisting largely of Foraminifera.

Nearer the land the growth of coral becomes less abundant and the water deepens, forming a shallow channel which at high water may be practicable for vessels of small draft.* The width of the reef is largely dependent on the slope of the sea bottom, becoming narrowed in proportion as this is steep; and also, as Semper has shown, on the set of the currents, which, when they are strong, hinder the extension of the reef past which they flow.

The edge of a fringing reef is often seen to be strewn with large masses of dead coral, which have been torn from the outer slopes by the breakers and

FIG. 144a.—A barrier reef (from Darwin), seen from within, from one of the high peaks of Bolabola, one of the Society Islands.

thrown upon the reef. Many such masses must, on the other hand, fall down the slope, and here, by their gradual accumulation, together with the shells of Pteropods, Foraminifera, and other pelagic organisms borne by ocean currents, form a basis on which the living margin of the reef may extend outward.

The structure and growth of a fringing reef do not appear to offer problems which are very difficult to solve, but the case is different when we come to the other two classes of reef, the barrier reef and atoll.

FIG. 144b.—A small atoll (from Darwin), being a sketch of Whitsunday Island in the S. Pacific, taken from Captain Beechey's Voyage. The whole circle has been converted into land, which is a comparatively rare occurrence.

A **barrier reef** resembles a fringing reef except in one important particular, namely, that it is separated from the shore by a channel, which is often of great width and which may attain a depth of 50–60 fathoms.

The great barrier reef of Australia which runs along the east coast of Queensland is over 1100 miles in length, and encloses a channel which is in many places

* Opposite the mouths of streams there are wide openings in the reef where the coral does not grow, by which the fresh waters reach the sea.

more than 30 miles wide, and 10-25 fathoms deep. A similar reef, though of less extent, borders the western shore of New Caledonia. In many cases, as in the Society Islands and elsewhere, a similar barrier or, as it is here called, an encircling reef surrounds an island or group of islands, having at intervals breaks in it, which lead into the deep lagoon channels and are often situated opposite the valleys on the land. Low islands (*coral islands*) formed of coral rock and sand thrown up by the sea, and supporting a littoral vegetation, are often situated on the reef.

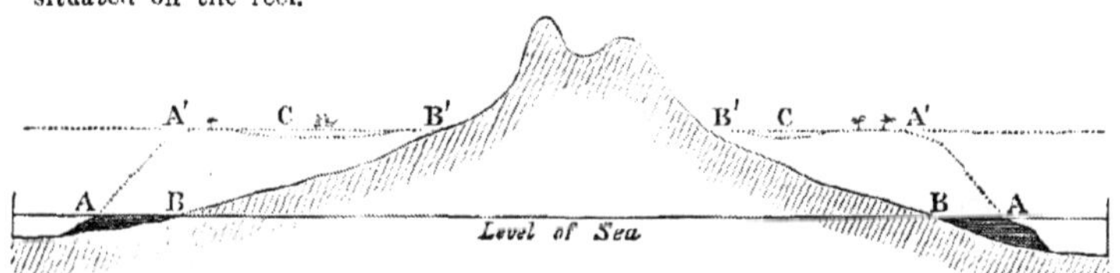

FIG. 144*d*.—Diagram illustrating Darwin's view of the formation of a barrier reef from a fringing reef (from Darwin). AA outer edge of the fringing reef at the level of the sea; BB shores of the island; A'A' outer edge of reef (now a barrier reef) after its upward growth during a period of subsidence; CC the lagoon channel between the reef and the island; B'B' the shore of the island.

The **lagoon islands** or **atolls** consist of a similar ring-shaped reef, enclosing an open lagoon in which there is no island (Fig. 144*b*). There may be gaps leading into the lagoon, or the ring may be so complete, that, as in the case of Fakaofu in the Union Islands, it is left at low tide brim full of water standing some feet above sea level. The floor of the lagoon is generally nearly flat, shelving slightly towards the reef, and is often from 20 to 35 fathoms below

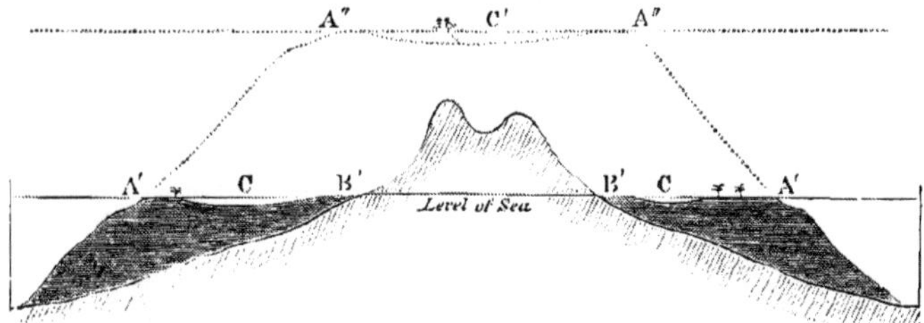

FIG. 144*e*.—Diagram illustrating Darwin's view of the formation of an atoll from a barrier reef by subsidence. A'A' outer edges of the barrier reef at the level of the sea. The cocoanut trees represent coral islets formed on the reef. CC the lagoon channel; B'B' the shores of the island generally formed of low alluvial land and of coral detritus from the lagoon channel; A"A" the lagoon of the newly formed atoll. According to the scale the depth of the lagoon and of the lagoon channel is exaggerated.

the surface—whereas the sea outside the reef rapidly deepens, so that it is not uncommon to find a depth of more than 1000 fathoms within a mile of the reef. The slope of the seaward side of the reef below the region of growing coral in many cases exceeds an angle of 45°. Low islands (*coral islands*) with glittering white sand and characteristic vegetation are often situated on the reef, scattered or united into a continuous belt of land.

The mode of the formation of the atoll has been much discussed. Whatever the foundation of an atoll may be, it is requisite (1) that it should at one time, at any rate, have reached so near the surface that coral polyps could form a settlement on it, and (2), as whole groups of islands have the atoll form, that it should in the great majority of cases not rise above that level.

Formerly the view was held that atolls were formed on submarine mountains (volcanic) which reached to within the limit of growing coral distance (40 fathoms), but this was objected to on the ground that it was exceedingly unlikely that there should have been so many mountains reaching to just within the right distance from the surface. But it has recently been pointed out* that a large number of banks exist in the ocean, springing from deep water and rising to a wide level top at a depth of about 30 fathoms from the surface. On many of these banks a rim of growing coral has been found raised a few fathoms above the general level. Such banks have probably been produced by a submarine volcanic eruption which has formed a mound, sometimes reaching to the surface, sometimes reaching above it, but composed of loose scoriae in a more or less finely divided state. Within the last few years the formation of such mounds has been observed in different parts of the sea, and they have since been, or are now in process of being reduced, by the action of the waves on the loose material, to the condition of submarine banks such as are known to exist, and such as are required for the foundation of atolls.

Darwin, to whom the existence of these banks was unknown, promulgated his well-known theory of coral reefs, which was as follows: fringing reefs are formed within the coral-reef zone on the shores of places where the conditions are favourable for coral growth. Barrier reefs and atolls are derived from fringing reefs by the subsidence of the land on which the reef is placed.

If we suppose an island bordered by a fringing reef to begin to subside (Fig. 144*d*), the corals, which as we have seen, are in the most vigorous condition on the edge of the reef, no longer limited in their growth by the level of low water, will grow upward; the width of the lagoon will be increased by the extent to which the sea now encroaches on the shore, and its depth by the amount of the subsidence. Continue the subsidence and the reef will rise like a wall round the land from which it is separated by the deepening lagoon, and the land itself, lessened in area, will be reduced to the higher parts of the original mass, forming an island or a group of islands in the centre of the wide lagoon (Fig. 144*d*). The breaks in the fringing reef at the mouths of rivers still remain in the barrier corresponding to the valleys, these parts of the shore line having been originally free from coral growth. Although the lagoon deepens as the island subsides, its depth will not be increased by the whole amount of that subsidence, being partially filled up by detrital matter carried down by streams on the one hand, and on the other by the material broken off by the wear and tear of the breakers on the outside of the reef.

Continue the subsidence further till the highest mountain peak disappears below the waters of the lagoon and an atoll remains, "like a monument, marking the place of the burial," of the island (Fig. 144*e*).

In confirmation of his view as to the conditions under which fringing reefs are formed, Darwin pointed out that their distribution is in many cases coincident with lines of recent volcanic activity, with which elevation of the earth's crust is often associated; as well as with raised shore lines giving direct evidence of elevation. Although not ignoring the possibility of atoll-shaped islands being formed without the aid of subsidence, Darwin regarded this as

* See "Foundations of Coral Atolls," by Admiral W. J. L. Wharton, F.R.S. *Nature* 1426, Feb. 25, 1897, p. 390.

exceptional, and, on the whole, the areas in which barrier reefs, encircling reefs, and atolls occur were considered by him to be areas of subsidence. To this view he was led in part by the difficulty of finding any other foundation for atolls than an island which had subsided. Direct indications of subsidence are, from the nature of the case, difficult to obtain, but such evidence as we have is not altogether in harmony with Darwin's theory. Thus Dana, although an adherent of the view, was of the opinion that the movement now going on in many of the Paumotu Islands is one of elevation; and similar evidence is forthcoming from elsewhere (Solomon Islands, Tonga Islands, etc.). Such evidence is, on the other hand, in keeping with the view that atolls rest, at any rate in some cases, on such banks as those described by Admiral Wharton, and referred to above (p. 174). It has been urged by Murray that an atoll on such a basis would increase in size by the extension of the reef to seaward on its own talus, while the lagoon would be widened by the solution of the dead portions of the reef. A fringing reef would be converted into a barrier reef by the same process.

The part which the *Anthozoa* take in the alteration of the earth's surface is considerable. In the present time they protect the coast from the consequences of the breaking of the waves, and assist in the formation of islands and rocks by producing immense masses of calcareous matter. In earlier geological epochs they have played a still more important part, judging from the great thickness of the coral formations of the Palaeozoic period and of the Jurassic formation.

Order 1. Rugosa = Tetracoralla.

Palaeozoic Corals with numerous symmetrically arranged septa grouped in multiples of four.

To these belong the families of the *Cyathophyllidae, Stauridae, etc.*

Order 2. Alcyonaria.*

Polyps and polyp colonies with eight pinnate tentacles and eight mesenteric folds.

The Alcyonaria are all marine and, with the exception of the *Haimeidae*, colonial. The buds are formed as a rule, not from the bodies of the polyps themselves, but from stolons which originate as tubular processes of the body-wall of the polyps at the base of the colony (Fig. 145, *A*), or from the small canals which ramify in the common jelly or coenenchyma and connect together the polyps (Fig. 145, *D*). In the simplest colonies the polyps arise directly from the basal stolon (*Cornularia*); an advance upon this occurs when the basal part of the polyp acquires a greatly developed jelly,

* Wright and Studer, "Report on the Alcyonaria," *Challenger Reports*, 1889. S. J. Hickson, "A Revision of the genera of the Alcyonaria Stolonifera," *Trans. Zool. Soc.*, vol. 13, 1894. A. Kölliker, *Anat.-syst. Beschreibung d. Alcyonarien*, 1872.

and spreads out so as to form a plate-like expansion of coenenchyma containing endodermal canals from which the new polyps bud (*Clavularia rosea*). By the increase of this basal coenenchyme we get the massive colonies of *Alcyonium*, with long polyp tubes and canal system (Fig. 145, *D*). Another variation in the colonial form may be deduced by supposing the plate-like expansion to be invaginated so as to bring its lower surface to the inside, and the polyps to the outside. Such a process would lead to the production of colonies with an ectodermal axial rod (*Gorgonia*, Fig. 145, *B*, ? *Pennatula*). Calcareous spicules are very generally present. They may occur either in the coenenchyma, or in the walls and tentacles of the polyps, and they may either be loose or coalesced to form a

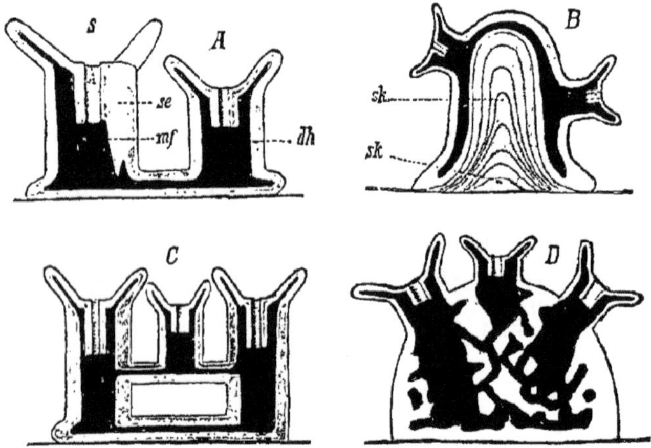

Fig. 145.—Diagrams to show the budding and mode of formation of the colonies of various *Alcyonaria*. *A*, general diagram. *B*, *Gorgonia*. *C*, *Tubipora*. *D*, *Alcyonium*. The enteric space and canals are black. *S* oesophagus; *se* mesenteries; *mf* mesenterial filaments; *dh* enteron; *sk* axial skeletal rods with lines to show the mode of growth.

continuous corallum. When this coalescence occurs in the polyp-walls we get the thecal tubes of *Tubipora;* when it occurs in the coenenchyma, the axial rod of the *Scleraxonia* (*Corallium rubrum*). Sometimes the spicules are embedded in a horny material, sometimes in a calcareous cement. In the *Pennatularea* and the *Holaxonia* (*Gorgonia*) there is an axial rod secreted by an epithelial layer contained in the stem or axis of the colony. This epithelial layer is of ectodermal origin in the *Holaxonia* (Fig. 145, *B*), but its origin in the *Pennatularea* is doubtful (?endodermal). In all other cases the spicules arise in the jelly of the coenenchyma.

The polyps (autozooids) have eight pinnate tentacles and eight mesenteries (Fig. 146). The gonidial groove (siphonoglyphe) when present is single, and the side on which it is placed is called **ventral.*** The longitudinal muscles are placed on the ventral faces of the mesenteries (Fig. 146). The dorsal mesenteries are often longer than the others, and are developed earlier in the bud, though later in the egg. The polyps vary considerably in their power of contractility. In the genera, with a good development of spicules in the body-wall, there is hardly any retractility, but the tentacles are simply folded over the mouth on irritation. In *Stereosoma*, which has a horny layer beneath the ectoderm, the polyps are non-retractile. The polyps are often dimorphic. There are the *autozooids* with tentacles and generative organs, and the *siphonozooids* without these structures, and with filaments only on the dorsal mesenteries. A gonidial groove (siphonoglyphe) is present in the *Alcyonidae* and in *Sarcophyton*; it is absent in the autozooids of *Pennatulacea*, *Heteroxenia*, and *Paragorgia*, but present in the siphonozooids, in which it is always specially developed. The *Pennatulacea* are phosphorescent.

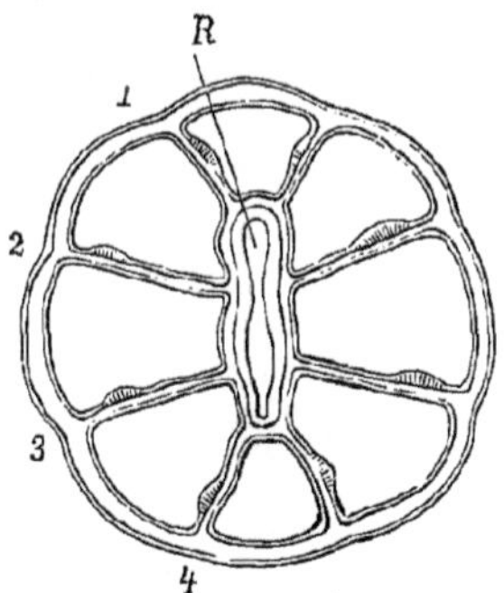

FIG. 146.—Transverse section through *Alcyonium* (after Hertwig). *R* gonidial groove; *1*, *2*, *3*, *4*, the four pairs of septa with their muscles.

The enteric cavities of the polyps are connected with the fine canal-system of the coenenchyma, when such exists, and are continued as main canals for a longer or shorter distance towards the base of the colony. In the *Stolonifera* they all, of course, open into the basal stolon.

Development. The ova develop inside (as far as the planula) and outside the parent. They have yolk, and the segmentation is on the centro-lecithal type, and is often delayed until the nucleus has undergone many divisions. There is usually a free-swimming planula-larva.

Extinct forms.† The genera *Heliolites* (palaeozoic) and *Polytremacis* (chalk, greensand, eocene) were probably *Helioporidae*, and

* This is a special use of the term ventral, and does not imply any homology with the ventral surface of bilateral animals.

† K. A. Zittel, *Handbuch der Palaeontologie*, Bd. 1, p. 208. Munich and Leipzig, 1876–80.

Heliopora itself is found in the cretaceous formation. *Syringopora* (Silurian, Devonian, Carboniferous) was probably allied to *Tubipora*. The *Favositidae* were probably Alcyonarian.

Sub-order 1. **PROTOALCYONARIA.**

Polyps solitary; with or without spicules.

Fam. 1. **Haimeidae.** *Haimea* M. Edw.; *Hartea* P. Wright; *Monoxenia* H.

Sub-order 2. **STOLONIFERA.**

Colonial Alcyonaria with a membranous or ribbon-like stolon. Jelly poorly developed. Polyps either entirely free from one another except at their bases, or connected by horizontal platforms (Tubipora) or connecting tubes (Clavularia viridis). Skeleton absent, or composed of calcareous spicules which may be joined together or be isolated. In some cases the body-wall is supported by a horny secretion.

Fam. 2. **Cornularidae.** Polyps are not united in bundles, but either spring from a plate-like expansion or a creeping stolon; or are branched and bear lateral buds. *Cornularia* Lam., no spicules, horny secretion on polyp-walls and stolon; *Rhizoxenia* Ehrb.; *Clavularia* Q. and G.; *Sarcodictyon* Forbes; *Anthelia* Savigny; *Gymnosarca* S. Kent; *Cornulariella* Verrill; *Telesto* Lamouroux, polyps rise from a flat base, or from stolons, and bear buds; *Coelogorgia* M. Edwards, colony arborescent, an axial polyp with buds; *Cyathopodium* Verrill, stolons calcified connecting the short cup-shaped polyps; *Scleranthelia* Studer: *Anthopodium* Verrill; *Sympodium* Ehrb.; *Stereosoma* Hickson, with spicules, with non-contractile polyps and tentacles, with a horny layer between the ectoderm and supporting lamella; *Erythropodium* Köll.; *Callipodium* Verr.: *Pseudogorgia* Köll., axial polyp with lateral polyps budded from the upper part.

Fam. 3. **Tubiporidae.** Colonies consist of tubular polyps parallel to one another, and united by horizontal platforms containing endodermal canals (Fig. 145, *C*). The platforms are formed as outgrowths of the lips of the polyps, into which prolongations of the enteric cavity pass to form the endodermal canals; they are at first without or with only a few spicules. The platforms and the greater part of the walls of the polyp-tubes contain a skeleton formed of coalesced spicules, so that the dry corallum has the form of parallel tubes united by lamellae. The first layer of platforms constitutes the plate-like stolon of origin. The tubes are divided at intervals by partitions called tabulae which may be funnel-shaped. *Tubipora* L., the organ-pipe coral.

Sub-order 3. **ALCYONACEA.**

Colonial Alcyonaria with a well-developed canaliferous coenenchyma and loose spicules. Without axial skeletal rod. The buds are formed from the coenenchymal canals (Fig. 145, *D*).

Fam. 4. **Xeniidae.** Colonies of long polyps, united in their lower portion by a canal system, ramifying in a connecting coenenchyma with feebly calcareous spicules. *Xenia* (*Heteroxenia* Köll.) Savigny.

Fam. 5. **Organidae.** Elongated polyps united together so as to form a short upright stem. Polyps and tentacles provided with spicules. *Organidus* Danielssen.

Fam. 6. **Alcyonidae.** Massive coenenchyma containing the polyp tubes, which are united by endodermal canals from which the buds are formed. Isolated spicules in the jelly of the coenenchyma. *Crystallophanes* Dan.;

Bellonella Gray; *Nidalia* Gray; *Paralcyonium* M. Edw.; *Sarakka* Dan.; *Alcyonium* L., *A. digitatum*, dead men's fingers; *Lobularia* Savigny; *Sarcophyton* Lesson, with dimorphic polyps; *Lobophytum* Marenzeller, with dimorphic polyps; *Anthomastus* Verr., dimorphic polyps; *Nannodendron* Dan., dimorphic polyps.

Fam. 7. **Nephthyidae.** A branched coenenchyma with sterile base and terminal polyps. The latter do not exhibit separate calycine and tentacular regions, and there is no invagination of the latter; when at rest tentacles folded over oral disc. Buds arise from fine endodermal canals between the polyps. *Voeringia* Dan.; *Fulla* Dan.; *Barathrobius* Dan.; *Gersemia* Marenzeller; *Gersemiopsis* Dan.; *Drifa* Dan.; *Duva* Kor. and Dan.; *Eunephthya* Verr.; *Ammothea* Sav.; *Nephthya* Sav.; *Spongodes* Lesson; *Paranephthya* Wright and Studer; *Scleronephthya* Wr. and St.; *Chironephthya* Wr. and St.; *Siphonogorgia* Köll.

Fam. 8. **Helioporidae.*** Compact corallum formed in the jelly of the coenenchyma. The corallum is traversed by tubes closed above, called coenenchymal tubes (possibly modified siphonozooids), and by tubes continued from the polyp calycles. Both systems of tubes are divided by tabulae and are united by endodermal canals in the superficial coenenchyma. False septa formed by denticulations of the margins of the calycles. *Heliopora* Blainville.

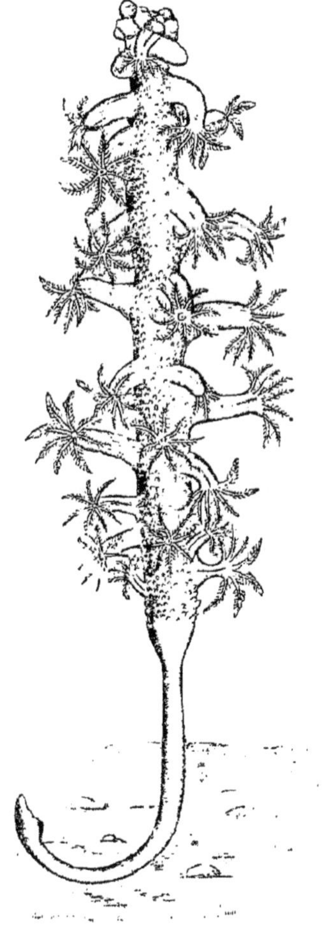

Fig. 147.—*Kophobelemnon Leuckartii.*

Sub-order 4. **PENNATULACEA.**†

Unattached polyp colonies with a stalk embedded in mud or sand, and a rachis bearing polyps. The stalk generally has an axial rod.

The stalk is without polyps and is embedded in mud or sand. The rachis is a continuation of the stalk and carries the polyps, which are arranged

* H. N. Moseley, "The Structure and Relations of the Alcyonarian Heliopora Coerulea," *Phil. Trans.*, 1876.

† A. Kölliker, *Anat.-syst. Beschreib. d. Alcyonarien Abt.* 1. *Die Pennatuliden*, Frankfurt-a-M., 1872. A. Kölliker, "Report on the Pennatulidae," *Challenger Reports*, 1880.

upon it in different ways. In *Veretillum* they are distributed all round the rachis; in *Kophobelemnon* (Fig. 147) they are absent on a streak of one side—the so-called *ventral side;* in most genera they are arranged bilaterally, there being a *dorsal* as well as a ventral streak free from them. Further, in some genera they are sessile on the rachis, but in the *Pennatulea*, or Sea-Feathers proper, the polyps are borne only on lateral processes of the coenenchyma, called the pinnules. The pinnules are broad triangular leaf-shaped structures attached by their base to the rachis and carrying the polyps on their dorsal edges (Fig. 148). The polyp-tubes project in some cases to form the so-called *cells;* the cells may have spines or tufts of spicules.

The polyps are dimorphic; the *autozooids* have tentacles and generative organs, and are without a gonidial groove; the *siphonozooids* possess a gonidial groove but are without tentacles and gonads, also they have filaments on the two dorsal mesenteries only. The siphonozooids are distributed over the whole rachis in *Renilla* and *Veretillum;* and in the *Pennatulea* they are on the rachis or on the pinnules.

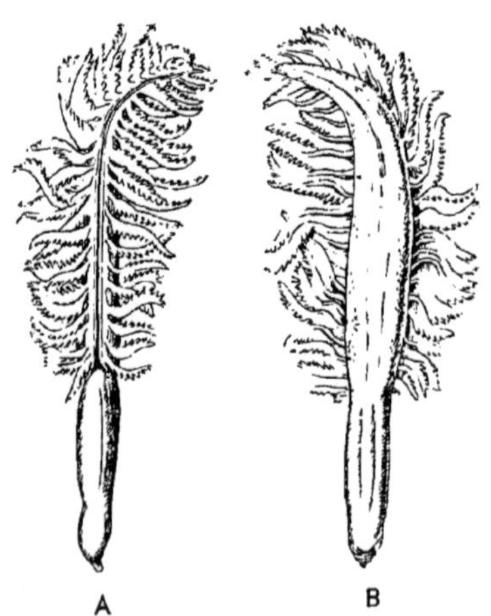

FIG. 148.—*Pennatula Sulcata* Köll. (after Kolliker). *A*, from the dorsal; *B*, from the ventral side.

The stalk generally contains an axial calcareous or horny rod surrounded by a sheath of epithelial cells; and the coenenchyma and bodies of the polyps may contain isolated spicules. The polyps are continued into tubes which join the canal system of the coenenchyma. This canal-system consists of large canals continued down from the polyps and opening after a longer or shorter course into a few "*main canals,*" which run in the stalk. There are generally four of the latter (two in *Renilla*), of which two are lateral, one dorsal, and one ventral. At the lower end of the stalk the lateral canals cease, leaving only the dorsal and ventral. They fuse at the end of the stalk and are said to open there. In addition to the large canals there are minute canals uniting them. Some of the polyps are closed below, and open at their base by narrow openings into the general canal system of the colony.

The *Pennatulacea* are not distributed uniformly over all seas. They are mainly littoral, but deep water forms are known, and these, it is important to notice, belong principally to the simpler families, *e.g.*, the *Protoptilidae* and *Umbellulidae*. They appear to be absent, or nearly so, in the deeper parts of the Pacific and Atlantic Oceans and the South Polar Sea at a certain distance from the shore.

Section 1. **Pennatulea.**

The Sea-feathers. With pinnules; rachis with a bilateral arrangement of the polyps, elongated, cylindrical.

Fam. 1. **Pteroeididae.** Pinnules well developed, with siphonozooids on the pinnules. *Pteroeides* Herklot; *Godefroyia* Köll.; *Sarcophyllum* Köll.

Fam. 2. **Pennatulidae.** Pinnules well developed; siphonozooids on the ventral and lateral sides of the rachis. *Pennatula* Lam.; *Leioptilum* Verr.; *Ptilosarcus* Gray; *Halisceptrum* Herklot.

Fam. 3. **Virgularidae.** Pinnules small, without a calcareous plate. *Virgularia* Lam.; *Scytalium* Herkl.; *Pavonaria* Köll.

Fam. 4. **Stylatulidae.** Pinnules small, with a calcareous plate. *Stylatula* Verr.; *Dubenia* Kor. and Dan.; *Acanthoptilum* Köll.

Section 2. **Spicata.**

Rachis elongated, cylindrical, with a bilateral arrangement of the polyps; without pinnules; polyps sessile.

Fam. 5. **Funiculinidae.** Polyps on both sides of the rachis in distinct rows, with cells; ventral siphonozooids absent. *Funiculina* Lamarck; *Halipteris* Köll.

Fam. 6. **Stachyptilidae.** Polyps (with cells) on both sides of the rachis in distinct rows. Ventral siphonozooids present. *Stachyptilum* Köll.

Fam. 7. **Anthoptilidae.** Polyps on both sides of the rachis in distinct rows, without cells. *Anthoptilum* Köll.

Fam. 8. **Kophobelemnonidae** (Fig. 147). Polyps on both sides of the rachis in a single series, or in indistinct rows, large and without cells; rachis elongated, cylindrical; ventral streak of rachis without polyps. *Kophobelemnon* Asbjörnsen, *Sclerobelemnon* Köll.; *Bathyptilum* Köll.

Fam. 9. **Umbellulidae.** Polyps on both sides of the rachis in a single series, or in indistinct rows, large and without cells; rachis short (*i.e.*, the polyps are placed at the end of the central stem). *Umbellula* Lam.

Fam. 10. **Protocaulidae.** Polyps on both sides of the rachis in a single series, or in indistinct rows, small and without cells. *Protocaulon* Köll; *Cladiscus* Kor. and Dan.

Fam. 11. **Protoptilidae.** Polyps on both sides of the rachis in a single series or in indistinct rows, with cells. *Protoptilum* Köll.; *Lygomorpha* Kor. and Dan.; *Microptilum* Köll.; *Leptoptilum* Köll.; *Trichoptilum* Köll.; *Scleroptilum* Köll.

Section 3. **Renillea.**

Rachis expanded in the form of a leaf, with bilateral arrangement of the polyps on one side of the expansion; without pinnules. A single large siphonozooid (exhalent zooid) terminates the end of the central stem.

Fam. 12. **Renillidae.** *Renilla* Lam.

Section 4. **Veretillea.**

Club-shaped colonies, without pinnules. Polyps arranged all round the rachis.

Fam. 13. **Cavernularidae.** Spicules long. *Cavernularia* Valenciennes; *Stylobelemnon* Köll.

Fam. 14. **Lituaridae.** Spicules short. *Lituaria* Val.; *Veretillum* Cuv.; *Policella* Gray; *Clavella* Gray.

Sub-order 5. **GORGONACEA.**

Fixed colonial Alcyonaria with a horny or calcareous axial rod, which is covered by a coenenchyma from which the polyps arise.

Section 1. **Scleraxonia (Pseudaxonia).**

Fixed upright branched colonies. The coenenchyma consists of a canaliferous cortical layer (with spicules) in which the polyps are placed, and of a medullary substance. The latter contains spicules (different in form from the cortical spicules) which are generally tightly packed, and sometimes fastened together by a horny substance, or cemented into a strong axis by calcareous matter. Without epithelial layer round the central rod.

Fam. 1. **Briareidae.** Coenenchyma consists of a polyp-bearing cortex and a medullary substance of closely packed spicules. These are either developed on the surface of an upright shrubby colony, or the medullary substance is relegated to the interior of a cylindrical stem, over which is spread the cortex. In the latter case there is a more or less well-defined axis, which may be permeated by nutritive canals. *Leucorlla* Gray; *Solenocaulon* Gray; *Semperina* Köll.; *Suberia* Studer; *Anthothela* Verr.; *Paragorgia* M.-Edw.; *Briareum* Blainville; *Titanideum* Ag.; *Iciligorgia* Ridley; *Spongioderma* Köll.

Fam. 2. **Sclerogorgidae.** An axis consisting of closely intercalated elongated spicules with dense horny sheaths. The axis is surrounded by longitudinal canals, into which there open the reticulated coenenchymatous canals, uniting the polyps. *Suberogorgia* Gray; *Keroeides* Wr. and St.

Fam. 3. **Melitodidae.** Axis jointed, consisting of alternate portions of calcareous and soft horny substance. *Melitodes* Verr.; *Mopsella* Gray; *Acabaria* Gray; *Psilacabaria* Ridley; *Wrightella* Gray; *Clathraria* Gray; *Parisis* Verr.

Fam. 4. **Corallidae.** Axis of a dense calcareous mass of fused spicules; polyps dimorphic; the siphonozooids are said to grow into autozooids. *Corallium* Lam.; *C. rubrum*, the red coral (Fig. 138); *Pleurocorallium* Gray.

Section 2. **Holaxonia (Axifera)**

Coenenchyma branched or simple, with cortical canaliferous layer and axial rod, which is either horny, or of calcified horn, or of alternating joints of calcareous matter and horn. Axial rod derived from a layer of ectoderm cells invaginated at the base of the colony, and surrounding it as an epithelium (Fig. 145).

Fam. 5. **Dasygorgidae.** Simple or branched, coenenchyma thin, polyps large; both polyps and coenenchyma contain spicules. When at rest the tentacles are folded over the oral disc. *Strophogorgia* Wright; *Chrysogorgia* Duch. and Mich.; *Herophila* Steenstrup; *Dasygorgia* Verr.; *Iridogorgia* Verr.

Fam. 6. **Isidae.** Axis consists of alternating horny and calcareous portions. *Bathygorgia* Wright; *Ceratoisis* Wright; *Callisis* Verr.; *Acanella* Gray; *Isidella* Gray; *Sclerisis* Studer; *Primnoisis* Wr. and St.; *Mopsea* Lamouroux; *Acanthoisis* Wr. and St.; *Isis* L.

Fam. 7. **Primnoidae.** Axis calcareous and horny, basal attachment calcareous. Polyps with club-shaped calycine portion. Operculum calycine formed by some of the scale-like spicules of the calycine region, which shut over the tentacular region. *Callozostron* Wright; *Calyptrophora* Gray; *Primnoa* Lamouroux; *Stachyodes* Wr. and St.; *Calypterinus* Wr. and St.; *Stenella* Gray; *Thouarella* Gray; *Amphilaphis* Wr. and St.; *Plumarella* Gray; *Primnoella* Gray; *Caligorgia* Gray.

Fam. 8. **Muriceidae.** Axis horny; spicules project beyond the surface of coenenchyma; operculum tentacular, formed by the spicules at the base of the tentacles which close over the calyx when the oral region is retracted. *Acanthogorgia* Gray; *Paramuricea* Köll; *Hypnogorgia* Duch. and Mich.; *Muriceides* Wr. and St.; *Clematissa* Wr. and St.; *Villogorgia* Duch. and Mich.; *Anthogorgia*

Verr.; *Menella* Gray; *Acis* Duch. and Mich.; *Thesea* Duch. and Mich.; *Bebryce* Philippi.

Fam. 9. **Plexauridae.** Colony branched, axis horny; polyps occur all over the thick coenenchyma; spicules large; cortical club-shaped and deeper spindle-shaped spicules. *Eunicea* Lamouroux; *Plexaura* Lamouroux; *Psammogorgia* Verr.; *Platygorgia* Studer.

Fam. 10. **Gorgonidae.** Colonies upright and branched usually in one plane; axis horny, rarely horny and calcareous; polyps arise from stem and twigs in a bilateral and biradiate manner. Coenenchyma smooth, spicules small. *Platycaulos* Wr. and St.; *Lophogorgia* M.-Edw.; *Leptogorgia* M.-Edw.; *Stenogorgia* Verr.; *Callistephanus* Wr. and St.; *Swiftia* Duch. and Mich.; *Gorgonia* L.; *Eugorgia* Verr.

Fam. 11. **Gorgonellidae.**

Order 3. Zoantharia = Hexactinia.

Polyps and polyp-colonies, usually with simple unbranched tentacles. There are usually incomplete as well as complete mesenteries, and the tentacles usually alternate in several circles.

This order includes the Sea-Anemones and Corals. The order owes its name *Hexactinia* to the fact that in some of the best-known forms the mesenteries and tentacles are arranged in some multiple of the number six. There is, however, the greatest variation in this respect, and with the progress of research it has become clear that the number six is by no means universally characteristic; indeed we may go further, and say that it is not even typical; and it appears probable that, when this matter has been more fully looked into, the hexactinian arrangement will be found to be only one of many mesenterial arrangements found in the group. In some cases the number of mesenteries increases with the growth of the animal.

Development. Our knowledge is not very complete. An invaginate gastrula has been observed, with a blastopore persisting as mouth. A ciliated, free-swimming larva is usually formed.

There are three sub-orders.

Sub-order 1. ACTINIARIA.* MALACODERMATA.

Solitary, rarely colonial polyps with mesenteries, the number of which is usually a multiple of six; without skeleton. Body moving freely, or adherent by means of the pedal disc; rarely firmly fixed.

* P. H. Gosse, *A History of the British Sea-Anemones and Corals*, London, 1860. R. Hertwig, "Report on the Actiniaria," *Challenger Reports*, Pt. 15, 1882. R. Hertwig, Supplement to the above, *Challenger Reports*, Pt. 73, 1888. A. Andres, "Le Attinie," *Fauna and Flora des Golfes von Neapel*, 1884. A. C. Haddon, "A Revision of the British Actiniae," Pts. 1 and 2, *Sci. Trans. Roy. Dublin Soc.* (2), 4, 1889–91. J. Playfair McMurrich, "Report on the Actiniae collected by the United States Fish Commission Steamer *Albatross*," *Proc. U. S. National Museum*, vol. xvi, p. 119, 1893.

The Zoantheae alone are colonial. As a rule there is a single corona of tentacles at the edge of the oral disc; these are the primary (marginal) tentacles. When they are in several rows the inner are the oldest. They arise both from intra- and intermesenterial spaces (see below, p. 184). In addition to these there are in some forms (*Corallimorphidae*) secondary or accessory tentacles arising from the disc midway between the mouth and margin, and these are always intramesenterial. There is nearly always a sphincter muscle at the peristomial margin, which may be endodermal or mesodermal (*i.e.* in the jelly). It closes the peristomial margin over the mouth and tentacles during retraction. The mouth is usually slit-like, and the oesophagus has two gonidial grooves (Gosse). Sometimes there is only one gonidial groove (*Peachia*, *Zoantheae*, *Cerianthus*). The thickened walls of these grooves are prolonged beyond the lip at the two ends of the long axis of the mouth as two tubercles (called in the *Siphonactinidae* the *conchula*).

The **oesophageal lappets** are processes of the oesophageal wall at each end of the long axis which hang down into the coelenteron beyond the rest of the oesophagus; the gonidial grooves are continued on to them.

FIG. 149.—Transverse section through *Adamsia* (*Sagartidae*) (after R. Hertwig). *Hf* the chambers between the directive mesenteries: *R* gonidial grooves.

The gonidial grooves are, in this work, defined as dorsal and ventral. The ventral groove is the most conspicuous, and the only one present in *Peachia* and the Zoantheae. When there is only one groove, it is not always possible to determine whether it is dorsal or ventral. In the terminology which has more recently been introduced by Haddon the ventral groove is termed the *sulcus*, and the dorsal the *sulculus*. It will, of course, be understood that the above use of the words dorsal and ventral is special to the group, and implies no homologies with the dorsal and ventral surfaces of other animals.

The mesenterial arrangement presents the greatest variation. The arrangement* generally described as typical is that of the *Hexactiniae*, in which the mesenteries are arranged in pairs (Fig. 149), and in the simplest cases in six pairs. These are the *primary mesenteries*, and they all reach and are inserted into the oesophagus. The mesenteries of each pair are usually provided with longitudinal muscles on those faces which are turned towards one another, except on the two pairs of directive mesenteries; these carry the longitudinal muscles on the faces turned from one another. The portion of coelenteron enclosed in each pair of mesenteries is called an *intramesenterial* space or **endocoele**, the portion between the pairs being *intermesenterial* (**exocoele**). There are, therefore, six of the former and six of the latter. There are in many cases *secondary*

* For variations see descriptions of the different sections.

mesenteries (formed later than the primary) in addition to the primary; these are complete in most *Hexactiniae* (incomplete in some *Sagartidae*, *Peachia*, etc.); there are six pairs of them, and they are placed in the intermesenterial spaces, the longitudinal muscular faces of each mesentery being turned towards the corresponding face of its fellow of a pair, as in the case of the lateral primaries. There may be, in addition, twelve pairs of *tertiary* mesenteries and twenty-four *quaternaries*, and so on; the mesenteries of each order being always in pairs and placed in intermesenterial spaces. The mesenteries of each order are smaller than those of the preceding order, and except in the case of the primaries and secondaries do not, as a rule, reach the oesophagus. The space between the mesenteries of a primary pair is a primary intramesenterial space, that between the mesenteries of the next cycle a secondary intramesenterial space, and so on.

The tentacles, like the mesenteries, are in cycles of different age, so that we can distinguish tentacles of the first, second, etc. order. There is often a corresponding distinction in size (*Corallimorphidae*), and the arrangement of the mesenteries is reflected in that of the tentacles. In such cases the six largest tentacles are over the primary mesenterial spaces; the next six, which are a little smaller, belong to the secondary intramesenterial chambers; then follow the twelve tentacles, still smaller, of the tertiary intramesenterial spaces; while the twenty-four last tentacles communicate with the intermesenterial spaces.

Acontia are present in the *Sagartidae*. The epithelial cells always carry flagella or cilia.

The sexes are usually separate, but a few are hermaphrodite. The generative cells in some cases, if not all, escape through the gonidial grooves.

Budding takes place in the colonial *Zoantheae*, and both budding and fission are occasionally observed in some solitary forms (*Anthea cereus* = *Anemonia sulcata*, *Actinoloba*, *Actinia*, and other genera).

The **development of mesenteries** in some larval *Actiniae* is interesting in view of permanent arrangements in the various tribes of the *Actiniaria*. It is as follows:—

Stage 1.—The two first mesenteries are at right angles to the long axis of the oesophagus, and divide the coelenteron into two, generally unequal, chambers.

Stage 2.—The mesenteries of the second pair are in the larger of the two chambers so formed.

Stage 3.—The mesenteries of the third pair develop in the smaller of the two primitive chambers.

Stage 4.—The mesenteries of the fourth pair are within the unpaired chamber enclosed by the mesenteries No. 2 (Fig. 149*a*).

Stage 5.—Two pairs now arise simultaneously, and for some time remain incomplete; they are respectively between the first and second, and the first and third; their longitudinal muscles face the longitudinal muscles of the first and second respectively. When these are completed we get the typical Hexactinian arrangement of the above mesenterial pairs; the third and second are directive. Moreover, it is to be noted that Stage 4 exists permanently in the *Edwardsiae*, while Stage 5 is found in *Gonactinia* and in a modified form in the *Zoantheae*.

Stage 6.—A pair of small mesenteries with their muscles facing each other appears in each intermesenterial chamber (exocoele), and so with subsequent cycles. In *Peachia* only four pairs of secondaries are formed, those of the dorsal intermesenterial chamber (opposite to the gonidial groove) being absent.

The deep-sea forms are very commonly distinguished by the reduction of their tentacles even to vanishing point and by the enlargement of their terminal openings to large slit-like stomidia*; also by variations from the type of mesenterial arrangement.

The *Actiniaria* comprise the Sea-Anemones; they live in the sea, attached to rocks or other bodies, or embedded in sand, or sessile, as commensals, on hermit crab shells.

Section 1. **Hexactiniae.**

With paired mesenteries. The mesenteries of each pair are usually provided with longitudinal muscular fibres on those faces which are turned towards one another, except on the two pairs of directive mesenteries, which carry the longitudinal muscles on the faces turned from one another (Fig. 149). Six or more pairs of mesenteries, increasing in multiples of six.† Mouth slit-like, oesophagus usually with two gonidial grooves.

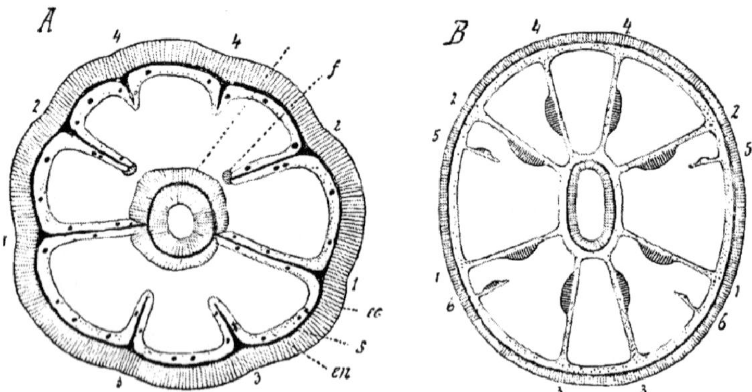

Fig. 149*a*.—Diagram of the growth of the mesenteries in Hexactinians (from Korschelt and Heider). *A*, stage of *Manicina areolata*, with eight primary mesenteries, in transverse section (after H. V. Wilson); *B*, stage of *Aulactinia stelloides* with twelve primary mesenteries (after McMurrich). The mesenteries are numbered in the order of their appearance. *ec* ectoderm; *en* endoderm; *s* supporting lamella; *f* mesenterial filaments.

Sub-tribe 1. **STICHODACTYLINAE.** Tentacles arranged radially, some or all of the intramesenterial chambers communicating with more than one tentacle.

Fam. 1. **Corallimorphidae.** With a double or multiple corona of tentacles (marginal, principal, and intermediate accessory), more than one tentacle communicating with each intramesenterial chamber. Tentacles various; pedal disc present; gonads on all the septa; muscular system weak; sphincter muscle various; acontia absent. *Corallimorphus* Moseley, from the deep-sea: *Corynactis* Allman; *Capnea* Forbes; *Discosoma* Leuck.; *Aureliana* Gosse; *Rhodactis*

* Recent researches render it probable that in these cases the tentacles have dropped off.

† There is, however, variation in this character, even within the section Hexactiniae.

M.-Ed. and H.; *Phymanthus* M.-Ed. and H.; *Crambactis* Haeckel; *Cryptodendrum* Klunzinger; *Actinothrix* D. and M.; *Heterodactyla* Ehr.

Fam. 2. **Minyadidae.** Pedal disc transformed into an apparatus for floating. Tentacles in some as in the *Corallimorphidae*. *Minyas* Cuv.; *Dactylominyas* And.; *Aceronimyas* And.; *Phyllominyas* And.

Sub-tribe 2. **ACTININAE.** Tentacles arranged in cycles, only a single tentacle communicating with each intramesenterial chamber.

Fam. 3. **Antheomorphidae.** Tentacles digitate; pedal disc present; accessory tentacles absent; gonads on all the mesenteries; numerous complete mesenteries; muscular system weak; without sphincter muscle or acontia. *Antheomorphe* R. Hertwig, deep-sea form.

Fam. 4. **Actiniidae.** Tentacles digitate, in a single corona; pedal disc present; acontia absent, sphincter muscle endodermal, weak; with numerous mesenteries. *Actinia* Browne; *Anemonia* Risso; *Condylactis* D. and M.; *Actinioides* Hadd. and Shack.; *Bolocera* Gosse.

Fam. 5. **Aliciidae.** With large flat base. Lateral body-wall with simple or compound hollow processes or vesicles, mostly in vertical rows. No cinclides. Sphincter variable, diffuse, endodermal. Acontia absent. *Alicia* Johnson; *Cystiactis* M.-Edw.; *Thaumactis* Fowler; *Bunodeopsis* And.; *Phyllactis* M.-Ed. and H.

Fam. 6. **Bunodidae.** Tentacles digitate; pedal disc present; acontia absent; sphincter well developed, circumscribed, endodermal; with numerous perfect mesenteries. *Tealia* Gosse; *Leiotealia* R. Hertwig.; *Bunodes* Gosse; *Phymactis* M.-Edw. and H.; *Aulactinia* Verr.; *Anthopleura* Duch.; *Evactis* Verr.; *Thelactis* Klunzinger; *Cereactis* Andres.

Fam. 7. **Paractidae.** Tentacles digitate; pedal disc present; acontia absent; sphincter strong, mesodermal; with numerous perfect mesenteries. *Paractis* M.-Edw.; *Dysactis* M.-Edw.; *Tealidium* Hertwig: *Antholoba* Hertwig; *Ophiodiscus* Hertwig; *Paranthus* Andres; *Paractinia* And.; *Actinernus* Verr.; *Actinostola* Verr.; *Pycnanthus* McM.; *Cymbactis* McM.: *Stomphia* Gosse.

Fam. 8. **Amphianthidae.** Tentacles digitate; pedal disc present; acontia absent; mesodermal sphincter present; transverse axis of body and mouth elongated, so that the two gonidial grooves almost touch; principal mesenteries sterile; secondary mesenteries incomplete. Attached to the axial skeletons of *Gorgonidae*. *Stephanactis* Hertw.; *Amphianthus* Hertw.

Fam. 9. **Sagartidae.** With mesodermal sphincter muscle, usually with only a few complete mesenteries, with acontia.

Sub-fam. 1. **Sagartinae.** With naked ectoderm, the acontia emitted through the mouth and through cinclides. The mesenteries of the second and subsequent cycles may, in a more irregular manner, reach the oesophagus. *Sagartia* Gosse; *Cereus* Oken; *Actinoloba* Blainv.; *Adamsia* Forbes, sessile upon Gastropod shells containing a Hermit crab; *Cylista* Gosse; *Mitactis* Haddon and Duerden; *Aiptasia* Gosse; *Gephyra* V. Koch (*Morph. Jahrb.* 4) solitary or colonial, sessile on zoophytes, to which it is glued by a cuticular matter secreted by the ectoderm, with more than 24 tentacles.

Sub-fam. 2. **Chondractininae.** With thick body-wall, upper portion different in character from the lower, which is provided with a cuticle. The 12 primary mesenteries alone are complete and without gonads. Acontia emitted by the mouth, cinclides absent. *Actinauge* Verrill; *Chitonanthus* McM.; *Hormathia* Gosse; *Chitonactis* Fischer; *Chondractinia* Lütken; *Paraphellia* Haddon.

Sub-fam. 3. **Phellinae.** Sagartidae with a cuticular covering; primary mesenteries alone fertile. *Phellia* Gosse; *Octophellia* And.; *Ilyactis* And.; *Ammonactis* Verr.

Fam. 10. **Heteractidae.** Tentacles clavate, knobbed. *Eloactis* And.; *Rhopalactis* And.; *Ragactis* And.; *Heteractis* M.-Edw. and H.; *Stauractis* And.

Fam. 11. **Sideractidae.** With sixteen pairs of perfect mesenteries, and three series of non-retractile tentacles, of which the innermost contains eight. *Sideractis* Dan.

Fam. 12. **Madoniactidae.** With a few principal mesenteries, acontia, and a prominent endodermal circular muscular system. Intermediate between *Bunodidae* and *Sagartidae*. *Madoniactis* Dan.

Fam. 13. **Andvakiadae.** Elongated, without any real pedal disc, seated loose in the sand, the greater part of the body encrusted. The uppermost bare part of the body, the oral disc, and the tentacles completely retractile. Few mesenteries. Endodermal circular muscular system. Near *Sagartidae*.

Fam. 14. **Liponemidae.*** Tentacles reduced to short tubes or stomidia: with numerous perfect mesenteries. Deep-sea forms. *Polysiphonia* R. Hert., tentacles short tubes with large terminal opening, allied to *Paractidae*; *Polystomidium* R. Hert., with stomidia, allied to *Actinidae*; *Liponema* R. Hertwig, stomidia very numerous; *Aulorchis* R. Hert., with gonads modified into a tube opening through the mouth.

Fam. 15. **Sarcophianthidae.** *Sarcophianthus* Lesson.

Fam. 16. **Thalassianthidae.** Tentacles replaced by bushy excrescences of the disc. *Thalassianthus* Leuck.; *Actineria* Blain.; *Megalactis* Ehrb.; *Actinodendron* Blain.

Fam. 17. **Ilyanthidae.** With single corona of tentacles; pedal disc absent, gonidial grooves and sphincter obscure. *Ilyanthus* Forbes; *Mesacmaea* And.; *Halcampa* Gosse; *Halcampella* And.

Fam. 18. **Siphonactinidae.** Like the last, but with a gonidial groove, the lips of which project beyond the mouth (*conchula*). *Peachia* Gosse; *Siphonactinia* K. and D.; *Philomedusa* Müller; *Actinopsis* K. and D.

Danielssen's genera *Aegir* and *Fraja*† do not exist. They were founded upon mutilated specimens.

Section 2. **Paractinia.**

With paired mesenteries. There are two pairs of directives, and the longitudinal muscles are arranged as in *Hexactinia*. Number of mesenteries has no relation to the number 6. With two gonidial grooves and two oesophageal lappets.

Fam. 1. **Sicyonidae.** Sessile, with tetramerous arrangement of the mesenteries, sphincter muscle mesodermal, tentacles as short knob-like stumps. Possibly related to the *Tetracorallia*. *Sicyonis* R. Hert. Deep-sea form.

Fam. 2. **Polyopidae.** Without pedal disc, tentacles transformed into stomidia.‡ *Polyopis* R. Hert. Deep-sea form, probably tetramerous.

* *Vide* note on p. 186.

† D. E. Danielssen, "Actinida," *The Norwegian North Atlantic Expedition* 1876-8. 1890.

‡ See note on p. 186.

Section 3. **Protactiniae (Protantheae).**

With 12 primary and 1 or 2 pairs of secondary mesenteries, which are dorsal rather than ventral. Body-wall and oesophagus with ectodermal ganglionic and muscular layers.

Fam. **Gonactinidae.** *Scytophorus* R. H.; *Gonactinia* Sars, with the power of transverse fission; *Oractis* McM.; *Protanthea* Calg.

Section 4. **Edwardsiae.**

Without pedal disc, with 8 mesenteries, including two pairs of directive mesenteries and 4 unpaired mesenteries (Fig. 150). All mesenteries with gonads. Tentacles usually more numerous than the mesenteries. The muscular (longitudinal) faces of the 4 unpaired mesenteries are all turned the same way. Live in sand. *Edwardsia* Quatrefages.

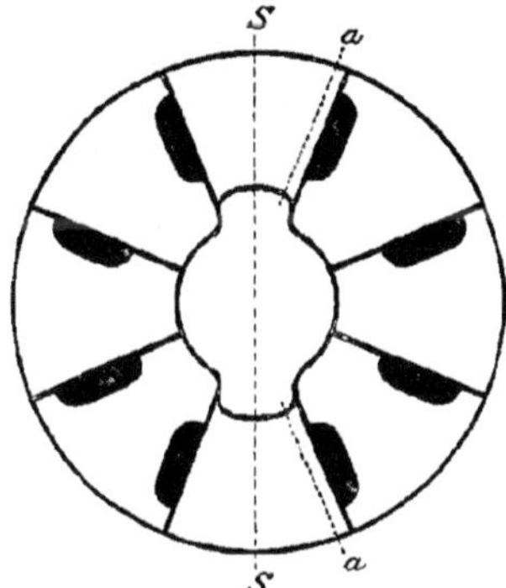

FIG. 150.—Diagram of the arrangement of the muscles and mesenteries of *Edwardsia* (from Chun after Boveri). *S*, *S* sagittal plane; *a*, *a* gonidial grooves.

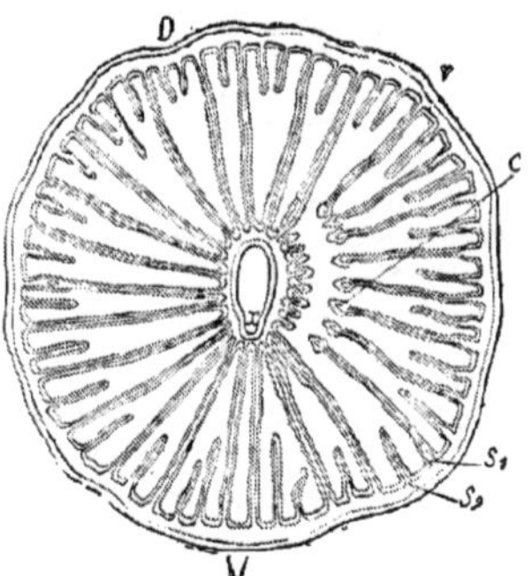

FIG. 151.—Diagram of a transverse section through a young *Zoanthus*. The section is a little oblique. *D* dorsal, *V* ventral side; *r* gonidial groove; *c* mesenteric filaments; s^1 macro-, s^2 micro-mesenteries (from Perrier).

Section 5. **Zoantheae.**

With numerous mesenteries of two kinds, (*a*) imperfect sterile micro-mesenteries, (*b*) larger perfect macro-mesenteries with gonads and filaments: the two kinds (Fig. 151) usually placed alternately, so that each pair is composed of a larger and a smaller mesentery; two pairs of directive mesenteries, one pair consisting of macro-, the other pair (dorsal) of micro-mesenteries; one gonidial groove ventral (near large directives). Usually colonial; wall of body usually traversed by ectodermal canals, and encrusted with foreign bodies which may even be embedded in the wall. The longitudinal muscular faces of the mesenteries of each pair are arranged as in *Hexactiniae*. New mesenteries are formed in the inter-mesenterial space on either side of the ventral directives. The colonial forms arise either from a branched stolon, or from a broad basal plate containing anastomosing endodermal canals.

Fam. **Zoanthidae.** *Zoanthus* Cuvier; *Gemmaria* D. and M.; *Isaurus* Gray; *Palythoa* Lamx.; *Sphenopus* Stenstr., with rounded aboral end, embedded in sand; *Epizoanthus* Gray, often on hermit-crab gastropod shells; *Parazoanthus* Haddon and Shack.

Section 6. **Ceriantheae.**

With numerous unpaired mesenteries (Fig. 152), and a single gonidial groove (ventral). The two mesenteries attached to the gonidial groove (directive) are very small; the mesentery on either side of these is large, and reaches to the aboral end; the remaining mesenteries diminish in size towards the dorsal region where new mesenteries are added (not ventrally as in *Zoantheae*).

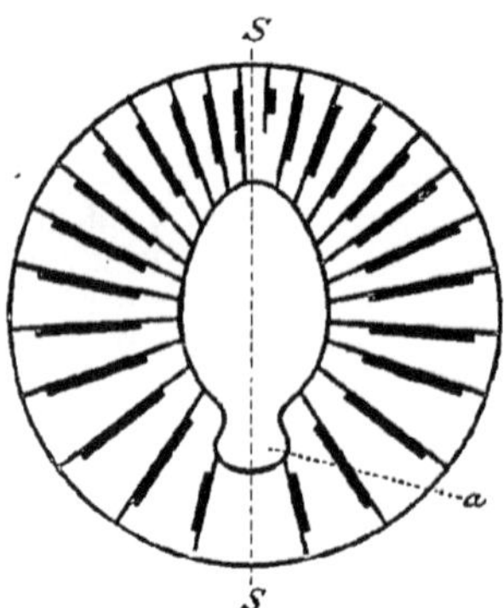

FIG. 152.—Diagram of the arrangement of the mesenteries of *Cerianthus*. *a* gonidial groove.

Fam. **Cerianthidae.** With a double corona of tentacles, marginal principal and circumoral accessory; aboral end rounded; without sphincter. With an aboral pore and a sheath of mud, sand grains, and nematocysts, in which the aboral end of the body lies as in a case. *Cerianthus* D. Chiaje; *Bathyanthus* Moseley; *Arachnactis* Sars, pelagic (possibly a larval form).

Sub-order 2. **ANTIPATHARIA.***

Colonial Zoantharia with a tendency to hexamery; with a usually branched, axial, hollow, horny skeletal rod contained in an epithelial sheath.

The coenenchyma consists of the fused bases of the polyps; it is always thin and without spicules. Except in one genus there is a central horny rod, round which the coenenchyma is disposed. The origin of the epithelial sheath which surrounds the rod is unknown. The polyps have generally six tentacles and six primary mesenteries, four of which are directives, and the other two transverse. The transverse mesenteries bear the gonads. The 4 or 6 secondary mesenteries fade away in the lower part of the polyp. The polyps are always much elongated in the transverse axis (*i.e.* at right angles to the elongation of the mouth), and in the *Schizopathinae* the body is actually constricted into three divisions, two lateral containing the gonads and one central with the mouth. Each division has two of the tentacles. To this phenomenon the name pseudo-dimorphism rather than dimorphism (gonozooids and gastrozooids) should be applied.

Fam. 1. **Savagliidae.** With 24 mesenteries and tentacles. The colonies are without an axial rod, but form a sheath round Gorgonid skeletons. Polyps with typical Actinian structure. Probably Actiniarians. *Savaglia* Nardo (*Gerardia* L. Duth.).

Fam. 2. **Antipathidae.** With 6 tentacles, 6 primary mesenteries, and with or without 4 or 6 secondary mesenteries. The two lateral primary mesenteries bear the gonads. The axial skeleton is spiny and has a central canal.

Sub-fam. 1. **Antipathinae.** Polyps not pseudo-dimorphic, each with 6 tentacles; transverse axis of the polyp more elongated than the axis (sagittal) which is marked by the long axis of the mouth. *Cirripathes* Blainv.; *Stichopathes* Brook; *Leiopathes* Gray; *Antipathes* Pall.; *Antipathella* Brook; *Aphanipathes* Brook; *Tylopathes* Brook; *Pteropathes* Brook; *Parantipathes* Brook.

* G. Brook, "Report on Antipatharia," *Challenger Reports*, Pt. 80, 1889.

Sub-fam. 2. **Schizopathinae.** Polyps exhibit pseudo-dimorphism; each with two tentacles (*i.e.* 6 to each polyp). *Schizopathes* Brook; *Bathypathes* Brook; *Taxipathes* Brook; *Cladopathes* Brook.

Fam. 3. **Dendrobrachiidae.** With branched retractile tentacles. Axial rod without central canal. Anatomy not known, possibly Alcyonarian. *Dendrobrachia* Brook.

Sub-order 3. MADREPORARIA.*

Colonial, rarely solitary, zoantharian polyps, which secrete by the ectoderm a continuous and complicated calcareous corallum.

This old and apparently well-established division of the *Zoantharia* is still, from a structural point of view, very imperfectly known. The less important features of structure, viz. the arrangement of the hard parts—the corallum—has been minutely examined, but the soft parts have been neglected, and had it not been for the investigations of the Oxford School, and of v. Koch, we should still know very little more about them than we do of the soft parts of extinct forms. These investigations which will, we may hope, soon lead to the possibility of a satisfactory classification of the sub-order, have established the following important points: (1) the complete disestablishment of the *Tabulate* division, which, on examination of the soft parts, has been found to comprise forms belonging to *Hydromedusae*, *Alcyonaria*, as well as to *Madreporaria*; (2) that the corallum is entirely a product of the epithelial ectoderm, and lies wholly outside the animal; (3) that the structure of the polyps varies in the different groups, though the Hexactinian type seems on the whole to prevail. The most important deviations from that type, so far known, are presented by those forms, in which the directive mesenteries (if indeed they can be called so) present the same arrangement of their muscles as do the other pairs (*Lophohelia*, *Mussa*, *Euphyllia*, *Heteropsammia*), and the number of mesenteries and tentacles is not a multiple of six.

Acontia appear to be absent; but peristomial cinclides are said to be present, allowing of the emission of the much convoluted mesenteric filaments. The colonies are generally dioecious; and the gonads are borne upon all or certain of the mesenteries. Pores at the apex of the tentacles seem to be absent.

Asexual reproduction by budding, or by fission, is always present. In *Fungia* and its allies there is formed from the egg a fixed nurse-stock, which has the property of nipping off its disc-shaped apical portion, and of forming in its place a new disc. The fixed nurse-stock is a typical polyp with theca and septa, and at first it does not terminate in an expanded disc. When the walls of the theca, which are at first vertical, have widened out into a disc, the lower part of it forms a stalk. It is this stalk which is left after the fission, and which produces a new disc. The new disc is not a bud, but is a product of the growth of the structures already existing in the base of its predecessor (Lister, *Q. J. M. S.*, 29). It is not certain whether the *Fungia* stock increases by budding.

* Martin Duncan, "A Revision of the Families and Genera of the Sclerodermic Zoantharia," *Journal of the Linnean Society*, vol. 18, 1885. H. N. Moseley, "Report on the Corals," *Challenger Reports*, 7, 1881. G. C. Bourne, "Anatomy of Mussa and Euphyllia," etc., *Q. J. M. S.*, 27, 1887, p. 21. G. H. Fowler, "Anatomy of the Madreporaria," I.–V., *Q. J. M. S.*, vol. 25 to vol. 28. V. Koch, "Üb. d. Verhältniss v. Skelet u. Weichtheile b. d. Madreporen," *Morph. Jahrb.*, 12, 1886. M. M. Ogilvie, "Microscopic and systematic study of Madreporarian types of Corals," *Phil. Trans.*, 187, 1896.

In the colonial forms, fission by division of the polyps into two may occur (*Oculinae*, *Astraea*), or the division may be confined to the oral disc, so that a complex polyp is formed, with several mouths and oesophaguses opening into a common coelenteron (*Maeandrina*, Fig. 144). In many cases, perhaps the majority, the colony is increased by budding from the extra-thecal coenosark.

The hard structures or corallum follows more or less closely the shape of the polyp, and were at one time thought to be actually contained in the tissue of the polyp. They consist of a cup or theca, from which the polyp projects and into which it can shrink. and in some forms of a connecting substance, which may or may not be porous, connecting the cups—this is the coenenchyma. The cup has, projecting inwards from its walls, a number of radiately and vertically arranged calcareous plates, which suggest calcified mesenteries. These are the *septa*: they are not mesenteries, but occur between mesenteries. The cup presents a *basal plate* below, from which rise the *walls*.

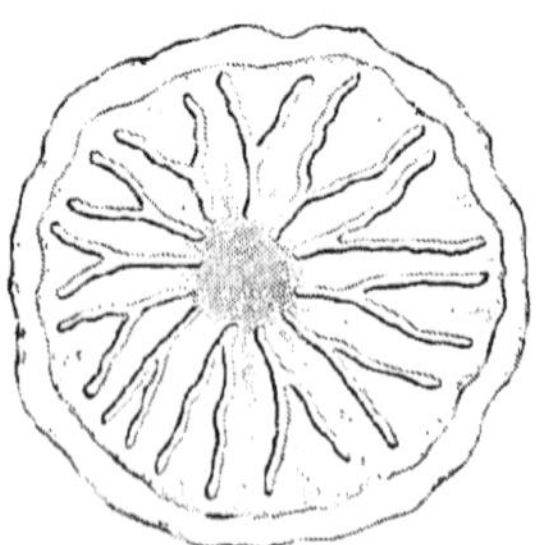

Fig. 153.—Basal plate of a larva of *Astroides calycularis*, soon after attachment. With 12 radial ridges (after Lacaze Duthiers, from Balfour).

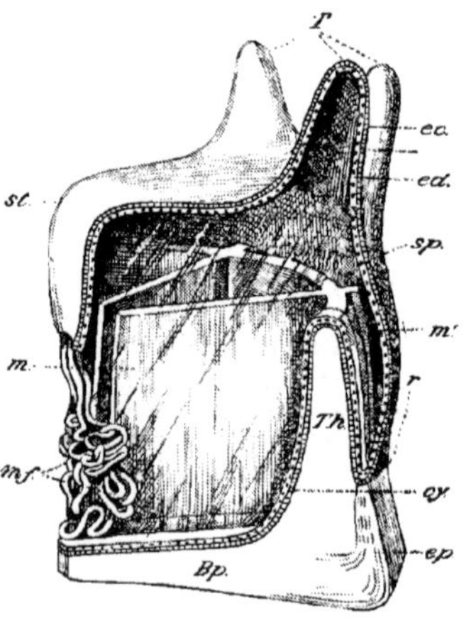

Fig. 154.—Diagram to exhibit the relations of the polyp to the corallum. *T* tentacles; *ec* ectoderm; *ed* endoderm; *st* oesophagus; *mf* mesenterial filaments; *r* extra-thecal coelenteron; *m* mesentery; *m'* extra-thecal portion of mesentery; *Bp* basal plate; *ep* epitheca; *Th* theca; *cy* calicoblasts (after G. C. Bourne).

The hard structures or corallum are secreted by, and on the outer side of, the ectoderm.

The first part of the skeleton to appear (*Astroides calycularis*) is an annular basal plate (Fig. 153), incomplete at first in its central part, between the basal ectoderm and the surface to which the young polyp is attached. Next twelve radially arranged folds of the basal body-wall rise up and project into the enteron. The ectoderm of these folds secrete calcareous deposits, which constitute the first trace of the septa (Fig. 153). The folds of the septa differ from those of the mesenteries (between which they are placed) in being folds of the whole body-wall, and not of the endoderm alone.

The septa, therefore, arise as rod-shaped structures in continuity with the basal plate. They increase in thickness and height as the polyp grows, and their outer ends, which do not reach to the body-wall of the polyp, become forked. The theca is formed by the junction (complete in the *Aporosa*, incomplete in the *Porosa*) of these forked extremities of the septa.

From this account it is obvious that the theca must not only project into the cavity of the polyp in exactly the same way as do the septa, and divide it into an extra- and intra-thecal portion (Fig. 154), but also must divide the mesenteries in a similar manner. This is actually found to occur in adult polyps, in which the body-wall projects over the lip of the calycle and lies on the outer side of the theca (Fig. 154, *r*).

It appears that when the polyps of a colony are connected by soft tissues (coenosark), the connection is effected by this extra-thecal portion of the polyp, and that the so-called coenenchyme or hard matter filling up the valleys between adjacent polyps is secreted by the ectoderm on the lower side of this connecting coenosark. The coenosark is generally broken up into canals which, in the *Porosa*, communicate with the coelenteron of the polyps by apertures left in the theca, and may in some cases, at any rate, be embedded in the superficial layer of the hard coenenchyma. The extra-thecal coelenteron is confined to the upper part of the thecae, and the coenosarkal continuation of it over the coenenchyme (when present) may or may not be broken up by continuations of the mesenteries. When there is no coenenchyme, and the thecae are isolated from one another except at their base, the living tissues appear to have died away round the basal parts of the thecae.

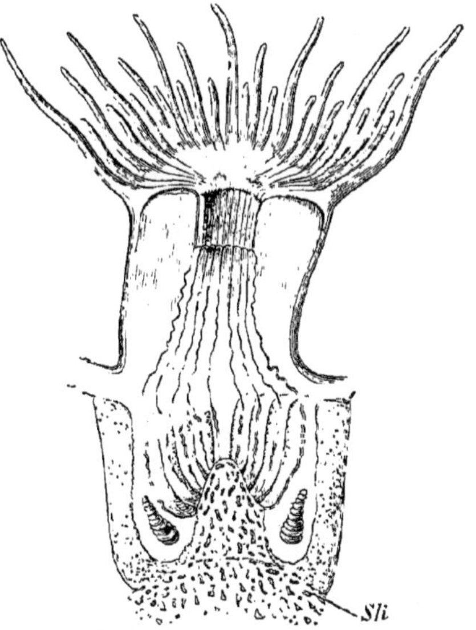

Fig. 155.—Vertical section through a polyp of *Astroides calycularis* (after Lacaze-Duthiers). The mouth-opening, oesophageal tube and mesenteries are seen; also the calcareous septa between the mesenteries, and the columella *Sk* (the line from *Sk* should be produced to the middle of the cup).

As may be gathered from the last statement, the polyps ascend as the thecae grow and forsake the lower older parts of the cup. In their ascent the ectoderm of their basal walls secretes calcareous laminae, which may either completely occlude the cup—as the *tabulae* of *Seriatopora* and *Pocillopora*, or merely stretch as imperfect plates between the septa as the so-called *dissepiments*.

Synapticula are more rod-like calcareous structures passing from septum to septum through the mesenteries. The *columella* (Fig. 155) is a central calcareous projection into the theca rising up from the basal plate, and *pali* (Fig. 156) are accessory columellae arranged in a circle round the central columella, and sometimes joined to the edges of the septa: they are sometimes looked upon as projections of the septa. The *costae* are vertical ridges along the outside of the theca: they are extra-thecal projections of the septa.

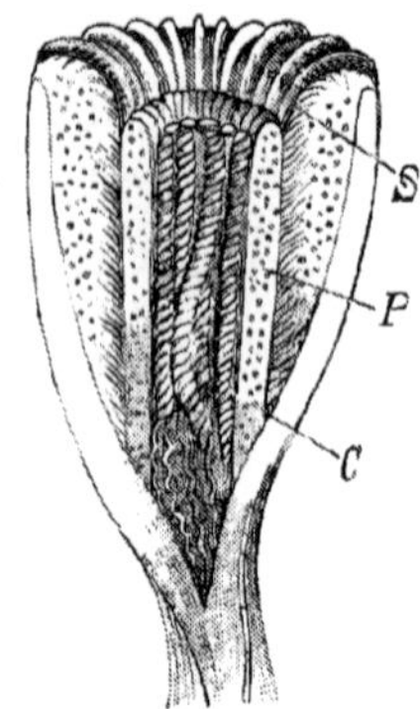

FIG. 156. — Vertical section through the cup of *Cyathina cyathus* E. and H. = *Caryophyllia cyathus* Lamk. (after M. Edwards). *S* septa; *P* pali; *C* columella.

The septa may be confined to the intramesenterial spaces, or they may occur in the intermesenterial as well. They are generally present in cycles of different sizes like the mesenteries, but the number is not always a multiple of six.

The *epitheca* when present is outside the theca: it is attached to the edge of the basal plate (Fig. 154).

Section 1. **Aporosa.**

Solitary or colonial forms. Hard parts usually solid and imperforate. Theca or wall solid, may be epithecate. Septa solid near the wall, and usually, but not invariably, solid at the further part. Interseptal loculi (*i.e.*, the chambers between the septa) open throughout, or closed more or less by endotheca in the form of dissepiments and tabulae.

One or more rows of tentacles in relation to the septa and interseptal loculi. The disc with one or more mouths; a mesentery usually in each interseptal loculus. Mesenteries usually in multiples of six.

Fam. 1. **Turbinolidae.** Corallum simple (solitary), or in colonies. Gemmation from the wall or from an expansion of the basal structures. Wall solid. Septal loculi open to the base. Endotheca rarely present.

(*a*) Corallum simple, rarely producing deciduous buds. *Smilotrochus* Ed. and H.; *Onchotrochus* Duncan; *Desmophyllum* Ehrb.; *Schizocyathus* Pourtales; *Flabellum* Lesson; *Rhizotrochus* Ed. and H.; *Thysanus* Dunc.; *Placotrochus* Ed. and H.; *Sphenotrochus* Ed. and H.; *Nototrochus* Dunc.; *Placocyathus* Ed. and H.; *Platytrochus* Ed. and H.; *Turbinolia* Ed. and H.: *Stylocyathus* d'Orb.; *Conocyathus* d'Orb.; *Bistylia* T. Woods; *Trematotrochus* T. Woods; *Trochocyathus* Ed. and H.; *Deltocyathus* Ed. and H.; *Odontocyathus* Moseley; *Caryophyllia* Lmk. (Fig. 156); *Ceratotrochus* Ed. and H.; *Discocyathus* Ed. and H.; *Brachytrochus* Duncan; *Sabinotrochus* Dunc.; *Stephanotrochus* Moseley; *Anthemiphyllia* Pourt.; *Fungiacyathus* Sars; *Guynia* Dunc.; *Duncania* Pourt.; *Haplophyllia* Pourt.

(*b*) Colonial; buds free above their origin; no exotheca uniting the corallites. *Cocnocyathus* Ed. and H.; *Gemmulatrochus* Dunc.

(*c*) Colony growing from basal expansions; exotheca absent. *Polycyathus* Dunc.

Fam. 2. **Oculinidae.** Colonial, in the form of branches, espaliers, irregular ramifications on a thick stem; or massive, or incrusting. Interseptal loculi usually open to the base, but dissepiments or tabulae sometimes occur. Walls of corallites often increasing in thickness exogenously with age, and becoming a solid mass by union with others. Solid intercalicular coenenchyma usually present. Polyps when expanded rising above the wall, or long and exsert, the mouth protruding; the tentacles 10 to 48 or more, elongated, tips usually swollen or capitate.

(*a*) Massive or incrusting colonies. Columella and pali absent, or a false columella may be present. Coenenchyma well developed between the calices. *Baryhelia*, Ed. and H.; *Neohelia* Moseley; *Diblasus* Lonsdale.

(*b*) Dendroid or bunch-shaped colonies. Corallites often coalescing; gemmation alternate. Columella absent or rudimentary. Tabulae or dissepiments present or not. *Lophohelia* Ed. and H.; *Amphihelia* Ed. and H.; *Acrohelia* Ed. and H.

(*c*) Arborescent or tufted colonies. Gemmation rarely from one side only. Columella various. *Oculina* Ed. and H.; *Cyathohelia* Ed. and H.; *Trymohelia* Ed. and H.; *Sclerohelia* Ed. and H.; *Bathelia* Moseley.

(*d*) Branched espalier-like colonies. Corallites projecting or twisted. Columella styliform. No pali. Coenenchyma well developed. *Prohelia* E. de Fro. Jurassic and cretaceous.

(*c*) Arborescent, palmate, or incrusting colonies. Septa few, unequal. Columella styliform. Costae short or absent. *Stylophora* Ed. and H.; *Madracis* From.

Fam. 3. **Pocilloporidae.** Colonial, with tabulae; septa small; columella well or ill developed. Inter-corallite structure coenenchymal and solid. Polyps with disc, tentacles, and one pair of long mesenterial filaments. *Pocillopora* Lamarck; *Seriatopora* Lamk.

Fam. 4. **Astraeidae.** Solitary or colonial, rarely reproducing by deciduous buds. Colonies increase by gemmation and fissiparous division. Interseptal loculi with dissepimental endotheca, rarely tabulae. Soft parts resembling those of Turbinolidae; the long serial calices have several mouths in the limited disc which is surrounded by tentacles. Corallites may unite by their walls, but true intermural solid coenenchyma is rarely seen. Includes the so-called brain corals.

Sub-fam. 1. **Astraeidae simplices.** Simple solitary forms. Propagation rarely by deciduous buds. Pali present or absent. Endotheca always present, but variable in amount. *Lophosmilia* Ed. and H.; *Sphenophyllia* Moseley; *Parasmilia* Ed. and H.; *Dasmosmilia* P.; *Lithophyllia* Ed. and H.; *Asterosmilia* Duncan.

Sub-fam. 2. **Astraeidae reptantes.** Colonies composed of short corallites, which arise by gemmation from stolons or basal expansions. *Cylicia* Ed. and H.; *Astrangia* Ed. and H.; *Ulangia* Ed. and H.; *Colangia* Pourt.

Sub-fam. 3. **Astraeidae gemmantes.** Colonies increasing by gemmation from the wall below the calicular margin. Endotheca dissepimental. *Cladocora* Ed. and H.; *Pourtalosmilia* Duncan.

Sub-fam. 4. **Astraeidae caespitosae.** Corallites isolated terminally, being free at their sides, springing from a common parent; increasing by fissiparity, separation occurring rapidly or serial growth persisting. Gemmation rare. *Eusmilia* Ed. and H.; *Solenosmilia* Dunc.; *Dasyphyllia* Ed. and H.; *Dendrocora* Dunc; *Trachyphyllia* Ed. and H.; *Mussa* Oken.

Sub-fam. 5. **Astraeidae confluentes.** Increase by fissiparity, with excess of serial growth. Gemmation may occur. Corallites united by their walls, costae, or by intermediate tissue, or free. *Euphyllia* Ed. and H.; *Dendrogyra* Ehrbg.; *Pectinia* Oken; *Diploria* Ed. and H.; *Manicina* Ehrb.; *Maeandrina* Ed. and H.; *Coeloria* Ed. and H.; *Leptoria* Ed. and H.; *Symphyllia* Ed. and H.; *Mycetophyllia* Ed. and H.; *Ulophyllia* Ed. and H.; *Tridacophyllia* Blainv.; *Colpophyllia* Ed. and H.; *Scapophyllia* Ed. and H.; *Plerogyra* Ed. and H.; *Physogyra* Quelch; *Hydnophora* Ed. and H.

Sub-fam. 6. **Astraeidae agglomeratae fissiparantes.** Colonies massive or incrusting. Corallites increasing by fissiparity, and sometimes also by gemmation; united by costae or coenenchyma; not forming long series. *Dichocoenia* Ed. and H.; *Favia* Oken; *Goniastraea* Ed. and H.

Sub-fam. 7. **Astraeidae agglomeratae gemmantes.** Massive and foliaceous colonies. Colonies increasing by gemmation from the wall from within the calice, or from intercorallite tissue. Corallites joined by costae, exotheca, or peritheca, or fused by their walls. Endotheca vesicular, rarely tabulate. *Heliastraea* Ed. and H.; *Phymastraea* Ed. and H.; *Solenastraea* Ed. and H.; *Plesiastraea* Ed. and H.; *Echinopora* Dana; *Galaxea* Oken; *Acanthopora* Verrill; *Leptastraea* Ed. and H.; *Acanthastraea* Ed. and H.; *Astrocoenia* Ed. and H.; *Prionastraea* Ed. and H.; *Merulina* Ehrbg.; *Moseleya* Quelch.

Section 2. **Fungida.**

Solitary or colonial forms. Septa and septo-costae with synapticula, which cross the interseptal and intercostal loculi. An endotheca is present or absent. Basal structures perforate or imperforate. Soft structures with short, lobe-like, scattered, sometimes obsolete tentacles, not covered when contracted; discs not circumscribed, and in colonial forms confluent.

Fam. 1. **Plesiofungidae.** Transitional between the *Aporosa* and *Fungida*. Simple or colonial, with synapticula in the interseptal loculi, besides endothecal dissepiments. Septa solid and imperforate, occasionally perforate and trabeculate. *Epistreptophyllum* Milaschewitsch; *Siderastraea* Blainv.; *Polyaraea* Fritsch.

Fam. 2. **Fungidae.** Simple or colonial, usually depressed; septa solid or porous. With synapticula, without dissepimental endotheca; tentacles short; scattered, sometimes absent. Wall perforated and echinulate. *Fungia* Dana; *Diafungia* Duncan, both solitary. The following are colonial: *Halomitra* Dana; *Sandalolitha* Quelch; *Cryptabacia* Ed. and H.; *Herpolitha* Esch.; *Polyphyllia* Q. and G.; *Lithactinia* Lesson; *Zoopilus* Dana.

Fam. 3. **Lophoseridae.** Wall neither perforated nor echinulated. Simple forms. *Trochoseris* Ed. and H.; *Cycloseris* Ed. and H.; *Diaseris* Ed. and H.; *Bathyactis* Moseley; *Psammoseris* Ed. and H.; *Stephanoseris* Ed. and H. Colonial forms: *Cyathoseris* Ed. and H.; *Lophoseris* Ed. and H.; *Haloseris* Ed. and H.; *Tichoseris* Quelch; *Mycedium* Oken.; *Phyllastraea* Dana; *Trachypora* Verrill; *Leptoseris* Ed. and H.; *Stephanaria* Verrill; *Agaricia* Lamck.; *Plesioseris* Duncan; *Psammocora* Dana; *Pachyseris* Ed. and H.; *Coscinaraea* Ed. and H.

Fam. 4. **Anabaciadae.** Simple or colonial, septa trabeculate and fenestrated. Synapticula small. Dissepiments absent. Wall indistinct. *Anabacia* d'Orb.

Fam. 5. **Plesioporitidae.** Transitional group with regularly perforate septa. *Maandroseris* Rouss.

Section 3. **Perforata.**

Corallum entirely or almost entirely composed of porous or reticulate coenenchyma. Dissepiments and tabulae may be present or absent. Septa solid or much perforated, or represented by trabeculae only.

Fam. 1. **Eupsammidae.** Simple or colonial. Walls with costae and apertures in the intercostal spaces. Calices well developed. Increase by gemmation and fission. *Stephanophyllia* Michelin.; *Leptopenus* Moseley, deep-water, southern hemisphere; *Balanophyllia* S. Wood; *Thecopsammia* Pourt.; *Eupsammia* Ed. and H.; *Heteropsammia* Ed. and H.; *Dendrophyllia* Ed. and H.; *Pachypsammia*

Verrill; *Leptopsammia* Ed. and H.; *Endopsammia* Ed. and H.; *Astroides* Blainv. (Fig. 155); *Lobopsammia* Ed. and H.; *Rhodopsammia* Semper; *Rhizopsammia* Verr.

Fam. 2. **Madreporidae.** Colonial, arising by gemmation from the sides of the parent polyp: coenenchyma more or less abundant, spongy and reticulate, slightly or not distinct from the porous corallite-walls. *Madrepora* L.; *Turbinaria* Oken; *Astraeopora* Blainv.; *Montipora* Q. and G.; *Anacropora* Ridley.

Fam. 3. **Poritidae.** Sclerenchyma reticulate and perforate. Septa never completely lamellary. Walls very porose. Corallites increasing by gemmation, and united directly or by intervening porous sclerenchyma. *Porites* Ed. and H.; *Synaraea* Verr.; *Napopora* Quelch; *Rhodaraea* Ed. and H.; *Alveopora* Q. and G.; *Dichoraea* T. Woods.

Sub-phylum II. CTENOPHORA.*

Free-swimming, transparent pelagic coelenterata, with eight meridional rows of vibratile plates formed of fused cilia. They possess an oesophageal tube—called the stomach—lined by ectoderm, and a gastrovascular canal system. Nematocysts are almost always absent.

The fundamental form of the *Ctenophora* is a gelatinous, spherical, or ovoid body, which swims in the sea by the activity of its ciliated plates. It has two poles—the oral pole marked by the mouth, and the aboral pole marked by the sense organ. The line connecting these two poles is the main axis, and in describing the structure of the body it is important to recognise two planes which pass through this axis at right angles to one another. The mouth leads into a tube called the *stomach* (sometimes called oesophagus or stomodaeum, because it is lined by ectoderm), and the stomach opens into the central part of the gastrovascular apparatus called the *funnel* (infundibulum). The stomach is furnished with two hepatic bands. Both stomach and funnel are flattened sacs, and both lie in the main axis—the funnel of course above the stomach—but with their long diameters

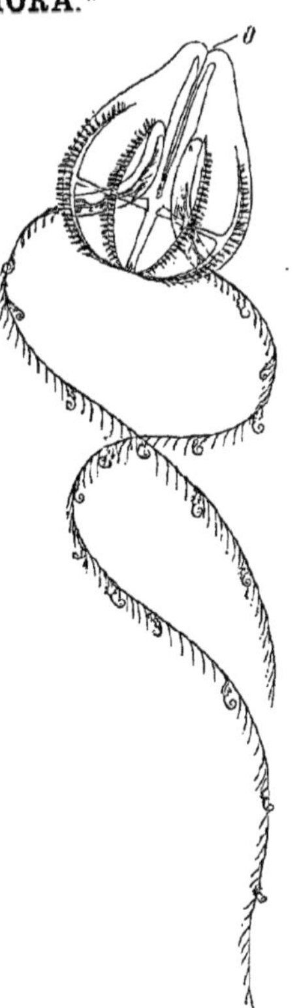

FIG. 157. — *Hormiphora* (*Cydippe*) *plumosa* (after Chun). *O* mouth.

* C. Chun, "Die Ctenophoren des Golfes von Neapel," *Fauna und Flora des Golfes von Neapel.* 1880.

in different planes. The long diameter of the stomach lies in one of the two planes mentioned above, while that of the funnel lies in the other plane, which is at right angles to the first. These two planes are called the **stomach-plane*** (Fig. 158, *M*, *M*.) and **funnel-plane** (*T*, *T*.) respectively.

The stomach-plane divides the body into a right and left half; but it is impossible to speak of the two parts of the body marked off by the funnel-plane as dorsal and ventral, or as anterior and posterior, as is done by some authors, because these two parts are identical and

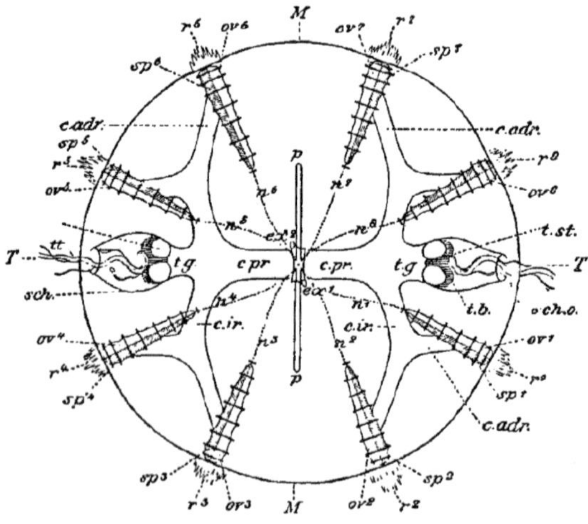

FIG. 158.—Diagram of a *Cydippe* seen from the aboral pole (after Chun). *M-M* stomach (sagittal) plane; *T-T* funnel (transverse) plane; r^1 to r^8 the eight rows of vibratile plates; *p* polar plates; n^1 to n^8 the eight ciliated grooves; *t.b* base of tentacle; *t.st* stalk of tentacle; *tt* branch of tentacle; *sch* sheath of tentacle; *sch.o* opening of tentacle-sheath; *c.pr* perradial vessel; *c.ir* interradial vessel; *c.adr* adradial vessel; sp^1 to sp^8 the sperm producing, ov^1 to ov^8 the ova-producing sides of the eight meridional vessels; *t.g* tentacle vessel; ex^1 and ex^2 the two aboral openings of the gastrovascular system.

not distinguishable from one another by any differential character. The tentacles when present are two in number: they arise from the sides of the body in the funnel-plane (Fig. 158, *T*, *T*). Further, we may speak of two transverse axes—a stomachal axis passing through the long diameter of the stomach, and a funnel axis passing through the long diameter of the funnel.

The body (Fig. 157) carries eight meridional rows of vibratile

* The plane of the stomach is sometimes called the sagittal plane, and that of the funnel the transverse plane.

plates—the ribs. These rows begin close to the aboral pole, and pass in the meridians of the animal towards the oral pole. Four of these rows lie in one half of the body, and four in the other. Further it is to be noted, that of these eight rows we may distinguish those which lie on each side of the tentacle (or funnel-plane)—these are the sub-tentacular rows*—and those which lie on each side of the stomach-plane, which we may call the sub-stomachal† rows (Fig. 158). There are therefore four sub-tentacular rows and four sub-stomachal rows of meridional plates.

The central nervous system and sense organ are placed at the aboral pole (Fig. 159). It has the form of a flat depression formed of ciliated sensory ectoderm, and covered over by a bell-shaped structure (the *bell*) formed of fused cilia (*gl*). Some of the cilia of the sensory area are very long and fused together to form four large triangular plates. These are the *springs* (*f*). Their tips are attached to, and carry a small mass of otoliths (*ot*), which is placed over the centre of the sensory area. The sensory plate is drawn out in the stomach plane into two lobes—the *polar plates* (Fig. 158, *p*). Beginning at the base of each of the four springs, or otolith-bearers, is a ciliated groove (*pl*), which, passing outward through a hole in the bell-like cover, divides into two grooves (n^{1-b}). These are continuous with the aboral ends of two rows of vibratile plates, and are sometimes called nerves because they seem to transmit any movement of the otolith-bearer to the row of vibratile plates, and so set the latter in motion. The movement of the vibratile plates begins at the aboral pole and passes oralwards, each plate successively bending energetically towards the aboral pole, and then slowly regaining its original position. The movement of the animal is thus with the oral end forwards.

The vibratile plates have the appearance of consisting of long cilia fused together at their bases. They arise from specially long ectoderm cells.

The gastrovascular apparatus (Fig. 158) consists of a central space, the funnel, which gives off two vessels—the *perradial* vessels (*c.pr*); these pass outwards in the funnel-plane in opposite directions and divide dichotomously into the *interradial* vessels (*c.ir*), of which there are four. These again divide, and give rise to eight *adradial* vessels (*c.adr*), which enter the *meridional* vessels. The meridional vessels underlie the rows of vibratile plates and end blindly above

* Sometimes called sub-transversal.

† Sometimes called sub-sagittal and sub-ventral.

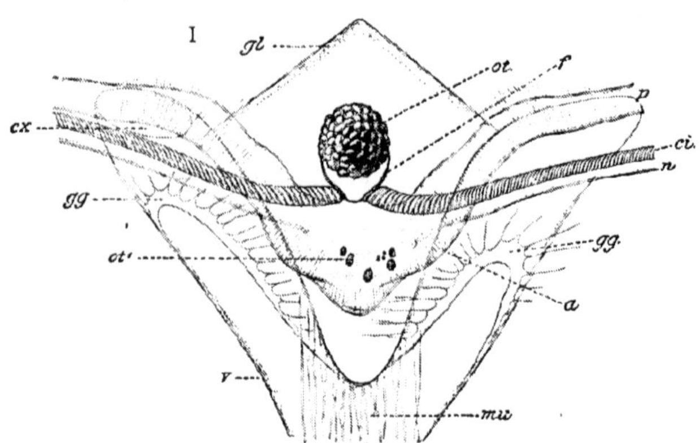

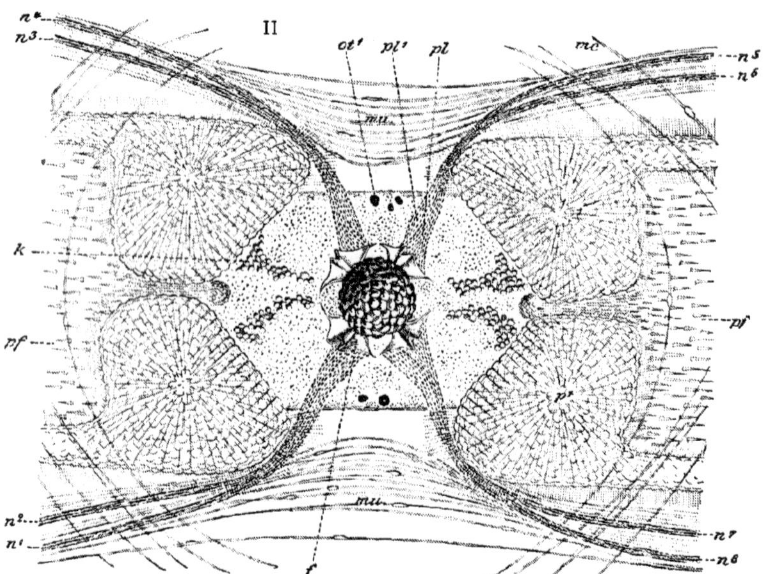

FIG. 159.—Sense organs and adjacent organs, I of *Cestus veneris* in side view (from the stomach plane) × 100, II of *Eucharis multicornis* from above × 120 (after Chun). *a* sense organ; *gl* bell; *f* the four springs; n^1 to n^8 the eight ciliated grooves with their cilia *ci*, and their plate-like expansions at the base of the springs *pl* and *pl'*; *k* granules; *ot* otoliths; ot^1 otoliths in process of formation; *p* edge of polar plate; *pf* middle part of polar plate (polar field); *mu* (in II) muscular fibres running beneath the ciliated grooves; *mc* circular muscular fibres; *mu* (in I) longitudinal muscles of the funnel-vessels; *v* funnel-vessel; *gg* branching of a funnel-vessel into the ampullae; *ex* aboral excretion pores.

and below. The perradial vessels give rise, close to their origin, to the *paragastric* canals, which run oralwards on each side of the stomach (Fig. 157). They end blindly, and are absent only in one genus (*Euchlora*). The tentacular vessels are direct continuations of the perradial vessels (Fig. 158, *tg*). The funnel is continued upwards as the funnel-vessel, which divides beneath the sense-organ into two limbs, each of which again divides into two ampullae, one situated in each quadrant of the body. Two of these ampullae (and two which lie in diagonally opposite quadrants) open to the exterior by small pores placed just outside the sensory plate (Fig. 159 I, *ex*).

There are two extensile tentacles (Fig. 157) in the funnel-plane. They consist of a stout base contained in a depression of the body-wall, which constitutes the tentacular sheath and into which the tentacle can be withdrawn. The tentacle vessel is not prolonged along the tentacle, but ends in two ampullae at its base. The tentacles carry a row of branches which are provided with the peculiar adhesive cells (Fig. 161).

The muscular tissue is feebly developed. It has the form of fibres, lying in the jelly and branched at both ends.

There is a sub-epithelial nervous plexus with scattered ganglion cells. It extends on to the stomach, but no connection has been observed between it and the sense-organ and ciliated grooves and ribs.

The description given above applies to one group of the *Ctenophora*, the *Cydippidae*, the structure of which may be taken as typical. There are, besides, the *Cestidae*, the *Lobatae*, and the *Beroidae*.

The *Cydippidae* include, besides the spherical form just described, forms in which the main axis is elongated (cylindrical *Pleurobrachiadae*), and forms in which the body is compressed in the stomach-plane, *i.e.*, the stomach-axis is much reduced (*Euchlora*, *Callianira*). In *Callianira* (Fig. 165) there are two wing-like processes of the aboral end of the body; they lie in the funnel-plane, and the meridional vessels are prolonged into them. In *Euchlora* the sub-tentacular ribs are longer than the sub-stomachal.

The larvae of the *Cestidae* and *Lobatae* closely resemble the above described typical form (particularly the *Mertensid* variety of it) and acquire the adult condition by a complicated metamorphosis (see below).

The *Cestidae* are ribbon-shaped (Fig. 160), and the body is enormously elongated in the stomach-plane and much compressed in the funnel-plane (*cf.* Fig. 158); *i.e.*, the stomach-axis of the body is very long, and the funnel-axis very short. Further, the sub-tentacular ribs (r^1, r^4, r^5, r^8) are very short, and reduced to a few plates at the aboral pole, while the sub-stomachal ribs (r^2, r^3, r^6, r^7) extend close together along the whole length of each side of the aboral surface of the body.

The perradial vessels are absent, as the four interradial vessels arise directly from the funnel (in correspondence with the compression of the body). The sub-stomachal meridional vessels run horizontally beneath the long ribs to the ends of the body, while the sub-tentacular (g^1, g^4, g^5, g^8) pass along the middle of each side of the body, also to the extremity, where they are connected with the sub-stomachal vessels (r^2), and with the paragastric vessels which are continued along the flattened oral surface of the body. In the *Cestidae* generative cells are only produced in the sub-stomachal meridional vessels.

The tentacular apparatus is hidden in a sheath. There is no projecting tentacle, but a very large tentacle base which is in the sheath. Each tentacle-sheath is continued right and left as a furrow, which extends along its own side of the oral edge of the body to the extremity. The numerous lateral fibres which arise from the tentacle base lie in these tentacular furrows and hang down from them all along like a fringe (Fig. 160), being held in the tentacular furrows by hooks.

In the *Lobatae* the body is compressed in the same way, but not to the same extent as in the *Cestidae*, *i.e.* the funnel axis is shortened. There are two large buccal lobes, one on each side of the mouth in the stomach-plane. The sub-tentacular ribs are shorter than the sub-stomachal, and at their oral ends arise the four *auricles*. The auricles are processes of the body, and each carries one row of vibratile plates. The sense-organ lies in a deep pit at the aboral pole, and the ciliated furrows (the so-called nerves), which pass from it, extend all along the ribs. The interradial vessels arise directly from the funnel, and the sub-tentacular vessels anastomose with the paragastric vessels at the base of the auricle. From the point of junction there arises a vessel which forms a loop in the lobes and anastomoses with its fellow of the neighbouring quadrant. The sub-stomachal vessels also form loops and anastomose in the oral lobes.

The *Beroidae* are without any tentacular apparatus in both larvae and adults. The body is compressed in the funnel-plane, and the main axis is elongated. The mouth and stomach are enormous, and the lower third of the stomach-wall is closely beset with sabre-like

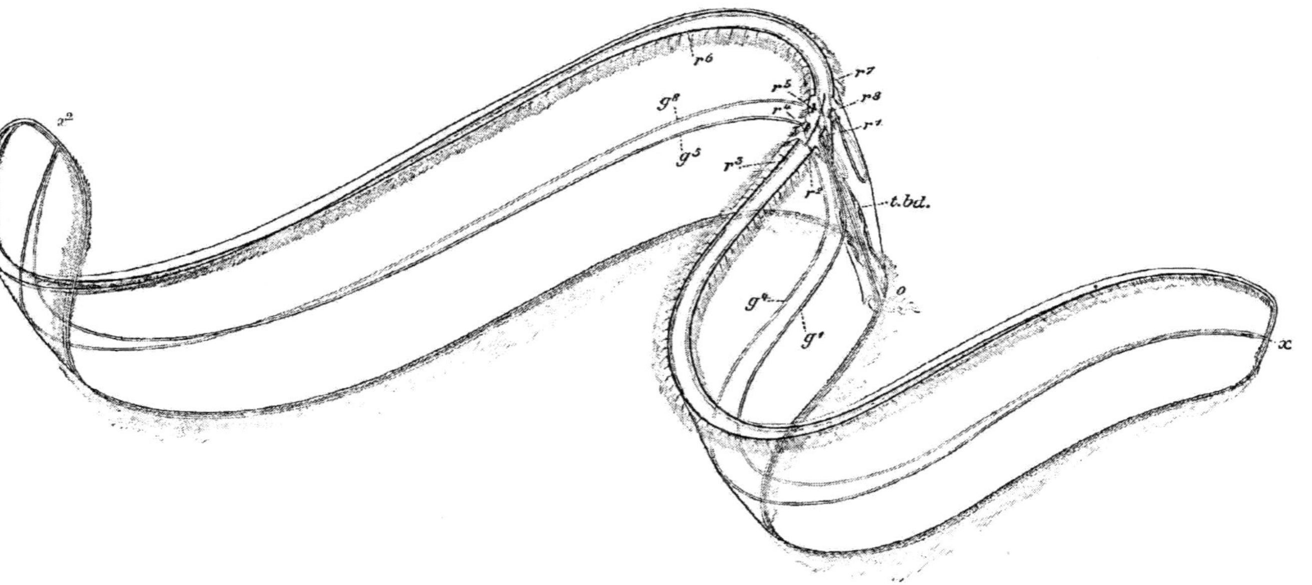

Fig. 160.—*Cestus veneris*, median-sized specimen (after Chun). r^1, r^4, r^5, r^8 the small sub-tentacular (sub-transversal) ribs; r^2, r^3, r^6, r^7 the large sub-stomachal (sub-sagittal) ribs; *o* mouth; *t.bd* tentacular band; g^1, g^4, g^5, g^8 sub-tentacular vessels which, at x^1 and x^2, communicate with the sub-stomachal vessels (which run beneath the long ribs).

hooks for the retention of food. The hepatic bands are absent. The edges of the polar plates of the sense-organ are dendritically branched. The interradial vessels and the ampullae spring directly from the funnel. The musculature is well developed, and the meridional and paragastric vessels give numerous branches which anastomose.

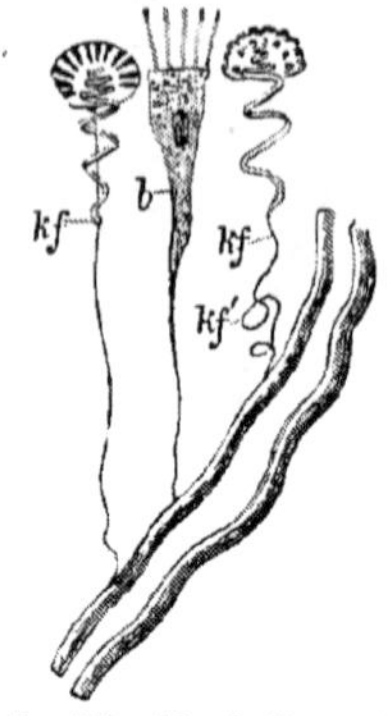

Fig. 161.—Muscle-fibres, adhesive cells (*kf*) and tactile cells (*b*) from the lateral filaments of the tentacle of *Euplocamis stationis* (after R. Hertwig). *kf'* prolongation of the contractile thread of a prehensile cell.

Thread-cells are almost universally absent; but the ectoderm of the tentacles contains peculiar "*adhesive* cells," the base of which is prolonged into a spirally coiled thread (Fig. 161), while the convex free end is soft and glutinous and becomes readily attached to any object which touches it. Thread-cells are found in the ectoderm of *Euchlora* alone, the tentacles of which are characterized by the absence of adhesive cells and the presence of amœboid ectodermal prominences.

The nervous system and sense-organs have already been described (p. 199).

The *Ctenophora* are hermaphrodite, and sexual reproduction alone is known. The generative cells arise on the walls of the meridional vessel, or of diverticula of the same. Sometimes they are localized (*Cestus*); sometimes they arise along the whole length of the canals, one side being beset by egg-follicles, the other by sperm-sacs. Ova and spermatozoa when ripe pass into the gastro-vascular space, and are ejected through the apertures of the same.

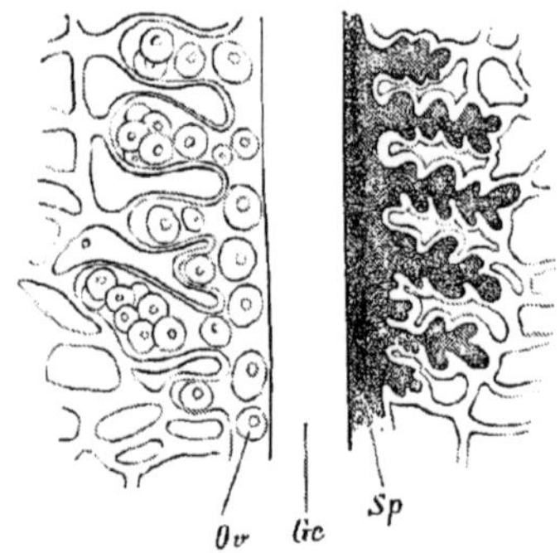

Fig. 162.—Meridional vessel (*Gc*) of *Beroe ovata* with ova (*Ov*) and spermatozoa (*Sp*) in their walls (after Will).

The early development takes place within the egg-membranes. The segmentation is complete. A cap of small ectoderm cells is soon formed (Fig. 163), which grow round the larger endoderm cells; but before this overgrowth is completed a number of cells are separated from the lower ends of the large endoderm cells, and constitute an *embryonic mesoderm* (Fig. 163, *5*, *Ms*) (Metschnikoff). The mesoderm gives rise to the protoplasmic network (wandering cells) of the jelly. The mouth is a new formation, and the stomodaeum (stomach) arises as an invagination of ectoderm at the lower pole (Fig. 163, *7*, *8*, *9*).

The larva when hatched differs more or less from the sexually mature animal in the simpler and usually more spherical form of the body, in the small size of

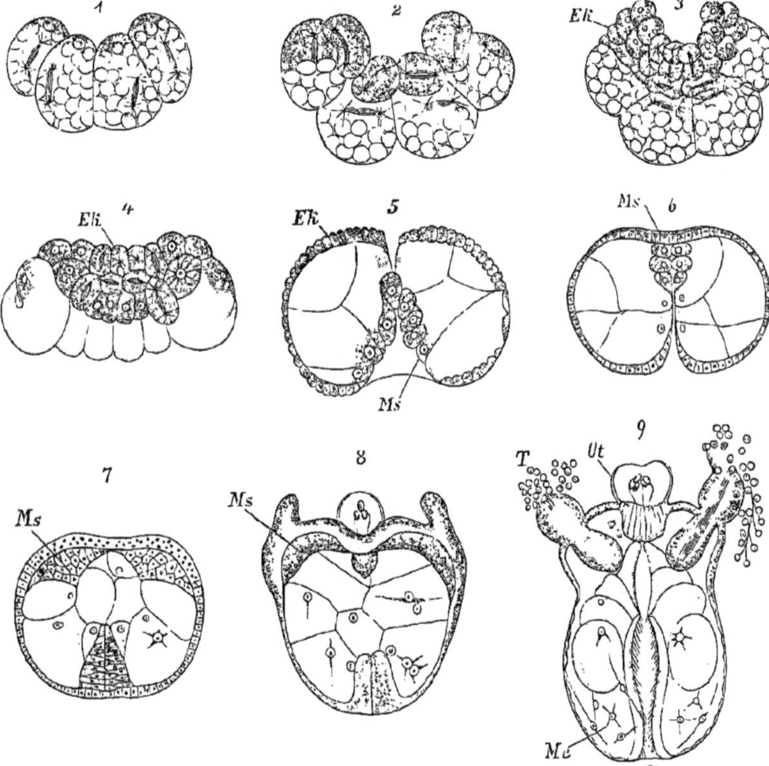

FIG. 163.—Development of *Callianira bialata* (after Metschnikoff). *1*, stage with eight; *2*, stage with sixteen blastomeres; *3*, the eight upper cells of the previous stage have divided into 48 micromeres (ectoderm), which form a cap on the 8 macromeres (endoderm); *4*, side view of an older stage; *5*, embryo in the stage of formation and invagination of mesoderm (*Ms*); *6*, more advanced stage in sagittal section; *7*, stage with developing stomodaeum; *8*, later stage with commencing formation of tentacles; *9*, ripe embryo; *T* tentacles; *Ot* sense-organ; *O* mouth; *Me* mesoderm (jelly).

the tentacles and swimming plates, and in the difference in the relative size of the oesophageal tube, infundibulum, and vascular canals. The differences, putting *Cestus* on one side, are most striking in the lobed *Ctenophora*, the embryos of which have a great similarity to the young of *Cydippe*. It is only after a longer period of larval life that the completely mature form is attained by the unequal growth of the swimming plates and their canals, the outgrowth of the tentacle-like processes, and the formation of two lobe-like projections round the mouth from those halves of the body which correspond to the longer rows of swimming plates. The *Lobatae* present a peculiar phenomenon, which has been called by Chun, **dissogony.** The Cydippe-like larva develops sexual cells on its four sub-stomach vessels, and becomes sexually mature during the hot period of the year. The eggs are fertilised and develop normally into larvae smaller than those produced from adults. After the sexual activity has continued for some days, the larva loses its generative organs, undergoes a complicated metamorphosis, and develops into the adult, with generative organs in all of its eight meridional vessels. The phenomenon of dissogony is therefore characterised by the fact that sexual maturity occurs twice in the same individual, and that the two sexual periods are separated by a sterile period in which there are no generative cells, and in which a complicated metamorphosis occurs.

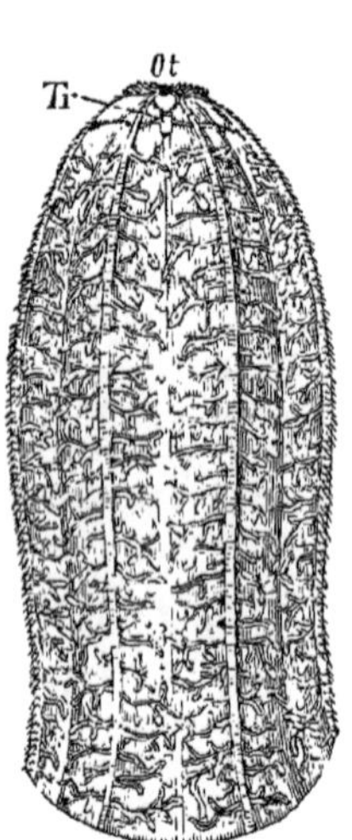

FIG. 164.—*Beroe ovatus.* *Ot* sense-organ, at its sides are the small tentacles of the polar areas; *Tr* funnel.

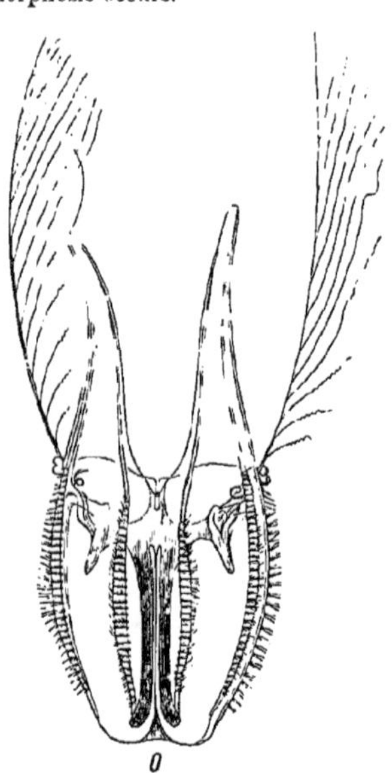

FIG. 165.—*Callianira bialata* (after Chun). *O* mouth.

The *Ctenophora* live in the warmer seas, and, under favourable conditions, often appear in great quantities at the surface. They feed on marine animals of various size, which they capture with their tentacles. Many, as the *Beroïdae*, which do not possess tentacles, are compensated for this deficiency by the possession of an unusually large mouth (Fig. 164), by means of which they are able to receive relatively large bodies, even fishes, into the wide oesophageal tube, and to digest them. Although the average size is small, some of them, as *Cestus*, *Eucharis*, reach the length of a foot.

Order 1. TENTACULATA.

Ctenophora with tentacles.

Section 1. **Cydippidae.**

Spherical or cylindrical Ctenophora, with two simple or pinnate tentacles, retractile into a sheath. The meridional and paragastric vessels end blindly.

Fam. 1. **Mertensidae.** Body compressed in the stomachal-plane; sub-tentacular ribs longer than the sub-stomachal. No wing-like appendages at the sensory pole. *Euchlora* Chun (*Haeckelia* Car. and Gerst., *Owenia* Köll., *Mertensia* Geg.); *Charistephane* Chun.

Fam. 2. **Callianiridae.** Body compressed in the stomach-plane: sub-tentacular ribs longer than the sub-stomachal. Wing-like appendages at the sensory pole. *Callianira* Péron (Fig. 165).

Fam. 3. **Pleurobrachiadae.** Body round in section. Sub-tentacular and sub-stomachal ribs equal in length. *Hormiphora* L. Ag. (*Cydippe* Geg.); *Pleurobrachia* Fleming; *Lampetia* Chun; *Euplokamis* Chun.

Section 2. **Lobatae.**

Body laterally compressed; stomach-axis longer than the funnel-axis. With two lateral lobes in the oral region, and four auricles. Lateral tentacles lie in a tentacular furrow; tentacle-sheath absent.

The auricles are provided with swimming plates, and are placed at the end of the sub-tentacular ribs. The central nervous system is sunk in a pit. The stomachal ribs are longer than the sub-tentacular. Mouth-opening wide, and extending into a buccal furrow reaching to the base of the lobes. The four interradial vessels arise direct from the funnel. The meridional vessels are continued on to the lobes in a sinuous course, and anastomose. The larvae are *Mertensia*-like forms which, in *Eucharis*, become sexually mature, and reproduce themselves.

Fam. 1. **Lesueuridae.** Lobes and lobe-windings of vessels rudimentary. Auricles long and ribbon-shaped. *Lesueuria* M.-Edw.

Fam. 2. **Bolinidae.** Lobes of medium size. Lobe-windings of vessels simple. Adradial vessels pass directly into the aboral ends of the meridional vessels. Auricles short. *Bolina* Mertens; *Bolinopsis* L. Ag.; *Hapalia* Esch.

Fam. 3. **Deïopeidae.** Body strongly compressed. Lobes of medium size. Windings of lobe-vessels more complicated than in the *Bolinidae*. Auricle short. Ribs consist of few but enormous plates. Sub-tentacular vessels possess short aboral blind processes. *Deïopea* Chun.

Fam. 4. **Eurhamphaeïdae.** Two aliform processes in the tentacular-plane at the aboral pole, on which the sub-tentacular ribs are continued. *Eurhamphaea* Gegenb.

Fam. 5. **Eucharidae.** Lobes of considerable size, with complicated vessel-windings. Auricles vermiform. Body beset with papillae. Aboral blind ends of sub-tentacular vessels long. A long main tentacular fibre as well as the lateral tentacular fibres. *Eucharis* Esch.

Fam. 6. **Mnemiidae.** Lobes very large. Origin of auricles and lobes placed almost at the same height as the funnel. Auricles long and ribbon-shaped. *Mnemia* Esch.; *Alcinoe* Rang; *Mnemiopsis* L. Ag.

Fam. 7. **Calymmidae.** Body strongly compressed. The great lobes arise almost at the height of the funnel. Ribs nearly horizontal. *Calymma* Esch.

Fam. 8. **Ocyroidae.** Lobes enormous, almost independent of the body. *Ocyroë* Rang.

Section 3. **Cestidae.**

Body much compressed in the funnel-plane, i.e., with the broad faces parallel to the stomach-plane.

The sub-tentacular ribs much shorter than the sub-stomachal, which extend all along the aboral edges of the body. The interradial vessels arise direct from the funnel. The sub-tentacular vessels run along the middle of the band to unite at the ends of the body with the long sub-stomachal and paragastric vessels. Gonads in the sub-stomachal vessels only. The larvae are *Mertensia*-like forms.

Fam. **Cestidae.** *Cestus* Les.; *Vexillum* Fol.

Order 2. Nontentaculata.

Without tentacles.

Fam. **Beroïdae.** With large mouth and stomach. Body conical or oviform, compressed in the funnel-plane. The vessels give off branches which anastomose in the jelly. *Beroë* Brown.

*Ctenoplana** Korotneff (*Z. f. w. Z.*, 43, 1886) and *Coeloplana* Kowalewsky (*Nachrichten der Liebhaber der Naturwiss.*, 1882, Russian) should probably be included amongst the Ctenophora.

Pemmatodiscus Monticelli,† a gastrula-like form living in the jelly of *Rhizostoma*, may be mentioned amongst the Coelenterates.

* A. Willey, "On Ctenoplana," *Q. J. M. S.*, vol. 39, p. 323.

† Monticelli, *Naples Mitth.*, 12, 1897.

CHAPTER V.

Phylum PLATYHELMINTHES.

Vermiform, bilateral, more or less elongated, and usually dorso-ventrally flattened animals, with an anteriorly-placed central nervous mass (cerebral ganglion), and an excretory system of ramified canals containing flame-cells. Enteron when present aproctous.

The forms included in this phylum are mostly parasitic *Entozoa*, but some live freely in water or on land. Suckers and hooks for attachment are often present, especially in the parasitic forms. The enteron is absent in one class (*Cestoda*), and when present is without an anus. Vascular system and body cavity are not found in the group, and the organs are embedded in a plasmatic mass—the parenchyma—in which muscular and connective tissue elements are differentiated. This is the so-called mesoderm. The excretory system is distributed as a system of branching and often anastomosing canaliculi throughout the parenchyma, and in the absence of a vascular system collects the excretory products in all parts of the body. Projecting from the walls of the canals at intervals are thick flame-shaped cilia, and sometimes finer cilia. The canaliculi are supposed to end blindly in hollow cells of the parenchyma, from which cells there often projects into the blind end of the canal a flame-shaped cilium. Such terminal cells are called flame-cells.

In the terms of the cell-theory this system may be described as consisting of a number of branched and hollow anastomosing cells, the spaces or vacuoles within which form a continuous system opening externally, and constituting the cavities of the excretory canaliculi. Such canals are often distinguished as intracellular, and contrasted with intercellular canals, in the walls of which cell limits can be made out. It is extremely doubtful if this distinction has the reality or morphological importance which is often attributed to it.

The external openings of the excretory canals vary much, both in number and position.

The animals are usually hermaphrodite, and the reproductive organs are complex. A vitellarium or yolk-gland is very generally present, and may be part of the ovary or distinct from it. Such glands are to be regarded as parts of the ovary, the cells of which are without the capacity of becoming ova, but store up yolk matter and are deposited round the ovum in the cocoon or egg-shell, and consumed by the embryo during its development. Asexual reproduction is very often found, and may take place at different stages of the development. The life history is in such cases very complicated. So far as is known parthenogenesis is not met with in the group, unless the germ-cells of the sporocysts and rediae of the Trematoda be regarded as parthenogenetic ova.

Class I. TURBELLARIA.*

Free-living Platyhelminthes with delicate, soft, and often leaf-shaped bodies, and with a ciliated ectoderm containing rhabdites and sometimes thread-cells, and with muscular protrusible pharynx.

The *Turbellaria* include fresh-water, marine, and terrestrial forms. They are distinguished by the possession of a ciliated ectoderm and soft delicate tissues. They usually have an oval, flattened body, and they reach only a small size. It is exceptional to find organs for adhering, viz., small hooks and suckers. The anterior end of the body is especially sensitive, and generally bears eyes and sometimes a pair of tentacles, which in a few cases (*Euryleptidae* and *Pseudoceridae*) contain prolongations of the enteron. In some fórms the dorsal surface is covered with papillae, and these also sometimes contain prolongations of the enteron. A pair of ciliated pits, not unlike those of Nemertines, are occasionally present on the front of the body, around which there may be present a ciliated marginal groove. A sucker is found in some *Polyclada* on the ventral surface.

The mouth is on the front, middle, or hinder part of the ventral surface. In the American Triclad *Phagocata* there are eight or nine pairs of mouths and pharynges in addition to the main one, and the same peculiarity has been observed as an exception in the genera *Planaria* and *Polycelis*. The generative openings are on the ventral surface behind the mouth.

The skin consists of a single layer of cells, or of a finely granular

* L. v. Graff, *Monographie der Turbellarien*, Leipzig, 1882. L. v. Graff, *Die Organisation der Turbellaria Acoela*, Leipzig, 1891. P. Hallez, "Catalogue des Turbellariés du Nord de la France," etc., *Revue Biologique du Nord de la France*, Ts. ii., iv., and v., 1892, 3. A. Lang, "Die Polycladen," *Fauna und Flora des Golfes von Neapel*, 1884. F. W. Gamble, "British Marine Turbellaria," *Q. J. M. S.*, 34, 1893, p. 433; and Article on Turbellaria, in the *Cambridge Natural History*, vol. 2, 1896.

nucleated material, which bears cilia and rests upon a stratified basement membrane (Fig. 166). It is covered externally by a homogeneous membrane bearing tactile hairs and comparable to a cuticle. Peculiar rod-like structures, called *rhabdites*, are found in the ectoderm cells and in cells lying deeper in the parenchyma, but connected with the ectoderm layer by processes. The rhabdites are homogeneous, highly refractile structures of unknown function, and can be extruded from the ectoderm. Other structures probably of a similar character, but differing slightly, are also found in the ectoderm—these are the *pseudorhabdites* and *sagittocysts*. Nematocysts are found in some members of the class (*Microstoma*, *Anonymus*, *Stylochoplana*). Various pigments are often present in the epidermis or in the parenchyma; and green vesicles containing chlorophyll and starch grains (*Vortex*), or yellow cells (*Convoluta*), are found in the parenchyma in some forms (?symbiotic *Algae*). The dermis, or outer layer of the parenchyma, lies beneath the basement membrane, and contains well-developed muscular layers (usually outer circular and inner longitudinal).

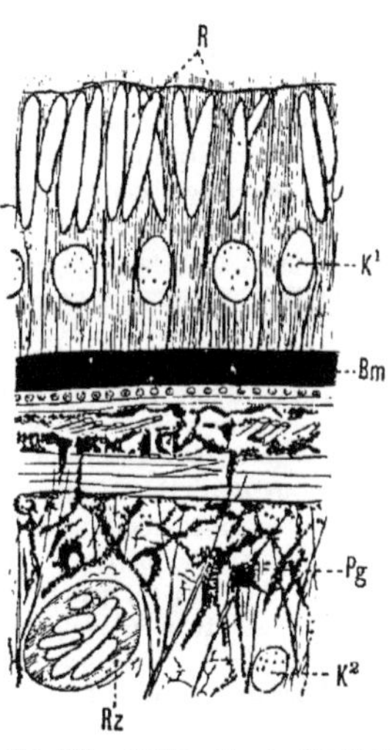

FIG. 166.—Portion of a longitudinal section of *Planaria polychroa*, showing the ectoderm and outer part of the parenchyma. *R* rhabdites in ectoderm; K^1 nuclei of ectoderm; *Bm* basement membrane; *Pg* pigmented connective tissue cell; K^2 nucleus of parenchyma cell; *Rz* deep rhabdite-forming cell (after Leuckart and Nitsche).

The structure of the **parenchyma** is difficult to make out. In the *Acoela* the whole of the tissues within the basement membrane may be described as consisting of a plasmatic mass containing nuclei, vacuoles, and fibres, both muscular and connective. The centre of this mass is of a softer consistency than the outer parts, and constitutes the solid digestive endoderm into which the food passes to be digested. The outer part of the parenchyma, which is not at all marked off from the central, contains dermal muscles externally, networks of connective tissue fibres, and muscular fibres running dorso-ventrally through the body. Further, the central nervous mass and the nerves and the gonads are all parts of this parenchyma, differentiated from it indeed, but not in any sense marked off from it (except in the case of the ripe generative cells), and passing perfectly gradually into it. Cell limits, save for the ripening generative cells, and possibly here and there a wandering cell, are entirely absent from it. It should also be mentioned that spaces or vacuoles filled with fluid are present in all parts of the parenchyma. The parenchyma of the other *Turbellaria*, generally speaking, resembles that of the *Acoela* in structure, but differs in the fact that the organs—enteron, cerebral ganglion, gonads—are more completely delimited from it.

It cannot be said that there is a perivisceral cavity in the *Turbellaria*. There aré vacuoles in the parenchyma, and these, in some *Rhabdocoelida*, are so much developed round the enteron as to suggest a body cavity. According to v. Graff indeed, there is in such cases a lining of endothelium.

As to the homologies of the parenchyma tissue, it appears probable, from its condition in the *Acoela*, that it is a part of the endoderm in which the enteric cavity is either not developed or has collapsed, or become filled with plasmodial extensions of its protoplasmic walls, of a nature similar to the filling up of the enteron described by Bourne in a coral polyp (*Q. J. M. S.*, 28, p. 29).

In the *Acoela* this absence of enteron extends to the whole of the endoderm. The close relation of the parenchyma to the endoderm is not only suggested by the condition in the *Acoela*, in which they are actually continuous, but is indicated by an observation of Lang's that the endodermal lining of the enteron contains extensions of the excretory system. This suggests that possibly the continuity between the parenchyma and the endoderm in the *Tricladа*, at any rate, is closer than is supposed. The origin of the generative cells in the parenchyma, and their passage through the vacuolated spongy mass to the genital opening is exactly what must happen in Bourne's coral polyp with a spongy enteron. At the same time I do not wish to suggest that the parenchyma ever had a continuous digestive space. I would rather suggest that the *Acoela* are connecting forms between large Infusoria and the higher animals, in which the endodermal protoplasm, though without a continuous cavity, is partly differentiated into a number of important organs.

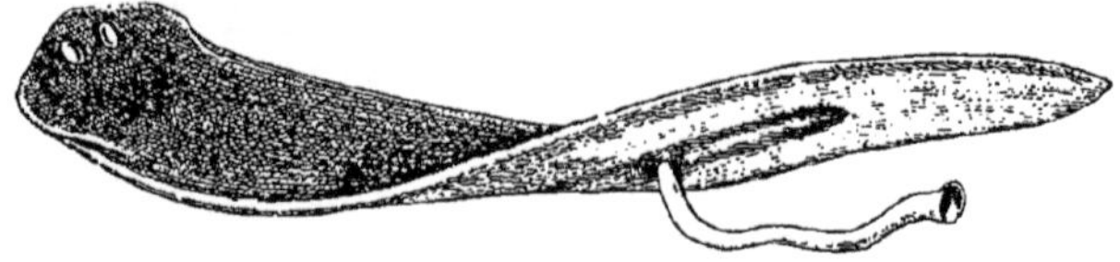

FIG. 167.—*Planaria polychroa* creeping with outstretched pharynx.

The mouth leads into a muscular pharynx, which is contained in a sheath, and can usually be protruded like a proboscis (Fig. 167). The enteron, into which the pharynx opens, consists only of a simple chamber (Fig. 168), or of a central chamber prolonged into branches, which may themselves branch and even anastomose. In some cases

(*Yungia, Cycloporus*) the enteric diverticula open to the exterior. The internal wall of the enteron is sometimes ciliated.

In the *Acoela* the pharynx leads into the central parenchyma, in the vacuoles of which digestion takes place.

The **excretory** organs are the draining system of the parenchyma. They also sometimes extend into the endoderm, which is really the central part of the parenchyma. A general account of what is known of these organs has already been given. To that may now be added that there is generally a main trunk opening externally by one or several openings (Fig. 172) on the dorsal, or on the ventral surface, or even into the pharynx-sheath; that this main canal gives off secondary canals which may branch and even anastomose; that the secondary canals receive the finest canaliculi, which come direct from the terminal cells—the flame-cells as they are called. Flame-like cilia and smaller cilia may project into these canals at different parts of their course.

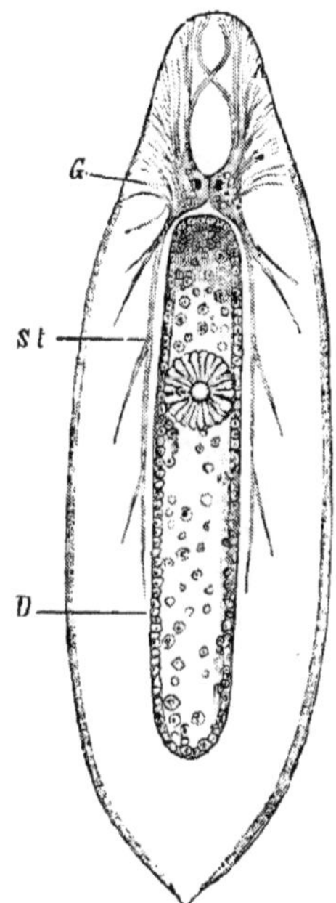

FIG. 168.—Alimentary canal and nervous system of *Mesostomum Ehrenbergii* (after Graff). *G* the two cerebral ganglia with two eye-spots; *St* the two lateral nerve-trunks; *D* alimentary canal with mouth and pharynx.

The whole system is said to be intracellular, and without well-marked walls. It can only be made out in the living

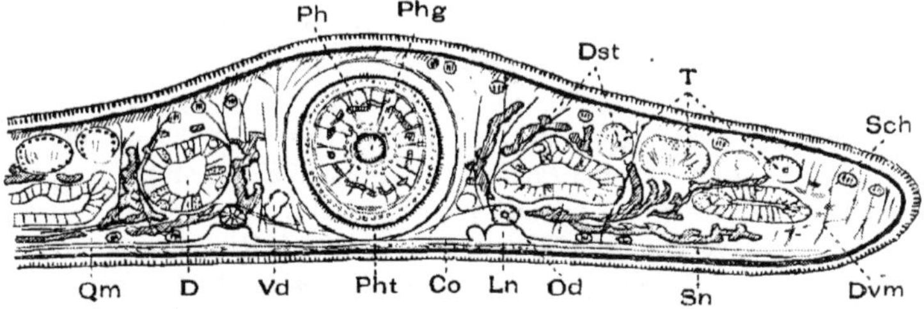

FIG. 169.—Transverse section of *Planaria polychroa*, passing through the pharynx (after Leuckart and Nitsche). *D* intestine; *Ph* pharynx; *Phg* cavity of pharynx; *Pht* sheath of pharynx; *Ln* lateral nerve-trunks with commissures *Co* connecting them, and lateral branches *Sn*; *T* testes; *Od* oviduct; *Dst* yolk-glands; *Dvm* dorsoventrally running muscular fibres; *Qm* transverse muscular fibres; *Sch* mucous glands opening externally at the edge of the ventral surface; *Vd* vas deferens.

animals. No doubt what we really have to do with here is a continuous system of tubular vacuoles in the plasmatic parenchyma—similar in nature to the tubes leading from the contractile vacuole of an Infusorian.

The excretory system has not yet been observed in the *Acoela*.

The **nervous system** (Fig. 168) consists of a bilobed ganglion at the anterior end of the body. It is embedded in the parenchyma, and gives off nerve trunks in all directions; of these, two especially large lateral trunks run backwards, one on either side (Fig. 168). The latter are sometimes connected at regular intervals by delicate transverse cords, and in the *Polyclada* and *Triclada* all the peripheral nerve-trunks anastomose. In a number of the two last-named groups a diverticulum of the enteron runs forward above the transverse commissure in a groove between the two cerebral lobes; and in some genera this enteric prolongation is encircled by a nervous commissure.

Sense-organs. Eye-spots are very generally present. In the lower forms (*Acoela*) they are simply pigment-spots in the ectoderm. As a rule, however, they are placed in the parenchyma or in the brain, and consist of many retinal rods contained in a pigmented sheath. The outer ends of these rods, *i.e.*, the ends turned towards the external surface, are continuous with the nerves passing to the brain.

In the *Polyclada* the eyes have been observed to increase by division, and in the *Triclada* they have been seen to fuse, so as to give rise to more complex eyes. The eyes, which may be present in any number, from one to a hundred or more, are placed over the brain, or on the tentacles, or on the margin of the body.

Auditory organs, as otocysts, are sometimes found. In *Acoela* and in some *Rhabdocoela* (*Monotus*), and more rarely in *Polyclada*, a single otocyst is found in the neighbourhood of the brain.

The integument is endowed with a highly developed tactile sense; the large hairs and stiff bristles which project between the cilia have probably this function. Lateral ciliated pits, which may also be explained as sense-organs, are in rare cases present at the anterior end of the body. In the *Proboscidae* the anterior end of the body is retractile into a sheath, and probably highly sensitive.

Reproductive organs. With the exception of *Microstomum* and *Stenostomum*, and possibly *Plagiostomum dioicum*, the *Turbellaria* are hermaphrodite; but steps between the hermaphrodite and dioecious condition are not wanting, for according to Metschnikoff in *Gyrator hermaphroditus* (*Prostomum lineare*), the male generative organs are sometimes developed, while the female organs remain rudimentary, or *vice versa.* This is probably only a case of successive hermaphroditism common amongst the *Turbellaria*, a condition in which the male organs attain maturity before the female.

The generative organs are always complex, and the gonads are either compact, or follicular and scattered. As a type of the first condition we may take *Mesostomum Ehrenbergii* (Fig. 170). Here there is a single external opening leading into an *atrium genitale*, into which opens the vas deferens (*Vd*) through the penis (*P*), the

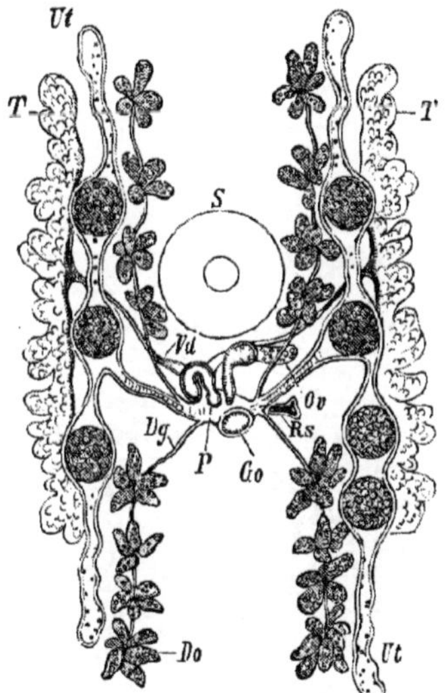

Fig. 170.—Generative apparatus of *Mesostomum Ehrenbergii* (combined from Graff and Schneider). *S* pharynx; *Go* sexual opening; *Ov* ovary; *Ut* uterus with winter eggs; *Do* yolk-gland; *Dg* duct of yolk-gland; *T* testes; *Vd* vas deferens; *P* penis; *Rs* receptaculum seminis.

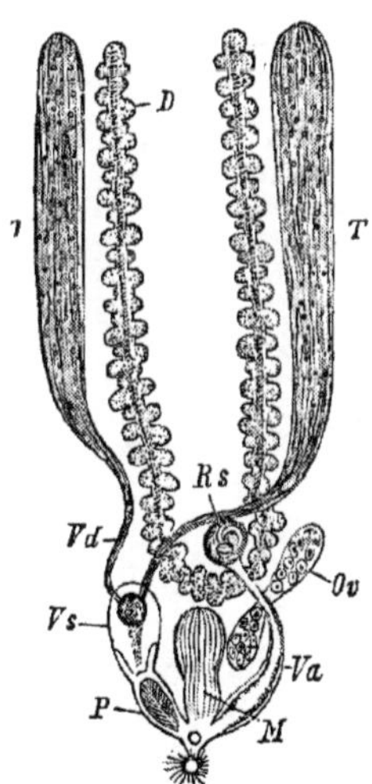

Fig. 171.—Generative organs of *Vortex viridis* (after M. Schultze). *T* testes; *Vd* vas deferens; *Vs* vesicula seminalis; *P* penis; *Ov* ovary; *Va* vagina; *M* uterus; *D* yolk-gland; *Rs* receptaculum seminis.

oviduct, the two uteruses (*Ut*), the receptaculum seminis (*Rs*), and the ducts of the yolk-glands (*Dg*). The testes (*T*) are paired and tubular, their ducts join in the penis, which projects into the atrium. The ovary (*Ov*) is single, and its duct opens directly into the *atrium genitale.* The receptaculum seminis stores up the sperm received in copulation, and the eggs remain in the uterus for a shorter or longer time. As a type of the second condition we may take the fresh-water Triclad *Planaria lactea.* Here (Fig. 172) also there is a single external opening (*Go*) leading into an *atrium genitale*, into which open the vasa deferentia through the penis, the oviduct (*Od*), the uterus (*Ut*), and a muscular sac of unknown function (*X*). There are two ovaries (*Ov*) placed between the third and fourth pairs of intestinal caeca or thereabouts. The two oviducts (*Od*) pass backwards, receiving as they go the yolk-glands which are placed between the caeca of the intestine; they join to open by a single tube into the genital atrium. The testes are numerous vesicles placed irregularly between the gut caeca, and communicating with a longitudinal tube on each side—

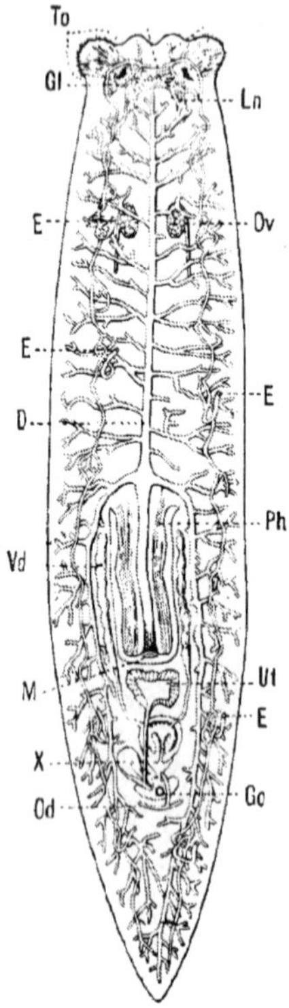

Fig. 172.—Diagrammatic representation of the anatomy of *Planaria lactea* (after Leuckart and Nitsche). *D* intestine; *Ph* pharynx; *M* opening of sheath of pharynx; *E* external openings of excretory vessels; *To* tactile organs; *Gl* cerebral ganglion; *Ln* origin of lateral nerve; *Ov* ovary; *Od* oviduct (anterior and posterior part only shown); *Ut* uterus; *X* muscular piriform organ; *Go* external genital opening; *Vd* vesiculae seminales opening into the base of the penis. The yolk-gland and testes are omitted, and the vasa deferentia are not shown.

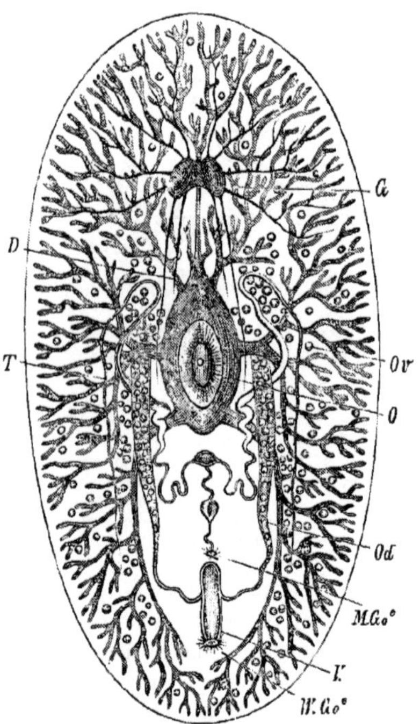

Fig. 173.—Anatomy of *Leptoplana pallida* (after Quatrefages). *G* cerebral ganglion with the nerves given off from it: *O* mouth; *D* branches of intestine; *Od* oviduct; *V* vagina; *W. Goe* female generative opening; *M. Goe* male generative opening; *T* vas deferens; *Ov* ova.

the vasa deferentia:* these open into glandular sacs—the vesiculae seminales (*Vd*), which themselves open into the penis. The penis is an eversible organ, which can be protruded through the genital pore.

In the *Polyclada* the ovaries as well as the testes are follicular and scattered, and there are no yolk-glands. Moreover, the external openings are separate, the male being in front of the female. In some *Rhabdocoela* also the generative openings are separate. In the *Acoela* the ovaries are compact, but the testes are follicular, and there are no generative ducts, so that the generative cells have to make their way through the parenchyma to the external generative opening or openings.

The penis is protrusible and very commonly armed with hooks, which are said to assist the animal in retaining hold in copulation, and also to serve as weapons of offence. In *Prorhynchus* and *Stylostomum* the penis opens into the mouth.

A curious instance of vegetative repetition is presented by some *Polyclada*, in which there may be (*Anonymus*) several pairs of penes and of male openings placed in two rows on the ventral surface.

In some cases it is said that true copulation takes place; but in many forms (*e.g.*, *Polyclada*) the sperm masses are deposited on the body of another worm, or even in a wound in the body-wall made by the impact of the penis. In such cases of sperm injection the spermatozoa must make their way to the receptaculum or uterus by travelling through the tissues of their host. These creatures do not always confine their attentions to individuals of their own species, for Lang observed a *Pseudoceros* crawl over a *Thysanozoon* and make several wounds, in each of which a sperm mass was deposited.

Some *Rhabdocoela* form two kinds of eggs—summer eggs, which have thin shells, and are retained in the uterus till hatching; and winter eggs, which have thick brown shells, are laid, and last through the winter to be hatched out in the spring. The eggs are either enclosed singly in egg-shells, or several of them are included with a number of yolk-cells in one shell or cocoon. The *Polyclada* agglutinate a number of eggs together in an albuminous mass.

The fresh-water *Turbellaria* as well as many marine forms undergo a simple direct development, and in the young state are often difficult to distinguish from *Infusoria*. Some of the *Polyclada* undergo a

* It is possible that there are no vasa deferentia, but that the spermatozoa make their way through the parenchyma to the vesiculae seminales.

metamorphosis, the larvae, known as Müller's larva, possessing finger-shaped ciliated lobes (Fig. 174).

Asexual reproduction by transverse fission takes place in the *Microstomidae* (Fig. 175). The posterior third of the body becomes separated from the rest by a septum, and develops its own mouth and organs. This process may continue both in the offspring and parent before separation takes place, so that a complex chain of individuals arises (Fig. 175).

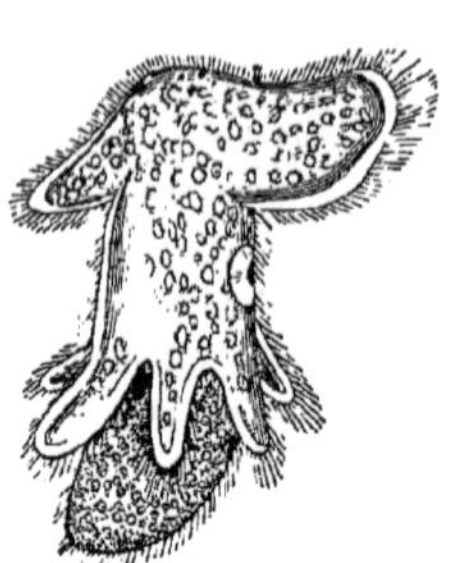

Fig. 174.—Larva of *Eurylepta auriculata* (after Hallez).

Planaria alpina is also known to undergo fission. The power of regenerating lost parts is great.

Order 1. Acoela.*

Small marine Turbellaria without digestive cavity, but with a spongy digestive parenchyma not differentiated into endoderm and mesoderm. Definite excretory organs have not been observed. A single otocyst is always present. Mouth ventral, leading into a short pharynx.

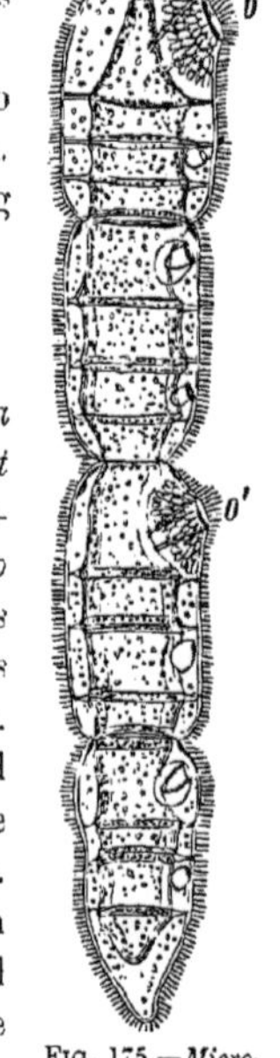

Fig. 175.—*Microstomum lineare* (after Graff). A chain produced by fission; *O*, *O'* mouth openings.

The central nervous system, generative organs, and digestive organs, are not sharply separated from the parenchyma, but appear rather to be a part of it. Generative ducts are not specially developed, though in the male organs there appear to be fairly-well specialized tracts along which the sperm passes. The genital opening is double or single; a penis is always, and a bursa seminalis sometimes present. There are no yolk-glands, and the excretory system has not been made out. There is always a pharynx, which ends in the central mass. At the front end of the body there is an organ of a glandular or sensory nature, which was formerly mistaken for the mouth. It is the "frontal organ."

The *Acoela* are marine. Most are carnivorous, but it is said that some species of *Convoluta* can live like a green plant by means of the green cells (? *symbiotic algae*), which live symbiotically in their tissues.

* Y. Delage, "Études histologiques sur les Planaires Rhabdocoeles Acoeles," *Arch. zool. exp. et gén.* (2) T. iv. 1886.

Fam. 1. **Proporidae.** With one generative opening, without accessory female apparatus; with soft penis. *Proporus* v. Graff, *Proporus venenosus* O. Schm., Plymouth; *Monoporus* v. Graff, *M. rubropunctatus* O. Schm., Plymouth. *Haplodiscus* Weldon.

Fam. 2. **Aphanostomidae.** With two sexual openings, the female being anterior to the male. With bursa seminalis and soft penis. *Aphanostomum* Oersted, *A. diversicolor* Oe., *A. elegans* Jen., Plymouth; *Nadina* Uljanin; *Cyrtomorpha* v. Graff; *Convoluta* Oersted, *C. paradoxa* Oerst., English coast, etc.; *Amphichoerus* v. Gr.; *Polychoerus.*

Order 2. Rhabdocoela.

With straight, rod-shaped enteron, and protrusible pharynx.

This order includes the smallest forms in the class, but not the most simply organised. On the contrary, the structure of the parenchyma lends support to the view that they are the most highly organised of the class. The enteron is unbranched, and lies in a space simulating a body-cavity. The vitellarium is sometimes not separated from the ovary. They live on the juices of small worms and of the larvae of *Entomostraca* and *Insecta*, which they envelop with a cutaneous secretion containing rhabdites, and afterwards suck. They are mostly fresh-water, but a few of them are to be met with in the sea, and some upon land (*Prorhynchus sphyrocephalus*). Parasitic forms are also known (*Graffilla muricicola* in the kidney of *Murex brandaris*, *Fecampia erythrocephala* in gut of *Carcinus maenas*, etc.).

Fam. 1. **Macrostomidae.** With two generative openings, the female being in front of the male. Without accessory female apparatus; with simple pharynx. *Mecynostomum* E. v. Ben., marine; *Macrostomum* E. v. Ben., f.w. and mar., *M. hystrix* Oe., f.w.; *Omalostomum* E. v. Ben., marine.

Fam. 2. **Microstomidae.** With sexual and asexual reproduction; with simple ovary, without accessory female apparatus; with simple pharynx. *Microstomum* O. Schmidt, dioecious, f.w. and mar., *M. lineare* Oe., f.w.; *Stenostomum* O. Schm., dioecious, f.w. and mar., *S. lemnae* Dug., *S. leucops* O. Schm., f.w.; *Alaurina* Busch, hermaphrodite with follicular testes, *A Claparedii* v. Graff, coast of Skye.

Fam. 3. **Prorhynchidae.** With separate generative openings, the female being ventral, the male combined with the mouth. With simple ovary divided into germarium and vitellarium; with variable pharynx. The so-called proboscis is the copulatory organ. Damp earth or fresh water. *Prorhynchus* M. Schultze, *P. stagnalis* M. Sch.

Fam. 4. **Mesostomidae.** With one or two generative openings; vitellarium and germarium either distinct organs or separate parts of the same organ. Usually with accessory female apparatus, and always with compact paired testes. With a ventrally-placed rosette-shaped pharynx. F.w. and mar. *Otomesostomum* v. Graff; *Mesostomum* Dugès (Fig. 168); *Bothromesostomum* M. Braun; *Castrada* O. Schm.; *Promesostomum* v. Gr., *P. marmoratum* Schultze, coasts of Britain;

P. solea O. Sch. Plymouth, *P. lenticulatum* O. Schm. Isle of Man; *Byrsophlebs* Jen., *B. Graffi* Jen. Plymouth, Millport, *B. intermedia* v. Gr. Isle of Man, Millport; *Proxenetes* Jen., *P. cochlear* v. Gr., Millport.

Fam. 5. **Proboscidae.** With a tactile proboscis, one or two generative openings, separate germarium and vitellarium, bursa seminalis and compact testes. Mouth ventral; pharynx rosette-shaped. The continuity of the enteron is interrupted at the period of sexual maturity. All marine except *Gyrator* Ehbg. *Pseudorhynchus* v. Gr., Irish Channel; *Acrorhynchus* v. Gr., Brit. coast; *Macrorhynchus* v. Gr., Brit. coast; *Hyporhynchus* v. Gr., Brit. coast; *Gyrator* Ehrb., St. Andrews.

Fam. 6. **Vorticidae.** With one generative opening; germarium and vitellarium combined or separate; with accessory female apparatus, always simple uterus, and compact paired testes. Mouth ventral, and usually near the anterior end; pharynx with a single exception cask-shaped. The chitinous penis very various. *Schultzia* v. Gr.; *Provortex* v. Gr., Brit. coast; *Vortex* Ehbg.; *Jensenia* v. Gr.; *Opistomum* O. Schm.; *Derostomum* Oers.; *Graffilla* v. Jhering, parasitic; *Anoplodium* Schneider; *Fecampia erythrocephala* Giard, in gut of *Carcinus maenas*.

Fam. 7. **Solenopharyngidae.** With one gen. opening, one germarium (ovary), paired compact elongated testes. The pharynx elongated, tubular, directed backwards. *Solenopharynx* v. Gr.

Order 3. Alloiocoela.

Enteron lobed or an irregularly widened sac. Testes follicular. Paired ovaries and vitellaria, combined or separate; vitellaria irregularly lobed or partially branched.

The parenchyma is less differentiated than in the *Rhabdocoela*, and more in the condition it has in the *Acoela*. The generative openings are separate or united, as in the latter group. With few exceptions they are marine (*Plagiostomum lemani*, Lake of Geneva, *Bothrioplana*).

Fam. 1. **Plagiostomidae.** With one gen. opening, and without accessory female apparatus (except *Cylindrostomum*); with testicular follicles in front of, alongside, and behind the brain. Pharynx varying in position and size. Otocyst absent. Usually small, cylindrical, or plano-convex forms with narrow hind end. *Acmostomum* v. Gr.; *Plagiostomum* O. Schm.; *Vorticeros* O. Schm.; *Enterostomum* Clap.; *Allostomum* P. J. v. Ben.; *Cylindrostomum* Oers. All recorded from British coast except *Acmostomum*.

Fam. 2. **Monotidae.** With two gen. openings, and with bursa seminalis; testicular follicles closely aggregated between brain and pharynx. Pharynx always elongated with opening directed backwards. With one otocyst. Elongated forms with narrow front end and broad hind end. *Monotus* Diesing; *Automolos* v. Gr. Both recorded from British coast.

Fam. 3. **Bothrioplanidae.*** Intestine with three branches, of which the two posterior unite behind the pharynx. With one genital opening. Allied to the Triclada. Genital glands with tunica propria. *Bothrioplana* Braun., fresh water.

* F. Vejdovsky, *Z. f. w. Z.*, 60, 1895, p. 198.

Order 4. TRICLADA.

Enteron with three branches, one forward and two backward; pharynx cylindrical, inserted at the junction of the three branches. Genital pore single, behind the mouth.

The *Triclada* comprises marine, fresh-water, and terrestrial forms. A description of a typical member of the order has already been given. The testes are numerous and follicular; the vitellarium consists of scattered follicles, and is only exceptionally compact (*Otoplana*); the ovaries are two in number, placed anteriorly. The mouth is usually behind the middle of the body, and the body is more or less flattened. The skin is often provided with glands, the secretion of which in certain land forms (*Bipalium*, *Rhynchodemus*) hardens to a fibrous web. The *Triclada* have great power of repairing lost parts; fission occurs in *Planaria alpina* and *cornuta*.

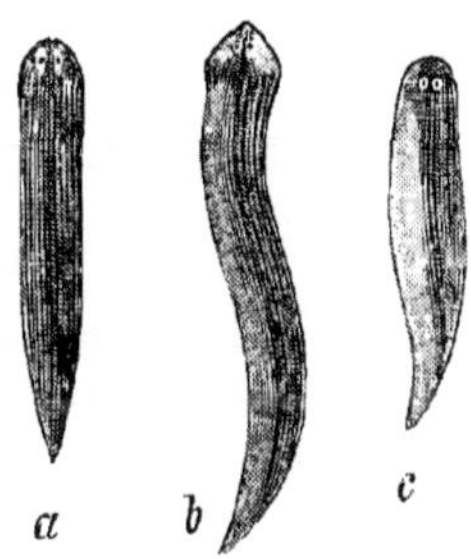

FIG. 176.—*Planaria polychroa* (*a*), *P. lugubris* (*b*), *P. torva* (*c*), about twice natural size (after O. Schmidt).

Marine forms. Enteric branches but little ramified, sometimes simply lobed. Mouth placed in the posterior half of the body (except *Bdellura*). Body flattened. Uterus placed behind the genital opening.

Fam. 1. **Otoplanidae.** With otocyst and ciliated pits, without eyes. *Otoplana* Du Plessis.

Fam. 2. **Procerodidae.** Without otocyst and ciliated pits. *Cercyra* O. Schm.; *Procerodes* Girard (including *Gunda*, *Fovia*, *Haga*); *Uteriporus* Bergendal. *Gunda* and *Fovia* recorded from British coast.

Fam. 3. **Bdelluridae.** Ectoparasitic on *Limulus*, provided with a caudal apparatus for fixation. *Bdellura* Leidy.

Fresh-water forms. Enteric branches much ramified. Mouth placed in the posterior half of the body. Body flattened. Uterus placed between the pharynx and the penis.

Fam. 1. **Planaridae.** Head without differentiated organ of fixation. *Planaria* O. F. Müll. (including *Dugesia*), with two eyes; *P. lactea* O. F. M.; *P. punctata* Pall.; *P. polychroa* Schm.; *P. torva* M. Sch.; *P. alpina* Dana are British. *Phagocata* Leidy, with 8 or 9 pairs of pharynges additional to the main one, Pennsylvania; *Anocelis* Stimpson, no eyes; *Polycelis* Hemprick and Ehrenb., many marginal eyes; *P. nigra* Ehr., *P. cornuta* are British.

Fam. 2. **Dendrocoelidae.** Head with several differentiated organs of fixation. *Dendrocoelum* Oerst. (incl. *Bdellocephala* and *Galeocephala*), two eyes; *Oligocelis* Stimpson, six eyes; *Procotyla* Leidy; *Sorocelis* Grube, and *Dicotylus* Grube, from Lake Baikal.

Terrestrial forms. Enteric branches generally simply lobed. Position of mouth and form of body variable. Uterus slightly developed and behind the genital pore. Ventral muscular system much developed.

Fam. 1. **Leimacopsidae.** Dorsal face very convex; mouth in the anterior part of body. *Leimacopsis* Diesing, 2 frontal tentacles with eyes at base, tropical America.

Fam. 2. **Geoplanidae.** Body sub-cylindrical; mouth almost median (except *Microplana* and *Dolichoplana*). *Geoplana* Stimpson (with *Geobia* and *Coenoplana*) S. America, Australia, N. Zealand, etc.; *Sphyrocephalus* Kuhl and v. Hasselt (with *Bipalium*), Ceylon, India, N. Zealand, etc.; *Geodesmus* Metschnikoff; *Rhynchodemus* Leidy, world-wide, mouth a little behind the middle of body. *Dolichoplana* Moseley, Philippines; *Microplana* Vejdovsky.

Fam. 3. **Polycladidae.** Body flattened; mouth in posterior part of body. *Polycladus* Blanchard.

Order 5. Polyclada.

Enteron with many branches, which ramify or anastomose. Male and female openings distinct (exceptionally united in Stylochoplana, Discocelis). Ovaries follicular, without vitelline glands.

The *Polyclada* are marine animals. Both testes and ovaries are numerous and follicular. A vitellarium is absent. The female genital opening is behind the male. In some forms the young are hatched out as larvae, called after their discoverer, Müller's larvae (Fig. 174).

ACOTYLEA.

Without suckers. Mouth in the middle of the body or behind it. Copulating apparatus in the posterior half of body. Without tentacles, or with nuchal tentacles.

Fam. 1. **Planoceridae.** With nuchal tentacles; mouth about median; male copulating apparatus directed backwards. *Planocera* de Blainville; *P. folium* Grube, Berwick Bay. *Imogine* Girard; *Conoceros* Lang; *Stylochus* Ehrb.; *Stylochoplana* Stimps., body widened in front, attenuated behind, male and female openings united, with two tentacles and eyes at their base; *St. maculata* Quatr., Brit. coast. *Diplonchus* Stimps.

Fam. 2. **Leptoplanidae.** Without tentacles; mouth about median; male copulating apparatus directed backwards. *Cryptocelis* Lang, with oval, consistent body, mouth median, genital openings separate, eyes small in indistinct groups between the brain and front body end, and round the whole edge of the body; *Discocelis* Ehrb., broadly oval, tolerably consistent body, mouth median or in front of middle, a single genital opening, eyes in two groups, one on each side of brain, and on the anterior edge of the body; *Leptoplana* Ehrb. (Fig. 173), body elongated, genital openings more or less removed from hind end of body, eyes in two sometimes indistinct tentacular groups, and in the brain area, small eyes on the anterior edge of body; *L. tremellaris* O. F. M., Brit. coast. *Trigonoporus* Lang.

Fam. 3. **Cestoplanidae.** Without tentacles, body ribbon-like. Mouth not far from hind end of body. Copulating apparatus directed forwards; eyes scattered over the whole head. *Cestoplana* Lang.

COTYLEA.

With central or sub-central ventral sucker always placed behind the openings of the body. Mouth in middle of body, or in front of it. Copulating apparatus (except *Anonymus*) in the anterior half of the body. Without tentacles, or with marginal tentacles.

Fam. 1. **Anonymidae.** Body broad, oval, without tentacles. Numerous penes in two lateral rows. Single female opening between mouth and sucker. Eyes in brain area and on whole edge of body. *Anonymus* Lang.

Fam. 2. **Pseudoceridae.** Body oval or elliptical, with fold-like marginal tentacles. Mouth in middle of anterior half of body. Eyes on brain areas, and on the tentacles. *Thysanozoon* Grube, with dorsal villi and double penis. *Pseudoceros* Lang, and *Yungia* Lang, without villi. In *Yungia* the gut diverticula open by numerous pores on the dorsal surface of the body.

Fam. 3. **Euryleptidae.** Body oval or elliptical, with or without pointed frontal tentacles. Mouth near the front end of the body. Pharynx tubular. Eyes in the brain area and on the tentacles (or at the two sides of front edge of body). *Prostheceraeus* Schmarda, Brit. coast: *Cycloporus* Lang, gut branches open along the whole of the edge of the body, *C. papillosus* Lang, Plymouth, Port Erin; *Eurylepta* Ehrb.; *Eu. cornuta* O. F. M., Brit. coast. *Oligocladus* Lang, Brit. coast; *Stylostomum* Lang, Brit. coast; *Aceros* Lang.

Fam. 4. **Prosthiostomidae.** Body elongated, without tentacles. Mouth immediately behind brain. Pharynx long and tubular. *Prosthiostomum* Quatrefages.

Class II. TREMATODA.*

Parasitic Platyhelminthes with unsegmented, usually flattened, rarely cylindrical, body. They possess a mouth and ventrally-placed organ for attachment. The intestine is forked and without an anus.

The Trematodes, in the main features of their organisation, are most nearly related to the Turbellaria. They differ from them in their parasitic habit, and in the absence in the adult of a ciliated ectoderm, though the *Temnocephalidae*, in possessing a partially ciliated ectoderm and in other respects, are intermediate between the two classes. In connection with their parasitic habit they possess special organs for adhering, such as suckers and hooks, which are stronger and better developed in those which are external parasites than in those which infest the internal organs of animals.

The mouth is invariably placed at the anterior end of the animal, usually in the middle of a small sucker. It usually leads into a

* M. Braun, "Trematoda" in Bronn's *Klassen u. Ordnungen*, Bd. iv., 1887–93. A. v. Nordmann, "*Mikrographische Beiträge z. Kenntniss der wirbellosen Thiere*, Berlin, 1832. P. J. v. Beneden and Hesse, *Mém. de l'Acad. Roy. Belgique*, 34, 1864, p. 60. P. J. v. Beneden, *Mém. s. les Vers intestinaux*, Paris, 1858–61. v. Linstow, *Compendium der Helminthologie*, Hanover, 1878. A. P. Thomas, "The Life-History of the Liver-Fluke," *Q. J. M. S.*, 23, 1883, p. 99. R. Leuckart, "Die Parasiten des Menschen," Ed. 2, Leipzig, 1879–1894. S. Goto, "Ectopar. Trematodes of Japan," *Journ. Coll. Sci. Imp. Univ. Japan*, 8.

muscular pharynx with a more or less elongated oesophagus (Fig. 177), which is prolonged into a forked intestine, ending blindly and often beset with coeca. The organs are embedded in a parenchyma of muscle and connective tissue.

The **excretory** apparatus consists of two large lateral trunks, and of a network of fine vessels permeating the tissues and ending in small ciliated lobules (flame-cells). The two large trunks usually open into a common contractile vesicle, which opens to the exterior at the hind end of the body.

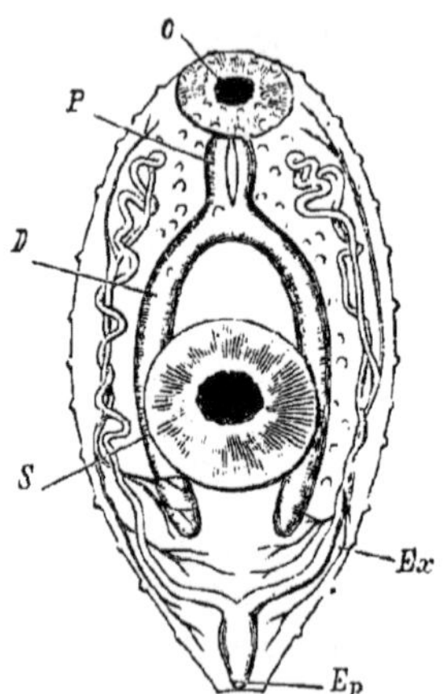

FIG. 177.—Young *Distomum* (after la Valette). *Ex* trunk of the excretory system; *Ep* excretory pore; *O* mouth with sucker; *S* ventral sucker; *P* pharynx; *D* forked intestine.

The **nervous system** consists of a bilobed ganglion lying above the oesophagus. From it there pass out, in addition to some small nerves, two strong backwardly directed ventral nerve-trunks. These are connected by transverse anastomoses with each other, and with two weaker lateral longitudinal cords, which are themselves connected with two dorsal trunks (Fig. 178). **Eye-spots** with refractive bodies are sometimes present on the larvae during their migration, and in many of the ectoparasitic forms.

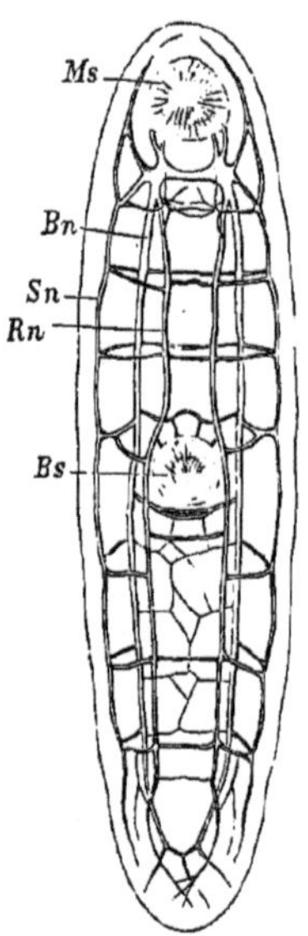

FIG. 178.—Nervous system of *Distomum isostomum* (after E. Gaffron). *Ms* oral sucker; *Bs* ventral sucker; *Sn* lateral nerve; *Rn* dorsal nerve; *Bn* ventral nerve.

Locomotion is effected by the dermo-muscular system, with the co-operation of the organs of adhesion (suckers and hooks) which present numerous modifications in number, form, and arrangement.

The **organs of adhesion**, in size and development, are related to the endoparasitic or ectoparasitic mode of life. In the endoparasitic Trematodes they are less developed, and usually consist of the oral sucker and of a second larger sucker on the ventral surface, either near the mouth as in *Distomum*, or at the opposite end of the body (*Amphistomum*). The large sucker may however be absent (*Monostomum*). The ectoparasitic forms are, on the other hand, distinguished by a much more powerful adhesive armature: for, besides two small suckers at the sides of the mouth, they possess one or more large suckers at the posterior end of the body (Fig. 185), which, moreover, may be supported by a chitinous framework. There are often in addition chitinous hooks, and very frequently two larger hooks among the posterior suckers in the middle line (Fig. 185). In the *Temnocephalidae* the anterior suckers are replaced by a group of tentacles, and in the *Aspidocotylea* the adhesive apparatus is a kind of large ventral foot bearing numerous suckers.

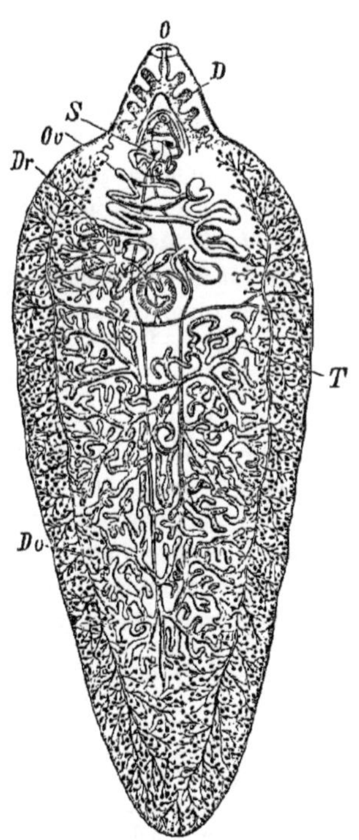

Fig. 179.—*Distomum hepaticum* (after Sommer). *O* mouth; *D* caecum of intestine (hinder part of intestine not shown); *S* sucker; *T* testes; *Do* yolk-glands; *Dr* ovary; *Ov* uterus.

Reproductive organs. The *Trematoda* are mostly hermaphrodite.* As a rule, the male generative opening and the uterine opening of the female are side by side or one behind the other, not far from the middle line of the ventral surface, near the anterior end of the body (Fig. 179). The openings may be separate, or they may lead into a genital cloaca which opens by the single *genital pore*. The male opening leads into a sac, the *cirrus sac*, which encloses the protrusible terminal part (*penis* or *cirrus*) of the vas deferens. The vas deferens soon divides into two, which lead back to the two large simple or multilobed—in *Distomum hepaticum* much branched—testes. Both testes and ovaries are, as a rule, placed between the two limbs of the intestine.

The female organs consist of a convoluted uterus and of an ovary and paired yolk-glands, to which may be added a special shell-gland.

* *Bilharzia*, *Distomum Okenii*, and perhaps one or two other *Digenea* are dioecious.

The ovary may be rounded (Fig. 185) or branched (Fig. 179); the yolk-glands, which secrete vitelline matter, are much branched tubular glands occupying the sides of the body (Fig. 179). The oviduct, after receiving the ducts from the yolk-gland, is continued as the convoluted, somewhat dilated, uterus which opens near the male opening. The shell-gland is placed round the junction of the oviduct and vitelline ducts, or round the first part of the uterus. Here the egg is fertilised, and the vitelline particles come into contact with the ova. In the *Heterocotylea* (*Polystomeae*) the part of the uterus into which the oviduct opens is dilated and called the **ootype.** Here, or in the beginning of the uterus if there is no ootype, each ovum acquires its investment of yolk, and is surrounded by a strong shell; it is then ready to be passed out through the opening of the uterus.

In a great many Trematodes, if not in all, there is a special paired or unpaired canal, called the canal of Laurer,* which opens externally in the dorsal middle line (*Distomum*), or laterally through two warts on the sides of the body (*Polystomum*, Fig. 185), and internally into the oviduct where it joins the yolk-ducts, or into the yolk-ducts near this point. This canal, certainly in some cases, probably in all, is functionally a vagina, and serves for the entrance of the spermatozoa received in copulation, or otherwise, from another individual. In some genera (*Polystomum*, *Diplozoon*, *Octobothrium*) there is a duct (vitello-intestinal) connecting the oviduct with the intestine. It appears to serve the purpose of carrying the superfluous yolk into the intestine.

The so-called *third vas deferens*, as a tube connecting the vas deferens with the oviduct, appears not to exist. It is probable that the vitello-intestinal canal (and possibly the canal of Laurer) have been mistaken for a tube connecting the oviduct with the male-duct in such a way as to permit of self-fertilization.

The function of Laurer's canal is in some doubt. It has very generally been interpreted as a vagina, but, excepting in *Polystomum integerrimum*, in which two worms have been detected in reciprocal copulation, with their male generative openings applied to one of its openings in the other worm, the apposition of the penis to it has never been observed. On the other hand, in the Distomic (digenetic) division reciprocal copulation with the penis of one worm inserted into the uterine opening of the other has been observed in some forms,† and in one case‡ the penis of one individual was found to be inserted into the adjacent opening of its own uterus. These latter cases point to the view that the uterus functions as vagina. Then the question arises, what is the function of the canal of Laurer? It has been suggested that its function is to carry off the superfluous

* J. Fr. Laurer, *Disquisitiones anatomicae de Amphistomo conico*, 1830, 4°.

† *Holostomum serpens*, *Monostomum faba*, *Distomum clavigerum*, *D. cylindraceum*.

‡ *Distomum cirrigerum*.

yolk. This view is not, however, borne out by an examination of its contents, which are often spermatozoa, though sometimes yolk, and even ova. Some observers hold that the single canal of the digenetic forms is not homologous with the two canals of many monogenetic forms, but with the vitello-intestinal canal, which is not present in the *Digenea.*

On the whole we incline to the view that the canal of Laurer, whether double or single, is in all cases homologous, and serves for the entrance of spermatozoa to the female ducts, when, as is often the case, the uterus is full of ova. It may be that in some cases the penis is used, like that of the *Turbellaria*, for hypodermic injection of spermatozoa; but as a general rule the thickness of the cuticle and the isolation of the individuals will prevent this; again, sometimes the penis may, more or less accidentally, find its way into the opening of its own uterus, or into that of another worm; but, as a general rule, the parasites being not contiguous, and having but little power of locomotion, the spermatozoa are discharged into the cavity in which the parasite is lodged. The testis being very large, the number of spermatozoa so discharged is probably very great, so that they spread over the whole surface of the infected organ, to which other parasites of the same species are affixed. Some of them, therefore, are sure to reach the body of other individuals, and to pass, by their own movements, into the opening of the canal in question.

That some kind of haphazard impregnation of this kind is the rule in Trematoda is rendered almost certain by the relatively enormous size of the testis. On this view the penis must be regarded as a vestigial organ, with little or no function under the present conditions of life.

Although the so-called internal or third vas deferens does not exist, self-impregnation* seems in some cases to occur. For in some species solitary specimens have been found with spermatozoa in the female passages (*e.g.*, *Polystomum integerrimum*, *Distomum agamos* encysted in *Gammarus pulex*). This self-fertilization may have been effected by the insertion of the penis into the opening of the uterus, but probably it has more often been effected in the manner above suggested.

The egg-shell is usually in two parts, a small piece being separated by a suture from a larger, and constituting the *operculum.* In the ectoparasitic forms it is often prolonged into a filament at one or both poles of the egg, which filament serves for the attachment of the egg (Fig. 188). Filaments are only rarely found in the endoparasitic forms.

The eggs of the endoparasitic forms are smaller and more numerous than those of the ectoparasitic division. The early stages of development are sometimes passed through in the uterus, but in some cases the segmentation does not begin till after oviposition. Most Trematodes lay their eggs, but a few are viviparous (*Gyrodactylus*).

The just hatched young either possess (in most ectoparasitic forms, *Heterocotylea*) the form and organisation of the parent (Fig. 186),

* A. Looss, "Beiträge z. Kenntniss der Trematoden," *Z. f. w. Z.*, 41, 1885, p. 420–427.

in which case there is no intermediate host and no alternation of generations: such forms are **monogenetic,** and are mostly ectoparasites; or they have the form of ciliated larvae (*Miracidium*) which pass into an intermediate host, and eventually become transformed into the adult without alternation of generations: such forms are monogenetic and **metastatic,** and are endoparasites; or they undergo a change of host and present the phenomenon of a complicated alternation of generations, in which the miracidium larva develops into an asexual form which reproduces itself: such forms are called **digenetic,** and are all endoparasites. In the monogenetic forms the relatively large eggs become, in most cases, attached on the place

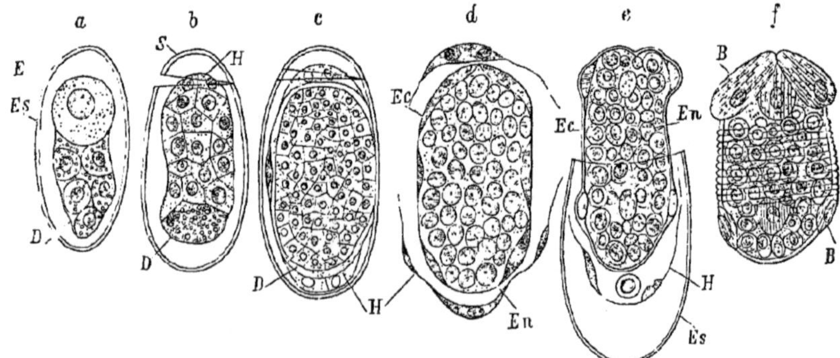

FIG. 180.—Embryonic development of *Distomum tereticolle* (after H. Schauinsland). *a*, egg after hardening in picric acid; *Es* shell; *E* ovum; *D* yolk-cells. *b*, the yolk has been largely used up by the embryonic cells; *H* the first-formed cells of the cellular membrane of the shell; *S* operculum. *c*, later stage, the shell membrane *H* surrounds the embryo, the yolk *D* almost gone. *d*, later stage showing ectoblast *Ec*, the internal mass being endoblast. *e*, later stage. *f*, a ripe embryo just before the hatching; *B* setigerous plates with their nuclei.

where the mother lives, and the young are hatched out as non-ciliated or partially ciliated forms (*Polystomum integerrimum* has transverse rows of cilia at hatching, Fig. 186). In the case of the metastatic and digenetic forms the eggs (relatively small in the *Digenea*) are deposited in a damp place, usually in the water.

In the *Digenea** the ovum is placed next the operculum—the pole of the egg where the anterior end of the future embryo will be formed (Fig. 180, *E*). It undergoes a total cleavage and gives rise to a solid cell-mass, from the surface of which is differentiated a cellular membrane (Fig. 180, *H*), which lines the shell and is

* H. Schauinsland, "Beitrag z. k. d. embryonalen Entwicklung der Trematoden," *Jena, Zeitschrift*, 16, 1883.

left behind at hatching. Beneath this there is formed a second membrane (*Ec*), which gives rise to the ectoderm of the just-hatched larva; this is ciliated in some forms (*D. hepaticum*, Fig. 182), but in others it develops a stout cuticular layer (*D. tereticolle*, Fig. 180, *e*, *Ec*).

The peripheral cells of the contained cell-mass now differentiate a third membrane of flattened epithelial cells, while of the remaining cells those at the head end give rise to the rudiment of the alimentary canal—the others becoming the germ cells of the larva. The embryo is now hatched, and becomes a small free larva—the *Miracidium*. This, which may be ciliated (Fig. 182) or non-ciliated, or even provided with stiff bristles (*D. tereticolle*, *D. ovocaudatum*), has a contractile body and often an X-shaped eye-spot; it is also provided with *excretory* canals in its walls, and in cases in which the structure has been fully examined, a sucker with a mouth opening, intestine, and a ganglion can be made out (Fig. 181). Further, between the intestine and the body-wall there are some ova-like cells—the germ cells.

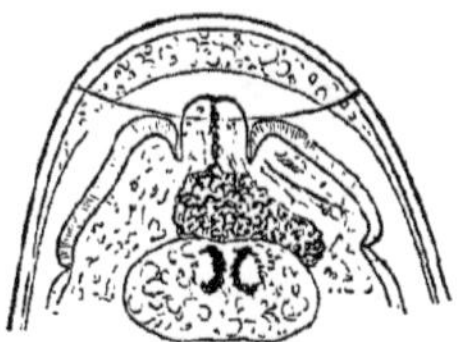

FIG. 181.—Anterior pole of an egg of the liver-fluke with fully developed embryo, in optical section. The proboscis, with mouth, oesophagus, and intestine, the ganglion and eyes are visible (after Leuckart).

This larva leaves the egg and, if ciliated, wanders about independently in search of a host, through the body-wall of which it bores its way; if non-ciliated it is taken up by the new host in its food.

Sometimes the unhatched egg is swallowed (*D. lanceolatum*) and the larva is set free in the alimentary canal. In any case the larva makes its way into the tissues of its host, which is a mollusc, usually a water-snail, and, casting its cuticularized or ciliated skin, becomes a *sporocyst* or a *redia*.* The sporocyst (Fig. 182, *b*) is a hollow sac with excretory canals in its wall, and containing in its cavity a number of *germ-cells*; a *redia* (Fig. 182, *c*) is like a sporocyst except that it contains a mouth at one end and an intestine, and two lateral processes near its hind end; it likewise contains a cavity with germ-cells. The germ-cells of these organisms develop into more sporocysts or rediae, or into *Cercariae*.* The Cercariae are young Trematodes, which eventually reach (often only after two migrations, an active

* The miracidium generally becomes a sporocyst, rarely a redia (*Monostomum flavum* and *mutabile*). The sporocyst may produce other sporocysts (*D. cygnoides*), or Cercariae direct, but generally gives rise to rediae (*D. hepaticum*, *Diplodiscus subclavatus*), which produce the Cercariae.

and a passive one) the final host, where they become sexually mature. They are furnished (Fig. 182, *d*) with an exceedingly motile caudal appendage, frequently with a buccal spine, and occasionally with eyes, and they present in the rest of their organization great resemblances to the adult, excepting that the generative organs are not developed. In this form they make their way out of the body of the *redia* or sporocyst, and of their host, and move about in the water, partly creeping and partly swimming. Here they either perish or find a new host (snail, worm, insect larva, crustacean, fish,

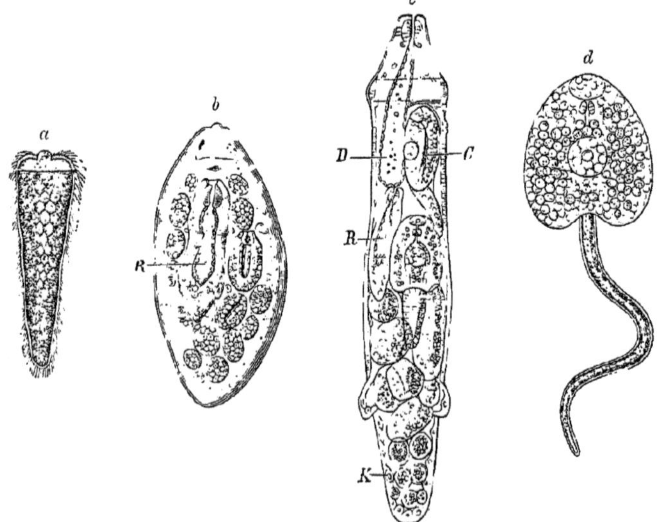

FIG. 182.—Stages in the life-history of *Distomum hepaticum*. *a*, Miracidium (ciliated embryo). *b*, Sporocyst with rediae *R* (after Leuckart). *c*, redia (after Thomas); *D* gut, *C* Cercaria, *R* redia, *K* germ-cells. *d*, Cercaria (after Thomas).

batrachian), into which they penetrate aided by the powerful vibrations of their tails; they then lose the tail and encyst.

The *Cercariae* thus become distributed amongst a number of hosts, and in each case give rise to an encysted form, which only differs from the adult in being without generative organs. This young Trematode migrates passively with the flesh of its host into the stomach of another animal, and thence, freed from its cyst, into the organ (intestine, liver, etc.) in which it becomes sexually mature.

There are then, as a rule, three different hosts, in the organs of which the different stages (miracidium, sporocyst, redia, encysted form, sexually mature animal) of the digenetic Trematoda are buried.

The transitions from one host to another are effected partly by independent migration (*Miracidium*, *Cercaria*), partly by passive migration (encysted sexless adult). Modifications of the ordinary course of development may, however, take place; these may be either complications or simplifications. The embryo at hatching may contain a single *redia* (as in *Monostomum flavum* and *mutabile*), which it carries about till it enters the first host (Fig. 184, *b*). In some cases the course of development is simplified by the omission of the second intermediate host, viz., that which contains the encysted immature Trematode (*Cercaria macrocerca* of *Distomum cygnoides*; also *Leucochloridium*, the sporocyst of *Distomum macrostomum*, in the tentacles of *Succinea amphibia*; *D. hepaticum*, in which the Cercaria encysts upon a blade of grass). In other cases the sporocysts produce rediae, which produce Cercariae or more rediae (*D. hepaticum*).

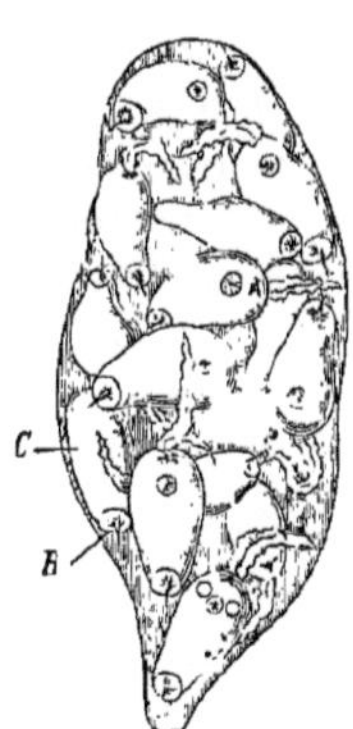

FIG. 183.—A sporocyst with contained Cercariae *C*, *B* boring spine of Cercariae.

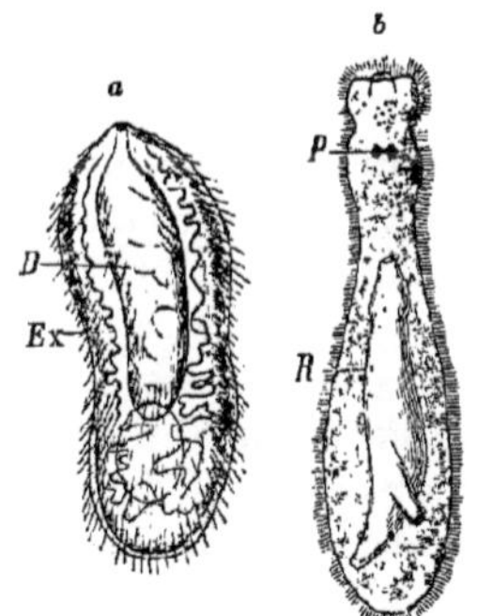

FIG. 184.—*a*, Miracidium (ciliated embryo) of *Diplodiscus* (*Amphistomum*) *subclavatus* (after G. Wagener). *D* intestine; *Ex* excretory vessels. *b*, Miracidium of *Monostomum mutabile* (after v. Siebold). *p* eye-spot; *R* redia.

Further, there are unencysted young Distoma which never become sexually mature in their host (*e.g.* in the lens and vitreous humour of the vertebrate eye, in the jelly of Coelenterates). On the other hand encysted forms are known (*Distomum agamos* of Gammarids) which are sexually ripe and produce eggs within the cyst (?self-impregnation). The sporocyst may increase by division (*e.g.* the sporocyst of *Cercaria minuta*). The tail of the *Cercaria* may become transformed into a sporocyst, and after detachment produce a brood.

Many forms possess great adaptability to changed conditions of life, *e.g.* *D. echinatum*, which proceeds from *Cercaria echinata* of *Paludina vivipara* and normally infests the intestine of the Duck and of water birds; it can, however, attain maturity in the intestine of the Dog, Mouse, and Rat.

If we summarise the development of the endoparasitic Trematodes we get the following results:

(1) Development with a non-ciliated larva, which changes direct into the sexual animal. No intermediate host, but transference to another individual. *Aspidogaster* and its allies.

(2) Development with ciliated larva, which enters other animals (Molluscs, Hirudinea, Fishes, Amphibians, Mammals), there encysts, and takes on a second larval form (metastatic Trematodes). *Holostomidae.*

(3) Development with ciliated or non-ciliated larva (Miracidium), which passes actively or passively into another animal (Mollusc), in which it becomes transformed into an asexual form (sporocyst or redia), which produces Cercariae either directly, or after the formation by internal budding and rarely by fission of more asexual forms. These are the larvae of the adult sexual animal, and they usually migrate into another host (Mollusc, or Crustacean, Insect, Fish, Amphibian, and even Mammal), where they lose their larval organs and encyst. Here they remain until they are passively transferred into the alimentary canal of their permanent host (almost always a Vertebrate), in which they become sexually mature (digenetic Trematodes). Most *Malacotylea.*

The life-cycle of the digenetic Trematoda must be looked upon as an alternation of generations of the variety called heterogamy. That is to say, the alternation is between a sexual generation (the so-called adult) and one or more parthenogenetically producing generations (sporocyst, redia). The *Miracidium* is a larva which, becoming parasitic, loses its gut, nervous system, etc., and degenerates into a sporocyst, or more rarely a redia. The sporocyst or redia possesses a kind of diffuse ovary in the germ-cells, which develop either into rediae or Cercariae (see above, p. 229). The redia is very similar to the sporocyst, but is not so degenerate. The Cercaria is simply the larva of the final stage; it possesses special larval organs, *e.g.* tail and spine, which however may be absent in cases in which they are not required; *e.g. Macrostomum*, in which the migration of the Cercaria into its final host is a passive one.

On this view the germ-cells of the sporocysts and rediae are ova, and from what is known of the development (see p. 229), they seem to be set apart at an early stage (the end of cleavage). Some observers hold the view that we have to do here with a kind of embryonic fission at a very early stage, the divided off cells remaining latent for some time, and only developing later into sporocysts or rediae. But whatever view be taken of the reproduction of the sporocysts and rediae, there are two facts to be noted with regard to it; (1) it starts, as does sexual reproduction, from a one-celled stage, and (2) the phases of the development of the so-called germ-cells, whether into rediae or Cercariae, seem closely to resemble what is known of the development of the fertilised ovum into the Miracidium.

Order 1. HETEROCOTYLEA = MONOGENEA.

Body variously shaped. Anterior end with or without suckers; hind end always with a suctorial organ. The ectodermal epithelium is transformed into a cuticular layer except in the *Temnocephalidae* and in the lateral suckers of the *Tristomidae*. Eyes are present. The openings of the excretory organs are usually paired, dorsal and anterior; rarely single and posterior. Always hermaphrodite; male and female openings separate or united, ventral and usually anterior. A paired or unpaired vagina is generally present, opening ventrally or laterally, or more rarely dorsally. Reproduction sexual; development direct, without metagenesis or heterogamy. For the most part external parasites, on the integument, or in the mouth, nasal passages, branchial cavity, in some cases in the urinary bladder, of Fishes, Amphibia, Reptiles, and Crustacea. They are usually hatched in the locality inhabited by the mother. Sometimes the development is a metamorphosis, and the young live in another place.

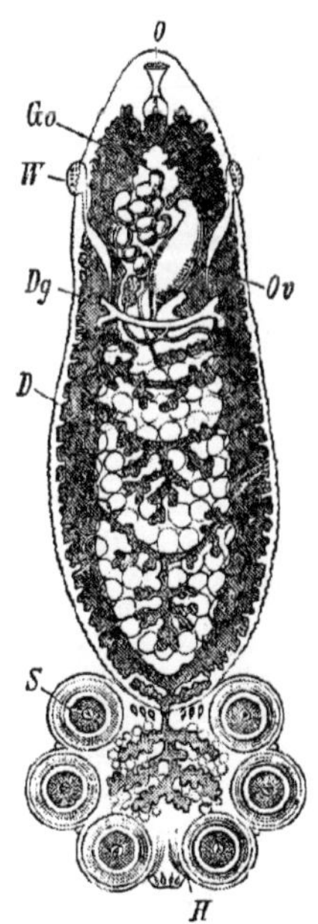

FIG. 185.—*Polystomum integerrimum* (after E. Zeller). *O* mouth; *Go* genital opening; *D* intestine; *W* lateral papillae bearing openings of vaginae (Laurer's canal); *Dg* yolk-gland duct; *S* sucker; *Ov* ovary; *H* hooks.

The development of *Polystomum integerrimum*, parasitic in the bladder of the frog, is the best known, owing to the researches of E. Zeller* (Fig. 186). The production of eggs begins in the spring, when the frog awakens from its winter sleep and proceeds to pair. It lasts from three to four weeks. It is easy then to observe the *Polystoma* in the process of reciprocal copulation. When the eggs are being laid, the parasite forces the anterior end of the body with the genital opening through the mouth of the bladder nearly as far as the anus. The development of the embryo takes place in water, and occupies a period of some weeks, so that the young larvae are not hatched until the tadpoles have acquired internal gills. The larvae (Fig. 186) are Gyrodactylus-like, and possess four eyes, a pharynx and alimentary canal, as well as a posterior disc for attachment, which is surrounded by sixteen hooks. They possess five transverse rows of cilia, three are ventral and anterior, two dorsal and posterior. There is also a ciliated cell upon the anterior extremity. The larvae now migrate into the branchial cavity of the tadpole, lose their cilia, and are transformed into young *Polystoma* by the formation of the two median

* *Z. f. w. Z.*, 22, 1872.

hooks, and of the three pairs of suckers upon the posterior disc. The young Polystomum, eight weeks after the migration into the branchial cavity, at the time when the latter begins to abort, passes through the stomach and intestine into the bladder, and there only becomes sexually mature after three or more years. In some exceptional cases, and always when the larva has passed on to the gills of a very young tadpole, it becomes sexually mature in the branchial cavity of the latter. The forms then remain very small, are without the vaginae and uterus, and die after the production of a single egg, without ever getting to the bladder.

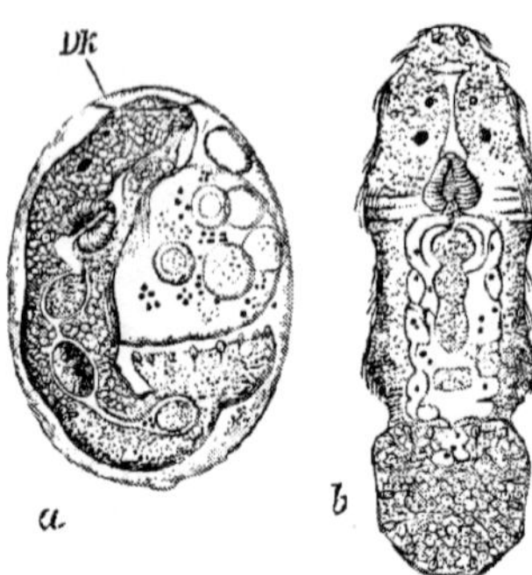

FIG. 186.—*a*, egg with embryo of *Polystomum integerrimum*; *b*, larva of same. *Dk* operculum (after Zeller).

Fam. 1. **Temnocephalidae.** With 4 to 12 tentacles at the anterior end, and a posterior sucker without hooks or marginal membrane. The integument consists of a cuticle, epidermis, and basement membrane; in some cases it is, in part, ciliated. There are rhabdites in the tentacles. Mouth subterminal, a muscular pharynx (absent in *Craspedella*), and an unbranched intestine. A pair of contractile excretory sacs opening anteriorly and dorsally. Genital pore single, ventral, and posterior; it leads into an atrium, which receives the opening of the uterus and the penis. Eggs with stalk-like appendages, by means of which they are sometimes attached. They live as external parasites on *Crustacea*, *Chelonia*, and *Mollusca* of fresh waters, and feed on Infusoria, small insect-larvae, Rotifers, and Crustacea. They have been found in Australia, New Zealand, Malay Archipelago, S. America, India, and may be regarded as connecting links between the *Turbellaria* and *Trematoda*. *Temnocephala* Blanchard; *Craspedella* Haswell.

Fam. 2. **Tristomatidae.** With flattened, discoidal, or elongated body; with two lateral anteriorly-placed suckers, and one large posterior sucker, often provided with radiations and hooks. Parasitic on the skin or the gills of marine fishes, or on the bodies of parasitic marine Crustacea.

Sub-fam. 1. **Tristomidae.** With flat body, two suckers, and a large ventral sucker; genital opening and opening of vagina usually on the left. On the skin and gills of marine fishes. *Nitzschia* v. Baer; *Epibdella* Blainv.; *Phyllonella* v. Ben. and Hesse; *Trochopus* Dies.; *Placunella* v. Ben. and Hesse; *Tristomum* Cuv.; *Acanthocotyle* Montic.; *Encotyllabe* Dies.

Sub-fam. 2. **Monocotylidae.** With flat body. Without lateral suckers, with small ventral sucker. Genital opening median, vagina paired. On skin, gills, or in cloaca of marine fishes. *Pseudocotyle* Ben. and Hesse; *Calicotyle* Dies.; *Monocotyle* Tschbg.

Sub-fam. 3. **Udonellidae.** With cylindrical body, with lateral suckers, and large simple ventral sucker. On parasitic Crustacea. *Udonella* Johnst.; *Echinella* v. Ben. and Hesse; *Pteronella* Ben. and Hesse.

Fam. 3. **Polystomatidae.** With a more or less distinctly marked-off adhesive organ at the hind end, bearing suckers and hooks; usually with two oral suckers. On the gills of fishes, in Amphibia and Reptiles on the skin, nasal passages, or bladder.

Sub-fam. 1. **Octocotylidae**. With two oral suckers and genital hooks; adhesive organ with 4, usually 8 small suckers. On the gills of marine and f.w. fish. *Octobothrium* Leuck.; *Pleurocotyle* Gerv. and Ben.; *Diplozoon* v. Nordm. (Fig. 187). The animal is double, two individuals being fused to form an x-shaped double animal, the posterior ends of which are provided

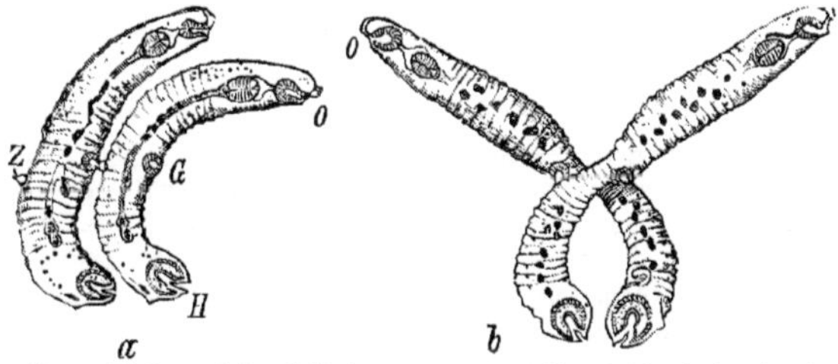

FIG. 187.—Young *Diplozoon* (after Zeller). *a*, two young *Diporpae* beginning to attach themselves together; *b*, after complete attachment. *O* mouth; *H* posterior suckers; *Z* dorsal papilla; *G* ventral sucker.

with two large suckers divided into four pits. In the young state they live solitarily as *Diporpa*; they then possess a ventral sucker and a dorsal papilla (Fig. 187, *a*). The *Diporpa* is without generative organs; these are not formed until after it has fused with another *Diporpa* to form the x-shaped *Diplozoon* (Fig. 187, *b*). The production of ova is confined to a definite period of the year, usually the spring. The eggs are laid singly after the formation of the thread by which they are attached, and two weeks later the embryo (Fig. 188), which only differs from *Diporpa* in the possession of two eye-spots and a ciliated apparatus upon the sides and on the posterior extremity of the body, is hatched. When an opportunity of fixing itself on the gills of a fresh-water fish occurs, the young animal loses its cilia and becomes a *Diporpa*, which possesses the characteristic apparatus for attachment, and sucks the branchial blood. The junction of two Diporpae can now take place; this is not effected as was formerly believed, by the fusion of the two ventral suckers, but in such a manner that the ventral sucker of each animal affixes itself to the dorsal papilla of the other, and fuses with it (Fig. 187, *b*). In this fusion the vas deferens of the one animal is directly connected with the opening of the vagina of the other. *D. paradoxum* v. Nord., on the gills of many fresh-water fish. *Anthocotyle* Ben. and Hesse; *Vallisia* Perugia-Parona;

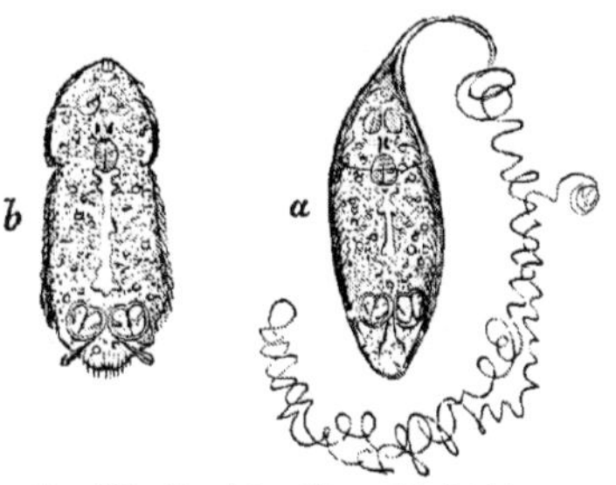

FIG. 188.—Egg (*a*) and larva (*b*) of *Diplozoon* (after Zeller).

Phyllocotyle Ben. and Hesse; *Hexacotyle* Blainv.; *Platycotyle* Ben. and Hesse; *Plectanocotyle* Dies.

Sub-fam. 2. **Polystomidae.** Without oral suckers and larger genital hooks. Adhesive disc usually with 6 suckers, and with hooks. On the gills of m. fishes, on the skin, gills, or in the bladder of Amphibia or Reptiles. *Polystomum* (Fig. 185) Zeder; *Onchocotyle* Dies.; *Erpocotyle* Ben. and Hesse; *Diplobothrium* Leuck.; *Sphyranura* R. Wright.

Sub-fam. 3. **Microcotylidae.** With two oral suckers, and with genital hooks. Adhesive disc with numerous suckers. On gills of m. fishes. *Microcotyle* Ben. and Hesse; *Gastrocotyle* Ben. and Hesse; *Axine* Abildgaard; *Pseudaxine* Par. and Per.

Sub-fam. 4. **Gyrodactylidae.** Usually without oral suckers. Anterior end with 2 or 4 cephalic lappets, or with sucker-like membrane; excretory organs opening at the hind end; adhesive disc usually with small radially arranged hooks, and 2 or 4 larger central hooks, without suckers. Reproduction by eggs which are either laid, or develop within the mother into a young form which may itself produce a second generation while still in the parent. v. Siebold believed that he had observed a young animal developing from a germ-cell of *Gyrodactylus*, and that this became pregnant during its development. He regarded the *Gyrodactylus* as an asexual form since he failed to find organs for the production of sperm. G. Wagener, however, showed that the reproduction is sexual, and conceived the idea that the germs from which the second and third generations are formed are derived from the remains of the fertilized ovum, from which the first generation is formed. Metschnikoff too is of the opinion that the individuals of the first and second generations are formed at the same time from a common mass of similar embryonic cells. On gills or integument of fishes. *Calceostoma* Ben. and Hesse; *Gyrodactylus* v. Nordm., *G. elegans* v. Nordm., from the gills of Cyprinoids and f.w. fishes; *Dactylogyrus* Dies.; *Tetraonchus* Dies.; *Amphibdella* Chatin; *Diplectanum* Dies.

Order 2. Aspidocotylea.

Body very variously shaped. Adhesive apparatus ventral, large, round, oval, or elongated, more or less distinctly marked off from the body, and possessing numerous suckers arranged in one or several rows, without armature. Mouth terminal or sub-terminal without oral sucker; oesophagus short, with a more or less developed pharynx; intestine saccular. Genital pore median, ventral, anterior to adhesive organ; penis sheath opens into widened end of uterus, which is much coiled; usually one testis; Laurer's canal absent; yolk-gland paired. Eggs without filaments. Excretory organs open posteriorly by one, somewhat dorsally placed, pore. Development direct with simple metamorphosis. A ciliated covering is not developed during the embryonic development. Parasitic in the alimentary canal and gall bladder in Chelonia and fishes, and in different organs in Molluscs.

Fam. **Aspidobothridae.** With the characters of the order. *Aspidogaster* v. Baer; *A. conchicola* v. Baer, from the pericardial cavity of *Anodonta* and *Unio*. *Platyaspis* Mont.; *Cotylogaster* Mcatic; *Macraspis* Olss.

Order 3. Malacotylea = Digenea.

Body variously shaped. Being internal parasites the adhesive apparatus is, as a rule, feebly developed, consisting of two suckers, one—the oral sucker—at the anterior end, and the other (less frequently present) on the ventral surface, or at the hind end. An armature of chitinous hooks is never present on the posterior sucker, and only exceptionally on the anterior sucker. Additional adhesive organs are occasionally present (*Holostomatidae*, *Amphistomatidae*). Mouth terminal or sub-terminal, exceptionally on the middle of the ventral surface (*Gasterostomum*), almost always surrounded by a sucker; eyes exceptionally present. Excretory organs open posteriorly by a terminal or dorsal pore. Hermaphrodite, except *Bilharzia* and possibly one or two other forms.* Genital organs open close to one another, or into an atrium. Genital pores usually on the ventral surface anteriorly, rarely placed posteriorly, or laterally, or at the front end. One ovary and usually two testes; yolk-gland paired, rarely unpaired. Laurer's canal usually present. Eggs usually without filaments, with operculum. Life-history with alternation of generations (digenetic, except *Holostomidae*) and more than one host. The sexual form is parasitic in the alimentary canal, and its appendages, of Vertebrates †; the asexual generations in Mollusca, and the encapsuled larva in Invertebrata and lower Vertebrata. Insectivorous birds are particularly infested by them. It may be of interest to *gourmets* to know that the trail of a woodcock largely consists of distomic Trematodes.

Fam. 1. **Holostomatidae (Metastatica).** With two suckers, and a peculiar adhesive apparatus behind the ventral sucker on the anterior region of the body. Body divided into an anterior flattened and a posterior cylindrical region (Fig. 189). Genital organs in the posterior region, and genital opening at the hind end. In the alimentary canal of birds, reptiles, and mammals, rarely in amphibia and fishes. The eggs are large and not very numerous, and hatch out

* Dimorphic forms are found in certain species of *Monostomum* and *Distomum* in connection with the division of labour of the sexual functions; one individual develops only male organs, and the other only female, the former producing spermatozoa, and the latter ova. The vestige of the functionless generative gland undergoes in these cases a more or less complete degeneration. Such forms are morphologically hermaphrodite, but practically of separate sexes.

† There are a few exceptions to this rule, *e.g.*, *Distomum Echiuri* in the nephridia of Echiurus pallasii, and *D. rhizophysae* in the gastrovascular apparatus of the siphonophore *Rhizophysa conifera*.

as ciliated larvae (of the miracidium type), which probably make their way into an intermediate host (fish, amphibian, mollusc, mammal), in which they develop into a larva known as *Tetracotyle* (usually with four suckers). The *Tetracotyle* probably becomes directly transformed into the adult when the intermediate host is eaten by the permanent host. This kind of development, in which there is a change of host but no alternation of generations, is called **metastatic.** *Diplostomum* Brds.; *Polycotyle* Will. and Schm.; *Hemistomum* Brds. (Fig. 189); *Holostomum* Nitzsch.

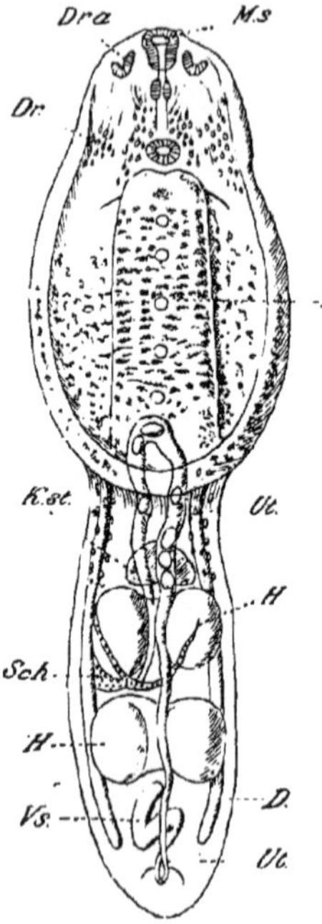

FIG. 189.—*Hemistomum clathratum* Dies., from the gut of *Lutra brasiliensis*, ventral view (from Bronn, after Brandes). *M.s* oral sucker; *K.st.* ovary; *Ut.* uterus, the opening of the uterus is rather indistinctly rendered, just below the lower *Ut.*; *H* testis; *V.s.* vesicula seminalis; *Dr.a* openings of glands *Dr.*; *z* ventral adhesive apparatus, the ventral sucker is just in front of this; *Sch* shell-glands.

Fam. 2. **Amphistomatidae.** Digenetic forms with two suckers; the posterior (ventral) sucker is terminal, and on it or just in front of it there are sometimes numerous papillae or pits for attachment. Genital opening median, ventral, in anterior third of body. Eggs with operculum, without filament. In alimentary canal of Vertebrata. *Amphistomum* Rud.; *Diplodiscus* Dies.; *Gastrodiscus* Cobb.; *Homalogaster* Poir.; *Gastrothylax* Poir.; *Aspidocotyle* Dies.

Fam. 3. **Distomatidae.** Digenetic forms with two suckers; the posterior sucker is on ventral surface. Genital opening median, in anterior third of body, anterior to ventral sucker, rarely behind the latter or lateral. Eggs usually with operculum, without filament. Laurer's canal usually present. In alimentary canal and its appendages of Vertebrates. *Distomum* Retz., *D. hepaticum* L. liver-fluke (Fig. 179); with conical anterior end and numerous spine-like prominences on the surface of the broad, leaf-shaped body, which is about 33 mm. long. In the bile-ducts of sheep and of other domestic animals, causing sheep-rot. It is occasionally found in man, and bores its way into the portal vein and into the system of the vena cava The embryo is hatched after the egg has been some time in water as a ciliated larva (Fig. 182), with an ×-shaped eye-spot; this passes into the water-snail *Limnaea truncatula*, and there casts its ciliated skin and emerges as a *sporocyst*. The sporocyst produces *rediae*, which produce more rediae or *Cercariae.* The Cercariae, which are provided with long tails, leave the host, swim about for a short time in water, and then encyst upon foreign objects, *e.g.*, blades of grass. In this condition they are eaten by the sheep. *D. crassum* Busk, perhaps identical with *D. rathouisi* Poir. (Fig. 190) in the intestine of Chinese. *D. lanceolatum* Mehlis, body lancet-shaped, 8-9 mm. long, lives in same place as *D. hepaticum*; the miracidium is pear-shaped, ciliated on anterior half of body only, and bears a styliform spine on the projecting apex. *D. conjunctum* Cobb., lancet-shaped, 12 mm. long, in the liver of dog, rarely on man, East Indies. *D. spathulatum* R. Leuck. =*D. sinense* Cobb., in liver of man and cat in Japan and China; *D. pulmonale* Bölz, in the lungs of man in Japan and China. *D. ophthalmobium* Dies., a doubtful species, 4 specimens only found, in the lens-capsule of a nine-months child. *D. heterophyes* in the intestine of man in Egypt. *D. macrostomum* Rud. (Fig. 191), in the intestine of insectivorous birds, with genital pore at hind end. The eggs are consumed by the snail, *Succinea amphibia*, in the gut of which the miracidium is set free; this makes its way through the gut-wall into the tissues, and becomes a branched sporocyst, known as *Leucochloridium paradoxum*. Some of the branches extend into the tentacles, to which they give a peculiar appearance by their colouring of green and white bands and red tip. A bird, attracted by this, pecks off the tentacle, and so swallows a branch of the sporocyst, in which have been formed tailless Cercariae. The latter are thus transferred to the intestine of the final host. The remarkable feature about this life-history is that the Cercaria is never free, and is without a

FIG. 190.—*Distomum rathouisi* Poir., after Leuckart.

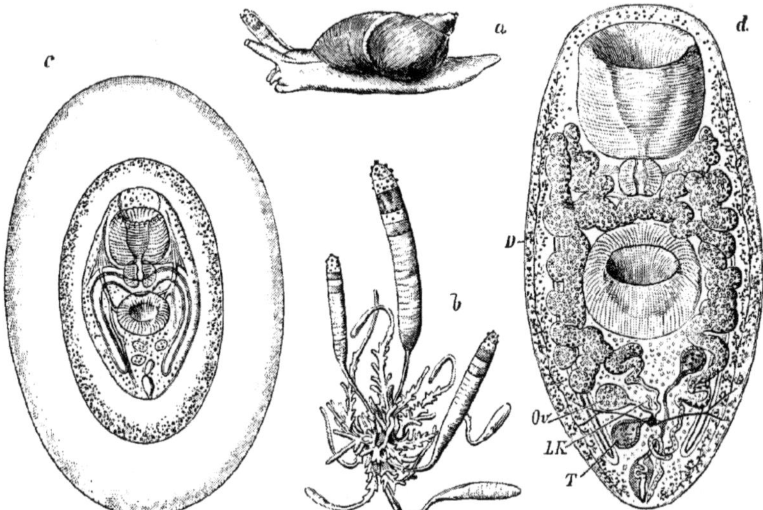

FIG. 191.—Life-history of *Distomum macrostomum*, after Heckert. *a*, *Succinea amphibia* containing a ripe sporocyst of a *Leucochloridium* in its right tentacle. *b*, *Leucochloridium paradoxum* isolated. *c*, Cercaria (tailless) ready for transference in a double membrane. *d*, sexual *Distomum macrostomum*; *D* yolk-glands; *T* testis; *Ov* ovary; *LK* canal of Laurer. The openings of the vas deferens and uterus are at the hind end.

tail. *Rhopalophorus* Dies.; *Koellikeria* Cobb., dioecious, living in pairs contained in cysts in the mucous membrane of the mouth and branchial cavity of fishes; the one individual is cylindrical and narrow, and produces spermatozoa; the other is swollen in the middle and posterior region of the body, and is filled with eggs. The dissimilar condition of the two individuals is probably due to the fact that copulation only leads to the fertilization of one of them, which alone is able to perform the female sexual functions. *K. filicolle* Rud. (*D. Okenii* Köll.) in *Brama Raji*. *Bilharzia** Cobb. (Fig. 192), dioecious; body elongated, nematode-like; the female is slender and cylindrical; the male has powerful suckers and the margins of its body are bent round so as to form a groove—the *canalis gynaecophorus*—for the reception of the female; they live in pairs in the blood-vessels of mammals. *B. haematobia* Cobb., portal veins and veins of bladder of man in most parts of Africa. The embryos, which are ciliated, escape by the urethra. By the deposition of masses of eggs in the vessels of the mucous membrane of the ureter, bladder, and great intestine, inflammation is set up which may cause haematuria and stone. *B. crassa* Sons., in *Bos taurus domesticus*; *B. magna* Cobb., in *Cercopithecus fuliginosus*.

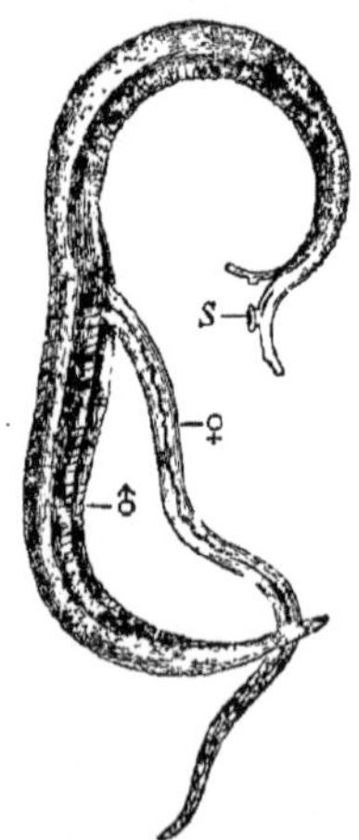

Fig. 192.—*Bilharzia haematobia.* Male and female, the latter being in the *canalis gynaecophorus* of the former. *S* sucker.

Fam. 4. **Gasterostomatidae.** With the anterior sucker only; mouth on the ventral surface, not in the sucker; eggs with operculum, without filament. Genital opening terminal, at hind end. The Cercaria is known as *Bucephalus*, and has a bifid tail. In alimentary canal of fishes. *Gasterostomum* v. Sieb.

Fam. 5. **Didymozoonidae.** With the anterior sucker only; mouth in the sucker; gut present or absent. Hermaphrodite, but live in pairs in cysts. Eggs with operculum without filament. Genital pore in front of the oral sucker, terminal. On the integument, or in the mouth and branchial cavity of marine fishes. *Didymozoon* Tschbg.; *Nematobothrium* v. Ben.

Fam. 6. **Monostomatidae.** With only one anterior sucker. Mouth in the sucker. Genital pore usually in front, median, ventral. Eggs often with two filaments. Laurer's canal usually absent. In alimentary canal of most vertebrate classes. *Monostomum* Zed.; *M. lentis* v. Nord., the young form without generative organs is found in the lens of the human eye; *M. bipartitum* Wedl., living in pairs enclosed in a common cyst, the one individual surrounded by the posterior lobe of the other, branchiae of Tunny-fish; *Notocotyle* Dies.; *Ogmogaster* Jaegerskiöld; *Opisthotrema* Leuck.

THE DICYEMIDAE AND ORTHONECTIDAE.†

The systematic position of these forms is obscure: it will, however, be convenient to place them provisionally near the *Trematoda*. They all consist of an

* G. Fritsch, "Zur Anat. d. Bilharzia haematobia Cobb.," *Arch. f. Mic. Anat.*, 31, 1888.

† C. O. Whitman, "A contribution to the embryology, life-history, and classification of the Dicyemids," *Naples Mittheilungen*, Bd. 4. 1882. Julin C., "Contribution à l'histoire des Mesozoaires," *Arch. de Biologie*, Bd. 3, 1882.

outer layer of ciliated cells surrounding an inner mass, which in the *Dicyemidae* consists of a protoplasmic mass with many nuclei, and in the *Orthonectidae* (single genus *Rhopalura*) of a mass of cells compacted together. None of them have a digestive cavity. The *Dicyemidae* are parasitic in the kidneys of *Cephalopoda*, the *Orthonectidae* in the gut of Turbellarians, in the body-wall and tissue-spaces of Nemertines, and in the body-cavity and brood pouches of Ophiurids.

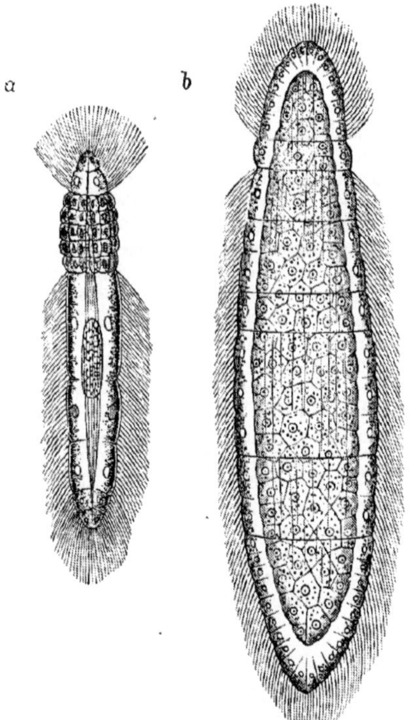

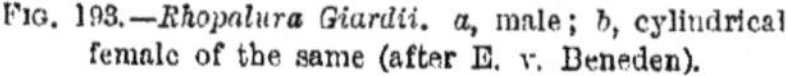

FIG. 193.—*Rhopalura Giardii.* *a*, male; *b*, cylindrical female of the same (after E. v. Beneden).

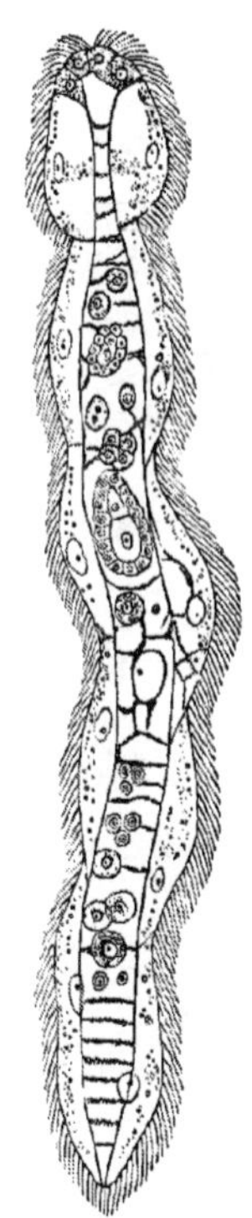

FIG. 194.—*Dicyemopsis macrocephalus* (after v. Beneden).

In the *Dicyemidae* egg-like germs are found in the central plasmic mass which give rise to embryos: no spermatozoa have been found (Fig. 194). In *Orthonectidae* both spermatozoa and ova are found in the central mass, and in different individuals. No light is thrown upon their affinities by their development. The *Orthonectidae* possess a layer of fibres, presumably muscular, between the outer cells and the inner mass. The males are smaller than the females (Fig. 193); and there are two kinds of females—the cylindrical forms (Fig. 193, *b*) and the flattened forms. The history of these forms is obscure. Both forms produce eggs, and are supposed to leave the host to wander into a new host, where the cylindrical female ends her existence in the act of expelling her eggs, while the flat female breaks up into a number of sacs, each enclosing a number of ova. The eggs of the cylindrical female are probably fertilised, while those of the flattened form are supposed to develop parthenogenetically.

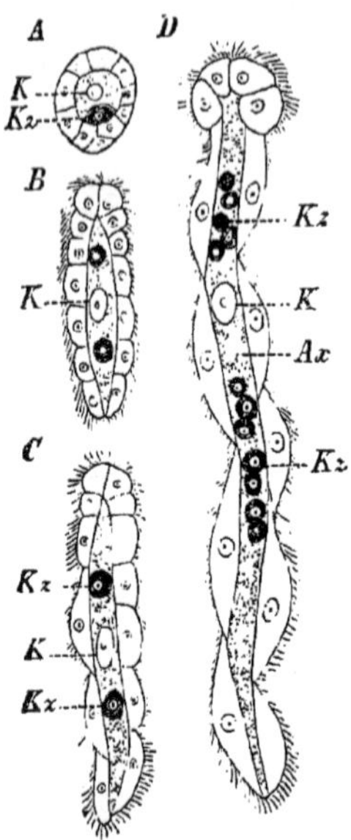

FIG. 195.—*A–D*, stages in the development of the vermiform embryos of *Dicyema*; *A*, of *Dicyemennea eledones* (after Whitman); *B–D*, of *Dicyema typus* (after E. van Beneden). *Ax* axial cell; *K* nucleus of the axial cell; *Kz* germ cells (from Korschelt and Heider).

The *Dicyemidae* have two kinds of embryos (hence the name of the group)—the *infusoriform* (Fig. 196) and the *vermiform* (Fig. 195)—which arise from the egg-like germs in the central plasmic mass. They arise in individuals of slightly different form—the individuals producing vermiform embryo are called *nematogens* and are longer and thinner than those producing infusoriform embryos, which are called *rhombogens*. The rhombogens, however, after producing a certain number of infusoriform embryos become nematogens and produce vermiform embryos.

The vermiform embryo changes directly into the parent form. The infusoriform embryo is very different from the parent, and its fate is unknown. Very possibly it has the power of making its way out of its host, and so distributing the species. No males or spermatozoa have been observed in the *Dicyemidae*, and it has been suggested that the infusoriform embryos are immature males, or that they contain the male elements. Whitman indeed states that he has found them in a modified form within nematogenic adults. The origin of the egg-like germs within the central plasmic mass has been described as a case of endogenous cell-formation; *i.e.* a nucleus gathers round itself a certain amount of protoplasm, which becomes delimited, to

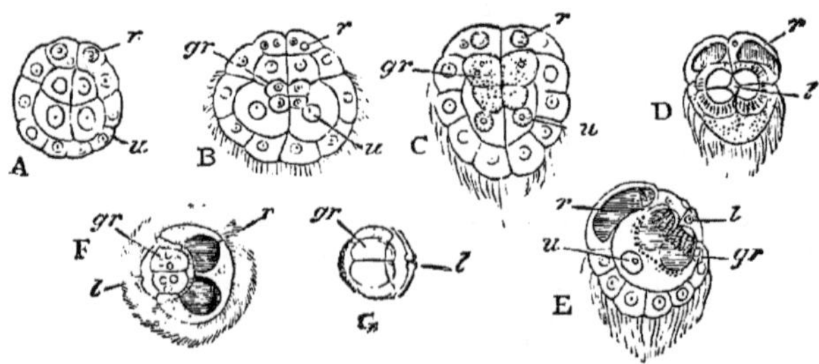

FIG. 196.—Infusoriform embryos and their development. *A–D*, of *Dicyema typus*; *E–G*, of *Dicyemella Wagneri* (after van Beneden from Balfour). *A–C*, developing embryos; *D*, embryo from the ventral side, *E* from the right side, *F* from the front; *G* side view of the urn isolated. *gr* granular bodies in the urn; *l* bed of urn; *u* floor of the urn; *r* refractive bodies.

an extent and in a manner which has not been ascertained, from the surrounding plasma.

By some naturalists the *Dicyemidae* and the *Orthonectidae* have been regarded as survivals of a most primitive Metazoan group—of a group possibly intermediate between the Protozoa and the Metazoa, and they have been grouped together as *Mesozoa*. There does not, however, appear to us to be any sufficient reason for this view, especially when we remember their parasitic habit. We are inclined to regard them as allied to the *Trematoda*, to the miracidium larva of which they do present some considerable resemblance.

Trichoplax, which may be mentioned here, has been found in salt-water aquaria (F. E. Schulze, *Zool. Anzeiger* 6). It is a small, flattened organism (1–3 mm. in diameter), and consists of a sponge-work of protoplasm, with nuclei at the nodes, and continuous with processes of the surface epithelial cells. The latter are ciliated on the lower surface of the animal. Nothing is known of the reproduction.

Salinella is another form which may be mentioned here (Frenzel, *Arch. f Naturg.*, 58, 1891). It has only been found in aquaria.

Order 3. Cestoda.*

Elongated and usually segmented Platyhelminthes without mouth or alimentary canal, with organs for attachment at the anterior extremity.

The tape-worms (Fig. 196), which may easily be recognised by their band-shaped, usually segmented bodies, are parasitic in the alimentary canal of Vertebrata, and were formerly taken for single animals. Steenstrup was the first to introduce a different view, according to which the tape-worm is a colonial animal (*Strobila*), a chain of single animals, each segment or *proglottis* being an individual. There are, however, *Cestoda*, like *Caryophyllaeus* (Fig. 212), which are destitute both of external segmentation and of

* Besides the older works and papers of Pallas, Zeder, Bremser, Rudolphi, Diesing, and others, compare van Beneden, "Les vers cestoïdes ou acotyles," Brussels, 1850. Küchenmeister, *Ueber Cestoden im Allgemeinen und die des Menschen insbesondere*, Dresden, 1853. V. Siebold, *Ueber die Band- und Blasenwürmer*, Leipzig, 1854. G. Wagener, "Die Entwickelung. der Cestoden," *Nov. Act. Leop.-Car.*, tom. 24, Suppl., 1854. G. Wagener, *Beitrag zur Entwickelungsgeschichte der Eingeweidewürmer*, Haarlem, 1857. R. Leuckart, *Die Blasenbandwürmer und ihre Entwickelung*, Giessen, 1856. R. Leuckart, *The Parasites of Man*, vol. 1, 1886, London. F. Sommer and L. Landois, "Ueber den Bau der geschlechtsreifen Glieder von Bothriocephalus latus," *Zeitschr. f. wiss. Zool.*, 1872. F. Sommer, "Ueber den Bau und die Entwickelungsgeschichte der Geschlechtsorgane von Taenia mediocanellata und Taenia solium," *Ibid.*, tom. 24, 1874. M. Braun, *Zur Entwick. gesch. des breiten Bandwürmes* (*Bothriocephalus latus*), Wurzburg, 1883. H. Shauinsland, "Die embryonalen Entwick. d. Bothriocephalen," *Jena, Zeitschr.*, 19, 1885. L. Niemiec, "Ueb. d. Nervensystem d. Cestoden," *Arb. a. d. Zool. Inst. Wien.*, 7, 1887. Fr. Zschokke, *La Structure anat. et hist. des Cestoïdes*, Geneva. 1888. B. Grassi u. G. Rovelli, "Embryol. Forsch. an Cestoden," *Centralbl. f. Bakteriol.*, 5, 1889. M. Braun, "Vermes" in Bronn's *Thierreich*, 4, 1895. O. v. Linstow, *Compendium der Helminthologie*, Hannover, 1878; and *Nachtrag* to the same, 1889.

segmentation of the generative organs; while in other cases the segments of the body are clearly differentiated, and each is provided with a set of generative organs, but they do not attain individual independence (*Ligula*). The *proglottides*, however, usually become separated off, and after their separation from the body of the tape-worm continue to live for some time independently, and in some cases even increase considerably in size, if they remain within the intestine of their host (*Echeneibothrium*, *Calliobothrium*, etc.). The proglottis after separation may remain for a certain time in the intestine, but eventually makes its way either passively in the faeces (*e.g.* *T. solium*), or actively by its own movement (*T. saginata*) to the exterior per anum. Here it retains its vital power for a short time, and crawling away from the faeces ascends the stalks of plants. It soon dies, and the body decomposes and the eggs are scattered; or in some cases the eggs escape through a rupture in the body-wall, and are left as a trail on the objects over which the proglottis crawls. The eggs soon lose their vitality in a dry atmosphere. The proglottides of *T. saginata*, which have considerable powers of movement, have been found on the wall a yard above the bed of their quondam host, and they frequently creep over his warm body.

These facts seem to be sufficient to justify the view that the tape-worm is a colonial or polyzoic animal, the individual members of which have the power of separate and independent life. At the same time the existence of monozoic forms like *Archigetes* (Fig. 211) and *Caryophyllaeus* (Fig. 212) must not be forgotten: in these—the *Cestodariidae*—there is only one set of generative organs, and the body is unsegmented; the head and body not being sharply distinguished. It would appear that these monozoic forms of *Cestoda* have the same relation to the proglottis that *Lucernaria* has to *Aurelia* in the *Acalephae*. Just as the *Lucernaria* may be compared to a *Scyphistoma*, which develops generative organs and does not strobilate, so an *Archigetes* may be looked upon as a scolex, which becomes sexual but does not bud off proglottides (*i.e.* does not develop into a strobila).

This is the only satisfactory mode of regarding the *Cestoda*; especially as the entire tape-worm, and not the proglottis alone, corresponds to the Trematode, and is to be regarded as being derived from the latter by a simplification of organization and loss of the alimentary canal.

The anterior part of the tape-worm is narrow, and presents a

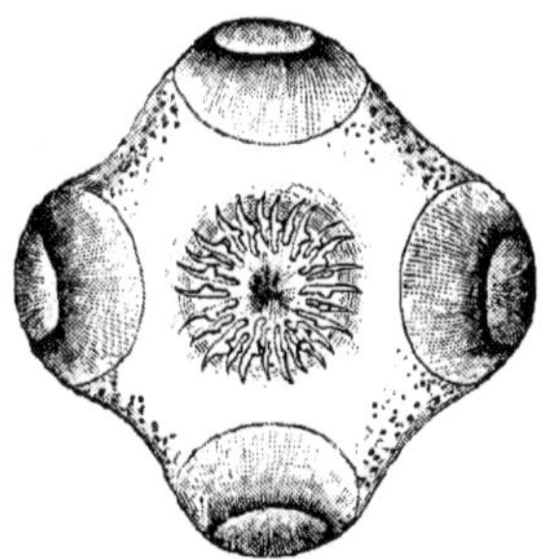

FIG. 196*a*.—Head of *Taenia solium*, viewed from the front (apical surface), with rostellum and double circle of hooks. The four suckers are visible (from Claus).

terminal swelling by which it attaches itself. This anterior swollen part is distinguished as the head of the tape-worm, but it is mainly its external form which entitles it to this name. In *Caryophyllaeus* the head armature is very weak, and consists of a lobed fringed expansion. The apex of the head often ends in a conical projection, the *rostellum*, which is armed with a circle of hooks, while the lateral surfaces of the head are furnished with four suckers (*Taenia*, Fig. 196*a*). In other cases only two suckers are present (*Bothriocephalus*); or we find suckers of more complicated structure and beset with hooks (*Acanthobothrium*), or four protrusible proboscises beset with recurved hooks (*Tetrarhynchus*, Fig. 198); while in other genera the head armature presents various special forms.

That portion of the animal, which follows the head and is distinguished as the neck, shows, as a rule, the first traces of commencing segmentation. The rings, which are at first faintly marked and very narrow, become more and more distinct and gradually larger the further they are removed from the head. At the posterior extremity the segments or proglottides are largest, and have the power of becoming detached.

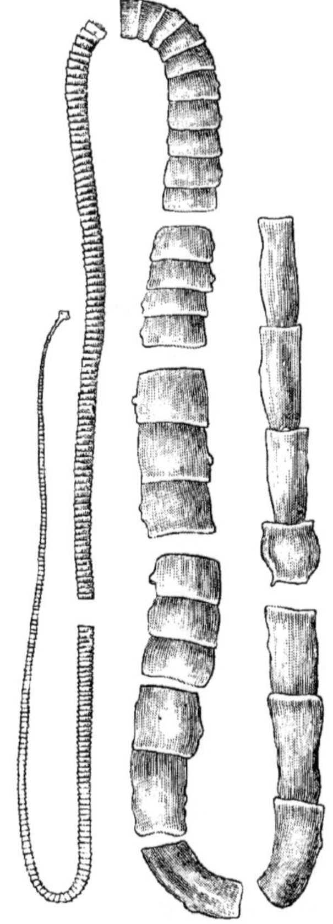

FIG. 197.—*Taenia saginata* (*mediocanellata*), natural size (after R. Leuckart).

FIG. 198. — Young *Tetrarhynchus* with beginning segmentation. The four excretory canals with the connecting loops in the head, and the terminal vesicle *B* are visible (from Claus).

The simplicity of the internal organization corresponds with the simple appearance of the external structure. Like the *Trematoda*, the *Cestoda* are said to possess no epithelial ectoderm. Beneath the cuticle-like outer membrane is a layer of spindle-shaped cells lying at right angles to the surface; their external ends abut upon the cuticle, and their inner ends are prolonged as fibres into the parenchyma. Beneath this layer there is a delicate superficial layer of longitudinal muscular fibres, and next a parenchyma of connective tissue, in which strongly-developed bundles of longitudinal muscular fibres, as well as an inner layer of circular muscles, are embedded; both these muscular layers are traversed, principally at the sides of the body, by groups of dorso-ventral muscular fibres. The power which the proglottis possesses of altering its form is due to the interaction of all these muscles. By means of them it is able to shorten itself considerably, at the same time becoming much broader and thicker, or to elongate to double its normal length, becoming much thinner. In the connective tissue parenchyma of the body, not only the muscles, but all the other organs are embedded. In its peripheral portion, especially in the neighbourhood of the head, we find small densely packed **calcareous concretions**, which are probably contained in connective tissue cells.

The **nervous system** consists of two lateral longitudinal cords passing externally to the main trunks of the excretory system. They are somewhat swollen in the head, where they are connected by a transverse commissure: these anterior swellings and the commissure may represent a cephalic ganglion. In *Moniezia* the lateral nerves are connected by two transverse commissures at the hind end of each proglottis. Distinct sense-organs are wanting, but the tactile sense may be ascribed to the skin, especially to that of

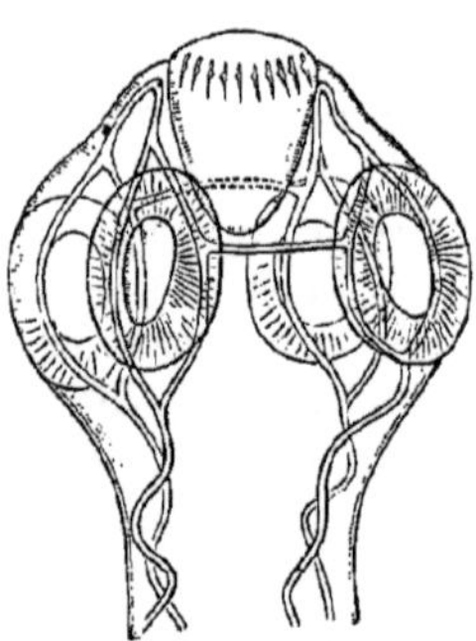

FIG. 199.—Head of a *Taenia* with the four suckers, and the connecting loops of the excretory canal (after Pintner).

the head and the suckers. An alimentary canal is also wanting. The nutritive fluid, already prepared for absorption by the host, passes endosmotically through the body-wall into the parenchyma.

The **excretory apparatus**, on the contrary, attains a considerable development as a system of much ramified canals which are distributed throughout the whole body.* It consists primarily of two longitudinal canals (a dorsal and a ventral †), running along each side of the body and connected in the head and in each segment by transverse trunks (Figs. 198, 199). According to the state of contraction of the muscular system these longitudinal trunks and cross branches appear sometimes straight and sometimes bent in a wavy or zigzag manner: their breadth also presents considerable variation, so that the power of contraction has been ascribed to their walls. The longitudinal trunks only serve as the efferent ducts of a system of very fine vessels which ramify throughout the whole parenchyma and receive numerous long tubes: the latter begin in the parenchyma with closed funnels, which contain a vibratile ciliated lappet (Fig. 200). The larger vessels are said to contain valves. In many cases, as in *Ligula* and *Caryophyllaeus*, these longitudinal trunks are broken up into numerous longitudinal vessels, which are connected by transverse anastomoses. In other cases, on the other hand, the two ventral vessels are enlarged at the cost of the two dorsal, which may entirely atrophy. The external opening of the excretory system is, as a rule, placed at the posterior end of the body, *i.e.*, at the hind end of the last segment, in which a small vesicle with an external opening receives the longitudinal trunks. According to the observations of Leuckart on *Taenia cucumerina*, the posterior transverse canals in the segments immediately preceding

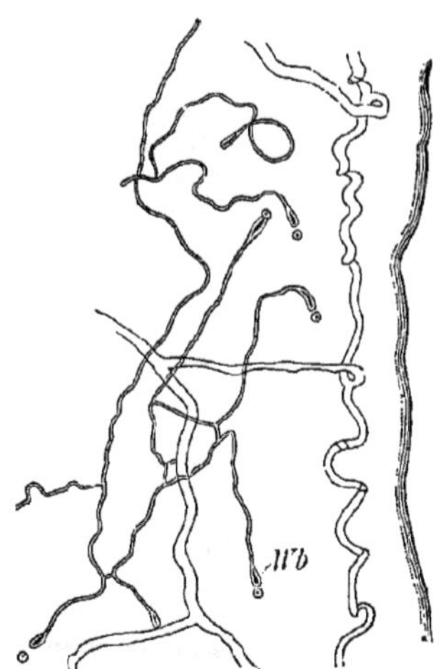

FIG. 200.—A portion of the excretory system of *Caryophyllaeus mutabilis* (after Pintner). *Wb* ciliated funnels with the nucleus of the cell belonging to them.

* Compare Th. Pintner, "Untersuchungen über den Bau des Bandwurmkörpers," Wien, 1880.

† These surfaces are distinguished by the generative apparatus (see below).

the last become, by their gradual shortening and the approach of the longitudinal trunks, transformed into the vesicle, which acquires an external opening when the segment behind it is detached. In rare cases the excretory system possesses additional openings in the anterior part of the body behind the suckers, and elsewhere.

The **generative apparatus** is repeated in each proglottis. The male apparatus consists of numerous small vesicles, the testes (Fig. 201, *T*), which are situated nearer one surface of the proglottis than the other; this surface is distinguished as the **dorsal** surface. Delicate ducts proceed from the testes to open into a common efferent

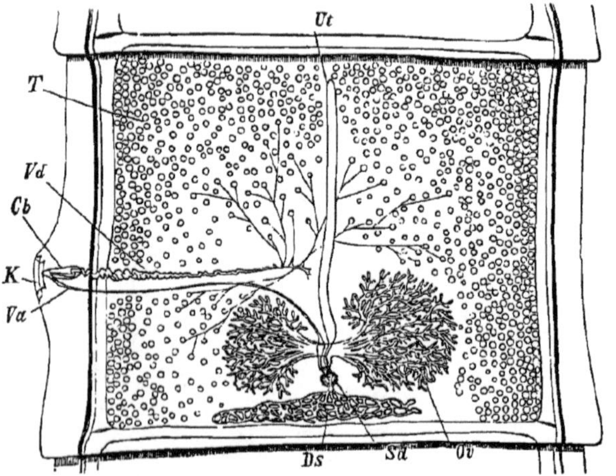

FIG. 201.—Proglottis of *Taenia saginata*, with male and female organs (after Sommer). *Ov* ovary; *Ds* yolk-gland (vitellarium); *Sd* shell-gland; *Ut* uterus; *T* testes; *Vd* vas deferens; *Cb* pouch of the cirrus; *K* generative cloaca; *Va* vagina.

duct (*vas deferens*). The coiled end of this duct lies in a muscular pouch (*cirrus sheath*), whence it can be protruded through the genital opening as the so-called cirrus. This cirrus is frequently beset with spines which are directed backwards, and serves as a copulatory organ. The female generative organs consist of *ovary*, *yolk-gland*, *shell-gland*, *uterus*, *receptaculum*, and *vagina*. The vagina and vas deferens usually open into a common genital cloaca, which lies either on the ventral surface of the segment (*Bothriocephalus*), or on the lateral margin (*Taenia*) (Fig. 201). In the last case it may be placed alternately on the right and on the left side. Sometimes the two genital openings are widely separate, the male opening being placed at the side, the female on the surface of the segment. In other cases there are two sets of generative organs in each segment, opening on the right and left margins (*Dipylidium*). In the *Bothriocephalidae* and other forms (possibly in all Cestodes except the *Taeniadae*) the uterus has a special opening of its own to the exterior in addition to the vagina, which opens close to the vas deferens (Fig. 216). This opening is comparable to the uterine opening in *Trematoda*, differing from the latter in being remote from the male opening. The vagina of Cestodes, which is not used for the exit of eggs, even when there is no uterine opening (see above, p. 244), must be compared to the canal of Laurer in the *Trematoda*. In forms with a special uterine opening, eggs are deposited through it while the proglottis is part of the chain in the intestine. When there is no uterine opening, the eggs are only set free by the rupture of the proglottis after it has broken away from the chain and reached the exterior.

As the segments increase in size and become further removed from the head, the contained generative organs gradually reach maturity in such a way that the male generative organs arrive at maturity rather earlier than the female. As soon as the male elements are mature, copulation is said to take place, and the receptaculum seminis is filled with sperm, and then only do the female generative organs reach maturity.

The method of sperm-transference is not fully understood. The penis is sometimes found projecting from the genital opening (Fig. 207), and Leuckart states that he has seen it inserted into the vagina of the same proglottis, thus being in a position to effect self-fertilization. In other cases it has been found (Pagenstecher) inserted into the vagina of another proglottis of the same chain. There is also the possibility of copulation between the proglottides of different chains in the same host, though this has not been observed; and it may be that in some cases the penis is used, like that of the *Turbellaria*, for hypodermic injection of spermatozoa. Finally we must admit the possibility of the sperm passing out into the fluids of the intestine in which the body of the tape-worm is bathed, and of the spermatozoa so set free migrating in sufficient numbers into the vagina of other proglottides.

The ova are fertilized and pass into the uterus, which then assumes its characteristic form and size. As the uterus becomes distended, the testes and then the ovaries and vitellaria are more or less completely absorbed (Fig. 202). The posterior proglottides, viz., those which are ready for separation, have alone undergone full development, and the eggs in their uterus often contain completely developed embryos. Accordingly we can recognize in a continuous series of the segments the course of development passed through by the sexual organs and products in their origin and gradual progress towards maturity. An examination of the segments between that with the first trace of the generative organs and the first proglottis with fully developed organs will give us an idea of the stages of structure through which each segment has to pass. The tape-worms are oviparous; either the embryo develops within the egg-shell in the body of the mother, or the development takes place outside the proglottis, for example, in water (*Bothriocephalus*).

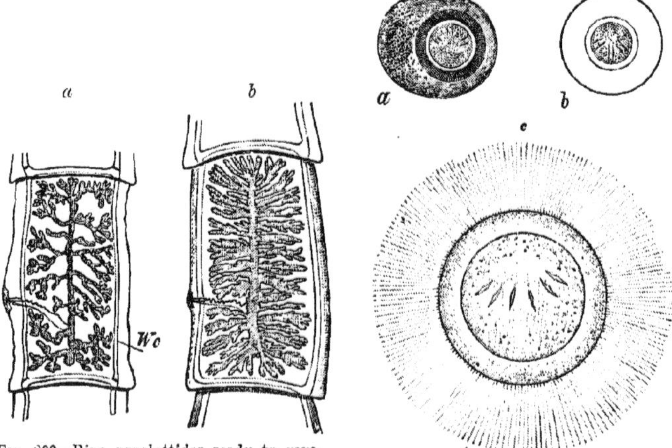

FIG. 202—Ripe proglottides ready to separate. *a*, of *Taenia solium*; *b*, of *Taenia saginata*. *Wc* water-vascular (excretory) canal (from Claus).

FIG. 203.—Egg with embryo: *a*, of *Taenia solium*; *b*, of *Microtaenia*; *c*, larva of *Bothriocephalus latus* (after R. Leuckart).

The eggs are round or oval and of small size (Fig. 203): they consist of the minute ovum embedded in yolk-cells and surrounded by a membrane, which is thin when the development takes place in the uterus, and thick and provided with an operculum when it occurs only after oviposition (*Bothriocephalidae*, etc.). The early development appears to be closely similar to that of Trematodes; the segmentation is complete, and is followed by the epibolic formation of two membranes, of which the outer, lining the inner surface of the shell in the *Bothriocephalidae* (Fig. 204), is called the *shell* or *enveloping membrane*, the inner being the so-called ectoderm or *outer layer*. These two membranes surround the rest of the embryo, which we shall call the *inner mass*. The enveloping membrane is

left in the shell at hatching, and the outer layer either becomes thick and ciliated, as in some *Bothriocephalidae* (Fig. 203, *c*), or thin and not ciliated as in other *Bothriocephalidae* and Taenias which inhabit aquatic animals, or finally it becomes cuticularised as a thick, radially striated layer (Fig. 203, *a*), as in the Taenias which infest land animals (protection against desiccation). Meanwhile, either while the embryo is free (*Bothriocephalidae*), or while it is within the uterus of the mother, the inner mass develops six hooks, and becomes the *six-hooked embryo* or **Onchosphere** (Fig. 205, *b*). In the *Bothriocephalidae* the further history of the embryo is unknown; it loses its ciliated coat and probably soon dies, unless it migrates into its next host, which is unknown. In other cases, in which the life-history has been followed, the development of the embryo into the asexual scolex rarely takes place in the intestine of the original host. It is said to do so in *Taenia* (*Hymenolepis*) *nana* (see p. 261), and it has been suggested that it might occur in the stomach of the same animal, if by reversed peristalsis a ripe proglottis was passed back into the stomach, and there digested (*T. saginata*).

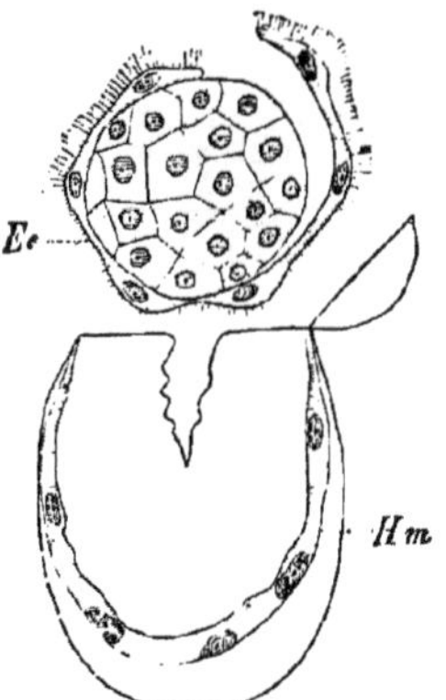

FIG. 204.—Embryo of *Bothriocephalus latus* pressed out of the egg. *Ec* outer layer (so-called ectoderm); *Hm* the shell (enveloping) membrane.

As a rule the **Scolex**, *i.e.* the head and neck of the tape-worm, is developed from the six-hooked embryo in another host, and in some cases (*Coenurus*, *Echinococcus*) more than one scolex arises from a single embryo (Fig. 206).

The eggs usually leave the intestine of the host in the proglottis, either by active migration or in the faeces. The proglottides are deposited on the ground or in water. Here they crawl about and deposit their eggs as described on page 244, but they soon die, especially if the temperature is unfavourable and the air dry. The embryo, which in the case of the land forms is protected against desiccation by the thick cuticle described above, retains its vitality for a time, which depends on the external conditions. Eventually it dies, unless it passes into the stomach of a suitable host. As a rule this host is an herbivorous or omnivorous animal, but it may be a carnivorous animal. The embryos are usually taken up in the food, or in drinking water, but occasionally they enter accidentally in consequence of dirty habits.* As soon as the egg membranes are digested or burst by the action of the juices of the stomach of the new host, the embryos, or onchospheres as they are called, which have been thus set free, bore their way into the gastric or intestinal vessels by means of their six (rarely four) hooks, the points of which can be approached and removed from one another over the periphery of the small globular embryonic body (Fig. 205, *b*). When they are once within the vascular system, they are no doubt carried along passively by the current of blood, and transported by a longer or shorter route into the capillaries of the different organs, as the liver, lungs, muscles, brain, etc. After losing their hooks, they usually become enveloped by a cyst of the connective tissue of their host,

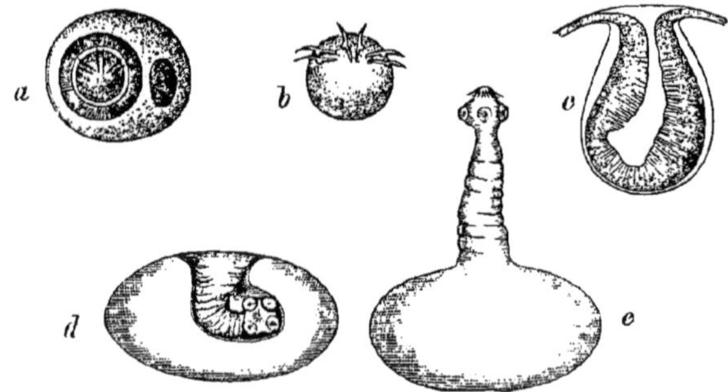

FIG. 205.—Stages in the development of *Taenia solium* to the *Cysticercus* stage (partly after R. Leuckart). *a*, egg with embryo; *b*, free embryo; *c*, rudiment of the head as a hollow papilla on the wall of the vesicle; *d*, bladder-worm with retracted head; *e*, the same with protruded head, magnified about four times.

and grow into large vesicles with liquid contents and a contractile wall. The vesicle gradually becomes a *cystic* or *bladder worm* by the formation of one (*Cysticercus*,† Fig. 205, *e*) or several (*Coenurus*) hollow buds, which are developed from the walls and project into the interior of the vesicle (Fig. 205, *c*). The armature of the tape-worm head (suckers and double circle of hooks) is formed on the inside and at the bottom of this invagination of the wall of the vesicle (Fig. 205, *d*). When these hollow buds are evaginated so as to form external appendages of the vesicle, they present the form and armature of the Cestode head, as well as a more or less developed

* The habit of allowing dogs to lick the face and to feed off plates which their owners use is, to say the least of it, an unclean one, and should be avoided.

† Exceptionally two or more heads are found in some Cysticercus forms.

neck, which presents even at this stage traces of segments (Fig. 205, *e*). The head and neck together constitute the **scolex.** In some cases (*Echinococcus*) the irregularly shaped maternal vesicle produces from its internal walls one or two generations * of secondary vesicles which project into it; and the Cestode heads originate in special small brood-capsules on these secondary vesicles (Fig. 206, *a*). In such cases the number of tape-worms which arise from one embryo

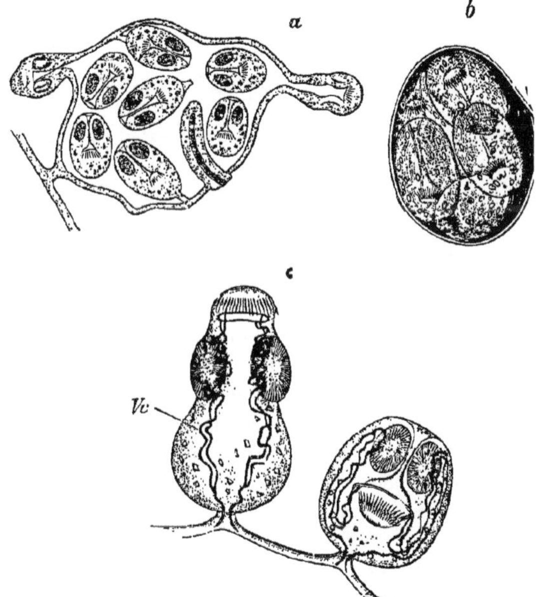

FIG. 206.—*a*, brood-capsule of *Echinococcus* with developing heads (after R. Leuckart). *b*, brood-capsule of *Echinococcus* (after G. Wagener). *c*, heads of *Echinococcus* still connected with the wall of the brood-capsule—one is evaginated; *Vc* excretory canals.

is naturally enormous, and the parent vesicle may reach a very considerable size, being sometimes as large as a man's head. In consequence of this enormous growth the vesicles frequently obtain an irregular shape; while on the other hand, the tape-worms which are developed from them remain very small, and carry, as a rule, only one ripe proglottis (Fig. 207). The cyst in which the bladder-worm lies, and which is caused by it, is called an **hydatid cyst.**

* In *Cysticerci* (*C. longicollis, tenuicollis*) also, sterile daughter vesicles are sometimes budded off.

So long as the tape-worm head (*scolex*) remains attached to the body of the bladder-worm and in the host of the latter, it never develops into a sexually mature tape-worm; although in many cases it grows to a considerable length (*Cysticercus fasciolaris* of the house-mouse). The bladder-worm must enter the alimentary canal of another animal before the head (*scolex*) can, after separation from the body of the bladder-worm, develop into the sexually mature tape-worm. This transportation is effected passively, the new host eating the flesh or organs of the animal infected with Cysticerci. The tape-worms, therefore, are principally found in the *Carnivora*, the *Insectivora*, and the *Omnivora*, which receive the bladder-worms in the flesh of the animals on which they feed. The vesicles are digested in the stomach, and the cestode head becomes free as a *scolex*. The latter is, perhaps, protected from the too intense action of the gastric juice by its calcareous concretions, and at once enters the small intestine, fastens itself to the intestinal wall, and grows by gradual segmentation into a tape-worm. From the *scolex* the chain of proglottides proceeds as the result of a growth in length accompanied by segmentation, a process which is to be looked upon as a form of asexual reproduction (budding in the direction of the long axis). The development of the scolex is then to be explained as a metamorphosis, characterized by the individualization of certain stages of the development. But the whole life-history is a case of metagenesis, inasmuch as the sexual proglottides alternate with the asexual scolex.

FIG. 207.—*Taenia Echinococcus* (after R. Leuckart), magnified 12 to 15 times.

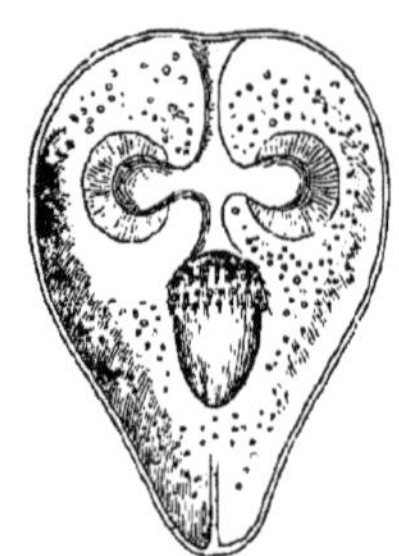

FIG. 208.—*Cysticercoid* of *Taenia cucumerina*, magnified 60 times (after R. Leuckart).

The development of some tape-worms (*Microtaeniinae*) presents considerable simplifications. In the cysticercus stage the vesicle is represented by a small appendage (Fig. 209, *b*), in which the cavity is much reduced or absent. Such cysticercus forms are called *Cysticercoids*, in which an appendage bearing the embryonic hooks is distinct from a larger part which represents the scolex (Fig. 210). Cysticercoids are found principally in Invertebrates (Gammarids, Cyclops, Insects, Slugs, Oligochaetes).

The tape-worms found in herbivorous animals are probably derived from the Cysticercoids of Invertebrates, but the intermediate host is generally unknown. The same may be said of the tape-worms of birds, only in this case the intermediate host is more often known.

In some cases the caudal appendage, which is the homologue of the vesicle of the *Cysticerci*, is elongated (Fig. 210), and the Cysticercoids then present a considerable resemblance to the *Cercaria* of a Trematode. If this comparison is a just one, as it probably is, the Cysticercoid must be regarded as a more primitive larval form than the Cysticercus. Moreover *Caryophyllaeus* (Fig. 212) with its single set of generative organs and unsegmented body is probably the most primitive member of the group, and may be compared to the ordinary sexual Trematode.

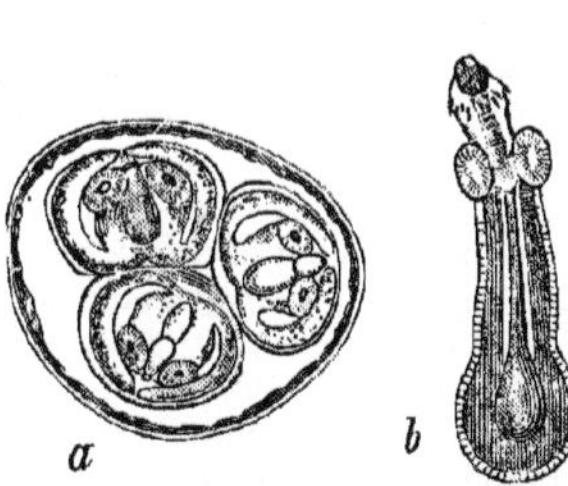

FIG. 209.—*a*, *Echinococcus*-like Cyst from the body-cavity of the earth-worm (after * E. Metschnikoff), containing three Cysticercoids; *b*, Cysticercoid with evaginated head.

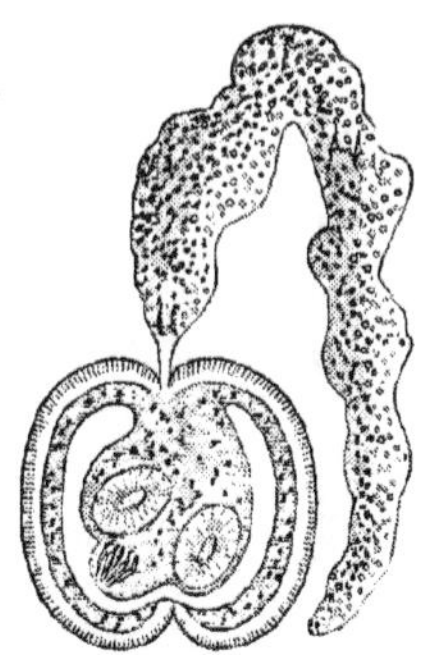

FIG. 210.—Cysticercoid of *Taenia sinuosa* from Gammarus pulex (after Hamann).

Amphilina (Fig. 213) and *Amphiptyches* are forms intermediate between the Trematodes and Cestodes, while *Archigetes* (Fig. 211) is either the most primitive Cestode, or a larval form which has become sexually mature. All the other Cestodes differ from the primitive *Caryophyllaeus* in the fissive reproduction which the body undergoes in the process of strobilization. The only part of the body which is not reproduced in this asexual increase is the organ of attachment; just as in the *Scyphistoma* (Fig. 132, *g*) all the organs of the body participate in the fission by which the *ephyrae* arise except the stalk of attachment. Finally it must be pointed out, that on this view *Ligula* is not a primitive, but a highly

* *Verh. d. Petersburger Naturforsch.*, 1868, Zool., p. 263.

specialized form, in which the sexual persons produced by division have lost their distinctness and do not separate from the colony. In other words, *Ligula* bears the same relation to a sharply segmented tape-worm that a hydroid colony with medusoids bears to a colony which buds off free-swimming medusae. *Bothriocephalus*, in which the segments separate off in groups (Fig. 214), may be regarded as a stage on the road to the condition found in *Ligula*.

Fam. 1. **Cestodariidae.** The body is unsegmented and the generative organs are not repeated. *Archigetes* Lkt. *A. Sieboldii* Lkt. (Fig. 211), in the body-cavity of the generative segments of Tubifex rivulorum, about 3 mm. long, with caudal appendage (? Onchosphere) which carries three pairs of small hooks at its free end. The anterior end of the body has two weak suckers. Life-history unknown. The only Cestode which attains sexual maturity outside the *Vertebrata*. It is to be regarded as a Cestode retaining the Onchosphere, and still fixed by the embryonic hooks. *Caryophyllaeus* Müll., elongated, no external distinction between head and body, without suckers; excretory opening at hind end; genital pore on ventral surface of hind end, receives the vas deferens, uterus, and vagina. *C. mutabilis* Rud. (Fig. 212), intestine of Cyprinoids, asexual form probably in Tubifex rivulorum. *Amphilina* Wagener, body-cavity of sturgeons, *A. foliacea* Rud. (Fig. 213). Body flattened, Trematode-like, 60 mm. long, pointed at one end which carries a sucker near by the opening of the uterus (*Ut.m*); the vagina (*Vg*) and vas deferens (*C*) open at the other end; body-cavity of sturgeon; produces a partially ciliated hooked embryo, life-history unknown. *Gyrocotyle* Dies. (*Amphiptyches* Wagener), alimentary canal of *Holocephali*, a sucker at the anterior end, edge of body folded. Uterus, vagina, vas deferens open separately; *G. urna* W., alimentary canal of *Chimaera*. The Onchosphere probably passes into bivalves. *Wageneria* Monticelli, in *Scymnus nicaeensis*.

FIG 211.—*Archigetes Sieboldii* Lkt. (after Leuckart).

Fam. 2. **Bothriocephalidae.** With only two suckers, which are weak and flat. Generative organs usually open on the flat surface of the proglottis. Proglottides are often detached in groups. Hydatid stage represented by an encysted scolex (Fig. 215), which is usually found in fishes. *Bothridium* Blainv. (*Solenophorus* Creplin) intestine of pythons and boas; *Diplocotyle* Krabbe, intestine of fishes; *Diphyllobothrium* Cobbold, intestine of dolphin; *Ptychobothrium* Lönnberg; *Duthiersia* Perrier, from *Varanus*.

Bothriocephalus Brems. Segmented body. Head with two pits, without hooks. The genital openings are on the middle of the ventral surface. The young stage usually in fishes. *B. latus* Brems. (Fig. 214), the largest of the tape-worms parasitic in man, twenty-four to thirty feet in length, principally found in Russia, Poland, Switzerland, and South France. The sexually mature segments are broader than they are long (about 10–12 mm. broad and 3–5 mm. long). They do not become detached singly, but in groups (Fig. 214). The segments of the hindermost portion of the body are, however, narrower and longer. The head is club-shaped, and is provided with two slit-like pits. The cortical parts of the lateral regions of the body contain a number of round masses of granules, the *yolk-glands* (Fig. 216, *Dst*), the contents of which are poured into the shell glands (coiled glands) through the so-called yellow ducts.

The genital openings lie close together, one behind the other, in the midst of the segment (Fig. 216, *a*). The anterior and larger belongs to the male generative apparatus, and leads into the muscular terminal portion of the vas deferens, which is enclosed in the cirrus sheath (Fig. 216, *Cb*), and can be evaginated as the cirrus. The vas deferens just before its entrance into the cirrus pouch

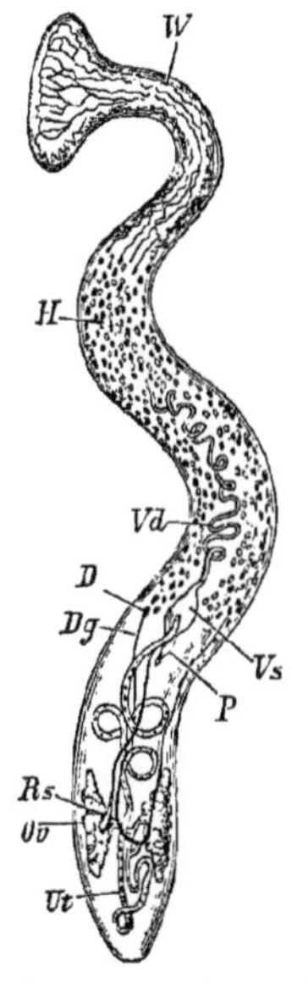

Fig. 212.—*Caryophyllaeus mutabilis* (after V. Carus). *W* excretory canal; *H* testes; *Vd* vas deferens; *Vs* vesicula seminalis; *P* penis; *Ov* ovary; *D* yolk-gland; *Dg* duct of yolk gland; *Ut* uterus; *Rs* receptaculum seminis.

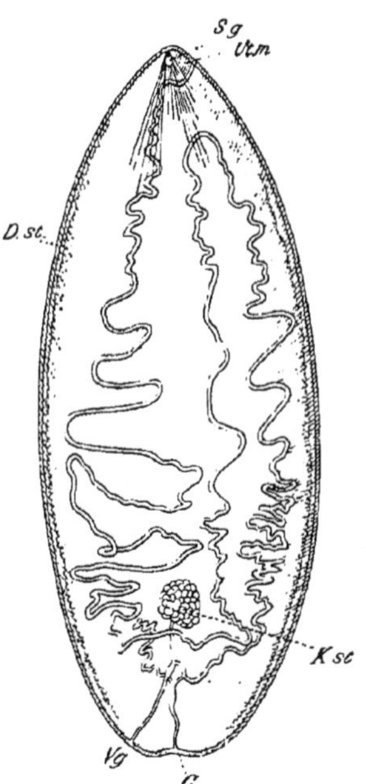

Fig. 213.—*Amphilina foliacea*, showing the generative ducts. *S.g* sucker; *Ut.m* opening of uterus; *D.st* yolk-gland; *K.st* ovary; *Vg* opening of vagina; *C* opening of vas deferens (after Wagener).

is dilated (Fig. 216, *b*) to form a large muscular swelling (the vesicula seminalis?). It then becomes coiled, and passes in the direction of the long axis of the segment on the dorsal surface and divides into two side branches. These receive the efferent canals of the delicate testicular sacs, which occupy the lateral parts of the middle layer (*T*). The female genital opening (Fig. 216) leads into a vagina (*Va*) situated behind the pouch of the cirrus, and frequently filled with semen. This vagina runs as a tolerably straight median canal on the ventral surface, and opens by a short, narrow tube into the oviduct. The vagina also functions as a *receptaculum seminis*. There is yet a third opening

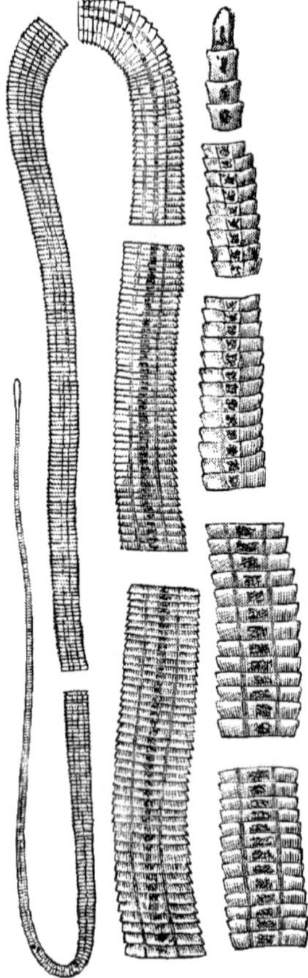

FIG. 214.—*Bothriocephalus latus* (after Leuckart).

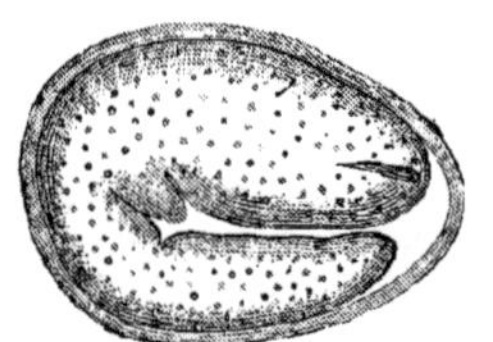

FIG. 215.—Encysted larva of *Bothriocephalus* from the Smelt (after Leuckart).

(Fig. 216, *a*), situated at some distance behind the other two; this is the opening of the tubular uterus (*Ut*), the convolutions of which give rise to a peculiar rosette-shaped figure in the midst of the segment. Close to the hind end of the segment the ducts of the yolk-glands (*Dst*) and of the ovaries (*Ov*) unite with each other and open into the uterus; the cells of the shell-gland (*Sd*) surround and open into the point of junction of these structures.

The ova are for the most part developed in water, and escape from the upper pole of the egg-shell through a lid-like valve (Fig. 204). The escaped embryo is covered with cilia (Fig. 203), by means of which it swims about for a long time. The encysted larval form (scolex without a bladder) is found between the muscles or in the viscera of the pike, turbot (*Lota vulgaris*), and possibly of other fresh-water fish (Fig. 215). How they become infected is not known, as experiment tends to show that the ciliated larva does not enter them directly, and no intermediate host is known. The scolex enters the body of its final host in the flesh of the fish. *B. cordatus* Lkt. With large, heart-shaped head, without a filiform neck; with deposits of numerous calcareous bodies in the parenchyma. It attains a length of about three feet, and lives in the intestines of man and of the dog in Greenland. *B. liguloides* Lkt. Young form about 20 cm. in the subperitoneal tissue of man in China and Japan.

Schistocephalus Crepl. Head split, with a sucker on each side. The body of the cestoid form is segmented. *S. solidus* Crepl. Lives in the body cavity of the stickleback, escapes into the water, and becomes sexually adult in the intestine of water-birds. *Triaenophorus* Rud. Head not distinct, with two weak suckers and with two pairs of tridentate hooks. The body has no external segmentation. The generative openings are marginal. *T. nodulosus* Rud. In the intestine of the pike. Asexual encysted form in the liver of *Cyprinus*. *Bothrimonus* Duvernoy, intestine of sturgeon.

Ligula Bloch. Body band-shaped and unsegmented. Without real suckers. Hooks may be present or absent. The Cestoid has no segmentation, but the generative organs are repeated. They live in the body cavity of Teleosteans and become sexually mature in the intestine of birds. *L. simplicissima* Rud., in the body cavity of fishes and in the intestine of aquatic birds. *L. tuba* v. Sieb., in the intestine of the tench.

Fam. 3. **Tetrarhynchidae.** Head with two or four suckers, and with four protractile proboscises armed with hooks (Fig. 217); genital openings marginal. The scolices live encysted in bony fishes, and the sexual worms in the intestine of skates and rays. *Tetrarhynchus* Cuv., with four suckers; *Rhynchobothrium* Blainv., with two suckers.

Fam. 4. **Echinobothridae.** With two suckers and two armed proboscises; genital openings marginal. *Echinobothrium* v. Ben.

Fam. 5. **Tetraphyllidae.** Head with four very mobile suckers often armed with hooks. Proglottides detached singly. Genital pores marginal. In the intestines of Selachians.

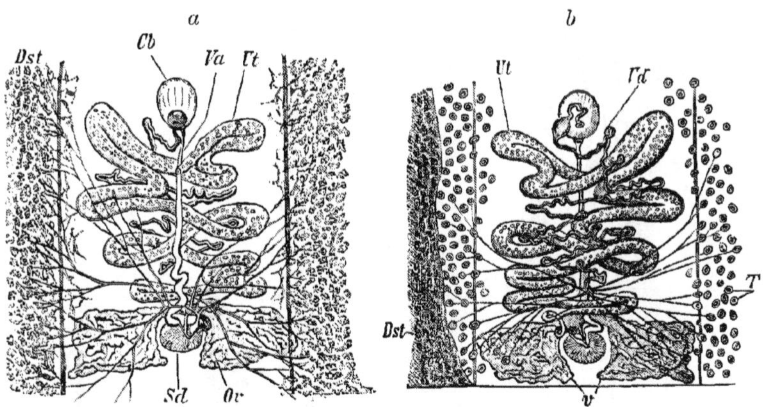

FIG. 216.—Generative organs of a sexually mature proglottis of *Bothriocephalus latus* (after Sommer and Landois); *a*, from the ventral surface; *b*, from the dorsal surface. *Ov* and *v* ovary; *Ut* uterus opening to the exterior independently of the vagina; *Sd* shell-gland; *Dst* vitellarium (yolk-gland); *Va* vagina with opening; *T* testes; *Cb* pouch of the cirrus; *Vd* vas deferens.

Sub-fam. 1. **Phyllobothrinae.** Suckers without hooks, more or less stalked. *Echeneibothrium* v. Ben.; *Phyllobothrium* v. Ben., suckers sessile; *Anthobothrium* v. Ben.

Sub-fam. 2. **Phyllacanthinae.** Suckers armed, each with 2 or 4 hooks. *Acanthobothrium* v. Ben.; *Calliobothrium* v. Ben.; *Onchobothrium* Blainv.

Fam. 6. **Taeniadae.** The armature of the head consists of four muscular suckers, to which is frequently added a single or double circle of hooks on the rostellum. The proglottides have a marginal sexual opening. The vagina is usually long, and enlarged at the internal end to form a receptaculum seminis (Fig. 201). The uterus is without a special opening to the exterior. The young stages are Cysticerci or Cysticercoids, rarely quite without caudal vesicle. Parasitic in warm and cold-blooded animals.

Sub-fam. 1. **Cystotaeniinae.** Rostellum usually with double row of hooks. Development by means of Cysticerci (bladder-worms). Both bladder-worm and tape-worm stage in mammals. *Taenia* L. (*Cystotaenia* R. Lkt). The vesicles of the Cysticerci are large. The heads arise from the bladder of the Cysticercus.

T. solium. L. 2–3 metres long. A double circle of 26 hooks. The ripe proglottides are 8–10 mm. long and 6–7 mm. broad; the uterus has 7–10 dendritic branches (Fig. 202). It lives in the human intestine. The bladder-worms belonging to it (*Cysticercus cellulosae*) live principally in the dermal cellular tissue and in the muscles of pigs, but also in the human body (muscles, eyes, brain), in which self-infection with them is possible if a *Taenia* is present in the digestive canal; more rarely in the muscles of the roe-deer, the dog, and the cat. In the human brain the *Cysticercus* acquires an elongated form, and sometimes does not produce a head.

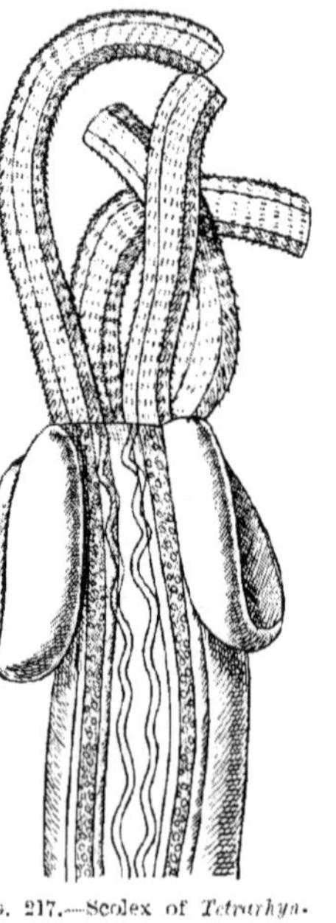

Fig. 217.—Scolex of *Tetrarhynchus rufícollis*, showing the four hooked proboscises protruded and two of the suckers (after van Beneden).

T. saginata Goeze = *mediocanellata* Kuchenm., in the intestine of man, distinguished by the older helminthologists as a variety of *T. solium*. Head without circle of hooks or rostellum, but with four more powerful suckers. The tape-worm reaches a length of four metres, and becomes much stronger and thicker. The mature proglottides are about 18 mm. long and 7–9 mm. broad. The uterus forms 20–35 dichotomous side branches (Fig. 202). The *Cysticercus* lives in the muscles of the ox (Fig. 218). It appears to be principally distributed in the warmer parts of the Old World, but is often found in great numbers in many places in the north. It is the common tape-worm of the Abyssinians.

Fig. 218. — *Cysticercus* of *Taenia saginata* (*mediocanellata*), magnified about eight times. The head is protruded (from Claus).

T. serrata Goeze, in the intestinal canal of the dog. The Cysticercus is known as *Cysticercus pisiformis* in the liver of the hare and rabbit. *T. crassicollis* Rud. in the cat, with *Cysticercus fasciolaris* of the common mouse. *T. marginata* Batsch. of the dog (butcher's dog) and wolf with *Cysticercus tenuicollis* from ruminants and pigs, and occasionally in man (*Cyst. visceralis*). *T. crassiceps* Rud. in the fox with *Cysticercus longicollis* from the thoracic cavity of the field-mouse. *T. coenurus* v. Sieb. in the intestine of the sheep-dog with *Coenurus cerebralis* in the brain of one-year-old sheep causing staggers. The presence of *Coenurus* in other places has been stated, as for instance, in the body cavity of the rabbit. *T. tenuicollis* Rud. in the intestine of the weasel and the pole-cat, with a *Cysticercus* which, according to Kuchenmeister, lives in the hepatic ducts of the field-mouse.

Echinococcifer Weinl. The heads bud on special brood-capsules in such a way that their invagination is turned towards the lumen of the vesicle (Fig. 206). *T. echinococcus* v. Sieb. (Fig. 207) in the intestine of the dog, 3–4 mm. long, forming but few proglottides. The hooks on the head are numerous but small. Its bladder-worm is distinguished by the great thickness of the stratified cuticula. It lives as *Echinococcus* principally in the liver and the lungs of man (*E. hominis*), and of domestic animals (*E. veterinorum*). The first form is also distinguished as *E. altricipariens* on account of the frequent production of primary and secondary vesicles; it usually reaches a very considerable size, and has a very irregular shape; while that form which inhabits domestic animals, *E. scolicipariens*, more frequently retains the form of the simple vesicle. Finally these echinococcus cysts frequently remain sterile, in which case they are called *Acephalocysts*. Another, and indeed pathological form is the so-called multilocular *Echinococcus*, which was for a long time taken for a colloid cancer. It is also found in mammalia (in cattle), and here presents a confusing resemblance to a mass of tubercles. The echinococcus disease (*hydatid plague*) was widely spread in Iceland. This disease likewise seems endemic in many places in Australia.

Sub-fam. 2. **Microtaeniinae.** The rostellum is frequently absent, or unarmed, or beset with small hooks. Development by means of *Cysticercoids*, the vesicle having but little fluid, or being absent. The Cysticercoid lives principally in invertebrates (slugs, insects, &c.), more rarely in cold-blooded vertebrates (the tench).

Sub-genus *Dipylidium* Leuck. With two sets of generative organs in each segment. *T. cucumerina* Bloch, in the intestine of dogs (house dogs). The Cysticercoid is entirely without the caudal vesicle, and lives (according to Melnikoff and R. Leuckart) in the body cavity of the dog-louse (*Trichodectes canis*). The infection with the Cysticercoids takes place when the dog swallows the parasites which are annoying him, while the parasites swallow the eggs contained in faeces adherent to the hair of the dog. Nearly allied is *T. elliptica* Batsch. in the intestine of the cat, occasionally in that of man.

Sub-genus *Hymenolepis* Weinland. One set of generative organs in each segment, opening on one side; eggs with two smooth shells. *T. nana* Bilh.-v. Sieb., in the intestine of the Abyssinians and in Sicily, hardly an inch long; probably identical with *T. murina* of the rat, the Cysticercoid of which, according to Grassi and Rovelli, is able to develop in the intestinal villi of its host, and then emerge to form a tape-worm in the intestine (self-infection). *T. flavopunctata* Weinl., in the human intestine (North America). Also found by Grassi in Italy, and regarded as identical with *T. diminuta* Rud. = *leptocephala* Crepl., of the rat. The Cysticercoids of the meal-worm are probably developed into tape-worms in the intestines of mice and rats.

In other partially unarmed *Taenias** the generative organs and development are as yet not accurately known; such are—*T.* (*Anoplocephala*

* See note on following page.

R. Bl.) *perfoliata* Goeze, and *T. plicata* Rud., in the horse; *T.* (*Moniezia* R. Bl.) *pectinata* Goeze, in the hare; *T. dispar* Rud., in the frog; *T.* (*Ctenotaenia* R.) *expansa* Rud., in the ox.

There is a special section of Taenias,* in which the genital opening is on the broad surface of the segment: such are *T. litterata* Batsch.; *T. lineata* Goeze, in the intestine of the dog. To another group belong Taenias from the gut of birds, *e.g.* *T. sinuosa* Zed., in the goose and duck; *T. tenuirostris* from *Merganser* and *Anas*, both with tailed Cysticercoids in Gammarus.

* A number of new genera have been recently created for the reception of species formerly united under *Taenia*: *vide* A. Railliet, "Notices parasitologiques," *Bull. Soc. Zool. France*, 1892; and R. Blanchard, *Mém. de la Soc. Zool. de France*, 4, 1891, p. 420.

CHAPTER VI.

PHYLUM NEMERTEA.*

Elongated, vermiform animals with a ciliated ectoderm, and an eversible proboscis lying in a sheath on the dorsal side of the enteron. With mouth and anus, simple gonads, and a vascular system. Dioecious.

The position of the *Nemertea* is difficult to settle. Formerly they were united with the *Platyhelminthes*, but the presence of an anus, and of a vascular system, and the higher organization of the organs and tissues in general, renders it advisable, for the present at any rate, to place them in a separate phylum. allied to the Platyhelminth phylum but not part of it. It has been suggested that the proboscis of Nemertines is homologous with the retractile anterior part of the body found in some *Turbellaria*, particularly *Prorhynchus*. The general structure of the Nemertine tissues strongly recalls that of *Balanoglossus*.

The *Nemertea* are elongated worms, some of them attaining an immense length, and are found in the sea, in fresh water, and on land. The marine forms are, however, by far the most numerous. They are often brilliantly coloured, and many of them have the power of forming a tube around their bodies by the mucous secretion of the skin. The body is excessively contractile, so much so that a worm which is measured by yards when extended may shrink to a length of as many inches. The mouth is large and placed on the ventral surface of the anterior end of the body. It leads into a straight alimentary canal consisting of oesophagus and intestine. The latter opens posteriorly by a terminal anus, and possesses lateral caeca which are generally, but not always, regularly arranged in pairs; it also gives off from its front end an anteriorly-directed caecum.

The most characteristic organ of the group is the **proboscis**. This

* W. C. McIntosh, *A Monograph of British Annelida*, Pt. 1, "Nemerteans," Ray Society, 1873-4. A. A. W. Hubrecht, "Unters. üb. Nemertinen a. d. Golf v. Neapel," *Niederl. Arch. f. Zoologie*, 2, 1874. Id., "The Genera of European Nemertines critically revised," *Notes from the Leiden Museum*, 1879-80. Id., "Report on the Nemertea," *Challenger Reports*, vol. 19, 1887. O. Bürger, "Die Enden des exkretorischen Apparates bei den Nemertinen," *Zeit. f. w. Zoologie*, 53, 1891. O. Bürger, "Die Nemertinen," *Fauna u. Flora des Golfes von Neapel*, 22, 1895.

lies in a sheath (Fig. 219 *bis*) which is placed on the dorsal side of the gut, and extends in most forms along the whole length of the body. It is a hollow tubular organ, opening nearly always at the front end* of the body, but closed behind. In the retracted state

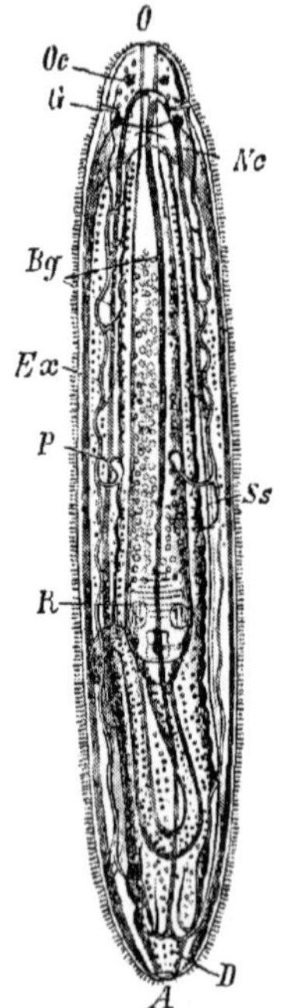

Fig. 219.—*Tetrastemma obscurum* (after M. Schulze). Young specimen about 3 lines in length; *O* mouth; *D* intestine; *A* anus; *Bg* blood-vessels; *R* proboscis armed with stylet; *Ex* lateral trunks of the excretory system; *P* excretory pore; *G* lateral organ; *Nc* nerve centre; *Ss* lateral nerve cords; *Oc* eyes.

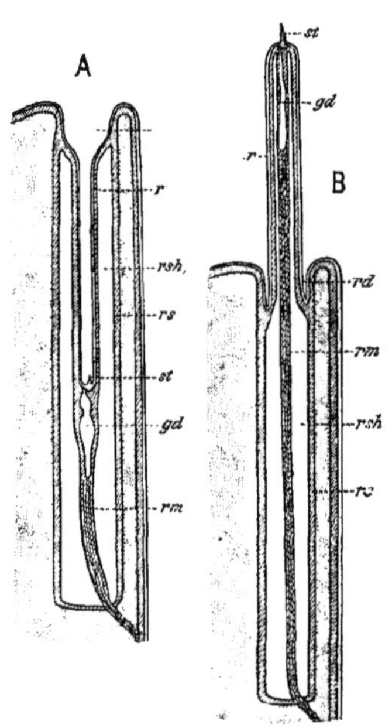

Fig. 219 *bis*.—Diagram showing the proboscis of an armed Nemertine in its relation to its sheath (from Lang). *A*, in the retracted condition; *B*, in the protruded state. *r* proboscis; *rs* proboscis sheath; *rsh* rhyncocoelom; *st* spine; *gd* cavity of the posterior non-eversible part of the proboscis; *rm* retractor muscle.

the proboscis lies in its sheath, to the hind end of which its blind end is fastened by a muscular band (Fig. 219 *bis*, *A*); its walls contain both transverse and longitudinal muscles, the latter of which are continued into the band. The proboscis sheath also

* In *Akrostomum*, *Malacobdella*, etc., the opening of the proboscis is within the mouth.

has muscular walls, and contains a corpusculated fluid; it contains a closed cavity (*rhynchocoelom*), and its epithelioid lining is continuous anteriorly with that covering the outer side (in the retracted state) of the proboscis, *i.e.* with the epithelium on that surface of the retracted proboscis which is in contact with the proboscis-sheath fluid. The proboscis and its sheath may therefore

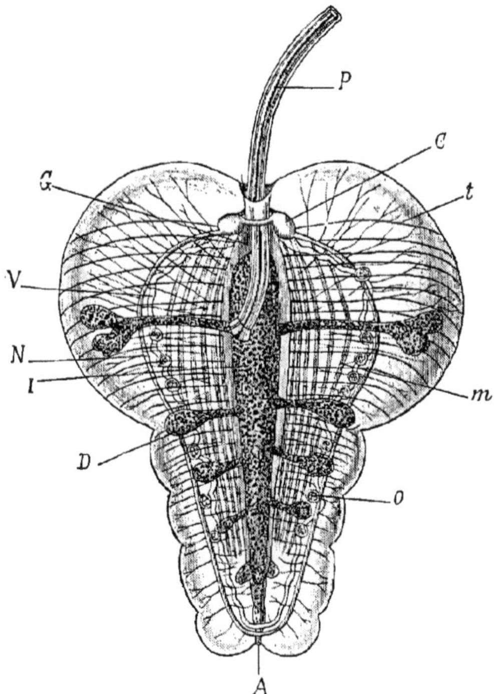

FIG. 220.—A young *Pelagonemertes Rollestoni*, dorsal view (after Hubrecht, from Perrier). *P* proboscis; *G* sheath of proboscis; *I* intestine; *D* diverticula of intestine; *C* cerebral ganglia; *N* lateral nerve cords; *V* lateral vessels; *O* ovaries; *A* anus; *m* longitudinal muscles; *t* transverse muscles.

be compared to the finger of a glove, of which the free end is pulled in upon the portion next the hand, the pulling in being effected by a string attached to the inner side of the tip of the glove, and lying inside the finger. From the above description it will also be clear that the epithelium lining the inner surface of the retracted proboscis is continuous with the surface ectoderm of the body at the anterior opening, and becomes external when the proboscis is projected by eversion (Fig. 219 *bis*, *B*). The eversion is no doubt caused by the contraction of the muscular wall of the sheath bringing pressure to bear upon the contained fluid, and, when completed, the middle portion of the retracted proboscis is at the front end of the protruded organ. That is to say, the hinder part of the proboscis is not eversible: it lies within the protruded part of the proboscis in the projected state, and in the hinder part of the sheath in the retracted state. This non-eversible portion has glandular walls and contains a fluid. In the armed forms (*Hoplonemertini*) it is separated from the eversible part by a contraction of its cavity, which almost closes up the eversible part, and a stylet, to which may be added two groups of small accessory reserve stylets, is placed at this point. The posterior part of the proboscis opens by a narrow aperture at the base of this stylet, and allows the contents of the glandular non-eversible part, which is very possibly poisonous, to exude. When the proboscis is completely everted the stylet projects freely at its apex.

In the unarmed forms (*Anopla*) the stylet is absent, but the surface of the proboscis which becomes external on eversion possesses numerous nematocysts. The proboscis is therefore clearly offensive, but very possibly it also possesses a tactile function.

The body-wall varies in structure in the different orders. It always possesses an external layer of ciliated ectoderm cells containing mucus-secreting gland-cells. Within this there is in the *Heteronemertini* (1) a thin basement membrane, (2) a cutis containing connective tissue and some longitudinal muscular fibres, (3) a layer of longitudinal muscular fibres, (4) a thin layer of a plexiform nervous matter containing at two points the lateral nerve-cords, (5) a layer of circular muscular fibres, and (6) of longitudinal muscles; this is followed by a compact mass of reticular connective tissue which passes into the muscular coats of the intestine and proboscis sheath. In other *Nemertini* there are only an external circular and internal longitudinal muscular layer. All the layers above mentioned are embedded in a gelatinous albuminoid material, which is especially developed in the transparent pelagic form, *Pelagonemertes*, and is of the same nature as the jelly of *Medusae*. There is therefore no **body cavity** in Nemertines, though, as in Rhabdocoeles, spaces in the connective tissue may in some cases be so large as to simulate one.

The **nervous system** consists of two cerebral ganglia connected above and below the proboscis sheath and each divided into two lobes, a dorsal and ventral. The ventral lobes are continued backwards as the lateral nerve-cords (Fig. 221), which in some cases unite posteriorly above the intestine by a supra-anal commissure. The dorsal lobe (Fig. 221, *C*) is partly or completely divided into an anterior and posterior lobe. In the armed forms (Fig. 221, *L*) this third lobe is completely separated off and united with the anterior only by nerves. This posterior part of the dorsal lobe contains in nearly all forms an epithelium-lined cavity which communicates with the exterior by a fine ciliated tube. The latter opens, in the forms with head-slits, into the slits, in the other forms on the side of the head. The whole apparatus including the brain lobe constitutes the **lateral organs**, or as it is sometimes called, the cerebral organs. The central nervous system lies in *Carinella* outside the muscular layers; in the *Heteronemertini* between the external longitudinal and the circular layer; and in the *Metanemertini* completely within the muscles. In all, except the *Metanemertini*, in which the central nervous system gives off nerve-cords, there is a complete nerve-sheath of a reticulate nervous substance, occupying the same position with regard to the muscles as do the central

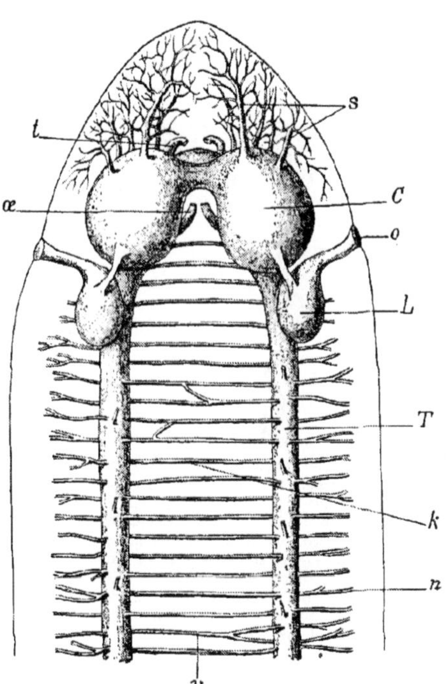

FIG. 221.—Diagram of the nervous system of *Drepanophorus Lankesteri* (from Perrier after Hubrecht). *C* dorsal lobes of the cerebral ganglia; *t* nerves of the proboscis; *s* sensory nerves of head; *œ* stomatogastric nerves; *L* posterior part of the dorsal lobes of the brain in which are the lateral organs; *O* opening of lateral organ; *n* peripheral nerves; *T* lateral nerve-cords; *v*, *k* transverse commissures.

organs. The lateral nerve-cords are thickenings of this sheath, and there is a median dorsal thickening constituting the proboscis nerve.

In *Heteronemertini* the central organs are impregnated with haemoglobin. As **sense organs** there are the lateral organs above mentioned, the tactile hairs of the ectoderm cells, cephalic eyes more or less numerous, and sometimes provided with refractive bodies. Exceptionally, as in *Oerstedia pallida*, two otolithic vesicles are found on the brain.

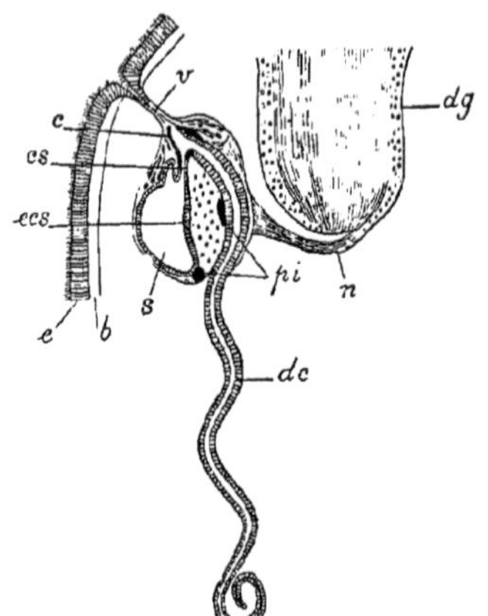

FIG. 222.—Lateral organ of *Drepanophorus rubinus* (after Bürger, from Perrier). *e* ectoderm; *b* basement membrane; *v* canal of the organ; *dg* dorsal lobe of cerebral ganglion; *n* nerve connecting *dg* to its posterior lobe; *c* point of division of the canal into a glandular tube *dc*, which projects back behind the brain and into a canal *cs* which passes into the dilatation *s* in the posterior lobe; *ecs* epithelium of the sac; *pi* pigment.

The **vascular system** consists of three longitudinal vessels with contractile walls placed just outside the intestinal wall; one of them is straight and dorso-median, and the other two are sinuous and lateral. They communicate in front and behind; and anteriorly the lateral vessels often dissolve themselves in a network of vascular spaces on the oesophagus. The longitudinal vessels give off lateral branches, which no doubt open into a system of lacunae in the tissues. The blood is usually colourless, but in some species it is red. In *Amphiporus splendens* and *Borlasia splendida*, the red colour (haemoglobin) is contained in the blood corpuscles.

The **excretory organs** consist of two lateral tubes, which lie close to the lateral vessel, and open externally by one, or rarely by several openings on each side. They are confined to the anterior region of the body, usually not extending further back than the oesophagus (fore-gut). These tubes have a glandular ciliated lining, and in some forms, possibly in all, give off branches which themselves branch and finally end in small swellings, containing a long flame-shaped cilium. These flame-cell ends are said to differ from the corresponding structure in *Turbellaria* by consisting of many cells. As already pointed out, this distinction is probably unimportant, depending merely on the size of the organ. The flame-cell knobs of Nemertines sometimes project into the blood-vessels, but do not open into them. In the forms in which flame-cells have not been observed, the excretory canals are said to open into the lateral blood-vessels. This will probably turn out to be an error. The animals are generally dioecious, and the **gonads** quite simple, consisting of paired sacs regularly placed between the gut caeca, or in some cases more numerous than the caeca and irregularly arranged. They open by simple openings on the dorso-lateral surface of the body.

Some genera are viviparous (*Prosorochmus Claparedii*, *Tetrastemma obscurum*), but most of them lay eggs in albuminous strings. The development is usually direct, but in many *Heteronemertini* the young are hatched as helmet-shaped, free-swimming larvae (Fig. 223), known as *Pilidium*.

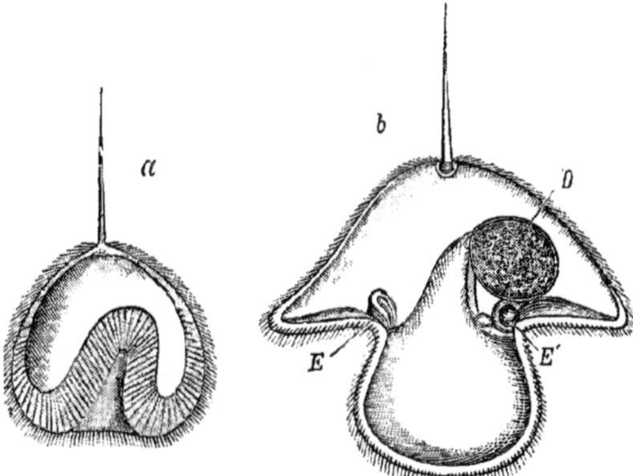

Fig. 223.—*Pilidium* (after E. Metschnikoff). *a*, free-swimming larva. *b*, later stage, helmet-shaped. *E*, *E′* the two pairs of ectodermal invaginations; *D* alimentary canal.

The Pilidium was formerly described as a species of a supposed independent genus. The enteron is formed by the invagination of the wall of a hollow blastosphere; the blastopore forms the larval mouth; at the aboral pole a long flagellum is developed, and a broad lobe grows out on each side of the mouth, on to the edges of which the circumoral ciliated band extends. By an elaborate metamorphosis this larva turns into the young worm.

Asexual reproduction is unknown, but the power of repair is great. In some cases the body readily breaks up into pieces when touched, and the pieces have the power of developing the whole.

The Nemertines live principally in the sea, under stones in the mud, but the smaller species swim freely. There are also forms which live on land (*Geonemertes*), fresh-water forms (species of *Tetrastemma*), and a pelagic form (*Pelagonemertes*). Certain species form tubes and passages which are lined by a slimy secretion. The food of the larger species principally consists of tubicolous worms, which they extract by their proboscis. There are, however, parasitic (or commensal) Nemertines, which infest *Crustacea*, or live on the mantle and gills of *Mollusca*. In this case they are, like the *Hirudinea*, furnished with a posterior sucker (*Malacobdella*).

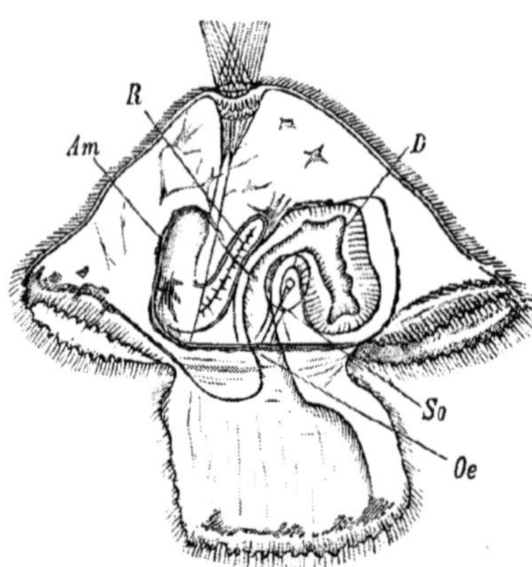

Fig. 224.—Later stage of *Pilidium*, with tuft of cilia and enclosed Nemertine (after O. Bütschli); *Oe* oesophagus; *D* alimentary canal; *Am* amnion; *R* rudimentary proboscis of the Nemertine; *So* lateral pit.

Order 1.
PROTONEMERTINI.

The cerebral ganglia and lateral nerves are outside the dermal muscles, either in the ectoderm or beneath the dermis. The body-wall consists of the ectoderm, a dermis, an external circular and an internal longitudinal layer of muscles, between which is usually interposed a diagonal layer. The mouth is behind the brain. There is no caecum. The proboscis is without stylets.

Fam. 1. **Carinellidae.** The lateral organs have the form of epithelial pits or canals, which only exceptionally perforate the dermis and penetrate into the brain; they have no relation to the lateral vessels. There is no dorsal vessel. The brain and lateral nerves lie in the epithelium or beneath the dermis. The dermis is homogeneous and has a gelatinous appearance. *Carinella* McIntosh.

Fam. 2. **Hubrechtidae.** The lateral organs are spherical structures which lie deep within the body-wall and project into the lateral vessels. Brain and lateral nerve-cords lie beneath the dermis, which is reticular. There is a dorsal vessel. The excretory organs constitute a richly-branched canal-system. *Hubrechtia* Bürger.

Order 2. MESONEMERTINI.

The lateral nerves lie in the dermal muscular layer. Body-wall as in Order 1. Mouth behind the brain. No caecum. The proboscis is without stylets.

Fam. 1. **Cephalothricidae.** The lateral nerves are in the longitudinal muscles. Lateral organs, and cephalic slits absent. *Carinoma* Oudemans; *Cephalothrix* Oersted.

Order 3. METANEMERTINI.

Brain and lateral nerves lie within the dermal muscles in the body-parenchyma. The body-wall is as in previous orders. Mouth in front of the ganglion. The mouth and the proboscis usually open together. The proboscis as a rule has stylets, and there is almost always a caecum given off from the anterior end of the intestine, and projecting forwards.

A. **PRORHYNCHOCOELOMIA.** With long and thin body, which coils itself in complicated windings. They do not swim. The proboscis is much shorter than the body.

Fam. 1. **Eunemertidae.** Usually several small eyes. No otocysts. Only one stylet. Slow in movement. *Eunemertes* Vaillant; *Nemertopsis* Bürger.

Fam. 2. **Ototyphlonemertidae.** Eyes absent. One, rarely two pairs of otocysts, ventral to brain. The body is more cylindrical than flat. The worms belonging to this family are small (1–3 cm.). *Ototyphlonemertes* Diesing.

B. **HOLORHYNCHOCOELOMIA.** Usually with short body. Some of them swim. The proboscis is at least as long as the body. The proboscis sheath always reaches into the hinder third of the body.

Fam. 3. **Prosorhochmidae.** With four eyes. Gut-pouches and gonads alternate with one another. The lateral organs are very small, in front of the brain. Usually hermaphrodite. *Prosorhochmus* Keferstein; *Prosadenoporus* Bürger; *Geonemertes* Semper, terrestrial form.

Fam. 4. **Amphiporidae.** Body when extended comparatively short and wide. Extensile part of proboscis thick and covered with adhesive papillae. The gonads and gut-pouches do not, as a rule, strictly alternate. The gut-pouches are branched. There are almost always numerous eyes. *Amphiporus* Ehrbg.; *A. lactifloreus* Johnst. Lives under stones and is distributed from the North Sea to the Mediterranean. *Drepanophorus* Hubrecht.

Fam. 5. **Tetrastemmidae.** The body is, as a rule, short (1–3 cm.). There are almost always four eyes. The gut-pouches are not branched, and the gonads alternate with them. The lateral organs are in front of the brain. For the most part dioecious. *Tetrastemma* Ehrbg.; *T. obscurum* M. Sch., viviparous; Baltic; *T. agricola* Will. Suhm., terrestrial; *T. clepsinoides* Dugès, fresh-water, North America, probably Europe, *e.g.*, Cherwell at Oxford, *T. lacustre* Du Plessis, Lake of Geneva. *Oerstedia* Quatref.

Fam. 6. **Nectonemertidae.** Deep-sea forms, with short, broad bodies and hind end flattened into a fin. They can swim. The mouth and proboscis openings are separate. The presence of stylets has not been certainly shown. Without eyes. *Nectonemertes* Verrill; *Hyalonemertes* Verrill.

Fam 7. **Pelagonemertidae.** Body leaf-shaped, gelatinous, hyaline. Anterior extremity broad and abrupt, posterior narrowed to a point. Digestive canal with 13 pairs of lateral ramifications. Integument thin and hyaline, with a thin muscular tunic immediately beneath it consisting of external circular and internal longitudinal fibres. The viscera and tissues are embedded in the hyaline gelatinous matter. Gonads open on the ventral surface. Nerve-cords internal to the muscular coats. Free-swimming, pelagic. *Pelagonemertes* Moseley (Fig. 220).

Fam. 8. **Malacobdellidae.** Parasitic forms. The body is short and provided with a sucker at the hind end. Intestine coiled, without pouches. The proboscis is unarmed and opens with the mouth. There is a dorsal vessel, and two lateral. The proboscis-sheath reaches to the anus. *Malacobdella* Blainv., in the mantle-cavity of various marine Lamellibranchs.

Order 4. HETERONEMERTINI.

The lateral nerves are in the dermal muscles; they lie outside the circular muscles. The body-wall consists of ectoderm, dermis, an outer longitudinal layer of muscles (which is not present in the other orders), a circular and an internal longitudinal muscular layer. The diagonal muscles, if present, lie between the circular and outer longitudinal. The mouth is behind the brain. There is no caecum. The proboscis is unarmed.

Fam. 1. **Eupoliidae.** There are usually no lateral cephalic slits. The canal of the lateral organ opens either directly to the exterior, or into a shallow ventral slit. *Eupolia* Hubrecht; *Poliopsis* Joubin.; *Valencinia* Quatrefages.

Fam. 2. **Lineidae.** The body is more or less flattened. There is a deep longitudinal lateral fissure on each side of the head. A ciliated groove leads from the bottom of the fissure into the posterior lobe of the ganglion. The nervous tissue is tinged with haemaglobin. Development often by ciliated larvae. *Lineus* Sow. (Fig. 225); *L. marinus* Mont.; *L. longissimus* Sim., sea-long-worm; *Borlasia* Oken; *Euborlasia* Vaill.; *Micrura* Ehrbg.; *Cerebratulus* Renier; *Langia* Hubrecht, the margins of the body slightly frilled.

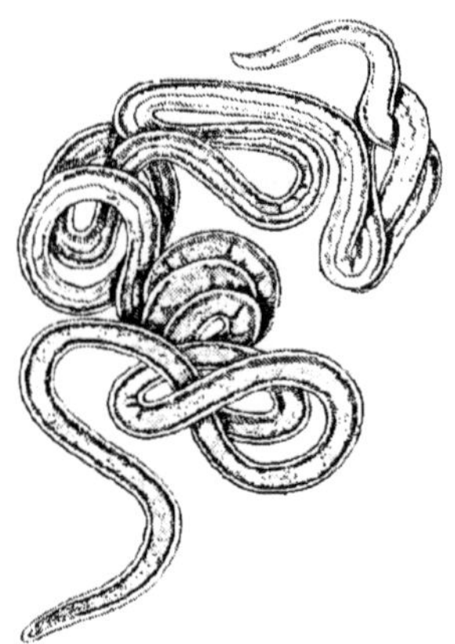

Fig. 225.—*Lineus sanguineus* (after McIntosh).

CHAPTER VII.

PHYLUM NEMATHELMINTHES.

THE Nemathelminthes include three orders—the *Nematoda*, the *Nematomorpha*, and the *Acanthocephala*—which have little else in common than the round form of body and the parasitic habit.

The body is unsegmented, rounded, more or less elongated, tubular or filiform, and both ends are, as a rule, tapered off. Appendages are always wanting, as are, with few exceptions, movable bristles. On the other hand, special organs for attack and attachment, such as teeth and hooks, are not unfrequently present on the anterior end of the body: and in some cases small suckers, which serve for attachment during copulation, may be developed near the hind end. As a rule the integument possesses a cuticular layer of relatively considerable thickness, and the ectoderm is very generally reduced to a nucleated granular layer, in which cell outlines are absent. These features of the skin are probably correlated with the endoparasitic habit, for we find them in the *Trematoda*, and the ectoderm of the *Cestoda* is, to say the least of it, much modified. In these latter groups the ectoderm is even less conspicuous than in the *Nemathelminthes*. There is a well-developed muscular layer, which generally consists of longitudinal fibres only, but circular fibres are also present in the *Acanthocephala*. A *body-cavity* is always present, but it appears generally to be without an epithelial lining, and in *Nematoda*, at least, is bounded on one side by the endodermal wall of the alimentary canal. As to the nature of this body-cavity we have little evidence except in the *Acanthocephala*, in which, from its relation to the generative organs and duct in the female we may infer it to be a coelom. It contains a vascular fluid, and in the other orders it is probably a haemocoele. *Blood vessels* and special *respiratory organs* are wanting. A *nervous system* is always present, but it presents very different features in the three orders. Of *sense organs* there are often sensory papillae in the neighbourhood of the mouth and genital openings, and eye spots are often present in the free-living forms. While in the *Acanthocephala* mouth and *alimentary canal* are completely absent, the *Nematoda* and *Nematomorpha* possess a mouth at the anterior end of the body, an oesophagus, and a straight digestive canal, which usually opens by the anus on the ventral surface near the hind end of the body. The *excretory organs* have various forms, and their nature is not understood: there are no flame-cells. In the *Nematoda* they consist of paired canals in the ectoderm, which open by a common pore on the same surface (ventral) as the anus. In the *Acanthocephala* there is a pair of organs which appear to be of the nature of nephridia. The absence of cilia may be stated as a general characteristic of the group; but they are said to be present in the supposed nephridia of the *Acanthocephala*.

With a few exceptions the *Nemathelminthes* have separated sexes, and the male organs often open into the rectum and are provided with copulatory spicula. The larvae and sexual animals are not unfrequently distributed in two different hosts.

The majority of the *Nemathelminthes* are parasites either during the whole period of their life or at certain stages. There are, however, also free-living forms which often show the closest relationship to the parasitic members of the group.

Class I. NEMATODA (THREAD-WORMS).*

Nemathelminthes, with mouth and alimentary canal. With longitudinal muscles only, with lateral lines, without cilia. The vas deferens opens into the rectum. They are principally parasites. Dioecious.

The Nematodes possess an extremely elongated thread-like body, which may be provided with papillae at the anterior pole in the region of the mouth, or with hooks and spines within the oral cavity. The mouth leads into a narrow oesophagus, which usually has thick muscular walls, a chitinous lining, and a triangular lumen,

* Besides the older writings of Rudolphi, Bremser, Cloquet, Dujardin, compare Diesing, "*Systema helminthum*," 2 Bde. Wien, 1850-51. Diesing, "Revision der Nematoden," *Wiener Sitzungsberichte*, 42, 1860. Claparède, "*De la formation et de la fécondation des œufs chez les vers Nematodes*," Genève, 1856. A. Schneider, "*Monographie der Nematoden*," Berlin, 1866. R. Leuckart. "*Untersuchungen über Trichina spiralis*," Leipzig and Heidelberg, 1866, 2nd edition; also "*Die menschlichen Parasiten*," etc., tom. 2., Leipzig and Heidelberg, 1876. C. Claus, "*Ueber Leptodera appendiculata*," Marburg, 1868. O. Butschli, "Untersuchungen über die beiden Nematoden der Periplaneta orientalis," *Zeitschr. für Wiss. Zool.*, tom. 21., 1871. And "Beiträge zur Kenntniss des Nervensystems der Nematoden," *Archiv. für Mikr. Anatomie*, tom. 10. A. Goette, "*Unters. z. Entwick. d. Würmer*," Leipzig, 1882. R. Leuckart, "*Neue Beiträge z. Kent. d. Bau u. d. Lebensgeschichte d. Nematoden*," Leipzig, 1887. A. E. Shipley, "Nemathelminthes," *Cambridge Natural History*, vol. 2, May, 1896.

and is frequently dilated behind to a muscular bulb (pharynx). In certain genera (*Rhabditis*, *Oxyuris*), the chitinous lining of the pharynx is raised into ridges or tooth-like prominences, to which the radial muscles converge in the form of conical bundles. According to its function, the oesophagus is essentially a suctorial tube, which pumps in fluids, and by peristaltic action passes them on to the intestine. The intestine follows the pharynx, and opens by the anus not far from the hind end of the body on the ventral surface (Fig. 226). Its walls are formed of cells and are non-muscular, and are coated with cuticle both inside and outside; it may be reduced to a row of perforated cells (*Filaria*). The terminal portion, or rectum, has a special investment of muscular fibres which render it contractile. Muscular fibres passing from the body-wall to the wall of the rectum are also frequently present. In certain Nematodes the anus may be wanting (*Mermis*); and in some genera (see p. 290) even the alimentary canal undergoes degeneration.

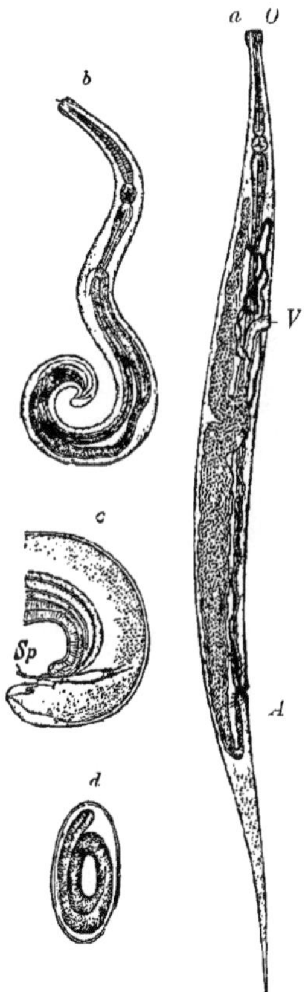

Fig. 226. — *Oxyuris vermicularis* (after R. Leuckart). *a*, female; *O* mouth; *A* anus; *V* genital opening. *b*, male with curved posterior end. *c*, the latter enlarged; *Sp* spiculum. *d*, egg with enclosed embryo.

Beneath the stiff cuticle, which is often transversely ringed and is composed of several layers, lies a soft granular nucleated sub-cuticular layer (*hypodermis*), which is without cell limits and is to be regarded as the matrix of the former. Beneath this lies the highly-developed muscular layer, which consists of band-shaped or fusiform longitudinal muscles. The surface of the body may present markings, as for instance polyhedric spaces and longitudinal ribs, also processes in the form of tubercles, spines,* and hairs. Ecdyses,

* There may also be prominences of various kinds, and even in some cases a complete covering of spines (*Cheiracanthus* Dies. = *Gnathostoma* Ow., *Ch. hispidum* Fedsch.).

i.e., sheddings of the cuticular layer, seem only to occur in the young forms. The muscles are each composed of a single cell, in which two parts are distinguishable,—a clear, sometimes a granular protoplasmic portion (medullary substance), which projects into the body-cavity and is often prolonged into processes; and an external fibrillated layer (Fig. 227).

The Nematodes may be distinguished as *Meromyaria* or *Polymyaria*, according to the arrangement of their muscular system. In the *Meromyaria* the number of muscle-cells (which are arranged according to definite laws) in the cross section is small (eight), while in the *Polymyaria* their number is considerable. In the latter the muscle-cells are often connected together by transverse processes of the medullary substance, which unite on the so-called median lines to form a longitudinal cord.

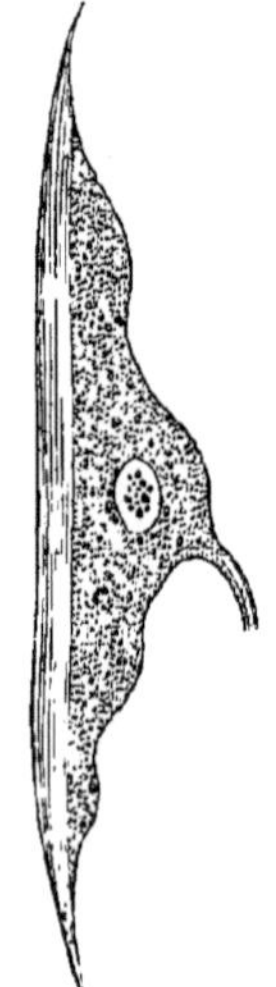

FIG. 227.—Muscle-cell of a *Nematode*.

In almost every case two lateral regions remain free from muscles, and form the so-called *lateral lines* or *regions*, which may equal in breadth the neighbouring muscular regions. These lateral regions are internal projections of the hypodermis, and are formed of a finely granular nucleated substance, and enclose a clear vessel containing granules. This vessel is connected with that of the opposite side in the anterior part of the body, and the two open by a common transverse slit, the *excretory pore*, on the ventral surface in the median line. The lateral lines are regarded as the *excretory organs*.* Median lines (*dorsal* and *ventral*), accessory median lines (sub-median lines), the latter being placed between the principal median line and the lateral line, are also to be distinguished. Cutaneous glands, in the form of unicellular glands, have been observed principally in the region of the oesophagus and in the tail.

The **nervous system**, owing to the difficulty which its investigation offers, has only been satisfactorily recognized in a few forms. It consists of a nerve ring (Fig. 228) surrounding the oesophagus, near the anterior end of the body, in *Asc. megalocephala* just in front of

* Hamann (Sitz. Ber. Akad. Berlin, 1891, p. 57) asserts that in *Lecanocephalus*, in which the right canal only is present, the excretory canal is coiled, and ends posteriorly in a small opening into the body-cavity. This statement needs confirmation.

the excretory pore, and sending off posteriorly six and anteriorly six nerve trunks (*Ascaris megalocephala*). The two largest posterior trunks run in the dorsal and ventral lines (*N. dorsalis, ventralis*), the four smaller, two in each lateral line, to the extremity of the tail; while of the six anterior nerves, two run in the lateral lines (*N. laterales*), and four in the interspaces between the lateral and median lines (*N. submediani*); they supply the papillae around the mouth. The ganglion cells lie partly near, in front of and behind the nerve ring, partly on the fibrous cords themselves, and are arranged in groups which can be distinguished as ventral, dorsal, and lateral ganglia. There are in addition groups of ganglion cells in the median lines and in the lateral lines in the caudal region.

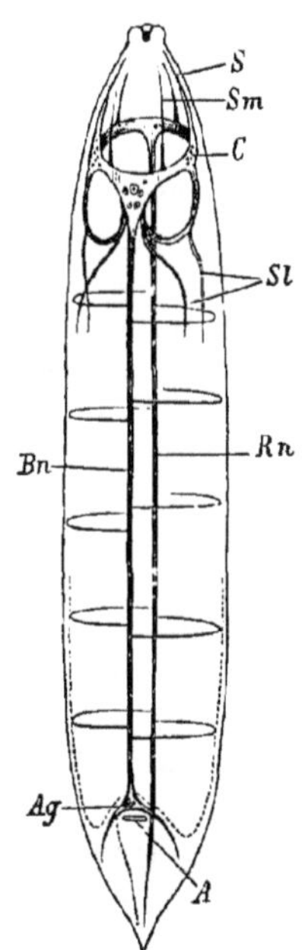

Fig. 228.—Nervous system of Nematodes, diagrammatic after Butschli. *C* lateral ganglion on the nerve ring; *S* anterior lateral nerve; *Sm* submedian nerve; *Sl* sublateral nerve; *Bn* ventral nerve; *Rn* dorsal nerve; *Ag* anal ganglion; *A* anus.

As **sense organs** we must mention the eyes found in the free-living *Nematoda*, and the papillae and tactile hairs found principally in the neighbourhood of the mouth. Each papilla is supplied by one nerve fibre, which is swollen to a knob and forms the axis of the papilla.

The **body-cavity** is a continuous space between the longitudinal muscles of the body-wall and the outer cuticle of the gut-wall; it contains a corpusculated and colourless fluid. The homology of the cavity is doubtful; it occurs between the muscles and the endoderm, and it has no relation to the gonads or excretory organs. There is no **vascular system.**

Generative organs. The Nematodes are dioecious (with exception of the hermaphrodite *Pelodytes*, and of *Rhabdonema* (*Ascaris*) *nigrovenosum* and *Allantonema mirabile*, which produce first spermatozoa and later ova). The males are characterised by their smaller size, and by the posterior end of the body being generally curved. Both kinds of generative organs consist of a single tube or of paired and often much coiled tubes, at the upper end of which the generative products are developed, the lower ends representing the efferent ducts and receptacula of the generative products. The usually paired ovarian tubes are distinguishable into the following regions: the *ovary* where the eggs are developed; the *oviduct* along which they pass to reach the *uterus*, a more dilated region where they are fertilised and often pass through a part or the whole (viviparous forms) of their development; the two uteruses lead into the single *vagina*, which opens to the exterior on the ventral surface somewhat in the anterior region of the body or near the middle, rarely near the hind end. The male generative apparatus, which contains amoeboid spermatozoa without flagellum or vibratile appendage, is almost invariably represented by an unpaired tube, and usually opens on the ventral surface near the posterior end of the body in a common opening with the intestine. As a rule, the common cloacal portion contains two pointed chitinous rods, the so-called *spicula*, in a pouch-like invagination. These spicula can be protruded and retracted by a special muscular apparatus, and serve to fasten the male body to the female during copulation. In many cases (*Strongylidae*) an umbrella-like bursa is added, or the terminal portion of the cloaca can be protruded like a penis (*Trichina*); in such cases the cloacal aperture lies almost at the extreme end but is still ventral (*Acrophalli*). In the male, papillae are almost always present in the region of the posterior end of the body, and their number and arrangement afford important specific characters. The upper ends of the generative organs in both sexes (ovary and testis) consist of a multinucleated cord in which cell limits are not discernible; lower down cell limits become discernible, but all the cells are attached by a stalk to a central protoplasmic cord—the rachis. In the oviduct and vas deferens the germ cells (progametes) become entirely free and lie loose in the cavity of the duct, and divide into the definite genital cells or gametes (ova or spermatozoa).

Development. The *Nematoda* for the most part lay eggs; it is only in rare cases that they bear living young. The eggs usually possess a hard shell and may be laid at different stages of the embryonic development or before it has begun. In the viviparous Nematodes the eggs lose their delicate membranes in the uterus of the mother (*Trichina*, *Filaria*). Fertilization takes place by the entry of a spermatozoon into the ovum, which is still without a membrane. The segmentation is equal, and leads to the formation of a kind of invaginate gastrula, from the two cell-layers of which are developed the body-wall and the alimentary canal. The embryo gradually assumes an elongated cylindrical form, and comes to lie rolled up in several coils within the shell. The excretory pore and the rudiments of generative organs, as well as a nerve ring, are present in the embryo, which is also provided with mouth and anus. The free development is a metamorphosis, usually complicated by the circumstance that it is not undergone in the habitat of the mother. The young stages or larvae, probably of most Nematodes, have a different habitat to that of the sexually adult animal, being contained in different organs of the same or even of different hosts. The larvae live for the most part in parenchymatous organs, either free or encysted in a connective tissue capsule; the adults, on the contrary, live principally in the alimentary canal.

The embryo is almost invariably characterised by the special form of the oral and caudal extremities, but sometimes also by the possession of a boring tooth. Sooner or later the skin is shed, and the animal enters its second stage, which may often still be considered as a larval stage; repeated ecdyses precede the sexually adult stage.

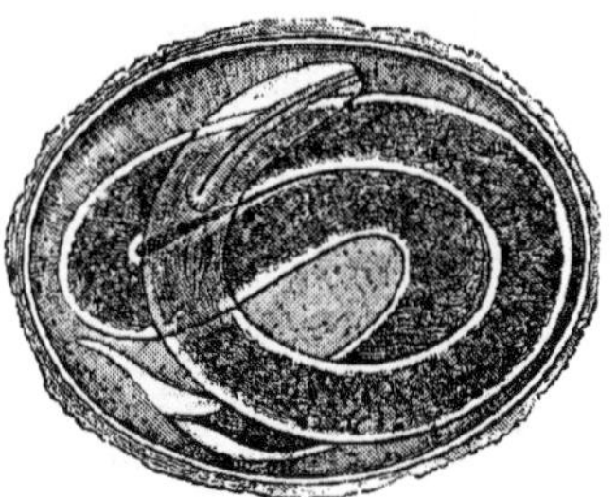

FIG. 229.—*Sclerostomum tetracanthum*, encysted (after R. Leuckart).

The post-embryonic development of the Nematodes presents numerous modifications. In the simplest cases the embryo, while still enveloped in the egg membranes, is transported passively in the food (*Oxyuris vermicularis* and *Trichocephalus*). In many *Ascaridae* the embryos, which are provided with a boring tooth, first make their way into an intermediate host, by which they are transported into the intestine of the second host with the food or water.

More frequently the young forms encyst within the intermediate host, and, enclosed in the cyst, are transferred into the stomach and intestine of the permanent host (Fig. 229). For example, the embryos of *Spiroptera obtusa* of the mouse, while still in the egg membranes, are taken with the food by the meal-worm, in the body cavity of which they encyst. In the viviparous *Trichina spiralis* there is a modification of this mode of development, inasmuch as the migration of the embryos and their development, to the stage found encysted in the muscles (muscle-trichina), takes place in the same animal which contains the sexually mature intestinal *Trichinas*.

The development of the Nematode larvae often makes a considerable advance within the intermediate host into which they have migrated. Thus, for instance, in *Cucullanus elegans*, the embryos migrate into the *Cyclops*, and in the body cavity of these small *Crustacea* undergo two ecdyses and essential alterations of form, obtaining at this early stage the characteristic oral capsule of the sexually adult stage, to which they only develop in the intestine of the perch. According to Fedschenko,* a similar mode of development occurs in *Filaria medinensis*. The embryos pass into puddles of water, and migrate thence into the body cavity of the *Cyclopidae;* and after casting their skin assume a form which, except for the absence of the oral capsule, resembles that of the larva of *Cucullanus*. After the expiration of two weeks there is another ecdysis, with which is connected the loss of the long tail. The later history is unknown. It has not yet been discovered whether the migration of the Filarian larva into the permanent host (man, see p. 289) takes place within the body of the Cyclops, or independently after copulating in the free state.

The embryos of some *Nematoda* develop in damp muddy earth, after casting their skin, to small so-called *Rhabditis* forms with a double enlargement of the oesophagus, and with a pharyngeal armature. They lead an independent life in this habitat, and finally migrate to lead a parasitic life within the permanent host, where, after several ecdyses and alterations of form, they attain the sexually mature condition. This mode of development occurs in *Dochmius trigonocephalus* from the intestine of the dog, and very probably in the nearly allied *D.* (*Ancylostomum*) *duodenalis* of man, and also in *Sclerostomum*.

The offspring of parasitic Nematodes may, however, attain sexual maturity in damp earth, as free *Rhabditis* forms, and represent a special generation of forms whose offspring again migrate and become parasites. Such a life-history is a case of heterogamy. It occurs in *Rhabdonema nigrovenosum*, a parasite in the lungs of Batrachians. These parasites, which are about half to three-quarters of an inch long, all have the structure of females, but contain spermatozoa, which are produced (as in the viviparous *Pelodytes*) in the same tubes as, but earlier than the ova. They are viviparous. The embryos make their way into the intestine of their host, and accumulate in

* Compare Fedschenko, "Ueber den Bau und Entwicklung der Filaria medinensis," in the *Berichten der Freunde der Naturwissenschaften in Moskau*, tom. 8 and 10.

the rectum, but finally pass to the exterior in the faeces, and so reach the damp earth or muddy water, where they develop in a short time into the *Rhabditis*-like forms, which have separate sexes and are barely 1 mm. in length (Fig. 230, *a* and *b*). The impregnated females of the latter produce only from two to four embryos, which become free inside the body of the mother, pass into her body cavity, and there feed on her organs, which disintegrate to form a granular detritus. They finally migrate as slender, already tolerably large

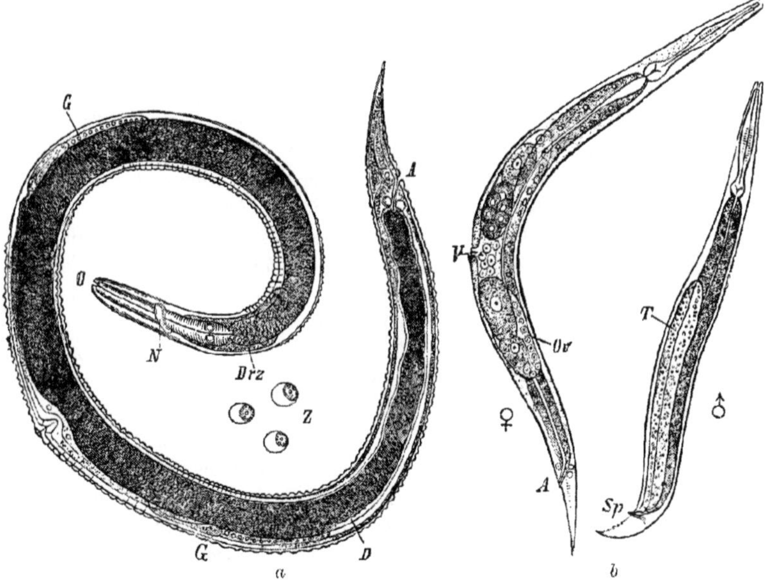

Fig. 230.—*a*, *Rhabdonema (Ascaris) nigrovenosum*, about 3·5 mm. in length, in the stage of maturity of the male products; *G* genital glands; *O* mouth; *D* intestine; *A* anus; *N* nerve-ring; *Drz* glandular cells; *Z* isolated spermatozoa. *b*, male and female *Rhabditis* forms from about 1·5 mm. to 2 mm. long; *Ov* ovary; *T* testis; *V* female genital opening; *Sp* spicula.

Nematodes into the lungs of the *Batrachia*, passing through the buccal cavity and glottis. A similar alternation of parasitic forms with free-living Rhabditis generations is presented by *Rhabdonema strongyloides* (*Anguillula stercoralis*), parasitic in man (p. 290), and by *Bradynema mirabile* (this is probably the real explanation of Zur Strasse's observation, see p. 290). *Rhabdonema* (*Leptodera*) *appendiculata*, which lives in the slug *Arion empiricorum*, also presents in its development a like alternation of heteromorphic generations, which, however, are not strictly alternating, inasmuch as numerous generations of the free *Rhabditis* form may succeed one another, before there is a return to the parasitic condition. The *Rhabdonema appendiculata* is also peculiar, in that the form parasitic in the slug is a larva characterized by the absence of a mouth and by the possession of two long band-shaped caudal appendages; it quickly attains maturity, but only after a migration into damp earth and after losing the caudal appendages and casting the skin.

The *Nematoda* feed on organic juices, some of them also on blood, and are enabled by their armed mouth to inflict wounds and to gnaw tissues. They move by bending their body with a rapid undulatory movement towards the ventral and dorsal surfaces. Although most *Nematoda* are parasitic, they usually lead an independent life in certain stages of their life-history. Numerous small *Nematoda* (*Anguillulidae*, *Enoplidae*), however, are never parasitic, but live freely in fresh and salt water, and in the earth. Some Nematodes are parasitic in plants, *e.g.*, *Anguillula tritici*, *dipsaci*, etc., and may even produce gall-like deformities (*Tylenchus*), others live in decaying vegetable matter, *e.g.*, the vinegar worm in fermenting vinegar and paste. In many cases the migration of the parasite is a condition necessary to the attainment of sexual maturity, *e.g.*, in *Mermis*, where sexual organs are not developed till the worm leaves its host and becomes free in damp earth, in which copulation of the sexes is effected. Finally there are certain small Nematodes the females of which alone are parasitic. These, after copulation in the free state with the small males, migrate into insects, and under the favourable conditions of parasitism not only increase enormously in size, but also undergo structural modifications favourable for the production of a large number of embryos. In *Attractonema gibbosum* and in *Sphaerularia bombi*, the remarkable parasite of the humble-bee, the females, after copulating in the free state, migrate, the former into the larva of the gall-fly *Cecidomyia pini*, the latter into the queen-bees, which live through the winter. Here the gut degenerates, and a kind of hernia of the body-wall, containing the generative organs, is formed, while the body of the worm shrinks to a small appendage (Fig. 231). The eggs develop in the body of the insect into larvae, which pass out of the body, become free, and, either after a few days (*Attractonema*) or after some months (*Sphaerularia*), become sexually mature.

The power possessed by small Nematodes of resisting the effects of prolonged desiccation and of coming to life again (so to speak) on being moistened is remarkable.

Fam. 1. **Ascaridae.** Body tolerably stout. Mouth with three lips furnished with papillae, one of which is dorsal, while the other two meet together in the ventral line. Buccal cavity rarely furnished with pieces of chitin. Posterior portion of oesophagus often forming a distinct pharyngeal bulb. Posterior end of male ventrally curved, and usually furnished with two horny spicula. Parasitic in the alimentary canal of animals.

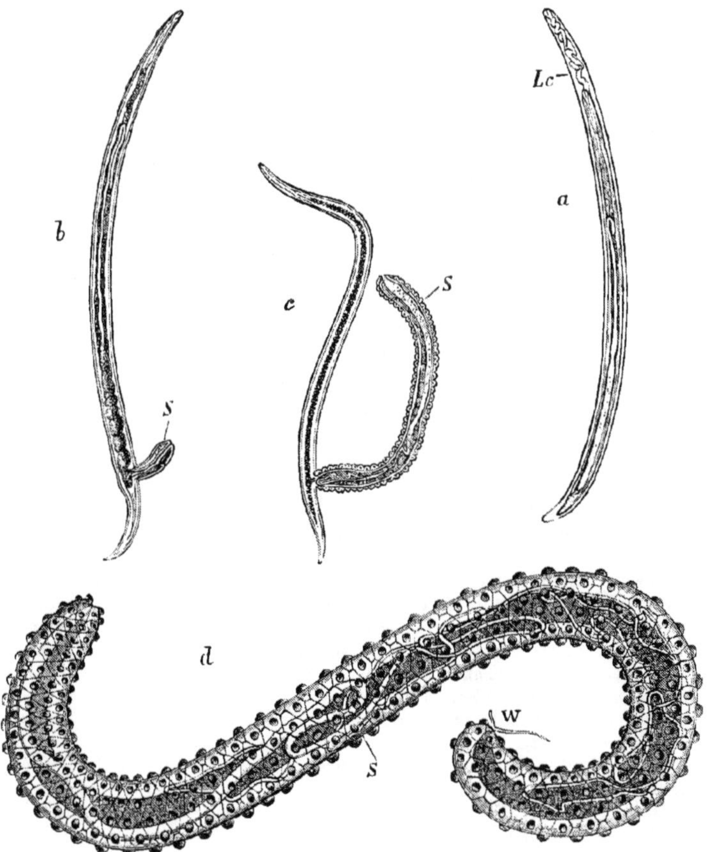

Fig. 231.—*a*, male *Sphaerularia* in its larval skin (*Lc*) × about 75. *b*, female with partially protruded vagina *S*. *c*, the same with still more developed uterine outgrowth *S*. *d*, uterine outgrowth fully formed, containing ovary, oviduct, and uterus × 10; *W* the relatively minute body of the worm.

Ascaris L. Polymyarian, edges of lips dentated in the larger species. Pharyngeal bulb not distinct; hind end usually short and conical, and in the male always provided with two spicula (Fig. 232); in *Vertebrata*. *A. lumbricoides*,

Cloquet, the human round worm, 4 to 14 in. in length, a smaller variety in the pig (*A. suilla* Duj.). The eggs pass out into water or damp earth, and remain there for some months, until the embryonic development is completed. They are probably introduced direct into the alimentary canal of their host (Grassi and Ebstein). The smallest worm found in the human intestine is 3 mm. in length. *A. megalocephala* Cloq., horse and ox, may attain a length of 17 inches. *A. mystax* Zed., dog and cat, sometimes man. There is a large number of species found in all classes of Vertebrates.

Heterakis Duj. Polymyarian, hind end of male with a preanal sucker and two lateral thickenings, spicules unequal, intestine of Vertebrates.

Oxyuris Rud. Meromyarian; pharyngeal bulb with teeth, hind end of female thin and pointed, male with one spiculum, in Vertebrates and Insects (cockroach and beetles); *O. vermicularis* L., large intestine of man, introduced directly as eggs; *O. curvula*, caecum of horse; *Nematoxys* Schn., in Amphibia; *Oxysoma* Schn.; *Isakis* Lespés, Arthropods and terrestrial Molluscs; *Labidurus* Schn.; *Aspidocephalus* Dies.; *Heterocheilus* Dies.; *Peritrachelius* Dies.

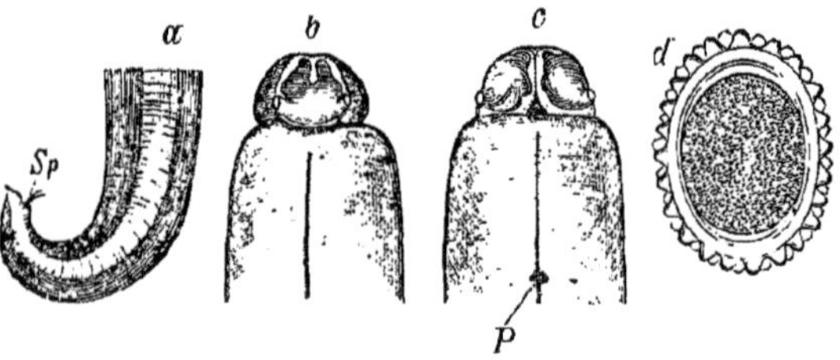

FIG. 232.—*Ascaris lumbricoides* (after Leuckart). *a*, hind end of male with the two spicula; *b*, anterior end from the dorsal side, showing the dorsal lip with its two papillae; *c*, the same from the ventral side with the two lateral ventral lips and the excretory pore (*P*); *d*, egg with the external membrane formed of small clear spherules.

Fam. 2. **Strongylidae.** The male genital opening is at the posterior end of the body, at the bottom of a bell-shaped bursa, the margin of which carries a variable number of papillae; no pharyngeal bulb. *Eustrongylus* Dies., polymyarian, with six oral papillae and a row of papillae on either lateral line. The bursa is bell-shaped and completely closed, with regular muscular walls and numerous marginal papillae. There is only one spiculum. The female genital opening is far forward. The larvae live encysted in fishes (*Filaria cystica* from *Symbranchus*). *E. gigas* Rud., the body of the female is three feet in length, and only twelve mm. thick. It lives singly in the pelvis of the kidney of the seal and otter, etc., and very rarely in man; got by eating raw fish.

Strongylus Rud. With six oral papillae and small mouth. Two conical cervical papillae upon the lateral lines. The posterior end of the male has an umbrella-like incompletely closed bursa. Two equal spicula, usually with unpaired supporting organ. The female sexual opening is sometimes approached to the posterior end of the body. They live for the most part in the lungs and bronchial tubes. The larvae live in damp earth, and probably pass into the host directly in the food. *St. longevaginatus* Dies. Body 26 mm. long, 5 to 7 mm. thick. The female sexual opening lies directly in front of the anus, and leads into a simple ovarian tube. Only once found in the lung of a six-year old boy, in Klausenburg. *St. paradoxus* Mehlis, in the bronchial tubes of the pig. *St. filaria* Rud., in the bronchial tubes of the sheep. *St. commutatus* Dies., in the trachea and bronchial tubes of the hare and rabbit. *St. auricularis* Rud., in the small intestine of *Batrachia*. *St. micrurus* in aneurisms on the arteries of the ox.

Dochmius Duj. With wide mouth and horny oral capsule, the edge of which is strongly toothed. Two ventrally placed teeth project at the bottom of the oral capsule, while on the dorsal wall a conical spine projects obliquely forwards. *D. duodenalis* Dub. (*Ancylostomum duodenale* Dub.), 10 to 18 mm. long, in the small intestine of man, discovered in Italy (Fig. 233); very widely distributed in the countries of the Nile (Bilharz and Griesinger). By aid of its strongly armed mouth it wounds the intestinal mucous membrane, and sucks the blood from the vessels. The frequent haemorrhages occasioned by these *Dochmii* are the cause of the illness known by the name of Egyptian chlorosis. It has lately been established that this worm occurs in Brazil, and that, like *D. trigonocephalus*, it develops in puddles of water (Wucherer). *D. trigonocephalus* Rud., in the dog. *Sclerostomum* Rud. With characters of Dochmius, but with a different oral capsule, into which two long glanular sacs open. *Sc. equinum* Duj. = *armatum* Dies. In the intestine and the mesenteric arteries of the horse. Lives freely in *Rhabditis* form. Bollinger* has shown that the phenomena of colic in the horse may be referred to embolic processes proceeding from aneurism of the intestinal artery. Each aneurism contains about nine worms. *Sc. tetracanthum* Mehlis, also in the intestine of the horse. The embryos, after migrating into the intestine, become encysted in the walls of the rectum and caecum, assume within the cyst their definite form, break out from the cyst, and escape again into the intestine. *Pseudalius* Duj.; *Physaloptera* Rud.; *Cucullanus* Müll.; *C. elegans* Zed., in the perch, buccal capsule well developed; viviparous, the young pass into water and enter *Cyclops*.

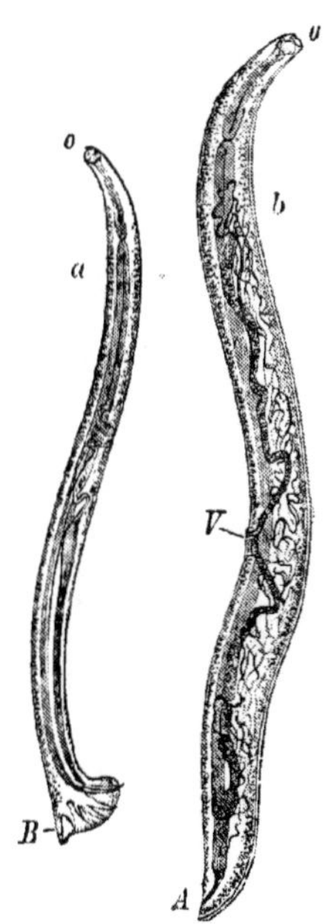

FIG. 233.—*Dochmius duodenalis* (after R. Leuckart). *a*, male; *O* mouth; *B* bursa. *b*, female; *O* mouth; *A* anus; *V* vulva.

Olullanus Lkt.; *O. tricuspis* Lkt., in the cat, sexless young encysted in the mouse. *Syngamus* Siebold; *S. trachealis* Sieb., in trachea and bronchi of poultry—cause of *gapes*, the male is permanently attached to the female by the application of the bursa to the female opening. The eggs hatch on the ground or in water. No second host, though the embryos may be eaten by earthworms and remain alive in their alimentary canal.

Fam. 3. **Trichotrachelidae**, with long, thin anterior portion of the body. Mouth small, without papillae. Oesophagus very long, traversing a peculiar cord of cells.

Trichocephalus Goeze. Anterior part (Fig. 234) of the body elongated and

* Bollinger, "Die Kolik der Pferde und das Wurmaneurysma der Eingeweidearterien," München, 1870.

whip-shaped: posterior part cylindrical and sharply distinct, enclosing the generative organs, curved in the male. Lateral lines absent. Main median lines present. The penis is slender and furnished with a sheath, which is turned inside out when the former is protruded. The hard-shelled, lemon-shaped eggs undergo the first part of their development in water, without intermediate host. *Tr. dispar* Rud. In the human colon: these worms do not live free in the intestine, but bury their filiform anterior extremity in the mucous membrane (Fig. 234). The eggs pass out of the host with the faeces, as yet without a sign of beginning development, which only takes place after a prolonged sojourn in the water or in a damp place. According to the experiments of Leuckart, performed with *Tr. affinis* of the sheep and *Tr. crenatus* of the pig, embryos with the egg membranes, if introduced into the intestine, develop into the adult *Trichocephalus;* and we may therefore conclude that the human *Tr. dispar* is introduced directly, and without an intermediate host, either in the drinking water or in uncleaned food. The young *Tr. dispar* is at first hair-like, and resembles a *Trichina*, and only gradually acquires the considerable thickness of the hind end of the body.

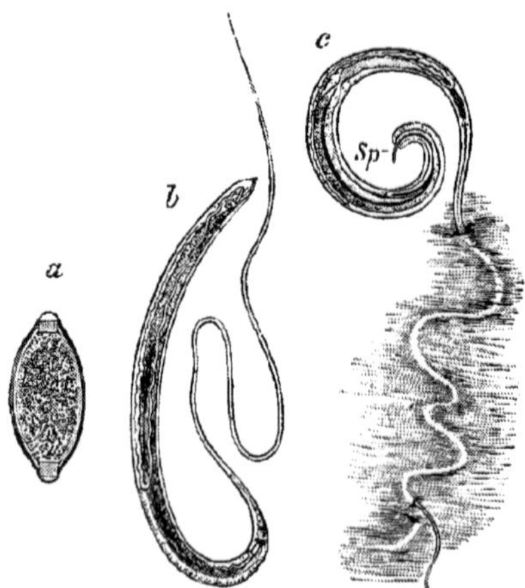

Fig. 234.—*Trichocephalus dispar* (after R. Leuckart). *a*, egg. *b*, female. *c*, male with the anterior part of the body buried in the mucous membrane; *Sp* spiculum.

Trichosomum Rud. Body thin, hair-like, but the posterior end of the body in the female is swollen. Lateral lines and the principal median lines are present. The male caudal extremity has a cutaneous fold and a simple penis (spiculum) and sheath. *Tr. muris* Creplin., in the large intestine of the house-mouse. *Tr. crassicauda* Bellingh., in the bladder of the rat. According to Leuckart, the dwarfed male lives in the uterus of the female, and there are usually two or three, more rarely four or five males in a single female. There is also a second species of *Trichosomum* found in the bladder of the rat, *Tr. Schmidtii* v. Linst., the larger male of which was formerly taken for that of *Tr. crassicauda*.

Trichina Owen.* Body thin, hair-like. Principal median lines and lateral lines are present. The female generative opening well forward. The hind end of the body without spicule; with two small conical papillae, between which the cloaca projects.

Tr. spiralis Owen, in the alimentary canal of man and numerous, principally carnivorous, mammalia; hardly two lines in length. The viviparous females begin to bring forth embryos about eight days after their entrance into the alimentary canal. These embryos traverse the intestinal walls and body cavity of the host, and migrate, partly by their own movements in the bundles of

* Compare the writings of R. Leuckart, Zenker, R. Virchow, Pagenstecher, etc.

connective tissue, partly with the aid of the currents of blood, into the striped muscles of the body. They pierce the sarcolemima and penetrate into the

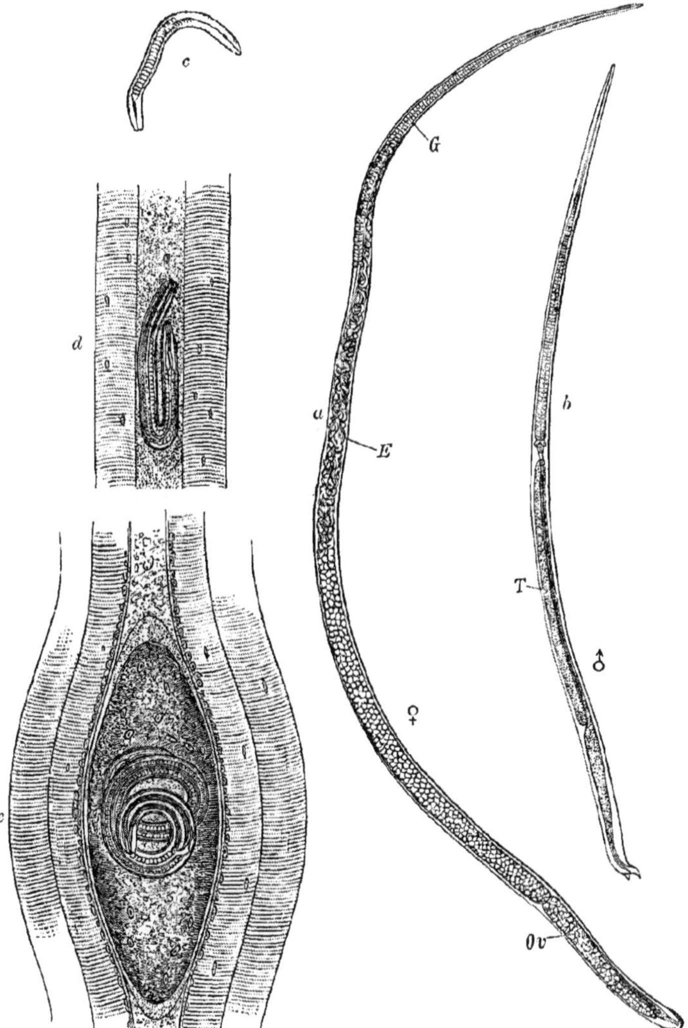

Fig. 235.—*Trichina spiralis.* *a*, mature female *Trichina* from the alimentary canal; *G* genital opening; *E* embryos; *Ov* ovary. *b*, male; *T* testis. *c*, embryo. *d*, embryo which has migrated into a muscle fibre, already considerably enlarged. *e*, the same developed into a coiled Muscle Trichina, and encysted.

primitive bundles, the substance of which degenerates, the degeneration being accompanied by an active multiplication of the nuclei. In a space of fourteen days they develop, within a sac-like swelling of the muscle fibres, into spirally coiled worms, around which and within the sarcolemma and its connective tissue investment a clear lemon-shaped capsule is excreted from the degenerated muscle substance. The young Muscle-Trichina can remain living for years within this capsule, which at first very delicate, gradually becomes thickened and hard by the formation of other layers and by the gradual deposition of calcareous matter. If the encysted animal is transferred into the intestine of some warm-blooded animal in the flesh of its first host, it is freed from its cyst by the action of the gastric juice, and the rudimentary generative organs, which are already tolerably far developed, quickly attain maturity. In from three to four days after their introduction the asexual Muscle-Trichinas become sexual Trichinas. These copulate and produce a brood of embryos which migrate into the tissues of the host (one female may produce as many as 1000 embryos) (Fig. 235). The house rat is especially to be mentioned as the natural host of the *Trichinae*. This animal does not hesitate to eat the carcase of its own species, and so the Trichina infection is passed on from generation to generation. Carcases infected with Trichinas are sometimes eaten by the omnivorous pig, in whose flesh the encysted *Trichinas* are introduced into the intestine of man, and occasion the well-known disease, Trichinosis, which when the migration takes place in numbers, often has a fatal result.

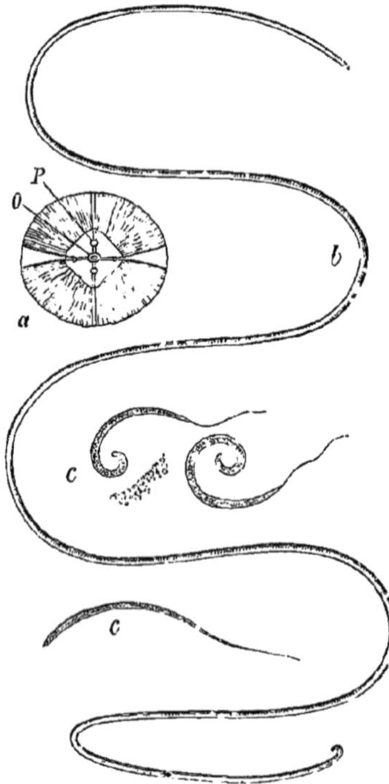

Fig. 236.—*Filaria medinensis* (after Bastian and Leuckart). *a*, anterior end seen from the oral surface; *O* mouth; *P* papilla. *b*, pregnant female (size reduced more than half). *c*, embryos strongly magnified.

Fam. 4. **Filariidae.** Body filiform, elongated, often with six oral papillae, sometimes with a horny oral capsule, with four praeanal pairs of papillae, to which an unpaired papilla may be added, with two unequal spicula or with simple spiculum.

Filaria O. Fr. Müll. (Fig. 236). With small mouth and narrow oesophagus. This genus, which is sometimes destitute of papillae, lives outside the viscera, usually in connective tissue, frequently beneath the skin, and is divided by

Diesing into numerous genera. *F.* (*Dracunculus*) *medinensis** Gmel. the Guinea worm, in the subcutaneous cellular tissue of man in the Tropics of the Old World, reaches a length of two feet or more. The head is provided with two small and two larger papillae. Alimentary canal degenerate. The female is viviparous, and without sexual opening. The male form is unknown. The worm lives in the connective tissue between the muscles and beneath the skin, and after reaching sexual maturity, occasions the formation of an abscess, with the contents of which the embryos escape to the exterior (Fig. 236). It has lately been proved (Fedschenko) that the embryos of *Filaria* migrate into a Cyclops and there undergo an ecdysis. Whether they are then (in the body of the Cyclops) introduced into man in his drinking water, or whether they first escape and copulate in a free state, is not known. *F. immitis* lives in the right ventricle of the dog, and is very abundant in East Asia. It is viviparous. The embryos pass directly into the blood, where, however, they do not undergo their further development. Similar immature *Filariae* are also found in the blood of man in the Tropics of the New and Old Worlds, and have been described as *F. sanguinis hominis nocturna*, the sexual form living in the lymphatic glands as *F. Bancrofti* Cobb. The intermediate host is probably the mosquito, on the death of which the larvae make their way into the water in which they, presumably, enter the alimentary canal of man. It is supposed that they make their way from the alimentary canal, by active migration, to the lymphatics where they mature and pair † The larvae enter the blood and probably escape by means of abscesses, etc., and by the kidneys, since they are also found in the urine; their appearance seems to have an aetiological connection with chyluria and elephantiasis. *F. sanguinis hominis nocturna* is periodic, appearing only in the blood at night. Two other varieties of *Filaria* are also found in the human blood, viz. *diurna* which appears in the day, and *perstans*. *Filaria perstans* is supposed to have a connection with the sleeping sickness of negroes. The sexual forms of *F. diurna* and *perstans* are unknown. In the East Indies, young Filaria also live in the blood of the street dog, and would seem to be related to the brood of *Filaria sanguinolenta*, since, according to Lewis, knotty swellings on the aorta and oesophagus are invariably found with these *Filaria*. *F. papillosa* Rud. in the peritoneum of the horse. *F. loa* Guyot., in the conjunctiva of negroes on the Congo. *F. labialis* Pane, only once observed at Naples. An immature *Filaria* described as *Filaria lentis* (*oculi humani*) has been found in the human capsula lentis. *Ichthyonema* Dies., anus and vulva absent, male minute, in fishes. *Spiroptera* Rud., *S. obtusa* Rud., stomach of mouse, asexual larva in meal-worm. *Spiroxys* Schn., *Hystrichis* Molin; *Tetrameres* Crepl. (*Tropidocera* Dies.); *Hedruris* Crepl.; *Ancyracanthus* Dies.

Fam. 5. **Mermithidae.** Aproctous Nematodes, with very long filiform body, and six oral papillae. The male caudal region is broad, and is provided with two spicula and three rows of numerous papillae. They live in the body-cavity of insects, and escape into the damp earth, where they attain sexual maturity and copulate. *Mermis nigrescens* Duj. was the occasion of the fable of the rain-worm. *M. albicans* v. Sieb. v. Siebold established by experiment the migration of the embryos into the caterpillars of *Tinea evonymella*.

* Compare H. C. Bastian, "On the Structure and Nature of the Dracunculus," *Trans. Linn. Society*, vol. xxiv., 1863. Fedschenko *l.c.*

† P. Manson, *China Customs Report*, No. xiv., 1878. T. Lewis, *Q.J.M.S.*, 19, 1879. See also Manson, in Davidson's "*Hygiene and Diseases in Warm Climates*," London, 1893.

Fam. 6. **Anguillulidae.*** Free-living Nematodes of small size, usually with a double swelling on oesophagus. Caudal glands are sometimes present. The lateral canals are often replaced by the so-called ventral glands; males with two equal spicules. Some species either live on or are parasitic in plants; others live in fermenting or decaying matter. The greater number, however, live free in earth or water. *Tylenchus* Bast. Buccal cavity small, and containing a small spine. The female genital opening lies far back. *T. scandens* Schn. = *tritici* Needham, in mildewed wheat grains. When the grains of wheat fall the dried embryos grow in the damp earth, bore through the softened membranes, and make their way on to the growing wheat plant. Here they remain some time, perhaps a whole winter without alteration, until the ears begin to be formed. They then pass into the latter, grow, and become sexually mature, while the ear is ripening. They copulate and deposit their eggs, from which the embryos creep out, and at length constitute the sole contents of the wheat grains (ear-cockles). *T. dipsaci* Kuhn, in heads of thistles (Carduus); *T. Davainii* Bast., on roots of moss and grass. *Heterodera Schachtii* Schmidt., roots of the beet-root, also of the cabbage, of wheat, barley, etc.

Rhabditis Duj. (divided by Schneider into *Leptodera* Duj. and *Pelodera* Schn.), with two strongly-developed oesophageal swellings, of which the hinder has a dental armature. *Rh. flexilis* Duj., head very sharply pointed, mouth with two lips; in the salivary glands of *Limax cinereus*. *Rh. angiostoma* Duj.

Rhabditis nigrovenosa, the free dioecious generation of *Rhabdonema* (*Ascaris*) *nigrovenosum*, which is hermaphrodite, and infests the lungs of frogs. *Rhabdonema strongyloides* Lkt. (*Anguillula intestinalis*) in intestine of man in Lombardy and Cochin China, causing diarrhoea; *Anguillula stercoralis* is the free Rhabditis generation of this.† *Rhabdonema* (*Leptodera*) *appendiculata* Schn., in damp earth, 3 mm. long. The parasitic form, which is without a mouth and has two caudal bands, lives in *Arion empiricorum*, and does not attain sexual maturity until after its escape from its host. It is dioecious, and gives rise to free-living, also dioecious, *Rhabditis* forms, many generations of which may succeed one another.

Bradynema Zur Strasse, *Br. rigidum* v. Sieb. in the body-cavity of the beetle Aphodius fimentarius, without mouth, intestine, anus, excretory and nervous systems. Mainly consisting of uterus full of embryos; viviparous; larvae, male and female, set free in body-cavity, bore through into intestine, and pass out by anus of host. According to Zur Strasse‡ the female larvae die without producing eggs, while the male larvae become protandrous hermaphrodites, and enter in an unknown way the body-cavity of their host. This account requires confirmation.

Allantonema mirabile Lkt. in the beetle *Hylobius pini*, 3 mm. long, sausage-shaped, surrounded by a membrane, and attached to tracheae in the body-cavity; without mouth, intestine, anus; are protandrous hermaphrodites, with a free-living dioecious Rhabditis generation.

* Davaine, "Recherches sur l'Anguillule du blé niellé," Paris, 1857. Kühn, "Ueber das Vorkommen von Anguilluleu in erkrankten Bluthenköpfen von Dipsacus fullonum," *Zeitschr. für wiss Zool.*, tom 9., 1859. Bastian, "Monograph of the Anguillulidae or free Nematoids, marine, land, and fresh water," London, 1864. O. Butschli, "Beiträge zur Kentniss der freilebenden Nematoden," *Nov. Acta*, tom. 36, 1873. J. G. de Man, "*Die frei in der reinen Erde und im süssen Wasser lebenden Nematoden der Nied. Fauna*," Leiden, 1884.

† Leuckart, *Ber. d. k. Sächs. Gesel. d. Wiss*, 1882.

‡ *Z. f. w. Z.*, 54, 1892, p. 700.

Attractonema gibbosum Lkt. in body-cavity of *Cecidomyia pini*, without mouth and anus; with alimentary canal reduced to a cell-cord; with prolapsed uterus and vagina, full of eggs, projecting from the body. Males and females are found in the free state; they copulate; the female alone enters the host.

*Sphaerularia bombi** Léon. Duf., in the body-cavity of the humble bees which have survived the winter (Fig. 231); the life-history is similar to that of *At. gibbosum*. The form in the body-cavity of the bee is the female, and consists of a 15 mm. vermiform body carrying a small Nematode-like worm. The former is the enormously grown protruded female generative apparatus, and the latter is the body of the female worm, which alone enters the host. The larvae become free in the bee, and leave it by the anus. When free and 1 mm. long they become sexual as males and females, and copulate. After copulation the female enters the bee.

Anguillula Ehrbg. Small buccal cavity, oesophagus with a posterior bulb and a valvular apparatus. Male without bursa. Usually two circular lateral organs. No anal gland. *A. aceti* = *glutinis* O. Fr. M., the vinegar- and paste-worm 1–2 mm. long. *Chromadora* Bast.; *Spilophora* Bast.; *Odontophora* Bast.

Fam. 7. **Enoplidae.**† Small, usually free-living, marine forms, without the posterior pharyngeal bulb, often with eyes and a buccal armature, often with fine hairs and bristles round the mouth. *Dorylaimus* Duj. (*Urolabes* Carter); *D. palustris* Cart., supposed by Carter to be a non-parasitic stage in the development of *Filaria medinensis*; *D. stagnalis* found in mud everywhere in Europe. *Trilobus* Bast.; *Monhystera* Bast.; *Comesoma* Bast.; *Enchelidium* Ehrbg., a large eye on the oesophagus, in the sea. *Enoplus* Duj., marine; *Symplocostoma* Bast.; *Oncholaimus* Duj.; *Odontobius* Roussel.

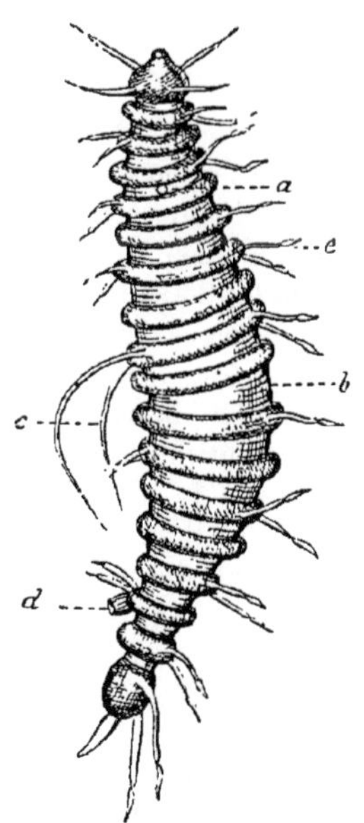

FIG. 237.—*Desmoscolex minutus* × 200. *a* eye; *b* ventral surface; *c* long dorsal bristles, which exist only in the female; *d* anus; *e* ventral bristles (after Greef, from Perrier).

The **Chaetosomatidae** and the **Desmoscolecidae** may be mentioned here. They are minute non-parasitic, marine organisms, presenting in the arrangement of their generative organs resemblances to the Nematoda. Their anatomy, however, is imperfectly known.

* A. Schneider, *Zool. Beiträge*, Breslau, vol. i. R. Leuckart, "Zool. Anzeiger," 1885. *Id.*, *Neue Beiträge z. Kennt. d. Nematoden*, Leipzig, 1887.

† Eberth, *Unters. üb. Nematoden*, Leipzig, 1863. Marion, "Rech. anat. et physiol. sur les Nématoides non-parasites marins," *Ann. Sc. Nat.*, XIII., 1870. O. Bütschli, "Üb. freileb. Nematoden, insb. d. Kieler Hafens," *Abh. Senk. Nat. Gesel. Frankfurt*, 9, 1874. De Man, "Onderz. o. v. in der Aarde lev. Nematoden," *Tyds. d. Nederlund. Tierkund. Vereenig*, 1875. *Id.* "*Contributions à la connaissance des Nématodes du golfe de Naples*," Leiden, 1876.

Chaetosomatidae.* The anterior end of the body is swollen into a head, with a semicircle of moveable hooks. The body is covered with fine hair. There is a terminal mouth, an oesophagus, intestine and ventral anus. Two rows of knobbed processes on the ventral surface of hind end of body. Dioecious. Testis single, vas deferens opens with the anus, and is provided with two spicules. Two ovaries, single vagina, opening ventrally about middle of body. Minute. Marine. *Chaetosoma* Clap., *Rhabdogaster* Metschnikoff.

Desmoscolecidae.† Minute, vermiform marine animals. Body marked with transverse ridges, which carry bristles. Mouth anterior and terminal, anus dorsal. Dioecious. Testis and ovary as unpaired tubes. Male with two spicula, vas deferens opens with the anus; ovary opens on ventral surface. They move by looping. Nervous system unknown. *Desmoscolex* Clap. (Fig. 237), *Trichoderma* Greef.

Class II. NEMATOMORPHA.‡

Elongated, unsegmented round worms without lateral lines, with ventral nerve-cord and circumoesophageal ring.

The position of these worms in the system is obscure. The name chosen for the order expresses the fact that in general form of body they resemble the *Nematoda*. Their anatomy, however, is very different from that of the latter group. The body is elongated and filiform, but is without oral papillae and lateral lines. There is a marked cuticle, beneath which is a cellular epidermis. The body-muscles are in one layer of longitudinal fibres. The nervous system consists of three cords closely approximated in the middle ventral line, within the muscular layer. This ventral cord divides in front into the two cords which embrace, and unite with one another dorsal to, the oesophagus. The mouth (which is anterior) and oesophagus are occluded in the adult state, and the anus is at the hind end of the body.

Between the body-wall and the gut there lie polygonal cells, which in the adult seem to differentiate into various organs, *e.g.* genital glands, genital ducts; and a space, which may or may not be a coelomic space, appears among them.

* E. Metschnikoff, *Z. f. w. Z.*, 17, 1867. Panceri, *Atti Accad. delle Scienze*, Naples 7, 1878, p. 7.

† Greef. Unters. üb. einige merkwürdige Thiergruppen der Arthropoden u. Wurmtypus. *Arch. Naturg.* 35 (1) 1869. Panceri, *Op. cit.*

‡ F. Vejdovsky, "Zur Morphologie der Gordiiden," *Z. f. w. Z.*, 43. 1886, p. 369, and *Ibid.*, 46, 1888, p. 188. v. Linstow, "Ueb. d. Entwick. und d. Anat. v. Gordius," *Arch. f. Mic. Anat.*, 34, 1889, p. 248. *Id.*, "Weitere Beobachtungen an Gordius," *Arch. f. Mic. Anat.*, 37, 1891, p. 239. *Id.*, "Über d. Entwick. v. Gordius," *Centralbl. Bact. Parasitk.*, 11, p. 475. O. Bürger, "Zur Kenntniss v. *Nectonema agile* Verr.," *Zool. Jahrb. Morph. Abth.*, 4. 1891, p. 631. H. B. Ward, "On Nectonema agile," *Bull. Mus. Harvard Coll.*, 23, 1892, p. 135.

The sexes are separate, and the generative organs are paired; they open to the exterior, with the anus near the hind end of the body. The ovaries are metamerically repeated masses of cells on each side. The ova are collected into two tubes—the egg-sacs, which behind become the oviducts. The oviducts open into a median atrium, which also receives the opening of the single receptaculum seminis. The male organs are much the same, except that the two vasa deferentia open separately into the rectum (cloaca). The male caudal region is forked, and is devoid of spicula.

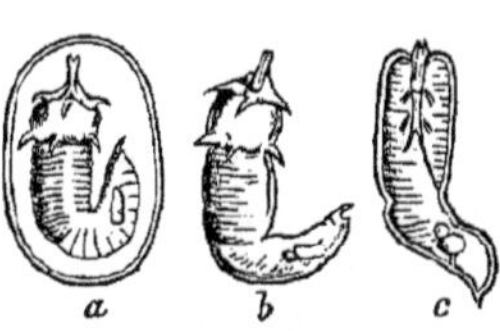

FIG. 238.—Larvae of *Gordius subbifurcus* (after Meissner). *a*, in the egg-membrane with protruded proboscis; *b*, out of the egg-membrane; *c*, with invaginated proboscis.

In the adult stage these animals are free and sexually mature, but the alimentary canal is generally degenerate, being occluded in front, and presumably functionless. They are found in fresh-water ponds and streams, and move in an undulating manner. The eggs give rise in the water to an embryo, with a circle of boring-spines, which leaves the egg and migrates into insect larvae (*Ephemerid larvae*, etc., they have been found even in fishes, molluscs, and oligochaetes) and there encysts. Water beetles and other aquatic predatory insects eat with the flesh of their prey the encysted young forms, which then develop in the body-cavity of their new and larger host to young *Gordiidae*, which make their way into the water. *Gordius* L.; *Nectonema* Verr., a marine form, life-history unknown.

Class III. ACANTHOCEPHALA.*

Elongated round worms, with protrusible proboscis furnished with hooks; without mouth and alimentary canal; with a body-cavity, into which the ova are dehisced and the oviducts open; parasitic.

The saccular, often transversely wrinkled body begins with a proboscis, which is furnished with recurved hooks, and can be retracted into a tube projecting into the body-cavity (sheath of the

* Besides Dujardin, Diesing, l. c., compare: R. Leuckart, "Parasiten des Menschen," tom 2, 1876. Greeff, "Untersuchungen über Echinorhynchus miliaris," *Arch. für Naturgesch*, 1864. A. Schneider, "Ueber den Bau der Acanthocephalen," *Müller's Archiv.*, 1868. Also the *Sitzungsberichte der Oberhessischen Gesellschaft für Natur- und Heilkunde*, 1871. B. Grassi and S. Calandruccio, "Ueb. einen Echinorhynchus, welcher auch in Menschen parasitirt u. dessen Zwischenwirth ein Blaps ist," *Centralbl. für Bakt. u. Parasitenkund.* 3, 1888. O. Hamann, *Die Nemathelminthen*, Heft 1, Monographie der Acanthocephalen, Jena 1891; Heft 2, 1895.

proboscis) (Fig. 239, *R* and *Rs*). The posterior end of this sheath is fastened to the body-wall by a ligament, and by retractor muscles (*retinacula*).

The **nervous system** (Fig. 239, *G*) is placed at the base of the proboscis, and consists of a simple ganglion formed of large cells. Nerves are given off from the ganglion anteriorly to the proboscis, and through the lateral retractors to the body-wall (Fig. 239, *R*). The latter supply partly the muscular system of the body, and partly the genital apparatus, in which there are, principally in the male animal, special nerve-centres, consisting of ganglionic enlargements. **Sense organs** are entirely wanting, as also are mouth, alimentary canal, and anus.

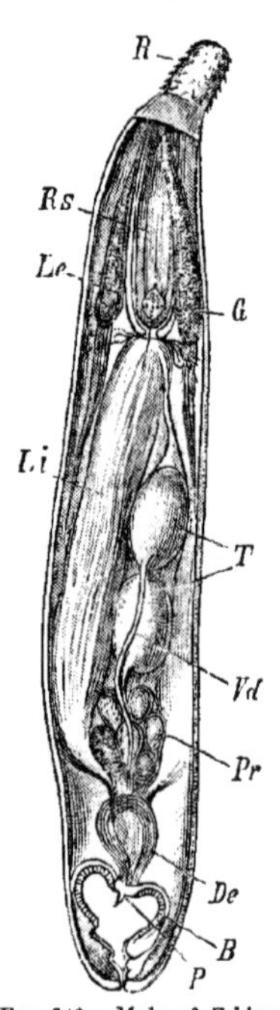

FIG. 240.—Male of *Echinorhynchus angustatus* (after R. Leuckart). *R* proboscis; *Rs* sheath of the proboscis; *Li* ligament; *G* ganglion; *Le* lemnisci; *T* testes; *Vd* vasa deferentia; *Pr* prostatic sacs; *De* ductus ejaculatorius; *P* penis; *B* retracted bursa.

The nutritive juices are taken in through the whole outer surface of the body. In the soft granular subcuticular layer of the integument (*epidermis*), in which, as in Nematodes, cell-limits are not discernible, lies a complicated system of canals, filled with a clear fluid containing granules. Beneath the subcuticular layer of the integument, which layer is often very extensive and of a yellow colour, is placed the powerful muscular tunic; it is composed of external transverse and internal longitudinal fibres, which latter bound the body-cavity. The structure of the muscular fibres is not unlike that of Nematodes. The complicated ramified system of the epidermal canals, of which two principal longitudinal trunks may be recognized, is filled with juices, and probably functions as a nutritive apparatus. The epidermis and canal system of the proboscis is entirely cut off from the epidermis and vessels of the trunk by a thin ingrowth of cuticle. The *lemnisci* are two bodies which

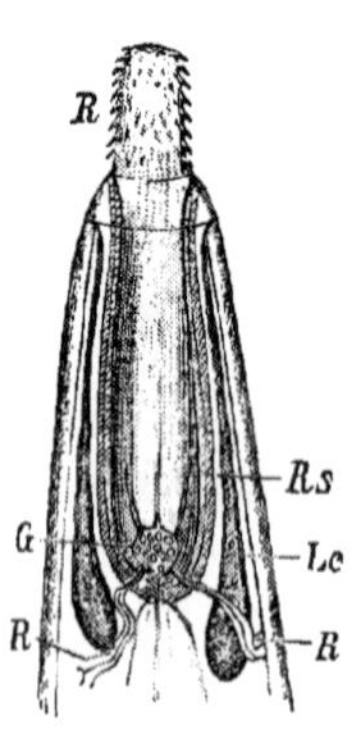

FIG. 239.—Anterior part of an *Echinorhynchus*. *R* Proboscis; *Rs* sheath of proboscis; *G* ganglion; *Le* lemnisci; *R* retinacula.

project through the muscular tunic into the body-cavity at the base of the proboscis (Fig. 240, *Le*). They consist of epidermis, continuous with that of the proboscis, and a thin muscular coat, and they contain a great number of anastomosing canals. It has been suggested that the epidermal spaces of the lemnisci and proboscis are organs for the protrusion of the proboscis; the fluid which in the retraction of the proboscis enters the epidermal spaces of the lemnisci being by the contraction of these latter organs forced back into the walls of the proboscis, so bringing about its protrusion. According to Schneider, the vessels of the lemnisci open into a circular vessel in the integument, and only communicate with the net-work of canals in the proboscis, while the other dermal vessels (nutritive apparatus), the contents of which differs from that of the vessels of the lemnisci, are, as above stated, completely shut off from the latter.

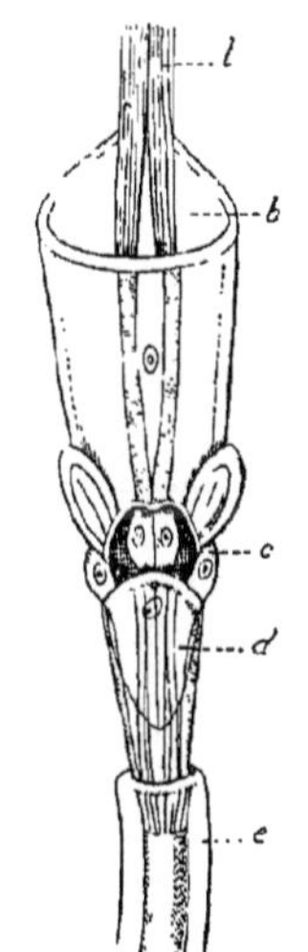

FIG. 241. — Generative ducts of a female *Echinorhynchus trichocephalus* from the so-called dorsal side, *i.e.*, the side of the posterior opening of the bell. The opening of the uterine bell *b* looking forward and its relation to the ligament *l*. *c* posterior opening of the bell, by which the unripe eggs pass back into the body-cavity; *d* uterus; *e* vagina (after Kaiser).

It has recently* been asserted that there is in *Echinorhynchus gigas* a pair of nephridia. They consist of a bunch of fine tubes, which terminate in a sieve-plate, the pores of which open into the body-cavity. The sieve-plate carries on its inner side a bunch of flagella. The fine tubes unite to form a duct, which joins its fellow. The single duct so formed opens into the generative duct.

Generative organs. The body-cavity through which fluids circulate encloses the greatly-developed generative organs, which are attached to the end of the sheath of the proboscis by the *ligament* (Figs. 240 *Li* and 241 *l*). The sexes are separate. The male (Fig. 240) has two testes (*T*), and the same number of efferent ducts (*Vd*). The latter unite behind to form a ductus ejaculatorius (*De*), which is often furnished with six or eight glandular sacs (*Pr*), and a conical penis (*P*), at the bottom of a bell-shaped protrusible bursa (*B*), situated at the posterior end of the body (Fig. 240). The generative organs of the larger females (Fig. 241) consist of the ovary developed in the ligament; of

* J. E. Kaiser "Die Acanthocephalen u. ihre Entwick." *Bibl. Zool.* Bd. 2, Heft 7, 2 Theile, 1892.

a complicated "uterine bell," which opens into the body-cavity (*b*), and to the base of which the ligament is attached; of two short oviducts connecting the bell with the uterus, which leads into the vagina; the vagina opens at the hind end. There is at the hind end of the bell a second opening into the body-cavity on the so-called dorsal side (*c*). The female generative ducts consist of a few very large cells, like the cells constituting the skin of the embryo, and the muscles of the ducts are fibrous differentiations of the outer parts of these cells.

It is only in the young stage that the ovary is a simple body enclosed by the membrane of the above-mentioned ligament. As the animal increases in size, the ovary grows, and becomes divided into numerous spherical masses of eggs, the pressure of which bursts the membrane of the ligament; the masses of ova, as well as the ripe elliptical eggs, which gradually become free from them, fall into the body-cavity. The egg membranes are not formed till after segmentation, and ought perhaps to be interpreted as embryonic membranes. The eggs, which already contain embryos, pass out of the body-cavity into the uterine bell, which is continually dilating and contracting, thence into the oviduct, and through the genital opening to the exterior; while the round still unripe eggs pass from the uterine bell through the dorsal posterior opening back into the body-cavity.

Fig. 242. — Embryo of *Echinorhynchus gigas* enclosed in the egg membranes (after Leuckart).

Development. Segmentation is irregular and complete, and results in the formation of an embryo, which is enclosed in three egg-membranes. The embryo has a small, somewhat long body, armed with small spines at the anterior pole, and consists of a central mass of small cells and a peripheral layer containing a few large nuclei and without cell-limits (Fig. 242). The peripheral layer gives rise to the ectoderm and the lemnisci; the central cells to the other organs. The large nuclei are said to break up into the small nuclei of the ectoderm of the adult except in *Neorhynchus* and *Arhynchus*. The body cavity arises in the central mass, and the cells on the outside of it form the muscles of the body-wall. In this development there can be no talk of layers in the ordinary sense. The embryo passes into the intestine of Amphipods (*Ech. proteus*, *polymorphus*), or of Isopods (*Ech. angustatus*), and there becomes free, bores through the wall of the intestine, and after losing the embryonic spines, develops to a small elongated larva, which, like a pupa, lies in the body-cavity of the small Crustacean, with its proboscis retracted and surrounded by its firm external skin as by a cyst (Fig. 243, *b*). As stated above, the skin of the larva gives rise only to the integument, the vessels and the lemnisci of the adult; while all the other organs enclosed within the dermal muscular envelope, viz., the nervous system, the sheath of the proboscis, and the generative organs, are developed from the so-called embryonic nucleus. It is only after their introduction into the intestine of fishes (*Ech. proteus*) or of aquatic birds (*Ech. polymorphus*), which feed on these *Crustacea*, that the larvae attain to sexual maturity, copulate, and reach their full size.

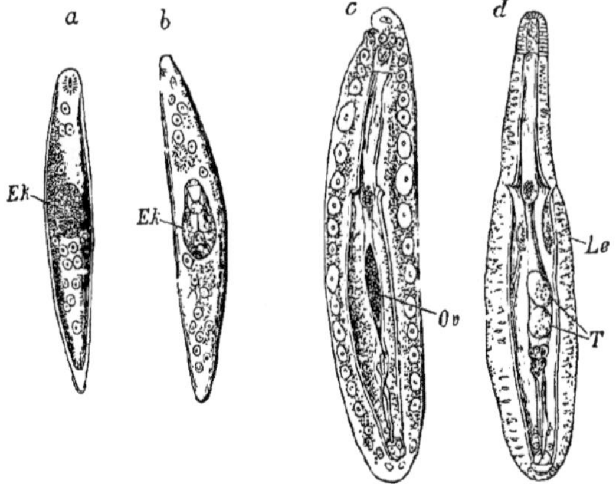

FIG. 243.—Larvae of *Echinorhynchus proteus* from *Gammarus* (after Leuckart). *a*, free embryo; *Ek* embryonic nucleus. *b*, older stage, with more differentiated embryonic nucleus. *c*, young female worm; *Ov* ovary. *d*, a young male worm; *T* testes; *Le* lemnisci.

The numerous species of the genus *Echinorhynchus* O. F. Müller, live in the alimentary canal of different Vertebrates, principally fishes, the gut wall of which may be as it were sown with these animals; the asexual larvae are found in small Crustacea (Amphipods, Isopods). Lambl found a small sexually immature *Echinorhynchus* in the small intestine of a child which died of leukhaemia. *Ech. polymorphus* Brems, in the intestine of the duck and other birds, larva in the crayfish; *E. proteus* Westrumb., larval form in Gammarus and in the body-cavity and liver of Phoxinus, sexual form in the trout. *Gigantorhynchus* Hamann, with ringed and flattened body, intestine of ant-eaters, birds, swine; larval form in grubs of insects; *G. gigas* Goeze, as large as an *Ascaris lumbricoides*, in the small intestine of the pig, larval form in grubs of beetles. *Neorhynchus* Hamann, the nuclei of the epidermis and lemnisci are large and few in number, as in the larvae of other forms; in the carp, larval form in larvae of Neuroptera, etc. *Arhynchus* Shipley, without eversible proboscis and hooks; epidermis and lemnisci with giant nuclei; inside the skin of a bird *Hemignathus procerus*.

Following the usual custom, the *Acanthocephala* are placed in the phylum Nemathelminthes; but it must be borne in mind that the relationship to the Nematoda thus expressed is more than doubtful. Indeed, there is nothing in common between the two groups, and there are many remarkable differences. The absence of an alimentary canal, the presence of a coelom with relations to the generative organs and ducts, the form of the nervous system, the presence of cilia in the so-called nephridia, are all characters of great importance and quite different from anything in the *Nematoda*. Further, the canals in the skin, though they may be compared to the lateral excretory organs of Nematodes, are, in reality, totally different, inasmuch as they form anastomosing net-works, and are found throughout the whole of the epidermis. Finally, the development cannot be brought into relation with anything else in the animal kingdom. It is peculiar in the fact of the giant nuclei of the outer layer and in their behaviour, and in the absence of a third layer or group of cells in the early stages.

Our inclination is to create a separate phylum—the *Acanthocephala*—for this strange group of parasites.

CHAPTER VIII.

PHYLUM ROTIFERA.*

Minute animals with a ciliated trochal disc, an anterior mouth, and a dorsal posterior anus. Perivisceral cavity well developed; excretory system with flame-cells; vascular system absent.

The *Rotifera* are small aquatic organisms which swim in the water by means of a ciliary apparatus at the front end of the body, called the **trochal disc.** This consists of the anterior end which generally has a somewhat discoidal form, and of the cilia which are disposed in one or two rows round its margin. The name "Wheel-animalcules," sometimes applied to this group, is due to the co-ordinated movement of these cilia, which produces the appearance of a rotating wheel, or in some cases, when the ciliary band is indented or interrupted at its median dorsal point and its median ventral, of two wheels (*Philodina, Limnias*). It will be useful to use the word **corona** for the discoidal anterior end of the body, and **velum** for the ciliary apparatus of its margin—the whole being the trochal disc. The posterior end of the body tapers, and is called the *foot* or *pseudopodium;* it may be jointed, and the joints are often telescopically retractile. It terminates either in a sucker-like surface for adhesion, on which the secretion of a cement-gland may be poured, or in two (or more) styles, which can be used as pincers for anchoring the animal (Fig. 244). It is used both for locomotion and attachment.

The trochal disc is generally retractile, and the foot is, in rare cases, absent.

The mouth is on the ventral side of the trochal disc, and the anus is on the dorsal surface, usually near the foot.

* Ehrenberg, "*Die Infusionsthierchen als vollkommene Organismen,*" Leipzig, 1838. C. T. Hudson and P. H. Gosse, "*The Rotifera or Wheel-animalcules,*" 2 vols., London, 1886. L. H. Plate, "Üb. d. Rotatorienfauna d. bottnischen Meerbusens," etc., *Z. f. w. Z.*, 49, 1890. C. Zelinka, "Studien ub. Rotatorien," *Z. f. w. Z.*, 44, 1886; 47, 1888; 53, 1892.

There is often a spur-like process, carrying a tuft of setae and projecting from the dorsal middle line close to the trochal disc: this is the **calcar** or **antenna**. There may be, in addition, a pair of ventral setigerous processes or antennae of a similar character. The single nerve-ganglion is placed on the dorsal side of the body, close to the trochal disc.

The velum varies in its arrangement. It may consist of a simple circle of cilia at the margin of the trochal disc (*Microcodon*): in this case it surrounds the mouth, and is a circum-oral ring. In the *Floscularidae* (Fig. 245, *3*), this ring is reduced to a half-ring on the ventral side of the mouth, and the edges of the disc may be produced into long processes (*Stephanoceros*). In other forms the ring encircles the disc twice by bending on itself (Fig. 245, *1*, *2*); thus enclosing the mouth, and having a dorsal gap between the points of flexure (*Rhizota*). In this form of velum a fusion of the two points of flexure would result in the formation of two rings, one pre-oral (above the mouth), and one post-oral (circum-oral) below the mouth. In the *Bdelloidea* the ring is doubled on itself, but there are two gaps, one dorsal, at the point of flexure, and the other ventral in the upper ring just above the mouth (Fig. 245, *4*). Very frequently, especially in the parasitic forms, the trochal disc is reduced, and in certain cases entirely aborted (*Apsilus*). In *Notommata tardigrada* the trochal disc is reduced to a ciliated disc round the mouth. In some cases the ciliated edge of the disc projects over the head and forms the double wheel above referred to (*Philodina*,

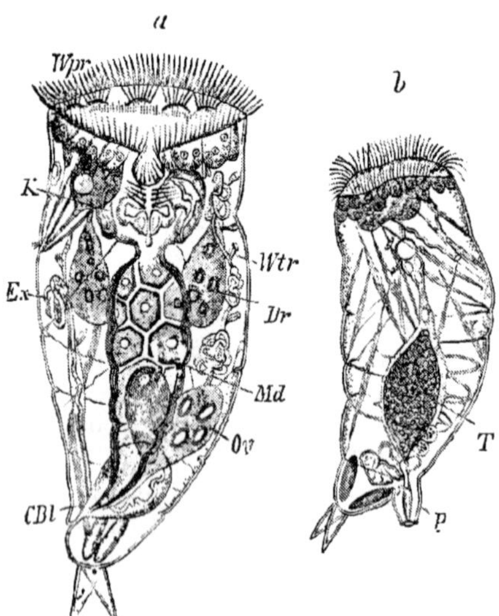

Fig. 244.—*Hydatina senta* (after F. Cohn). *a*, female. *b*, male. *Wpr* ciliated ring (velum); *CBl* contractile vesicle of the excretory system; *Wtr* ciliated lobule of the excretory organs (*Ex*); *K* mastax; *Dr* salivary glands; *Md* stomach; *Ov* ovary; *T* testis; *P* penis.

Brachionus), or becomes a ciliated cephalic shield (*Megalotrocha*, *Tubicolaria*). Finally it may be produced into ciliated processes of various form (*Floscularia*, *Stephanoceros*). The cilia are con-

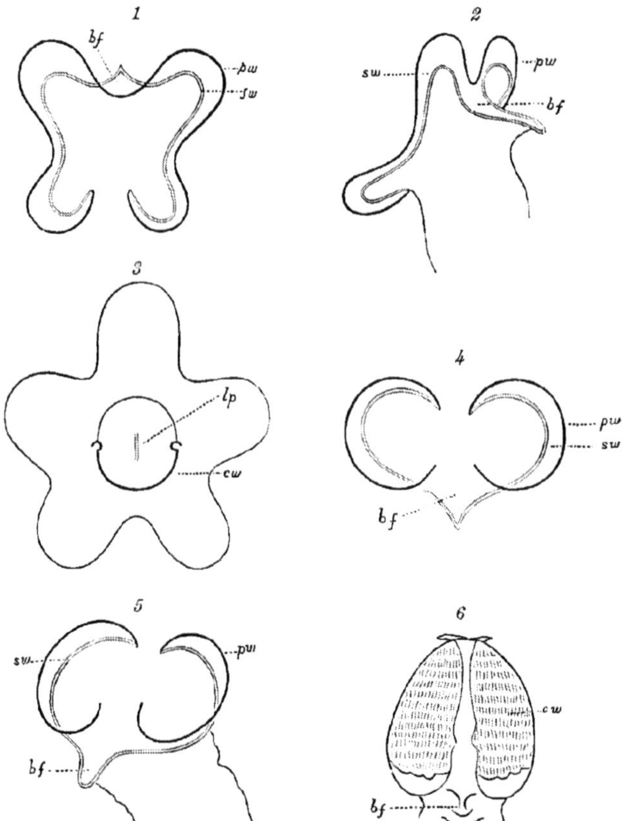

Fig. 245.—Vela of various Rotifera (after Hudson and Gosse). *1*, Rhizotic velum, front view (*Melicerta ringens*). *2*, Rhizotic velum, side view (*M. ringens*). *3*, Velum of *Floscularia campanulata*. *4*, Bdelloidic velum (*Rotifer citrinus*). *5*, The same, side view. *6*, Bdelloidic velum of *Adineta vaga*. *bf* the buccal funnel; *pw* the principal ring of cilia; *sw* the secondary ring; *lp* the lips; *cw* the velum.

cerned with locomotion, but in addition they play an important part in procuring food.

Locomotive organs. In addition to the foot and trochal disc, there are often styliform processes of the body, into the bases of which special muscles are inserted. Such are found in *Triarthra*, *Polyarthra*, etc. In one family—the *Pedalionidae*—the body possesses hollow appendages containing muscles, and recalling very closely the limbs of Arthropoda (Fig. 246).

The body-wall consists of a cuticle with a subjacent protoplasmic layer with scattered nuclei, but without cell-limits. This doubtless represents the ectoderm. There does not appear to be any con-

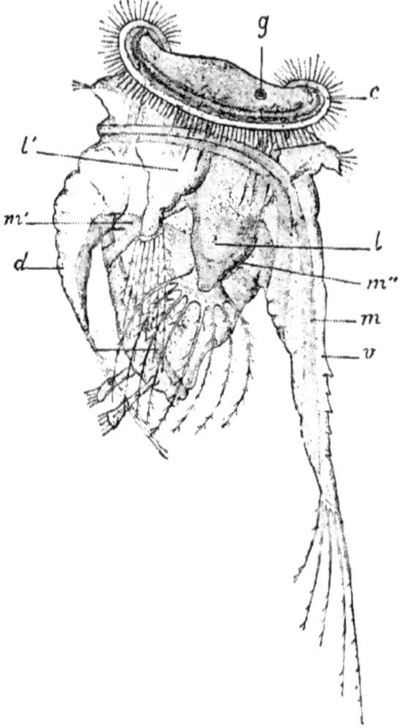

FIG. 246.—*Pedalion mirum*, side view (from Perrier, after Gosse). *g* eyes; *c* velum; *m*, *m'*, *m''* muscular bands; *v* ventral appendage; *d* dorsal appendage; *l*, *l'* lateral appendages.

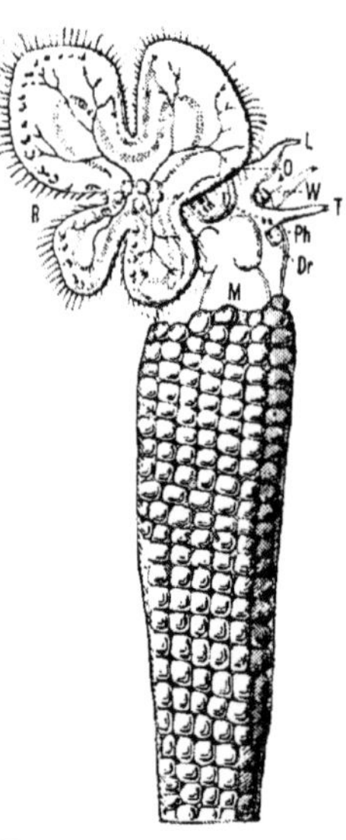

FIG. 247.—*Melicerta ringens* (after Leuckart and Nitsch, modified from Ehrenberg and Joliet). showing the tube formed of pellets cemented together. *O* mouth; *Ph* pharynx; *L* ciliated lobe overhanging *W* the ciliated cup; *Dr* cement gland; *M* stomach; *T* tentacle; *R* trochal disc.

tinuous muscular layer in the body-wall, but there may be a small amount of connective tissue, inasmuch as connective tissue fibres can be seen passing between the various organs across the body-cavity.

The cuticle is sometimes soft and thin, sometimes hardened into a kind of shell, called the lorica (the *Ploima*); there are sometimes joints in it, which give the body a segmented appearance. The skin is not ciliated except in certain spots, of which the trochal disc is the chief. The principal integumentary glands are the cement glands of the foot, which open at the extremity of the pincers and secrete the substance by which the animal is fixed. The tube or *urceolus*, in which some Rotifera live, is probably a secretion of the skin; it is often gelatinous, but in *Melicerta* it consists of pellets. These pellets are formed of foreign or faecal particles which are worked up and cemented together by a ciliated cup on the ventral side close to the trochal disc, the cementing substance being a secretion of a gland (*Dr*) near the ciliated cup (Fig. 247).

The **body-cavity** is a well-marked space, traversed as stated above by a few fine connective tissue strands and filled with a clear vascular fluid. As to its nature we know nothing. The generative and, apparently, the renal organs are in no way connected with it, so that presumably it is not coelomic. It may be a haemocoele; if it is, it is the only representative of the **vascular system**, for nothing in the shape of heart or blood-vessels are known.

The **alimentary canal** (Fig. 244) is ciliated throughout. It consists of a mouth leading into a muscular pharynx called the *mastax*, and provided with a special armature—the *trophi*. The parts of the armature (see below) are in continual movement, and serve for mastication. In the more predaceous forms they can be protruded so as to act as jaws. Following the pharynx is a short oesophageal tube; this leads into the digestive sac or stomach. The stomach is lined with large ciliated cells, which often contain green or yellow pigment grains. The anterior or gastric part of this cavity is wide and receives two large glandular tubes, which may be explained as salivary or pancreatic glands. The posterior narrow intestinal part leads into a cloacal chamber, which opens on the dorsal surface at the point where the foot-like posterior region joins the anterior part of the body. In some cases, *e.g. Asplanchna*, the intestine ends blindly, and the faeces are rejected through the mouth.

Respiration is carried on by the general surface of the body; special organs are wanting.

The muscular system is complex, being composed mainly of muscular bands passing between definite points. The muscular tissue is in part, at any rate, cross-striped.

Excretory organs. The so-called respiratory canals are excretory. They consist of two sinuous longitudinal canals, with fluid contents; they open into the cloaca either directly or by means of a contractile vesicle, and are beset by short appendages called the vibratile tags (Fig. 244). The latter appear to be simply flame-cells, which open into the main trunks: whether they open into the perivisceral cavity or not is disputed.

The **nervous system** consists of a simple or bilobed cerebral ganglion placed dorsal to the oesophagus, and giving off nerves to peculiar cutaneous sense organs and to the muscles. In *Callidina* and *Discopus* there is in addition a small ventral ganglion connected with the dorsal ganglion round the oesophagus. Eyes are often present, and lie upon the brain either as an X-shaped, unpaired pigment body, or as paired pigment spots provided with refractile spheres. There is often a small mass of calcareous granules in connection with the ganglion: its function is unknown. The above-mentioned cutaneous sense-organs, which have the form of prominences beset with hairs and setae, *e.g.* calcar, so-called antennae, etc., are probably tactile.

Generative organs. The sexes are separate, and are distinguished, except in *Seison*, by a strongly marked dimorphism. The males (Fig. 244, *b*) are much smaller than the females, and often have a very different form; they possess excretory tubes and ganglion, etc., but are without either oesophagus or intestinal canal—these organs being reduced to a cord of cells: they leave the egg completely developed. Their generative organs are reduced to a testicular sac, the muscular duct of which opens at the hind end of the body, sometimes on a papilliform protuberance. This is the penis, which either introduces sperm into the female's cloaca in true copulation, or perforates her body-wall and deposits the spermatozoa in the body-cavity.

The males of many species are unknown, and when they are known they often appear only in the autumn, when the winter-eggs are formed; but they are also found in summer. The generative organs of the female consist of a roundish ovary, with which a yolk-gland is associated, and of a short oviduct which usually opens into the cloaca. The oviduct is sometimes absent (*Rotifer*), and in the *Philodinidae* and *Seisonidae* the ovaries are double. Almost all Rotifera are oviparous, and their eggs are distinguishable into thin-shelled *summer eggs*, which develop immediately without being fertilized, and thick-shelled *winter eggs*, which last through the winter and are probably fertilized. They carry both kinds of eggs about on their body in their tube, but the summer eggs not unfrequently develop in the oviduct. The summer eggs are of two kinds, large and small. The large develop into females, and the small into males; they are produced by different individuals. Asexual reproduction is unknown.

It has been established by Maupas* that in certain Rotifera (*Hydatina senta*) the same animal produces only eggs of one sex; thus, a female which has once laid the small male egg never produces a female egg, and *vice versa*. He has further shown that while the general rule is that female eggs should be produced—a rise in temperature, if applied at the earliest stages of their short life, will ensure the production of the small male eggs. Further, he has shown that the layers of female eggs are devoted to parthenogenesis, and only produce the thin-shelled *immediate* eggs. Copulation of such with a male has no effect. On the other hand copulation between a young female, which would, if left alone, lay only male eggs, and a male brings about a complete change in the character of the eggs; in this case the egg acquires a thick, hard, often ornamented shell, and develops immediately to a certain stage; but the development soon ceases and the egg becomes a resting—so-called winter—egg. Such eggs may be produced at any season of the year, and their peculiar property is to be able to resist drought and other adverse influences; they give rise only to parthenogenetic females, which lay thin-shelled immediate eggs. Finally, copulation between a male-producing female, in which the eggs have been formed, and a male has no effect; in such a case the female lays small *immediate* eggs, which develop at once into males.

It thus appears that, as in the bees, an egg which would, if fertilized, produce a female, will, if not fertilized, produce a male. Should these most important and interesting observations be confirmed—and here it may be observed is a subject of the most fascinating and far-reaching character for research—it would appear that in the *Rotifera* the problem of how to control the production of the sexes has been solved.

The development has been studied in the parthenogenetic *immediate* eggs. It throws no light upon the affinities of the group. The young are hatched with practically the form of the adult. There is said to be no mesoderm in the ordinary sense of the word, but the rudiments of the organs are laid down independently.

The *Rotifera* principally inhabit fresh water, but marine forms are known. Some species live in gelatinous tubes which they secrete, or in a case consisting of pellets manufactured by the animal

* *Comptes rendus Acad. Sci.*, 109, 1889, 90, 91.

(*Melicerta*, Fig. 247). A small number are parasitic. Many of them possess to a remarkable degree the property of resisting the effects of drought and a high temperature (200° F.). This power is particularly developed in the *Bdelloidea*. On the approach of the adverse influence they protect themselves by a gelatinous secretion

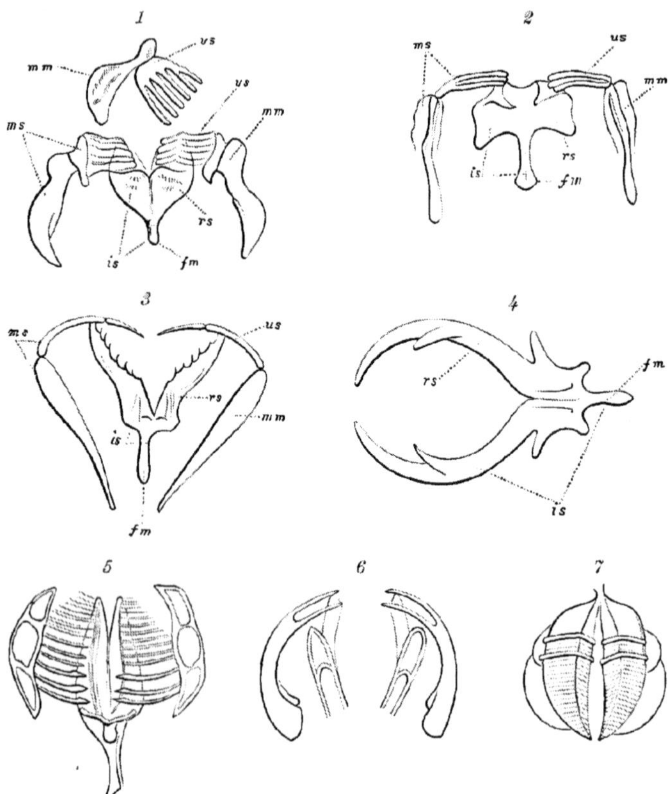

FIG. 248.—Jaws of *Rotifera*.—*1*, Malleate; *2*, Sub-malleate; *3*, Forcipate; *4*, Incudate; *5*, Malleo-ramate; *6*, Uncinate; *7*, Ramate (from Hudson and Gosse). *ms* malleus; *is* incus; *rs* rami; *fm* fulcrum; *mm* manubrium; *us* uncus.

which protects them against evaporation. The duration of life is often very short. The female of *Hydatina senta* attains maturity in three days and lives only fourteen days; and the males live only three days.

The position of the *Rotifera* in the system is obscure. Apparently they form an isolated phylum with affinities through *Pedalion* (Fig. 246) to the *Arthropoda;* and through the Trochosphere larva to Annelids and Molluscs. As in the Trochosphere, it is impossible to say whether the perivisceral space is a coelom or a haemocoele; probably it is a haemocoele. There is a further important feature of resemblance to the Trochosphere in the form of the excretory organs, which also recall those of Platyhelminthes and Nemertines. The *Rotifera* are highly specialized animals, and perhaps the most remarkable general fact to note about them is the rarity of marine forms. World-wide in distribution, and extremely rich in species, still they are mainly confined to fresh waters. One cannot help feeling that there must be an equally important marine branch of the phylum which has so far been overlooked, and which when discovered will throw light upon the affinities of the group.

The jaws or trophi are three in number; two *mallei* and one *incus* (Fig. 248, *1*). The malleus *ms* consists of a head or *uncus* (*us*), and of a handle or *manubrium* (*mm*). The incus (*is*) consists of two *rami* (*rs*), against which the mallei work, and of a basal piece—the *fulcrum* (*fm*). The modifications (Fig. 248, *1-7*) of their jaws are as follows:—1. *Malleate*. Mallei stout, manubria and unci of equal length. 2. *Sub-malleate*. Mallei slender; manubria about twice as long as unci. 3. *Forcipate*. Mallei rod-like; manubria and fulcrum long; unci pointed or absent; rami used as forceps. 4. *Incudate*. Mallei absent; rami as curved forceps. 5. *Malleo-ramate*. Mallei fastened by unci to rami. Unci 3-toothed; rami large, fulcrum slender. 6. *Uncinate*. Unci 2-toothed; incus slender. 7. *Ramate*. Rami, crossed by two or three teeth, fulcrum rudimentary; manubria absent.

Order 1. RHIZOTA.

Fixed when adult, usually inhabiting a gelatinous tube excreted from the skin; foot transversely wrinkled, not retractile within the body, ending in an adhesive disc or cup.

The foot of the *Rhizota* is unlike the foot of all other *Rotifera*, in that it is a prolongation of the dorsal, and not of the ventral, region of the body.

Fam. 1. **Flosculariidae.** Corona* produced into setigerous lobes: mouth central; velum* a single half circle ventral to the buccal orifice (*i.e.*, the gap in velum is dorsal, Fig. 245, *3*); trophi uncinate. *Floscularia* Oken; *Acyclus* Leidy, without setae; *Apsilus* Metschnikoff, without setae, velum, and foot; *Stephanoceros* Ehrbg., lobes long, convergent; *St. eichhornii* Ehbg.

Fam. 2. **Melicertidae.**† Corona without setigerous lobes; mouth lateral; velum a marginal continuous band, bent on itself at the dorsal surface, so as to encircle the corona twice; with the mouth between its upper and lower curves, and having also a dorsal gap between its points of flexure (Fig. 245, *1*); trophi malleo-ramate. Social, the tubes are often adherent to each other. *Melicerta* Schrank, corona of four lobes, dorsal antenna minute, ventral antennae

* **Corona** is used for trochal disc, and **velum** for ciliary ring or rings.

† This family must not be confused with the Leptomedusan sub-family of the same name (p. 134).

obvious, *Melicerta ringens* L. (Fig. 247); *Limnias* Schrank, tube without pellets; *Cephalosiphon* Ehrb., vent. antennae absent; *Oecistes* Ehrb., dors. ant. absent; *Lacinularia* Schweigger, adherent gelatinous tubes in clusters, antennae absent; *Megalotrocha* Ehrb., clustered, without tubes, ant. absent; *Trochosphaera* Semper (Fig. 249), solitary, free-swimming, spherical, velum as preoral ring broken dorsally, ventral antennae minute, trophi malleo-ramate, lateral canals end in cloaca, nerve ganglion close to mastax, male unknown, ovary opens into cloaca. Fresh water of Philippine Islands. Interesting from its resemblance in form, gut, velum, and appearance to a trochosphere larva; *Conochilus* Ehrb., gap in velum ventral.

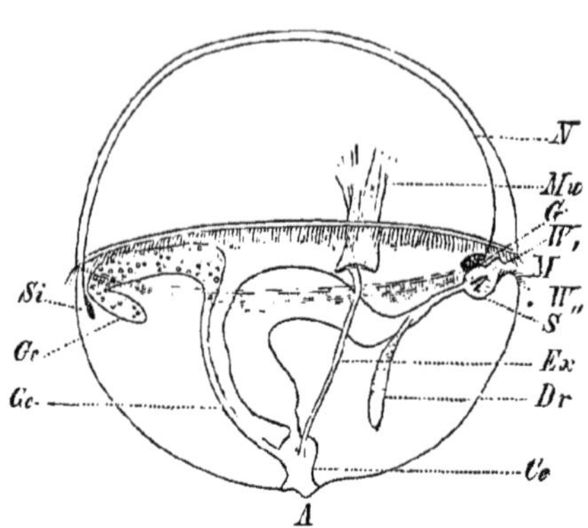

Fig. 249.—*Trochosphaera aequatorialis* (after Semper, from Korschelt and Heider). *Ce* cloaca; *Dr* glands of the foregut; *Ex* duct of the excretory organs; *G* brain; *Ge* ovary and oviduct; *M* mouth; *Mu* muscles; *N* nerve; *S* mastax; *Si* sense-organ; *W*, preoral, *W*,, postoral ciliated band; *A* cloacal orifice.

Order 2. Bdelloida.

Swimming with their velum, and creeping like a leech. Foot telescopic, wholly retractile within the body, usually ending in three toes.

Fam. 3. **Philodinidae.** Corona a pair of circular lobes, transversely placed. Velum a continuous marginal curve, bent on itself at the dorsal surface so as to encircle the corona twice, with the mouth between its upper and lower curves, and having also two gaps, the one dorsal between its points of flexure, and the other ventral in the upper curve opposite to the mouth (Fig. 245, *4*); trophi ramate. *Philodina* Ehrb., eyes two, cervical; *Rotifer* Schrank, eyes two, within the frontal column; in both these genera no oviduct has been seen, but the eggs develop in the body-cavity, and both eggs and young have been seen to leave the cloaca; males not known. *Actinurus* Ehrb.; *Callidina* Ehrb.

Fam. 4. **Adinetidae.** Corona a flat prone surface; velum as the furred (with cilia) ventral surface of the corona; trophi ramate; frontal column soldered to dorsal surface, and ending in two hooks. *Adineta* Hudson.

Order 3. Ploima.

Swimming with their velum, and (in some cases) creeping with their toes.

The mastax can be protruded so as to seize the prey; the ganglion is well developed and the eyes often have lenses.

Sub-order 1. ILLORICATA.

Integument flexible, not stiffened to an enclosing shell; foot when present almost always furcate, but not transversely wrinkled, rarely more than feebly telescopic, and partially retractile.

Fam. 5. **Microcodidae.** Corona obliquely transverse, flat, circular; buccal orifice central; velum a marginal continuous band encircling the corona and two curves of larger cilia, one on each side of the mouth; trophi forcipate; foot stylate. *Microcodon* Ehrb.

Fam. 6. **Asplanchnidae.** Corona sub-conical with one or two apices; velum single, edging the corona; with large stomach which ends blindly, intestine and cloaca being absent. *Asplanchna* Gosse; *Sacculus* Gosse.

Fam. 7. **Synchaetidae.** Corona a transverse spheroidal segment, sometimes much flattened, with styligerous prominences; velum a single interrupted or continuous marginal band; mastax large, piriform; trophi forcipate; foot minute, furcate. *Synchaeta.*

Fam. 8. **Triarthridae.** Body furnished with skipping appendages; corona transverse; velum single, marginal; foot absent. *Polyarthra* Ehrb.; *Pteroessa* Gosse; *Triarthra* Ehrb.; *Pedetes* Gosse.

Fam. 9. **Hydatinidae.** Corona truncate with styligerous prominences; velum two parallel bands, the one marginal fringing the corona and mouth, the other lying within the first, the styligerous prominences being between the two; trophi malleate; foot furcate. Both rings are continued into the mouth. *Hydatina* Ehrb., eye absent, *H. senta* O. F. M. (Fig. 244); *Rhinops* Hudson, eyes two; *Notops* Hudson, eye single. In *Hydatina senta* there is a small styliferous pit on the dorsal surface with a strong nerve from the ganglion.

Fam. 10. **Notommatidae.** Corona obliquely transverse; velum of interrupted curves and clusters, usually with a marginal band surrounding the mouth; trophi forcipate; foot furcate. *Albertia* Duj., vermiform, entozoically parasitic in Annelida; *Taphrocampa* Gosse, *Pleurotrocha* Ehrb., without eyes; *Notommata* Gosse, not annulose, cylindrical, with projecting tail; auricles on head; brain contains opaque chalk masses, trophi virgate; *Copeus* Gosse; *Proales* Gosse; *Furcularia* Ehrb.; *Eosphora* Ehrb.; *Diglena* Ehrb.; *Distemma* Ehrb.

Sub-order 2. **LORICATA.**

With a stiffened, wholly or partially enclosing shell; foot various.

Fam. 11. **Rattulidae.** Body cylindric or fusiform, smooth without plicae or angles; contained in a lorica closed all round, but open at each end, often ridged; trophi long asymmetric; eye single, cervical. Generally subject to abnormal conditions. With tendency to asymmetry, in mastax, antennae, toes. *Mastigocerca* Ehrb., toe as single style with accessory stylets; *Rattulus* Ehrb., toes two; *Coelopus* Gosse, body curved, toes one broad plate with another laid upon it in a different plane.

Fam. 12. **Dinocharidae.** Lorica entire, vase-shaped or depressed; sometimes facetted, often spinous; foot and toes often greatly developed; trophi symmetrical. *Dinocharis* Ehrb., foot and toes long, foot with spurs; *Scaridium* Ehrb., foot without spurs, toes long; *Stephanops* Ehrb., head with a wide circular shield.

Fam. 13. **Salpinidae.** Body more or less completely enclosed in a firm lorica which is open at each end and divided down the back by a fissure whose sides are united by membrane; two furcate toes always exposed. *Diaschiza* Gosse, *Diplax* Gosse, *Salpina* Ehrb., lorica with spines, trophi sub-malleate, eye single, cervical; *Diplois* Gosse.

Fam. 14. **Euchlanidae.** Lorica of two dissimilar plates, one dorsal and one ventral, united so as to form two confluent cavities of which the upper is much the larger; foot jointed, furcate. *Euchlanis* Ehrb.; *Cathypna* Gosse; *Distyla* Eckstein; *Monostyla* Ehbg.

Fam. 16. **Coluridae.** Body enclosed in a lorica, open at both ends, closed dorsally, usually open or wanting ventrally; head surmounted by a chitinous hood; toes two, rarely one, always exposed. *Colurus* Ehbg.; *Metopidia* Ehbg.; *Monura* Ehbg.; *Mytilia* Gosse; *Cochleare* Gosse.

Fam. 17. **Pterodinidae.** Lorica entire, various; corona and velum those of the *Philodinidae;* trophi malleo-ramate; foot wholly retractile, transversely wrinkled, jointless, toeless, ending in a ciliated cup; or foot absent; (foot and trophi Rhizotic, corona Bdelloidic). *Pterodina* Ehbg.; *Pompholyx* Gosse.

Fam. 18. **Brachionidae.** Lorica box-like, open at each end, and generally armed with anterior and posterior spines; foot long, very flexible, wholly retractile, wrinkled, ending in two toes. *Brachionus* Ehrb., lacustrine and marine; *Noteus, Notholca* Gosse; *Eretmia* Gosse.

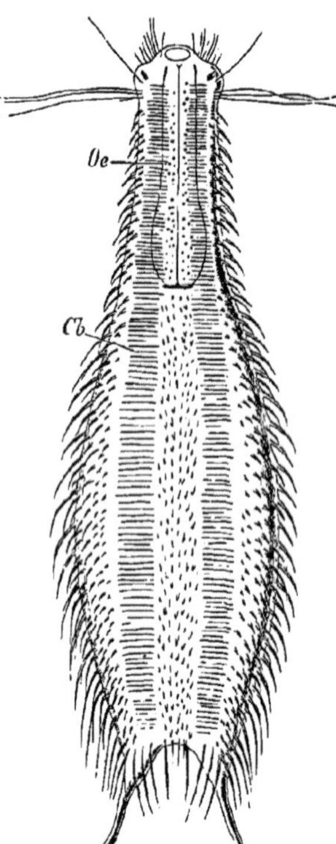

FIG. 250.—*Chaetonotus maximus* (after Bütschli), ventral view. *Cb* row of cilia; *Oe* oesophagus.

Order 4. SCIRTOPODA.

Swimming with velum, and skipping with arthropodous limbs; foot absent.

Fam. 20. **Pedalionidae.** Six hollow limbs; head truncate, corona of two concave lobes, and velum as in *Philodinidae;* trophi malleo-ramate. *Pedalion* Hudson (Fig. 246), one limb (the longest) ventro-median, another dorso-median, and two pairs of lateral limbs; and two stylate ciliated appendages on its posterior dorsal surface; *Hexarthra* Schmarda, with three pairs of limbs attached to the ventral surface.

Order 5. SEISONACEA.*

Marine forms parasitic on Nebalia.

The cloaca opens at base of neck in male, at hind end of body in female; male otherwise like female. Intestine complete or blind.

Fam. 21. **Seisonidae.** *Seison* Grube; *Paraseison* Plate; *Saccobdella* v. Ben. and Hesse.

GASTROTRICHA.†

Small aquatic organisms with a double ventral row of cilia, a cuticle

* C. Claus, "Ueb. die Organisation, etc., der Gattung Seison," *Festschrift z. F. d. 25 j. Best. d. K. k. z. b. Gesellschaft*, Wien, 1876. L. Plate, "Ectopar. Rotatorien d. Golfes v. Neapel," *Naples Mit.*, 7, 1886

† C. Zelinka, "Die Gastrotrichen," *Z. f. w. Z.*, 49, 1890.

often prolonged into spines and hairs, a cerebral ganglion connected with the ectoderm and with mouth and anus.

The Gastrotricha (Fig. 250) are small fresh-water organisms of unknown affinities. They have a ventral mouth in front, a dorsal anus behind, and an alimentary canal presenting a muscular pharynx, stomach lined with a few large cells, and a rectum. The body cavity is indistinct, and if present is without epithelioid lining. There is no vascular system, and the excretory organs consist of two coiled tubes opening on the ventral surface, and internally ending in a ciliated portion. The ovaries are paired, and no oviducts have been observed. It is doubtful if they are hermaphrodite, and if a small median organ behind the ovaries is a testis. No male is known. There is only one kind of egg, found in summer as well as winter. The muscular system is highly developed, and in part in specialized muscular bands. There are no transversely arranged muscular fibres. The cilia are in two ventral rows, and a patch is found on the head in the neighbourhood of the mouth. There is a dorsal ganglion in front which extends a little distance backwards on each side.

The marine forms, *Hemidasys* Clap. and *Turbanella* M. Sch., appear not to be *Gastrotricha*. The former has a parenchyma, a well-developed testis, vas deferens, and chitinous penis, and two small ciliated pits in front like Nemertines and some Turbellarians; the ovary, nervous system, and excretory organs are unknown. In *Turbanella* the whole ventral surface is ciliated, and there is no chitinous cuticle; and there are two ciliated pits.

Section 1. **Euichthydina.**

With forked tail, and adhesive glandular apparatus.

Fam. 1. **Ichthydiidae.** Skin either naked or beset with scales or papillae, never with spines. *Ichthydium* Ehbg.; *Lepidoderma* Zelinka.

Fam. 2. **Chaetonotidae.** Skin either with spines or with spines placed on scales. *Chaetonotus* Ehbg.; *Chaetura* Metschn.

Section 2. **Apodina.**

Without caudal fork; hind end either simply rounded, or lobed and provided with tufts of hairs. *Dasydytes* Gosse; *Gossea* Zelinka.

ECHINODERIDAE.*

These are minute marine animals with an elongated body, covered with a strong cuticle which is divided into segments, and an anterior protrusible spiniferous proboscis, at the apex of which the mouth opens. The anus opens at the hind end of the body. The cuticle is produced into bristles and spines. The central nervous system, which lies in the ectoderm, consists of a nerve-ring surrounding the pharynx and clothed with nerve-cells, and of a ventral cord with segmentally arranged groups of ganglia. The alimentary canal consists of a muscular pharynx, a non-muscular intestine, and a rectum. The excretory organs are represented by blind ciliated tubes, which open separately on the dorsal surface. The animals are dioecious; the testes and ovaries are paired tubes opening ventrally on either side of the anus. They occur in mud on the sea-bottom, and on marine Algae.

* W. Reinhard, "Kinorhyncha (Echinoderes), anat. Bau u. Stellung im System," *Z. f. w. Z.*, 45, 1887. C. Zelinka, "Üb. Echinoderes," *Verh. d. deutschen Zool. Gesel.*, 1894.

The systematic position of these animals is very uncertain ; indeed, our knowledge of their anatomy hardly justifies us in coming to a decision on the matter. In the published accounts of their anatomy there is no description of the vascular system, nor is the arrangement of the body-cavity made clear. It would appear, however, that the kidneys and generative organs are not connected with any other space, and it is possible that they are to be regarded as coelomic, as the products of two pairs of coelomic chambers (mesoblastic somites). If this should turn out to be a correct supposition, it would support the view that they are related to Annelids—a view which is prompted by the relations and segmentation of the central nervous system. *Echinoderes* Duj.

CHAPTER IX.

THE COELOMATA.

THE coelom, the possession of which characterizes all the remaining groups of the animal kingdom, is an organ of the greatest importance, and one the real nature of which has only recently been appreciated by anatomists.

Formerly the word coelom was used as synonymous with body-cavity or perivisceral cavity, and no distinction was recognized between the body-cavity of the *Arthropoda* and the same structure in such forms as the *Vertebrata*. In fact, there are works on Zoology now in use in which the term coelom is applied alike to the blood-containing space of the Arthropods, and to the body-cavity of Annelids and Vertebrates, which is free of blood, and into which the generative ducts and kidney tubes open and the generative cells are dehisced. We now know that the organ called coelom, so far from being necessarily or primarily a body-cavity, may in some cases have nothing of that mechanical relation to the viscera which is implied in the conception of a perivisceral space. It is true that in the majority of cases a portion of the coelom does enlarge, acquire a thin smooth wall, and enter into a relation with some of the more important viscera, a relation the purpose of which is apparently to enable the functional movements of the internal organs to take place with the least amount of friction and of resistance from surrounding structures. But there is one group, the *Arthropoda*, in which the coelom, though present and discharging important functions, develops no such perivisceral portion, and another, the *Mollusca*, in which the perivisceral portion is sometimes much limited, not extending to any other viscus than the heart.

The functions which the coelom discharges in such cases, and the real nature and relations of the organ will be dealt with more fully in the general part of this work; here we shall only consider the subject in its main outlines, and not being in a position to place all the facts before the reader, our treatment of it must be more or less dogmatic.

The coelom presents the fullest and, what we may call most typical development in the dibranchiate *Cephalopoda*. Here it is divided into three parts, all of which communicate. These parts are, (1) the genital sac or gonadial coelom, from the epithelial walls of which the reproductive cells arise; (2) the viscero-pericardial sac or perivisceral coelom, which has the functions of a typical body-cavity for the heart and certain other viscera; (3) the renal sacs or nephridial coelom, the epithelial walls of which secrete the nitrogenous waste, thus performing the functions of kidneys.

As stated above all these sections of the coelom are in communication, the gonadial with the viscero-pericardial, and the viscero-pericardial with the nephridial. Moreover, the gonadial and nephridial open directly to the exterior, the former through the generative ducts, and the latter through the nephridiopores. From this it would appear that the coelom, in addition to its mechanical relations, has two most important functions: the one of these is to bud out the reproductive cells, and the other to secrete the nitrogenous waste. These functions are always discharged by the coelom in the *Coelomata*. In certain forms, *e.g.*, *Peripatus*, elasmobranch fishes, *Cephalopoda*, some *Gastropoda* and the *Annelida*, this is perfectly obvious; but in most Arthropods, and in some Vertebrates and Molluscs it is a matter of inference, and might possibly be disputed by those zoologists who have not followed recent morphological research. We, however, have no hesitation in laying it down as a general law that in all *Coelomata* the functions of producing the reproductive cells and secreting the nitrogenous waste are discharged by this organ which we now call Coelom.

With regard to the origin of the coelom and its relation to other organs, we have here only the following statements to make.

The coelom is derived from the enteron; it is a part of the enteric cavity, which has, in all *Coelomata*, lost its connection with that portion which constitutes the alimentary canal in the adult, though it has retained its connection with the enteron, thus showing its enteric origin, in the young stages of certain groups, *e.g.*, the *Echinodermata*, *Chaetognatha*, *Brachiopoda*, *Enteropneusta*, and in *Amphioxus*. In the one division of the *Coelenterata*, viz., the *Actinozoa*, there is an incipient coelom in the enteric pouches, on the walls of which the reproductive cells are formed. The *Coelenterata*, however, are not *Coelomata*; they have no organ distinct from the enteron concerned with the renal and with the reproductive functions.

The *Platyhelminthes*, *Nemathelminthes*, and *Rotifera* are also not *Coelomata;* or perhaps it would be better to say that they cannot be ranked as *Coelomata* in the present state of our knowledge. It may be that in these phyla also, or in some of them, the progress of research will show that the renal and reproductive organs are enteric in origin, and are to be regarded as homologous with the coelom.

It has already (p. 312) been pointed out that in the lower *Turbellaria* there is reason to believe, from a study of the adult structure, that the whole of the mesodermal parenchyma, of which the genital and renal organs are part, is in reality a portion of the enteric wall.

Finally, before leaving this subject, there is one more point to notice with regard to the coelom. In all metamerically segmented animals the coelom is the first organ to show segmentation in the embryo: it is the first organ to show that repetition along the long axis which constitutes metameric segmentation. In fact, we may lay it down as a general law that in metamerically segmented animals the coelom is not only the first organ to present segmentation, but is segmented from its very first appearance, and that the subsequent segmentation or repetition of other organs is evolved out of this primal segmentation.

There are a few apparent exceptions to this law amongst certain groups in which the number of segments is very small, being, in fact, limited to three. These exceptions are the *Brachiopoda*, *Chaetognatha*, and the *Echinodermata*. In the first two of these groups the coelom is said to be separated off as a *single sac* on each side, which, presumably, subsequently becomes divided into the three sacs characteristic of the adult (for qualification with regard to Brachiopods see chapter on *Brachiopoda*). In the Echinoderms there is some variation with regard to this point.

CHAPTER X.

PHYLUM MOLLUSCA.*

Bilaterally symmetrical unsegmented animals, with a ventral foot, and usually with a radula, a mantle fold, and a univalve or bivalve calcareous shell. The central nervous system consists of a circum-oesophageal ring with various ganglionic developments, and both the haemocoel and coelom contribute to the formation of the perivisceral cavities.

The group *Mollusca*† with somewhat of its present limits was established by Cuvier (1798). Linnaeus had used the word before, but his group *Mollusca* bore very little resemblance to that of Cuvier, and even Cuvier himself included in the "embranchement" a number of forms which are now known to have nothing to do with it. Cuvier's *Mollusca* included the *Cephalopoda*, the *Gastropoda*, the *Pteropoda*, the *Acephala*, and the following groups now removed from them: the *Brachiopoda*, the *Tunicata*, and the *Cirripedia*.

In many points the Mollusca resemble the Annelida, with which phylum they are clearly closely allied. The arrangement of the

* J. Poli, "Testacea utriusque Siciliae, eorumque historia et anatome," 3 vols., Parma, 1791–1795. G. Cuvier, *Mémoires pour servir à l'histoire et à l'anatomie des Mollusques*, Paris, 1817. G. B. Deshayes, "*Histoire naturelle des Mollusques* (*Exploration de l'Algérie*)," Paris, 1844–48. R. Leuckart, "*Ueb. d. Morphologie u. d. Verwand. Verhält. der Wirbellosen Thiere*," Braunschweig, 1848. Eydoux and Souleyet, "*Voyage autour du monde sur la corvette la Bonite. Histoire naturelle Zoologie*," Paris, 1852. T. H. Huxley, "On the Morphology of the Cephalous Mollusca as illustrated by the Anatomy of certain Heteropoda and Pteropoda, etc.," *Phil. Trans.*, 1853. Bronn and Keferstein, "*Die Klassen u. Ordnungen der Weich-thiere*," Leipzig and Heidelberg, 1862–66. E. Ray Lankester, "Contributions to the developmental history of the Mollusca," *Phil. Trans.*, 1875. P. Fischer, "*Manual de Conchyliologie et de Paléontologie conchyliologique, Histoire nat. des Mollusques vivants et fossiles*," 2 vols., Paris, 1887. P. Pelseneer, "Introduction à l'étude des Mollusques," Bruxelles, 1894. K. A. Zittel, *Handbuch der Palaeontologie*, Abth. 1, Palaeozoologie, Bd. 2, Mollusca and Arthropoda, Munich and Leipzig, 1881–85. S. P. Woodward, *A Manual of the Mollusca*, 3rd Ed., London, 1875.

† For a short literary history of the Mollusca, *vide* E. Ray Lankester's Article on *Mollusca* in the last edition of the *Encyclopaedia Britannica*, and republished in "*Zoological Articles*," London, 1891.

central nervous system as a circum-oesophageal ring, of which the sub-oesophageal portion may be pulled out into ventral cords; the presence of a more or less developed perivisceral portion of the coelom; the dorsal position of the mainvascular trunk; and, finally, the very general presence of a trochosphere larva, are all important characters, which point to a connection more close than that which ordinarily exists between groups of phyletic rank.

On the other hand, the differences, which are also of great importance, must be noted. The *Mollusca* are almost entirely without the phenomenon of metamerism, which is so generally characteristic of *Annelida.* It is true that signs of this phenomenon are not altogether absent, for we find in *Nautilus* a twofold repetition of the gills, kidneys, and auricles, and in *Chiton* multifold repetition of the shell-plates and gills. But these are isolated instances, and not generally characteristic of the group. Moreover, so far as is known there is nothing in the embryo comparable to the repetition of the mesoblastic somites of the Annelids—the phenomenon by which the adult metamerism is preceded, and on which it is based. Finally, whereas in Annelids the perivisceral cavity is entirely furnished by coelom, and the haemocoel is almost completely canalicular; in Molluscs vascular sinuses are well developed, and in some parts of the body so large that they constitute perivisceral cavities to the organs in relation with them. Thus in Molluscs part of the body-cavity is coelomic—the portion called pericardial or viscero-pericardial—and part is haemocoelic; but it must be carefully borne in mind that the coelomic portion of the body-cavity is entirely separate from the haemocoelic, and that, however much they may appear to resemble each other in the adult, their relations and development are totally different.

The asymmetry and distortion found in the class *Gastropoda* is a very interesting phenomenon, and in some respects comparable to the still more remarkable distortion from the original symmetry found in the *Echinodermata.* But the two phenomena differ in their mode of occurrence in certain important respects, which it may be desirable to point out. In *Echinodermata* not only is the distortion completely carried out in every living member of the group, but a new symmetry has in all cases been obtained, and sometimes (certain *Echinoidea*) a second time lost. Therefore, in order to arrive at an understanding of the course which the distortion has taken, we are entirely dependent upon embryology; indeed, were it not for the knowledge gained by a study of the development, we should not know that any departure from the original symmetry had occurred. In the *Gastropoda*, on the other hand, not only is there one form in which there is no asymmetry at all, but the other members of the group present the distortion in various degrees, and in no living Gastropod has a new symmetry been completely attained. It is possible, therefore, by a comparison of adult forms to arrive at an understanding of the course by which the distortion in the most extreme members has been brought about.

The Mollusca are unsegmented animals without jointed appendages. The body is covered by a soft, slimy integument, bounded externally by an epithelium frequently ciliated and containing a considerable number of gland-cells. They lack both an internal and external locomotory skeleton, and are therefore especially suited for life in water. But few of them are terrestrial, and when this is the case the movements are always limited and slow; while the aquatic forms may be endowed with the power of rapid swimming.

Beneath the epithelium is the dermal connective tissue, which consists of cells of various forms, branched cells, elongated fibre-cells, and especially characteristic are the vesicular cells, which sometimes secrete and contain calcareous concretions or spicules (*Pleurobranchs*, and some *Nudibranchs*). This tissue contains blood-spaces, the filling of which with blood may cause a turgescent swollen condition of that part of the integument. Sometimes it is condensed, and forms the cartilage of the Cephalopoda, the skeletal tissue of the branchial filaments of Lamellibranchs, the shell of the *Cymbuliidae*.

The muscular fibres are generally non-striped. There is often an appearance of striation in consequence of the arrangement of the granules, *e.g.*, muscles of buccal mass, of heart, columellar muscles of the larvae of some Nudibranchs, etc., and in some cases the fibres are more distinctly cross-striped, *e.g.*, portion of adductor muscles of *Pecten*.

The dermal muscular system plays an important part in the locomotion of these animals, especially that part of it which is placed on the ventral surface of the body. In this region it is greatly developed, and gives rise to a more or less projecting locomotory organ of very various shape, the **foot** (*Fig. 254*).

The foot always consists of an unpaired median structure, which is sometimes divided into several parts, and may possess, in addition, paired lateral lobes—the **epipodia**.

In most *Mollusca*—the *Cephalophora* as opposed to the *Acephala* or *Lamellibranchiata*—the anterior part of the body is marked off as a distinct head, and bears the mouth, tentacles, and special organs of sense. The dorsal part of the body behind the head constitutes the main mass of the animal, and is called the **visceral sac.** The integument of the visceral sac constitutes the *mantle* or *pallium*, and is nearly always folded near the junction of the visceral sac with the head and foot. This fold is the *mantle-fold*, or pallial-fold, and the groove enclosed by it is the *mantle-groove*, or pallial-groove. The mantle-groove contains the gills, anus, and renal openings, and is nearly always deeper in the neighbourhood of these structures than elsewhere, thus constituting the *mantle-* or *pallial-cavity* proper. The outer surface of the mantle generally secretes a calcareous *shell*—though this structure may, by secondary modification, be absent in the adult. The head and foot are attached to the shell by paired and symmetrical muscles in the *Placophora*, *Scaphopoda*, *Lamellibranchiata* (retractors of the foot), and *Cephalopoda* (retractors of the head and funnel); and in *Gastropoda* by a single muscle—the columellar or spindle-muscle.

The **central nervous system** consists of a circum-oesophageal ring with a uniform coating of nerve-cells, or the latter may be specially aggregated into ganglia—the *cerebral* or supra-oesophageal, and the *pedal* or sub-oesophageal. In some forms the pedal or ventral portion of this nerve-ring may extend as a pair of long cords into the foot (*Chiton*, *Haliotis*, etc.). Sometimes there is a second circum-oesophageal commissure connecting the cerebral and pedal, and provided with a ganglion—the *pleural.* These various ganglia give off nerves—the cerebral to the head, cephalic sense organs and lips, the pedal to the foot, and the pleural to the mantle and gills.

In addition to this somatic system, as we may call it, there is the visceral system, which generally consists of two commissures with ganglia in their course. Of these there is (1) the *stomato-gastric commissure*, which starts on each side from the cerebral, and is completed ventrally to the oesophagus; the ganglia upon it are called the *buccal*, and supply the buccal mass and stomach; and (2) the *visceral commissure*, which starts on each side from the pleural, and is completed posteriorly ventral to the intestine. The ganglia upon this commissure vary in number in different forms, and are called the *visceral.* They innervate the intestine, vascular, renal, and reproductive organs. The stomatogastric and visceral commissures very frequently anastomose (*Cephalopoda*, *Gastropoda*).

Tactile organs. In addition to the general sensibility of the skin there are tentacles on the head, or in the *Acephala* on the mantle-edge. The **olfactory sense** is partly discharged by the tentacles, and partly by the **osphradia,** which are sensory patches of the mantle epithelium near the base of the gills, and are innervated by nerves from the visceral commissure. The **eyes** have generally a complicated structure; there are usually two on the head, and in rare cases they are found on the mantle-edge (*Acephala*), and even on the dorsal surface (*Chiton*, *Onchidium*).

Auditory organs, or organs which, from their structure, are supposed to have that function are very generally present. They have the form of otocysts, provided with sense-hairs on their internal walls, and containing one or more otoliths. They are usually paired, and lie either on the cerebral or pedal ganglia. They are, however, said to be always innervated from the former.

The **alimentary canal** (absent in two parasitic forms, *Entoconcha* and *Entocolax*) is always provided with a mouth and anus. The mouth is always at the front end of the body, on the head in the *Cephalophora*, and the anus opens into the mantle-cavity behind, excepting in Gastropods, in which the mantle-cavity is on the front side in consequence of the torsion of the visceral sac. Three parts can be distinguished—(1) the stomodaeum, consisting of buccal cavity and oesophagus; (2) the mesenteron, consisting of stomach and intestine; and (3) the proctodaeum, which is very short in most forms, even if it can be distinguished at all.

The buccal cavity is absent in *Acephala* (Lamellibranchs); but in the other forms, which may be called *Odontophora*, it forms a chamber with muscular and cartilaginous walls, which receives the salivary ducts, and has on its floor an apparatus called the **odontophore.** This consists of a toothed chitinous ribbon—the **radula,** a pair of cartilaginous pieces beneath the buccal floor, and a sheath in which the ribbon is formed (for detailed description see under Gastropoda, p. 356). There is usually an extensive liver opening into some part of the mesenteron.

The relations of the **perivisceral** or **body-cavity,** in the formation of which the coelom and haemocoele take varying parts, have already been described, and will be dealt with again in detail under the various classes.

The **vascular system** forms a *completely closed* system. Even in cases when some of the vessels open out into large spaces, which surround some of the viscera and form a true perivisceral cavity, there is no communication with the coelom, or with any other organ, or with the exterior. The so-called *aquiferous* canals, when present, do not lead into the blood system, but merely into spaces in the integument.

The heart consists of a median ventricle, and of two lateral auricles (four in *Nautilus*, one in most *Gastropoda*); it is placed on the dorsal surface in the pericardium (except in *Anomia* and the *Octopoda*). As already hinted the ventricle is merely a portion of a longitudinal dorsal vessel (the rest being formed by the aortae) comparable to the dorsal vessel of Annelids.

The arteries branch, but, except in *Cephalopoda*, in which capillaries are found, they end in sinuses amongst the organs, from which the blood passes into the large veins; these conduct the blood to the gills, part or all of it passing through the kidney on the way. In many cases the blood in the mantle passes directly to the heart without traversing the gills.

The blood is either colourless or of a slightly blue tinge, due to the presence of haemocyanin (an albuminoid pigment containing copper). In rare cases it is red owing to the presence of haemoglobin in the plasma (*Planorbis*), or in the corpuscles (*Solenogastres*, some Lamellibranchs).

A **lymphatic gland** consisting of a framework of connective tissue, in which blood corpuscles are formed, is often present on the course of the aorta.

Respiration is in all cases carried on through the general outer surface of the body, but in addition special respiratory organs in the form of *branchiae* or of *lungs* are generally present.

The *branchiae* are projections (often ciliated) of the body surface and are usually placed in the mantle-cavity. Very generally they have a certain form designated by the term **Ctenidium.** A ctenidium is a branchia consisting of an axis attached to the body and bearing two rows of projecting lamellae. Although not always found, the ctenidium is the typical Molluscan gill, and is always contained to the number of one, two, or four (*Nautilus*) in the mantle-cavity. In the *Chitonidae* only are the ctenidia more numerous than four. The gills may have other forms (*Nudibranchiata*), and in the terrestrial forms are absent altogether, the mantle-cavity having assumed a form and structure suitable for aerial respiration.

The **excretory organs**, often called the *organs of Bojanus*, are special portions of the coelom, and as such are nearly always in free communication with other parts of the coelom. In Nautilus alone are they isolated. The glandular tissue of the excretory organs is arranged in various ways in different forms (for which see accounts of separate groups). The forms in which the nitrogenous waste is got rid of is said to be guanin in Cephalopods, uric acid in many Opisthobranchs, and urea in Lamellibranchs.

The pericardial or viscero-pericardial division of the coelom also often contains glandular tissue, which constitutes the *pericardial gland*.

Reproduction is always sexual, and no cases of parthenogenesis are known in the phylum. The power of reproducing lost parts, whether cast off voluntarily or lost accidentally, is considerable—parts of the foot and its appendages, siphons of some bivalves, dorsal papillae, etc., of some Nudibranchs, cephalic tentacles, arms of Cephalopoda; and very often the part reproduced may bear organs of a complicated kind, *e.g.*, eyes, suckers, etc. But this power of reproducing lost parts is never so great—so far as is known—as to lead to the complete formation of a new individual from the part removed from the body, *i.e.*, asexual reproduction is unknown in the group. The hermaphrodite condition is fairly common.

The genital glands, however much they may be modified, are always to be regarded as portions of the coelom. This fact is perfectly obvious in the *Scaphopoda*, and in some Lamellibranchs and *Gastropoda*, in which they open into the renal division of the coelom; and in the *Solenogastres* and *Cephalopoda*, in which they open into the perivisceral part of the coelom. But in the majority of Lamellibranchs and Gastropods it is not so obvious; but there can be no doubt of this being the real, if modified and concealed, relation, when such forms are compared with others in the same class in which the two organs communicate; and further, it must not be forgotten that in some Lamellibranchs, and in *Paludina* amongst Gastropods, the genital cells can, in their origin in the embryo, be directly related to the coelom.

Development. We cannot here enter into details of the development; the greatest variety prevails from the large meroblastic eggs of *Cephalopoda*, through viviparous forms such as *Paludina*, to the immense majority of forms in which the eggs are small, and the young hatched at an early stage as a larva. The larva always has the trochosphere form, and soon acquires a shell-gland on its dorsal, and a rudiment of the foot on its ventral surface (Fig. 268).

The velum at the same time becomes drawn out into lobes, and we get the well-known veliger larva. A point to notice about the development is that the blastopore often assumes the form of a slit which may completely close, or close behind and remain open in front as the mouth, or less often close in front and remain open behind as the anus.

The *Mollusca* are essentially aquatic animals, and especially marine; only a few live on land, and these seek damp situations. They are one of the largest and most diversified of animal groups, and apparently have always been so since the earliest fossiliferous periods, for all of the great groups are represented in the Palaeozoic period, and a *Helix*, or a form allied to *Helix* and belonging to the most specialized of *Mollusca* is now known even from the Carboniferous.

At the present day about 25,000 species are known; they are distributed over the whole surface of the earth, and are found in the sea to a depth of nearly 3000 fathoms. Their habits of life are most various; there are parasitic forms, *Entoconcha*, *Eulima*, *Entocolax*, *Entovalva*, *Stilifer*, *Thyca* (all on Echinoderms); commensals (*e.g.*, *Montacuta*); fixed forms (*Vermetus*, *Ostrea*); pelagic forms; and creeping forms, the latter constituting the majority.

Their duration of life, where known, varies from one to thirty years; the Pulmonates generally live two years, but the garden-snail has been known to live five years. The oyster is adult at about five years, and lives to ten years. The *Anodonta* do not arrive at sexual maturity till five years, and live for twenty or thirty years.

The following is the classification of the Mollusca adopted in this work:—

CLASS I. LAMELLIBRANCHIATA.
,, II. SCAPHOPODA.
,, III. SOLENOGASTRES.
,, IV. GASTROPODA.
Sub-class 1. ISOPLEURA.
,, 2. ANISOPLEURA.
Order 1. Streptoneura.
Sub-order 1. *Aspidobranchiata.*
Tribe 1. DOCOGLOSSA.
,, 2. RHIPIDOGLOSSA.
Section *A.* Zygobranchiata.
,, *B.* Azygobranchiata.
Sub-order 2. *Pectinobranchiata.*
Tribe 1. PTENOGLOSSA.
,, 2. RACHIGLOSSA.

CLASS IV. GASTROPODA—*Continued.*
Tribe 3. TOXOGLOSSA.
,, 4. TAENIOGLOSSA.
Section *A.* Platypoda.
,, *B.* Heteropoda.
Order 2. Euthyneura.
Sub-order 1. *Opisthobranchiata.*
Tribe 1. TECTIBRANCHIATA (Includes Pteropoda).
,, 2. NUDIBRANCHIATA.
Sub-order 2. *Pulmonata.*
Tribe 1. BASOMMATOPHORA.
,, 2. STYLOMMATOPHORA.
CLASS V. CEPHALOPODA.
Order 1. Dibranchiata.
,, 2. Tetrabranchiata.

Class I.

LAMELLIBRANCHIATA* (PELECYPODA. LIPOCEPHALA).

Symmetrical Mollusca without head or odontophore; with bilobed mantle, bivalve shell, and usually lamellate gills.

The name of this class† is taken from the form which the gills present in most members of the group. The fresh-water mussel *Anodonta cygnea* is a typical member of the class.

The mantle-fold is in reality a continuous annular fold of the dorsal integument, but in connection with the lateral compression of the body it is especially developed on the two sides of the animal into a right and left lobe. It is, therefore, usually described as being bilobed. The mantle secretes over the whole of its outer surface a cuticular covering, which becomes calcified everywhere except in the dorsal middle line, forming the two pieces, or *valves*, of which the shell is composed. In the dorsal middle line it remains as an uncalcified elastic membrane—the *hinge-ligament*—which connects together the two valves of the shell. The latter are rarely exactly alike; nevertheless, the term unequivalve is only applied to those shells in which the asymmetry is very marked and the valves can be distinguished as an upper and lower. In such forms (*Ostreidae*) the lower valve is the larger and more arched, while the upper is smaller and flatter, closing up the concavity of the lower after the manner of an operculum. The two valves of the shell are in contact dorsally, where they are also connected by the ligament. This line of contact may be complicated by the presence of interlocking hinge-teeth. The hinge-edge with the ligament is, therefore, to be distinguished from the free edge of the shell, which is divided into an anterior, ventral, and posterior (siphonal) edge. The anterior and posterior edge

* G. Cuvier, "*L'histoire et l'anatomie des Mollusques*," Paris, 1817. Bojanus, "Ueber die Athem—und Kreislaufswerkzeuge der zweischaligen Muscheln," *Isis*, 1817, 1820, 1827. Deshayes, "Histoire Naturelle des Mollusques" (*Exploration de l'Algérie*), 1844–1848. S. Lovén, *K. Vet. Akad. Handlgr.* Stockholm, 1848. Translated in the *Arch. f. Naturgesch.*, 1849. Lacaze Duthiers, *Ann. des Sc. Nat.* 1854–1861. H. and A. Adams, "*The Genera of recent Mollusca*," London, 1853–1858. L. Reeve, "*Conchologica iconica*," London, 1846–1858. R. H. Peck "The Minute Structure of the Gills of Lamellibranch Mollusca," *Quart. J. Mic Sci.*, vol. xvii., 1877. K. Mitsukuri, "Structure and Significance of some aberrant forms of Lamellibranchiate Gill," *Quart. J. Mic. Sci.*, xxi., 1881. P. Pelseneer, "Contribution à l'étude des Lamellibranches," *Arch. d. Biologie.* xi., 1891. P. Pelseneer, "*Introduction a l'étude des Mollusques*," Bruxelles, 1894.

† There does not appear to us to be any sufficient reason for altering this well-established name in favour of the less suitable, but perhaps more symmetrical term, *Pelecypoda* (πέλεκυς, an axe).

may generally be determined by the fact that the hinge-ligament is posterior to the two *umbones* (*nates*), which have the form of two beak-like prominences projecting over the dorsal edge of the shell (Fig. 250 *bis*, *u*), and indicate the point (*apex*) where the development of the valves began. The *area* (*c*) is behind the apex, and includes the dorsal posterior side of the shell. The part of the dorsal edge in front of the apex is usually shorter, and contains, at least in the equivalve species, an excavation, the *lunula* (Fig. 250 *bis*, *l*), by means of which the anterior edge can at once be recognized. The apex in some forms (*Isocardia*, *Diceras*) is spirally twisted.

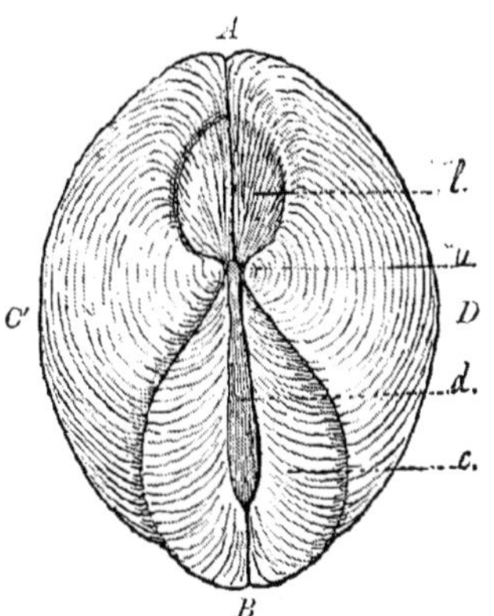

Fig. 250 *bis*.—Dorsal view of shell of *Lucina pennsylvanica*. *AB* anteroposterior axis; *CD* transverse (right and left) axis; *l* lunula; *u* umbo; *d* ligament; *c* area (from Perrier, after Fischer).

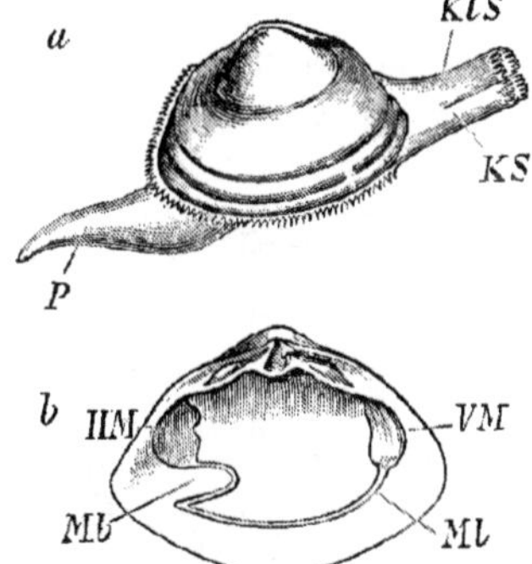

Fig. 251.—*a*, *Mactra elliptica* in its shell; *b*, inner surface of left valve of *Mactra solida*. *KlS* dorsal siphon (cloacal or exhalent); *KS* ventral siphon (branchial or inhalent); *P* foot; *VM* impression of anterior adductor muscle; *HM* impression of posterior adductor muscle; *Ml* pallial line; *Mb* pallial indentation.

The hinge-teeth are placed ventral to the umbo. They generally consist of *cardinal* teeth placed below the umbo, and of anterior lateral teeth in front of, and posterior lateral behind, the umbo. Hinge-teeth are entirely absent in some genera (*e.g.*, *Anodonta*, *Ostrea*).

The ligament is an elastic structure, and is so arranged that it is either stretched or compressed when the shell is closed by the action of the adductor muscles. In the former case it connects the two valves on the dorsal side of the hinge-line (Fig. 250 *bis*), and is said to be external; in the latter it is on the ventral side of the hinge, and is called internal.

While the outer surface of the shell presents various sculpture markings, the inner surface is smooth and shines with the lustre of mother-of-pearl. On a closer examination impressions and pits become visible on the inner surface. A narrow line, the so-called *mantle* or pallial line (the line of attachment of the *pallial muscle* of the mantle-edge to the shell), is placed near and fairly parallel to the ventral edge of the shell (Fig. 251). In the siphoned forms this presents posteriorly a bend directed forwards and upwards (*Mb*)—the *pallial bay* or *indentation* (*Sinupalliatae*).

This curve in the attachment of the mantle-edge permits of the siphons being contained within the shell when retracted. Impressions are usually caused by the insertion of an anterior and posterior *adductor muscle* which pass through the body transversely from one side to the other, and are attached to the inner surface of the shell (Fig. 251).

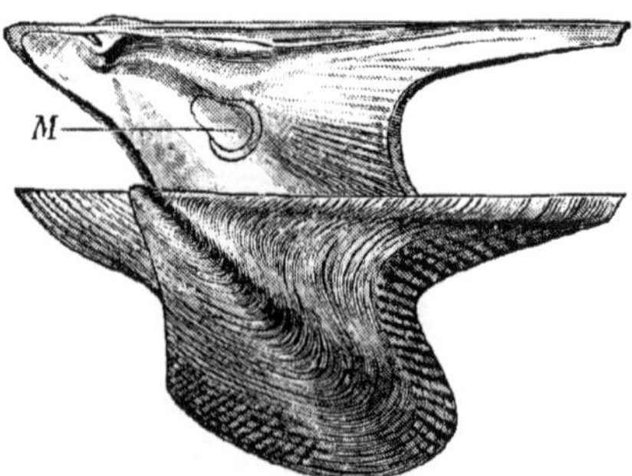

FIG. 252.—*Avicula semisagitta*, the valves are shifted over one another. *M* impression of the single adductor muscle.

While in the equivalve mussels (*Orthoconcha*) the two impressions are usually of equal size, in the unequivalve forms (*Pleuroconcha*) the anterior adductor may be reduced, and even completely vanish, in which case the posterior adductor has a much larger size and shifts forwards to the middle of the shell. Hence the names *Dimyaria* and *Monomyaria*.

Chemically the shell consists of carbonate of lime and an organic matrix (*conchyolin*), which usually has a laminated texture. In addition to this laminated layer there is also a thick external calcareous layer composed of large, pallisade-like prisms, which are placed side by side and may be compared to the enamel of teeth (Fig. 253). Finally, on the outer surface of the shell there is a horny cuticle, the so-called epidermis or *periostracum*.

The internal laminated layer is called the *nacreous* or *mother-of-pearl* layer; it has an iridescent lustre, and is secreted by the whole surface of the mantle. The middle layer is called the prismatic layer. It and the periostracum are secreted only by the free edge of the mantle.

The whole of the shell is a cuticular formation of the epidermis, and its growth is effected in two ways: (1) by additions to the nacreous layer whereby the shell increases in thickness; (2) by additions to the prismatic and horny layers, whereby it increases in superficial extent. Accordingly the outer coloured part of the shell, which is composed of vertical prisms and the horny cuticle, when once formed does not increase in thickness; while new concentric layers are continually being added to the colourless nacreous layer during the whole life of the animal. It is this nacreous secretion of the mantle surface which, when thrown down round foreign objects which have worked their way in between the mantle and the shell, in the so-called pearl oysters (*Meleagrina*) and to a less extent in other forms (*Unio, Margaritana*), gives rise to pearls.

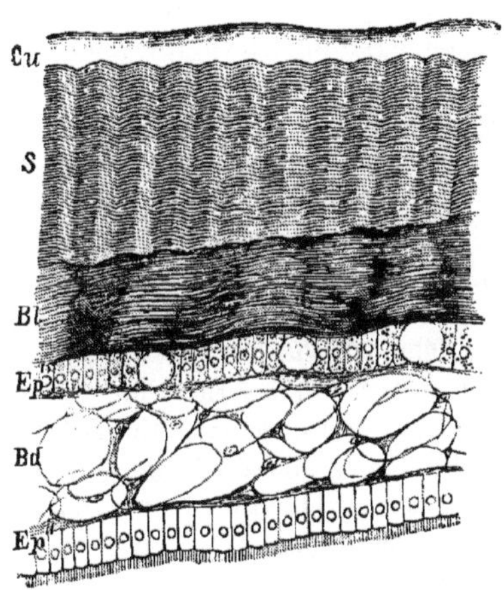

FIG. 253.—Vertical section through the shell and mantle of *Anodonta* (after Leydig). *Cu* cuticle or horny layer (periostracum); *S* prismatic layer; *Bl* laminated nacreous layer (mother-of-pearl); *Ep'* external epithelium of mantle; *Bd* connective tissue of mantle; *Ep''* internal epithelium of mantle, ciliated.

In some cases the valves do not meet ventrally, but always gape (*Pholadidae*, *Gastrochaenidae*, etc.). In exceptional cases the valves are fused dorsally (*Pinna*). The edges of the mantle are folded back over the shell in the *Galeommidae* and in *Entovalva*. In some gaping forms the parts of the body projecting beyond the shell secrete accessory pieces. Such pieces may be independent of the shell, as in the case of the dorsal pieces of *Pholas* (Fig. 270), and the calcareous tube with which *Teredo* lines its burrow (Fig. 271), or fused to the shell, as the calcareous tube of *Aspergillum* (Fig. 272).

The skin consists of a slimy, one-layered epidermis, beneath which lies a highly vascular connective tissue traversed by abundant muscular fibres. The epidermis on the outer surface of the mantle consists of columnar cells; while on the inner surface the cells composing it are ciliated (Fig. 253). Pigments are present principally upon the edges of the mantle, which are thickened and frequently folded and beset with tentacles, papillae, and eyes.

A head, properly so-called, is absent in this class, the parts of the body in the neighbourhood of the mouth being devoid of sense-organs. These are placed mainly on the mantle edges, which are the parts of the body in closest relation with the external medium.

The edges of the mantle may be entirely free from each other (those forms in which the gill-filaments are not connected to the mantle, and in which there is no concrescence of the mantle-lobes, *e.g.*, *Nucula*, the *Anomiidae*, *Arcidae*, *Trigoniidae*, *Pectinidae*); or they may be united to one another indirectly by the attachment of the branchiae, *e.g.*, *Unionidae*, *Ostreidae*; or they may be fused with one another in one, two, or three places. When there is only one fusion it separates off the opening of the cloacal or supra-branchial chamber from the general mantle-opening (*Mytilidae*, Fig. 256, *Carditidae*, *Astartidae*, *Crassatellidae*, most of the *Lucinidae*, etc.). When there are two fusions—as there are in *Yoldia* and *Leda*, and in most of the Eulamellibranchs, and in Septibranchs—the one separates off the cloacal opening as in the forms with one fusion, while the second is near the first, and with it bounds an opening adjacent to the cloacal opening. This second opening is the branchial opening; it leads into the general mantle-chamber (*Cardium*, Fig. 255). In such forms there is a third opening—the pedal opening—in front of the second fusion, through which the foot can be protruded. The size of the foot-opening is in inverse proportion to the extent of the second fusion. When the second fusion is much elongated, there may be a fourth opening between the pedal and branchial orifices (*Solen*, *Lutraria*, and some *Anatinacea*). In some cases, at any rate, this fourth opening is in relation to the byssus, for in *Lyonsia* the byssus filaments project through it. The further forwards the fusion of the two mantle-lobes extends, the more marked becomes a peculiar elongation of the posterior mantle region round the inhalent (branchial) and exhalent (cloacal) openings — an elongation of such a nature that two contractile tubes or **siphons** (Fig. 251) become formed (especially in boring and burrowing bivalves). These may be so large that they cannot be drawn between the posterior edges of the gaping valves of the shell. The two siphons are often fused with one another, but the two canals, with their openings surrounded by tentacles, remain separate. In the most extreme cases the siphons are enormously enlarged and the posterior region of the body is peculiarly elongated and uncovered by the rudimentary shell; so that the whole animal acquires a vermiform appearance, the shell-bearing anterior part of the body constituting the head (*Teredo*, Fig. 271).

In cases in which there is no fusion the hind end of the edges of the mantle lobes often presents two slight contiguous excavations (*Anodonta*, Fig. 254), the ventral of which is bordered by numerous papillae. When the two halves of the mantle are applied together these excavations form, with the corresponding structures of the opposite side, two slit-like openings placed one above the other.

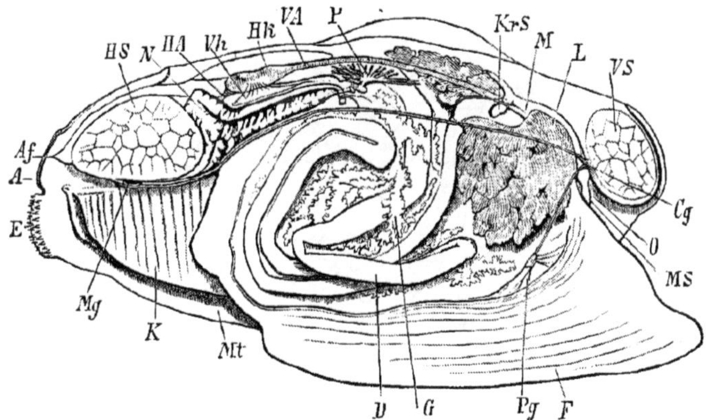

FIG. 254.—Anatomy of *Unio pictorum* (after C. Grobben). *A* region of mantle lobes bounding the cloacal or exhalent orifice; *Af* anus; *Cg* cerebral ganglion; *D* intestine; *E* region of mantle lobes bounding the inhalent or branchial orifice; *F* foot; *G* generative organs; *HA* posterior aorta; *Hk* ventricle; *HS* posterior adductor muscle; *K* branchiae; *KrS* crystalline style; *L* liver; *M* stomach; *Mg* splanchnic ganglion; *MS* labial palp; *Mt* mantle; *N* kidney; *O* mouth; *P* pericardial gland; *Pg* pedal ganglion; *Va* anterior aorta; *Vh* auricle; *VS* anterior adductor muscle.

The dorsal of these two openings functions as the cloacal (exhalent) opening, the ventral as the branchial (inhalent). So that in such cases, though there is no actual fusion, functionally two siphons are present as in the siphoned forms.

The most important of the **muscles** attached to the shell are (1) those of the edge of the mantle: these are attached to the pallial line of the shell, and serve to retract the edges of the mantle; the siphonal muscles which serve for the retraction of the siphons are a specialized portion of this system, and arise from the pallial indentation. (2) The adductors which pass from valve to valve of the shell. These are typically two in number (*Dimyaria*), but in some forms the anterior adductor is smaller than the posterior (*Mytilus*), and may be absent altogether (*Monomyaria*), as in *Ostrea*, *Pecten*, etc. (Fig. 252). The adductors sometimes consist of two parts of a different aspect: the fibres of one of these parts are said to be transversely striated, and capable of rapid contraction, those of the other being smooth, but this has been denied. (3) The anterior and posterior *retractors*, which are specialized portions of the adductors, serve for the retraction of the animal into the shell; and the *protractor* which passes from its attachment just behind the ventral portion of the anterior adductor.

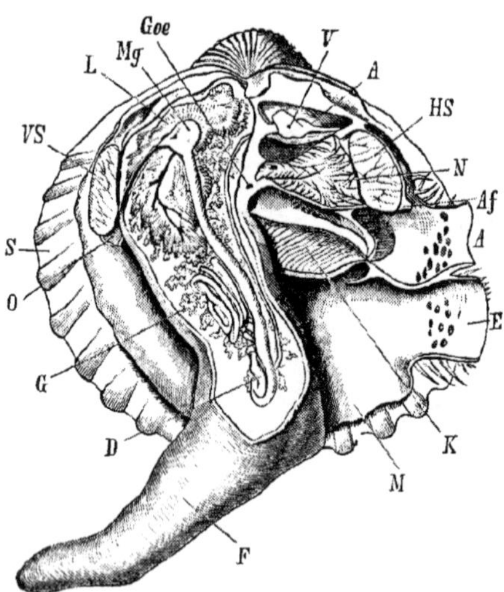

FIG. 255.—Anatomy of *Cardium tuberculatum* (after C. Grobben). *A* exhalent siphon; *A* auricle of heart; *Af* anus; *D* intestine; *E* inhalent siphon; *F* foot; *G* generative organ; *Goe* generative opening; *HS* posterior adductor; *K* gill of right side; *L* liver; *M* right lobe of mantle; *Mg* stomach; *N* kidney; *O* mouth; *S* right valve of shell; *V* ventricle; *VS* anterior adductor muscle.

The **foot**, which often contains some of the viscera (intestine, liver, and gonad), is completely absent in comparatively few forms, and only in those which have lost the power of locomotion (*Ostrea*, *Anomia*, *Teredo*, etc.). In many forms, principally in the larva, less frequently in the adult (*Mytilus*, Fig. 256), the foot possesses a **byssus** gland, which secretes silk-like fibres, by which a temporary or permanent attachment of the animal is effected. The form and size of the foot vary very considerably, in accordance with particular kinds of locomotion. The foot is most frequently used for creeping in sand, and is then hatchet-shaped; in other cases it is spread out laterally, and its creeping surface has the form of a disc. More rarely it is of a large size and bent, in which case it is used for springing movements in the water (*Cardium*, Fig. 255). Some Lamellibranchs possess a linear club-shaped or cylindrical foot (*Solen*, *Solenomya*), and move by rapidly retracting it and ejecting water through the siphons. Many use the foot for burying themselves in mud; others bore into wood (*Teredo*), or hard rock (*Pholas*, *Lithodomus*, *Saxicava*, etc.), for which purpose they push themselves against the rock with their short blunt foot, and use the hard and often serrated edge of their shell as a grater, giving it a rotatory movement. According to Hancock, the foot and edge of the mantle at the anterior edge of the gaping shell are beset with silicious crystals, and effect the excavation of the rock after the manner of a file.

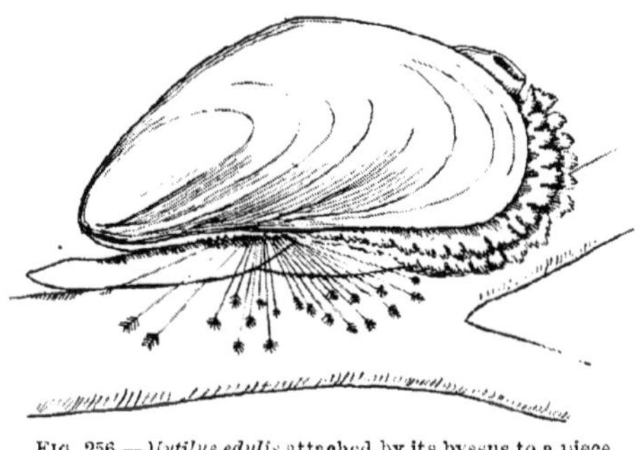

FIG. 256.—*Mytilus edulis* attached by its byssus to a piece of wood (after Meyer and Moebius).

The protraction of the foot is due to its turgescence by blood; its retraction to the retractor muscles. An aquiferous pore by which the vascular system communicates with the exterior does not exist.

The byssus-gland opens by a pore in the middle line of the foot. It is well-developed in the adult in *Anomia*, *Arca*, *Mytilus*, *Avicula*, *Pecten*, *Saxicava*, *Lyonsia*, *Tridacna*, *Dreissena*, etc. In *Anomia* the byssus passes through a hole in the right valve, and is calcified.

The **nervous system** presents three pairs of ganglia, the cerebral (supra-oesophageal), pedal (sub-oesophageal), and visceral ganglia (Fig. 257). In *Nucula* (Fig. 258) there is, in addition, close to the cerebral a pair of pleural ganglia; these are connected with the pedal by a pleuro-pedal connective which, however, joins the cerebro-pedal connective before it reaches the pedal. In *Solenomya*, in which there is also a pleural ganglion, the pleuro-pedal connective is fused with the cerebro-pedal throughout almost its whole extent. In all other Lamellibranchs a pleural ganglion is not present as a distinct ganglion, but is in all probability fused with the cerebral, which must, therefore, be regarded as a cerebro-pleural ganglion. The visceral ganglia are on the hinder part of the visceral loop; they are usually placed on the ventral side of the posterior adductor muscle.

Since there is never a distinct head and sense-organs are not present upon the anterior part of the body, the cerebral ganglia are but slightly developed They supply mainly the regions round the mouth and the anterior part of the mantle, to the margins of which two large nerves pass. They also send fibres along the cerebro-pedal connective to the otocysts. The two ganglia are frequently (*Unio*) far removed from one another laterally, and are sometimes approximated to the anteriorly placed pedal ganglia (*Pecten*), the nerves of which supply the ventral region of the body. The pedal ganglia are reduced when the foot is atrophied

The visceral ganglia innervate the viscera, the gills, the heart, and the

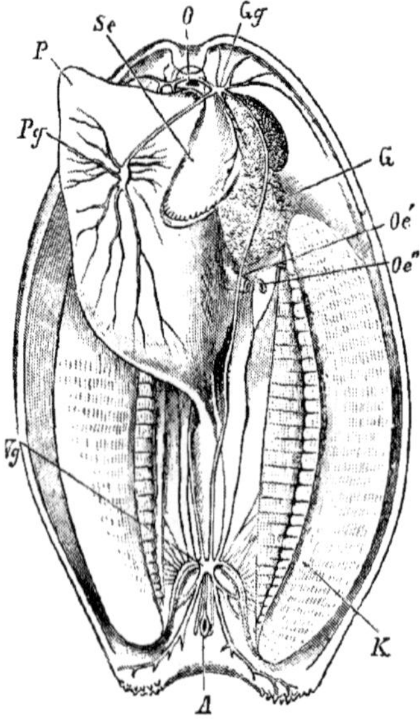

FIG. 257.—Nervous system of the pond mussel *Anodonta* (after Keber). *O* mouth; *A* anus; *K* gills; *P* foot; *Se* labial palps; *Gg* cerebral ganglion; *Pg* pedal ganglion; *Vg* splanchnic (visceral) ganglion; *G* generative gland; *Oe'* external opening of generative gland; *Oe"* opening of kidney.

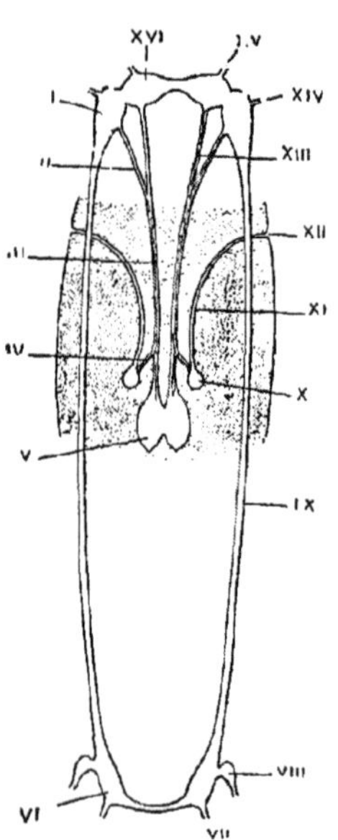

FIG. 258. — Dorsal view of nervous system of *Nucula*, the middle part of the foot is represented by dotted shading (after Pelseneer). *I* pleural ganglion; *II* pleuro-pedal connective; *III* combined pleuro-pedal and cerebro-pedal connective; *IV* nerve to the otocyst; *V* pedal ganglion; *VI* visceral ganglion; *VII* posterior pallial nerve; *VIII* osphradium; *IX* visceral commissure; *X* otocyst; *XI* canal from otocyst to exterior, and *XII* its external opening; *XIII* cerebro-pedal connective; *XIV* anterior pallial nerve; *XV* nerve to palps; *XVI* cerebral ganglion.

posterior part of the mantle. The nerves supplying the latter are two large trunks which run in the edge of the mantle, and anastomose with the mantle nerves from the cerebral ganglia. Large nerves also pass from the visceral ganglia to the siphons, at the base of which they form a pair of accessory ganglia.

Sense organs. Auditory, visual, tactile, and probably olfactory organs are present. The *auditory organs* have the form of paired otocysts placed in the foot. They are apparently innervated from the pedal ganglia, or from the cerebro-pedal connective close to the pedal ganglion, but the nerve fibres really arise in the cerebral ganglia. They contain one or more otoliths, and are lined by hair-cells and by ciliated cells. In the *Protobranchiata* (Fig. 258) they open to the exterior by a fine canal, and contain foreign bodies (fine grains of sand).

Eyes, when present, are placed on the edge of the mantle: they may either be simple pigment spots at the end of the respiratory tube (*Solen*, *Venus*), or they may be more highly developed and placed along the edges of the mantle in *Arca*, *Pectunculus*, *Tellina*, and especially in *Pecten* and *Spondylus*. In the two latter genera

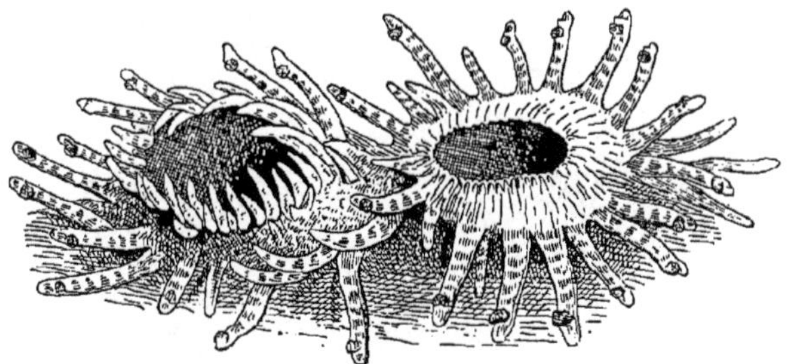

FIG. 259.—Openings of the siphons of *Cardium edule* (from Perrier after Möbius).

they are placed on stalks between the marginal tentacles, and have an emerald-green or brown-red colour: they consist (Fig. 260) of an eye-bulb with a corneal lens, choroid, iris, and a well-developed layer of rods, into the external ends of which the optic nerve (from the circumpallial nerve) passes. The *sense of touch* is specially localized on the exposed parts of the body, *i.e.* on the edges of the mantle and round the openings of the siphons. In these regions there are very generally present papillae, cirri, or even tentacles (Fig. 259).

The *olfactory sense* is supposed to reside in the so-called **osphradium,** which is a pigmented patch of epithelium placed one on each side close behind the gills (in the roof of the supra-branchial chamber near the visceral ganglia); it is innervated from a small ganglion placed close to the visceral on the visceral commissure, in which its nerve-fibres run from their origin in the cerebral ganglion.

Many bivalves, especially littoral forms, are highly sensitive to light. This is due to the presence of pigmented cells with a refractile cuticle on the edge of the mantle; these may be aggregated, as in *Arca*, into groups, and so assume the form of a compound eye. In *Arca* some of the eyes have the form of pits

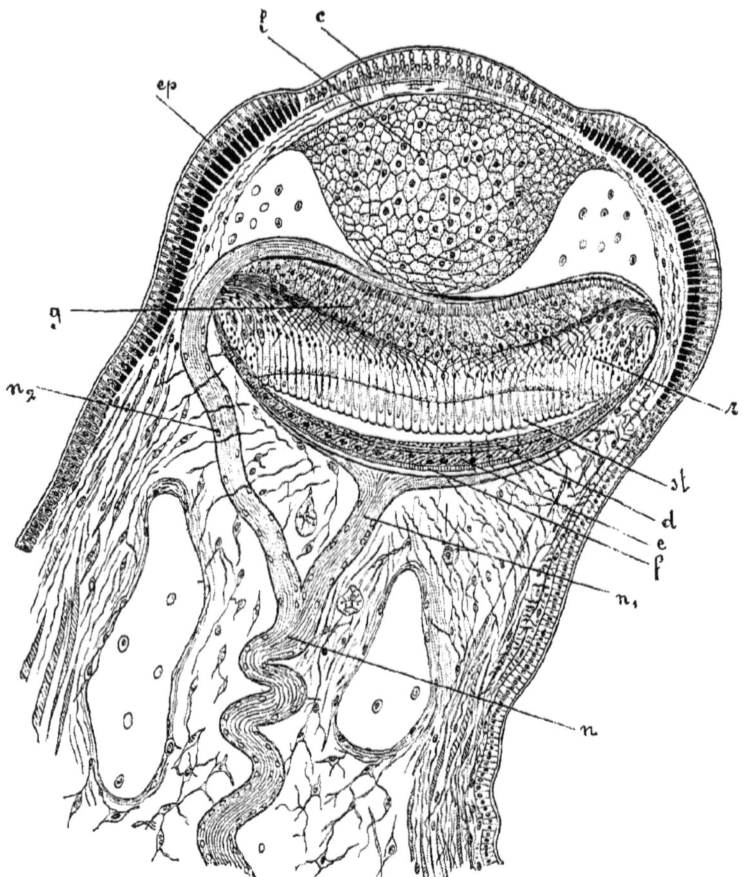

FIG. 260.—Axial section through the eye of *Pecten* (from Lang after Patten). *c* cornea; *l* lens; *ep* pigmented ectoderm; *g* layer of ganglion cells; *r* retina; *st* layer of rods of retina; *d* tapetum; *e* pigment; *f* sclerotic; *n* optic nerve, n_1 and n_2 its two branches.

of the skin, while in others the sensitive surface is convex towards the light, and the eye forms a slight projection on the surface: the latter are on the type of compound eyes. In *Pecten* and *Spondylus* the eyes are much more complex (Fig. 260). They are placed at the end of short tentacles at the edge of the mantle, and the sensitive surface is concave; the light being refracted on to it by a cellular lens. Further, these eyes are constructed on the so-called vertebrate type, the optic nerve entering the retina on the side turned towards the light and running back to the rods, which are on the inner side of the retina.

Alimentary canal. The mouth is at the front end of the body, and ventral to the anterior adductor when that is present. It is placed on the median bridge, which connects the two **labial palps** (Fig. 263). These latter structures are extensions, so to speak, of the margins of the mouth. They are usually bilobed at the end remote from the mouth, and marked by a median groove running from the cleft between the peripheral lobes to the mouth. Their surface on each side of the median groove is marked by transversely directed grooves leading into the median groove. The whole surface is richly ciliated, and they are to be looked upon as food-procuring organs which create currents of water, carrying the floating particles of which the food of these animals consists to the mouth. Jaws and tongue are always absent. The mouth leads by a short oesophagus into the stomach, at the pyloric end of which there is sometimes a blind sac. A rod-like transparent structure (*crystalline style*) is often present in this diverticulum of the stomach, or in the alimentary canal itself. The significance of this structure is doubtful, but it is a secretion of the alimentary epithelium, and is probably to be regarded as a reserve of nutriment, for it is periodically renewed. The *liver* or hepato-pancreas surrounds the stomach, and opens into it by a duct on either side; it also extends into the foot. The intestine is of considerable length, is much coiled, and is surrounded by the liver and gonads; it projects into the foot, and then ascends again behind the stomach to the dorsal surface, where it enters the pericardium and passes through the ventricle. After leaving the ventricle it passes dorsal to the posterior adductor muscle to open at the end of a short papilla into the cloacal chamber.

Vascular System. The heart, which is contained in the pericardium and lies in the dorsal region slightly in front of the posterior adductor, consists of a median ventricle and two lateral symmetrically placed auricles. The ventricle is continued as an anterior aorta dorsal to the intestine, and a posterior aorta ventral to it. The ramifications of the aortae lead the blood into a complicated system of lacunae in the mantle and in the interspaces between the viscera. These represent the capillaries and finer venous vessels. From them the blood passes into the large venous sinuses, the chief of which are two lateral sinuses placed at the bases of the gills and a pedal sinus leading into the large median sinus or vena cava in the floor of the pericardium. The course of the circulation, though it cannot be certainly determined, seems to be somewhat as follows: the blood from the mantle, which acts as a respiratory organ, returns directly to the heart; of the rest of the systemic blood, part is supposed to go direct to the gills, and thence to the auricles; the bulk of it, however, is probably collected into the vena cava, whence it passes through blood spaces in the adjacent kidneys to the gills, and thence to the heart. There is a valve at the junction of the main pedal sinus with the vena cava, which closes during the turgescence of the foot.

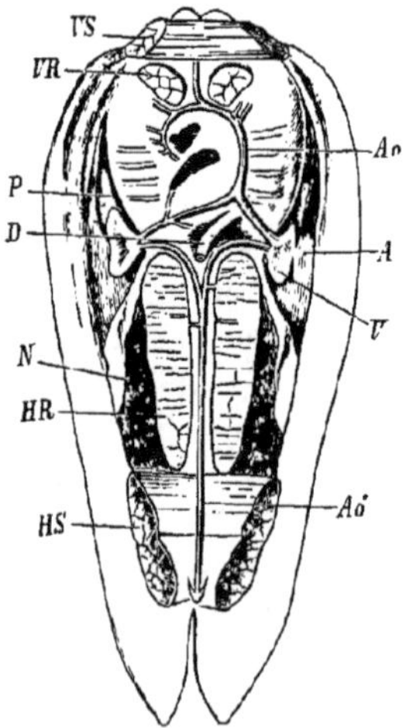

Fig. 261.—Dorsal view of *Arca noae* removed from the shell (after Grobben). The double pericardial cavities *P* are opened, and the rectum *D* turned forward. *VS* anterior, *HS* posterior adductor muscle; *VR* anterior, *HR* posterior retractor; *V* ventricle; *A* auricle; *Ao* anterior, *Ao'* posterior aorta; *N* kidney.

The blood is generally colourless, though in some forms it has a bluish tint owing to the presence of haemocyanin. It contains amoeboid cells, and in *Solen legumen* and *Arca noae* discoidal corpuscles charged with haemoglobin, which gives the blood a red colour.

The relation of the ventricle to the rectum varies in different forms. As a rule the rectum perforates the ventricle, but it passes dorsal to it in *Arca*, *Nucula*, and *Anomia*, ventral to it in *Teredo* and most species of *Ostrea*, and has a tendency to the latter arrangement in *Pinna*, *Perna*, and *Avicula*.

The vascular system does not communicate either with the exterior or with the pericardium.

In *Arca* (Fig. 261) there are two ventricles, and each gives off a single artery, which divides at once into an anterior and posterior vessel; the two anterior arteries fuse to form a single anterior aorta, and the two posterior unite into a posterior aorta.

In *Ostreidae* (Fig. 262) the auricles have partly coalesced, though their outer portions have remained distinct.

Organs of respiration. The Lamellibranchs possess two ctenidia attached to the ventral surface of the body, one on either side of the foot. In the simplest cases (*Protobranchiata*) the ctenidium has the typical Molluscan form (Figs. 263, 264 *A*); that is to say, it consists of a vascular axis bearing two rows of hollow lamellate processes or filaments. The axis projects freely posteriorly, and the filaments are directed transversely in *Nucula* and *Leda*, while in *Solenomya* one is directed dorsally and the other ventrally. In the *Filibranchiata* the

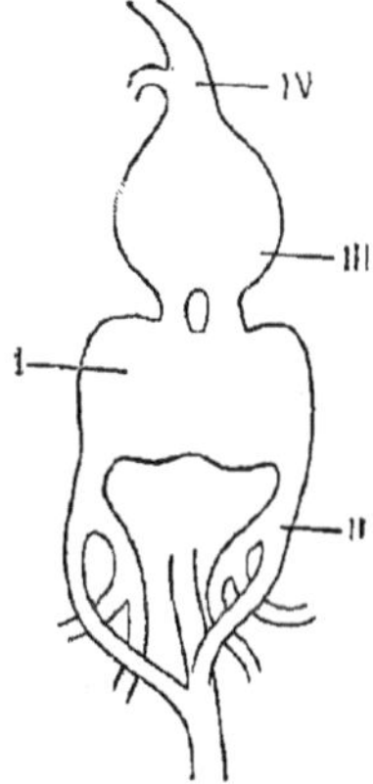

FIG. 262.—Heart of *Ostrea* magnified (after Poli, from Pelseneer). *I* fused auricles; *II* afferent veins; *III* ventricle; *IV* aorta.

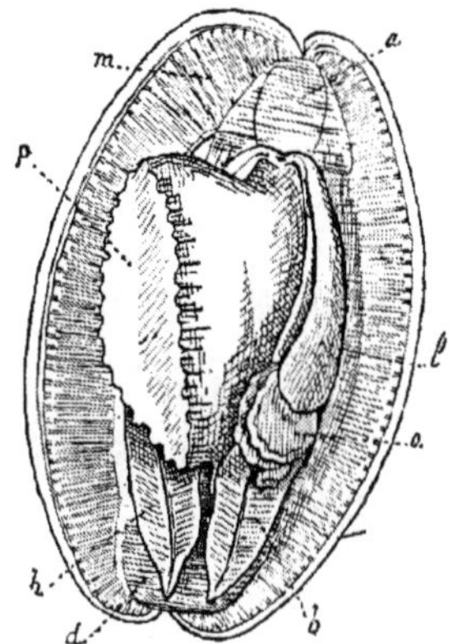

FIG. 263.—Ventral view of *Nucula* (after Deshayes, from Perrier). *a* anterior adductor muscle; *b* ctenidium; *d* posterior adductor; *l* labial palps; *m* internal surface of mantle; *o* posterior appendage of palps; *p* foot.

branchial processes of the ctenidial axis have the form of filaments (Fig. 264 *B*), which project ventrally into the mantle cavity; moreover, they are bent upon themselves in such a manner that the reflected portion, or ascending limb, is external in the case of the filaments of

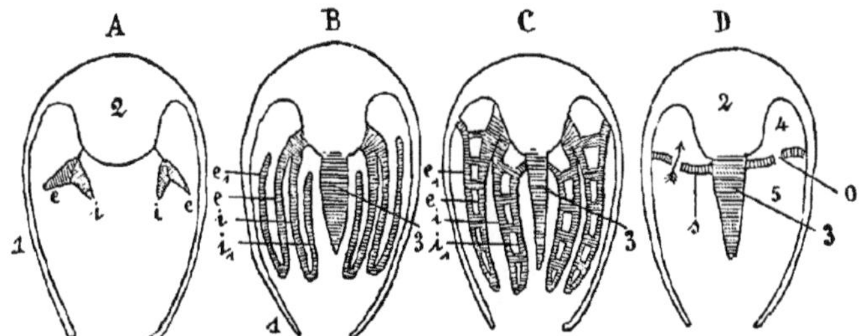

FIG. 264.—Series of diagrams of transverse sections, illustrating the different arrangement of the gills in Lamellibranchs. *A*, Protobranchiata. *B*, Filibranchiata. *C*, Eulamellibranchiata. *D*, Septibranchiata. *1* mantle; *2* body; *3* foot; *4* suprabranchial cavity; *5* branchial (mantle) cavity; *e* outer plate or filament of ctenidium—descending limb; e_1 ascending limb of *e*; *i*, i_1 the corresponding parts of the inner filaments of ctenidium; *o* pores in the branchial membrane of the *Septibranchiata* (from Lang).

the outer row, and internal in the case of the inner filaments. The adjacent filaments of the same row, both in their descending and ascending limbs, are held together by the interlocking of some especially long cilia (Fig. 265, *cj*); and the ascending and descending limbs of the same filament are in some cases united by fusion of tissue at certain points, giving rise to the so-called *interlamellar concrescences* (Fig. 266, *ilj*). Further, the ends of the ascending or reflected limb of the outer filaments may be united to the mantle, or may be free (Fig. 264 *B*, *C*), and similarly the ends of the ascending limbs of the inner filaments may be free, or those behind the foot may be united with their fellows across the middle line, thus forming a septum, which cuts off a suprabranchial chamber from the general mantle cavity.

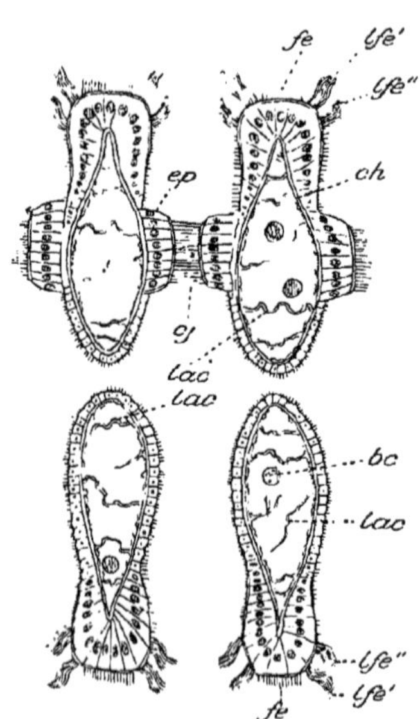

FIG. 265.—Transverse section through the ascending and descending limbs of two adjacent gill-filaments in *Mytilus* (after Peck). The two filaments are connected by a ciliary junction in the upper part of the figure. *bc* blood-corpuscle; *ch* chitinous layer; *cj* ciliary junction; *ep* epithelial prominence which carries the cilia of a ciliary junction; *fe* frontal epithelium; *lac* lacunar (vascular) tissue; *lfe'* latero-frontal epithelium with long cilia; *lfe''* second row of the same.

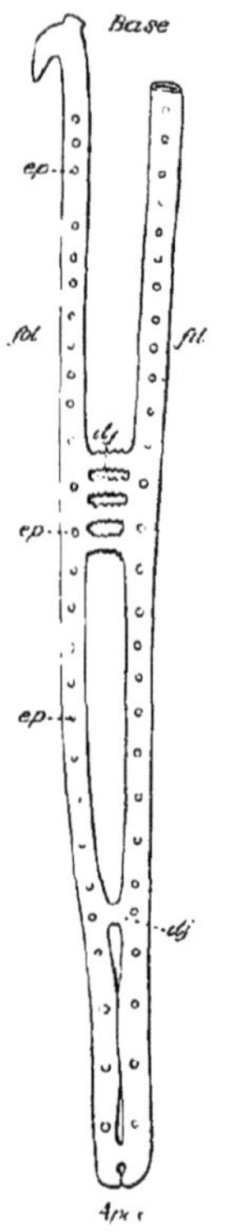

FIG. 266.—One filament of the gill of *Mytilus* (after Peck). *fil* descending limb, *fol* ascending limb of the filament; *ilj* interlamellar concrescence; *ep* position of interfilamentar ciliary junctions. The ascending limb ends freely in a hook-like manner. The *apex* or angle of the filament is grooved.

In all the *Eulamellibranchiata* and in a few of the Pseudolamellibranchs (*Ostreidae*, *Lima*, *Pinna*, etc.), the neighbouring filaments of the same row, in both their limbs, are united at certain points by interfilamentar vascular concrescences (Fig. 267). The filamentous character is still obvious, but both the descending and ascending limbs of adjacent filaments are connected so as to form a kind of trellis-like membrane, the ascending limbs into one membrane and the descending limbs into another. Between the porous membranes so formed is contained a space, which communicates with the general mantle-chamber by the pores left between the filaments where the latter have not undergone these interfilamentar concrescences.

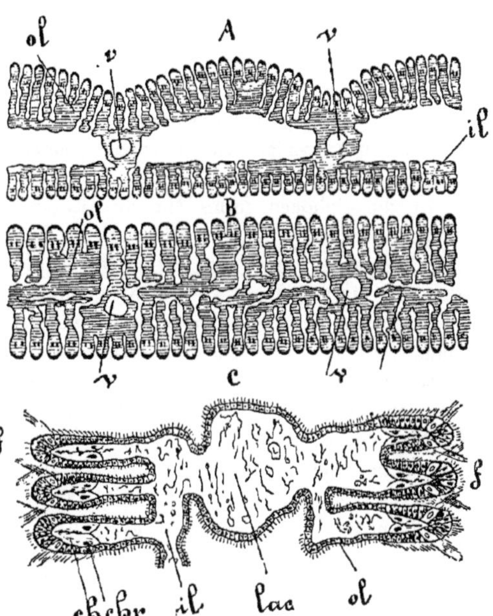

Fig. 207.—Pieces of transverse sections through the branchiae of *Anodonta* (from Lang, after Peck). *A* outer, *B* inner gill. In each section the two lamellae are shown as connected together by the interlamellar concrescences, which are formed by the sub-filamentar vascular tissues *of*, in *B*. The interfilamentar concrescences are effected by the same tissue. In *C*, which is a part of *B* more magnified, this tissue is marked *il*, *lac*, and *ol*; and the filaments with their frontal epithelium and their chitinous tissue *ch*, and rods *chr*. *ol* outer lamella; *il* inner lamella; *v* blood vessel; *of* subfilamentar tissue; *f* filaments; *ch* chitinous tissue of filaments; *chr* specially condensed chitinous rod in *ch*; *lac* subfilamentar tissue.

In this way the Eulamellibranch condition, in which there appear to be on each side two gills, each composed of two lamellae, is arrived at. A very similar condition is presented by the Filibranchs and some Pseudolamellibranchs, the difference being that in these groups the lamellae of the gill break up at the slightest touch into their constituent filaments, because the filaments are only held together by the interlocking of cilia; whereas in the former the lamellae are coherent, because the filaments (which are still perfectly obvious) are held together by continuity of tissue. Further, in the Eulamellibranchs the outer lamella of the outer gill has undergone concrescence with the mantle (Fig. 264), and the inner lamella of the inner gill with its fellow across the middle line behind the foot. In this way a septum is formed which divides the mantle cavity into a dorsal part —the suprabranchial cavity, and a ventral—the general mantle cavity. These two cavities communicate by the pores in the gill lamellae, and in many forms along the sides of the foot where the inner lamella of the inner gill may end freely, not having undergone concrescence with the foot. In some forms the inner lamella of the inner gill is joined to the base of the foot in front, but not behind The part of the suprabranchial cavity within the gill lamellae is broken up by vascular strands which connect the lamellae, and which result from the interlamellar concrescences already referred to. In the siphoned forms the median septum resulting from the fusion of the inner lamellae of the inner gills across the middle line is continued into the siphon, and causes the division of that tube into a dorsal or exhalent channel, which communicates with the suprabranchial chamber, and a ventral or inhalent which communicates with the general mantle-cavity.

In the *Septibranchiata* (Fig. 264 *D*) the gills are represented by a muscular septum which is perforated by a certain number of pores, and which, being fused with the mantle and foot and being continuous across the middle line behind the foot, completely divides the mantle-cavity into a dorsal and ventral part.

The filaments of the gills are always clothed with an epithelium, which is in part at any rate ciliated. The cilia are specially long on the so-called latero-frontal cells (Fig. 265), and in the Filibranchs and Pseudo-lamellibranchs on the ciliated discs of the ciliary junctions. The filaments are moreover stiffened by a dense chitin-like connective tissue, which in the Filibranchs forms a tube (Fig. 265) lying just beneath the epithelium and surrounding the central blood space (which is often divided by a septum); and in the Eulamellibranchs, on the other hand, this chitinous supporting substance has the form of two stout rods lying side by side in the filament part of the gill (Fig. 267 *C*), the vascular tissue being mainly contained in the internal outgrowths of the filaments, which have brought about the interfilamentous and interlamellar concrescences (Fig. 267, *of, il, lac,* and *ol*).

Amongst the Eulamellibranchs we find some peculiar modifications of the outer gill. In *Tellina* it is directed dorsally; in the *Anatinacea* it is directed dorsally and the outer lamella is absent; in *Lucina* it is absent altogether.

In Pseudolamellibranchs and some Eulamellibranchs the gills are folded along dorso-ventral lines, each fold including a certain number of filaments. The filament at the re-entering angle of each fold is stronger than the others.

Summary and additional details of gill-structure. The successive filaments of the same row are united (*a*) by the cilia of the lateral cells (*Nuculidae*, *Solenomyidae*, *Anomiidae*); (*b*) by the cilia of the ciliated disks of the ciliary junctions (*Arcidae*, *Trigoniidae*, *Mytilidae*, *Avicula*, *Pecten*, and *Meleagrina*); by vascular interfilamentar concrescences (*Lima*, *Ostrea*, *Pinna*, and all Eulamellibranchs). *Interlamellar concrescences*, *i.e.* fusions between the ascending and descending limbs of the same filament, or between the lamellae of the same gill, are absent in *Anomiidae*, *Arcidae*, *Trigoniidae*; they are present in a non-vascular form (*i.e.* consist only of epithelium and connective tissue) in *Mytilidae*, *Pectinidae*, and in a vascular form in Pseudolamellibranchs except *Pectinidae*, and in Eulamellibranchs. Finally there are forms like *Lima* with non-vascular interlamellar concrescences and vascular interfilamentar junctions.

Concrescence of the gills with the mantle and with those of the opposite side, is absent in *Nuculidae*, *Solenomyidae*, *Arcidae*, *Trigoniidae*, *Pectinidae*. In *Anomia* there is concrescence between the two branchiae, but none with the mantle. In all other Lamellibranchs they are fused to the mantle by the ascending limb of the outer filaments, and to their fellows across the middle line by the ascending limb of their inner filaments behind the foot.

Excretory system and pericardium. The pericardium is a dorsal median chamber enclosing the heart (except in *Anomia*). There are two nephridia—the so-called *Organs of Bojanus*. These usually have the form of twisted tubes, more or less dilated in certain parts, and opening at one end into the pericardium and at the other to the exterior on the ventral surface of the body on each side of the foot. The part of the tube which opens into the pericardium is generally lined with yellow or dark-coloured glandular tissue, which secretes concrements containing calcareous matter, uric acid, and guanin. The kidneys of the two sides communicate in some forms. In *Ostrea* the glandular part of the kidneys is a branched gland ramifying on the surface of the visceral mass.

The **pericardial glands** are differentiations of the epithelial lining of the pericardium. They may be placed on the auricles, to which they impart a yellow colour, or near the auricles at the anterior end of the pericardium.

Generative organs. The sexes are, with a few exceptions (some species of *Pecten*, *Ostrea*, *Cardium*, and the genera *Cyclas*, *Pisidium*, *Poromya*, *Entovalva*), separate, and except in some species of *Unio* do not present external differences. The generative glands lie amongst the viscera, and have the form of lobed or racemose glands, which are placed near the liver, surround the windings of the intestine, and extend into the base of the foot, and in some forms into the mantle lobes (*Mytilus*). The testis and ovary can often be distinguished from one another by their colour; the ovary being red, and the testis milk white or yellow. In the *Protobranchs* the generative glands open into the kidney near the pericardial opening; in the *Anomiidae* and *Pectinidae* also into the kidney, but nearer the external opening. In *Ostrea*, *Cyclas*, and some *Lucinidae* they open with the renal duct by a common opening. As a rule, however, the two organs (renal and genital) open separately, but close together, on the external side of the visceral commissure (Fig. 257).

In the hermaphrodite forms the whole gland may be hermaphrodite even to the ultimate acini, which produce ova and spermatozoa simultaneously or alternately (*Ostrea edulis* and *plicata*); or the male and female follicles may be separate, and open together by a common duct (species of *Pecten* and *Cyclas*), or by separate ducts leading to separate openings (*Poromya*, *Anatinacea*). Hermaphrodite individuals are sometimes met with in the fresh-water mussels—*Unio* and *Anodonta*.

Development. The fertilization of the eggs is sometimes effected in the branchial cavity, and the first part of the development often takes place in the mantle cavity, or between the lamellae of the inner (*Cyclas*) or outer gill (*Unio*, *Anodonta*) as in a brood-pouch; sometimes it takes place outside the mother (*Pecten*, dioecious *Ostreae*, *Mytilus*, in all of which artificial fertilization is possible). In *Ostrea edulis* it is effected in the oviduct. The egg is surrounded by a vitelline membrane with a micropylar aperture at one point.

The segmentation is unequal (Fig. 268), the formative pole being opposite to the micropyle. The gastrula is generally formed by epibole, rarely by invagination. The blastopore sometimes remains open, *e.g.* *Ostrea*, or closes, *e g.* *Cyclas*, *Teredo*, but the mouth is formed almost immediately by an ectodermal invagination at the same point. The stomach, liver, and intestine are formed from the endoderm, and the proctodaeum is established later as an ectodermal invagination after the shell has been formed.

The embryo, which is partially ciliated and often rotates within the egg-membranes, soon acquires a preoral ring of cilia and a shell-gland (Fig. 268). The latter is on the side opposite to the blastopore, and gives rise to a pellicle which is calcified from two symmetrical points, thus forming the rudiments of the two valves (Fig. 268 *c*). The cuticular shell remains uncalcified in the middle dorsal line, where it gives rise to the ligament. An ectodermal invagination, giving rise to the byssus gland, is almost always formed at the hind end of the foot, even in forms which are without a byssus in the adult. The anterior adductor appears before the posterior. Among the provisional arrangements the **velum**, as the preoral ciliated ring is called, is very generally present, and in the free-swimming larvae has the form of a large ciliated ring or collar; a pair of larval eyes provided with a lens may also exist within the velar area.

The gills appear as filaments, one by one from behind forwards in the posterior part of the larva between the mantle and the visceral mass. A pair of larval kidneys has been observed in some groups (Fig. 268 *d*, *N*). They consist of an internal part, having the form of a ciliated canal, and of a superficial part which opens externally on the posterior and ventral side of the cephalic region (*Cyclas*, *Teredo*).

The development of the fresh-water forms (*Cyclas*, *Unio*, *Anodonta*), in which the eggs and embryos are contained in well protected brood-pouches, may be

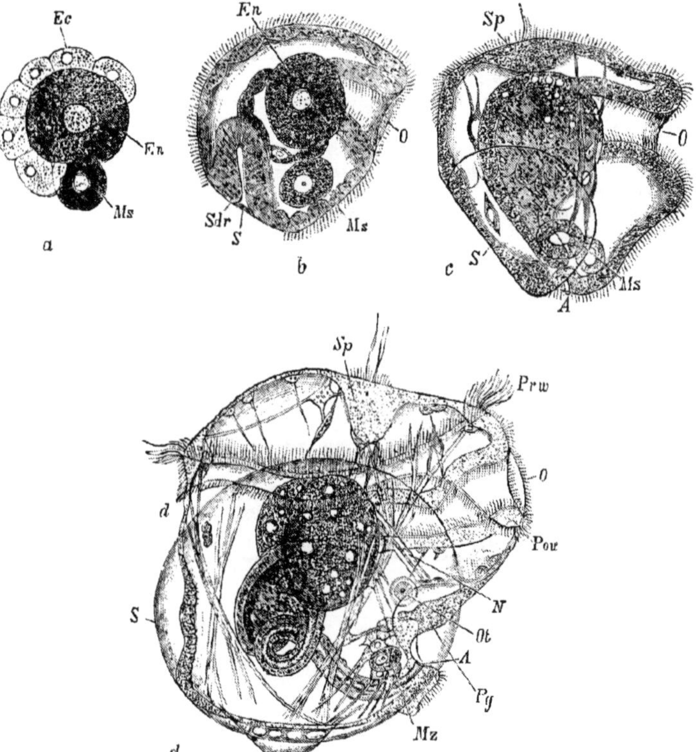

FIG. 268.—Stages in the development of the larva of *Teredo* (after Hatschek). *Ec* ectoderm; *En* endoderm; *Ms* mesoderm. *a*, median optical section of an embryo with two mesoderm cells *Ms* and two endoderm cells *En*. *b*, ciliated embryo with mouth *O*, stomach, intestine, shell-gland *Sdr* and shell *S*. *c*, later stage with anal invagination *A*, apical plate *Sp*, and more extensive shell *S*. *d*, Trochosphere larva of *Teredo*. *O* mouth; *A* anus; *Prw* preoral ciliated ring or velum; *Pow* postoral ciliated ring; *N* larval kidney or pronephros (head-kidney); *Ot* otocyst; *Pg* pedal ganglion; *Mz* mesoderm.

called direct. The marine Lamellibranchs on the other hand are set free at an early stage and swim about for a long time as larvae with large umbrella-like vela, from which the labial palps are developed.

The *Unionidae* have a somewhat complicated development.* The eggs are

* Schierholz, C. "Ueber die Entw. d. Unioniden," *Denkschr. k. Akad. Wiss. zu Wien, Math.-Naturw.* Bd. 45, 1889.

laid in spring or summer, and pass into the space between the lamellae of the inner gill, and thence into that between the lamellae of the outer gill at the posterior end of the gill, where these spaces communicate. In this space they remain as in a brood-pouch, and undergo the first stages of their development. These take two months. They then cease to develop, and pass the winter in the brood-pouch. In the following spring they pass out by the exhalent orifice under the larval form, known as *Glochidium* (Fig. 269). This possesses a bivalve shell with hooks in the middle of the ventral edge of the valves, a single adductor (the anterior), and a byssus thread which appears to arise just behind the adductor. They swim by snapping their valves, and attach themselves to the skin (gills or fin) of a fish, in which they become embedded. Here they remain as parasites from two to five weeks, and undergo further development; but they are not fully formed for some time after leaving their host. The gills grow, and the external lamina is not developed till the third year. Sexual maturity is not attained till the fifth year, and growth continues later.

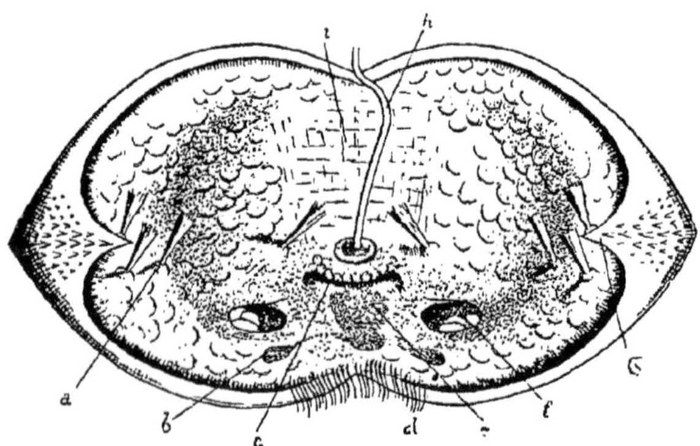

FIG. 269.—Ventral view of the Glochidium of *Anodonta* (from Pelseneer after Schierholtz). *a* bunch of setae; *b* visceral ganglia; *c* stomodeal invagination; *d* ciliated patch; *e* enteron; *f* lateral pits; *g* hooks on the edge of the valves; *h* byssus filament; *i* the single adductor.

The Lamellibranchs are all aquatic animals, for the most part marine, but a few are fresh-water. They feed on microscopic, mainly vegetable organisms (Diatoms, etc.), which float in the water, and are conveyed to them by currents set up by the cilia of the mantle-chamber. The Septibranchs alone are carnivorous. A great many forms live partly or wholly embedded in sand or mud, procuring food and water and getting rid of waste by their siphons which project at the surface. Many are sedentary, being attached to foreign objects by their byssus, or by one of the valves of their shell (*Ostrea*, *Spondylus*). Some are borers into wood (*Teredo*) or into stone (*Lithodomus*, *Pholas*, *Clavagella*). Some build nests by means of their byssus (*Lima*); while some are commensals—*Modio-*

laria marmorata in the test of Ascidians; *Vulsella* in sponges; *Montacuta* on Spatangids; *Entovalva* is parasitic in the oesophagus of *Synapta*. Some species are very active, leaping by aid of their foot (*Cardium*, *Tellina*, etc.); some are crawlers (*Cyclas*, *Lasaea*), and some swimmers, by flapping the valves of their shells (*Pecten*, *Lima*).

The *Lamellibranchs* are found in all parts of the world to the number of more than 5000 species. They have been found to a depth of 2900 fathoms. They have a wide distribution in the earlier periods of the earth's history, being known since the Silurian. Their fossil shells are most excellently preserved, and they are of the greatest importance as characteristic fossils for the determination of the age of formations.

Order 1. PROTOBRANCHIATA.

The gill-filaments are plate-like and not reflected, and the mantle-cavity is not divided into two parts.

The mantle has an hypo-branchial gland external to each gill; the foot has a ventral plantar surface, and the byssus gland is but slightly developed. The nervous system has a distinct pleural ganglion; the otocysts are connected with the surface by a tube. The sexes are separate, and the genital glands open into the inner end of the kidney-tube.

Fam. 1. **Nuculidae.** Palps large, with posterior appendage (Fig. 263). *Nucula* Lam., heart dorsal to rectum; *Leda* Schumacher, heart traversed by rectum, mantle with two siphons; *Yoldia* Möll.

Fam. 2. **Solenomyidae.** The palps are not bilobed; in each gill one row of filaments is directed dorsally, the other ventrally. Mantle lobes fused, having a single posterior opening, and one anteriorly for the foot. *Solenomya* Lam.

Order 2. FILIBRANCHIATA.

The gill-filaments are reflected, and united by ciliary junctions. The foot usually with a well-developed byssus apparatus.

Sub-order 1. ANOMIACEA.

Asymmetrical; posterior adductor large; heart dorsal to rectum, and causing projection into mantle-cavity; the reflected limbs of the inner filaments are fused across the middle line with their fellows; genital glands open into kidney, and that of right side extends into mantle.

Fam. 1. **Anomiidae.** *Anomia* L., byssus calcified, passing through a hole in the right valve. *Placuna* Bruguière.

Sub-order 2. ARCACEA.

Symmetrical; mantle open; both adductors well developed; gills without interlamellar junctions; renal and generative openings separate.

Fam. 2. **Arcidae.** Edges of mantle with compound pallial eyes. *Pectunculus* Lam., *Limopsis* Sassi; *Arca* L., heart dorsal to rectum (Fig. 261).

Fam. 3. **Trigoniidae.** *Trigonia* Bruguière.

Sub-order 3. **MYTILACEA.**

Symmetrical; mantle-lobes fused posteriorly; anterior adductor smaller than posterior; a single aorta; gill-filaments with interlamellar junctions; genital glands extending into the mantle, and opening by the side of the kidney openings.

Fam. 4. **Mytilidae.** Mussels. Attached by the byssus fibres of the tongue-shaped foot. *Mytilus* L.; *M. edulis* L. (Fig. 256), edible mussel of North Sea and Baltic; *Modiolaria* Lovèn; *Modiola* Lam.; *Lithodomus* Cuv.; *L. dactylus* Sow., Mediterranean (Temple of Serapis at Pozzuoli).

Order 3. PSEUDOLAMELLIBRANCHIATA.

Mantle entirely open; gills folded and filament at recntrant angle of fold modified; interfilamentar junctions effected by ciliated discs or vascular concrescence.

Foot feebly developed; anterior adductor usually absent; the auricles communicate with one another; the branchial filaments have interlamellar concrescences: genital glands open into or near the kidneys.

Fam. 1. **Aviculidae.** Byssus-apparatus well developed; shell usually unequivalve, dorsal margin straight and often long; ascending limbs of outer gill-filaments fused to mantle. *Avicula* Klein (Fig. 252); *Pinna* L.; *Perna* Bruguière; *Meleagrina* Lam.; *M. margaritifera* L., pearl-oysters, Indian Ocean, Persian Gulf, and Gulf of Mexico; *Malleus* Lam.: *Vulsella* Lam.

Fam. 2. **Ostreidae.** Byssus absent; fixed by shell; outer gill-filaments fused to mantle; shell unequivalve, fixed by left valve. *Ostrea* L. Oysters. Shell valves unequal, laminated, with weak hinge usually without teeth, single somewhat ventrally placed adductor muscle. In the true oysters the more arched left valve is firmly attached, while the right and upper valve, which is fastened by an internal ligament, lies as an operculum on the lower valve. Edges of mantle fringed, not fused; gill lamellae fused with mantle and across the middle line. Foot absent or rudimentary. They usually live together, and form banks of considerable extent. Found in the Jura and the Chalk. *Ostrea edulis* L., on the coasts of Europe on rocky ground; probably includes a number of different species according to the locality. According to Davaine, the oysters are said to produce only male sexual products towards the end of the first year, and it is not until later, from the third year onwards, that they become females and produce ova. Moebius, on the contrary, asserts that the sperm is the later formed, and not until after the pregnant beast has got rid of her eggs. The reproduction takes place especially in the months of June and July, at which time in spite of their extraordinary fertility the oysters should not be gathered. In the American oyster (*O. virginiana*) and the Portuguese oyster (*O. angulata*) the sexes are separate. The green oyster owes its colour to its food—a diatom *Navicula ostrearia*.* The colour is confined to its gills and labial palps, and is said to be due to selective absorption of the pigment from the blood. It is, however, possible that the diatoms may adhere to the surface of the gills and palps, and be consumed by phagocytes or other cells. This suggestion requires testing.

Fam. 3. **Pectinidae.** Byssus absent or feebly developed; shell ribbed; mantle-edge tentaculiferous; a duplication of mantle-edge folded internally;

* Ray Lankester, *Q. J. M. S.*, 26, p. 71.

generally with emerald-green pallial eyes; outer gill-filaments free. Some are fixed by one valve, and some can swim by opening and shutting their valves. *Pecten* Lam.; *Pedum* Brug.; *Hinnites* Defr.; *Spondylus* L.; *Lima* Bruguière. *Pectens*, the Scollops are edible; they swim by flapping their shells; *Spondylus* is attached by one valve.

The **Dimyidae** are allied here.

Order 4. EULAMELLIBRANCHIATA.

Interfilamentar and interlamellar (vascular) concrescences always present.

The mantle-lobes are connected together either by direct fusion or by the gills; the genital glands have independent external openings.

Sub-order 1. SUBMYTILACEA.

Mantle generally widely open, usually with one fusion; usually without siphons.

Fam. 1. **Carditidae.** A single mantle-fusion; foot keeled, often byssiferous. *Cardita* Bruguière. The families **Astartidae** and **Crassatellidae** are allied here.

Fam. 2. **Cyprinidae.** Foot long and bent: two mantle-fusions; papillose orifices; umbones often spiral. *Cyprina* Lam.; *Isocardia* Lam.

Fam. 3. **Lucinidae.** The external gill sometimes absent; foot vermiform without byssus. *Lucina* Bruguière (Fig. 250 *bis*); *Axinus* Sowerby; *Montacuta* Turton; *Corbis* Cuv.; *Diplodonta* Br.; *Ungulina* Daudet.

Fam. 4. **Erycinidae.** Foot byssiferous or with ventral surface enlarged. *Kellya* Turton; *Lepton* Turton; *Lasaea* Leach.

Fam. 5. **Galeommidae.** Foot with furrow, mantle edges reflected over the shell. *Galeomma* Turton; *Ephipodonta* Tate. *Chlamydoconcha* Dall without adductor muscles, and *Scioberetia* Bernard with only one gill on each side, have an internal shell and are allied here.

Fam. 6. **Cyrenidae.** Foot non-byssiferous; two siphons usually not united; hermaphrodite, viviparous, fluviatile. *Cyclas* Bruguière; *Pisidium* Pfeiffer, siphons united; *Galatea* Brug.; *Corbicula* Mulhf.

Fam. 7. **Unionidae.** Fresh-water mussels. Foot long, compressed, without byssus. *Anodonta* Lamarck (Fig. 257); *A. cygnea* Lam., in ponds; *A. anatina* L., in rivers and brooks; *Unio* Philipsson (Fig. 254); *Margaritana* Schum., river pearl-mussels; *Mycetopus* d'Orb., America. *Mutela* Scop., and *Pliodon* Conrad, Africa, are allied here.

Fam. 8. **Aetheriidae.** Fluviatile forms without foot, usually fixed by one valve. *Aetheria* Lamck., Africa.

Fam. 9. **Dreissenidae.** Foot cylindrical, byssiferous, two siphons. *Dreissena* v. Ben., fresh-water.

Sub-order 2. TELLINACEA.

Mantle well open; gills smooth; siphons well developed, separate; foot compressed, elongated; palps large.

Fam. 10. **Tellinidae.** External gill directed dorsally; siphons much elongated. *Tellina* L.; *Scrobicularia* Schumacher; *Semele* Schum.; *Gastrana* Schum.

Fam. 11. **Donacidae.** External gill directed ventrally; siphons separated. *Donax* L.

Fam. 12. **Mactridae.** External gill directed ventrally; siphons united. *Mactra* L., clams (Fig. 251).

Sub-order 3. VENERACEA.

Gills slightly folded ; foot compressed ; siphons generally short.

Fam. 13. **Veneridae.** Foot tongue-shaped ; siphons more or less united. *Venus* L. ; *Tapes* Megerle ; *Dosinia* Scop. ; *Cytherea* Lam. ; *Lucinopsis* Forb. ; *Venerupis* Lam. The **Petricolidae** (rock-borers) are allied here.

Sub-order 4. CARDIACEA.

Gills much folded ; foot cylindrical, more or less elongated ; generally without siphons.

Fam. 14. **Cardiidae.** Cockles. Foot long, bent, without byssus ; pallial orifices contiguous, with short siphons surrounded by papillae. *Cardium* L. (Figs. 255, 259) ; *Hemicardium* Cuv.

Fam. 15. **Tridacnidae.** Foot short, byssiferous ; pallial orifices apart ; posterior adductor only. *Hippopus* Lam. ; *Tridacna* Bruguière.

Fam. 16. **Chamidae.** Foot short, without byssus ; two adductors ; shell fixed, asymmetrical ; pallial orifices separated. *Chama* Bruguière.

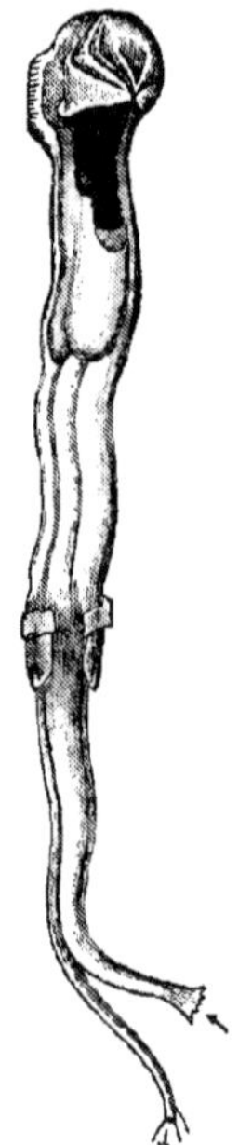

FIG. 271. — *Teredo navalis* removed from its calcareous tube, with elongated siphons (after Quatrefages).

Sub-order 5. MYACEA.

Branchiae much folded ; foot compressed, more or less reduced ; pedal orifice generally small ; siphons well developed.

Fam. 17. **Psammobiidae.** Siphons separated, elongated ; foot large, tongue-shaped. *Psammobia* Lamarck ; *Sanguinolaria* Lam.

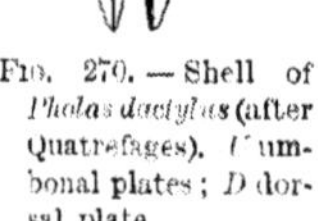

FIG. 270. — Shell of *Pholas dactylus* (after Quatrefages). *U* umbonal plates ; *D* dorsal plate.

Fam. 18. **Myidae.** Siphons united ; foot reduced, without byssus. *Mya* L. ; *Lutraria* Lamarck ; *Corbula* Brug.

Fam. 19. **Solenidae.** Foot strong, elongated, often cylindrical, without byssus ; siphons more or less short. *Solenocurtus* Blain. ; *Solen* L. (razor shell).

Fam. 20. **Saxicavidae.** Rock-borers. Foot small, byssiferous ; pedal orifice very short. *Saxicava* Fleuriau.

Fam. 21. **Gastrochaenidae.** Foot cylindrical, very small, without byssus ; gills narrow. The thin shell is sometimes contained in a calcareous tube secreted by the mantle. *Gastrochaena* Spengler.

Sub-order 6. PHOLADACEA (Boring Mussels).

Foot very short, truncated ; siphons long, united ; without shell-ligament.

Fam. 22. **Pholadidae.** Organs contained in the shell ; with one or several accessory shell-pieces. Gills prolonged into branchial siphon. *Pholas* L. (Fig. 270), *P. dactylus* L., mantle and siphons phosphorescent ; *Pholadidea*

Fam. 23. **Teredinidae.** The branchiae are to a large extent contained in the branchial siphon. Siphonal region vermiform, provided behind with two accessory shell-pieces. Shell very small, covering only anterior part of animal. *Teredo* L., *T. navalis* L. (Fig. 271), shipworm, was the cause of the famous dam-break in Holland at the beginning of last century.

Sub-order 7. ANATINACEA.

Hermaphrodite; ovary and testis with separate orifices; external gill directed dorsally, and without the reflected (outer) lamella.

Fam. 24. **Pandoridae.** Foot tongue-shaped, without byssus. Siphons very short. *Pandora* Bruguière; *Myochama* Stutchbury.

Fam. 25. **Lyonsiidae.** Foot cylindrical, byssiferous; siphons short. *Lyonsia* Sturton; *Lyonsiella* Sars.

Fam. 26. **Anatinidae.** Foot slender, without byssus; siphons long; a fourth pallial orifice. *Anatina* Lam.; *Thracia* Blainville; *Pholadomya* Sowerby.

Fam 27. **Clavagellidae.** Foot reduced, without byssus; siphons long, united; valves continued by a calcareous tube secreted by the siphons. *Clavagella* Lam.; *Aspergillum* Lam. (Fig. 272).

Fig. 272. — Shell of *Aspergillum javanum* (after Adams).

Order 5. Septibranchiata.

With branchial septum.

There are three pallial fusions, two siphons more or less elongated, and two adductors. The gills (Fig. 264 *D*) have the form of a muscular septum, extending from the anterior adductor to the junction of the two siphons and surrounding the foot, with which it is continuous. This septum presents symmetrical orifices.

Fam. 1. **Poromyidae.** Siphons short; foot elongated. On each half of the septum there are several groups of lamellae separated by orifices. Palps well developed. Hermaphrodite. *Poromya* Forbes; *Silenia* Smith.

Fam. 2. **Cuspidariidae.** Siphons elongated, united; foot reduced; palps reduced or absent; branchial septum pierced by isolated symmetrical orifices; sexes separate. *Cuspidaria* Nardo.

The genus *Entovalva* Voeltzkow, is not well enough known for its affinities to be determined. The mantle has a posterior orifice; the foot is large, with a posterior sucker. There is an hermaphrodite gland. It inhabits the oesophagus of a Holothurian from Madagascar.

Class II. SCAPHOPODA* (SOLENOCONCHAE).

Dioecious Mollusca without eyes or heart. The edges of the mantle are fused to form a tube which is open before and behind, and secretes a tubular calcareous shell.

* Lacaze Duthiers, "Histoire de l'organisation et du développement du Dentale," *Ann. Sc. Nat.* 1856-58. Plate, "Ueb. d. Bau u. d. Verwandt. d. Solenoconchen," *Zool. Jahrb. f. Morph.* Bd. 5, 1892. Kowalewsky, "Études sur l'Embryogénie du Dentale," *Ann. du Mus. de Marseille*, t. 1, 1883.

The body is elongated, slightly curved, with a dorsal concavity. The shell is shaped like an elephant's tusk, and is open at both ends. The animal, which is entirely contained within the shell, has a similar shape, and is enclosed by a tubular mantle also open at the two ends. It is attached to the shell by a muscle at the hind end. The mantle is to be regarded as paired folds of the dorso-lateral integument which have undergone concrescence ventrally.

A distinct head is not present, but there is an egg-shaped projection into the mantle-cavity at the front end, at the apex of which is the mouth, surrounded by leaf-like labial appendages. Springing from two lobes at the base of the cephalic prominence are a number of ciliated contractile filaments (*captacula*) which are swollen at their ends. These have been supposed to represent the ctenidia, which are otherwise not represented.

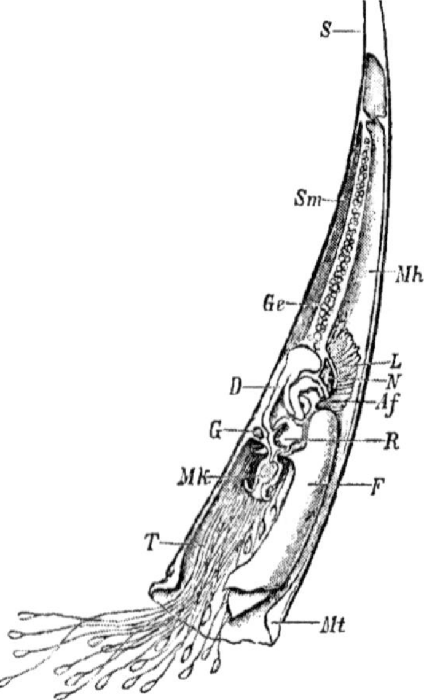

FIG. 273.—*Dentalium* as seen in longitudinal section (except the foot) after Grobben. *S* shell; *Mt* mantle; *Sm* shell muscle; *Mh* mantle-cavity; *F* foot; *Mk* cephalic prominence or oral cone; *T* captacula; *R* radula; *D* intestine; *L* liver; *Af* anus; *G* cerebral ganglion; *N* kidney; *Ge* generative gland.

The **foot** is cylindrical and directed forwards; it can be protruded through the anterior (larger) opening of the mantle. Its front end is trilobed (*Dentalium*), or carries a retractile disc with papillose margins (*Siphonodentalium*), from the centre of which a filiform tentacle arises in *Pulsellum*.

The **nervous system** consists of four groups of ganglia: a pair of *cerebral* with closely adjacent *pleural*, a pair of *pedal*, and a pair of *visceral* ganglia just in front of the anus. The visceral commissure arises from the pleural, and the pleuro-pedal commissure is completely fused with the cerebro-pedal. There is a system of *labial ganglia* (stomatogastric) in connection with the buccal mass: it is connected with the cerebral.

Sense organs. Eyes are absent. A pair of otocysts are placed on the pedal ganglia. The tentacles serve as tactile organs.

The **alimentary canal** is divided into a buccal cavity, oesophagus with two lateral pockets, stomach with large liver, and an intestine, which after several coils closely pressed together, opens behind the foot and the visceral commissure into the middle of the mantle cavity. The buccal cavity is placed in the body at the base of the cephalic projection, and contains a dorsal jaw and a ventral *radula*, which has a short sac and powerful muscles.

The **vascular system** is reduced to two mantle vessels and a complicated system of wall-less spaces throughout the body. There are no specialized **respiratory organs.**

The **excretory organs** are paired and lie in the middle region of the body. They open on either side of the anus.

The *Scaphopoda* are dioecious. The **ovaries** and **testes** are unpaired organs occupying the posterior part of the body behind the liver and intestine. They open into the right kidney.

The animals live buried in mud, and creep about slowly by means of their foot. There are about 100 species scattered in all seas, from the littoral to a depth of about 2000 fathoms. They are known since the Devonian.

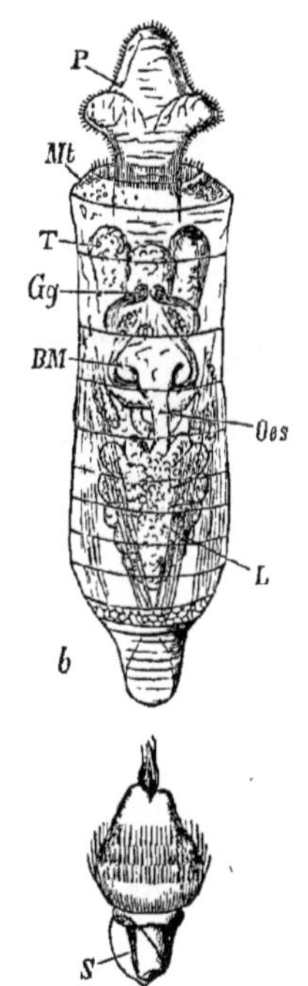

FIG. 274.—Larva of *Dentalium* (after Lacaze Duthiers). *a*, young larva with first rudiment of shell. *b*, older larva seen from the dorsal surface. *T* tentacle-collar; *MT* circular muscle of mantle; *P* foot; *Gg* cerebral ganglion; *Oes* oesophagus; *L* liver; *BM* buccal mass; *S* rudiment of shell.

Development. The eggs are laid singly. There is an invaginate gastrula with a large blastopore, which is at first at the hind end. The embryo elongates and the hinder part of the body grows out behind the blastopore, which becomes the mouth. The free-swimming larva has a preoral region with a ciliated tuft and several circles of cilia, which eventually consolidate into the velum. The mantle arises as two dorso-lateral folds which eventually coalesce ventrally. The shell also is at first bivalve, but subsequently becomes tubular.

The *Scaphopoda* are allied to the Lamellibranchs by their mantle and nervous system, but they possess an odontophore which approximates them to the *Cephalophora*. It is, however, impossible to say that they are more nearly allied to one than to the other.

There are three genera: *Dentalium* L.; *Siphonodentalium* Sars; *Pulsellum* Stoliczka.

In the preceding account the thin end of the animal has been spoken of as posterior; it is, however, possible to regard it as dorsal and as corresponding roughly to the visceral sac of a Gastropod.

Class III. SOLENOGASTRES* (APLACOPHORA).

Symmetrical vermiform animals without mantle-fold, distinct foot, or shell. The integument is provided with a cuticle and calcareous spicules or scales.

It appears that the Class *Amphineura*, which is established by many authors to include the *Solenogastres* and *Chitonidae* is quite unjustifiable; for, whereas the Chitons are clearly Gastropods, it is by no means certain that the Solenogastres are really Mollusca at all. Certainly they are not *Gastropoda*, for they differ from that class in numerous features, of which we may call special attention to the fact, of great morphological importance, that in them the gonad opens directly into the pericardium, a feature found in no Gastropod.

The *Solenogastres* comprise two families, the *Neomeniidae* and the *Chaetodermidae*. They are elongated, vermiform animals with a skin stiffened by the cuticle and spicules; and although in the *Neomeniidae* there is a ventral ciliated furrow, which is supposed to be homologous with the Molluscan foot, it may generally be said of them that they are without a foot. Further, there is neither mantle-fold nor shell; the respiratory structures, when present, cannot be certainly homologised with ctenidia; the alimentary canal is perfectly straight, passing between the anterior mouth and the posterior anus; the blood is red; and finally, the head is but ill marked off from the body, and entirely unprovided with special organs of sense. There is a haemocoelic body-cavity, a coelom consisting of a pericardium, gonad, and two nephridia; the gonad communicates with the pericardium. The reasons for regarding them as allied to the *Gastropoda* are to be found in the presence on the floor of the mouth of a chitinous structure, more or less closely resembling in its relations and appearance the radula; in the arrangement of the central nervous system; in the fact that

* A. A. W. Hubrecht, "*Proneomenia Sluiteri*," *Niederl. Arch. f. Zoologie*, Sup. Bd. 2, 1881. Kowalewsky and Marion, "Contributions à l'histoire des Solénogastres," *Ann. Mus. Marseille* (*Zoologie*), 3, 1889. Pruvot, "Sur l'organisation de quelques Néoméniens des côtes de France," *Arch. Zool. Exp.* (2), 9, 1891. Wiren, "Studien üb. die Solenogastres," I. and II., *Svenska Vet. Akad. Handl.*, 24 and 25, 1892–3.

the body-cavity is a haemocoele and contains blood; and in the presence of a pericardium which communicates with the exterior by a pair of nephridia. With regard to the latter point, however, it is not quite certain that the structure called pericardium is of that nature (it may be merely a conjoined portion of the generative ducts, of which the structures called nephridia constitute the remainder), and it has been maintained that in the *Neomeniidae* at least the organs called nephridia show no signs of being renal in function. And even if the cavity in question is pericardial it differs entirely from the pericardium of *Gastropoda* in the fact that it communicates with the gonad, and the generative cells pass through it on their way to the nephridia, through which they are ejected.

The *Solenogastres* are marine, but not littoral animals. Hitherto they have been found only at considerable depths—from 15 fathoms downwards as far as the abyss, often in association with colonies of *Hydrozoa* and *Actinozoa* on which they appear to feed.

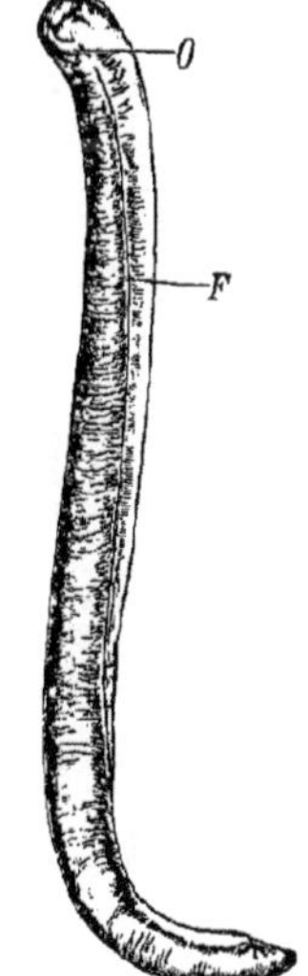

FIG. 275.—*Proneomenia sluiteri* (after Hubrecht). *O* mouth; *F* ventral furrow.

Fam. 1. **Neomeniidae.** *Hermaphrodite Solenogastres with ventral pedal groove, without differentiated liver, with paired nephridial tubes with a common opening.* The body is covered with spicules, which are embedded in the cuticle and in relation internally with epithelial papillae. There is a ventral furrow which is free from spicules; this structure begins in a rather marked ciliated pit, which is placed just behind the mouth and contains the openings of a large mucous gland, and ends behind by passing into the cloacal depression; further, it contains along its floor a ciliated projection, which is supposed to be homologous with the Molluscan foot.

The **mouth** is anterior and ventral, and leads into a buccal cavity which has muscular walls, and is sometimes protractile; it is lined by a thick cuticle, and the ducts of the salivary glands and the sheath of the radula open into it. The latter is absent in *Neomenia* and in certain species of *Proneomenia* and *Dondersia;* elsewhere it bears several transverse rows of chitinous teeth. The oesophagus is short, the stomach tubular and often provided with an anterior dorsal caecum; the intestine is straight, and opens into the cloaca, which also receives the openings of the nephridia and of a mucous gland. The liver is represented by several pairs of short lateral diverticula of the stomach.

The **nervous system** (Fig. 276) consists of a large cerebral ganglion in front of the buccal mass, giving origin to a stomatogastric commissure with two small ganglia and to two cords on each side—the pallial and pedal—passing backwards. The pedal cords are swollen into a ganglion below the oesophagus, and into smaller ganglia at intervals along their course. The swellings are united by commissures. The pallial cords have a ganglion near their origin from the cerebral, and are connected with the pedal by commissures; they unite in a supra-rectal ganglion behind. In *Neomenia* (Fig. 276) these cords are united for a short distance in front, and sometimes they join behind.

There are no special organs of sense.

The **blood corpuscles** contain haemoglobin, which gives the blood a red colour. There are no definite vessels, but there are two more or less marked sinuses—a

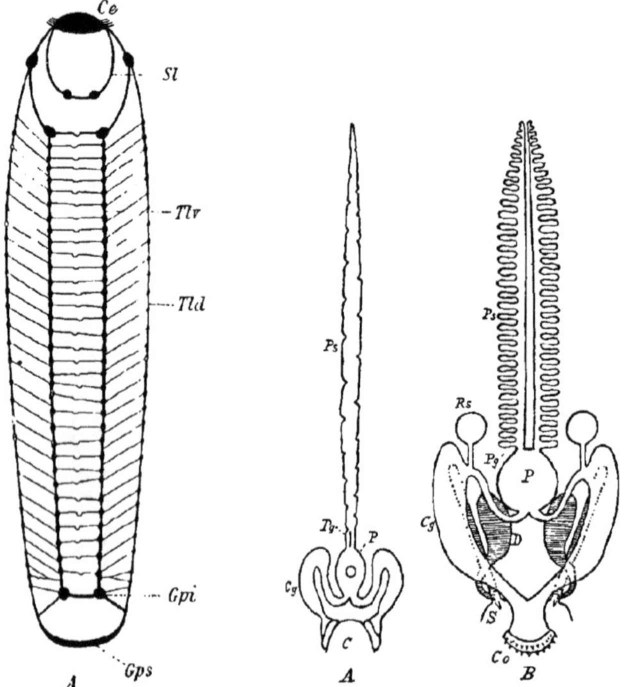

FIG. 276.—Diagram of the central nervous system of *Neomenia carinata* (after Wirén). *Ce* cerebral ganglion; *Sl* stomatogastric ganglia; *Tlv* pedal cord; *Gpi* posterior ganglion on the pedal cords; *Tld* pallial cords; *Gps* supra-rectal ganglion (from Bronn).

FIG. 277.—Diagram of the renal and generative organs and pericardium; *A*, of *Chaetoderma nitidulum*; *B*, of *Neomenia carinata* (after Wirén). *Ps* generative gland; *Pg* opening of generative gland into pericardium *P*; *Cg* nephridium; *Rs* receptaculum seminis; *Co* copulatory organ; *C* cloaca; *S*, *D* accessory glands.

ventral between the intestine and the ventral surface, and a dorsal sinus, the hinder part of which is contractile and supposed to represent the heart. These presumably communicate with the perivisceral cavity, which contains blood. Respiratory organs are supposed to be represented by some epithelial folds of he cloacal wall.

The cavity called **pericardial** (Fig. 277 *B*) is dorsal to the rectum. The heart projects into it dorsally; it communicates with the exterior by two tubes—the so-called *nephridia*—which are bent on themselves like those of so many Mollusca, and open by a common median opening into the cloaca ventral to the anus.

The animals are hermaphrodite, and the **genital glands** are paired and tubular; they open posteriorly into the pericardium.

It is extremely doubtful whether we ought to regard the nephridia as anything else than generative ducts, as they bear accessory organs on their course, and no signs of renal excretory products have been seen in them.

The development has been partly followed in *Dondersia*. There is an invaginate gastrula and a trochosphere with a velum, but no veliger stage nor shell-gland, nor foot.

The *Neomeniidae* have been taken in the North Atlantic and in the Mediterranean. There are 6 genera and about 20 species.

Neomenia Tullberg, with branchiae, without radula; *Paramenia* Pruvot, with branchiae and radula; *Proneomenia* Hubrecht, without branchiae, thick cuticle enclosing epithelial papillae; *Ismenia* Pruvot, cuticle thin, a pre-cloacal ventral prominence; *Lepidomenia* Kow. and Mar., cuticle thin, radula large; *Dondersia* Hubrecht, cuticle thin, radula rudimentary or absent.

Fam. 2. **Chaetodermidae.** *Dioecious Solenogastres without pedal groove, with radula as single horny tooth; nephridia with separate openings; branchiae paired and projecting; gonad median.* Body with uniform covering of short spicules embedded in the cuticle. Hind end of body bell-like, consisting of widely open cloaca, which contains two foliaceous branchiae, the anus and the two openings of the nephridia.

The **nervous system** consists of two closely apposed cerebral ganglia, from which arises a stomatogastric commissure with two ganglia, and on each side two lateral cords, the pallial and pedal, which are joined at intervals by commissures. Posteriorly the pedal joins the pallial, and the single cord thus formed joins its fellow dorsal to the rectum in the supra-rectal ganglion. There are no special organs of sense.

FIG. 278. — *Chaetoderma nitidulum* (from Perrier). *a* anterior end; *b* posterior end.

The mouth is anterior and terminal, the radula is represented by a single horny spine in a pit in the floor of the mouth, the alimentary canal is straight, and the liver is a single ventral caecum directed forwards and opening into the middle region of the alimentary canal.

The **vascular system** is much as in the Neomeniidae except that the haemoglobin is in the plasma.

The **nephridia** (Fig. 277) resemble those of *Neomenia*, but they are without accessory genital structures, are more obviously renal in function, and they open separately.

The **sexes** are separate. The gonad is a single median tube opening into the **pericardium**, and the genital products pass out by the kidneys (Fig. 277 *A*).

The development is unknown.

There is one genus with three species; from the North Atlantic, the Arctic, and the Pacific Oceans.

Chaetoderma Lovèn.

Class IV. GASTROPODA.*

Mollusca with a distinct head generally bearing tentacles, with a ventral muscular foot usually used for creeping, and typically with a continuous mantle-fold. The shell when present is composed of one or of more than two pieces.

The organization of the *Gastropoda* is with one exception asymmetrical, but the asymmetry is confined to the organs of the visceral mass, and rarely, if ever, affects the head. The head has the mouth at its anterior extremity, and usually bears one (*Streptoneura, Thecosomata, Phyllirhoë*, Elysiids, and some Pulmonates) or two (most Opisthobranchs and some Pulmonates) pairs of tentacles, and two eyes, which are placed at the base, sometimes at the apex of a pair of tentacles. The tentacles are contractile, and in the stylommatophorous Pulmonates they are invaginable.

Their form varies and they are often modified, and may even be absent (*Olivella, Homalogyra, Pterotrachea*, etc.); in most of the Bulloids both pairs are widened out and transformed into a quadrangular cephalic shield, the four corners of which correspond to the apices of the four tentacles. The single pair of the *Amphibolidae, Oncidiidae, Siphonariidae*, and *Gadiniidae* are reduced, and give to the head the aspect of a flattened disc. The anterior pair of *Pleurobranchidae, Tritoniidae*, etc., is transformed into a frontal velum. They may also be bifid, flattened or branched. The **labial palps**, which are processes of the lips, found in some forms, are not to be regarded as tentacles; nor are the small lobe-like processes (palmettes) sometimes found between the tentacles. The **pseudopallium** is a process of the cephalic integument which projects back over the shell.

The **foot** is a muscular organ, and projects from the ventral surface of the body. Typically it has a flat sole, and is used for creeping.

* Martini and Chemnitz, *Conchylien Cabinet*, 12 Bde., Nürnberg, 1837–1865. Sowerby, "*Thesaurus conchyliorum, or Figures and Descriptions of Shells*," London, 1832-62. Reeve, "*Conchologica iconica, etc.*," London, 1842-62. H. and A. Adams, "*The Genera of the recent Mollusca*," 3 vols, London, 1858. H. Troschel, "*Das Gebiss der Schnecken*," Berlin, 1856–78. S. P. Woodward, "*Manual of the Mollusca*," ed. 3, London, 1875. V. Hensen, "Ueb. das Auge einiger Cephalophoren," *Z. f. w. Z.*, 15, 1865. J. W. Spengel, "Die Geruchsorgane u. d. Nervensystem der Mollusken," *Z. f. w. Z.*, 35. Souleyet, "*Voyage de la Bonite*," Zoologie, T. 2., 1852. Hilger, "Beiträge zur Kenntniss des Gastropodenauges," *Morph. Jahrb.*, 10, 1885. Willem, "Observations sur la vision et les organes visuels de quelques Mollusques, etc.," *Arch. Biol.*, 12, 1892. Leydig, "Ueb. d. Gehörorgan der Gastropoden," *Arch. f. Mic. Anat.*, 7, 1871. Lacaze Duthiers, "Otocysts ou capsules auditives des Mollusques," *Arch. Zool. Exp. et Gén.* (1), 1, 1872. Houssay, "Recherches sur l'opercule et les glandes du pied des Gasteropodes," *Arch. Zool. Exp.* (2), 2, 1884. Grobben, "Die Pericardialdrüse der Gastropoden," *Arb. Zool. Inst. Wien*, 9, 1890. Baudelot, "Recherches sur l'appareil générateur des Mollusques Gast.," *Ann. Sci. Nat.* (*Zoologie*) (4), 19, 1863. P. Pelseneer, "*Introduction à l'Étude des Mollusques*," Paris, 1897. A. H. Cooke, "Molluscs," *Cambridge Natural History*, London, 1895.

It very generally bears on its hind end a horny or calcareous *operculum* for closing the shell-aperture when the animal is retracted. The form and size of the foot present various modifications according to the condition of life.

In the sedentary Gastropods it is reduced—in *Vermetus* and *Magilus*, which are fixed, to a small discoidal projection; and in *Thyca* and *Stilifer*, which are parasitic, to a small appendage. In the pelagic forms it is flattened laterally as a fin (*Heteropoda*) or may even be absent (*Phyllirhoë*). In the leaping forms (*Strombidae*) it is also flattened.

The creeping surface is sometimes divided by a longitudinal furrow, and the two halves may move alternately (*Cyclostoma*). The two anterior corners of the foot may be prolonged as tentacles.

FIG. 279.—*Helix pomatia.* O eyes at the extremities of the long tentacles; *Pe* foot.

The anterior part of the foot may project beneath the head, forming the *propodium* (burrowing forms); in *Natica* the propodium projects back on to the cephalic region.

The **epipodia** (sometimes called parapodia) are fin-like, produced lateral portions of the foot (in many Opisthobranchs); in *Notarchus* the epipodia fuse over the dorsal surface. In the *Rhipidoglossa* the epipodia are present, and carry papillae. The posterior part of the foot is often marked off from the rest as *metapodium*; in such cases when a propodium is present as well, the middle part is called mesopodium.

The **pedal glands** secrete a mucous substance, which lubricates the surface on which the animal moves, or hardens on contact with air or water into a thread by which the animal suspends itself (*Limax*, *Litiopa*, *Cerithidea*, etc.), or, as in *Ianthina*, becomes entangled with bubbles of air and forms a float to which, in the female, the eggs are attached. The secreting cells are distributed as unicellular glands in the epithelium of the foot, and are often aggregated in special invaginations of the skin. Such are the supra-pedal gland which opens at the front edge of the foot (*Pulmonata*, etc.); the labial glands which open into a furrow at the anterior end of the foot (creeping *Streptoneura* and Opisthobranchs); and the median pedal gland which opens on the ventral surface, and is comparable to the byssus gland of Lamellibranchs (*Cyclostoma*, *Cypraea*, etc.); it was formerly mistaken for an "aquiferous pore." Finally, in some forms there are glands at the hind end of the foot, either dorsal or ventral.

The integument consists of a superficial layer of cylindrical cells which are frequently ciliated; and of a connective-tissue dermis which is inseparably connected with the dermal muscles. Calcareous and pigment glands are placed in the integument; they are specially numerous at the edge of the mantle-fold, where they contribute to the growth and peculiar colouring of the shell.

The **mantle** (pallium) or integument of the visceral sac is thrown into a continuous fold, which completely encircles the body at the junction of the visceral sac with the head and foot. This fold is the mantle-fold; it encloses between itself and the body a space called the mantle- or pallial-groove. The mantle-groove is quite shallow, and groove-like over the greater part of its extent, and in the *Placophora* (Fig. 305 *B*) over its whole extent; but in all other forms it is especially deepened at one point. This specially deep part of the mantle-groove is the mantle-cavity proper. The mantle-cavity contains the ctenidium (in the *Zygobranchiata* the two ctenidia) when there is a ctenidium, and generally the anus, renal orifice, and generative opening.

The free edge of the mantle-fold is thickened and may be provided with short tentacles, pigment spots, and glands. In the *Zygobranchiata*, and in one or two other forms, there is a slit along the roof of the mantle-cavity in the mantle-fold. This slit, the edges of which may be fused at one or more points, allows of the exit of the spent water from the mantle-cavity.

The **siphon** (Fig. 280, *22*) is a kind of spout-like continuation of the mantle-edge on the left side of the mantle-cavity. It occurs in many, principally carnivorous, Streptoneura, and it conducts water into the mantle-cavity.

The **hypobranchial gland** (Fig. 280, *17*) is a highly glandular and generally folded portion of the mantle-lining between the two gills and the rectum (Zygobranchs), or between the single gill and the rectum (most other Gastropods).

In the typical *Gastropoda*, such as the whelk (*Buccinum*) or the garden-snail (*Helix*), the visceral sac is covered by a univalve shell. This shell has, to a certain extent, the same shape as the visceral sac (Fig. 312), and can usually completely receive and protect the head and foot when the animal is retracted. As a rule it is hard and calcareous, consisting of the three layers, an outer periostracum, a middle prismatic or porcellanous, and an internal nacreous layer: the nacreous layer is not, however, always present in Gastropods, and the porcellanous is of complex structure, consisting often of three layers of laminae, which are themselves composed of prisms. Sometimes it is a delicate structure, horny and flexible, or it may have a gelatinous or cartilaginous consistence, as in the *Cymbuliidae*, where, however, it is not a mantle-shell, but a subepidermic dermal product.

The shell is secreted by the epidermis of the mantle; the epidermic (periostracum) and porcellanous layers by the edge, and the nacreous layer by the whole surface. In cases in which the nacreous layer is absent, the mantle surface still has a shell-forming

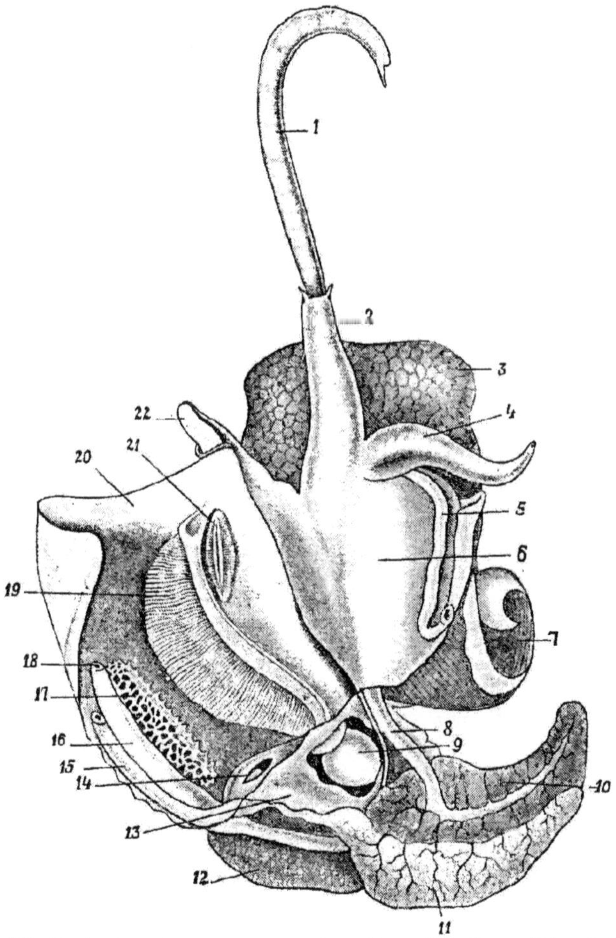

Fig. 280.—*Pyrula tuba*, male, removed from the shell. The mantle has been cut through along the right side of the mantle-cavity and turned over to the left. The pallial organs are accordingly reversed (after Souleyet from Lang). *1* proboscis; *2* head; *3* foot; *4* penis; *5* vas deferens cut and continued in the roof of mantle-cavity at *15*; *6* floor of the mantle-cavity; *7* spindle muscle; *8* intestine; *9* heart in the opened pericardium; *10* liver; *11* testis; *12* and *13* kidney; *14* kidney opening; *15* vas deferens; *16* rectum; *17* hypobranchial gland; *18* anus; *19* ctenidium; *20* roof of mantle-cavity; *21* osphradium; *22* siphon.

power, for lesions are repaired by a cement-like substance secreted by the mantle.

The shell consists of one piece, except in the *Chitonidae* (Fig. 305) in which there are eight pieces, and it usually has the same shape as the visceral sac. When that is flat or conical, the shell is also flat and conical (*Patella, Fissurella*); when the visceral sac is spirally coiled, the shell also is spirally coiled; and when the mantle-fold has a slit extending along the roof of the mantle-cavity, the shell has a corresponding slit or perforation (*Fissurella, Haliotis*). Finally, prolongations of the edge of the mantle, such as the siphon, leave their mark upon the lip of the shell-aperture (*Siphonostomata*).

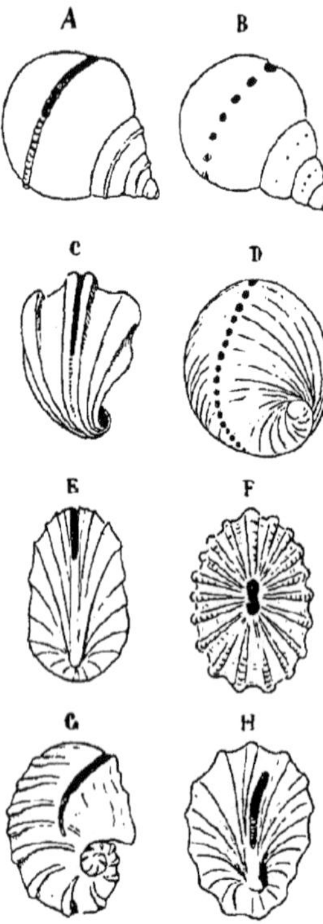

Fig. 281.—Shells of *A Pleurotomaria, B Polytremaria, C* and *E Emarginula, D Haliotis, F Fissurella, G* and *H* stages in the growth of the shell of *Fissurella* (from Lang).

The visceral sac of Chiton is not prominent and never coiled. The shell-pieces of this animal are partially covered by upgrowths of the mantle, and may be entirely covered by them (*Cryptochiton*). The shell-beds so formed have been compared by Lankester to the persisting shell-sac (in this case multiple) of the larva.

The visceral sac and shell are almost always coiled to a certain extent in the young or larva, even if not in the adult. The coils may be lost by *decollation*, as in *Caecum*, in which the spiral part drops off; or by modification during growth as in *Fissurella* (Fig. 281, *F, G, H*), or by subsequent addition during growth (by secretion from two reflected lobes of the mantle) of calcareous matter, which overlies the spiral shell and hides it, as in the cowries (*Cypraea*).

The spiral is generally *dextral* (leiotropic), *i.e.*, if the shell is placed with its spire uppermost and the aperture towards the observer, the aperture is to the right; or it may be *sinistral*. When it is sinistral the organization is sometimes affected, and the organs and apertures usually placed to the right are on the left side. This is always the case with sinistral monstrosities, but not always when the species is normally sinistral. In *Spirialis, Limacina*,

Lanistes, and a few other forms the shell is sinistral, but the organization is dextral; such shells are said to be *ultra-dextral* (Pelseneer). In some cases both sinistral and dextral species are known in the same genus.

As a general rule the whorls of the spire are closely applied together, the line or groove of apposition being called the *suture*. Sometimes, however, the whorls are more or less separate (*Scalaria*, *Cyclostoma*, *Valvata*, *Vermetus*). The shell is never multilocular, but sometimes the animal ceases to occupy the upper whorls, which drop off (*decollation*). The axis round which the whorls are coiled is called the *columella*; it may be hollow, in which case the shell is said to be *perforated* or *umbilicated* (*Solarium*) and the axial tube is called the *umbilicus*; the umbilicus may be shallow or deep. In *Natica* the umbilicus is filled up with calcareous matter. In other shells the whorls are closely coiled, and the *columella* is a central pillar; such shells are *imperforate*. In addition to the above the following terms are applied to the shell: the *apex* is the top of the spire—the first formed part of the shell; the *aperture* is the opening of the shell at the end of the last, or body-whorl. The aperture is entire, as in most vegetable feeders (*holostomatous*), or produced at its anterior end into a spout (*canal*), as in many carnivorous families (*siphonostomatous*). Sometimes there is a posterior canal as well, this is anal in function, and in some forms is represented by the so-called *slit* (*Emarginula*, *Pleurotomaria*, *Scissurella*). The slit lies over a slit in the mantle (see above). In *Fissurella* (Fig. 281) it becomes in the adult an apical pore, and in *Haliotis* a number of pores.

The margin of the aperture is called the *peristome*, the right side of which is the outer lip or *labrum*, the inner or columellar lip being the *labium*. The outer lip is sometimes thin, more often thickened, or reflected or inflected (*Cypraea*), or expanded (*Pteroceras*) or fringed with spines (*Murex*). When the fringes or expansions of the outer lip are formed periodically during the growth of the shell they are called *varices*.

Shells which are always covered by the mantle are colourless (*Limax*, *Parmophorus*). Those which are covered by the mantle when the animal expands acquire a glazed or enamelled surface like the cowries. When the shell is deeply immersed in the foot of the animal it becomes partly glazed (*Cymba*). In all other cases there is an epidermis or periostracum, though it is often thin and transparent.

In some cases the parts of the shell separating the successive whorls and the columella are absorbed in the adult, *e.g.*, *Nerita*, *Olivella*, *Cypraea*, *Auriculidae*; in such cases the visceral sac loses its spiral form.

The **operculum** is attached to the dorsal hind end of the foot. It consists of a horny basis which may be calcified. It grows with the shell, and the oldest part of it is called the *nucleus*. It is marked by lines of growth, which may be concentric or spiral (sinistral in dextral shells). Sometimes it is claw-shaped or *unguiculate*, in which case the nucleus is apical. It generally fits the mouth of the shell; but is sometimes too small, or even ridiculously inadequate for this purpose, as in *Bullia*, *Conus*, etc.; it also persists in some limpet-like forms which adhere by their flat foot, and in which it is not used (*Navicella*, *Concholepas*, *Sigaretus*). In such cases the operculum affords a good instance of an organ persisting after it has entirely lost its original function. The operculum is present in most adult *Streptoneura*, and in their larvae, if absent in the adult (except *Stilifer*); it is absent in the adults of most *Euthyneura* (except *Actaeon*, *Limacina*, *Amphibola*), but is present in their development except in some of the more specialized, *e.g.* *Pulmonata* (except *Auriculidae*, *Siphonaria*, *Gadinia*) and *Cavoliniidae*. It may be present or absent in the same genus (*Stomatella*, *Vermetus*, *Conus*), and may even be caducous in adults (*Limacina helicina*).

The edges of the mantle are often folded over the shell, so as to cover a part of it (many *Fissurellidae*, *Marsenina*, many *Cypraeidae* and *Marginellidae*, *Aplysia*, and some *Bullidae*, various Pulmonates), or the whole of it. In the latter case the edges fuse, and the shell is enclosed in a sac. Such an internal shell is found in *Cryptochiton*, *Pupilia*, most *Lamellariidae*, *Pustularia*, *Notarchus*, *Doridium*, *Gastropteron*, *Philine*, *Pleurobranchus*, and some species of *Limax*, and is very generally much reduced in size, and quite incapable of receiving the head and foot in retraction. Finally the shell and its sac may be absent, and the visceral sac become secondarily symmetrical (*Titiscaniidae*, *Pterotrachea*, *Runcina*, Gymnosomatous Pteropods and *Cymbuliidae*, *Pleurobranchaea*, *Nudibranchiata*, *Philomycidae*, *Onchidiidae*, *Vaginulidae*). In such cases the shell exists only in the larva (in some Pulmonates it is never formed at all), and disappears at the end of larval life. In most cases in which there is no shell the pallial chamber (and groove) and ctenidium are reduced or absent; *e.g.* in Nudibranchs where the ctenidium is said to be replaced by the papillae of the mantle (*i.e.* of the dorsal integument) called **cerata** (Fig. 328). In certain cases in which the larval shell is shed, another persistent and internal shell is formed (*Lamellaria*, the first spiny shell of which is called *Echinospira*). The attachment to the shell is effected by the *columellar* or *spindle muscle* (Fig. 296), which arises from the foot and is inserted into the columella in the spiral forms, or when there is no columella to the internal surface of the shell in a horse-shoe-shaped line (*Patella*), or over an oval area (*Haliotis*, Fig. 296 *S*).

The central **nervous system** consists typically of the three pairs of ganglia in the head (Fig. 282)—the cerebral supplying the head and sense-organs, the pedal supplying the foot, and the pleural which innervate the mantle and spindle muscle; and of two commissures which are completed ventrally to the gut and contain ganglia in their course; these are the anterior or stomatogastric commissure, on which are developed the buccal ganglia, and the posterior or visceral, on which may be developed the supra- and sub-intestinal ganglia and one or more abdominal ganglia. The stomatogastric commissure is connected with the cerebral ganglia, and supplies the buccal mass, and alimentary canal, while the visceral commissure is connected with the pleural ganglia, and innervates the vascular, excretory, and reproductive organs, and also gives off from the supra- and sub-intestinal ganglia nerves to the mantle, gills and osphradium. In all *Gastropoda* except the

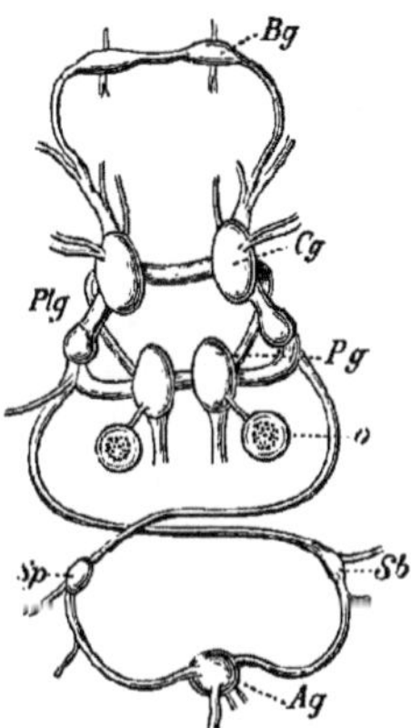

Fig. 282.—Nervous system of *Paludina* (after v. Jhering). *Cg* cerebral, *Plg* pleural, *Pg* pedal, *Sp* supra-intestinal, *Sb* sub-intestinal, *Ag* abdominal ganglion; *o* otocyst; *Bg* stomatogastric (buccal) ganglion.

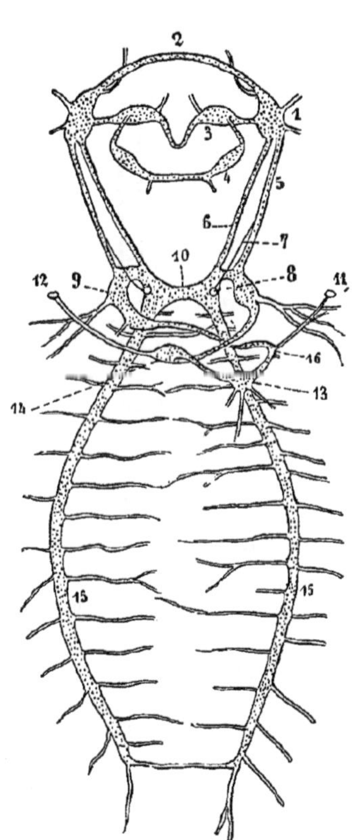

Fig. 283.—Nervous system of *Patella* (from Lang, combined after Pelseneer and Bouvier). *1* cerebral ganglion; *2* cerebral commissure; *3* labial ganglion; *4* buccal ganglion; *5* cerebro-pleural commissure; *6* cerebro-pedal commissure; *7* nerve to otocyst *8*; *9* pleural ganglion; *10* pedal commissure; *11* right, *12* left osphradium; *13* visceral ganglion; *14* supra-intestinal ganglion; *15* pedal cord; *16* indication of a sub-intestinal ganglion.

Chitonidae the visceral commissure is asymmetrical, in connection with the asymmetry of the organs which it supplies.

In the arrangement which is usually described as primitive the cerebral ganglia are placed at the sides of the oesophagus, and are connected by a long commissure; the pedal ganglia are drawn out into long ganglionic cords (Fig. 283, *15*, Fig. 284, *pe*) extending along the ventral surface (Aspidobranchs, *Paludina*, *Cyclophorus*, *Cypraea*), and connected in a ladder-like manner by numerous commissures; the pleural ganglia are approximated to the pedal; and the visceral commissure is of considerable

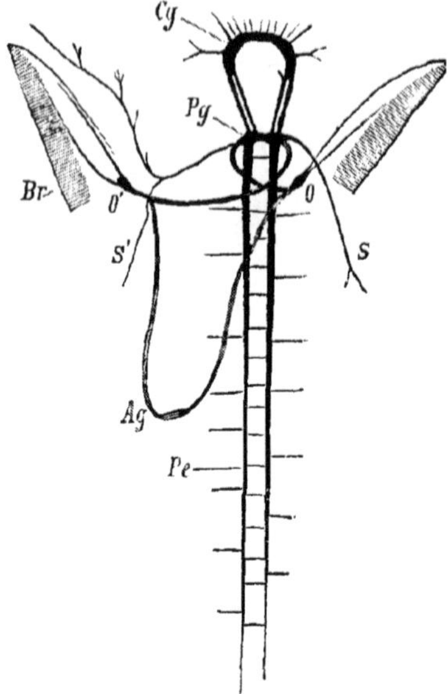

FIG. 284.—Nervous system of *Haliotis* (diagrammatic after Spengel). *Cg* cerebral ganglion; *Pg* fused pleural and pedal ganglia; *Ag* abdominal ganglion; *O* and *O'* osphradia; *Pe* pedal cord; *S* and *S'* pallial nerves; *Br* gills.

length, and its ganglia are widely separate (*Streptoneura* and the less specialized *Euthyneura*).

The principal variations in this arrangement are as follows: the cerebral ganglia are close together, and the pleural are approximated to them or may even fuse with them (most Pectinibranchs, *Pteropoda Thecosomata, Actaeon*); the pedal ganglia are concentrated and not drawn out into cords; the visceral commissure is short, and its ganglia approximated both to each other and to the pleural (Fig. 285, most *Euthyneura*); finally, all the ganglia (cerebral, pedal, pleural, and ganglia of visceral commissure) may be closely approximated on the dorsal side of the oesophagus (many Nudibranchs).

The visceral commissure is in the *Streptoneura* twisted into a figure of 8 in the following manner (Figs. 282, 283); the commissure from the right pleural ganglion passes dorsal to the alimentary canal to the left side, and there forms a ganglion—the *supra-intestinal* ganglion (*sp*, *14*)—which supplies the left osphradium and ctenidium and left side of the mantle,

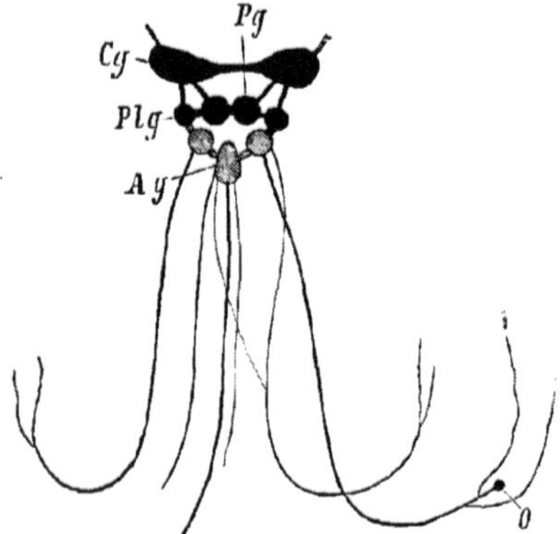

FIG. 285.—Nervous system of *Limnaea* (after L. Duthiers). *Cg* cerebral, *Pg* pedal, *Plg* pleural, *Ag* abdominal ganglion; *O* osphradium.

or without forming a ganglion gives off a strong branchial nerve, which passes to a ganglion —the branchial ganglion, close under the osphradium; while the commissure from the left pleural ganglion passes ventral to the alimentary canal to the right side, and there gives rise to the *sub-intestinal* ganglion (*sb*, *16*), which supplies the right side, or gives off a nerve to the right branchial ganglion. The part of the commissure connecting these two ganglia generally contains one or more purely visceral ganglia. The supra-intestinal ganglion supplies the left osphradium, which is generally present, while the sub-intestinal supplies the right osphradium, which is absent in all except the Zygobranchiate forms.

The nervous system has not been seen in the adult of *Entoconcha* and *Entocolax*.

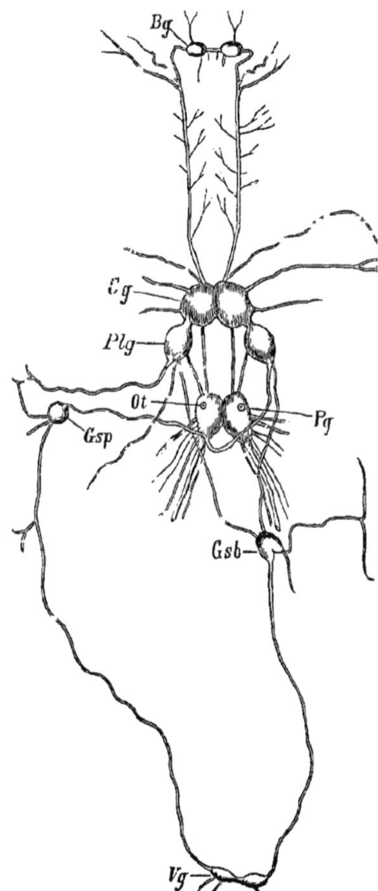

FIG. 286.—Nervous system of *Cassidaria* (after Haller), illustrating dialyneury on the left side, and zygoneury on the right. The pallial nerves from the left pleural ganglion *Plg* anastomose with the pallial nerves from the supra-intestinal ganglion *Gsp* (dialyneury), whereas the pallial nerve from the right pleural (not lettered) runs to the subintestinal ganglion *Gsb* (zygoneury), from which all the mantle-nerves of the right side arise. *Cg* cerebral; *Plg* left pleural; *Pg* right pedal; *Gsp* supra-intestinal; *Gsb* subintestinal; *Vg* abdominal ganglion; *Ot* otocyst.

In the *Streptoneura* each side of the mantle is usually innervated by a mantle-nerve from the pleural of the same side, and by a mantle-nerve from the supra- (left side of mantle) or sub-intestinal (right side of mantle) ganglia, or, if they are absent, from the corresponding parts of the visceral commissure. The nerves from these two sources generally anastomose in the mantle (**dialyneury**). This arrangement is sometimes modified (more often on the right side), in that the mantle-nerve of the pleural passes direct to the ganglion of the visceral commissure of its own side (supra- or subintestinal), from which ganglion all the mantle-nerves of that side appear to be given off (**zygoneury**, right side of Fig. 286).

In some *Streptoneura* there is, in addition to the stomatogastric commissure, a **labial commissure** (Fig. 283, *3*) given off from the cerebral ganglia and passing ventrally to the oesophagus.

Sense organs. Tactile organs are represented by the tentacles, the edges of the lips, which are often folded (labial palps), and tentacular and lobe-like prolongations which are found here and there on the head, mantle, and foot (see these organs).

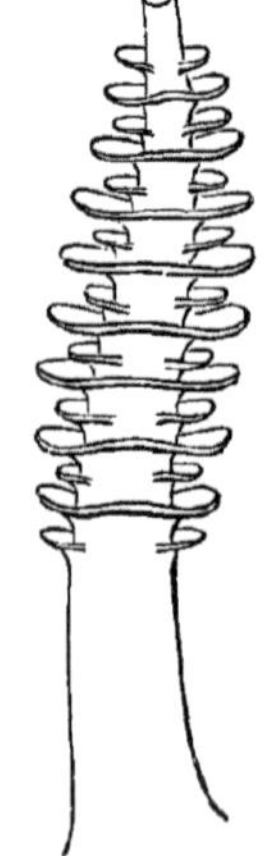

Fig. 287.—Lamellar rhinophore of *Eolis coronata* (from Perrier, after Alder and Hancock).

The **rhinophores** are the organs which subserve the olfactory sense; they are placed upon the cephalic tentacles, and on the posterior tentacles when there are two pairs. They are epithelial structures, and may be localised in a patch of high epithelium at the end of the tentacles, or in a pit, the walls of which may be even folded (some Opisthobranchs). The olfactory nerve arises from the cerebral ganglion, sometimes in common with the optic, and may have a ganglion on its course. The olfactory sense is well developed in most Gastropods.

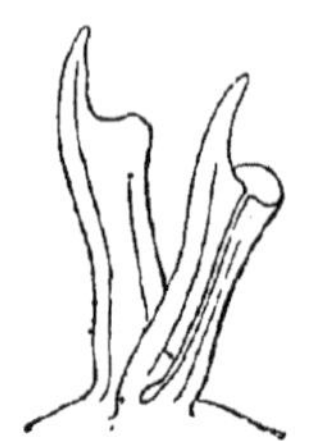

Fig. 288.—Rhinophore of *Hermaea bifida* (from Perrier, after Alder and Hancock).

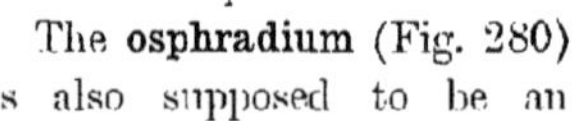

The **osphradium** (Fig. 280) is also supposed to be an olfactory organ. It lies in the mantle-cavity close to the ctenidium and consists of a special patch of epithelium generally columnar and ciliated, and frequently placed upon a special ganglion (Fig. 284, branchial ganglion, see above). It is innervated by a nerve from the visceral commissure (supra-intestinal ganglion—if it is present). In some forms its surface is much folded, and it has the appearance of a bipectinate ctenidium (Fig. 280), for which it was formerly mistaken (*e.g.*, *Natica*, *Buccinum*, *Cypraea*, *Strombidae*, etc.); in some *Euthyneura* it is invaginated into the ganglion, and has the form of a pit. In the Zygobranchiate forms there are two osphradia—the right one supplied by the sub-intestinal, and the left by the supra-intestinal ganglion, or from the part of the visceral commissure where these ganglia would be if present. In the *Chitonidae* there is an osphradium at the base of each ctenidium. In *Fissurella* it is only present in a diffuse and indistinct form. It is also found in some forms in which the ctenidium is absent (*Patella*, *Clione*, etc., *Basommatophora*).

Otocysts which are supposed to be auditory in function exist in most gastropods (absent in *Ianthina*, *Vermetus*) as a pair of closed sacs, which are lined by an epithelium bearing cilia and sensory hairs, and contain concretions and a fluid. They are usually placed in the foot near the pedal ganglia (though innervated from the cerebral by way of the cerebro-pedal commissure, Fig. 283), but in the pelagic forms (*Heteropoda*) and in most Nudibranchs they are near the cerebral ganglia (Fig. 315). They contain a fluid, and one large concretion—the *otolith* (as in the Ctenobranchiate *Streptoneura*, and a few Opisthobranchs), or numerous small ones—the *otoconia* (Aspidobranchs, most *Euthyneura* and dialyneuric *Taenioglossa*). It has not been proved by actual observation that any gastropod possesses the sense of hearing.

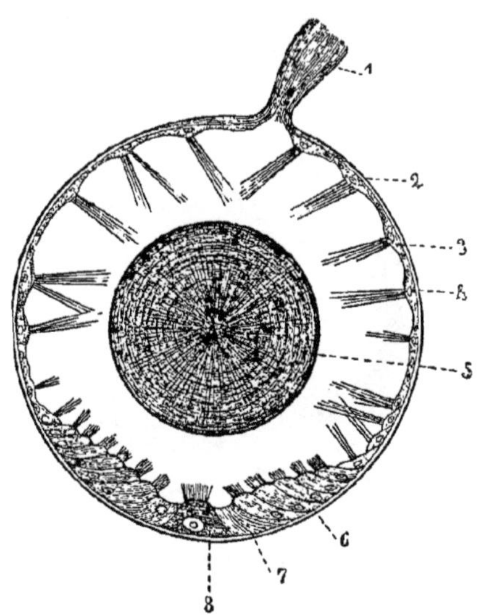

Fig. 289.—Otocyst of *Pterotrachea* (after Claus). *1* auditory nerve; *2* structureless membrane; *3* and *4* ciliated cells; *5* otolith; *6* hair cells; *8* central cell (from Lang).

Eyes. There is a pair of cephalic eyes in almost all gastropods. They are placed at the base of the tentacles (of the second pair in Opisthobranchs) or at the apex of the posterior tentacles (*Stylommatophora*), or half-way along the tentacles (some *Streptoneura*). They consist in their simplest form of a widely open pit of the skin (Fig. 290 *A*), the epithelium of which is pigmented and carries on its free surface a layer of rods (*Docoglossa*); in some of the *Rhipidoglossa*, *e.g.*, *Haliotis*, *Trochus*, the lips of this pit are approximated and its cavity filled with a cuticular lens (Fig. 290 *B*); finally, the pit may be closed and the retinal epithelium continued in front of the lens as a kind of internal cornea, thus lining a complete vesicle, while the outer epithelium, together with some connective tissue, is continued over the eyes as a cornea proper (Fig. 290 *C*). The optic nerve enters the retina on its internal side, and the percipient elements are on the side of the retina next the light.

The cephalic eyes are vestigial, being buried in the skin or absent, in most burrowing forms; absent in most abyssal and subterranean species, and in the *Chitonidae*; and, curiously enough, absent in

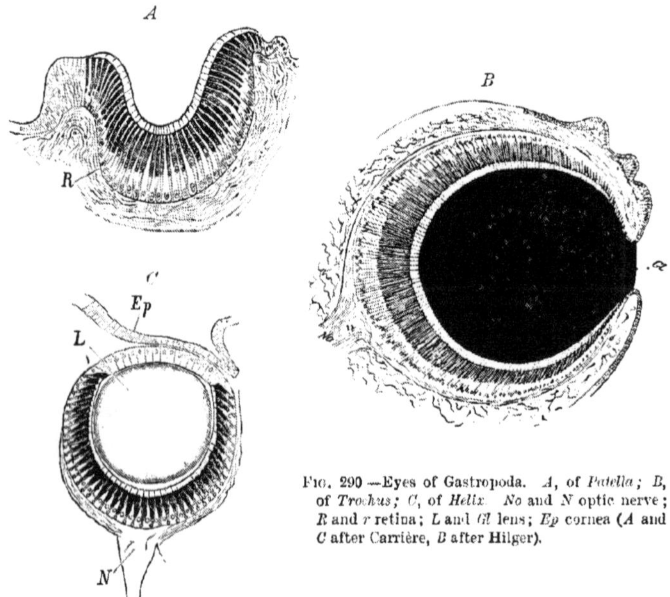

FIG. 290.—Eyes of Gastropoda. *A*, of *Patella*; *B*, of *Trochus*; *C*, of *Helix*. *No* and *N* optic nerve; *R* and *r* retina; *L* and *Gl* lens; *Ep* cornea (*A* and *C* after Carrière, *B* after Hilger).

some pelagic forms, *e.g.* *Ianthina*, many Pteropods; though the *Heteropoda* possess the best developed eyes of the class.

Eyes are found in considerable numbers on the dorsal surface in the *Onchidiidae*,* in addition to the cephalic eyes. These pallial eyes are constructed upon the so-called Vertebrate type, *i.e.* the optic nerve perforates the retina, spreads out on the surface turned towards the lens, and then passes inwards to the percipient elements which are on the side of the retina turned inwards. Moreover, they possess

* Semper C., "*Forms of Animal Life.*"

a cellular lens. Pallial eyes* are also found in certain genera of the *Chitonidae* on the shell pieces; they possess a calcareous cornea.

The **alimentary canal** is usually coiled, and, as a rule, bends forward to open in front on the right side, into the mantle-cavity if that is present. In *Chitonidae* it opens into the mantle-cavity in the middle line behind, and in some Nudibranchs the anus is median, dorsal, and posterior.

Most of the more specialized forms possess an invaginable proboscis (Fig. 280), the invagination generally beginning at the base; but some possess one which is retractile from the point.

In the *Naticidae* there is a glandular disc on the ventral face of the proboscis, the function of which is to assist the animal in perforating the shells of bivalves. In the *Pneumodermatidae* the proboscis bears some suckers, which are either isolated or placed upon two retractile lobes (Fig. 321).

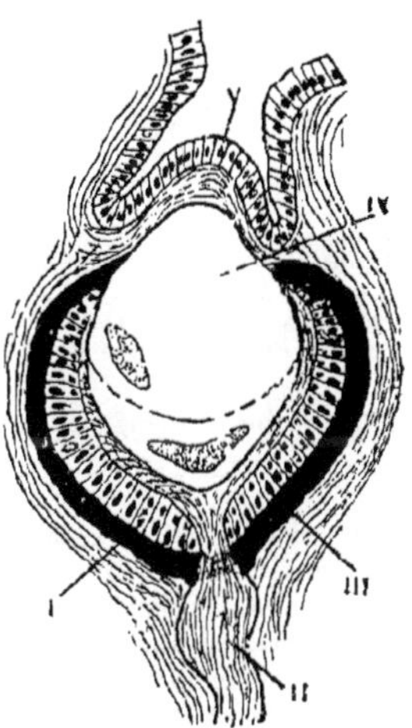

FIG. 291.—Axial section of the eye of *Onchidium* (from Pelseneer after Semper). *I* retina; *II* optic nerve; *III* pigment; *IV* lens; *V* cornea.

The buccal cavity is provided in its anterior part with a pair of horny jaws, which may be closely approximated dorsally (*Natica*), or even fused. In *Patella*, *Aegirus*, and all Pulmonates there is only a single dorsal jaw, and in many forms the jaws are absent entirely.

The walls of the buccal-cavity are much thickened owing to the muscles of the jaws and odontophore, and to the cartilages of the latter, and give rise to the structure called the **buccal mass**.

There is no lower jaw, but on the floor of the buccal-cavity there is a ridge, partly muscular, partly cartilaginous, which has received the same name, though it should rather be regarded as a tongue: this is the **odontophore** (Fig. 292). The surface of this tongue is covered by a chitinous or horny membrane, called the lingual ribbon, or *radula*, on which are carried horny teeth of various forms. Behind, the radula passes into a cylindrical pocket, the so-called *sheath* of the radula, which projects as a small, generally papilla-like prominence from the ventral and posterior end of the buccal mass. The radula and its teeth

* Moseley, H. N., *Q. J. M. S.*, 1884.

are secreted by the epithelium of the radula sheath, which may be compared to the bed of a nail, or to the persistent pulp of a tooth; for as the front part of the radula is continually being worn away by use, it is replaced from behind by the new growth in the sheath.

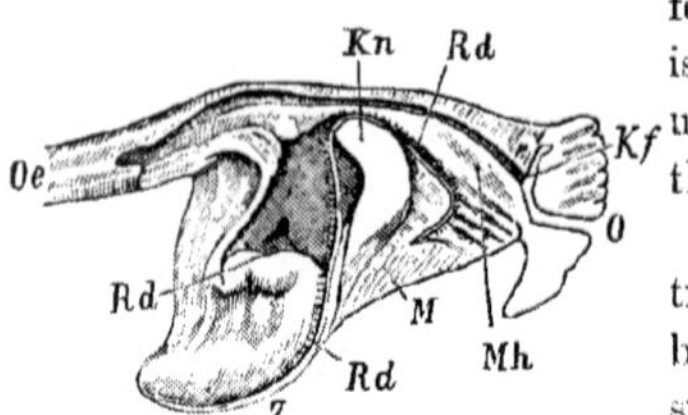

Fig. 292.—Longitudinal section through the buccal mass of *Helix* (after Keferstein). *O* mouth; *Mh* buccal cavity; *M* muscles; *Rd* radula; *Kn* lingual cartilage; *Oe* oesophagus; *Kf* jaw; *Z* sheath of radula.

The odontophore can be protruded and retracted as a whole by its muscles, and there is possibly a small power of movement of the ribbon itself on the subradular membrane.

The teeth are arranged in transverse rows on the surface of the radula; their form, size, and number vary in different species, and afford important systematic characters of species, genera, and families.

The middle tooth of each row is called the *central* or *rachidian* tooth; the teeth on each side of this are called *laterals*, while the outermost of all are the *marginals* or *uncini*. The laterals and marginals sometimes merge into one another, so that the distinction between them is lost. The varying arrangement in number of these different teeth is expressed by formulae: thus the typical formula of the *Toxoglossa* is 1.0.0.0.1., which means that one

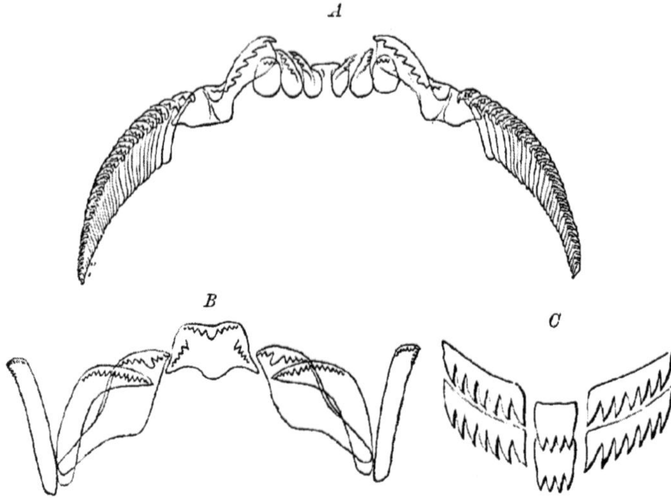

Fig. 293.—One row of teeth of the radula of *A*, *Helicina tropica*, Rhipidoglossa; *B*, *Bythinia tentaculata*, Taenioglossa; *C*, two rows of *Fasciolaria tarentina*, Rachiglossa (from Perrier).

marginal tooth alone is present, and that central and laterals are absent altogether; in the *Rachiglossa* (Fig. 293 *C*) it is 1.1.1., which means that a central and one lateral on each side are present; the teeth are strongly cusped in this group. In the *Taenioglossa* (ribbon-tongued) the typical formula is 2.1.1.1.2. (Fig. 293 *B*). In the *Ptenoglossa* there are an indefinite number of outer teeth, not distinguishable into marginals and laterals, and the formula is written ∞. 1. ∞. The *Rhipidoglossa* (fan-tongued) are characterized by an indefinite number of marginals, arranged like the ribs of a fan; typical formula ∞. 5. 1. 5. ∞. (Fig. 293 *A*). The *Docoglossa* have a few, but strong columnar teeth, and the special feature is the multiplication of the central; thus *Patella* has 3.1.4.1.3., or, as it may be written, 3.1. (2+0+2) 1.3.

In the *Euthyneura* there is generally no distinction into marginals and laterals. The length of the radula, and the number of rows of teeth, varies much in different species. In *Patella* it is very long, in the *Pulmonata* it is short and broad. The radula is entirely absent in the *Eulimidae*, *Pyramidellidae*, *Thyca*, *Entoconcha*, *Entocolax*, *Coralliophilidae*, certain species of *Terebra*, *Tornatinidae*, *Cymbuliopsis*, *Gleba*, *Dorididae*, *Doridopsis*, *Corambe*, *Phyllidia*, *Tethyidae* (*i.e.*, in parasitic and suctorial forms).

The subradular organ is a papilla in front of the radula on the floor of the mouth.

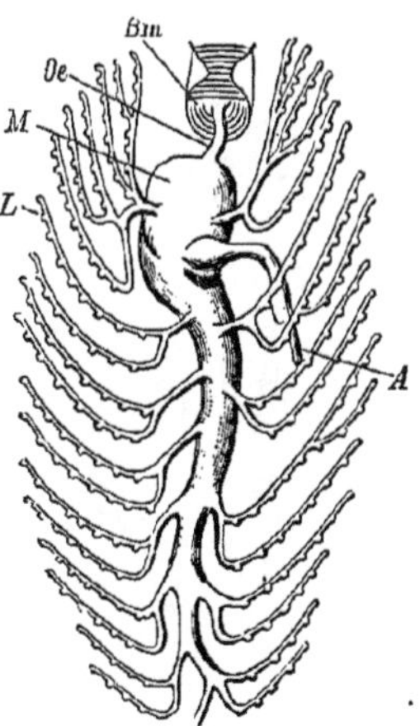

FIG. 294.—Alimentary canal of *Aeolis papillosa* (after Hancock). *Bm* buccal mass; *Oe* oesophagus; *M* stomach; *L* liver caeca which enter the dorsal appendages (cerata): *A* anus.

In addition to buccal glands, sometimes found round the buccal opening, there is always a pair of salivary glands opening into the buccal-cavity. The buccal-cavity leads into the oesophagus, which is followed by a dilated stomach, usually provided with a caecal appendage. The stomach opens into an intestine, which is usually long and coiled, and surrounded by a large multilobed liver. The liver occupies nearly all the upper coils of the visceral sac, and pours its secretion into the intestine and into the stomach (Fig. 295).

The arrangement of the alimentary canal and of the liver presents in detail many essential modifications; one of the most remarkable being that offered by the stomach with its diffuse hepatic caeca (often extending into the cerata) in so many of the *Nudibranchiata* (Fig. 294). The terminal part of the intestine is distinguished by its size, and may be called the rectum.

The oesophagus may have a crop-like swelling in its course, and sometimes there is a gland on about the middle of its length (gland of Leiblein, Fig. 295). The stomach may contain in its middle region masticatory plates (some Opisthobranchs), or be divided by constrictions into three regions (*Aplysia*).

The **vascular system** presents numerous and important modifications. The heart is enclosed in a special pericardium, and is usually placed on one side of the middle line, more or less anteriorly, near the respiratory organs. It usually consists of a conical ventricle (Fig. 280) which gives off the aorta, and of an auricle which is turned towards the respiratory organs, and into which the blood passes by the respiratory veins. In some Gastropods (*Rhipidoglossa*, except *Helicina*) there are two auricles (the right one being usually the smaller), and the ventricle is pierced by the rectum (Fig. 297 *b*, *c*). In the *Chitonidae* alone is the heart not only symmetrical (with two symmetrical auricles), but placed at the hind end of the animal.

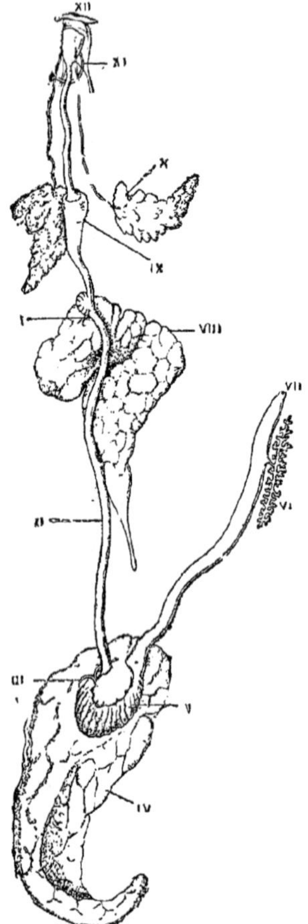

FIG. 295.—Enlarged dorsal view of the alimentary canal of *Murex* (from Pelseneer, after Haller). *I* duct of gland of Leiblein; *II* oesophagus; *III* duct of liver; *IV* liver; *V* stomach; *VI* anal gland; *VII* anus; *VIII* gland of Leiblein; *IX* crop or proventriculus; *X* salivary gland; *XI* radula; *XII* mouth.

When there is only one auricle it is the left, though in the majority of the *Streptoneura* and of the *Pulmonata*, and in *Actaeon*, *Limacina*, *Clio virgula* and *C. acicula* it lies in front of the ventricle, and in most Opisthobranchs, in *Testacellidae*, *Onchidiidae*, *Firolidae*, and some *Calyptraeidae* it is behind the ventricle.

The aorta, which is given off from the end of the ventricle opposite to the auricle, or from the hind end in the *Rhipidoglossa* with two auricles, usually divides into two arteries, of which one passes forward and branches to the head and foot, while the other passes dorsalwards to the viscera. There is a bulb, extra- or intrapericardial, on the root of the aorta in some forms (*Patella*, etc.). The arteries terminate by opening into irregular spaces in the tissues and amongst the organs, some of which spaces have coalesced or dilated to give rise to the **perivisceral cavity** (exclusive of the pericardium) of these animals. From these systemic spaces the blood passes through the branchial (pulmonary) vessels to the respiratory organs. In some cases there is, in addition to the ctenidial blood, a certain amount of blood returning to the auricle from other parts of the mantle, *e.g.*, *Acmaeidae*, *Heteropoda*, *Pleurobranchidae*, *Pneumodermatidae*. Of these the *Pleurobranchidae*, *Heteropoda*, and some *Acmaeidae* do not possess secondary branchial structures on the mantle; but in other *Acmaeidae* and in the *Pneumodermatidae* there are secondary branchiae in addition to the ctenidium.

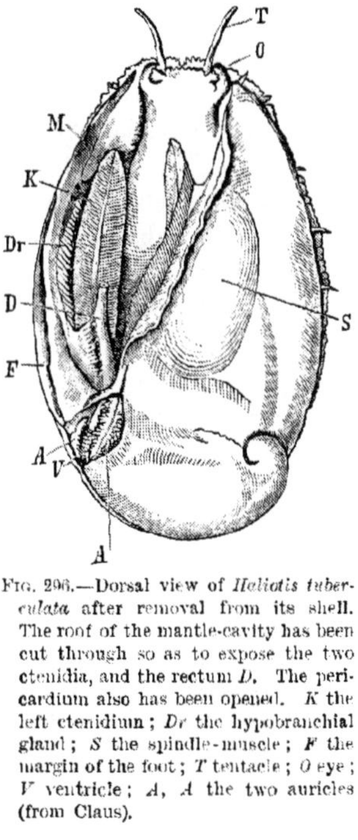

The blood is generally colourless; it is red in *Planorbis* (haemoglobin in the plasma), and in certain forms has a blue tinge (haemocyanin). There is a blood-gland on the aorta in some Opisthobranchs (Bulloids, *Pleurobranchus*, Dorids); and in certain streptoneurous *Platypoda* there is a corresponding organ near the kidney, communicating with the auricle.

FIG. 296.—Dorsal view of *Haliotis tuberculata* after removal from its shell. The roof of the mantle-cavity has been cut through so as to expose the two ctenidia, and the rectum *D*. The pericardium also has been opened. *K* the left ctenidium; *Dr* the hypobranchial gland; *S* the spindle-muscle; *F* the margin of the foot; *T* tentacle; *O* eye; *V* ventricle; *A*, *A* the two auricles (from Claus).

Respiratory organs. In only a small number of Gastropods is respiration effected exclusively through the general integument (some Nudibranchs). By far the greater number breathe through gills, and many through lungs; a few combine branchial and pulmonary respiration.

The gills are either typical ctenidia contained in the mantle-cavity, or freely-exposed branched processes of the dorsal integument (some Nudibranchs).

Excluding the *Chitonidae* (Fig. 305), which are unlike other Gastropods in having a number of pairs of ctenidia bearing two rows of lamellae (*i.e.*, bipennate) and placed in the hinder part of the mantle-groove, there are never more than two ctenidia (Zygobranchs, Fig. 296); but usually an asymmetrical development takes place, and only one gill (the left) remains (Fig. 280).

The Gastropod ctenidium is (except in *Chitonidae*) attached by the whole or greater part of its length to the mantle-wall, and carries typically (in the Aspidobranchs and Tectibranchs) two rows of plates arranged perpendicularly to the axis (Fig. 296), but in the more specialized *Streptoneura* (Fig. 280) one row of plates is absent (it is reduced in the monobranchiate Aspidobranchs). In the former case the gill is said to be *bipectinate*, in the latter *monopectinate*.

The respiration of air is confined to some Prosobranchs and to the *Pulmonata*. In this case the mantle-cavity serves as the respiratory cavity, but it differs from the branchial mantle-cavity by containing air, and by possessing instead of a gill a rich network of blood-vessels on the inner surface of its roof. The mantle-cavity communicates by a long slit with the external medium, but in some forms it is very widely open (Tectibranchs), and in others the opening is reduced to a small, round aperture capable of being closed (*Pulmonata*). Frequently the edge of the mantle is prolonged into a siphon (see above, p. 358).

The arrangement of the respiratory organs is of importance for the classification of the larger groups. According to their position with regard to the heart two great divisions can, as Milne-Edwards has pointed out, be established: (1) the *Opisthobranchiata*, in which the auricle and gill are behind the ventricle; (2) the *Prosobranchiata*, in which the auricle and gill are in front of the ventricle (Fig. 280). So far as this character is concerned, most *Pulmonata* are Prosobranchiate; but the *Pulmonata* in many features of their organization, particularly in their hermaphroditism, stand closer to the *Opisthobranchiata*.

In some aquatic forms the ctenidium is entirely absent, and respiration is carried on either by the mantle and integument generally, without any secondary branchiae (*Lepetidae*, *Dermatobranchus*, *Heterodoris*, many Elysiids, *Phyllirhoe*, *Clionidae*, *Halopsychidae*), or by secondary branchiae; these may be in the mantle-groove as the branchial plates of *Patella*, or on the external surface of the mantle, as in *Clionopsis*, *Notobranchaea*, and most Nudibranchs.

Some littoral forms (*Littorina*, etc.) are able to live for a long time out of water and habitually do so when the tide recedes; in such cases the internal surface of the mantle may be modified, and in *Cerithidea obtusa* the ctenidium even completely disappears. In *Ampullaria* one part of the mantle-chamber is modified as a lung, and is partly separated from the other, which contains the ctenidium; such forms are truly amphibious. Finally, there are the so-called land Operculates; these are streptoneurous forms in which the ctenidium has disappeared and the mantle-cavity is completely transformed into a lung. Of such forms we have the rhipidoglossate family—the *Helicinidae*, and three families of the *Taenioglossa*, viz., the *Cyclophoridae*, *Cyclostomatidae*, *Aciculidae*.

In the *Pulmonata* proper the ctenidium is entirely absent and the opening of the mantle-cavity (pulmonary cavity) much restricted: in the *Onchidiidae* the lung sac is said to be absent, but this is doubtful (Bergh. *Morph. Jahrb.* 10).

A few of the *Pulmonata* have partly or wholly reverted to an aquatic life (the so-called fresh-water *Pulmonata*); some of these forms (*Limnaeidae*) are really air-breathers, and make periodic visits to the surface to procure air, while others (*Amphibola, Siphonaria, Gadinia, Ancylus, Limnaeidae* of deep lakes) are truly aquatic, as the lung sac is always filled with water.

Body-cavity. In *Gastropoda* there is usually a well-developed perivisceral cavity in relation with the alimentary canal or with the anterior part of it. It is the cavity which is opened up in dissecting the alimentary canal of *Chitonidae*, and the anterior part at least of the alimentary canal of most other Gastropoda. We may call this cavity the *perivisceral.* There is also another cavity, which has no connection with the perivisceral, and is called the *pericardial* because it is related to the heart. Whereas it is quite certain that the pericardial cavity is a part of the coelom, the nature of the perivisceral cavity is doubtful. By most anatomists it is regarded as haemocoelic in nature, and this is probably the correct view, but recently it has been interpreted as a part of the coelom at least in the *Chitonidae.* This view is not, however, supported by any evidence anatomical or embryological, for the space in question has no connection with the nephridia or generative organs, and has not been traced from the paired cavity, which appears at an early stage in the development of *Chiton* and is probably coelomic. We may, therefore, safely say that so far as our knowledge at present goes the perivisceral cavity of Gastropoda is a part of the vascular system, and therefore haemocoelic.

It has been recently asserted* that in some of the *Docoglossa* the pericardium has a considerable visceral extension.

The **coelom** of the Gastropoda is in three sections: (1) the pericardium; (2) the nephridia; (3) the gonad.

(1) The pericardium is in relation with the heart; it normally communicates with the nephridial system, and part of its lining is generally glandular and forms the pericardial gland. It has no connection with the blood system.

(2) The nephridial part of the coelom will be described below under the head of excretory organs.

(3) The gonad retains its connection with the rest of the coelom only in the *Aspidobranchia*, in which it communicates with the right kidney. In *Chitonidae* it is separated both from the nephridia and

* B. Haller, "*Studien üb. Docoglosse u. Rhipidoglosse Prosobranchier,*" Leipzig, 1894.

pericardium, and in other Gastropods it has its own special duct, which has been interpreted as, and probably is, a persistent portion of the otherwise absent right (primitive left) nephridium. The developmental evidence in favour of this view is not very strong, but we may hold it at present as a provisional view. In no Gastropod is the gonadial section of the coelom in direct connection with the pericardial.

Excretory organs. The kidneys, or organs of Bojanus, of the *Gastropoda* are typically paired and symmetrical, somewhat dorsally placed near the pericardium, and open externally on each side into the mantle-cavity, generally not far from the anus, and internally into the pericardium by the reno-pericardial openings. This typical condition is found in the *Chitonidae* (Fig. 308) alone. In all other Gastropods the kidneys are of unequal size, and in all but the *Aspidobranchiata* there is only a single kidney, the right one (primitive left) being absent (Fig. 297).

In the Aspidobranchs there are two kidneys (except the *Neritidae*); of these the right is always larger than the left, and in addition to its proper function discharges that of generative duct, inasmuch as the genital gland opens into it (in *Haliotis* by a large slit, and in *Fissurella* on a papilla not far from the external opening, Fig. 297 *b* and *c*). They both open into the mantle-cavity, one on each side of the anus; but with regard to their pericardial opening there is some variation. In *Trochus*, *Turbo*, and *Haliotis* the small left sac opens into the pericardium, while the right kidney is without the pericardial opening. In *Fissurella*, *Emarginula*, *Patella*, etc., it is said by some authors that neither of the kidneys have the reno-pericardial opening; whereas in *Lottia*, and according to other authors *Patella* and *Fissurella*, the right kidney sac alone has a pericardial opening* (Fig. 297 *b*). In other Gastropods there is only one kidney, the left (primitive right); it opens externally near the anus, generally into the mantle-cavity, but sometimes it opens with the anus, with or without a long ureter, and there is a common cloaca (some Pulmonates, *Pteropoda Gymnosomata*); internally it opens by a ciliated orifice into the pericardium (Fig. 297 *e*). In *Paludina* there are two kidneys in the embryo, but the one on the right side disappears in the course of development.

In the simplest cases the kidney is a sac with glandular walls, but it may acquire a spongy texture, or it may consist of a non-glandular

* R. v. Erlanger, "On the paired nephridia of Prosobranchs," *Q.J.M.S.* 33, p. 587.

sac giving off branching glandular tubes (*Chitonidae*, some Nudibranchs).

Pericardial glands are present as differentiations of the lining of the pericardium.

Generative organs. The *Streptoneura* are dioecious, with the exception of *Valvata, Marsenina, Onchidiopsis, Odostomia*, and *Entoconcha*; the *Euthyneura* and these five genera are hermaphrodite.

In the dioecious forms there is usually but little to mark the sexes externally, unless a penis be present. The generative gland

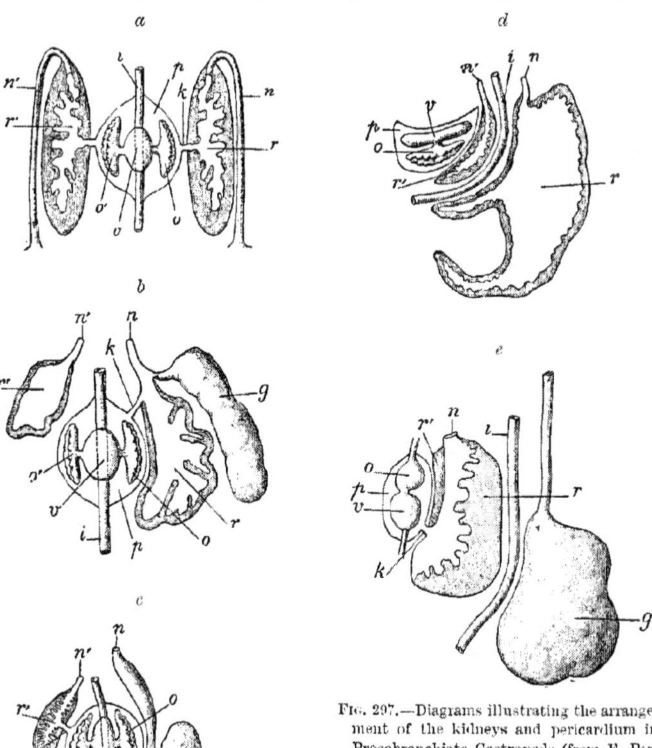

FIG. 297.—Diagrams illustrating the arrangement of the kidneys and pericardium in Prosobranchiate Gastropods (from E. Perrier, after R. Perrier). *a*, Lamellibranchs. *b*, *Fissurella*. *c*, *Haliotis*. *d*, *Patella*. *e*, Pectinibranchs. *g* gonad; *i* rectum; *k* reno-pericardial opening; *n* duct of right kidney (of left in *e*), *n'* duct of left; *o* right, *o'* left auricle (except in *d* and *e*, in which *o* is the left auricle); *p* pericardium; *r* right, *r'* left kidney (except in *e*, where *r* is the left kidney, and *r'* a blood gland); *v* ventricle.

is a racemose body placed at the dorsal summit of the visceral mass in the liver. In the Aspidobranchs as already mentioned it opens into the right kidney, and there is no penis. In the *Neritidae* and Pectinibranchs it has its own duct, which passes along the right side of the rectum (Figs. 280 and 297 *e*) and opens into the mantle-cavity not far from the anus.

The penis, which is very generally present, is a muscular, non-invaginable process of the neck of the animal (Fig. 280) on the right side; and there is a ciliated groove, for the conduct of the sperm, leading from the male opening in the mantle-cavity to the penis, and along the penis to its end (*Ampullaria*, *Littorinidae*, *Naticidae*, *Cypraeidae*, *Heteropoda*, *Murex*, *Dolium*, *Strombus*). In many forms, however, this groove is closed into a canal so that the vas deferens is continued as a tube to the end of the penis (Fig. 280).

There are not usually accessory glands in the dioecious Gastropods, but in some forms there is a glandular region in the oviduct, which secretes albumen, and there is a receptaculum seminis in *Neritidae*, *Paludinidae*, *Heteropoda*, etc.

In the hermaphrodite *Streptoneura* there may be an hermaphrodite duct leading to the external opening, or dividing into male and female portions, which open separately; and the accessory glands are more developed than in the dioecious forms.

In the *Euthyneura* the organs are more complicated. The generative gland, which has the same relations as in the *Streptoneura*, may be itself hermaphrodite to its ultimate follicles; or the male and female follicles may be separate, several female follicles being grouped round and opening into a central male follicle (Pleurobranchs, most Nudibranchs). In the streptoneurous form *Entoconcha* alone are the male and female parts quite separate.

There is always a single hermaphrodite duct leaving the ovo-testis. In the simplest cases it is undivided (**monaulic**) and passes to its opening on the right side; from the latter passes a seminal groove along the side of the body to the penis, which is placed in front (Bullids, Aplysids). In *Valvata*, *Pleurobranchidae*, most Nudibranchs and Pulmonates the duct is **diaulic**, *i.e.*, at a certain point in its course it divides into the male part—the *vas deferens*, which passes to the penis, and into a female part—the *oviduct*, which passes to the female opening. In the diaulic arrangement the male and female openings are either remote from each other (*Valvata*, most *Basommatophora*, *Onchidium*, *Vaginulus*), or close together (most Nudibranchs), or the two ducts join again to open into a common cloaca (*Stylommatophora*, *Siphonaria*) (Fig. 298).

There is a receptaculum seminis, which opens into the hermaphrodite duct in the monaulic forms, and into the oviduct (vaginal part) in the diaulic forms; in the Dorids and Elysids this structure not only opens into the oviduct but communicates directly with the surface of the body by a tube which may be called the vagina. The vagina opens close to the oviduct. This is the **triaulic** arrangement, and recalls that of the *Trematoda*.

The penis, which is always invaginable, is cephalic in Pulmonates, pedal in most Opisthobranchs.

The accessory glands are numerous in the hermaphrodite forms (Fig. 298). In addition to the receptaculum seminis already mentioned, there is generally an albuminous gland in connection with the hermaphrodite duct, and a mucous gland (multifid vesicles of *Helix*) opening into the vaginal part of the oviduct. In some *Helicidae* a peculiar sac—the *dart-sac*—which produces in its interior a dart-like calcareous rod, opens into the vagina between the two groups of multifid vesicles. This rod—the so-called love-dart—is attached to a papilla at the base of the sac; it is protruded during copulation, and seems to play the part of a stimulating organ. It is usually broken off during use, and often remains in the body of the other snail.

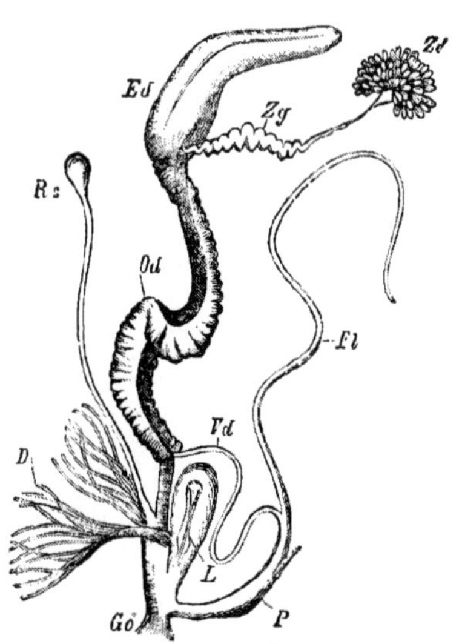

FIG. 298.—Sexual organs of the Roman Snail (*Helix pomatia*), illustrating the diaulic arrangement. *Zd* hermaphrodite gland; *Zg* its duct; *Ed* albumen gland; *Od* oviduct and seminal groove; *Vd* vas deferens; *P* penis; *Fl* flagellum; *Rs* receptaculum seminis; *D* mucous glands; *L* love-dart in dart-sac; *Go* genital opening (after Baasen).

The vas deferens often presents a prostatic gland, which in some forms is in close connection with the evaginable penial part, and is called the flagellum.

The hermaphrodite forms appear as a rule to be protandrous, and rarely, if ever, self-fertilizing. It is stated that specimens of *Arion* and *Limnaea*, which had been kept isolated all their lives, have been known to lay eggs, from which young were hatched; but these may have been cases of parthenogenesis, a phenomenon which has of late been observed in quite unexpected quarters.

In some Gastropods (*Paludina*, *Ampullaria*, and some species of *Murex*) two kinds of spermatozoa are produced in the same gonad. One of these is hair-like and the other vermiform. The meaning of the phenomenon is unknown, but it appears that the former alone conjugate with the ovum.

Fertilization is nearly always internal, and brought about by copulation, which in hermaphrodite forms with contiguous openings for male and female may be reciprocal; but in *Chitonidae*, *Patella*, *Haliotis*, and other forms without a penis, it is probable that it takes place externally in the sea (it certainly does so in *Chiton*). In such forms also the eggs are often laid singly and without any accessory protective capsule (*Patella*, *Haliotis*, *Chiton*), though in *Fissurella* they are embedded in jelly. In aquatic forms the eggs are generally united in masses of gelatinous matter, which may be rounded, elongated, or ribbon-shaped, and are attached to plants or other bodies (fresh-water Pulmonates, Opisthobranchs, *Bithynia*, *Valvata*, *Heteropoda*, etc.). In *Streptoneura* several eggs floating in albumen are generally enclosed in hard and coriaceous capsules or cocoons, and it often happens that all the eggs in a cocoon do not develop, some serving as food for the others. The eggs of *Natica* are embedded in albumen mixed with sand, and have the appearance of a sandy ribbon. In some *Streptoneura* the eggs are attached to a part of the body or shell, *e.g.* *Vermetus* on the inside of the shell, *Nerita* on the outside. In *Ianthina* they are attached to the float.

In *Stylommatophora* the eggs are laid singly, and are often remarkable for their size; in *Bulimus*, *Achatina*, and some species of *Helix* they are as large as the eggs of small birds, and provided with calcareous shells. In such cases, however, the ovum itself is small, the mass of the egg consisting of albumen. The young of *Helix Waltoni* when first hatched is about the size of a full-grown *Helix hortensis*. In *Helix hortensis* and *aspersa* the eggs are laid in hollows in the ground and covered with earth. In some species of *Paludina*, *Melania*, *Littorina*, *Cymba*, *Ianthina*, *Clausilia*, *Helix*, *Pupa*, and *Vitrina* the eggs develop in the uterus, and the young are born fully developed.

The **development** begins with an unequal segmentation leading to the formation of a blastula and gastrula. The blastopore varies in its fate: it often has a slit-like form, and as a rule it closes up from behind forwards either entirely or persists as the mouth-opening. In *Paludina*, however, it persists as the anus. There is a well-developed stomodaeum, giving rise to the buccal cavity and a short proctodaeum. The mesoderm is derived from the endoderm near the blastopore (or according to Tönniges from the ectoderm along the site of the blastopore in *Paludina*), and often gives rise to two mesoblastic bands, which in *Chiton* and some other forms early exhibit a cavity. In *Chiton*, however, the fate of this cavity is unknown; it appears to break up, at any rate it has not been traced. In some other forms (*Paludina, Bithynia*) v. Erlanger* has shown that it gives rise to the pericardium, kidney, and gonad. The same observer also asserts that in *Paludina*

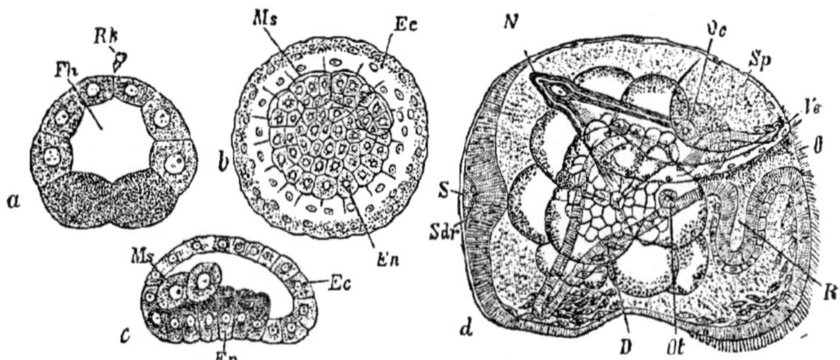

FIG. 299.—Some stages in the embryonic development of *Planorbis* (after C. Rabl). *a*, optical section through a segmenting ovum, with 24 segments. *b*, stage with four mesoderm cells, viewed from the lower pole. *c*, oblique optical section through *b*. *d*, older embryo, with shell gland (trochosphere stage). *D* alimentary canal; *Ec* ectoderm; *En* endoderm; *Fh* segmentation cavity; *Ms* mesoderm; *N* primitive kidney (pronephros); *O* mouth; *Oc* eyes; *Ot* otocyst; *R* rudiment of radula; *Rk* polar bodies; *S* shell; *Sdr* shell-gland: *Sp* apical plate (thickening of ectoderm of preoral lobe); *Ve* velum.

it is developed as a bilobed outgrowth of the primitive gut of the embryo in the neighbourhood of the blastopore.† In the land Pulmonates the young are born with the adult form, and there are but very slight traces of a velum in the development, but in most *Gastropoda* the young leave the egg-membranes as a trochosphere larva (Fig. 299 *d*), which very soon acquires, if it does not already possess them, a shell-gland as an ectodermal invagination of the extra-velar dorsal surface, and a rudiment of the foot as a projection of the ventral surface (Fig. 299 *d*). The shell-gland may or may not secrete a chitinous plug, but in the veliger stage (Fig. 300), which follows the tro-

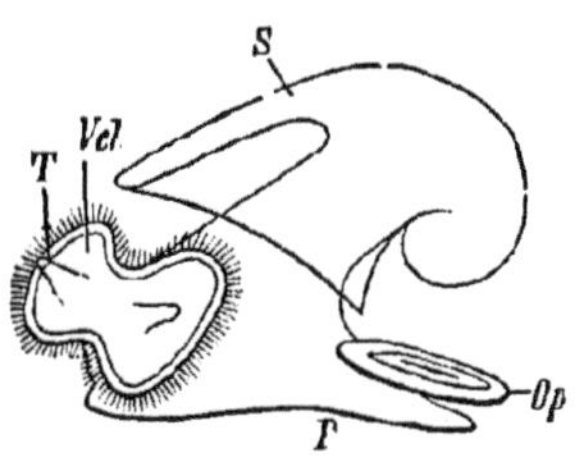

FIG. 300.—Young veliger larva of a Gastropod (after Gegenbaur). *S* shell; *P* foot; *Op* operculum; *Vel* velum; *T* tentacles.

* R. v. Erlanger, "Zur Entwick. d. Paludina vivipara," Pt. 1 and 2. *Morph. Jahrb.*, Bd. 17, 1891, and "Zur. Ent. v. *Bithynia tentaculata*," *Mith. Zool. Stat. Neapel.*, Bd. 10, 1892.

† This is denied by Tönniges, who asserts that the mesoderm in *Paludina* is derived from the ventral ectoderm along the line of the blastopore. *Vide Z. f. w. Z.*, 61, 1896.

chosphere, it flattens out and secretes a plate-like shell, which with the development of the visceral sac acquires a cap-like form. At the same time the dorsal integument at the edge of the shell is formed into an annular fold, encircling the visceral sac. This is the mantle-fold; it becomes especially deepened on the right side, and gives rise to the rudiment of the mantle-cavity. The preoral part of the body along the line of the ciliated ring is prolonged into 2, 4, or 6 processes, and forms the definite velum or sail of the veliger larva. The foot also becomes much more prominent, and develops in most cases an operculum on its hind end (Fig. 300).

While these changes are going on, the visceral sac and shell become coiled in a nautiloid fashion, and at the same time more developed on the left side than on the right; the beginning of this asymmetry has already been indicated by the development of the mantle-cavity somewhat on the right side of the visceral sac. It soon becomes more marked and results in the mantle-cavity coming to lie on the anterior side of the visceral sac, and in the commencement of that dextral spiral twist of the visceral sac and shell which is characteristic of so many Gastropods.

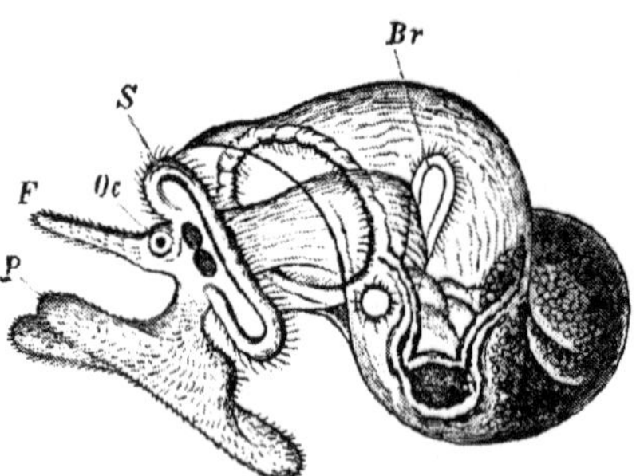

FIG. 301.—Larva of *Vermetus* (after L. Duthiers). Older veliger. *S* velum; *Br* gill; *F* tentacle; *P* foot; *Oc* eye.

The vascular system arises as a set of spaces in the mesoderm; it is customary to state that these spaces are persistent remnants of the segmentation-cavity, but it is very doubtful if there is any meaning in this statement.

The larva often possesses some other special organs which deserve mention: these are the *contractile sinuses* and the *larval kidneys*. The former are blister-like projections of the integument, with muscular walls; they are supposed to assist in the larval circulation, and are found in various parts of the body, *e.g.*, foot. The latter (Fig. 299 *d*) are a pair of glandular tubes which open externally just outside the velar area on each side. It has been stated that they have internal openings into the blastocoel, but the opposite has also been stated, and the balance of evidence is against the existence of such openings.

Metamorphosis of the larva into the adult. However symmetrical the adult may appear to be, the larva (except in the *Chitonidae*) always presents, after the trochosphere stage, an asymmetrical development of its visceral sac, and there is, in almost all, a shell more or less nautiloid in character, and generally, though less invariably, an operculum. In the great majority of *Euthyneura* the shell is lost in the adult, and the operculum, with a very few exceptions, also disappears in the same group. In some cases the larval shell is replaced by a permanent shell.

The asymmetry sometimes appears before the anus and mantle-cavity are formed, in which case the visceral sac becomes more developed on the left side, and the anus and mantle-cavity appear on the right side and then, by the relative growth of parts, shift forwards along the right side until they are placed on the anterior face of the visceral sac. In cases in which the anus appears before the asymmetry, *e.g.*, *Paludina*, it and the mantle-cavity are at first posterior and median; they then gradually pass round to the right side, and so to the anterior face of the visceral sac, this shifting being accompanied by a greater growth of the visceral sac towards the left. This movement of the anus and mantle-cavity, if viewed from the dorsal surface, has been in a direction opposite to that of the hands of a watch, and has consisted of a twisting of the visceral sac about an axis, which may roughly be described as dorso-ventral, through an angle of 180°, so that the original posterior face of the visceral sac becomes the actual anterior face, and the visceral organs of the original right side become placed on the left, and *vice versa*. But in all Gastropods, except the *Aspidobranchiata*, the organs of the original left side of the visceral sac are either not developed, or, if they are developed, disappear; consequently the anus, instead of opening into the middle of the mantle-cavity, is on its right side, and the actual right gill, osphradium, kidney, and auricle are not present. This torsion of the visceral sac round a dorso-ventral axis in a direction opposite to that of the hands of the watch when viewed from the dorsal pole, is not the only twisting the visceral sac undergoes. It also twists round a horizontal axis passing from right to left in such a manner as to form a spiral. If this twisting be supposed to have taken place before the torsion first described, the visceral sac must have fallen forwards, and so formed a coil like that of the shell of Nautilus on the anterior side: such a coil is called

exogastric (Fig. 303). But when the visceral sac has undergone the torsion just described, which resulted in its anterior face becoming posterior, this coil would naturally be on the hinder face of the visceral sac (Fig. 303). Such a coil is secondarily **endogastric** and is the permanent condition in the ordinary dextral visceral sac of most Gastropoda.

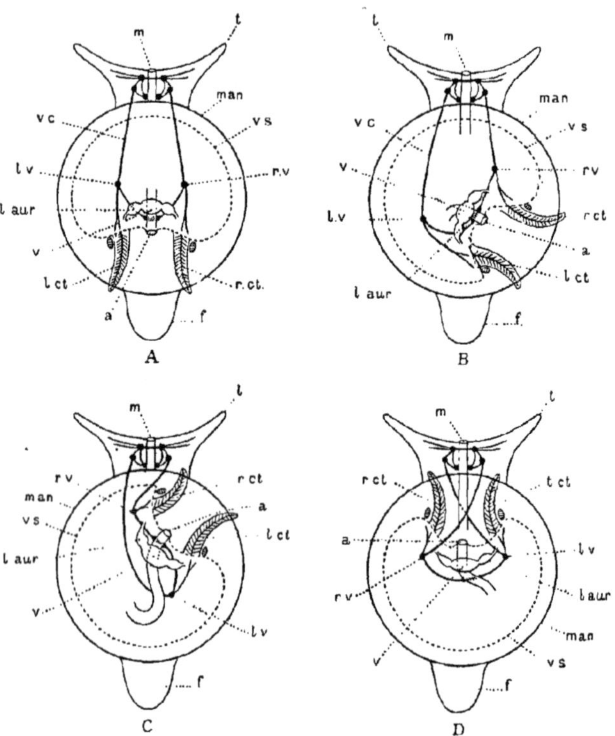

FIG. 302.—Diagrams illustrating the torsion of the visceral sac, which is supposed to have taken place in most *Gastropoda* (after Lang). *A* represents the hypothetical primitive form in which no torsion has taken place. In *B* the visceral sac has rotated through an angle of 45° in a direction opposite to that of the hands of a watch; the mantle-cavity with the gills and anus, etc. is now placed on the right side of the body. *C*, Intermediate stage. *D*, final stage, in which the visceral sac has rotated through an angle of 180° in a direction opposite to that of the hands of a watch, and about a dorso-ventral axis; the mantle-cavity with gills, anus, etc. are now on the anterior side of the visceral sac, and the primitive right gills, kidney, auricle, have become placed on the left side. The visceral commissure has become twisted into a figure of 8 loop. The outline of the visceral sac is indicated by the dotted line. *a* anus; *l.aur* left auricle; in *C*, *l.aur* points to the right auricle; *f* foot; *l.ct* primitive left gill; *l.v* primitive left visceral ganglion (sub-intestinal); *m* mouth; *man* edge of mantle-fold; *r.ct* primitive right gill; *r.v* primitive right visceral ganglion (supra-intestinal); *t* tentacle; *v* ventricle; *v.c* visceral commissure; *v.s* outline of visceral sac.

A glance at the accompanying diagram will explain the effect of the first-described torsion about the dorso-ventral axis upon the organs of the visceral sac, and the course of the peculiar figure-of-8 loop of the visceral commissure. Why the organs of the primitive left side disappear in so many Gastropods is not clear, but their disappearance is doubtless connected in some way with the torsion, and consequent asymmetry produced in the parts affected by the torsion.

FIG. 303.—*Ampullaria cornu arietis* (règne animal), illustrating the spiral torsion of the visceral sac about a right and left horizontal axis, in which the visceral sac has apparently fallen backwards (endogastric). It is, however, really an exogastric coil, because the visceral sac has been twisted so that its hinder surface is in front.

In some groups there is a tendency for the visceral sac to flatten out into a conical form, and for the nautiloid shell to become conical (*Patella*, *Fissurella*, *Aplysia*, *Pleurobranchus*, *Doris*, etc.). This is, however, a purely secondary feature, for the young possess the typical nautiloid shell, and in some cases the stages of the loss of the spire can be completely followed.

This acquisition of a secondary symmetry, which is found very commonly in the *Euthyneura*, must not, however, be confused with a tendency to **detorsion** of the visceral sac which is found in most *Euthyneura*. It shows itself in the shifting back of the anus well on to the right side and even to the hind end, and in the untwisted condition or only partially twisted condition of the visceral loop (see account of *Actaeon* and allies). Formerly this condition was looked upon as an arrested stage in the torsion, but having regard to the arrangement of the visceral commissure in *Actaeon* and its allies, and to the fact that in the suppression of the organs of the right (primitive left) side the *Euthyneura* are as specialized as the most specialized *Streptoneura*, it is more probable that the condition referred to is due to detorsion. Forms in which the anus is secondarily shifted backwards almost always present a reduction or disappearance of the mantle and shell (*Pterotrachea*, *Aplysia*, Dorids, *Janus*, *Alderia*, *Limapontia*, *Testacella*, *Onchidium*, *Vaginulus*).

By far the majority of the *Gastropoda* are marine. Almost the whole of one group of the *Pulmonata* (*Basommatophora*) and the following *Streptoneura* live in fresh water—some *Neritidae*, the *Ampullariidae*, *Paludinidae*, *Valvatidae*, *Bithyniidae*, *Hydrobiidae*, some *Cerithiidae*, the *Melaniidae*, *Cremnoconchus*, *Canidia*. The *Pulmonata Stylommatophora* and certain *Streptoneura* (*Helicinidae*, *Cyclophoridae*, *Cyclostomatidae*, *Aciculidae*) are terrestrial. Many *Littorina*, *Cerithium*, *Melania*, etc., live in brackish water. Many branchiate Gastropods are able to live for some time out of water in dry places; in such circumstances they are withdrawn into their shells, the opening of which is closed by the operculum. Almost all move by creeping; some, however, as *Strombus*, jump; others, as *Oliva* and *Ancillaria* swim excellently by the aid of the lobes of their foot. Some marine forms, *e.g.*, *Magilus*, *Vermetus*, etc., are fixed by their shells; a few only are parasitic, as *Stilifer* on Sea-urchins and Starfishes, *Entoconcha* in Synapta. The method of nutrition differs as much as does the habitat. Many, especially the siphonostomate forms, are voracious predatory animals, and prey on living animals; some, as *Murex* and *Natica*, with this object bore into the shells of Molluscs; several (*Strombus*, *Buccinum*) prefer dead animals. An equally large number, *viz.* almost all Pulmonates and holostomatous branchiate Gastropods, feed on plants. More than 17,000 species are known: they are found on all parts of the earth; in the deep sea to a depth of 2650 fathoms, and from the Himalayas above the snow line. Geologically they are found at the very commencement of the palaeozoic epoch in the Cambrian, and the genus *Helix* (*Zonites* Montf.) is known in the Carboniferous.

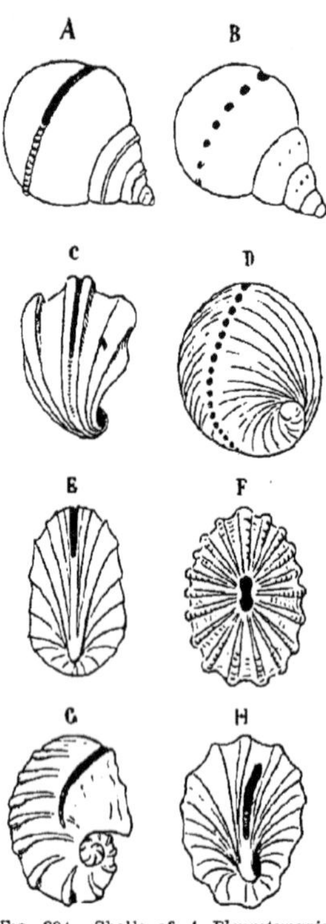

Fig. 304.—Shells of *A Pleurotomaria*, *B Polytremaria*, *C* and *E Emarginula*, *D Haliotis*, *F Fissurella*, *G* and *H* stages in the growth of the shell of *Fissurella*.

SUB-CLASS 1. ISOPLEURA* (PLACOPHORA).

Symmetrical Gastropoda, with metameric repetition of the ctenidia and shell-plates.

In this sub-class the head is without tentacles and sense-organs. The foot occupies the greater part of the ventral surface, and has a broad flat sole (in *Cryptoplax* the foot is narrow). The mantle occupies the whole dorsal surface, and completely hides the head. The visceral sac is flattened and not drawn out into a dome. The mantle-fold encloses a shallow groove which completely surrounds the body, and is roughly of uniform depth throughout. The ctenidia, which are bipectinate and projecting, are attached to the

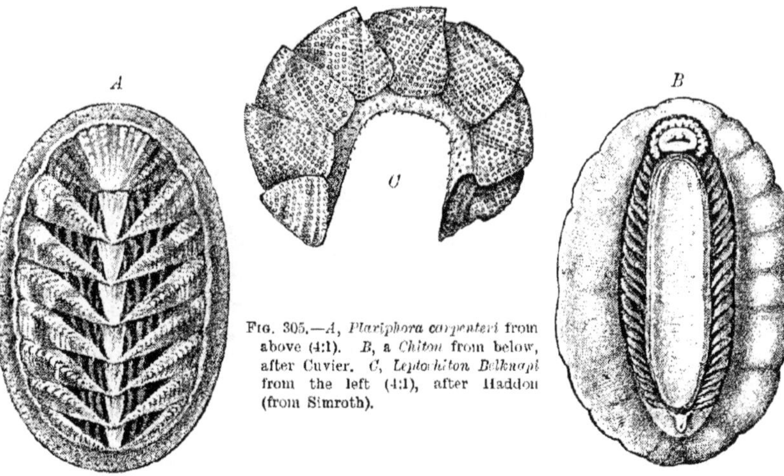

FIG. 305.—*A*, *Placiphora carpenteri* from above (4:1). *B*, a *Chiton* from below, after Cuvier. *C*, *Leptochiton Belknapi* from the left (4:1), after Haddon (from Simroth).

floor of this groove, generally in the hinder part, but in some forms they extend forwards nearly to the head; they vary in number from six to eighty pairs.

The shell-plates, of which there are eight, are partly, and in *Cryptochiton* wholly, embedded in the mantle; each plate is overlapped by the plate in front, except in *Cryptoplax* (*Chitonellus*), in which the plates do not touch (Fig. 306). The parts of the mantle

* A. Th. Middendorf, "Beiträge zu einer Malacozoologica rossica," *Mém. Acad. Imp. Petersburg*, 1848. A. Sedgwick, "On certain points in the anatomy of Chiton," *Proc. Roy. Soc.*, 1881. H. Simroth, "*Mollusca*" in Bronn's *Klassen u. Ordnungen*, 1893–94. I. Blumrich, "Das Integument der Chitonen," *Z. f. w. Z.*, 52, 1891. A. C. Haddon, "Report on the Polyplacophora," *Challenger Reports*, Pt. 43, 1886.

not covered by the plates bear spicules, which are inserted in ectodermal pits.

The mantle-groove contains, in addition to the gills, the anus on a short papilla in the middle line behind, and on each side the opening of the kidney, and in front of that the opening of the generative duct (Fig. 308).

In the **nervous system** (Fig. 307) the nerve-cells are not concentrated into ganglia, but are diffused along the main nerve trunks. There are two pairs of main longitudinal trunks—the *pedal* in the foot, and the *pallial* along the mantle-groove. The former are connected by a number of commissures throughout their extent. The latter supply the mantle, gills, and osphradia,

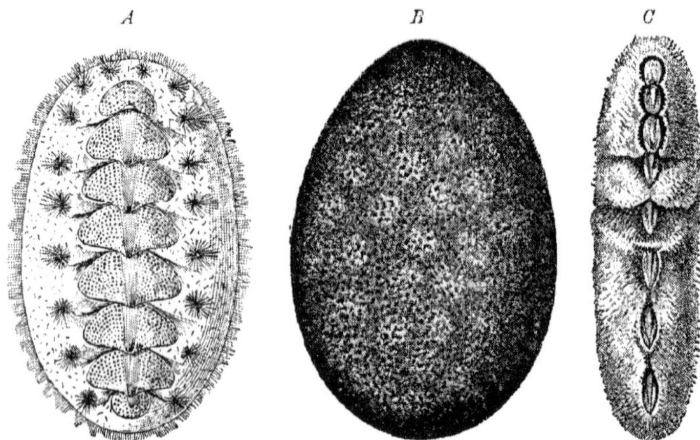

FIG. 306.—*A*, *Acanthochiton fascicularis* $\frac{4}{5}$ (after Blumrich). *B*, *Cryptochiton stelleri var. violacea* $\frac{1}{2}$ (after A. v. Nordmann). *C*, *Cryptoplax oculatus* $\frac{2}{3}$ (after Haddon). From Simroth.

and to a certain extent the viscera, and end behind by running into one another dorsal to the rectum: they are to be regarded as corresponding to the mantle-nerves, and partly as the much extended pleural ganglia of other forms. These two cords on each side join in front, and the single ganglionic cord passes round the oesophagus to unite with its fellow across the middle line dorsal to the oesophagus. This broad ganglionic band corresponds generally to the cerebral ganglia of other forms.

The visceral commissure (Fig. 307, *VII*) is given off from the pallial nerves close to where these join the pedal cords, and passes only a short distance backwards ventral to the alimentary canal.

It has a pair of small ganglia posteriorly near the anterior part of the stomach.

The stomatogastric commissure, with its contained buccal ganglia, arises from the point of division of the cerebral into pallial and pedal trunks, and passes between the buccal mass and oesophagus. Finally, there is another suboesophageal commissure with a pair of small ganglia arising from the labial commissure and supplying the subradular organ.

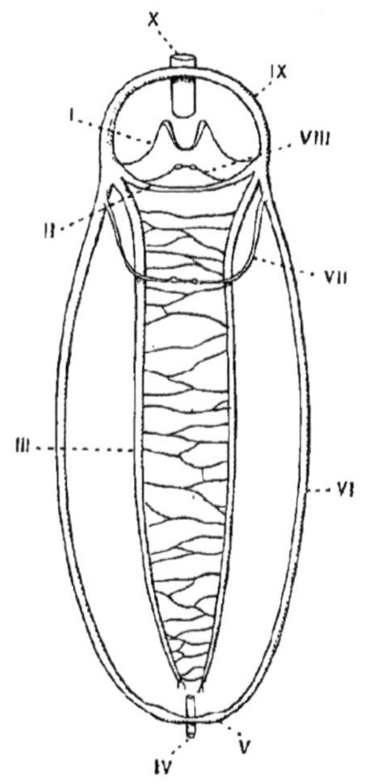

Fig. 307.—Central nervous system of *Chiton* seen from the dorsal side. The anterior and posterior ends of the alimentary canal are represented diagrammatically in their relation to the nervous system (modified from Pelseneer). *I* stomatogastric commissure (buccal); *II* labial commissure; *III* pedal cord; *IV* anus; *V* supra-rectal pallial commissure; *VI* pallial cord; *VII* visceral commissure; *VIII* subradular commissure; *IX* cerebral commissure; *X* mouth.

Sense organs. The edges of the snout or head are extended into short labial palps. On the floor of the mouth just in front of the radula is a small epithelial projection, which is supposed to be a sense-organ—the subradular organ. The shell-pieces are traversed by fine nerve-cords, which end in the megalaesthetes and the micraesthetes. These are sensory epithelial papillae, and in some forms the megalaesthetes are modified into eyes (Moseley), with a calcareous cornea, a lens, and a pigmented envelope. It has been asserted that the osphradia are multiple—one at the base of each ctenidium—but it is doubtful whether there are any structures corresponding to the osphradia of other Gastropods. Ridges of high epithelium containing sense-cells have been described in the mantle-groove.

The **alimentary canal** begins with the mouth, which is placed on a projection in front of the foot, the head or snout; it passes back to the anus, which is median and posterior. The radula is well developed, and its formula is $(3 + 1) \cdot (2 + 1) \cdot 3 \cdot (1 + 2)\ (1 + 3)$. There are two salivary glands with very short ducts. The oesophagus is short, and is extended into a glandular pocket on each side. The stomach is large and surrounded by the large liver, the ducts of which it receives. The intestine is long and coiled.

Body-cavity. There is a well-marked perivisceral-cavity in relation with the alimentary canal. Its nature is not certain, for its development is not known, and it is not connected with the pericardium. Provisionally it may be regarded as haemocoele.

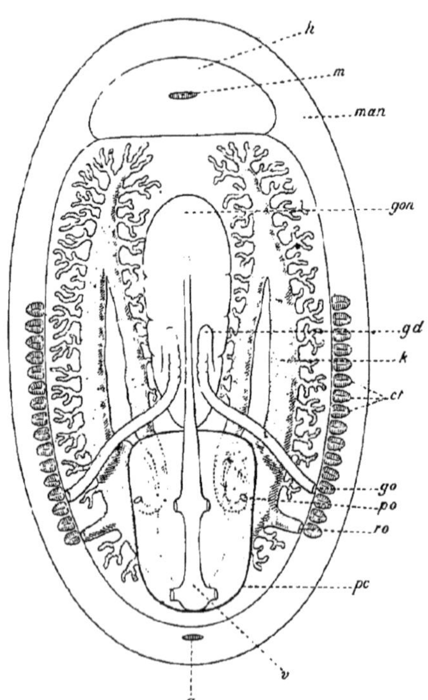

Fig. 308.—Dorsal view of the kidneys, generative organs, and heart of *Chiton discrepans* (modified after Sedgwick). *a* anus; *ct* ctenidia; *g.d* generative duct; *g.o* opening of generative duct into the mantle-groove; *gon* generative organ; *h* head; *k* kidney with its glandular diverticula; *m* mouth; *man* mantle-fold; *pc* wall of pericardium; *p.o* opening of kidney into the pericardium; *r.o* external opening of kidney into the mantle-groove; *v* ventricle of heart, the auricles are cut away, but their openings into the ventricle are shown.

Coelom. The pericardial cavity placed at the hind end on the dorsal surface is undoubtedly coelomic, for although it has no connection with the gonad, it opens into the nephridia.

The **heart** (Fig. 308) consists of a median ventricle giving off an anterior aorta, and of two symmetrical auricles, each of which has two openings into the ventricle (in *Chiton magnificus* it has recently been stated that there are four openings on each side*). The blood flows to the auricles from the gills.

Excretory organs. There are two kidneys (Fig. 308), each consisting of bent tubes of the typical Molluscan type with glandular diverticula. They open into the pericardium on each side, and to the exterior in the mantle-groove between two of the posterior ctenidia.

* B. Haller. Beiträge zur Kenntniss der Placophoren, *Morph. Jahrb.* 21.

Generative organs. The sexes are separate, and the generative gland is single and median, and occupies the dorsal part of the body just in front of the pericardium, with which, however, it has no connection. The generative ducts are paired, and arise from the hinder end of its ventral surface; they are curved on themselves near their origin, and in the female have a glandular enlargement on their course. They open into the mantle-groove a little in front of the renal openings (Fig. 308).

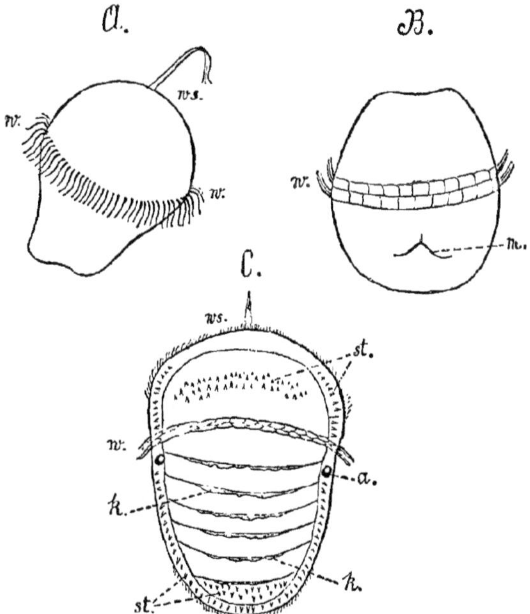

FIG. 309.—*A* larva of *Chiton marginatus* (after Lovén). *B* embryo, and *C* larva of *Chiton polii* (after Kowalevsky) from Korschelt and Heider. *a* eye; *k* rudiment of shell-plates; *m* mouth; *st* spines; *v* velum; *vs* apical tuft of cilia.

The eggs when laid have a spiny shell, and may be kept in the mantle-groove or expelled freely into the sea. Fertilization is external.

In the development the young are hatched as larvae and pass through a trochosphere stage. There is no veliger stage or nautiloid shell. The blastopore becomes the mouth.

The Chitons are marine animals which live like limpets attached to stones by the foot. They inhabit the littoral zone, and have been found to a depth of 2000 fathoms. They are known fossil since the Silurian.

There is only one family, the **Chitonidae**, but there is considerable variation in the group, both specific and individual, and consequently a large number of species has been described. These have been arranged in genera, of which there are a considerable number; but the attempt which has been made by some authors to arrange these genera in several families, and these again in sub-orders, seems to be going further than the facts warrant, for the uniformity in structure in the group is remarkable, and the differences, though numerous, are trivial and unimportant.

Fam. **Chitonidae.** With characters of sub-class. *Chiton* L. plates of shell much exposed; *Leptochiton* Gray, edge of mantle uniformly covered with scale-like spicules; *Callochiton* Gray, with branchiae only in the hinder part of the mantle-groove; *Acanthochiton* Leach, edges of mantle with spicules united in bundles corresponding to the plates of the shell; *Plaxiphora* Gray; *Cryptoplax* Blainville (*Chitonellus* Lam.), shell-plates largely concealed and not articulated together, foot narrow; *Cryptochiton* Middendorf, shell-plates completely embedded in the mantle.

SUB-CLASS 2. ANISOPLEURA.

Asymmetrical Gastropoda with shell of one piece, never more than two ctenidia, and with a well-marked visceral commissure. Generally with a veliger larva.

The above definition contains the main points in which the rest of the Gastropoda differ from the Isopleura. The *Anisopleura* are divided into two orders, the *Streptoneura* and the *Euthyneura*.

Order 1. STREPTONEURA.*

Dioecious Anisopleura with shell and generally with operculum; with gills in front of the heart. Visceral commissure twisted into a figure-of-8.

Behind the head, which is usually distinct and bears a single pair of tentacles, lies the mantle-cavity, into which the rectum, kidney, and oviduct open. In rare cases two gills are present, but as a rule the right gill is absent. The branchial veins enter the heart from the front. The visceral commissure is twisted into a figure-of-8, of which the right half passes dorsal to the alimentary canal,

* E. Claparède, "Anatomie u. Entwick. d. Neritina fluv.," *Z. f. w. Z.*, 2, 1850. H. Lacaze Duthiers, "Sur le système nerv. de l'Haliotide et sur la Pourpre," *Ann. Sci. Nat.* (4), 12, 1859. Id., "Sur l'anat. et l'embryog. des Vermets," *Ann. Sci. Nat.* (4), 13, 1860. E. L. Bouvier, "Système nerv., morph. gén. et classification des Gastéropodes prosobranches," *Ann. Sci. Nat.* (7), 3, 1887. Perrier, "Sur l'anat. et l'hist. du rein des Gastéropodes pros.," *Ann. Sci. Nat.* (7), 8, 1889. Wegmann, "Contributions a l'histoire naturelle des Haliotides," *Arch. Zool. Exp.* (2), 2, 1884. Boutan, "Sur l'anat. et le dével. de la Fissurella," *Arch. Zool. Exp.* (2), 3 bis, 1886.

the left half ventral to it. The pleural ganglia are sometimes joined to the opposite half of the visceral commissure by an anastomotic branch of the pallial nerve (*dialyneury*, *e.g.*, *Cassidaria*, on left side), or the pallial nerve directly joins the ganglion on the opposite half of the visceral commissure, *i.e.*, the pallial nerve from the right pleural joins the sub-intestinal ganglion, and that from the left pleural joins the supra-intestinal (*zygoneury*, more common on the right side, *e.g.*, *Cassidaria*, where the right pleural is directly joined to the sub-intestinal, see above, p. 365, Fig. 286).

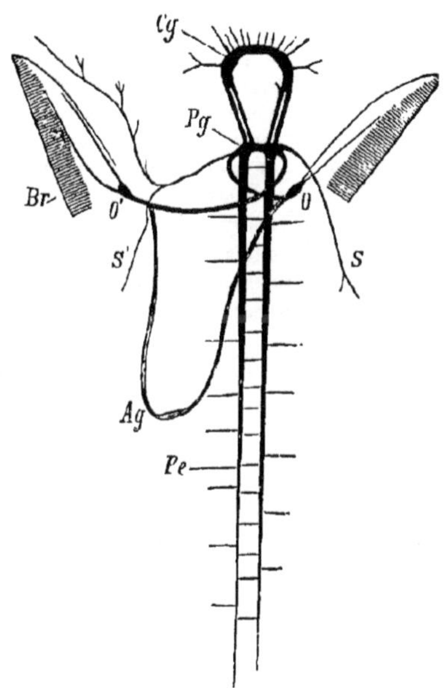

Fig. 310.—Nervous system of *Haliotis* (diagrammatic after Spengel). *Cg* cerebral ganglion; *Pg* fused pleural and pedal ganglia; *O* right and *O'* left osphradium; *Ag* abdominal ganglion; *Pe* pedal cord; *S* and *S'* pallial nerves; *Br* gills.

The males are often provided with a large penis placed on the right anterior side of the body. In the generative organs accessory glands are usually absent. The eggs are surrounded by albumen and laid in capsules, which are frequently fixed to foreign objects; sometimes they are attached to a raft of mucous bubbles, to which the animal itself adheres (*Ianthina*).

Sub-order 1. **Aspidobranchiata (Diotocardia, Scutibranchia).**

Ctenidium bipectinate or absent, labial commissure present.

The pleural ganglia are approximated to the pedal, which are prolonged posteriorly into long ganglionic cords (Figs. 283, 310). The cerebral ganglia are widely separate from each other, and the commissure connecting them passes in front of the buccal mass and salivary glands. There is a sub-oesophageal labial commissure with buccal ganglia (Fig. 283). The **osphradium** is small, and placed on the branchial nerve. The **otocyst** has numerous small concretions (*otoconia*). The **eye** is open (Fig. 290 *A*), or closed with a very small cornea. The **radula** has multiple centrals. The **ctenidia** are bipectinate and free anteriorly. Traces of bilateral symmetry are usually present. There are generally two auricles and two kidneys, which open on short papillae. The gonad (Fig. 297 *b*) opens into the right kidney (except in the *Neritidae*).

Tribe 1. **DOCOGLOSSA.**

Nervous system without dialyneury; eyes open without lens. Two osphradia. Operculum and hypobranchial glands absent. Jaw unpaired and dorsal. Radula has strong, pillar-shaped teeth, and never more than three marginals on each side. A single auricle; the rectum does not traverse the ventricle or the pericardium. The visceral sac is conical in the adult, and not twisted.

Fam. 1. **Acmaeidae.** Left ctenidium only; it is to a great extent free. *Acmaea* Esch.; *Scurria* Gray; *Addisonia* Dall.

Fam. 2. **Patellidae.** Limpets. Ctenidia absent, replaced by a circlet of plate-like structures in the mantle-groove. *Patella* L.; *Helcion* Gray.

Fam. 3. **Lepetidae.** Ctenidia, pallial branchiae, and eyes absent. *Lepeta* Gray.

Tribe 2. **RHIPIDOGLOSSA.**

Nervous system with dialyneury; eyes with lens; a single osphradium (the left), except when there are two ctenidia; one or two hypobranchial glands. Jaws paired; marginal teeth of radula numerous, and crowded together like the ribs of a fan (Fig. 293 *A*). A crop, oesophageal gland, and a stomach caecum (often spiral) present. Two auricles, ventricle traversed by rectum (except in *Helicinidae*). Often with an epipodial projection, which is frequently tentaculiferous, on each side of foot.

Section A. With two ctenidia (*Zygobranchiate*). Shell with marginal slit or holes corresponding to an anal opening in mantle.

Fam. 1. **Pleurotomaridae.** Visceral mass and shell twisted. Mantle slit in front in the middle line. Operculate. Cambrian onwards. *Pleurotomaria* Defrance (Fig. 304 *A*); *P. quoyana* Gulf of Mexico; *Scissurella* d'Orb.; *Trochotoma* Desh.

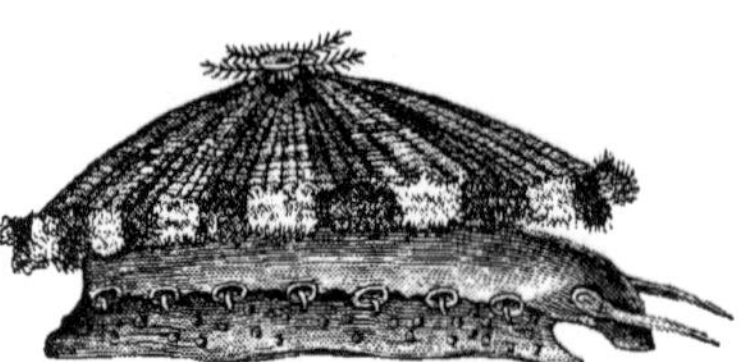

FIG. 311.—*Fissurella maxima* (from Bronn).

Fam. 2. **Haliotidae.** Spire of the visceral sac and shell much reduced. Shell flat, with wide aperture, with internal mother-of-pearl lustre and a row of holes on the left side over the mantle slit. The mantle-cavity is displaced to the left side by the enormous spindle-muscle. Epipodium tentaculiferous. Without operculum. *Haliotis* L., ear shell, ormer (Figs. 296 and 304 *D*), adherent to rocks like a limpet; found in large numbers in the Channel Islands.

Fam. 3. **Fissurellidae.** Visceral sac and shell conical, with an apical aperture or an anterior marginal excavation. Limpet-like. Without operculum. *Fissurella* Brug., key-hole limpet (Figs. 304 *F*, and 311); *Puncturella* Lowe; *Pupillia* Gray; *Emarginula* Lam., with anterior marginal excavation;

Scutum (*Scutus*) Montfort (*Parmophorus* Blainv.), Australia, shell partly covered by mantle; *Cemoria* Leech, *Rimula* Defrance. *Propilidium* Forbes and H., and *Cocculina* Dall, allied here.

Section B. With one (left) ctenidium (*Azygobranchiate*).

Fam. 4. **Trochidae.** Visceral sac and shell coiled in a spiral; eyes open; operculum horny; with head-lobes between the tentacles. *Trochus* L.; *Margarita* Leach (Fig. 312); *Monodonta* Lam.

Fam. 5. **Stomatiidae.** Spire of visceral sac and shell reduced. *Stomatella* Lam.; *Gena* Gray; *Broderipia* Gray; *Stomatia* Lam., without operculum; *Phaneta* Ad., fluviatile.

Fam. 6. **Delphinulidae.** Visceral mass and shell spiral; no cephalic lobes. *Delphinula* Lam.; *Cyclostrema* Marryat.

Fam. 7. **Liotiidae.** *Liotia* Gray.

Fam. 8. **Turbinidae.** Visceral sac and shell spirally coiled; epipodial tentacles; calcareous operculum. *Turbo* L., top-shell. *Phasianella* Lam., pheasant-shell; *Mölleria* Jeffreys.

Fam. 9. **Neritopsidae.** *Neritopsis* Grateloup.

Fam. 10. **Macluritidae.** *Maclurea* Lesueur.

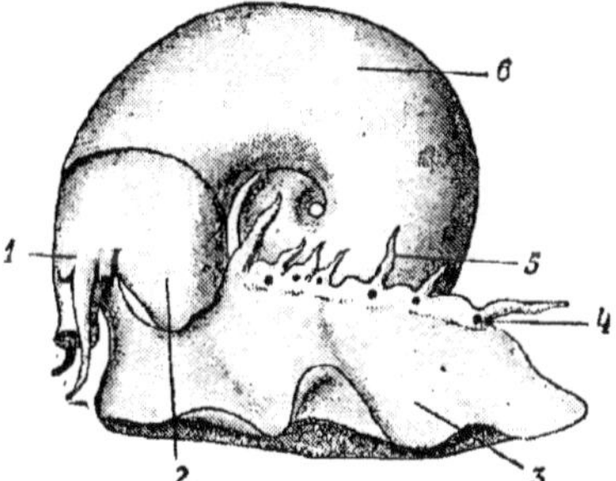

Fig. 312.—*Margarita groenlandica* (after Pelseneer, from Lang). *1* head; *2* anterior epipodal lobes; *3* foot; *4* pigmented eminences at the base of the epipodial tentacles *5*; *6* visceral sac.

Fam. 11. **Neritidae.** Epipodium little developed, without tentacles; a cephalic penis; calcareous operculum; eyes stalked. *Nerita* L.; *Neritina* Lam., chiefly brackish water and fluviatile; land species are known; *N. fluviatilis* Müller, British rivers. *Theodoxia* Mont.; *Smaragdia* Issel; *Septaria* Fér.

Fam. 12. **Titiscanidae.** Shell and operculum absent. *Titiscania* Bergh; *T. limacina* Bergh., Pacific Ocean.

Fam. 13. **Helicinidae.** Epipodium without tentacles; branchia absent, mantle cavity transformed into a pulmonary chamber; heart with a single auricle, not traversed by the rectum. Operculum without apophysis. Terrestrial. *Helicina* Lam., Central America and Bahamas; *Hydrocena* Parreys; *Georissa* Blanford.

Fam. 14. **Proserpinidae.** Branchia replaced by pulmonary cavity; mantle partly reflected over shell; operculum absent. Terrestrial. *Proserpina* Gray, Cuba, Jamaica, Venezuela; *Ceres* Gray, Mexico.

Sub-order 2. Pectinibranchiata (Ctenobranchia, Monotocardia).

Ctenidium monopectinate, generally without labial commissure.

The nerve-collar is behind the buccal mass, and there is no labial commissure (except in *Paludina* and *Ampullaria*). The single osphradium is often pectinate, and well differentiated. The eye is always closed. Otocysts with otolith (except *Paludina*, *Valvata*,

Ampullaria, *Cyclophorus*, *Acicula*, some *Cerithiidae*, etc.). Radula with central tooth single or absent.

The ctenidium, auricle, and kidney of one (right) side are absent, and the ventricle is not traversed by the rectum. The ctenidium is monopectinate (Fig. 280), and attached to the mantle by its whole length. The kidney opens by a slit, exceptionally by a ureter (*Paludina*, *Cyclophorus*, *Valvata*), and never receives the generative products. A penis is generally present, and the gonad has its own duct and opening.

Tribe 1. PTENOGLOSSA.

Without siphon. Peristome entire. Mouth with proboscis or snout. Penis absent. Radula without median tooth, with numerous lateral hooks ($\infty . 0 . \infty$).

Fam. 1. **Ianthinidae.** Snout prominent, blunt; tentacles bifid; no eyes; shell thin, spiral, without operculum. Foot small, secreting a float, to which the eggs are attached. Pelagic, carnivorous. *Ianthina* Lam.; *I. fragilis* L., Atlantic and Med.; *Recluzia* Pet.

Fam. 2. **Solariidae.** Shell flattened, with a large umbilicus which extends to summit of spire. Head short, proboscis long; foot small. Tentacles slit along their whole length. *Solarium* Lam. Stair-case shell.

Fam. 3. **Scalariidae.** Shell turriculated; a small siphon; foot and proboscis short. The animal secretes a purple liquid when molested. Carnivorous. *Scalaria* Lam., wentle-trap.

Tribe 2. RACHIGLOSSA.

With long proboscis, evaginable from the base. Radula with at most three teeth in each row, a cusped median tooth, and a cusped lateral on each side, which may be absent; marginals absent (Fig. 293 *C*). All possess a siphon and are predatory.

Fam. 1. **Muricidae.** Eyes at the base of the tentacles; foot truncated; an anal gland. *Murex* L. The Ancients obtained their purple from species of *Murex*; it is secreted by the anal gland. *M. crinaceus*, sting-winkle, English Channel: *M. trunculus* L. Heaps of shells of this species still to be seen on the Tyrian shore. *Purpura* Brug.; *P. lapillus* L., destructive to mussel beds; *Trophon* Montf.; *Urosalpinx* Stimpson.

Fam. 2. **Coralliophilidae.** Radula absent. Inhabits coral reefs. *Magilus* Montf.; *Leptoconchus* Rupp.; *Coralliophila* H. and A. Adams.

Fam. 3. **Columbellidae.** *Columbella*, Lam.

Fam. 4. **Nassidae.** Foot long, broad, often with terminal appendage; siphon long; eyes on outer base of tentacles; central tooth of radula arched, multicuspid, laterals bicuspid with small denticles between the cusps; shell buccinoid. *Nassa* Lam.; *N. reticulata* L., dog-whelk, common on English coasts at low water; *Canidia* Adams; *Cyclonassa* Swainson; *Bullia* Gray.

Fam. 5. **Buccinidae.** Eyes at base of tentacles; foot large; proboscis long. Central tooth of radula with 5–7 cusps, laterals bi- or tri-cusped. Shell thick, covered with periostracum; canal of varying length; operculum corneous. *Buccinum* L., whelk; *Chrysodomus* Swainson.

The **Haliidae** (*Halia* Risso) are allied here.

Fam. 6. **Fasciolaridae.** Shell fusiform; spire long; operculum corneous; head small; short tentacles, with eyes at base; siphon moderate. *Fasciolaria* Lam.; *Fusus* Lam.

The **Turbinellidae** (*Hemifusus* Swainson; *Fulgur* Montf.; *Latirus* Montf.; *Lagena* Schum.) are allied here.

Fam. 7. **Mitridae.** Tentacles carrying eyes laterally; proboscis long; shell fusiform, spire pointed; operculum absent. *Mitra* Lam., mitre-shell; *Thala* Ad.; *Turricula* Ad.

Fam. 8. **Volutidae.** Head flattened, and prolonged laterally into lobes, on which are the eyes. Proboscis short; siphon with interior appendages; operculum generally absent. *Voluta* L.; *Guivillea* Watson; *Cymba* Brod. and Sow., boat-shell; *Cymbium* Montf.

Fam. 9. **Olividae.** Anterior part of foot separated by a transverse furrow, mesopodium reflected laterally over the shell; eyes placed half-way up the tentacles: shell olive-shaped; operculum present or absent. *Oliva* Brug., olive-shell; *Olivella* Swainson; *Ancillaria* Lam.; *Olivancillaria* d'Orb.

The **Harpidae** (*Harpa* Lam., harp-shell) and **Marginellidae** (*Marginella* Lam.) are allied here.

Tribe 3. TOXOGLOSSA.

Radula without median and lateral teeth (except *Spirotropis*), 1.0.0.0.1. Oesophagus with a large poison gland; siphon and proboscis well developed; carnivorous.

Fam. 1. **Terebridae.** Eyes at end of tentacles; foot small; shell turriculated. *Terebra* Adanson, auger-shell.

Fam. 2. **Conidae.** Eyes on outer side of tentacles; siphon prominent; shell conical. *Conus* L., cone-shell (Fig. 313); *Bela* Leach; *Clavatula* Lam.; *Drillia* Gray; *Conorbis* Sow.

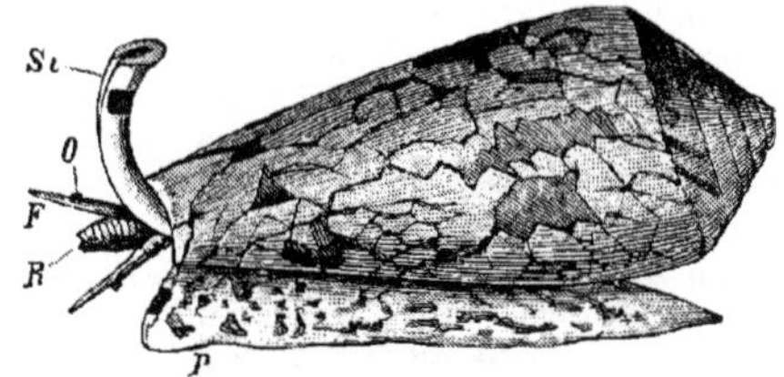

Fig. 313.—*Conus textilis* (règne animal). *R* proboscis; *S* siphon; *F* tentacle; *O* eye; *P* foot.

Fam. 3. **Cancellariidae.** Proboscis short, usually no radula; no operculum; vegetable feeders. *Cancellaria* Lam.

Tribe 4. TAENIOGLOSSA.

Radula with normal formula 2.1.1.1.2. (Fig. 293 *E*). Marginals sometimes numerous.

There are two sections, the *Platypoda* and *Heteropoda*.

Section 1. Platypoda.

Normal creeping Pectinibranchs, little modified. Foot more or less flattened ventrally. Rarely with accessory glands on the generative ducts (*Paludina, Cyclostoma, Naticidae*, etc.). Generally with jaws.

Fam. 1. **Naticidae.** Foot very large, partly covers shell; propodium reflected on the head; shell semi-globular, spire small; operculum. Eyes often absent. carnivorous, feeding on bivalves. *Natica* Lam.; *Sigaretus* Lam.

Fam. 2. **Lamellariidae.** Mantle more or less completely covering the shell. Without operculum. Jaws fused dorsally. *Lamellaria* Montagu; *Onchidiopsis* Beck; *Velutina* Blainv.; *Marsenina* Gray.

Fam. 3. **Trichotropidae.** Radula allied to that of *Velutina*. *Trichotropis* Broderip.

Fam. 4. **Naricidae.** Foot circular, carrying an epipodial lobe on each side; tentacles flattened; shell naticoid, with a velvety periostracum. *Narica* Recluz.

Fam. 5. **Xenophoridae.** Mineralogists, conchologists. Snout elongated: foot divided transversely into two parts, of which the posterior carries the operculum. Shell trochiform, attaching foreign bodies (stones, shells) externally. *Xenophora* Fischer.

Fam. 6. **Capulidae.** Visceral sac and shell conical (patelliform), slightly curved behind, usually with an internal plate; without operculum; columellar muscle horseshoe-shaped. *Capulus* Montf. (*Pileopsis* Lam.), bonnet limpet; *Crepidula* Lam.; *Calyptraea* Lam., cup and saucer limpet; *Thyca* Adams.

Fam. 7. **Hipponycidae.** Visceral mass and shell conical; foot reduced, may secrete a calcareous base. *Hipponyx* Defrance.

Fam. 8. **Littorinidae.** Proboscis short, broad; tentacles long, eyes at their outer bases; penis behind the right tentacle; oviparous or viviparous; radula long; shell turbinate, solid, operculum corneous. Marine or brackish water; mostly littoral. *Littorina* Férussac, periwinkle; *Lacuna* Turton; *Cremnoconchus* Blanford. *Fossarus* Philippi is allied here.

Fam. 9. **Cyclophoridae.** Mantle-cavity without ctenidium, transformed into a pulmonary sac. Tentacles long, filiform. Pedal ganglia elongated into cords. Otocysts with otoconia. Operculum circular. Terrestrial. *Pomatias* Hartmann; *P. obscurum* S. Europe; *Cyclophorus* Montf.; *Cyclosurus* Morelet; *Pupina* Vign.; *Cataulus* Pfr.; *Cyclotus* Guilding; *Pterocyclus* Benson; *Leptopoma* Pfr.; *Megalomastoma* Guild.; *Craspedopoma* Pfr.; *Diplommatina* Bens.

Fam. 10. **Cyclostomatidae.** Mantle-cavity pulmonary. Ctenidium absent. Otocysts with otolith; jaws absent; deep median longitudinal groove in foot. Operculum generally calcareous. Terrestrial. *Cyclostoma* Drap.; *C. elegans* Müller, Britain and temp. Europe.

The **Aciculidae** (*Acicula* Hartm.), also terrestrial and found in Britain, are allied here.

Fam. 11. **Truncatellidae.** Looping snails. A monopectinate gill; snout long, bilobed; foot very short. They walk by contracting the space between the lips and foot. Found between tide-marks, survive many weeks out of water, *Truncatella* Risso.

Fam. 12. **Rissoidae.** Epipodial filaments; operculigerous lobe with appendages; shell small, acuminate; marine and brackish water. *Rissoa* Freminville; *Litiopa* Rang.

Fam. 13. **Hydrobiidae.** Operculigerous lobe without filaments; shell small, acuminate. Brackish or fresh-water. *Hydrobia* Hartm.; *Bithynia* Gray; *Lithoglyphus* Mühlfeldt; *Pomatiopsis* Tyron; *Bithynella* Moquin; *Assiminea* Leach.

The **Skeneidae** (*Skenea* Fleming) and **Jeffreysiidae** (*Jeffreysia* Alder) are allied here; as also are probably the **Homalogyridae** (*Homalogyra* Jeffreys) and the **Choristidae** (*Choristes* Carpentier).

Fam. 14. **Adeorbidae.** *Adeorbis* S. Wood.

Fam. 15. **Paludinidae.** Shell conical or globular, with thick olive-green periostracum; operculum horny, normally concentric. Snout blunt, tentacles long, eyes on short pedicels outside tentacles. Pedal ganglia elongated into cords; kidney with ureter. Viviparous. Fresh-water. *Paludina* Lam., river-snail.

Fam. 16. **Valvatidae.** A bipectinate ctenidium projecting freely on the left side; a filiform appendage on the right side; eyes sessile behind the tentacles. Hermaphrodite. Fresh-water. *Valvata* Müller.

Fam. 17. **Ampullariidae.** Mantle-cavity divided into a right branchial division containing a monopectinate ctenidium, and into a left half functioning as a pulmonary chamber. Oesophageal nerve-collar in front of buccal mass. Fresh water. *Ampullaria* Lam. (Fig. 303). Apple-snail or idol-shell. Inhabit lakes and rivers in warmer parts of world, retiring into the mud in the dry season.

Fam. 18. **Cerithiidae.** Spire elongated; siphon short; operculum horny. Marine or brackish water. *Cerithium* Adanson; *Cerithidea* Swainson; *Triforis* Deshayes; *Laeocochlis* Dunker and Metzger; *Potamides* Brongniart.

The **Planaxidae** (*Planaxis* Lam.) and the **Modulidae** (*Modulus* Gray) are allied here.

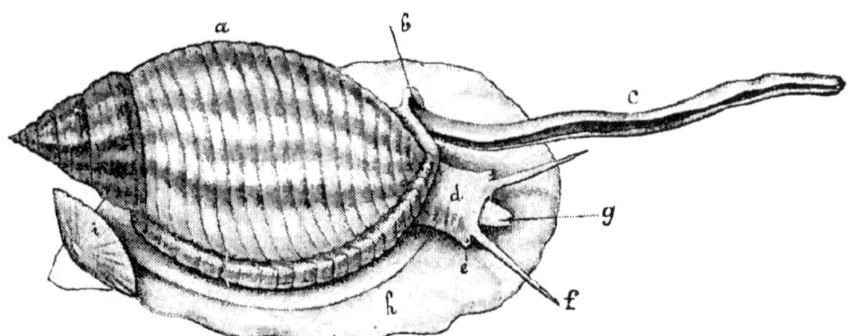

Fig. 314.—*Cassis sulcosa* (after Poli, from Lang). *a* shell; *b* anterior canal or spout for the siphon; *c* siphon; *d* head; *g* proboscis; *e* eye; *f* tentacle; *h* foot; *i* operculum.

Fam. 19. **Nerinaeidae.** *Nerinaea* Defrance.

Fam. 20. **Melaniidae.** Spire elongated; thick, dark periostracum; edge of mantle fringed. Viviparous. Fresh-water in warmer parts of world. *Melania* Lam.; *Melanopsis* Lam.; *Paludomus* Swainson.

Pleuroceridae (fresh-water), **Pseudomelaniidae** (marine) are allied here.

Fam. 21. **Turritelidae.** Shell multispiral or tubular; operculum horny; head large and projecting; edge of mantle fringed; foot broad. *Turritella* Lam.; *Mathilda* Semper; *Coecum* Fleming.

Fam. 22. **Vermetidae.** Shell tubular, attached, irregularly coiled, last coils not in contact; operculum circular. *Vermetus* Adans., worm-shell; *Siliquaria* Brug.

Fam. 23. **Strombidae.** Wing-shells. Foot narrow, arched, compressed laterally, without ventral sole; snout long; large and elaborate eyes placed on thick pedicels with slender tentacles arising from the middle of the pedicels. Very active, progressing by a sort of leaping movement. *Strombus* L.; *Pterocerus* Lam., scorpion-shell; *Terebellum* Klein; *Rostellaria* Lam.

Fam. 24. **Chenopodidae (Aporrhaidae).** Foot elongated, narrow; proboscis short, tentacles long; siphon short. Shell like that of *Strombus*. *Chenopus* Philippi (*Aporrhais* Aldrovandus).

Fam. 25. **Struthiolariidae.** Shell buccinoid; operculum claw-shaped, notched; foot broad and short. *Struthiolaria* Lam.

Fam. 26. **Cypraeidae.** Cowries. Mantle reflected over the shell, and provided with appendages; shell convolute, enamelled; spire concealed; outer lip thickened, inflected; operculum absent. The young shell has a prominent spire. *Cypraea* L., Cowry; *Erato* Risso; *Ovula* Brug.; *Pustularia* Swainson.

Fam. 27. **Doliidae.** Shell ventricose, spirally furrowed; no operculum; eyes stalked; foot large and expanded; siphon long. *Dolium* Lam., the Tun; *Pyrula* (*Pirula*) Lam., fig-shell.

Fam. 28. **Cassididae.** Shell ventricose with varices, spire short, outer lip reflected or thickened, operculum semilunar; eyes sessile; foot large, rounded in front; proboscis and siphon long. *Cassis* Lam., helmet-shell (Fig. 314); *Cassidaria* Lam.

Fam. 29. **Tritonidae.** A proboscis; siphon well developed but short; foot short; shell thick, varicose, outer lip inflected and thickened, operculum corneous; periostracum often thick and hairy. *Triton* Lam.; *Ranella* Lam. frog-shell. The genus *Oocorys* Fischer, is allied here.

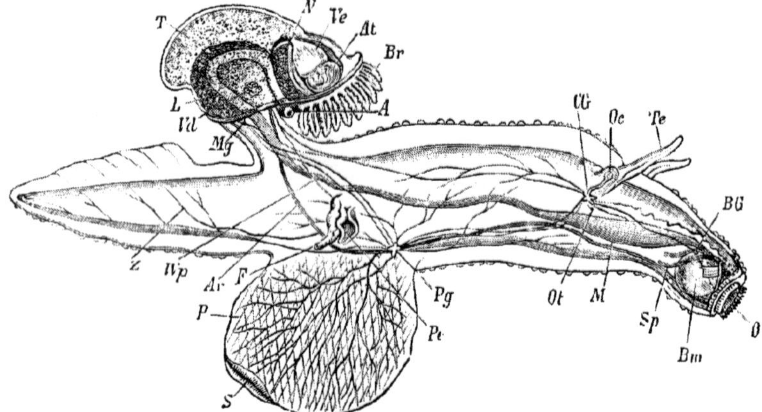

FIG. 315.—Male of *Carinaria mediterranea* (after Souleyet, Gegenbaur, and Keferstein). *A* anus; *Ar* anterior aorta; *At* auricle; *BG* buccal ganglion; *Bm* buccal mass; *Br* gills; *CG* cerebral ganglion; *F* flagellum; *L* liver; *M* crop; *Mg* mantle ganglion; *N* kidney; *O* mouth; *Oc* eye; *Ot* otocyst; *p* pro- and mesopodium; *Pe* penis; *Pg* pedal ganglion; *S* sucker; *Sp* salivary gland; *T* testis; *Te* tentacles; *Vd* vas deferens; *Ve* ventricle; *Wp* ciliated furrow; *Z* posterior branch of anterior aorta.

Section 2. **Heteropoda.***

The *Heteropoda* are free-swimming, pelagic Pectinibranchiates, with laterally-compressed foot, taenioglossate radula, and otocysts near the cerebral ganglia. They are without jaws.

* C. Gegenbaur, "*Untersuch. üb. Pteropoden u. Heteropoden*," Leipzig, 1854. T. H. Huxley, "On the Morphology of the Cephalous Mollusca," etc., *Phil. Trans.* 1853.

The main peculiarity of the body consists in the large size of the head and foot, and the small size of the visceral sac. The **foot** is divided into an anterior part—the so-called pro- and mesopodium (Fig. 315 *p*)—and a posterior part—the metapodium. The former is foliaceous, and often carries a sucker on its posterior part; while the latter, considerably elongated and extended far backwards, forms the caudal continuation of the body.

The visceral sac is either spirally twisted and enclosed by a mantle and spiral shell (*Atlanta*), or has the form of a projecting mass, which is placed over the hinder part of the foot, and is likewise covered by the mantle and a reduced hat-shaped shell (*Carinaria*); or finally, the visceral sac is reduced to a small, scarcely-projecting nucleus, which is covered on the front side by a membrane with a metallic lustre, and is completely without a shell.

The **nervous system** is well-developed. The cerebral ganglia are conjoined, and the pleural are fused with them (visibly in *Atlanta* and *Pterotrachea*). There are two pedal connectives on each side, which are partially free proximally in *Atlanta*. The pedal ganglia are situated at the base of the fin (pro-mesopodium). The visceral commissure is long, streptoneurous, and provided with several ganglia, but without dialyneury or zygoneury (p. 365). In the *Carinariidae* there are secondary viscero-pedal anastomoses which are not twisted. In the *Firolidae* the pedal connectives are fused with the anterior part of the visceral commissure; and behind the pedal ganglion the two branches of the visceral commissure are fused together for the greater part of their extent.

The **osphradium** has the form of a more or less elongated ciliated organ in the pallial cavity to the left of the ctenidium. The **otocysts** are close to the cerebral ganglia. The **eyes** are very large, contained in capsules, and moved by special muscles; they are highly elaborate and are placed at the base of the tentacles, which are entirely absent in *Pterotrachea*.

The **alimentary canal** has a protractile pharynx with radula, a long oesophagus slightly swollen towards its middle, and a stomach and liver placed behind. The intestine is short and is not curved forward in *Pterotrachea*.

The **heart** is near the stomach; it is in front of the gill in *Pterotrachea*, but is disposed as in the *Platypoda* in the less specialized forms. The arteries terminate abruptly in the blood-sinuses. The **ctenidium** is monopectinate, not covered by mantle-fold in *Pterotrachea*, and absent in *Firoloida*. The **kidney** is a transparent sac with contractile walls; it has the same relations as in other Gastropods, and opens not far from the anus.

The males are distinguished by the possession of a large penis, which projects freely on the right side of the foot. The males of *Pterotrachea* possess a sucker on the hinder part of the anterior division of the foot; in *Atlanta* and *Carinaria* the sucker is present in both sexes. The **ovaries** and **testes** fill the posterior part of the visceral sac, and are partially embedded in the liver. The ducts open on the right side of the body near the anus. There is a ciliated groove to conduct the sperm from the generative opening to the penis. The penis is provided with a glandular appendage or flagellum. The oviduct possesses a large albumen gland and a receptaculum seminis opening into it.

The *Heteropoda* are exclusively pelagic animals, and are often found in great numbers in the warmer seas. Their tissues and shell are transparent. They are somewhat clumsy in their movements, which are effected with the ventral surface uppermost, by oscillations of the whole body and fin. They are all carnivorous. When the tongue is protruded, the lateral teeth fly apart from one another like the limbs of a pair of forceps, and when retracted they again fall together; in this way small marine animals are seized and drawn into the mouth.

Fam. 1. **Atlantidae.** Visceral sac and shell coiled in a spiral. The metapodium carries an operculum. *Atlanta* Lesueur.

Fam. 2. **Carinariidae.** Visceral sac and shell conical, and small in relation to rest of body; foot elongated, without operculum. *Carinaria* Lam.

Fam. 3. **Pterotracheidae.** Visceral sac reduced; without mantle and shell; a sucker on foot in male only. *Pterotrachea* Forskal. No tentacles: a filiform appendage at posterior part of foot. *Firoloida* Lesueur. Tentacles present; gill and posterior pedal appendage absent.

Tribe 5. **GYMNOGLOSSA.**

Radula and jaws absent.

Fam. 1. **Eulimidae.** Proboscis elongated; tentacles without furrow: often parasitic on Echinoderms. Shell small, long, subulate, polished. *Eulima* Risso; *Stilifer* (*Stylifer*) Brod., a cephalic pseudopallium extending over shell, in body-wall of male star-fish.

It is convenient to place here the two genera *Entocolax* and *Entoconcha*, both parasitic on Echinoderms. In these genera there is only a vestige of the alimentary canal with a single opening. The visceral mass is surrounded by the pseudopallium, and the cavity so formed receives the generative duct and opens to the exterior by a small orifice. *Entocolax* Voigt. Fixed by the aboral extremity; sexes separate. *E. Ludwigi*, fixed to the body-wall in body-cavity of a Holothurian. *Entoconcha* Müller, vermiform, fixed by the oral end; hermaphrodite. Larva with a shell (resembling that of *Natica*), operculum and velum. *E. mirabilis* Müller, in body-cavity of *Synapta digitata* attached to intestinal wall.

Order 2. EUTHYNEURA.*

Hermaphrodite Anisopleura, in which the visceral commissure is not twisted (save in the **Actaeonidae**). *The primitive left gill, nephridium, and auricle are always aborted, and the operculum is generally absent.*

There are typically two pairs of tentacles on the head. The visceral commissure is generally very short, and its ganglia (Fig. 316)

* Alder and Hancock, "*A Monograph of the British Nudibranchiate Mollusca*," London, 1845-55. Hancock, "On the Anatomy of Doridopsis," *Trans. Linn. Soc. London*, 25, 1865. H. Lacaze Duthiers, "Anat. et Phys. du Pleurobranche orange," *Ann. Sci. Nat.* (4), 11, 1859. Id., "Du système nerveux des Mollusques gastéropodes pulmonés aquatiques," *Arch. Zool. Exp.* (1), 1, 1872. Id., "Histoire de la Testacelle," *ib.* (2), 5, 1887. Vayssière, "Recherches zoologiques et anatomiques sur les mollusques Opisthobranches du Golfe de Marseille," *Ann. Mus. Marseille* (*Zool.*), 2 and 3, 1885-88. P. Pelseneer, "Report on the Pteropoda," *Challenger Reports*, pt. 66, 1888. Bergh, "Die kladohepatischen Nudibranchier," *Zool. Jahrb.* (*Abth. f. System.*), 5, 1890. Leidy, "Special Anatomy of the terrestrial Gastropoda of the U.S." in Binney, *The terrestrial air-breathing Molluscs of the U.S.*, vol. i., Boston, 1851. P. Pelseneer, "Recherches sur divers Opisthobranches," *Mémoires Couronnés et Mémoires des Savants Etrangers, Acad. Bruxelles*, 53, 1894. L. Pfeiffer, "*Monographia Heliceorum Viventium*," Leipzig, 1848-69; and *Monographia Auriculaceorum Viventium*, Cassel, 1856. A. Rossmässler, "*Iconographie der Land-u. Süsswasser Mollusken Europas*, Leipzig, 1835-59. Férussac et Deshayes, *Hist. nat. gén. et partic. des Mollusques terrestres et fluviatiles*, Paris, 1829-51.

all approximated to the oesophagus (except in the Bulloids and in *Aplysia*). In the Bulloids it shows traces of the characteristic torsion of the Streptoneura; and in one form, *Actaeon*, it is actually streptoneurous (Fig. 318). Hermaphroditism is universal and the generative system is complicated by the presence of a number of accessory structures (albumen and mucous glands, spermathecae, etc.).

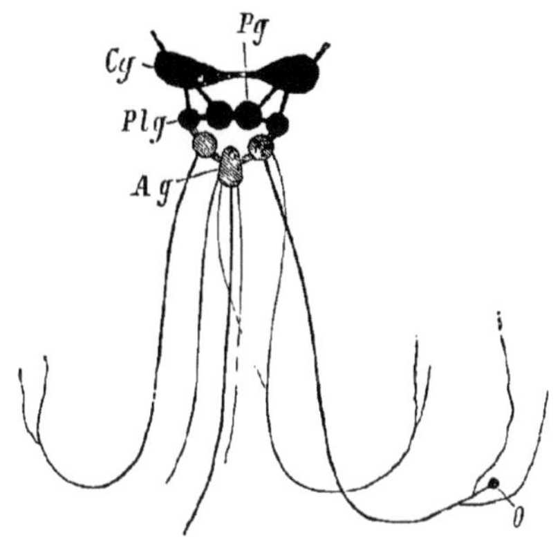

FIG. 316.—Nervous system of *Limnaea* (after L. Duthiers). *Cg* cerebral, *Pg* pedal, *Plg* pleural, *Ag* abdominal ganglion; *O* osphradium.

The otocysts generally contain otoconia. Further, there is in this order a tendency to the reduction of the pallial cavity, the shell, and the operculum. The operculum is absent in almost all forms; the shell is completely absent in Nudibranchs, and it tends to disappear in Pulmonates and Tectibranchs. The mantle-cavity is absent in Nudibranchs, and is but slightly marked off in most Tectibranchs; and in cases in which it is present its reduction is indicated by the fact that the genital ducts generally open outside it.

As explained on p. 385, the *Euthyneura* may be looked upon as streptoneurous forms which have undergone partial detorsion, the result of which is that the visceral commissure is generally untwisted, and the mantle-cavity and external openings are, as a rule, shifted off the anterior face of the visceral sac to the right side, and often displaced far backwards.

Sub-order 1. **Opisthobranchiata.**

Marine Euthyneura with aquatic respiration. The gills are generally behind the heart. The pallial-cavity when it exists is widely open.

The mantle-cavity never contains more than one ctenidium, placed on the right side of the body. Often there is no ctenidium (Nudibranchs).

Tribe 1. **TECTIBRANCHIATA.**

Opisthobranchs provided with a mantle, and a shell (except in *Runcina*, *Pleurobranchaea*, the *Pneumatodermatidae*, *Clionopsidae*, and *Clionidae*), with a ctenidium (except in the three above-named families), and with an osphradium.

The tribe includes three sections—the *Bulloidea*, the *Aplysioidea*, and the *Pleurobranchioidea*. All the families formerly united as Gymnosomatous *Pteropoda* are included with the *Aplysioidea*, while the thecosomatous Pteropods are placed with the *Bulloidea*. All are marine.

The leading features of the group are as follows:—

1. The displacement backwards of the pallial opening, the anus and the circumanal complex.
2. The disappearance of the visceral twist and of the operculum; the reduction of the mantle and shell; the exposure of the genital opening, and of the ctenidium.
3. The acquisition of external symmetry.
4. The absence of the spermatic groove, which is transformed into a vas deferens.
5. The reduction in the number of teeth in the transverse rows of the radula.
6. The concentration of the central nervous system, and the shortening of the visceral commissure.
7. The displacement of the oesophageal nerve collar towards the posterior part of the buccal mass.

In the following table the distribution of these characters amongst the families is shown:—

	1	2	3	4	5	6	7
Pleurobranchidae	—	—	—	—		—	—
Umbrellidae	—	—	—	—		—	—
Gymnosomata		—	—			—	—
Aplysiidae	—	—	—			—*	—
Lobiger	—			—	—	—	—
Pelta	—	—	—			—	—
Thecosomata			—*		—	—	—
Philinidae	—	—	—				
Bullidae	—				—*		—*
Actaeonidae				—			

* In some members of the type only.

From this table it is apparent that *Actaeon* most resembles the Streptoneura, inasmuch as it has its visceral commissure twisted, auricle anterior to ventricle, an operculum, genital opening (female) in mantle-cavity, anus forward; but it differs from that group in being hermaphrodite. With all these streptoneurous affinities one might expect the genital apparatus to be of the most unspecialized construction, *i.e.*, with an hermaphrodite opening and perhaps without even a spermatic groove, but this is not the case: in its ducts *Actaeon* is as specialized as any Tectibranch. If we attempt to draw up a genealogical scheme for the group, we cannot place *Actaeon* at the root because in this particular it is as highly specialized as any form. *Aplysia* on the other hand, while most highly specialized in many of its organs, presents the unspecialized features of having an open spermatic groove and a much elongated visceral commissure.

Section 1. **Bulloidea.**

Shell well developed, external or internal (absent only in *Runcina*). Head usually without tentacles, with a broad cephalic disc. Epipodia continuous with ventral face of foot. Stomach generally provided with masticatory plates. In *Actaeon* and the *Limacinidae* the foot carries a semilunar operculum. The pleural ganglia are close to or fused with the cerebral. Visceral commissure long except in the *Limacinidae*, *Cymbuliidae*, *Cavoliniidae*. The visceral commissure is twisted in *Actaeon* in the streptoneurous fashion, and shows indications of the same feature in many other *Bulloidea*. For instance, in *Scaphander* the left half of the commissure passes, ventral to the oesophagus, to the subintestinal ganglion; while the right half is placed mainly dorsal to the alimentary canal and passes to the supra-intestinal ganglion which supplies the gill and the osphradium. The nerve collar is anterior to the buccal mass in *Actaeon*, *Bulla striata*, *Philine*, *Doridium*, while in others (*Bulla hydatis*, *B. cornea*, *Acera*, etc.) it is behind. The male and female generative openings may be in common or separate. In the latter case the male opening is at the end of the penis on the right side in front of the mantle opening, while the female is within the mantle cavity (*Actaeon*). When there is a common opening (*e.g.*, *Cymbuliidae*, *Limacinidae*, *Cavoliniidae*, *Pelta*, etc.) there is a seminal groove to the penis.

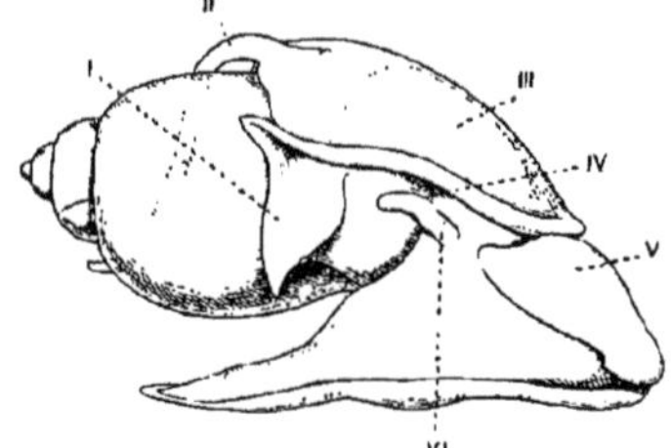

Fig. 317.—*Actaeon tornatilis* (after Pelseneer) × 5. *I* inferior lobe of mantle; *II* pallial gland; *III* hypobranchial gland; *IV* pallial aperture; *V* eye and cephalic shield; *VI* penis.

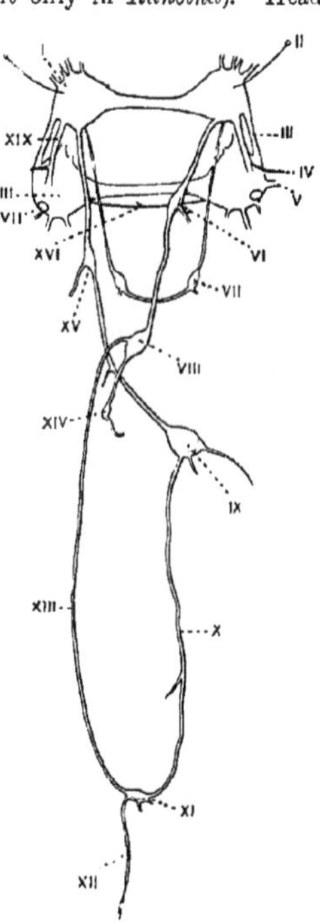

Fig. 318.—Nervous system of *Actaeon*, dorsal view, × 20 (after Pelseneer). *I* cerebropleural ganglion; *II* eye; *III* cerebropedal connective, *IV* nerve from pleuropedal connective, *V* penial nerve; *VI* right accessory pallial, *VII* stomatogastric, *VIII* supraintestinal, *IX* sub-intestinal, and *XI* abdominal ganglion; *X* sub-intestinal part of the visceral commissure, *XIII* supraintestinal part of same; *XII* genital nerve; *XIV* osphradial and *XV* left accessory pallial ganglion; *XVI* parapedal commissure; *XVII* otocyst; *XVIII* pedal ganglion; *XIX* pleuropedal connective.

Fam. 1. **Actaeonidae.** Cephalic disc bifid behind. Shell external with prominent spire, entirely covering animal; operculum horny: visceral loop streptoneurous; epipodia absent. *Actaeon* Montf. (Fig. 317).

Fam. 2. **Ringiculidae.** Cephalic disc forms behind an open tube. Shell external with prominent spire, without operculum. *Ringicula* Deshayes.

Fam. 3. **Tornatinidae.** Shell as before, but spire concealed. Radula absent. *Tornatina* Adams; *Volvula* Adams.

Fam. 4. **Scaphandridae.** Shell external, without projecting spire; cephalic disc short, truncated behind; radula with first lateral very large; epipodia well developed; stomach with three well-developed calcareous plates. *Scaphander* Montf.; *Cylichna* Lovén; *Amphisphyra* Lovén; *Atys* Montf.; *Smaragdinella* Ad.

Fam. 5. **Bullidae.** Shell external without projecting spire; cephalic disc bifurcated behind; epipodia large; radula usually multiserial. *Bulla* L., bubble-shell; *Acera* Müller.

The **Aplustridae** (*Aplustrum* Schumacher) are allied here.

Fam. 6. **Philinidae.** Shell internal; epipodia large; cephalic disc simple. *Philine* Ascanius, three calcareous stomach-plates; *Doridium* Meckel, no stomach-plates, two posterior pallial appendages; *Gastropteron* Meckel, mantle and shell much reduced, epipodia large, united behind.

Fam. 7. **Runcinidae.** Cephalic shield and mantle continuous; shell absent; four stomach-plates. *Runcina* Forbes (*Pelta* Quatrefages).

The following three families were formerly united as **Pteropoda Thecosomata,** characterized by their foot extending round the dorsal side of the head and being entirely transformed into two anterior lateral fins, by the existence of a mantle-fold, by the absence, in the adult, of a ctenidium (except in certain *Cavolinia*) and of eyes, by the presence of one pair of tentacles of which the right is often the larger, by the position of the cerebral ganglia at the sides of the oesophagus, by the radula having three teeth in each row (1:1:1), by the stomach containing horny plates, and by being pelagic in habit. By the operculum of *Limacina* they are allied to *Actaeon*, and by the approximation of the pleural to the cerebral ganglia and by the stomach-plates, they are allied to the *Bulloidea* generally. The heart has one auricle and one ventricle: the kidney is on the right side, and opens into the pericardium and into the mantle-cavity. The genital glands open on the right side, and there is a ciliated groove for the sperm leading from the opening to the penis, which is anterior and cephalo-dorsal.

Fam. 8. **Limacinidae.** Fins large, mantle-cavity dorsal; shell spiral, sinistral (ultradextral), operculate. Anus, etc., on right side. *Peraclis* Forbes: *Limacina* Cuvier.

Fam. 9. **Cymbuliidae.** Adult without shell; a cartilaginoid pseudo-shell, subepithelial and formed by the connective tissue; mantle-cavity ventral. Larva with a calcareous, spiral operculate shell. *Cymbulia* Pér. et Les.; *Tiedmannia* D. Chiaje; *Cymbuliopsis* Pelseneer; *Gleba* Forskal; *Desmopterus* Chun.

Fam. 10. **Cavoliniidae.** Visceral mass and shell not coiled, symmetrical: pallial-cavity ventral. *Cavolinia* Abildgaard (*Hyalea* Lam., Fig. 319); *Cuvierina* Boas; *Clio* L., with subgenera *Creseis* Rang (Fig. 320), *Hyalocylix* Fol, and *Styliola* Les.

The **Lophocercidae** are allied to the *Bulloidea*. They have an external shell, a long foot, epipodia separated from the ventral face of foot, and a short visceral commissure, and a ctenidium and branchial chamber on right side. *Lobiger* Krohn; *Lophocercus* Krohn.

Section 2. **Aplysioidea.**

The shell is reduced or absent. The head is without cephalic disc and has two pairs of tentacles. The epipodia arise from the sides of the body, and are not direct continuations of the ventral surface of the foot. The visceral commissure is much shortened (except in *Aplysia*), and the nerve ganglia are closely aggregated. There is a second commissure connecting the pedal ganglia called the parapedal commissure, in addition to the ordinary one; the aorta passes between the two. The osphradium and its ganglion (supplied by a nerve from the supraintestinal) is placed at the base of the gill, between the openings of the genital organ and the kidney. There is a seminal groove leading from the former to the penis, which is on the right side of the head.

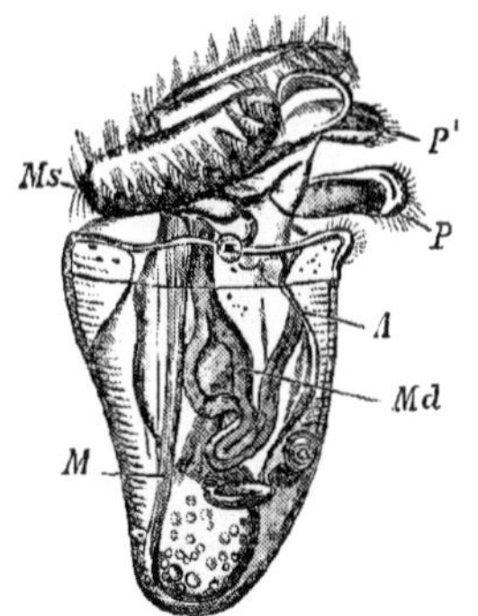

FIG. 319.—Larva of *Cavolinia* (*Hyalea*) *tridentata* (after Fol). *Ms* velum; *P* foot; *P'* the two lateral (epipodial) lobes of the foot; *A* anus; *M* retractor muscle; *Md* stomach.

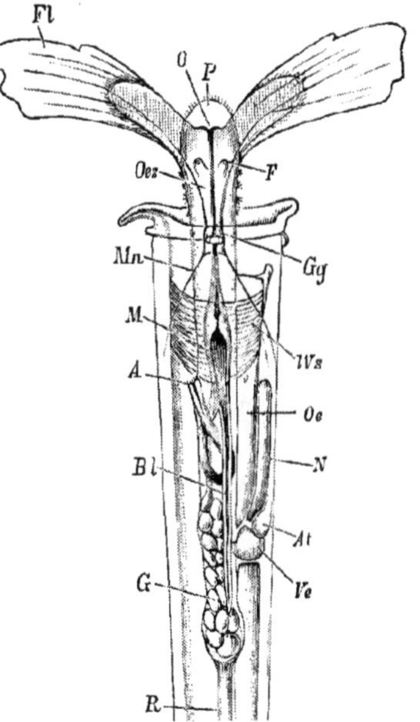

FIG. 320.—*Creseis acicula*, dorsal view, after Gegenbaur (hind end omitted). *A* anus; *At* auricle; *Bl* blind sac of stomach; *F* tentacle; *Fl* fins; *G* gonad; *Gg* cerebral ganglion; *M* stomach; *Mn* mantle-nerve; *N* kidney; *O* mouth; *Oe* opening of kidney into mantle-cavity; *Oes* oesophagus; *P* median lobe of foot; *R* retractor; *Ve* ventricle; *Ws* ciliated shield. The terminal part of intestine passes ventral to anterior, not dorsal as here represented.

The *Aplysiidae* and the **Pteropoda Gymnosomata** are included in this section. The *Aplysiidae* have a widely open mantle-cavity, ctenidium, and an internal shell, all of which are absent in the other families. *Pneumoderma*, however, is said to retain the ctenidium in the lateral gill.

Fam. 1. **Aplysiidae.** Shell internal, but the sac in which it lies opens by a pore in the centre of the dorsal surface. With a mantle-cavity and ctenidium. *Aplysia* L., Sea-hare; *Aplysiella* Fischer; *Phyllaplysia* Fischer, shell absent; *Notarchus* Cuvier, epipodia fused, dorsally to the visceral sac; *Dolabella* Lmk.

The following families, formerly included under **Pteropoda Gymnosomata**, are without mantle or shell. The head is well developed and bears two pairs of tentacles, the posterior pair with eyes. The median part of the foot consists of a posterior lobe and two antero-lateral lobes joined in front (horse-shoe-shaped): the epipodia or fins are quite separate from this median part of the foot and from the head. The penis is latero-ventral on the right side of the foot; a groove leads to it from the genital opening. There is an evaginable proboscis generally bearing suckers or appendages with suckers. These have nothing to do with the foot. The buccal cavity contains evaginable hook-sacs opening on each side of the radula. The jaws are united ventrally in the middle line. The stomach is without plates. The anus is on the right side. The cerebral ganglia are closely approximated on the dorsal side of the oesophagus, and the pleural are near the pedal; in this they differ from the so-called thecosomatous Pteropods, and resemble the *Aplysiidae*. There may be branchial processes of the integument—a lateral one on the right side—and a posterior gill.

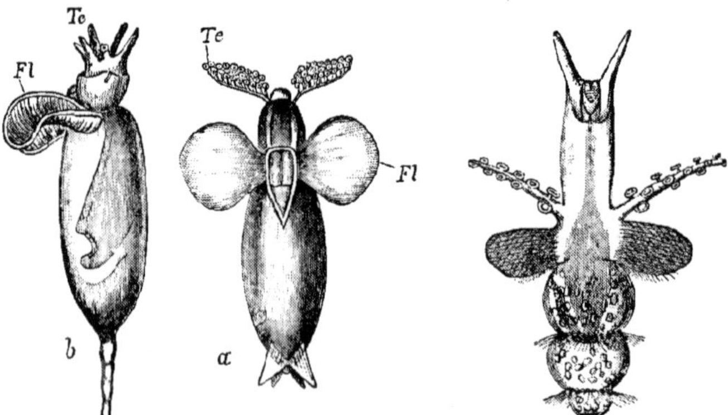

FIG. 321.—*a*, *Pneumoderma violaceum* (from Bronn) from the ventral surface. *b*, *Clione australis* (règne animal) from the side. *Fl* fins (epipodia); *Te* appendages of the proboscis, with suckers in *a*.

FIG. 322.—Larva of *Pneumoderma* (after Gegenbaur).

Fam. 2. **Pneumodermatidae.** With suckers on the proboscis. *Dexiobranchaea* Boas; *Pneumoderma* Cuvier (Fig. 321); *Spongiobranchaea* d'Orb.

Fam. 3. **Clionopsidae.** Without suckers and buccal appendages. Proboscis very long. Posterior gill tetraradiate. *Clionopsis* Troschel.

Fam. 4. **Clionidae.** With conical glandular buccal appendages (cephaloconesʼ). No gill. *Clione* Pallas (Fig. 321).

Fam. 5. **Notobranchaeidae.** With conical buccal appendages, without lateral gill. *Notobranchaea* Pelseneer.

Fam. 6. **Halopsychidae.** Body ovate, rounded behind; without gills or proboscis; fins broadened at the ends. *Halopsyche* Bronn.

Section 3. **Pleurobranchioidea.**

The head has two pairs of tentacles, of which the posterior are the rhinophores; the foot is without epipodia; the mantle-cavity is shallow and contains a large ctenidium on the right side. The male and female generative openings are near but separate, and there is no seminal groove. The pleural ganglia are close to, or fused with, the cerebral, which are themselves closely approximated. The visceral commissure is short and possesses but few ganglia. In *Tylodina* alone is there an osphradium and osphradial ganglion.

Fam. 1. **Umbrellidae.** Visceral sac and external shell in the form of a flat cone; foot very thick. *Umbrella* Lam.; *Tylodina* Rafinesque.

Fam. 2. **Pleurobranchidae.** Shell internal or absent; anterior tentacles forming a frontal velum; spicules in the mantle; foot flat. *Pleurobranchus* Cuv., internal shell; *Pleurobranchaea* Meckel (Fig. 323), no shell; *Neda* Adams.

Tribe 2. **NUDIBRANCHIATA.**

Opisthobranchs without shell in the adult state.

There is no ctenidium or osphradium; the nervous system is much concentrated, and the ganglia are generally aggregated, and more or less fused together, on the dorsal side of the oesophagus; but the different commissures—visceral, pedal, and buccal—are distinct. There are four suboesophageal commissures—the visceral, the pedal, the parapedal (a second thin commissure between the pedal ganglia), and the subcerebral (Fig. 324). In addition to these there is the stomatogastric system, which constitutes another commissure completed ventral to the oesophagus. The nerves springing from the visceral commissure are always unsymmetrical and are exclusively genital and reno-pericardial.

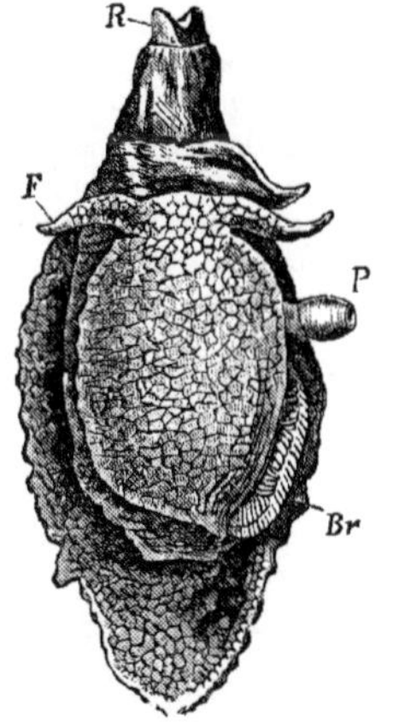

Fig. 323.—*Pleurobranchaea Meckelii* (règne animal). *Br* ctenidium; *F* posterior tentacle (rhinophore); *R* proboscis; *P* penis.

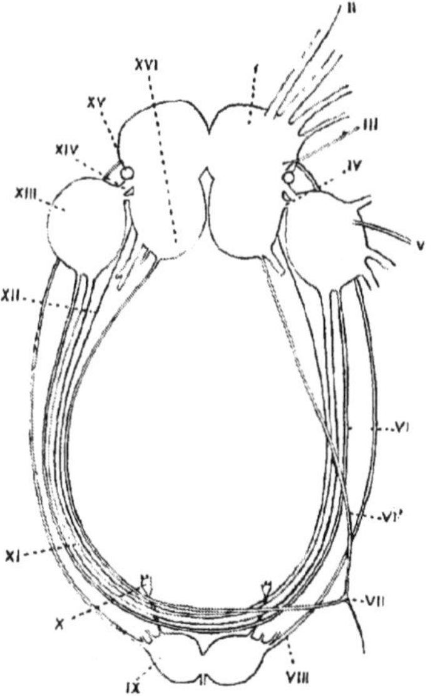

Fig. 324.—Nervous system of *Tritonia hombergi* (after Pelseneer), dorsal view × 15. *I* cerebral ganglion; *II* tentacular, *III* optic nerve; *IV* pleuropedal commissure; *V* penial nerve; *VI* parapedal commissure; *VI'* visceral commissure; *VII* abdominal ganglion and genital nerve; *VIII* stomatogastric commissure; *IX* buccal, *X* gastro-oesophageal ganglion; *XI* pedal, *XII* sub-cerebral commissure; *XIII* pedal ganglion; *XIV* cerebro-pedal commissure; *XV* otocyst; *XVI* pleural ganglion.

The pedal commissure is generally long, but it is short in *Fiona*, *Ancula*, and most *Elysioidea*, in which the pedal ganglia are approximated ventral to the oesophagus.

The parapedal commissure is homologous with the same structure in Tectibranchs, but it gives off no nerves. The relation of the aorta to these commissures varies, as it does in the Tectibranchs. The dorsal appendages or cerata are innervated from the pleural ganglia, showing that these structures are homologous and of a pallial nature. The penis is of a pedal nature, being innervated from the pedal ganglion as in the majority of *Streptoneura*.

The hermaphrodite gland has male and female acini, which however communicate. The animals are protandrous.

The characters of specialization of the Nudibranchs are as follows:

(1) Return to external symmetry. This was indicated in Tectibranchs, but it is carried further in Nudibranchs, where there may be complete external symmetry as in the *Doridioidea* with median anus, and even internal symmetry in many of the organs (heart, kidney, liver), but never in the generative apparatus. In the *Tritonioidea* this internal symmetry is not found.

(2) Displacement of the nervous system behind the buccal mass.

(3) Concentration of the central nervous system dorsal to the oesophagus. The pedal ganglia are approximated to the cerebral, and the pleural are fused with the cerebral except in *Tritoniidae*. But as already mentioned in *Fiona*, *Ancula*, and in most *Elysioidea*, we find the pedal more or less approximated ventrally to oesophagus, and distinct ganglia on the short visceral commissure. In the presence of distinct ganglia on the visceral commissure, often three in number, the *Elysioidea* are more primitive than other Nudibranchs.

(4) Reduction of the number of teeth of the radula (even to a single tooth for each transverse row).

(5) Diffusion of the liver.

(6) Triaulic condition of genital ducts.

The *Tritoniidae*, which are Nudibranchs because they are without ctenidium, osphradium, mantle-fold, shell, and because they have cerata, are without any of these features of specialization; except the dorsal grouping of the cerebral, pedal, and pleural ganglia, which are, however, distinct. But even in the

If we had another dimension we could connect the Elysid group and Janus to show the common presence of cerata.

nervous system they show the specialized feature of the reduction of ganglia on the visceral commissure.

The Dorids are the most specialized of all in possessing the median anus and the triaulic generative ducts; but they lack the diffusion of the liver.

The Eolids are highly specialized in their nervous system and cerata, but have the diaulic ducts. *Janus*, however, goes further in having a median anus, though the kidney is still lateral; but if we should be tempted to put *Janus* at the top of our genealogical tree we should be prevented by the fact that in its nervous system it approaches *Tritonia*, and that it has diaulic generative ducts.

The Elysiids retain primitive features of the nervous system, but in other respects are the most specialized of the group. But the nervous system drags them down, and would prevent us from putting them at the top of our genealogical tree. Indeed, by it we might feel inclined with Bergh to place them in a separate Tribe between the *Tecti-* and *Nudi-branchiata.*

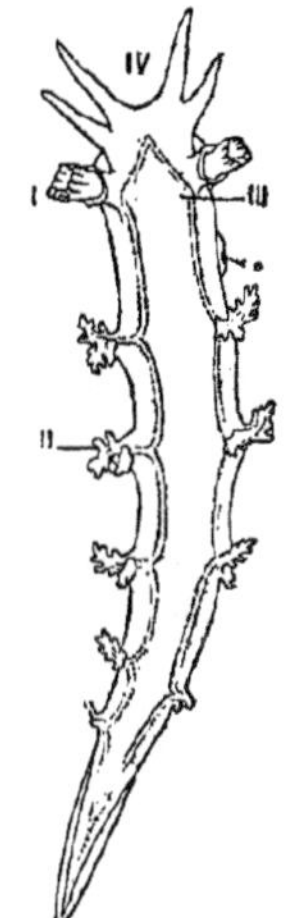

FIG. 325. — *Tritonia lineata*, dorsal view, × 7 (after Hancock, from Pelseneer). *I* posterior tentacle; *II* dorsal appendages (cerata); *III* eye; *IV* frontal velum; o genital opening.

Section 1. **Tritonioidea.**

The liver is contained entirely or principally in the visceral mass. The anus is on the right side. There are generally two rows of branched dorsal appendages. Male and female openings contiguous; genital ducts diaulic. The heart is asymmetrical, being on the right side.

Fam. 1. **Tritoniidae.** Anterior tentacles forming a frontal velum, foot large. Pleural ganglion distinct from cerebral. Kidney a simple sac opening externally on right side above the anus. Two rows of branched cerata without liver continuations. *Tritonia* Cuv. (Fig. 325), *Marionia* Vayssière.

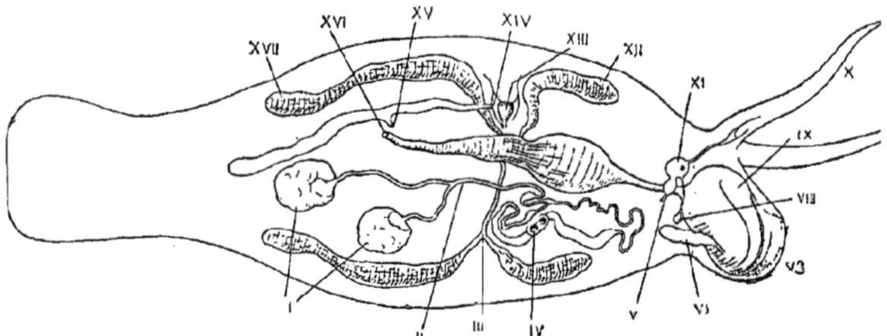

FIG. 326.—*Phyllirhoe bucephalum*, from the right side, × 3 (after Souleyet). The genital ducts are supposed to be a little unrolled. *I* genital gland; *II* sperm oviduct; *III* hepatic duct; *IV* female genital orifice; *V* pedal ganglion; *VI* salivary gland; *VII* mouth; *VIII* stomato-gastric ganglion; *IX* buccal mass; *X* tentacle; *XI* cerebral and pleural ganglion; *XII* lobe of liver; *XIII* heart in pericardium; *XIV* reno-pericardial opening; *XV* external opening of kidney; *XVI* anus; *XVII* liver.

Fam. 2. **Scyllaeidae.** Anterior tentacles absent; two pairs of large foliaceous cerata; foot narrow. *Scyllaea* L., on floating seaweed.

Fam. 3. **Phyllirhoidae.** Anterior tentacles and cerata absent. Body compressed laterally, without foot. Stomach with caeca. Orifices on the right side. With fin-like tail. Transparent. Pelagic. *Phyllirhoe* Per. et Les. (Fig. 326).

Fam. 4. **Tethyidae.** Anterior tentacles absent; rhinophores conical; head surrounded by a funnel-like velum; cerata foliaceous; foot large; radula absent. *Tethys* L. *Melibe* Rang. Cerata of *Tethys* are capable of independent movement after separation.

Fam. 5. **Dendronotidae.** Tentacles forming a fringed frontal velum; cerata branched; rhinophores arborescent; radula 10 : 1 : 10; liver extending into the cerata. *Dendronotus* A. and H.; *Hero* Lovén; *Lomanotus* Verany.

Tribe 2. DORIDIOIDEA.

Liver not branched; anus median posterior, generally dorsal and surrounded by ramified appendages—the cerata or branchiae. Genital ducts triaulic. Spicules in the mantle.

Fam. 1. **Polyceridae.** A frontal velum more or less projecting. Branchiae non-retractile. Rhinophores foliate. *Euplocamus* Philippi; *Triopa* Johnston; *Polycera* Cuv.; *Ancula* Lovén; *Goniodoris* Forbes; *Idalia* Leuckart; *Heterodoris* Verril and Emerton; *Aegirus* Lovén; *Acanthodoris* Grube.

Fam. 2. **Dorididae.** Mantle covering the head; anterior tentacles small; rhinophores foliate; branchiae retractile into a perianal pocket. *Doris* L.; *Chromodoris* A. and H.; *Hexabranchus* Ehrbg.

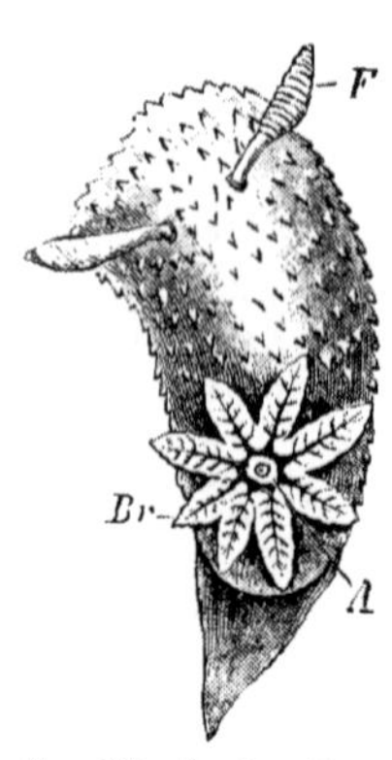

Fig. 327.—*Doris pilosa* (after Alder and Hancock). *Br* gills; *A* anus; *F* tentacle.

Fam. 3. **Doridopsidae.** Doris-like, but mouth suctorial, without radula. *Doridopsis* A. and H.

Fam. 4. **Corambidae.** Doris-like, but anus and branchiae behind, below the edge of the mantle. *Corambe* Bergh.

Fam. 5. **Phyllidiidae.** Mouth suctorial; no radula; branchiae all round the body between the mantle and the foot. *Phyllidia* Cuvier.

Tribe 3. AEOLIDIOIDEA.

The liver caeca extend into the cerata, which often contain nematocysts. Genital duct diaulic, with contiguous male and female openings. Jaws are present.

Fam. 1. **Aeolididae.** Cerata terminating in open sacs (Fig. 329), which communicate with the hepatic caeca and give origin by their lining epithelium to nematocysts (Fig. 330). *Aeolis* Cuvier (Fig. 328).

Fam. 2. **Glaucidae.** With three pairs of lateral lobes (cerata) carrying tegumentary papillae. Foot narrow. *Glaucus* Forster.

Fam. 3. **Pleurophyllidiidae.** Anterior tentacles constitute a burrowing shield. Branchiae beneath the edge of the mantle; radula 30 : 1 : 30. *Pleurophyllidia* Meekel (Fig. 331); *Dermatobranchus* Van Hasselt.

Fam. 4. **Dotonidae.** Cerata tuberculated, without nematocysts, and in one row on each side. *Doto* Oken.

Fam. 5. **Proctonotidae.** Anus posterior in the middle dorsal line. Anterior tentacles atrophied. *Janus* Verany ; *Proctonotus* A. and H.

Fam. 6. **Fionidae.** Liver as two longitudinal canals, into which open the caeca of the cerata. *Fiona* Hancock and Embleton.

Tribe 4. ELYSIOIDEA (SACCOGLOSSA).

Liver branched, the branches generally extending into cerata. Genital ducts usually triaulic, with widely separated genital openings. The vaginal orifice is, in some cases at any rate, developed later than the other two openings. Without jaws. Radula with one series of strong teeth, the worn-out teeth at the front end not dropping off, but preserved in a special sac (*Ascoglossa* Bergh). The pedal ganglia are near together, and the visceral commissure has three ganglia almost in contact.

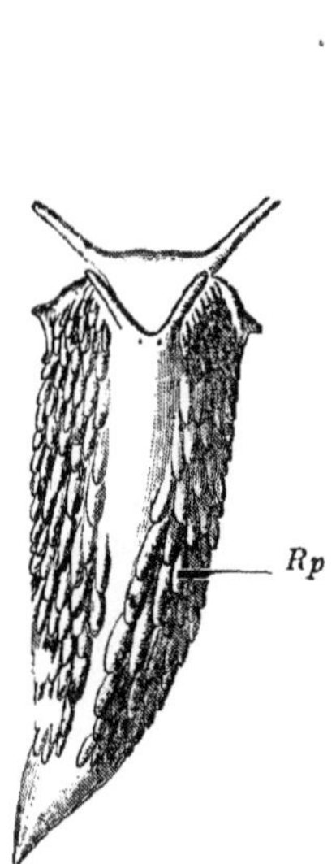

FIG. 328.—*Aeolis papillosa* (after Alder and Hancock). *Rp* dorsal papillae (cerata).

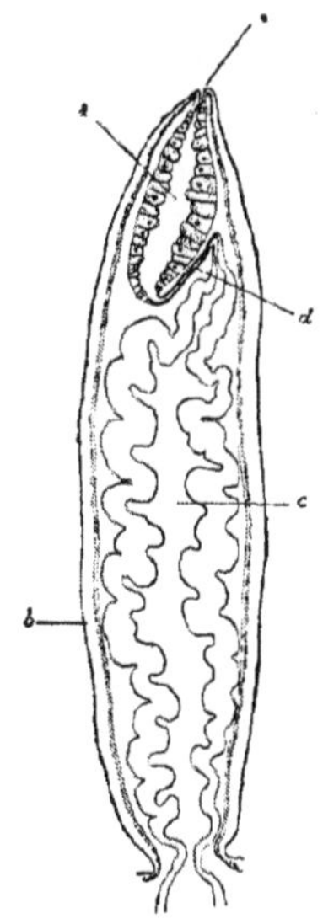

FIG. 329.—Section of a dorsal papilla of *Eolis* ×.40. *a* terminal sac which produces nematocysts; *b* ectoderm; *c* hepatic caecum; *d* tube connecting the caecum with the terminal sac; *e* opening of the sac (after Pelseneer).

FIG. 330.—A nematocyst with protruded thread of *Aeolis punctata* × 500 (from Pelseneer, after Vayssière).

Fam. 1. **Hermaeidae.** Cerata in several rows; anus dorsal. *Hermaea* Lovèn, *Stiliger* Ehrb., *Phyllobranchus* A. and H., *Cyerce* Bergh, hepatic caeca do not extend into cerata ; *Alderia* Allman.

Fam. 2. **Elysiidae.** Cerata absent; dorsal integument forms two lateral expansions containing hepatic caeca. Anus lateral. Reno-pericardial openings numerous. *Elysia* Risso.

Fam. 3. **Limapontiidae.** Without lateral expansions or cerata. Anus median postero-dorsal. Liver but little branched. *Limapontia* Johnston; *Actaeonia* Quatref.

The position of the **Rhodopidae** is uncertain: they are planarian-like naked forms, without branchiae, tentacles, buccal mass, or radula. Recently it has been shown* that the larva of this form is without a shell; thus destroying the main reason for regarding it as a nudibranch mollusc.

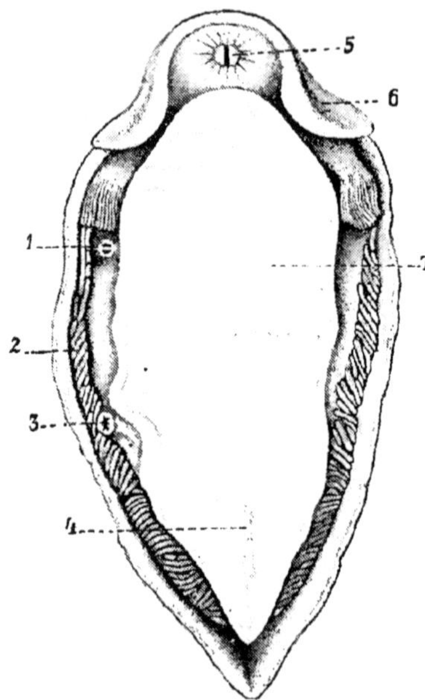

Fig. 331.—*Pleurophyllidia lineata*, ventral view (from Lang, after Souleyet). *1* genital opening; *2* branchial processes; *3* anus; *4* pedal gland; *5* mouth *6* tentacle shield; *7* foot.

Sub-order 2. Pulmonata.

Euthyneura in which the pallial-cavity is modified as a lung, and is without a ctenidium. The opening of the pallial-cavity is small.

The pallial-cavity is often reduced, as well as the shell. Sometimes the shell is internal or absent. There is never an operculum in the adult except in *Amphibola* (it is present in the young stages of *Auricula, Siphonaria, Gadinia*). The roof of the mantle-cavity is provided with a network of vessels for aerial respiration, and the mantle-cavity constitutes a lung. In *Vaginulus, Peronia*, and *Onchidium* the lung is absent in consequence of the almost complete disappearance of the mantle-cavity. In some cases the lung may be filled with water and serve for aquatic respiration (*Siphonaria*, some *Limnaea* and *Planorbis*). The auricle is usually in front of the ventricle. The kidney has usually a more or less elongated duct (ureter).

The opening of the pallial-cavity is reduced to a pore placed on the right side. The anus and renal openings are placed near the opening of the lung (in *Auricula* alone does the anus open in the

* S. Trichinese, "Nuove osservationi s. *Rhodope Veranii*," *Rend. Accad. Napoli* (2), vol. 1., p. 131. See also L. Böhmig in *Z. f. w. Z.*, 56.

lung sac). The generative organs open some way in front, on the right side.

The *Pulmonata* are generally terrestrial, sometimes fresh-water, rarely marine. They are cosmopolitan and include about 6000 species. They generally become torpid during part of the year—in warm countries in the dry season (aestivation), in cold in winter (hibernation). Hibernation in this country lasts about one quarter of the year, and during it the heart beats no more than twice a minute.

The mouth armature consists of an unpaired, horny upper jaw (which may be absent), and of a radula with a great number of teeth in the transverse row.

All are hermaphrodite. A few, *e.g.* species of *Clausilia* and *Pupa*, are viviparous. Most lay eggs, either, as in the fresh-water forms on water plants, united in tubular or flat masses, or as in the terrestrial forms in damp places, each one being surrounded by a protecting shell which may be calcareous. The ovum is always embedded in a large mass of albumen, which serves as nourishment for the developing embryo.

Tribe 1. **BASOMMATOPHORA.**

Shell always present and external. There is one pair of non-invaginable tentacles, at the base of which are the eyes. The penis is remote from the female orifice (except in *Amphibola* and *Siphonaria*). In *Auricula* alone among Pulmonates is the hermaphrodite duct not divided; and, passing from its opening which is anterior to the mantle-opening, there is a groove which is closed in front and leads to the penis. The closed part of this groove lies close beneath a superficial groove in the skin. In *Amphibola*, *Chilina*, and the *Siphonariidae* the duct splits, but the penial and vaginal openings are joined. There is generally a circular osphradium near the opening of the pulmonary cavity, between it and the renal opening. Radula multiserial. Fresh-water or quasi-marine.

Fam. 1. **Auriculidae.** Terrestrial, usually maritime. Respiratory aperture behind. Shell spiral. *Auricula* Lam., found in brackish-water swamps of tropical islands, on roots of mangroves, and by small streams within the influence of the tide; traces of small anterior tentacles; *Carychium* Müller; *Alexia* Lam.; *Pedipes* Adanson; *Melampus* Montf.; *Otina* Gray.

Fam 2. **Amphibolidae.** Visceral mass and shell spiral; operculum present. Aquatic, marine. *Amphibola* Schumacher, shores of New Zealand and Pacific Islands.

Fam. 3. **Siphonariidae.** Visceral mass and shell patelliform; tentacles atrophied; marine with aquatic respiration. *Siphonaria* Sow., secondary branchial lamellae on roof of mantle-cavity; *Gadinia* Gray, no branchia.

Fam. 4. **Limnaeidae.** Fresh-water animals with aerial respiration. Shell variable. *Limnaea* L.; *L. stagnalis* L., pond-snail; *Amphipeplea* Nillsson;

Physa Draparnaud; *Planorbis* Guettard, spire in same plane; *Ancylus* Geoffroy, river-limpet, visceral mass conical.

Fam. 5. **Chilinidae.** Pulmonary aperture larger than in any other pulmonate. Visceral commissure longer than usual. *Chilina* Gray, Chilian-snail.

Tribe 2. **STYLOMMATOPHORA.**

Two pairs of invaginable tentacles (except *Athoracophorus*), the posterior being oculiferous: male and female openings united (except in *Vaginulus, Onchidium*, and *Peronia*); no osphradium.

Fam. 1. **Succineidae.** Anterior tentacles reduced; male and female openings distinct, but contiguous. *Succinea* Draparnaud, amber-snail.

The **Athoracophoridae** (*Athoracophorus* Gould) are allied here.

Fam. 2. **Helicidae.** Land-snails. Shell external, spire short; genital organs generally with a dart and multifid vesicles; genital orifice under right tentacle. *Helix* L., over 1600 species; *H. aspersa* Müller, hedge-snail; *H. pomatia* L., Roman snail; *Vitrina* Drap., glass-snail; *Bulimus* Scopoli (1200 species); *Hemphillia* Binney and Bland.

The **Philomycidae** (*Philomycus*), without shell, allied here.

Fam. 3. **Arionidae.** Shell absent, or represented by calcareous granules in mantle. Slug-like. Genital opening just below pulmonary. *Arion* Fér., land-sole; eggs of *A. hortensis* phosphoresce for 15 days after deposition.

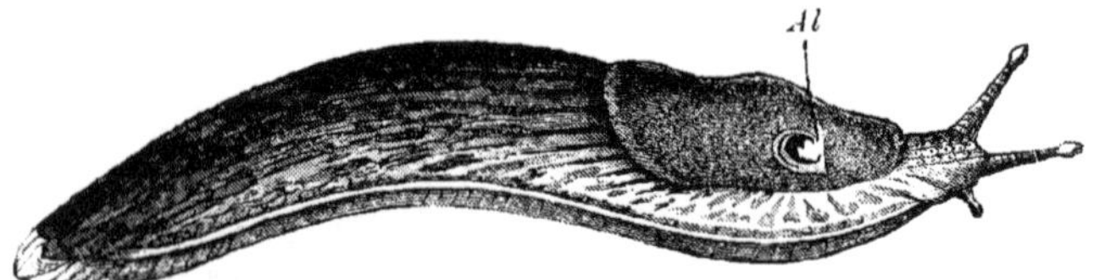

FIG. 332.—*Arion empiricorum* (règne animal). *Al* respiratory aperture.

Fam. 4. **Pupidae.** Shell external, spire long; male duct without multifid vesicles. *Pupa* Lam., chrysalis shell; *Clausilia* Drap.; *Vertigo* Müller; *Zospeum* Bourguignat; *Balea* Prideaux; *Ferussacia* Risso; *Caecilianella* Fér.

Fam. 5. **Limacidae.** Slugs. Shell small or absent, generally internal, may be spiral; tail often with mucous pore. Genital ducts without multifid vesicles; genital opening under right tentacle. *Limax* L., slug, mantle reduced, shell internal, 51 species; *Vitrina* Drap., glass-snail (here or with *Helix*); *Parmacella* Cuvier; *Zonites* Montf.; *Urocyclus* Gray; *Orpiella* Gould.

Fam. 6. **Testacellidae.** Pharynx protractile; jaw absent; cervical region elongated. Slug-like, or with visceral sac spirally coiled; carnivorous. *Glandina* Schumacher; *Daudebardia* Hartm.; *Testacella* Cuv., slug-like, subterranean, feeds on earthworms, S. Europe, Canaries, introduced into Britain.

Fam. 7. **Vaginulidae.** Without shell; male opening beneath the right tentacle, female midway beneath the mantle; respiratory and excretory orifices at hind end. *Vaginulus* Fér.; *Atopos* Simroth.

Fam. 8. **Onchidiidae.** Marine, without shell, mantle often warty, sometimes with eyes; genital openings widely separate; female opening posterior, near anus; anus and pulmonary opening as in *Vaginulus*. *Onchidium* Buchanan; *Peronia* Blainv.

Class V. CEPHALOPODA.*

With well-marked head, a circle of processes bearing suckers or tentacles round the mouth, and a funnel composed of two separate or fused halves. The genital coelom is continuous with the pericardial. Dioecious.

The *Cephalopoda* are symmetrical animals with a much-shortened antero-posterior axis, and with a strongly-developed visceral sac, which has undergone neither torsion nor asymmetrical development, and which is, except in *Nautilus*, unprotected by an external shell. The mantle-fold is circular and the mantle-cavity is especially developed on the posterior side of the visceral sac; in it are placed the ctenidia, either two (*Dibranchiata*) or four (*Tetrabranchiata*), and into it open the median anus, the ink-sac, the paired kidneys, and the genital duct. There are always processes round the mouth, which are either lobe-like and carry tentacles as in the Tetrabranchiates, or are arm-like and carry suckers as in the Dibranchiates. They are active, voracious animals, with a complex organization, highly developed sense-organs, and often possessed of considerable intelligence. The Cephalopods are marine animals, some frequenting the coast, others the high seas, and some the floor of the ocean to a depth of nearly 2000 fathoms. About 400 species are known. They feed on the flesh of animals, especially Crustacea, and some of them attain a great size.† The flesh is eaten, and the colouring matter of the ink-sac (sepia) and the dorsal shell (os sepiae or cuttle-bone) are used by man. The remains of Cephalopods occur in all formations from the Cambrian, and constitute important characteristic fossils (*Belemnites*, *Ammonites*).

* Férussac et d'Orbigny, "*Histoire naturelle générale et particulière des Céphalopodes acétabulifères vivants et fossiles,*" Paris, 1835–45. J. B. Verany, "*Mollusques méditerranés observés, etc., d'après le vivant,*" 1e Partie, Gênes, 1847–51. H. Müller, "Ueber das Männchen von Argonauta argo u. die Hectocotylen," *Z. f. w. Z.*, 1855. Jap. Steenstrup, "Hectocotylus dannelsen hos Octopodsl., etc.,"-*K. Danks. Vidensk. Selskabs Skrifter*, 1856; translated in *Archiv. f. Naturgesch.*, 1856. C. Grobben, "Morphologische Studien üb. den Harn-u. Geschlechtsapparat, etc., der Cephalopoden," *Arb. a. d. Zool. Inst. Wien*, 5, 1884. W. E. Hoyle, "Report on the Cephalopoda," *Challenger Reports*, vol. 16, 1886. J. Brock, "Zur Anat. u. Syst. d. Cephalopoden," *Z. f. w. Z.*, 36, 1882. P. Pelseneer, "Sur la value morphologique des bras et la composition du syst. nerv. cent. d. Cephalopodes," *Arch. Biol.* 8, 1888. Milne-Edwards et Valenciennes, "Obs. sur la circulation chez les Mollusques," *Mem. Acad. Sci. Paris*, 20, 1840. Vigelius, "Ueb. d. excretionssystem der Cephalopoden," *Niederl. Arch. f. Zool.*, 5, 1880. Milne-Edwards, "Sur les spermatophores des Céphalopodes," *Ann. Sci. Nat. Zool.* (2), 18, 1842.

† Specimens of *Architeuthis* have been taken measuring from apex of visceral sac to end of extended arms more than 50 feet, and with eyes 15 inches across.

The normal swimming position for a Cephalopod (Fig. 333) is horizontal, with the anterior surface of the visceral sac upward, and the posterior, *i.e.*, the surface on which lies the mantle-cavity and the mantle-opening, downward. When creeping the animals apply their arms to the substratum and the hump projects backward.

FIG. 333.—*Octopus vulgaris* (after Merculiano, from Lang), in swimming and sitting positions.

Bearing in mind these facts and the fact already mentioned that the antero-posterior axis is much shortened, the ordinary terminology, which is often confusing, used in describing *Cephalopoda* will be readily understood. The ventral surface is the region including the mouth and the anus. The funnel is therefore a purely ventral structure. The dorsal surface includes the visceral sac; roughly the apex of the visceral sac may be regarded as the middle of the dorsal surface. The anterior end is that surface of the head on the sides of which the eyes are placed; while the posterior end may be said to be marked by the anus. Inasmuch as the visceral sac is normally carried horizontally, the words "upper" and

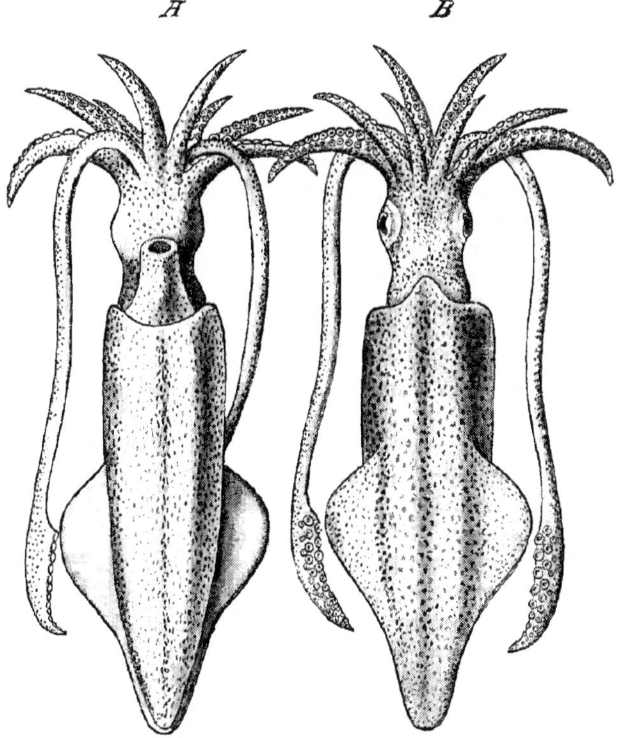

FIG. 334.—*Loligo vulgaris* (from Lang, after D'Orbigny). *A*, from the posterior side, showing the funnel just in front of the free edge of the mantle. *B*, from the front side showing the eyes on the head. There are also shown the ten arms, the fins, and the chromatophores in the skin.

"lower" are sometimes used in the sense of "anterior-dorsal" and "posterior-ventral" respectively; further, inasmuch as the visceral hump is sometimes shown in figures in its proper morphological position, as standing straight up above the head, the words upper and lower are sometimes used in the sense of dorsal and ventral.

In our descriptions the words upper and lower will be discarded, and we shall use only anterior and posterior, dorsal and ventral, in the sense described above.

The head is sometimes described as being placed in the centre of the foot, and the arms are looked upon as the frayed-out margins of the foot; on this view the name *Cephalopoda* is justified anatomically as well as functionally. On the other hand, this view is not taken by all anatomists, many of whom regard the funnel as the sole representative of the foot of other Gastropods. We shall consider this question further on.

The head is well-developed, and carries on its anterior sides a pair of conspicuous and generally elaborate eyes. Immediately behind (ventral to) the eyes there is often on each side a pit supposed to be olfactory. In certain *Oegopsida* the eyes are stalked; in *Nautilus* they are also stalked, and there are on each side two cephalic tentacles in relation with the eye, the preocular and the postocular tentacles (Fig. 338, *Pr.o*, *Pt.o*).

The margins of the head are produced into exceedingly muscular sucker-bearing arms in the Dibranchiates, and into tentacle-bearing lobes in *Nautilus*. Of the former there are always four pairs (*Octopoda*), and sometimes five (*Decapoda*); they are arranged in a circle round the head, and the anterior arm of each side is called the first, while the posterior arm, viz., that nearest the funnel, is called the fourth. The arms of the fifth pair, which are found in the *Decapoda*, are longer than, and attached somewhat internally to, the others between the third and fourth; they are often inserted into pits, into which they may be retractile (*Sepia*, *Sepiola*, *Rossia*), and are called the prehensile or *tentacular arms*. The tentacular arms have suckers only near their free, club-shaped end, but the other arms are suckered all along their internal (oral) surface.

The basal parts of some or all of the eight arms of the *Octopoda* may be united by a membrane; in *Tremoctopus* the four anterior, in *Histioteuthis* the six anterior, in *Alloposus* and *Cirroteuthis* all the arms are so united along their whole length. The *umbrella* so formed assists in locomotion by its alternate contraction and expansion.

In the female *Argonauta* (Fig. 335) the terminal parts of the two anterior (dorsal) arms are expanded into thin membranes (*vela*), which secrete the unilocular spiral shell.

Finally, in the males one of the arms is more or less modified for the purposes of sperm-transference in reproduction. The arm so modified is said to be *hectocotylized*, because in some species (*Argo-*

nauta, Tremoctopus, Philonexis) it is entirely detached from the male and left in the mantle-cavity of the female, where it was found by Cuvier, who mistook it for a parasitic worm and called it *Hectocotylus*.

The suckers are stalked in *Decapoda*, sessile in *Octopoda*. Further, in *Decapoda* they have a horny ring, which may be smooth or denticulated in some cases; on certain parts of the arms the suckers may be replaced by hooks, and on the tentacular arms of *Onychoteuthis* a large retractile hook arises from the centre of each sucker. The suckers consist of a disc-like surface, in the centre of which is a pit; the depth of this pit can be increased by the retraction of its bottom, to which muscle-fibres are attached.

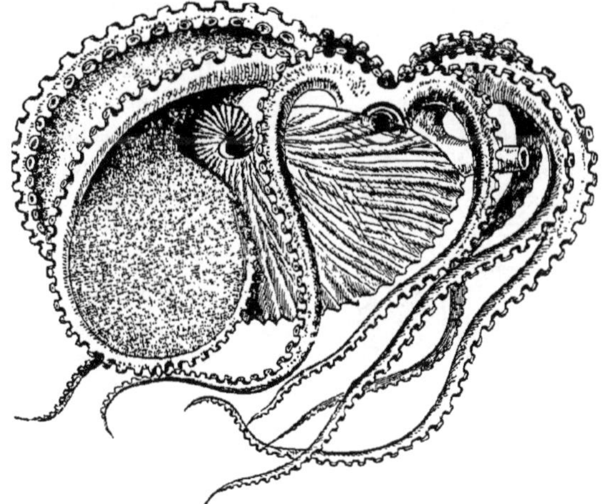

FIG. 335.—*Argonauta argo* (female), the paper Nautilus, swimming.

In the Tetrabranchiates (*Nautilus*) there are no arms, but in their place we find lobe-like prolongations of the margin of the head, which bear tentacles. The tentacles are retractile into sheaths, which are possibly comparable to the suckers of the Dibranchiates. The lobes are as follows (Fig. 336): an external annular lobe (*b*), the anterior part of which forms the *hood*, into which the coil of the shell fits, while the posterior part is much reduced; this lobe carries 19 tentacles on each side; within the annular lobe are in the female three tentaculiferous lobes, a posterior (ventral) and two lateral (*d*, *c*); in the male the posterior lobe (*d*) is reduced to a paired group of lamellae and bears no tentacles, and the right and left inner lobes are divided into two parts, a larger and a smaller. On the posterior inner lobe of the female there is a lamellated organ (*n*), and behind it, on the inner side of the annular lobe, another lamellated organ (*m*) to which the spermatophores of the male appear to be affixed (Kerr). In the male the smaller of the two parts into which the left inner lobe is divided ends in imbricated modified foliaceous tentacles (*p*); it is the *spadix*, and has been regarded as the lobe used in sperm transference and corresponding to the hectocotylized arm of Dibranchs. The corresponding part of the right inner lobe carries four tentacles, and is called the *antispadix* (*q*). In addition to these tentacles of the lobes, there are the four ocular tentacles already referred to.

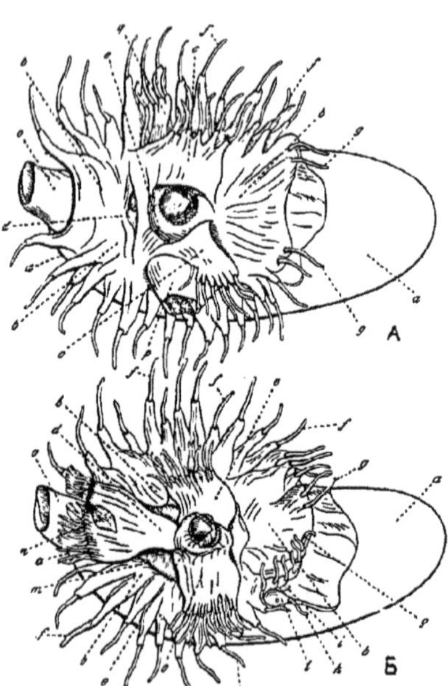

FIG. 336.—Oral surface of a male (*A*) and female (*B*) specimen of *Nautilus pompilius* in the expanded condition, one-third the natural size linear (from Lankester, after Bourne). *a* the shell; *b* the external annular lobe carrying 19 tentacles on each side, and anteriorly enlarged to form the hood (Fig. 337); *c* the right and left inner lobes, each carrying 12 tentacles in the female, and divided in the male into two parts; *d* the posterior inner lobe; *e* the oral cone; *f* the tentacles of the outer annular lobe projecting from their sheaths; *g* the two anterior tentacles of this lobe belonging to the hood; *i* superior, *k* inferior ophthalmic tentacle; *l* eye; *n* lamellated organ on the posterior inner lobe of the female; *m* paired laminated organ on each side of the posterior inner lobe of the female; *o* the funnel; *p* spadix; *q* antispadix.

The **funnel** is a tube with muscular walls leading into the mantle-cavity, and often containing (in *Nautilus* and most Decapods) a valve which admits of passage outwards only. It is placed on the ventral surface behind the head, and is to be regarded either as the entire foot of the animal, or as the epipodia, the rest of the foot having been wrapped round the head and frayed out peripherally into the arms or lobes.* The view that the funnel represents the whole foot is strongly suggested by the arrangement in *Nautilus*. In this animal it consists of two muscular lateral lobes of the ventral surface just behind the head, which are rolled round one another without fusion. When these lobes, which are of considerable size,

* The view that the arms represent a portion of the foot is not in our opinion satisfactorily established. It rests largely upon the fact that the nervous supply is derived from that portion of the central nervous system which is supposed to represent the pedal ganglion of other *Mollusca*. But this, as Graham Kerr has pointed out, is not a strong argument, for a main reason for regarding that part of the central nervous system, from which the brachial nerves arise, as pedal, is that the nerves to the arms arise from it. As pointed out below, separate and distinct ganglia are not distinguishable in the central nervous system of *Cephalopoda*.

are unrolled and flattened out, their resemblance to the foot of a Gastropod is considerable, and Graham Kerr has suggested the possibility of the animal being able actually to unroll them in life and to use them as a foot (Fig. 337). In the Dibranchs the edges of these folds are fused, so that the funnel is a complete tube (Fig. 334), and in both Dibranchs and *Nautilus* the broad hinder part of it is covered by the mantle-folds.

On each side of the funnel there is in Decapods a peculiar sucker-like arrangement by which the mantle-fold can be attached to the funnel so as to close the general mantle-opening; it consists of a smooth cartilaginous projection on the mantle, and a corresponding

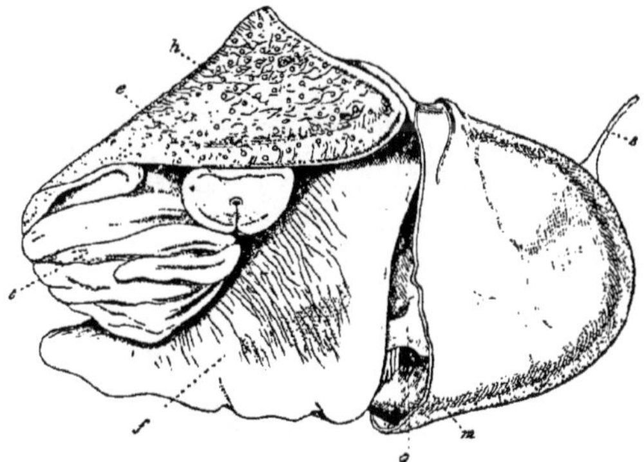

Fig. 337.—Side view of *Nautilus pompilius* extracted from the shell, the funnel has been opened out, and the mantle-flap partly cut away (after Graham Kerr). *f* funnel; *e* eye; *g* gill; *h* hood; *m* cut edge of mantle; *s* siphuncle; *t* tentaculiferous lobes.

depression on the funnel. In the *Octopoda* the mantle-fold is fused to the head anteriorly and laterally, so that the mantle opening is much restricted.

The deep part of the mantle-cavity is placed on the posterior surface of the visceral sac, which in the natural position is the under surface. It has in Dibranchs thick, muscular walls, and it contains one or two pairs of bipectinate ctenidia, the median anus, the paired renal openings, and the generative opening, which is sometimes single and sometimes paired. The regularly repeated contraction of the mantle-muscles causes the expulsion through the funnel of the respiratory water, which has been taken in through the mantle-opening, and with it of the excrementitious (renal and anal) and generative products. When the contraction is rapid and violent the jet from the funnel causes the animal to shoot rapidly backwards; on any alarm the ink-gland, which opens with the anus, pours out its black secretion into the water in the mantle-cavity, so that on the retreat of the animal backwards in the rapid manner just explained, the ejected water is black and produces a murky cloud which completely obscures its movements, and under cover of which it escapes.

Nautilus alone has an external **shell** secreted by the mantle. It is spirally coiled, the coil being directed on to the anterior face of the animal, *i.e.*, being of the kind called *exogastric;* moreover, it is divided by transverse septa into chambers. The bulk of the animal

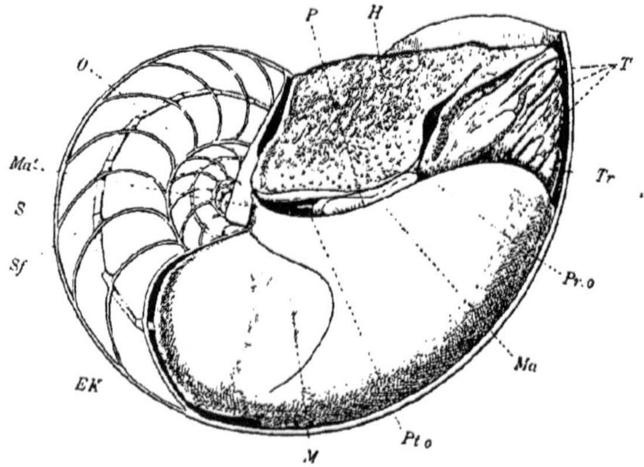

Fig. 338.—*Nautilus pompilius* (original drawing made under direction of Graham Kerr). *EK* penultimate chamber of the shell, which is in some specimens larger than the ante-penultimate; *H* hood; *G* groove in the hood; *M* shell-muscle; *Ma* edge of mantle-fold passing at *Ma'* on to the anterior face of the visceral sac, where it is in relation with the anterior convexity of the shell-coil; *P* eye partly covered by the mantle-fold; *Pr.o* preocular tentacle; *Pt.o* postocular tentacle; *O* part of the outer wall of shell which has not been cut away; *On* initial coil of the shell; *S* siphon; *Sf* postseptal neck; *T* tentaculiferous lobes; *Tr* funnel.

occupies only the last of these chambers, but it is incorrect to say that all the chambers except the last are untenanted by the animal, for there is a delicate process of the dorsal end of the hump—the **siphon** or siphuncle, which is coated with a thin layer of calcareous matter, and passes through all the septa right up to the first chamber. The chambers of the shell contain a gas which is said to have approximately the composition of atmospheric air.

The retractor muscles of the head and foot are inserted on to the internal wall of the shell, and a small anterior flap of the mantle-fold extends on to the convexity of the coil just dorsal to the hood (Fig. 338, *Ma*).

In the Dibranchs there is a spiral, chambered, mantle-shell in the Decapod *Spirula*, but it is coiled in the opposite way to that of *Nautilus* (endogastric), and does not enclose the visceral sac. It is in fact partly internal, being largely covered by lobes of the mantle. The septa are, however, perforated by a siphon, which contains a prolongation of the visceral sac.

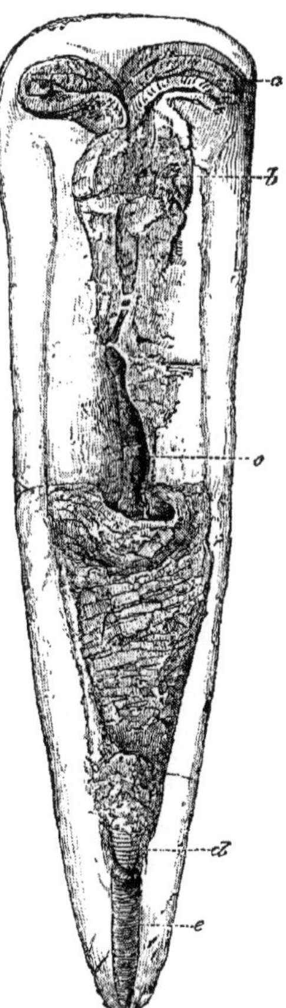

FIG. 340.—*Belemnites* with the remains of the body of the animal (after Huxley). *a* arms with hooks; *b* head; *c* ink-bag; *d* phragmocone; *e* guard.

In certain extinct Dibranchs there was a similar internal, chambered shell —either coiled (*Spirulirostra*) or straight (*Belemnitidae*). That these were internal or partly internal is shown by the fact that the chambered part of the shell or *phragmocone* is covered by a calcareous layer, often laminated, which forms the *rostrum* or *guard*. In the Belemnites the wall of the phragmocone (conatheca) is continued forwards into a proostracum, which must have been somewhere in the neighbourhood of the animal's body and head.

FIG. 339.—*Spirula peronii* (Bronn).

In all living Dibranchs except *Spirula* the shell is quite internal, being contained in a sac in the anterior wall of the visceral sac, and much reduced. In the *Sepiidae*, or cuttle-fishes, it is called the *cuttle-bone* or *sepiostaire*, and consists of a broad plate composed of laminated tissue containing air spaces and ending behind in a pointed rostrum. In the Squids it is a lamellar body composed of conchyolin, without calcareous matter, and is called the *pen*. In the *Octopoda* both the shell and its sac are absent. The shell of the female Argonaut, which has already been referred to, is not a mantle-shell. Fins of various forms are often present as lateral expansions of the mantle.

The dermis contains the remarkable **chromatophores** which cause the well-known play of colours. These consist of large cells filled with pigment (red, blue, yellow, or dark colours); to their walls, which are formed of a cellular membrane, numerous radiating muscular fibres (by some observers said to be connective tissue fibres) are attached. When the latter contract the cells are dilated and the pigment spreads over a larger area, and so gives colour to the skin. When the contraction ceases the cell returns, in virtue of the elasticity of its walls, to its original shape, and the pigment is again concentrated in a small space, and the skin becomes uncoloured; thus the animal changes its colour. The chromatophores are probably under the control of the will, and are connected with a special centre in the stalk of the optic ganglion. The eye seems to be the organ most intimately connected with them, for if the optic nerve be cut the power of voluntarily changing colour on that side is said to be lost. Nevertheless there seem to be peripheral centres through which these organs* can be brought into action, and the animals seem to have the power of changing their colour involuntarily according to the colour of their environment. In addition to the chromatophores there is a deeper layer of small shining spangles, which produce interference colours and thus give rise to the peculiar lustre and iridescence of the skin.

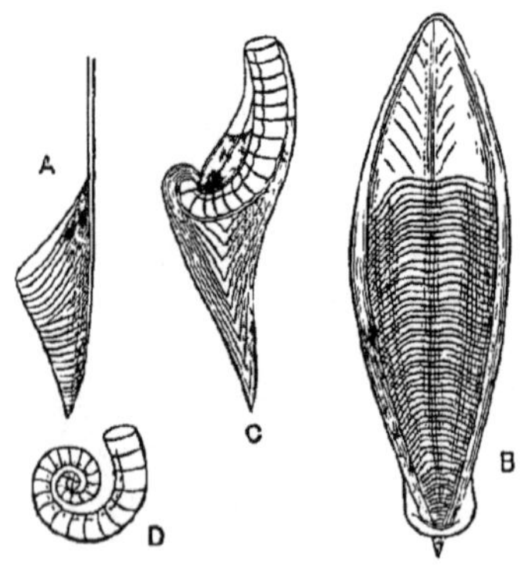

FIG. 341.—Internal shells of *Cephalopoda* (from Lankester). *A*, shell of *Conoteuthis dupiniana* from the Neocomian of France. *B*, shell of *Sepia Orbigniana*, Mediterranean. *C*, shell of *Spirulirostra Bellardii*, from the Miocene of Turin; the specimen is cut so as to show the chambered shell and the laminated guard deposited upon its surface. *D*, shell of *Spirula laevis*.

In certain abyssal Cephalopods there are cutaneous phosphorescent organs consisting of a superficial refractile structure and a deep photogenic layer; they

* Krukenberg, *Vergl. physiol. Studien an den Küsten der Adria*, Heidelberg, 1880.

are all directed towards the oral extremity (*Histioteuthis*). Aquiferous pores leading into spaces in the integument, but not communicating with the vascular system, are found in many Dibranchs.

There is an internal **cartilaginous skeleton** which serves for the protection of nerve-centres and sense-organs, and for the attachment of muscles. The cartilaginous tissue itself closely resembles the hyaline cartilage of vertebrates, differing in the fact that the cells are connected by their branching processes which traverse the matrix in all directions. In the head there is in Dibranchs a complete cartilaginous investment for the great ganglionic masses and otocysts, which furnishes lateral cup-like expansions for the eyes. In *Nautilus* there is a corresponding cephalic cartilage on the ventral side only of the nerve-centres, extending also into the tissue of the funnel. Cartilage is present also in other parts of the body, *e.g.* the branchial cartilage, fin cartilages, nuchal cartilage, dorsal cartilage.

Nervous system. There is a great concentration of ganglionic matter round the oesophagus, and it is difficult to trace exact homologies with the nerve-centres of other Molluscs.

In *Nautilus* there are two ganglionic rings round the oesophagus behind the buccal mass (Fig. 342). They are connected dorsally, above the oesophagus, in a common mass (*X*), which may be called the cerebral mass and compared to the cerebral ganglion of other Molluscs. The anterior ring innervates the funnel and cephalic lobes, and may be compared to the cerebro-pedal commissure and pedal ganglia of other types; the posterior ring innervates the mantle and viscera, and may be compared to the cerebro-pleural commissure, pleural ganglia, and visceral commissure of other Molluscs; it gives off a pair of stout nerves (*V*) which pass backward (dorsalwards), on either side of the vena cava; these supply the gills, osphradia, and viscera, but do not form dorsalwards in the visceral sac, visceral ganglia. The cerebral portion supplies the eyes, otocysts, ocular tentacles, lips, etc., and gives off on each side two nerves to form stomatogastric commissures (*XII*), which surround the oesophagus immediately behind

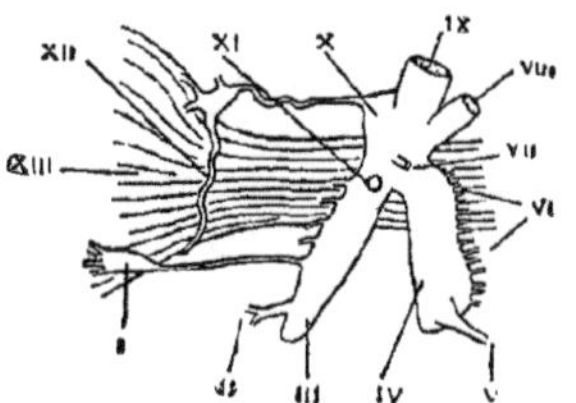

Fig. 342.—Central nervous system of a female *Nautilus* from the left side (from Pelseneer after Valenciennes). *I* accessory pedal ganglion; *II* nerve to funnel; *III* pedal; *IV* visceral ganglion; *V* visceral nerve; *VI* pallial nerves; *VII* tentacular nerves; *VIII* olfactory, *IX* optic nerve; *X* cerebral ganglion; *XI* otocyst; *XII* stomatogastric commissure (for detail see Fig. 348); *XIII* buccal mass.

the buccal mass. This stomatogastric system (Fig. 343) consists of the two above-mentioned nerves (*c.c*) on each side, which pass forward from the cerebral ganglion to end in the two pharyngeal ganglia (*ph. g*). These are connected by a long anterior commissure (*ant. com*) which passes ventral to the mouth, and by a shorter posterior commissure, the buccal commissure—(*buc. phar. com.* and *b. com.*)—which is also ventral to the oesophagus, and contains in its course two buccal ganglia (*buc.g.*). The ganglia and commissures of the stomatogastric system supply the buccal mass. In the female there is an accessory ganglion supplying the inner ventral cephalic lobe, and connected with what has been called above the pedal ganglion (Fig. 342, *I*).

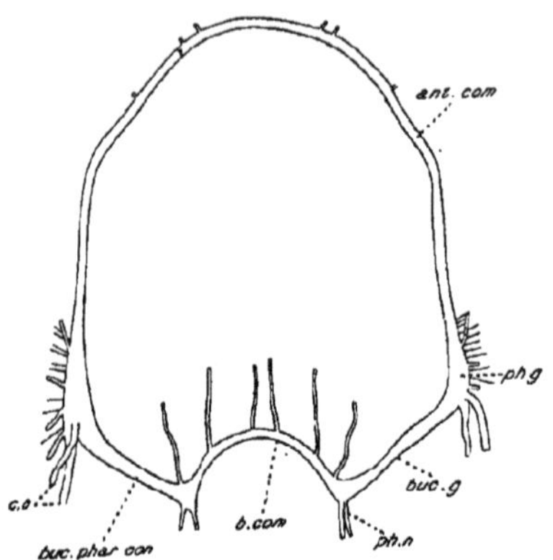

FIG. 343.—Buccal nervous system of *Nautilus pompilius* (after Graham Kerr). *ph.g* pharyngeal ganglion; *buc.g* buccal ganglion (*buc.g* ought to point to the swelling from which the nerves *ph.n* come off); *c.c* cerebro-pharyngeal connective; *buc.phar.con* bucco-pharyngeal connective; *ph.n* pharyngeal nerves; *b.com* buccal commissure (stomatogastric); *ant.com* anterior pharyngeal commissure.

It should be noted that the nerves to the cephalic arms come off rather high up on the anterior ring, and that the nerve to the funnel arises quite ventrally.

In the Dibranchs the great nerve centres are completely enclosed in the skull. As in *Nautilus* it is difficult to speak of special ganglia, for the whole mass is ganglionic. The part dorsal to the oesophagus we may call the cerebral ganglion (Fig. 344, *l*); this gives off later-

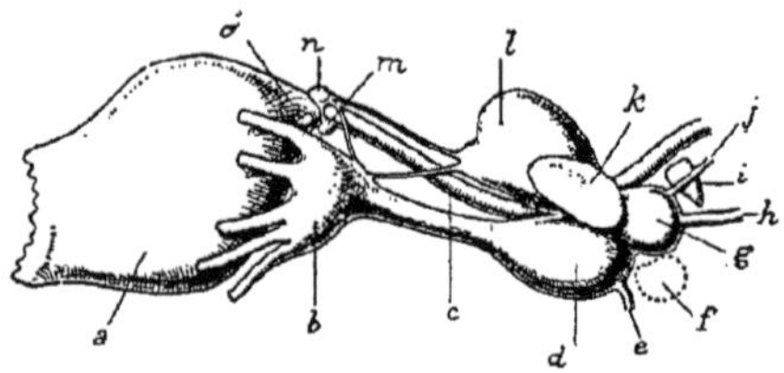

FIG. 344.—Central nervous system of *Ommatostrephes* from the left side, magnified (after Pelseneer). *a* buccal mass; *b* brachial ganglion; *c* oesophagus; *d* pedal ganglion; *e* nerve of the funnel; *f* position of the otocyst; *g* pleuro-visceral ganglion; *h* visceral nerve; *i* posterior salivary gland; *j* pallial nerve; *k* optic nerve, cut; *l* cerebral ganglion; *m* stomatogastric (buccal) ganglion; *n* anterior part of the cerebral ganglion (suprabuccal); *o* anterior salivary gland.

ally the great optic nerves (*k*), and is continued at the sides of the oesophagus as a broad commissure, which leads to the sub-oesophageal mass. The sub-oesophageal mass is indistinctly divided into a posterior part, which is supposed to consist of the pleural and visceral ganglia (*i.e.*, ganglia of the visceral commissure) fused (*g*), and an anterior comparable to the pedal of other types (*d*), the anterior aorta passing ventralwards between the two. The pedal portion is generally divided into two parts, a portion behind—the *pedal* proper—supplying the funnel, and a *brachial* part (*b*) in front, supplying the arms.

The cerebral ganglion is connected by two thin cords, or by one soon dividing into two, with a ganglion on the buccal mass called the *supra-buccal* ganglion (Fig. 344, *n*; Fig. 345, *b*). This is to be regarded as a detached portion of the cerebral; it gives off the stomatogastric commissure, which is completed ventrally to the oesophagus in the *infra-buccal ganglion* (Fig. 344, *m*; Fig. 345, *a*), or buccal ganglion proper. The cerebral and the supra-buccal are both connected with the brachial or anterior part of the pedal by separate commissures (Figs. 344 and 345). In *Octopus* the supra-buccal ganglion is not separated from the cerebral.

The pleuro-visceral portion of the sub-oesophageal gives off from the pleural portion two large pallial nerves (Fig. 344, *j*; Fig. 345, *m*)

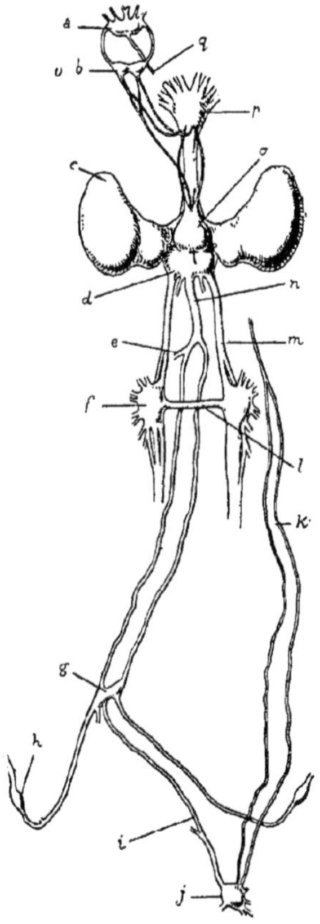

FIG. 345.—Central nervous system of *Ommatostrephes*, dorsal view (after Hancock, from Pelseneer). *a* stomatogastric (buccal) ganglion; *b* anterior part of cerebral ganglion (suprabuccal); *c* optic ganglion; *d* visceral ganglion; *e* rectal nerve; *f* stellate ganglion (mantle ganglion); *g* commissural ganglion of the visceral nerves; *h* brachial ganglion; *i* viscero-stomatogastric anastomosis; *j* stomach ganglion; *k* oesophageal stomatogastric nerves; *l* commissure of the mantle ganglia; *m* pallial nerve; *n* visceral nerve; *o* cerebral ganglion, beneath which a probe indicates the passage of the oesophagus, and of the stomatogastric nerve from *q* to *k*; *p* brachial ganglion; *q* oesophageal stomatogastric nerve.

to the *mantle ganglion* (*ganglion stellatum*), which are in some forms connected by a commissure (Fig. 345, *l*) dorsal to the oesophagus, and from the visceral portion two visceral nerves, fused at their origin, to the viscera (Fig. 345, *n*). The visceral nerves join behind in a ganglion (*g*), which gives off a right and left nerve to the two brachial ganglia (*p*). The infra-buccal ganglion gives off a nerve backwards (Fig. 345, *k*, and Fig. 342, *XI*), which runs along the oesophagus and ends in a large ganglion on the stomach (*j*). This ganglion gives off a nerve which anastomoses with the visceral nerves (*i*).

Organs of sense. Tactile sensibility is specially localized in the arms of Dibranchs and tentacles of *Nautilus*. There is an organ which is supposed to be *olfactory*, just ventral to the eye; it generally has the form of a pit (*Sepia*), but it may be tubercular. It is innervated from the cerebral ganglion. *Osphradia* are absent in Dibranchs, but are supposed to be present in *Nautilus* (see below). The *otocysts* in Nautilus are placed high up on the anterior ring close to the cerebral ganglion (Fig. 342, *XI*), and are sometimes described as being adjacent to the pedal centres; they contain numerous otoconia. In Dibranchs they are embedded in the floor of the skull, and they form a kind of labyrinth, their walls being drawn out into short diverticula; they each contain one large otolith, and are innervated from the pedal ganglion by a nerve which arises from the cerebral. The tube which connects them with the exterior in the embryo persists as a caecal process from the otocyst. Damage to the otocysts of *Cephalopoda* has been found to interfere with their power of maintaining equilibrium.

The **eyes** in most forms are extremely complicated in structure. In *Nautilus*, however, they are very simple, being altogether without refractive media, and consisting merely of a vesicle with an extremely narrow opening to the exterior. The lining of the vesicle constitutes the retina, and is continuous through the aperture with the external ectoderm; it consists of two layers separated by a layer of pigment. The eye of Nautilus is therefore constructed on the principle of the pin-hole camera, there being a dark chamber lined by the sensitive membrane, and a minute hole for the entrance of light.

In the Dibranchs (Fig. 346) the optic vesicle is closed, and its front wall secretes a cuticular biconvex lens (*L*); part of this lens is thrown down by the lining epithelium of the outer wall of the vesicle, and part of it by the outer ectoderm. The lens is therefore theoretically in two parts, and is traversed by the front wall of the vesicle.

The front wall of the vesicle at the point where it runs into the lens is thickened, and constitutes the *ciliary body* (*Ci*). The outside skin is thrown into folds over the eye: internally there is a fold which covers over the ciliary body, and extends as far as the periphery of the lens; it is highly pigmented and constitutes the *iris* (*Jk*). Outside this is another fold of much greater extent, which meets over the front of the lens (*C*), and bounds a space—the anterior optic chamber—the inner wall of which is formed by the lens, iridean folds, and choroid. This fold constitutes the sclerotic and cornea; it is in the *Oegopsida* perforated by a small pore over the lens. Outside the skin may be again folded to form an eyelid.

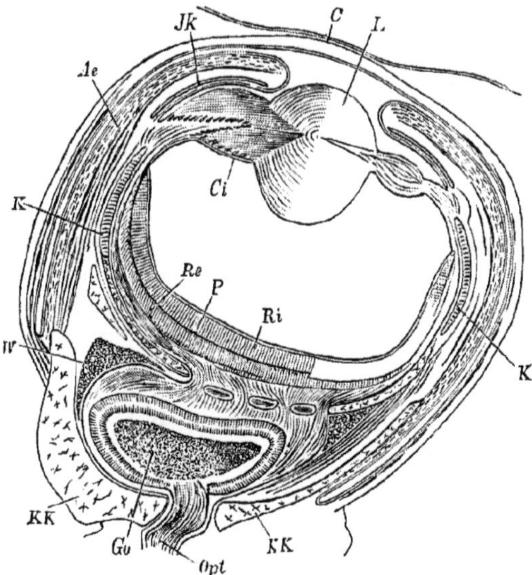

FIG. 346.—Section through the eye of *Sepia*, diagrammatic after Hensen. *Ae* argentea externa; *C* cornea; *Ci* ciliary body; *Go* optic ganglion; *Jk* iris cartilage; *K* cartilage of optic bulb; *KK* cephalic cartilage; *L* lens; *Opt* optic nerve; *P* pigment layer of retina; *Re* outer layer of retina; *Ri* inner layer of retina; *W* white body.

The retina consists of two layers (*Re* and *Ri*), as in *Nautilus*, divided by a layer of pigment (*P*), and the sensitive layer is that which lines the optic vesicle or posterior optic chamber (vitreous humour). The optic nerve comes from the large optic ganglion (*Go*), which is protected by the orbital expansion of the cephalic cartilage (*KK*). The *choroid* is the internal continuation of the iris, and forms the inner wall of the deeper parts of the anterior optic chamber; it is pigmented and contains from without inwards (1) a layer of epithelium,* (2) a layer of obliquely-placed plates called the argentea externa, (3) a layer of muscles, (4) the argentea interna, (5) a cartilaginous capsule (*K*), which lies next (6) the spreading out fibres of the optic nerve.

At the side of the optic ganglion within the orbit is a glandular body called the *white body* (*W*).

Alimentary canal. The mouth, which is placed within the circlet of arms, is surrounded by a circular fold forming a kind of lip (Fig. 347, *L*), and, in addition, in the Dibranchs by a membrane called the buccal membrane. The buccal membrane is in some Decapods divided into lobes alternating with the arms and bearing suckers. The entrance to the mouth is armed with two powerful horny jaws, an upper and a lower (*Mxi, Mxs*), which resemble in form a reversed parrot's beak, the lower jaw being the larger and overlapping the upper. The *radula* (*Ra*), which arises in a sheath, usually has in each row a median tooth and three teeth on each side, and there may be in addition, outside these, some flat, non-toothed plates (the radula is absent in *Cirroteuthis*). There is a sub-radular organ in front of the radula. The salivary glands (*Spd*) open into the buccal cavity; in *Octopoda* there are two pairs, of which the ventral is applied against the buccal mass, while the dorsal is dorsal to the skull. The ducts of the latter unite, and pass forwards with the oesophagus. In the Decapods the dorsal glands are alone present.

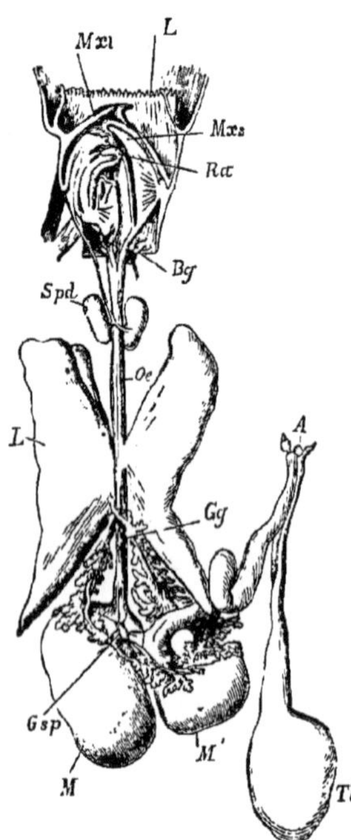

FIG. 347.—Digestive apparatus of *Sepia* (after W. Keferstein). *L* lip; *Mxi, Mxs* lower and upper jaws; *Ra* radula; *Bg* supra-buccal ganglion; *Spd* salivary gland; *Oe* oesophagus; *L* liver; *Gg* bile duct; *Gsp* splanchnic (stomach) ganglion; *M* stomach; *M'* caecum of stomach; *A* anus; *Tb* ink sac.

* Not very distinct in the figure; it is the inner lining of the space enclosed by the great corneal sclerotic folds.

The oesophagus (*Oe*), with or without a crop-like dilatation, is long and ends behind in the stomach (*M*), which is provided with a large, often spiral, caecum (*M′*). The intestine leaves the stomach close to the oesophageal entrance, and passes as a straight or slightly coiled tube (*Nautilus*, *Octopus*) to the anus (*A*).

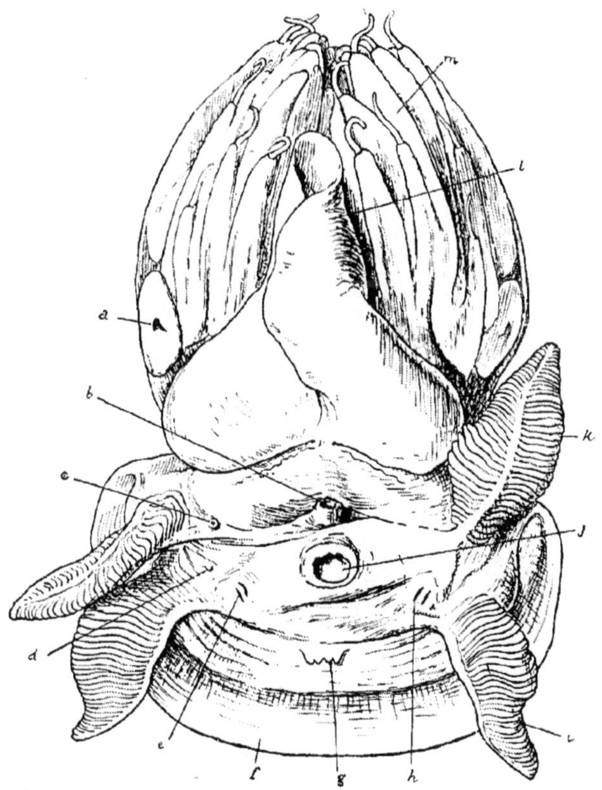

FIG. 348.—Male *Nautilus*, ventral view; the mantle is drawn back; reduced, after Keferstein (from Pelseneer). *a* eye; *b* genital opening; *c* opening of the anterior kidney; *d* inter-branchial papilla; *e* opening of the posterior kidney; *f* edge of mantle; *g* postanal papilla: *h* external opening of the pericardium; *i* posterior gill; *j* anus; *k* anterior gill; *l* funnel; *m* tentaculiferous appendages.

The liver (*L*) is a compact gland, and consists of two lobes, more or less united, one on each side of the oesophagus. The bile ducts (*Gg*) are two in number and covered by a glandular tissue called *pancreatic*, but really renal (see below); they open into the caecum of the stomach.

2 F

In *Nautilus* the dorsal salivary glands are not present and the liver is less compact and in four lobes, each with its own bile duct.

The **ink sac** (*Tb*), which is absent in *Nautilus* and *Cirroteuthis*, is a rectal gland. It is placed on the posterior side of the visceral sac close to the mantle-lining, and opens into the rectum.

The **perivisceral cavity** is, as in Gastropods, partly coelomic and partly haemocoelic, but the coelomic portion has a much greater extension than in other Molluscs.

The *haemocoelic* part of the body-cavity is best developed in *Nautilus*, where it forms a cavity, in relation with the crop, vena cava and one loop of the intestine; it occupies the anterior side of the visceral sac, and does not extend to the apex (which is occupied by genital coelom). It is traversed by connective-tissue strands, and communicates with the vena cava by numerous foramina in the wall of the latter. In Dibranchs the haemocoele is less developed; in Octopods it has the form of a large sinus surrounding the oesophagus, dorsal salivary glands, bile ducts, and anterior aorta; and communicates with the great vena cava. Also the cavity round the buccal mass is a blood space.

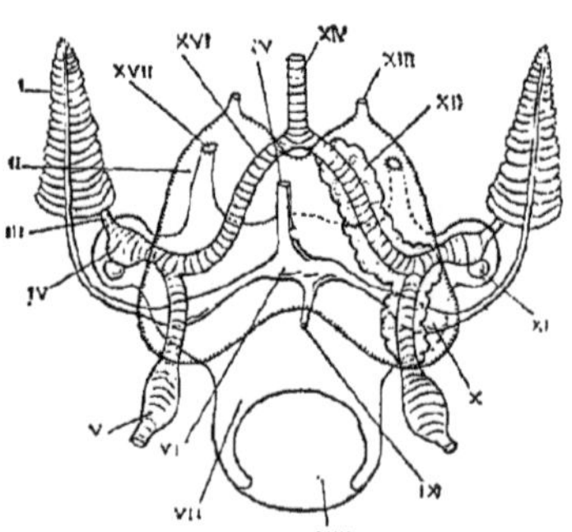

FIG. 349.—Diagram of the renal and circulatory apparatus of a Decapod, ventral view (after Pelseneer). *I* ctenidium; *II* renal sac; *III* afferent branchial vessel; *IV* branchial heart; *V* abdominal vein; *VI* ventricle; *VII* coelom; *VIII* genital gland projecting into the genital part of the coelom, which in front is in relation with the heart as the viscero-pericardial coelom, and opens into the renal sac at *XVII*; *IX* posterior aorta; *X* auricle; *XI* appendage of branchial heart (pericardial gland); *XII* glandular tissue of the kidney; *XIII* external opening of kidney; *XIV* vena cava; *XV* anterior aorta; *XVI* branch of vena cava; *XVII* reno-pericardial opening.

The perivisceral part of the *coelom* is divided into two parts, which however communicate—the so-called viscero-pericardial sac, corresponding to the pericardium of other types, and the genital portion. In *Nautilus* the viscero-pericardial cavity is in relation with the heart and pericardial gland, while the genital portion contains the gonad and has relations with the stomach and intestine. Both open to the exterior, the pericardium by two openings placed respectively just internally to the openings of the posterior nephridia (Fig. 348, *h*), while the genital opens by the single genital duct into the posterior region of the mantle-cavity (Fig. 348, *b*). In Dibranchs (Fig. 349) the perivisceral part of the coelom is also in two communicating parts: (1) the genital division, which opens to the exterior by the genital duct, and is related only to the gonad; and (2) the viscero-pericardial which has relations to the heart, stomach, intestine, branchial hearts, and pericardial glands (*Decapoda*), and does not open to the exterior. In *Octopoda* (Fig. 350) the viscero-pericardial coelom is practically absent, being reduced to a small space (*VI*) on each side round the glandular appendages of the branchial heart (pericardial glands), which opens (*VII*) into the nephridium and communicates by a narrow canal (*II*) with the apically-placed genital division of the coelom. In Dibranchs the perivisceral (viscero-pericardial) part of the coelom also communicates with the kidneys, and not directly with the exterior (Fig. 349, *XVII*), whereas in *Nautilus* it has no opening into the kidney.

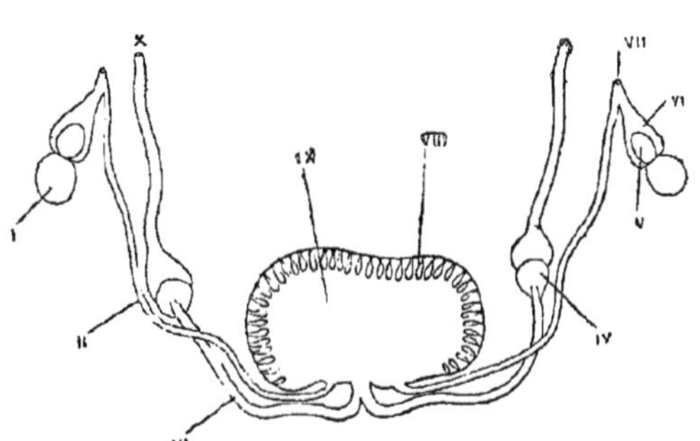

FIG. 350.—Diagram of the coelom of a female Octopus, ventral (posterior) view (from Pelseneer, after Brock). *I* branchial heart; *II* canal connecting the genital and visceral parts of the coelom; *III* oviducts; *IV* oviducal gland; *V* appendage of branchial heart; *VI* perivisceral (pericardial) coelom; *VII* reno-pericardial opening; *VIII* ovary; *IX* genital coelom.

The **branchiae** have the form of two (*Dibranchiata*) or four (*Tetrabranchiata*) bilamellate ctenidia, which are placed on the visceral sac in the mantle-cavity, and are not ciliated. In *Nautilus* they project freely and the posterior are a little larger than the anterior; in Dibranchs they are attached to the body along one side of the axis.

Osphradia are absent in Dibranchs, but in Nautilus there are two pairs of papillae in the mantle-cavity, which are supposed to represent them. The anterior pair is between the two pairs of gills (Fig. 348, *d*), while the posterior osphradia* are the so-called post-anal papillae placed near together on the mantle between the renal sacs and the nidamental glands in the female, and in the same

* *Vide* A. Willey, *Q. J. M. S.*, 40, 1897.

position in the male (Willey). They are innervated by branches of the visceral nerves.

Vascular system. The heart lies in the viscero-pericardial cavity (except in the *Octopoda*), and is placed at about the middle of the visceral sac. It consists of a median ventricle and of as many lateral auricles as there are gills (Fig. 351). A large anterior (ventral) aorta passes off from the ventricle, and gives in its course strong branches to the mantle, alimentary canal and funnel, and breaks up in the head into vessels to the cephalic organs. A posteriorly (dorsally) directed visceral artery also leaves the ventricle, supplying the viscera and gonad. The capillary network, which is richly developed in all the organs, passes partly into sinuses, partly into veins which are collected through lateral veins into a large anterior and a posterior vena cava. Each of these bifurcates into two or four trunks (according to the number of the gills) which carry the blood through the kidneys to the gills. Immediately before their entrance into the gills, the walls of these afferent branchial vessels are (except in *Nautilus*) especially muscular and rhythmically contractile, and constitute the *branchial hearts*. The glandular appendages of the branchial hearts are the *pericardial glands*.

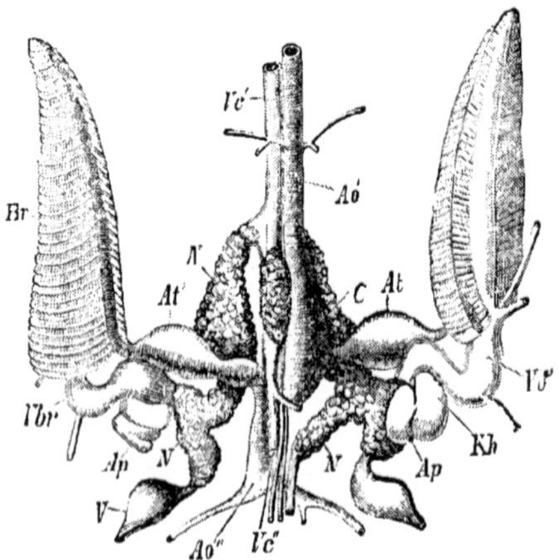

Fig. 351.—Circulatory and renal organs of *Sepia officinalis* from the dorsal (anterior) side (after Hunter). *Ao'* and *Ao''* anterior and posterior aorta; *Ap* appendage of the branchial heart: *At*, *At'* auricles; *Br* gills; *C* ventricle; *Kh* branchial heart; *N* renal appendages of the veins; *V* lateral vein; *Vbr* afferent branchial vessels; *Vc'*, *Vc''* anterior and posterior vena cava.

In *Nautilus* the vascular system is largely lacunar, but in Dibranchs capillaries are developed, though these may end in sinuses which open into the veins; in *Octopoda* especially there is a large sinus already described, which opens into the anterior vena cava.

The **excretory organs** lie on the posterior side of the visceral sac close to the mantle-cavity, into which they open. In Dibranchs they are two in number and are either completely separate (*Octopoda*), or they are connected together, generally through an anteriorly and dorsally-placed portion which is of considerable extent and lies close under the shell. This unpaired portion contains the so-called pancreatic tissue of the bile ducts, and is in relation with these structures. The anterior and posterior afferent branchial vessels run in the walls of these kidney sacs on their way to the gills, and give off blind diverticula which project into the cavities of the kidneys. The kidney epithelium, which is flattened over the rest of the sac, is especially glandular on these vascular diverticula (Fig. 349, *XII*) and secretes the waste matter in the form of concretions which largely consist of guanin.

Each kidney opens into the mantle-cavity through a papilla placed not far from the anus, and into the pericardial coelom by a pore not far from the external opening.

In *Nautilus* there are four kidneys which open to the exterior; one pair opens just anterior (*i.e.*, ventral) to the anterior gills, and the other pair just anterior to the posterior gills and close to the openings of the pericardium; they are without any internal opening into the coelom. The four afferent branchial vessels run in the wall separating the four kidneys from the pericardium, and give off diverticula, which are covered with glandular tissue, into the pericardium as well as into the renal sacs. The four tufts of glandular processes thus projecting into the pericardium constitute the pericardial gland, and are larger than the corresponding processes on the opposite side of the septa projecting into the renal sacs.

The kidneys and gills of *Nautilus* have been spoken of in the text as anterior and posterior, but it must be borne in mind that, owing to the fact that the part of the body in which they lie has been prolonged ventrally in the mantle-fold, the posterior kidney and gill are really ventral, *i.e.*, nearer to the mouth than the anterior, which are placed just at the point where the mantle-fold is given off from the body. The pericardium, which is also in the mantle-fold, is actually posterior to the kidneys, and the posterior walls of the kidney-sacs form the anterior wall of the pericardium; but it must be remembered that *morphologically* the pericardium is dorsal to, *i.e.*, nearer the apex of the visceral sac than, the kidneys.

The *Cephalopoda* are dioecious. The sexes present external differences, which are sometimes very marked. In many cases the males are much smaller than the females, as in *Argonauta*, in which genus the female is further distinguished from the male by possessing a shell. In *Nautilus* there are considerable differences in the number and arrangement of the cephalic tentacles (see above p. 421); and in all Dibranchs one of the arms of the male is different from the rest and said to be hectocotylized (see p. 420).

The **sexual gland** is single, and at the apex of the visceral sac. It is in both sexes a special development of a portion of the epithelium lining the genital division of the coelom. The sexual cells are dehisced into the coelom and pass out by the genital ducts, which open into the coelom and possess accessory glands on their course.

The generative duct of one side is usually suppressed or vestigial. In *Nautilus* in both sexes the right duct persists, and the left is vestigial, having lost its internal opening but retained the external. In all male Dibranchs the left duct is alone present; this is also the case in all females excepting the *Oegopsida* and *Octopoda* (except *Cirroteuthis*), in which both oviducts are present in functional development.

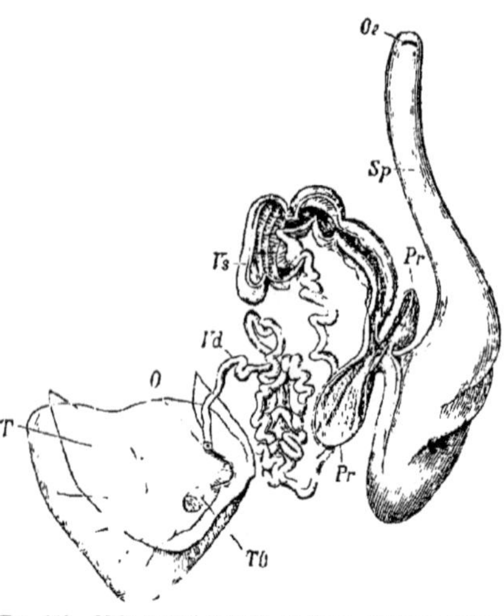

FIG. 312.—Male sexual organs of *Sepia officinalis* (after Duvernoy, modified from Grobben). *T* testis with a piece of peritoneum; *To* opening of the testicular tubes into the coelom; *Vd* vas deferens; *O* opening of the vas deferens into the body-cavity; *Vs* vesicula seminalis; *Pr* prostate; *Sp* Needham's sac; *Oe* generative opening.

In *Nautilus* the genital coelom communicates with the pericardial by three openings in the septum which separates them, and the stomach and intestine project into it. The genital gland is a hollow structure, and is to be regarded as a folded-off portion of the genital coelom, to the anterior wall of which it is attached; it opens into the genital coelom close to the openings of this latter structure into the pericardium. The reproductive cells are produced from the lining of the genital gland, and in the female the ripening ova, covered with their follicle cells and the flat cells of the peritoneum, project into its cavity. In the male the walls of the genital gland are folded so as to give rise to a number of branched tubes, which open into the central cavity of the organ. The genital duct in both sexes opens into the coelom close to the opening of the genital gland, and into the mantle-cavity to the right side of the middle line on the oral side of the anus; while the vestigial left duct opens to the left as the so-called *pyriform organ*. The vas deferens has an accessory gland and a spermatophore sac, and opens at the end of a papilla—the so-called penis.

In Dibranchs the genital glands and genital coelom are much as in *Nautilus*, except that the latter has no relation to the alimentary canal. The oviduct possesses an oviducal gland in its course, and opens into the mantle-cavity on the left side, or when two are present symmetrically on both sides.

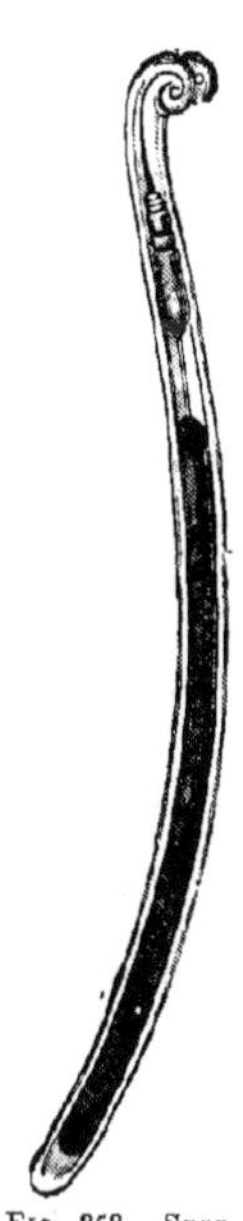

Fig. 353.—Spermatophore of *Sepia* (after M. Edwards).

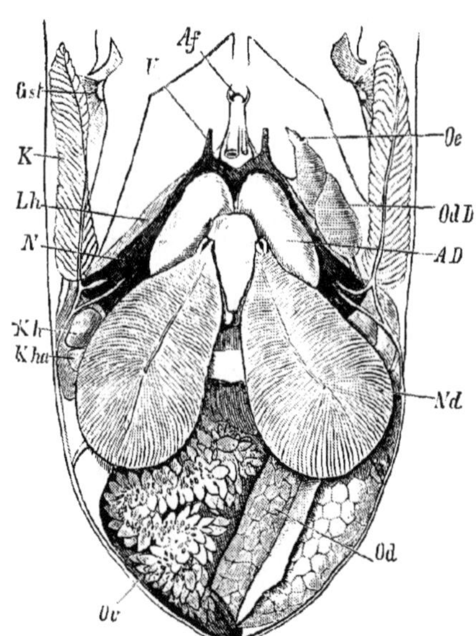

Fig. 354.—Posterior view of visceral sac of a female *Sepia* partly dissected (after Grobben). *Ad* accessory nidamental gland; *Af* anus; *Gst* stellate ganglion; *K* gills; *Kh* branchial heart; *Kha* pericardial gland (appendage of branchial heart); *Lk* tube from perivisceral coelom to the kidney (renal coelom), often called the aquiferous canal; *N* kidney; *Nd* nidamental gland; *Od* oviduct; *Od. D* oviducal gland; *Oe* external opening of oviduct; *Ov* ovary in the genital coelom which has been opened; *U* ureter.

The vas deferens opens at the same place as the oviduct, but has a much more complicated structure. In *Sepia* we get the following parts (Fig. 352): (1) a coiled narrow tube (*Vd*) which opens into the genital coelom, (2) a long dilated vesicula seminalis (*Vs*) with two prostatic glands opening into its terminal portion (*Pr*), (3) a spacious sac, known as Needham's sac, in which the spermatophores are stored, and which opens into the mantle-cavity at the apex of a papilla placed on the left side.

The *nidamental glands* are accessory glands of the female, which secrete the egg envelopes and open into the mantle-cavity. In *Nautilus* there is a single gland in the mantle-wall dorsally. In *Decapoda* they are in the anterior wall of the mantle-cavity superficial to the kidneys, and there are generally two pairs —the ventral being the smaller (Fig. 354, *Ad, Nd*).

The eggs are surrounded (*Argonauta, Octopus, Sepia*) by capsules with long stalks, which are united together in racemose masses (so-called sea-grapes), and fastened to foreign objects in the sea. In other cases the eggs are aggregated in gelatinous tubes which may be attached together in great numbers (*Loligo*). In *Argonauta* they are placed in the shell.

The *spermatophores* (Fig. 353) are manufactured in the male generative ducts. They are elongated, vermiform bodies containing spermatozoa, and often presenting a most elaborate structure. In Dibranchs they may be described as tubes containing spermatozoa at one end, and a piston and spiral elastic spring at the other. We may presume that under proper conditions the spiral spring elongates and drives down the piston, which then expels the spermatozoa.

The spermatophore are sometimes very long: in *Eledone* 8 centimetres, in *Octopoda* with autotomous hectocotylus they may attain a length of 50 centimetres; and in *Nautilus*, in which they are coiled on themselves, they may be more than 30 centimetres long.

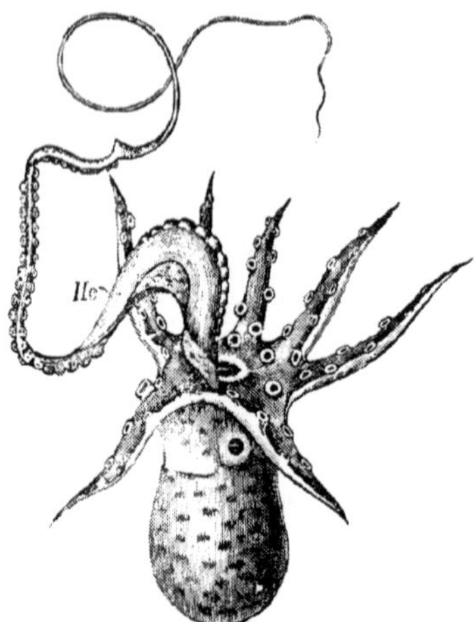

FIG. 355.—Male of *Argonauta argo* (after H. Müller). *Hc* hectocotylized arm.

In the Dibranchs one of the arms of the male is always modified, or *hectocotylized* as it is called, for purposes of copulation. In the *Decapoda* it is usually the fourth left arm; but in *Enoploteuthis* it is the fourth right, and it may vary from the fourth right to the fourth left in the same species; in *Idiosepius* and *Spirula* both right and left fourth arms are modified, and in *Spirula* enclosed in a common envelope; in *Octopoda* it is usually the third right, but in *Scaeurgus* and *Argonauta* it is the third left, and the second right in *Cirroteuthis*. In most cases the modification consists mainly in a reduction of the suckers, and the arm is not detached; it affects the extremity of the arm in *Enoploteuthis, Eledone*,

Octopus; its base in *Sepia,* its whole length in *Idiosepius* and *Rossia,* and *Loliolus.* But in three genera, *Philonexis, Tremoctopus,* and *Argonauta,* the modification is much more extensive, and the arm affected is charged with spermatozoa, cast off, and deposited in the mantle-cavity of the female (see above). This is called an autotomous hectocotylus. The modification is as follows: before sexual congress the arm in question is represented by a somewhat globular sac, which consists of flaps of the basal part of the arm wrapped round the distal part. Soon this sac bursts and allows the distal part and the suckers of the basal part to appear (Fig. 355). The distal part thus set free has at its end a small sac, which in its turn bursts and allows a long terminal filament to issue. The folds which formed the first-named sac, give rise to a receptacle on the aboral side of the arm, which becomes charged with spermatophores. This spermatophore receptacle communicates with a small vesicle in the base of the arm, which leads into a long canal extending along the arm and filament, and opening at the end of the latter. How this arm, which is deposited in the mantle-cavity of the female, is used in fertilization, and how it becomes charged with spermatophores is not known. After the arm is detached a new encysted arm is said to be formed on the scar.

Those *Cephalopoda* which are without this autotomous hectocotylus are said to copulate mouth to mouth. the modified arm being used to affix the spermatophores to the buccal membrane, or to transfer them to the mantle-cavity of the female. In *Sepia* and *Loligo* spermatophores are found in the pockets in the buccal membrane, and in *Nautilus* within the annular lobe of the head (*v.* p. 421).

Development. The egg of all *Cephalopoda* is large and heavily charged with yolk. This is the case even with *Nautilus*, as may be gathered from the ovarian egg, for the laid egg* of *Nautilus* has never been found. There is no larval stage, not even a trace of the velum, and the young are hatched with the form of the adult. The development is therefore very different from that of other molluscs.

In Dibranchs the cleavage is partial, and confined to one pole of the egg; as in the bird's egg it gives rise to a blastoderm or germinal disc, which in the subsequent development is raised more and more from the subjacent yolk. Soon several projections appear on the embryonic rudiment (Fig. 356). First in the centre of the germ a flattened circular ridge is formed round a central depression; the ridge is the mantle, and the depression is the shell-gland. In *Octopoda* (*Argonauta*) the shell-gland seems to shallow out and disappear, but in Decapods it becomes closed over and persists as a sac in which the shell is deposited.

On each side of the mantle the two parts of the funnel (*Tr*), which even in Dibranchs are separate in the embryo, appear; and between these and the mantle the gills (*Br*). Also laterally, but external to the folds of the funnel, the first traces of the head appear as two pairs of elongated lobes, of which the

* Since this was written, the eggs of *Nautilus* have been found by Dr. Arthur Willey, the Balfour student of the University of Cambridge (*Proc. Roy. Soc.*, vol. 60, 1897, p. 467). The eggs, which are laid singly, are enclosed in a double capsule of cartilaginous consistency. The ovum is very large (17 mm. in its longest diameter) and is surrounded by albumen.

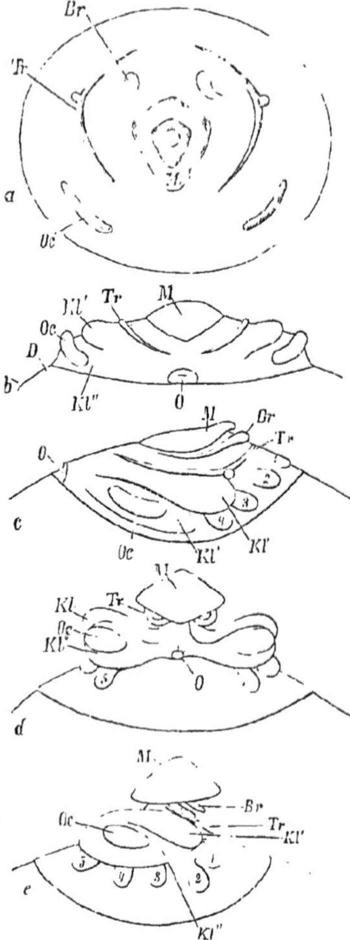

FIG. 356. — Embryonic development of *Sepia officinalis* (after Kolliker). *a*, view of germinal disc from above; the commencing embryo lying on the yolk; *Br* gills; *Tr* folds of the funnel; *Oc* eye; *M* mantle. *b*, somewhat later stage, from the front; *D* yolk; *Kl'* anterior, *Kl''* posterior cephalic lobe; *O* mouth. *c*, later stage, from the side; *1–4* first rudiments of arms. *d*, older stage from the front; *5* fifth pair of arms. *e*, still later stage in lateral view; the halves of the funnel have united.

anterior external pair bear the eyes (*Oc*). On the outer edge of the disc papilliform structures are formed, the first rudiments of the arms. In the later growth of this absolutely symmetrical embryo, the Cephalopod form becomes gradually more and more apparent: the mantle projects considerably, and grows over the gills and two parts of the funnel, which fuse to form the definite funnel. The cephalic

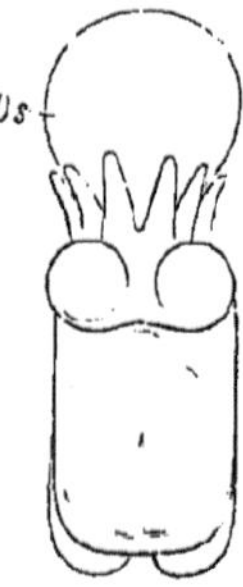

FIG. 357. — Almost ripe embryo of *Sepia officinalis* from the dorsal (anterior) face (after Kolliker). *Ds* yolk sac.

lobes grow together between the mouth and funnel, and on their oral sides become more sharply constricted off from the yolk, which, with a few exceptions, persists for some time as a yolk sac. The yolk sac is attached between the mouth and anus, and may be regarded in the later stages as a swollen up portion of the foot. It should be noted that the embryo is formed on the dorsal side of the yolk sac, and never completely surrounds it.

Order 1. TETRABRANCHIATA.*

Cephalopoda with four gills in the mantle-cavity.

The appendages of the head are peculiar, consisting of a number of lobes carrying sheathed tentacles; they are described on p. 421.

The cephalic cartilage, instead of forming a complete ring, consists of two horse-shoe-shaped limbs, on which the central parts of the nervous system lie. The eyes are without a lens or other refractive media. The funnel is formed of two lobes which are not fused. There is no ink sac. The ctenidia are four in number, as also are the branchial vessels, auricles, and kidneys, and probably the osphradia. The pericardium has a pair of external openings, and does not communicate with the kidneys. There is a multilocular external shell, in the last chamber of which the animal lies. It is secreted by the mantle and covers the visceral sac. Its chambers are filled with air and are traversed by a siphon. The shell consists of an external, frequently coloured, calcareous, porcellanous layer, and an internal mother-of-pearl layer.

The position and structure of the siphon, as well as the form of the septa, and the lines of fusion of the latter with the shell, afford important characters for the classification of the fossil Tetrabranchiates.

There is only one living genus confined to the Indian and Pacific Oceans, but the extinct members of the order are numerous and important.

The siphon (or siphuncle) in *Nautilus* may have a complete but thin nacreous investment, or only a partial one next one face of the septa, in which case it is called a *septal neck* (Fig. 338, *Sf*). In most *Nautilidae* the septal necks project from the sides of the septa which look towards the apex of the shell, while in the *Ammonitidae* they are, as a rule, on the other sides of the septa. The siphuncle varies in position; it may perforate the centre of the septa, in which case it is central or sub-central (*Nautilidae*), or it may be marginal, on the outer side of the septa (*Ammonitidae*). The *suture* is the name given to the line of union of the edge of the septum with the shell-wall. In the *Nautilidae* the sutures are uniformly curved or straight; in *Ammonitidae* they are exceedingly complex, being folded into a number of *lobes* with the concavity towards the aperture; the parts of the suture between the lobes are called the *saddles*,

* R. Owen, "*Memoir on the Pearly Nautilus*," London, 1832. J. Van der Hoeven, "Bydraagen tot de ontleedkundige Kennis aangaande Nautilus pompilius," *Verhandel. k. Akad. Amsterdam*, diel 3, 1856. T. H. Huxley, "On some points in the Anatomy of Nautilus pompilius," *Journ. Proc. Linn. Soc.*, London, 3, 1859. Keferstein, "Beitr. z. Anat. d. Nautilus pomp.," *Nachrichtsbl. k. Ges. wiss. Göttingen*, 1865. Lankester and Bourne, "On the existence of Spengel's olfactory organ and of paired genital ducts in the Pearly Nautilus," *Q. J. M. S.*, vol. 23, 1883. J. Graham Kerr, "On some points in the Anatomy of Nautilus pompilius," *Proc. Zool. Soc. London*, 1895. A. Willey, *Q. J. M. S.*, 39, 40.

and are convex towards the aperture. The lobes and saddles may be themselves secondarily folded or denticulated (*Ammonites*), or the lobes alone, the saddles being round (*Ceratites*).

The *Aptychus* of the *Ammonitidae* is a calcareous or horny plate, which was probably secreted by the hood and served as an operculum; it may be a single plate, or divided by a suture into two (*Synaptychus*).

Fam. 1. **Nautilidae.** Shell straight, or coiled; aperture simple. Septa concave towards aperture; suture simple; siphon usually central, or internal. From Cambrian to present day; attained their highest development in Devonian and Silurian. *Orthoceras* and *Nautilus* alone persist beyond the Palaeozoic epoch. *Nautilus* L., sole living genus, *N. pompilius* L.; *Orthoceras* Breynius, shell straight, lower Silurian to Lias.

Fam. 2. **Ammonitidae.** Shell of various forms, straight to spiral or turretted. Septa much folded, suture complex; siphon external. Extinct; Silurian to Eocene. *Goniatites* De Haan; *Ceratites* De Haan; *Ammonites* Bruguière.

Order 2. DIBRANCHIATA.*

Cephalopoda with two gills in the mantle-cavity.

The appendages are four pairs of arms with suckers on their oral faces; in the Decapoda there are, in addition, two long, prehensile, tentacle-like arms with suckers at their extremities only, placed between the third and fourth arms.

The cephalic cartilage constitutes a complete investment for the central nerve-organs. The eyes are elaborate, and have refractive media. The lobes of the funnel are fused. An ink-sac is generally present. The ctenidia are two in number, as are the branchial vessels, auricles, and kidneys. There are no osphradia. The kidneys open internally into the pericardium, and the latter has no external opening. The shell is in many forms completely absent; in the rest it is internal, or partly internal, and never protective to the visceral sac.

Sub-order 1. DECAPODA.

In addition to the eight arms there are two long tentacles between the third and fourth pairs of arms (ventral). The suckers are stalked and provided with a horny rim. The eyes are without a sphincter-like lid. The mantle bears two lateral fins, and there is a well-developed apparatus for closing the mantle-opening. An internal shell is present. The heart is contained in the coelom. Nidamental glands are generally present.

Fam. 1. **Ommatostrephidae.** Tentacular arms short and broad; suckers with toothed ring. *Ommatostrephes* Gray, sagittated calamary, is able to

* Hancock, "On certain points in the Anatomy and Physiology of the Dibranchiate Cephalopoda," *Nat. Hist. Review*, 1861. R. Owen, "Supplementary Observations on the Anatomy of Spirula Australis," *Ann. Mag. Nat. Hist.* (3), 3, 1879. T. H. Huxley and P. Pelseneer, "Report on *Spirula*," *Challenger Reports*, Pt. 83 (bound in at the end of the second volume of the "Summary of Results").

project itself for long distances from the water and is not unfrequently found on the decks of ships; *Ctenopteryx* Appellöf; *Chaunoteuthis* Appellöf; *Architeuthis* Steenstrup, includes the largest *Cephalopoda* known.

The **Thysanoteuthidae** are allied here.

Fam. 2. **Onychoteuthidae.** Tentacular arms long; suckers with hooks. *Onychoteuthis* Lichtenstein, the uncinated calamary, with hooks on the tentacular arms, and a small group of suckers at the base of each club, enabling them to act like forceps. *Enoploteuthis* d'Orb., armed calamary, with hooks on all the arms (as well as suckers). *Verania* Krohn, tentacular arms atrophied in the adult.

The **Gonatidae** (*Gonatus* Gray) are allied here.

Fam. 3. **Chiroteuthidae.** Tentacular arms very long, pen expanded at each end. *Chiroteuthis* d'Orb.; *Histioteuthis* d'Orb.; *Histiopsis* Hoyle; *Calliteuthis* Verr.; *Doratopsis* Rochebrune.

Fam. 4. **Cranchiidae.** Arms very short; fins terminal, small; eyes projecting. *Loligopsis* Lam.; *Histioteuthis* d'Orb.; *Cranchia* Leach; *Leachia* Les.; *Taonius* Steenstrup.

The above-named four families constitute the **Oegopsida**, which are characterized by their pelagic habit, by the possession of a perforated cornea, and by the presence of two oviducts.

Fam. 5. **Spirulidae.** Female with a single oviduct (right) and two nidamental glands. Shell spiral, coils not touching, multilocular, with internal siphon, partly internal, enclosed in two lobes of the mantle; it is coiled ventrally (Fig. 339), *i.e.* in the opposite way to that of *Nautilus*. Without rostrum and proostracum. *Spirula* Lam.; *S. Peronii* Lam., Pacific Ocean, probably abyssal.

Fam. 6. **Belemnitidae.** Arms with hooks, shell multilocular, straight, probably internal, with rostrum and proostracum. Extinct. Lias to Cretaceous. *Belemnites* Breyn.

Fam. 7. **Sepiolidae.** Body short, rounded dorsally; fins rounded, inserted somewhat on anterior surface of visceral sac, midway. *Sepiola* Leach; *Rossia* Owen; *Stoloteuthis* Verrill; *Inioteuthis* Verrill.

The **Idiosepiidae** (*Idiosepius* Steenstrup smallest cephalopod known) and **Sepiadariidae** are allied here.

Fam. 8. **Loliginidae.** Body elongated, conical; fins terminal, rhombic; pen lanceolate. *Loligo* Lam., calamary; the pens increase in number with age; several are found in old specimens (Owen, in Todd's *Cyclopaedia of Anatomy and Physiology*, vol. i. p. 546). *Sepioteuthis* Blainville; *Loliolus* Steenstrup.

Fam. 9. **Sepiidae.** Body flattened, broad; fins narrow, elongated, shell internal calcareous. *Sepia* L., cuttle-fish.

The families *Sepiolidae*, *Loliginidae*, *Sepiidae* constitute the group **Myopsida** characterized by their unperforated cornea and single oviduct (left).

Sub-order 2. OCTOPODA.

With eight arms, without tentacular arms. The suckers are sessile. Eyes relatively small, with sphincter-like lid. Shell absent. The heart does not project into coelom. Oviduct paired. Nidamental glands absent. Funnel without valve. Mantle without sucker-like apparatus for closing mantle-opening.

Fam. 1. **Cirrhoteuthidae.** Arms united by a membrane and carrying on each side of the suckers filamentous appendages; radula absent; deep water. *Cirrhoteuthis* Eschricht.

Fam. 2. **Amphitretidae.** Mantle fused with funnel in median line, leaving two openings into mantle-cavity. *Amphitretus* Hoyle.

Fam. 3. **Octopodidae.** Arms alike, more or less webbed; often without fins. *Octopus* Cuv. poulpe; *Scaeurgus* Trosch.; *Eledone* Leach; *Pinnoctopus* d'Orb., with fins; *Alloposus* Verrill.

Fam. 4. **Argonautidae.** Female with a unilocular spiral shell (Fig. 335); males very small; hectocotylized arm autotomous (Fig. 355). *Argonauta* L., the paper-nautilus.

Fam. 5. **Philonexidae.** Hectocotylized arm autotomous; other arms all alike in the two sexes; large aquiferous pores near the head and funnel. *Philonexis* d'Orb.; *Tremoctopus* Delle Chiaje.

CHAPTER XI.

ANNELIDA.

Segmented worms in which the perivisceral cavity is a part of the coelom. They almost all possess chitinous setae embedded in and secreted by pits of the skin.

The *Annelida* include the segmented worms. They differ from the *Arthropoda*, which in many respects they resemble, by the possession of a perivisceral division of the coelom. In other words the body-cavity is a part of the coelom. In this respect and in the arrangement of the central nervous system, and in the wide-spread occurrence of the trochosphere larva, they approach the *Mollusca*. They differ, however, from the *Mollusca* in the fact that the body is segmented. This segmentation, which is exhibited by a considerable number of organs, proceeds from and is based upon the mesoblastic somites in the embryo. The *Annelida* possess a dermo-muscular body-wall; that is to say, muscular tissue enters largely into the composition of the integument. In consequence of this fact the body-wall is always extremely contractile, and the body shrinks considerably when the animals are placed in spirit.

They all possess chitinous spines—the **setae**, which are secreted by the ectoderm and are embedded in pits of the skin. These setae are very conspicuous in the class *Chaetopoda*, less so in the classes *Hirudinea*, *Echiuroidea*, and *Archiannelida*.

The alimentary canal is tubular, and generally straight: it opens by a mouth which is placed on the ventral surface of the front end, and by an anus which is terminal, or subterminal and dorsal.

The **head** consists of the anterior part of the body, on the ventral side of which the mouth is placed. It is divided, very often by a mark, into a preoral portion—the **prestomium**—and a postoral portion, which is called the **peristomium**. The prestomium is sometimes called the first segment, but in the enumeration of segments the peristomium is counted as the first (Fig. 358).

Whether the prestomium should be regarded as a segment or not is a disputed question, and one which can only be settled by a study of development. Unfortunately Embryology speaks with an uncertain voice on this point in the *Annelida*; but from our knowledge of the *Arthropoda* we are inclined to the view that the preoral part of the body does contain at least one pair of mesoblastic somites (or their equivalent)—the only real test of a segment—which are serially homologous with the other somites. This is clearly shown in the embryo of *Peripatus*, in which the somites of the preoral region possess the rudiments of a nephridium, and send extensions into the preoral appendages called antennae; just as the posterior somites are prolonged into the legs.

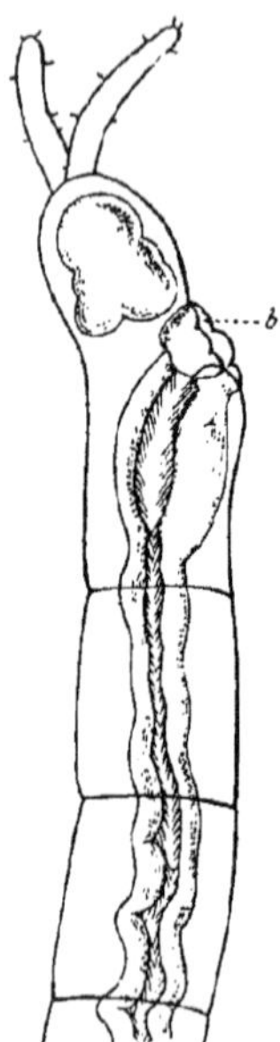

FIG 358.—Anterior end of *Polygordius neapolitanus*, seen in profile as a transparency. The distinction between the prestomium and peristomium, and between the peristomium and the next segment is clearly shown. *b* the mouth (after Fraipont).

The prestomium is sometimes quite small and inconspicuous: it may however be large and much elongated into a proboscis-like organ, as in *Nais lacustris* and the *Echiuroidea*. In some cases it bears special sensory appendages called tentacles and palps. It sometimes happens in the *Polychaeta* that a certain number of body segments are fused with the peristomium, forming a secondary composite head.

The central **nervous system** consists of two nervous tracts, mainly ventral and called the nerve cords. These are generally closely approximated in the middle ventral line over the greater part of their course, but sometimes they are widely separated (some *Polychaeta sedentaria*), and in the *Echiuroidea* the ventral parts of them are fused to form a single ventral cord. But in all forms, even in those in which there is complete union between them behind, they separate from one another in front, and embracing the anterior part of the alimentary canal become continuous with one another on the dorsal side at the front end of the body. The ventral portions of these cords are almost always swollen at segmental intervals into the so-called ganglia, from which the nerves generally proceed, and there is universally a single or bilobed swelling at the point where they are continuous with one another dorsal to the alimentary canal. This dorsal swelling (or swellings) is called the cerebral ganglion (or ganglia), or brain: the nerves which pass out from it (or rather which enter it) are the sensory nerves of the anterior end of the body, and of the important sense organs there located. The first swelling on the ventral cord is called the sub-oesophageal ganglion, and the subsequent swellings the ventral ganglia. The ganglionated condition is not due to the exclusive presence of nerve cells at certain points, for, as in the *Vertebrata*, nerve cells are found along the whole length of the central organ; but is rather due to the fact that there are more nerve cells and more nerve fibres in the ganglia than elsewhere, in consequence of the fact that the nerve fibres do not as a rule pass out all along the cords, but are gathered up into bundles, and leave the ventral cords at any rate at segmental intervals. The central nervous system is generally separated from the ectoderm and placed within the muscular layer, but in the *Archiannelida* and in one or two *Chaetopoda* it lies in and forms part of the ectoderm.

A **vascular system** is nearly always, probably always, present. It has a canalicular character, and the principal vessels are the dorsal and subintestinal longitudinal vessels.

The **perivisceral cavity** is a portion of the coelom. It is derived from the paired cavities of the mesoblastic somites of the embryo, which swell up and surround the intestine. Primitively therefore the body-cavity is divided into a series of paired cavities, the walls of which are in contact with the walls of the somites anterior and posterior to them in the series, and with the walls of their fellows of the opposite side dorsal and ventral to the alimentary canal. Two kinds of septa are thus formed, the one separating the cavities of somites adjacent to each other in the series: these are the transverse septa found running between the body-wall and the gut-wall in many *Chaetopoda*; and the other separating the cavities of the two somites of a pair on the dorsal and ventral side of the alimentary canal: these are the dorsal and ventral longitudinal mesenteries, which also run between the body-wall and the gut-wall, but in a longitudinal direction. These two kinds of coelomic septa are found coexisting in the *Archiannelida*, and possibly in one or two Chaetopods, but as a general rule the dorsal and ventral mesenteries break down in the adult, so that the two sides of the body-cavity become continuous with each other on the dorsal and ventral sides of the alimentary canal; while the transverse septa, though more often found in the adult (*Oligochaeta* and some *Polychaeta*), also generally break down either partially or completely, so that the perivisceral cavity becomes a continuous space from end to end of the animal (*Echiuroidea*, some *Polychaeta*). In the *Hirudinea* the coelom generally (*Acanthobdella* excepted) becomes more or less broken up into a system of communicating spaces.

The renal organs, called **nephridia**, are portions of the coelom, though this is not so obviously the case as in certain other groups (*e.g.*, *Vertebrata*, *Peripatus*, *Mollusca*). Still they almost always retain a communication with the rest of the coelom in the nephrostome or ciliated coelomic funnel (*Chaetogaster* and one or two other forms excepted). They almost always have a tubular form, and open externally on the ventro-lateral surface of the body. As might be surmised from the condition of the coelom in the embryo, they share in the segmentation of the body; and, moreover, with regard to them it should be noted that they very rarely (*Echiuroidea*, *Sternaspis*, *Chlorhaemidae*) lose their relation to the segmentation of the body. As a general rule there is a single pair of nephridia in each segment, but this rule is sometimes departed from in the *Chaetopoda*, where there may be more than one pair in a segment (*Perichaetidae*, *Acanthodrilidae*, *Capitellidae*.)

In a few cases some of the anterior and posterior nephridia shift their external openings from the outer surface of the body into the anterior and posterior end of the alimentary canal. This is seen in some *Oligochaeta* (so-called pepto-nephridia), and in the case of one posterior pair in the *Echiuroidea* (anal vesicles).

The **generative organs** are always products of the coelomic walls, being thickened patches of the coelomic epithelium. Occasionally, in the male more often than in the female, the generative section of the coelom is cut off from the rest by special membranes. This is seen in the sperm-reservoirs of some *Oligochaeta*.

The generative ducts open into the coelom, and have, of course, a tubular character with ciliated internal openings. Sometimes they are distinct from the nephridia (*Oligochaeta*), but, as a rule, some or all of the nephridia are used for the exit of the generative products, thus combining a renal and generative function. This association of the generative and excretory functions in the carrying apparatus is, of course, to be connected with the fact that these two functions are discharged by the same organ, viz., the coelom; it is a feature characteristic on the whole of all coelomate animals (*cf.* Vertebrata).

The *Annelida* are for the most part aquatic animals, but terrestrial forms are known (earth worms).

Their eggs are sometimes laid, several together, in a cocoon (*Hirudinea* and *Oligochaeta*), but, as a general rule they are deposited in the water, with but little more protection than is afforded by the vitelline membrane.

In the former case the development is direct and there is no larval stage, the development taking place at the expense of albuminous matter, which is included in the cocoon. But in the latter case the young undergo only a small part of their development in the egg-membranes, and are hatched out as larvae at an early stage, to undergo the rest of their development in the free-swimming state. The larval form, most commonly, one may almost say invariably, found, is that called the *trochosphere*, This larval form, which we have already met with in *Mollusca*, is fully described further on. The larva of *Polygordius*, which is known as the larva of Lovén, is a typical trochosphere (Fig. 360), and the larvae of many *Chaetopoda*, and of *Echiuroidea* (Fig. 431), are typical trochospheres.

The *Annelida* are classified as follows:—

CLASS 1. ARCHIANNELIDA.
,, 2. CHAETOPODA.
Order 1. Polychaeta.
,, 2. Oligochaeta.
CLASS 3. HIRUDINEA.
,, 4. ECHIUROIDEA (GEPHYREA ARMATA).

For reasons fully stated in the sequel, we have thought it necessary to break up the old group *Gephyrea*, and to exclude some of its divisions (*Sipunculoidea, Priapuloidea, Phoronis*) from the *Annelida*. The *Echiuroidea*, however, we retain as being obviously true Annelids.

Class I. ARCHIANNELIDA.*

Marine Chaetopoda without setae or parapodia.

There are two genera in this order, *Polygordius* Schn. and *Protodrilus* Hatschek, the former found in sand of European seas, and the latter in the sand of an inland sea-lake at Faro near Messina. They are elongated and thread-like, and are entirely without setae or parapodia. The head has two tentacles and two ciliated pits (Fig. 358). The segmentation is but faintly marked externally, by slight

* Fraipont, "Le Genre Polygordius," *Fauna and Flora d. Golfes v. Neapel* xiv., 1887. Hatschek, "Protodrilus," *Arb. Zool. Inst. Wien*, 3, 1881. Weldon, "Dinophilus Gigas," *Q. J. M. S.*, 1886, 27. Harmer, "Dinophilus," *Journ. Mar. Biol. Ass.*, New Series, 1, 1889. For a description of Polygordius see T. J. Parker, *Lessons in Elementary Biology*, London, 1891.

grooves in *Polygordius*, and by ciliated rings in *Protodrilus*. There is a median ciliated groove extending along the ventral surface of *Protodrilus*. The segmentation is homonomous and is marked by the coelomic septa (Fig. 361, *d*). There are no circular muscles in the body-wall.

The cerebral ganglion is in the preoral lobe, and the ventral cords lie in the epidermis (Fig. 359), and are without ganglionic swellings; in *Protodrilus* they are separated, one being placed on each

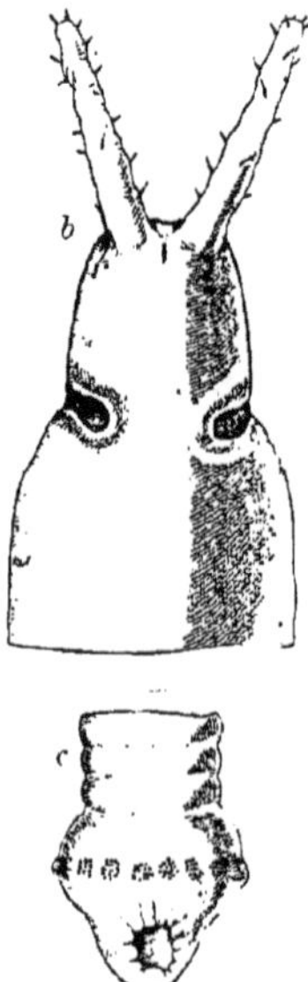

FIG. 359.—*Polygordius neapolitanus*. *b*, dorsal view of head and anterior part of body, showing the ciliated pits. *c*, hind end showing the last three segments, and the anal segment with a circle of papillae, and the anus (after Fraipont).

side of the ventral groove; while in *Polygordius** they are fused. There is in *Polygordius* a very short eversible buccal region followed by an oesophagus, which does not extend beyond the head (peristomial part). In *Protodrilus* there is a U-shaped muscular tube placed ventral to, and opening into the oesophagus.

There is a dorsal and ventral mesentery, and oblique longitudinal septa passing across the body-cavity on each side from the region

* In *Polygordius* there is a fine canal in the ventral cord, which is said to be a remnant of the ventral ciliated groove of the larva.

of the nerve cord to the lateral body-walls (Fig. 359), and in *Protodrilus* the body-cavity is traversed by a sparse reticulum.

There is a pair of simple nephridia in each segment, opening internally into the preceding segment; the first pair opens into the body-cavity of the hinder part of the head in *Protodrilus*.

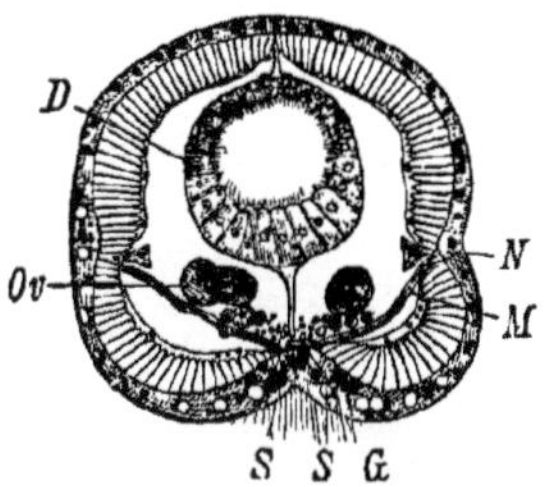

Fig. 360.—Transverse section through the body of *Protodrilus* (after Hatschek). *S*, *S* the two ventral nerve cords; *G* ganglionic layer of the same; *D* alimentary canal; *N* nephridium; *M* oblique muscular sheets (transverse muscles); *Ov* ova. The dorsal and ventral mesenteries are shown.

The vascular system consists of a dorsal vessel in the dorsal mesentery, and a ventral in the ventral mesentery, and of connecting vessels. The blood is red, yellow, or green, or colourless.

The generative organs are developed from the coelomic epithelium, and are discharged into the body-cavity. In *Polygordius** the ova probably escape by rupture of the body-wall, and the spermatozoa by the nephridia. *Protodrilus* is hermaphrodite.

The larva of *Polygordius* is known as Lovén's larva. It is a typical trochosphere (Fig. 361, *a*), and possesses a preoral ciliated ring (*Prw*), a weaker postoral ring (*Pow*), a thickening of ectoderm on the preoral lobe—the apical plate, a ventral mouth, and a terminal anus. There is a pair of mesoblastic bands at the hind end (*Ms*), terminating in front in the larval kidney. This posterior part of the body elongates, its mesoderm becomes segmented into somites; and it gives rise to the body of the worm (Fig. 361, *b*, *c*). Later the anterior end diminishes relatively in size, and becomes the head of the adult (Fig. 361, *d*).

Saccocirrus Bobr.,† should probably be placed here. It is found in the Black Sea and Mediterranean in shallow water on gravel, has a long (20 to 80 mm.) and narrow body with a groove on its ventral surface. The segments are numerous, and each of them, except the first, carries two dorso-lateral bundles of simple setae. The preoral lobe has two long tentacles and two eyes; between the preoral and postoral part (buccal segment) of the head are two ciliated pits. There are no parapodia, but the lower parts of the bundles of setae are enveloped in a cutaneous sheath, which can be protruded or retracted into the body. The setae are enlarged and grooved at their free ends. There are two characteristically marked appendages at the hind end, on each side of the anus, by which

* In *Protodrilus* it is said that the eggs pass backwards through the meshes of the body-cavity reticulum to a pore on the ventral side of the last segment (Repiachoff).

† Marion and Bobretzky, *Ann. Sci. Nat.* (6), 2, 1875, p. 69.

the animal fixes itself to the stones on which it creeps. The circular muscles of the body-wall are less developed than the longitudinal. There are also obliquely-placed muscular sheets passing from the ventral body-wall to the lateral, and dividing the body-cavity into two lateral and one median chamber. The insertion of these oblique muscles is associated with the division of the longitudinal muscle into four bands—one dorsal, two lateral, and one ventral.

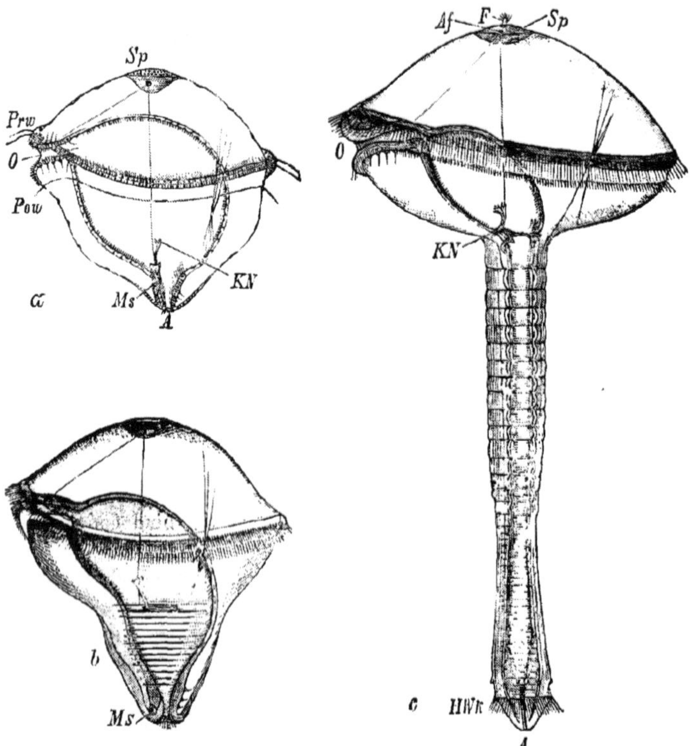

Fig. 361.—Development of *Polygordius* (after Hatschek). *a*, young trochosphere larva; *Sp* apical plate with pigment spots; *Prw* preoral, *Pow* postoral ring of cilia; *O* mouth; *A* anus; *Ms* mesoderm; *Kn* larval kidney (pronephros or head-kidney). *b*, older larva with commencing segmentation of the body. *c*, older stage; the body is elongated and vermiform, and segmented; *HWk* posterior circle of cilia; *Af* eye spot; *F* tentacle.

The ventral nerve cords are widely separate, and are placed, as is also the cerebral ganglion, in contact with the ectoderm immediately beneath the ventral insertions of these oblique septa. They are without ganglionic swellings. There are two contractile vesicles at the base of the tentacles, which drive a colourless fluid into the hollow tentacles and so distend them.

The oesophagus extends for thirteen or fourteen segments, and is followed by the intestine, which is dilated segmentally. Coelomic septa and dorsal and ventral mesenteries are present. Of the vascular system a dorsal vessel in the dorsal mesentery has been detected.

The sexes are separate, and the gonads are found in all segments behind the fourteenth. The germ-cells are developed from paired patches of the coelomic epithelium on the posterior faces of the septa, and are dehisced into the body-cavity, where they ripen. They escape by the nephridia of the part of the body in which they are contained.

The nephridia are simple tubes opening externally close to the bundles of setae, and internally into the preceding segment. In the male the nephridia of the testicular segments are swollen near their openings, and their terminal parts pass into small papillae which project on the surface as small penes. These penes can be retracted into penial sheaths. In the female there are, in the ovarian segments, receptacula seminis; they are yellowish vesicles lying in the central chamber of the body-cavity, and passing into a duct which traverses the oblique muscles and the lateral chamber to open on the ventral surface.

*Histriodrilus** (*Histriobdella* van Ben.) formerly classified with the leeches, is said to belong to the *Archiannelida*. It is a small worm (1·4 mm.) parasitic on the lobster, the eggs of which it devours. It has a distinct cephalic region, and an externally segmented body without setae. There is a pair of limbs on the head, and another on the last segment. There are cerebral ganglia and two ventral nerve cords which are continuous with the epidermis and united in each segment to form a ganglion; there are about eight segments. There is a muscular sub-oesophageal sac armed with three chitinous teeth, and the alimentary canal is ciliated. There are four or five pairs of nephridia, the internal openings of which have not been satisfactorily made out. The sexes are separate, and the generative organs complicated. The ovaries and testes are coelomic. A vascular system has not been observed. The body-cavity is present, and there are dorsal and ventral mesenteries. There are four longitudinal bands of muscular fibres in the body-wall.

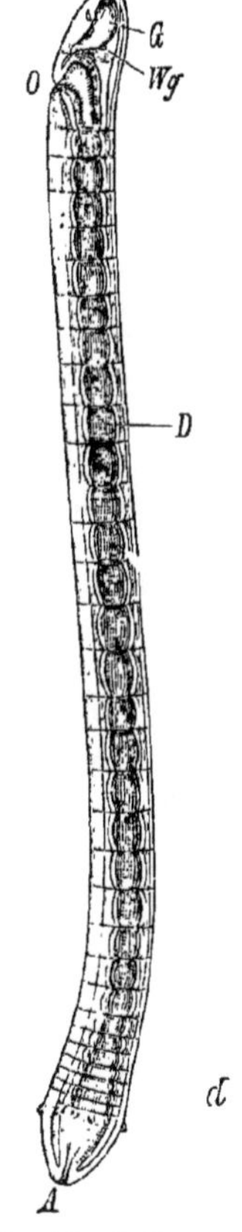

Fig. 361.—*d*, The young *Polygordius*; *G* cerebral ganglion; *Wg* ciliated pit; *D* alimentary canal.

Aeolosoma† Ehrb., often included amongst the *Oligochaeta* as the sole member of the sub-order *Aphaneura*, must be placed with the forms of uncertain position. It is a small (1–10 mm.), fresh-water transparent worm with a loose and uncertain outline. The prestomium is ciliated ventrally, and bears a pair of ciliated pits. The segmentation is ill-marked. There are four bundles of hair-like setae on each segment behind the peristomium. The ectoderm contains

* Foettinger, *Arch. Biol.*, 5, 1884, p. 435.
† Beddard, *Monograph of Oligochaeta*, Oxford, 1895.

coloured gland-cells. The circular and longitudinal muscular layers are both thin. Segmental coelomic septa are not present, except the septum between the peristomium and the next segment. Dorsal pores and head pores are not present. One pair of nephridia is present in each of the setigerous segments. The alimentary canal consists of pharynx, oesophagus, and intestine; and the anus is at the hind end of the body. The central nervous system consists of the cerebral ganglion, which is continuous with the ectoderm of the prestomium. It is said that there is no ventral nerve cord. Generative organs are only occasionally developed, reproduction generally taking place by fission. The testis is median and unpaired, and lies in the fifth segment (the peristomium being the first); the ovary is similarly placed in the sixth segment. There are no generative ducts. The sperm escapes by nephridia, the ova by a pore on the ventral side of the sixth segment. Spermathecae are small oval sacs in segments 3-5. At sexual maturity there is a feeble clitellum on the ventral surface of segments 5-7. The oviposition is unknown. One species has the habit of encysting in a chitinous capsule. There are several English species.

Aeolosoma differs from Oligochaets in the following characters, some of which point to Archiannelidan affinities; the continuity of the cerebral ganglion with the ectoderm, the absence of a ventral nerve cord, the absence of generative ducts, the cephalic ciliated pits, the absence of transverse coelomic septa; it resembles them in being hermaphrodite and in having spermathecae (which, however, are found in *Saccocirrus*).

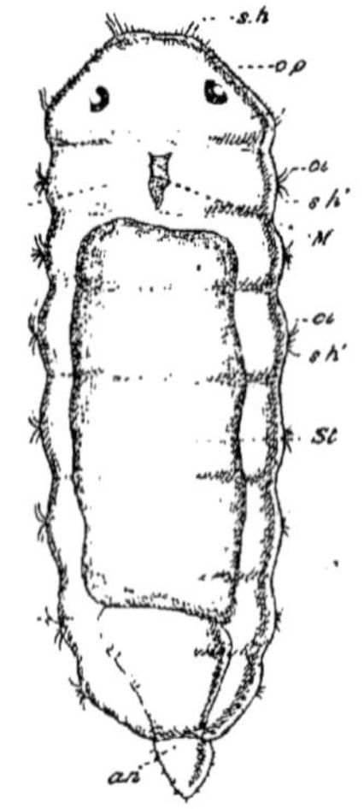

FIG. 362.—*Dinophilus gigas* (after Weldon). *an* anus; *cp* cephalic pits; *ci* transverse ciliated bands; *M* mouth; *St* stomach; *sh* cephalic sense hairs; *sh'* post-cephalic rings of sense hairs placed close behind the ciliary rings.

It is possible that the marine genus *Ctenodrilus* (Scharff. *Q. J. M. S.*, 27) should be placed with *Aeolosoma*. It has one row of comb-like setae on each segment, the prestomium is ciliated on its ventral surface, and bears two ciliated pits. There is only a single pair of nephridia in the peristomium, and only a single layer of muscles—longitudinal—in the body-wall. The nervous system, which consists of a cerebral ganglion and ventral cord, lies in the ectoderm. The sexual worm is unknown.

Dinophilus is a small marine worm found in the Channel in the spring on sea-weeds. There is a head consisting of preoral and postoral portions, and a body consisting of five or six segments, each marked externally by one or two ciliated rings. The ventral surface is ciliated. The anus is posterior, and projecting on the ventral side of it is a kind of tail-appendage. The preoral lobe either possesses two ciliated rings, or is uniformly ciliated: it also bears a pair of eyes and some groups of stiff sensory hairs, and a pair of ciliated pits.

The alimentary canal is straight and ciliated; there is a ventral muscular organ opening into the oesophagus like that of *Protodrilus*.

The nervous system consists of a ganglion in the preoral lobe, and a pair of separated lateral nerve cords placed close to, though not in, the ectoderm. The ventral cords are sometimes ganglionated segmentally.

The body-cavity is occupied (or traversed) by a network of connective tissue, in which the genital cells appear to arise. The spaces of this network are specially enlarged round the internal ends of the nephridia. There are no septa, no dorsal and ventral mesentery, and no vascular system.

The nephridia are in five pairs, of which the posterior in the male communicate by a ciliated funnel with the cavity of the testis. With the exception of this fifth pair, the nephridia open on the sides of the ventral surface: they are simple tubes which internally possess a large flame-shaped cilium, or a row of small cilia which give the appearance of a flame; it is uncertain whether they open internally or not; indeed it is uncertain whether we ought to speak of a body-cavity at all in these animals. The part of the nephridia next the external opening is ciliated, and no nuclei are distinguishable in connection with their walls.

The animals are dioecious. The nephridia of the fifth pair are connected with the testis internally and join together to open at the hind end of the body into a vestibule, which opens externally in the middle ventral line; they are dilated near the opening, and constitute a pair of vesiculae seminales. There is a median penis projecting into the vestibule. Spermatozoa are introduced into the female by means of the penis which perforates the skin at any point of the surface, and so introduces spermatozoa into the body-cavity. The ovaries like the testes appear to arise from cords of the parenchyma, and the eggs when ripe probably pass into a special median ventral passage in this parenchyma, which opens to the exterior by a median pore at the base of the caudal appendage.

There appears to be a thin layer of circular muscles, and a pair of longitudinal muscular bands just external to the lateral nerve cords.

Saccocirrus in many points of its anatomy closely resembles *Polygordius* and *Protodrilus*, near which it may be provisionally placed. The principal point of difference is the presence of setae—while the resemblances are numerous: the ciliated pits, the uniformly segmented body, the oblique septa, the dorsal and ventral muscles, the position of the ventral cords in the ectoderm. It clearly resembles the two Archiannelidan genera far more closely than does *Dinophilus* (see below), which is without coelomic septa, oblique septa, or vascular system, and has a body-cavity extensively occupied by a reticulum, and nephridia of doubtful relations. Moreover, the structure of the male and female generative organs of *Dinophilus* is quite different from that of Archiannelids, with the doubtful exception of the female *Protodrilus*, which needs reinvestigation on this point.

It must be remembered that the sub-oesophageal muscular organ, upon which so much stress has been laid as indicating affinity between *Dinophilus* and *Archiannelida*, though present in *Protodrilus*, is altogether absent from *Polygordius*.

The union of *Histriodrilus* with the Archiannelids is more fully justified, but before finally deciding the matter we require more knowledge of the generative organs and nephridia, and a re-examination of the vascular system.

In the present state of our knowledge the *Archiannelida* must be regarded as having very much the same relation to the *Chaetopoda* as *Chiton* has to the other *Gastropoda*; while, carrying on the same comparison, *Dinophilus* and *Histriodrilus* occupy, relatively to the main group, a similar position to that assigned above to *Neomenia* and *Chaetoderma* in the Molluscan phylum. They are isolated forms with Annelidan affinities. As to *Saccocirrus* it seems clear that it connects Archiannelids with such Polychaets as the *Opheliidae*, in which, though coelomic septa are absent, the oblique septa, predominance of the longitudinal muscles and in *Polyophthalmus* at least, if not in other genera, the cephalic ciliated pits, and contiguity of the ventral nerve cords to the ectoderm are noticeable characters.

Class 2. CHAETOPODA.

Annelids with conspicuous setae. The perivisceral cavity is in many cases divided by septa.

The *Chaetopoda* include *Polychaeta*, in which the setae are generally borne on parapodial processes of the body, which may be fairly compared to the appendages of Arthropods, and the *Oligochaeta*, in which there are no parapodia, the setae simply projecting from the body-wall.

Order 1. Polychaeta.*

Marine Chaetopoda with numerous setae embedded in parapodia: usually with distinct head, tentacles, cirri and branchiae. They are for the most part dioecious, and develop with a metamorphosis.

The *Polychaeta* are, with a few exceptions, marine Chaetopods, in which the setae are numerous and borne upon special processes of the body-wall called *parapodia*. The head, which is called the *prestomium*, very generally bears tentacular appendages, while the next segment, called the *peristomium*, is usually modified and may be fused with one or two of the following segments. The alimentary canal generally possesses an eversible buccal region, and a muscular protrusible (not eversible) pharynx. The coelomic septa may persist,

* Audouin et Milne-Edwards, *Ann. Sci. Nat.*, t. 27–30, 1832–38. Delle Chiaje, "*Descrizione e notomia degli animali senza vertebre d. Sicilia citeriore,*" Napoli, 1841. Quatrefages, "*Histoire Nat. d. Annelés,*" 1 and 2, 1865. Ed. Claparède, "*Annélides Chétopodes du Golf de Naples,*" 1868, and Supplement, 1870. Ehlers, "*Die Borstenwürmer,*" 1868. Johnston, "*Brit. Mus. Catalogue of non-parasitical worms,*" 1865. McIntosh, "British Annelida," *Trans. Zool. Soc.*, ix., 1877. Malmgren, "Nordiska Hafs-Annulater," *Ofversigt k. Vet. Akad. Fördhandlingar*, 1865 and 1867. St. Joseph, "Ann. Polychétes des côtes de Dinard," *Ann. Sci. Nat.* (7), 1, 5, 17, 20, 1886–94. Malaquin, *Recherches sur les Syllidiens*, 1893. Bobretzky et Marion, "Ét. s. les Annélides du Golf de Marseille, *Ann. Sci. Nat.* (6), 2, 1875. McIntosh, "Report on the Annelida Polychaeta," *Challenger Reports*, 1885, and numerous papers by E. Grube. W. B. Benham, "Polychaeta," *Cambridge Natural History*, 1896. Meyer, "Stud. üb. d. Körperbau der Anneliden," *Naples Mitth.*, 7 and 8, 1887–8. Pruvot, *Arch. Zool. Expér.* (2), 3, 1885.

but in a large number of cases partly or entirely break down and disappear. As a rule there are no special generative ducts, and the generative products, which are dehisced into the coelom, pass out by any or some of the nephridia. A vascular system is present, except in a few cases, in which its absence, though stated on good authority, must be regarded as doubtful until renewed investigations of the matter have been made. Some are free, but a large number inhabit tubes which they manufacture or construct for themselves.

The body is generally elongated, and the internal segmentation shown externally. The number of segments is usually considerable, and in some cases is variable, increasing with age. The segments may be all alike, or the body may be divided into two regions, often called **thorax** and **abdomen**, which differ from one another in the form of the parapodia, of the setae, and in other respects (*Sabellidae*, *Capitellidae*, etc.). In some forms the hind end of the body is much reduced, and may be without setae.

The **head** consists of the anterior part of the body, on the ventral side of which the mouth is placed. It is almost always divided by a mark into a preoral portion—the **prestomium**, and a postoral portion, which is miscalled the **peristomium**. The prestomium is often called the first segment, and the peristomium the second; but in this work the peristomium is reckoned as the first.

The prestomium may be a well-marked structure, or it may be much reduced, and hardly distinguishable from the peristomium (*Arenicola*). In the *Cryptocephala* the prestomium is hidden by the forward extension of the peristomium. The prestomium may be without appendages, but it usually bears two kinds of sensory appendages—the tentacles, which are attached dorsally and vary in number from two to five, and the palps which are two in number and ventrally placed (Fig. 363). It often bears one or two pairs of eyes. The peristomium is, in rare cases, provided with parapodia and setae, like the other segments (*Aphrodite*, *Hermione*, *Nephthys*); usually it is without setae and parapodia, though the cirri are often present as long, tentacle-like structures called **tentacular cirri** (Fig. 363, *Fc*). When there is more than one pair of tentacular cirri it is supposed that one or more of the hinder segments have become merged in the peristomium, and have lost their parapodia; in such cases the head is a compound or secondary head, one or two body segments having fused with the peristomeal.

The **parapodia** are segmental, hollow, lateral projections of the body. They carry the setae and are either biramous or uniramous.

When they are biramous the dorsal branch is called the *notopodium*, and the ventral the *neuropodium*.

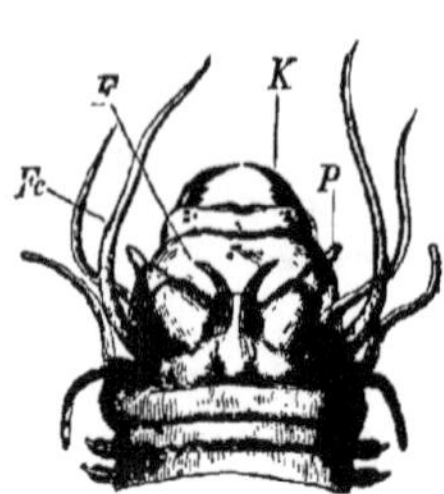

FIG. 363.—*Nereis margaritacea* (after M. Edwards). Head with everted mouth and protruded jaws (*K*). The small spines borne by the eversible wall of the mouth-cavity are shown. The prestomium carries at its front end two tentacles *F*, and ventrally two palps *P*. The peristomium carries four tentacular cirri *Fc* on each side.

The *acicula* are especially strong, dark-coloured setae, hardly projecting at all, and deeply embedded in the parapodia (Fig. 364). Typically there is one in the noto- and one in the neuropodium; they serve for the attachment of the muscles of the setae. The setae are chitinous, and project in groups from sacs on the parapodia; each seta is formed by a single large cell at the bottom of the sac.

The form of the setae varies extremely, and affords a good character for the classification of families and genera. According to the strength, form, and mode of ending the following forms may be distinguished: simple setae, which may be hair-like or flattened (*paleae*), or lance-shaped, or curved at the end (crotchets), etc.; jointed setae (composite), which carry a terminal articulated appendix (Fig. 365, *g*)—found in the *Nereidiformia; uncini* (*a*, *b*), setae with a sharply-

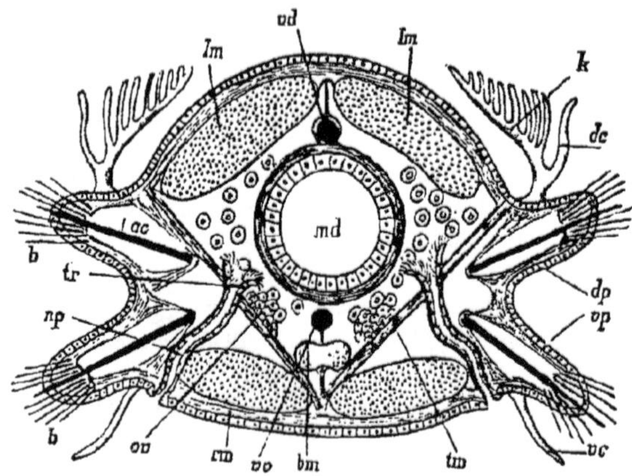

FIG. 364.—Section through a segment of a Polychaet, diagrammatic (from Lang). *ac* aciculum; *b* setae; *bm* ventral nerve cord; *dc* dorsal cirrus; *dp* notopodium; *k* gill; *lm* longitudinal muscles; *md* intestine; *np* nephridium; *ov* ovary; *rm* circular muscles; *tm* transverse muscle; *tr* funnel of nephridium; *vc* ventral cirrus; *vd* dorsal, *vo* ventral vessel; *vp* neuropodium. There are ova in the body-cavity.

curved hook (*Terebelliformia*, *Sabelliformia*). When the parapodia are completely wanting, the setae project from the body-wall (*Capitella*).

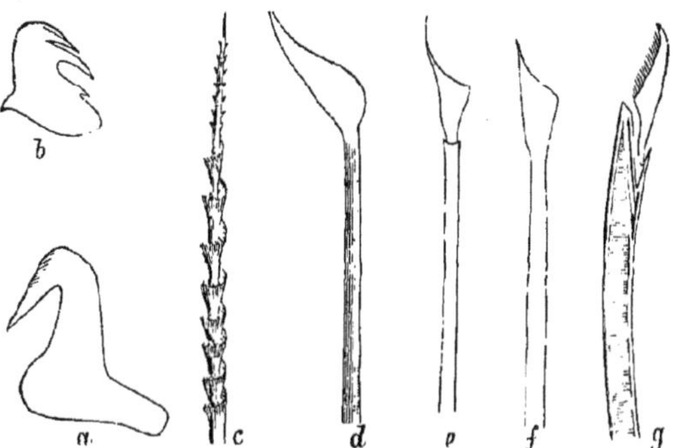

FIG. 365.—Setae of different *Polychaeta* (after Malmgren and Claparède) *a*, hooked setae of *Sabella crassicornis*; *b*, of *Terebella Danielsseni*; *c*, seta with spiral ridge from *Sthenelais*; *d*, lance-shaped seta of *Phyllochaetopterus*; *e*, of *Sabella crassicornis*; *f*, of *Sabella pavonis*; *g*, composite, sickle-shaped seta of *Nereis cultrifera*.

The appendages of the parapodia present a great variety of form, and not unfrequently vary in the different parts of the body. Most important are the *cirri*, which are attached to the dorsal and ventral surfaces of the parapodia. The cirri are for the most part filiform, and sometimes pointed; they may also be ringed. In some cases the dorsal cirri are flattened out as broad scale-like structures —the *elytra* (Fig. 366)— which constitute a protective covering to the back (*Aphroditidae*). In other cases the dorsal cirri* are modified as branchiae, which may be filiform, branched and antler-like, comb-shaped, or in tufts; sometimes they are confined to the middle region of the body (*Arenicola*, Fig. 397),

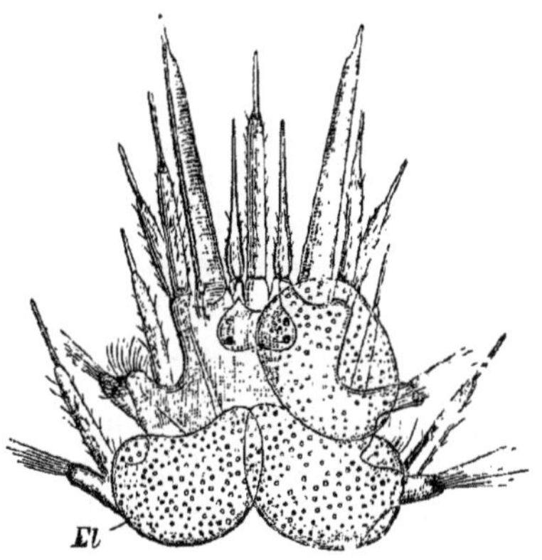

FIG. 366.—Anterior end of *Polynoe extenuata*, the first elytron on the left hand being removed (after Claparède). The two setae of the oral segment are visible. *El* elytron.

* It sometimes happens that cirri are found on branchiferous segments, in which case the gills must be regarded as additional to cirri.

or are extended almost across the whole dorsal surface (*Euphrosyne*); sometimes they are confined to the anterior segments immediately following the oral segment (*Terebella*).* Both cirri, or one of them, may be altogether absent.

The last or anal segment is always modified: the parapodia and setae are absent, but one or two pairs of cirri—the anal cirri—are present. The anus is usually terminal; in *Notopygos* it is dorsal, and in *Caobangia* it is anterior and ventral.

The **alimentary canal** is usually straight (coiled in *Pectinaria*, *Sternaspis*, *Siphonostoma*, etc.); it is divided into buccal cavity, pharynx, oesophagus, and intestine. The buccal cavity has muscular walls, and is often eversible, while the pharynx which also has muscular walls is often protrusible (Figs. 363 and 367). The buccal cavity has a chitinous lining, which may be thickened at certain

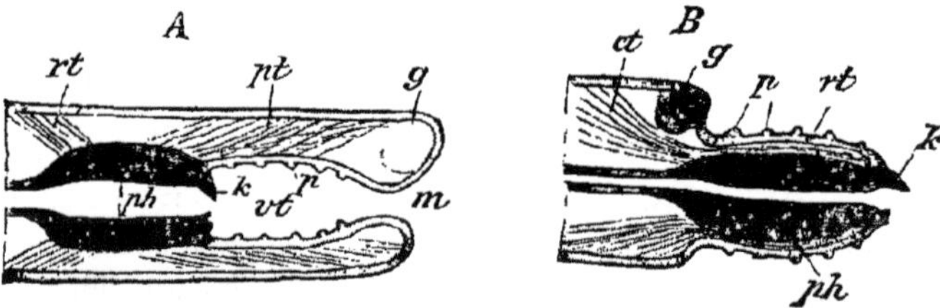

FIG. 367.—Diagrammatic representation of the pharyngeal apparatus of a predatory Chaetopod. *g* brain; *k* jaw; *m* mouth; *ph* pharynx; *pt* protractor muscles; *rt*, *ct* retractors; *vt* wall of buccal cavity (eversible) with papillae or denticles *p*. In *A* the apparatus is retracted, in *B* protruded. (After Lang.)

spots into denticles, and the pharynx is often armed with powerful and movable chitinous jaws or plates. These structures, however, are not always present (*e.g.* Terebellids and Sabellids); they are well developed in the *Nereidiformia*. The oesophagus is often provided with a pair of diverticula, which in some *Syllidae* and *Hesionidae* contain air. The intestine reaches to the anus, and is constricted segmentally. In the *Aphroditidae* the sacculations are elongated into caeca (Fig. 368). In Syllids and some Terebellids there is a muscular gizzard with a hard chitinous lining. In the *Capitellidae* there is a siphonal tube leaving the intestine in front, and entering it again behind.

The **nervous system** is constructed on the typical Annelidan plan. It is sometimes in contact with the epidermis, though more often separated from the skin. The cerebral ganglion is contained in

* The cephalic branchiae of the *Sabelliformia* are palps not cirri.

the prestomium, and the eyes, when present, are often sessile upon it. The ventral cord is double, and the two halves are usually closely approximated, though in some *Sabelliformia* (*Serpulidae*) they are widely separated (Fig. 369), especially in front, and connected together by transverse commissures which pass between the ganglionic swellings. These latter are segmental, the first or subpharyngeal being often contained in the third segment, and the nerves are given off from them. In some forms, *e.g.* *Arenicola*, the ganglionic enlargements are very inconspicuous. A stomatogastric or visceral system is given off from the cerebral ganglion, or from the circum-pharyngeal commissures, or from both. Giant fibres are often present on the dorsal side of the ventral cords. The nerves to the palps arise from a special section of the cerebral ganglion; the nerves to the tentacular cirri come off from the subpharyngeal ganglion, and sometimes from the circum-pharyngeal commissures as well. The subpharyngeal ganglion, which also supplies the first-parapodium, seems, as a rule, to consist of the ganglia of two or more segments fused.

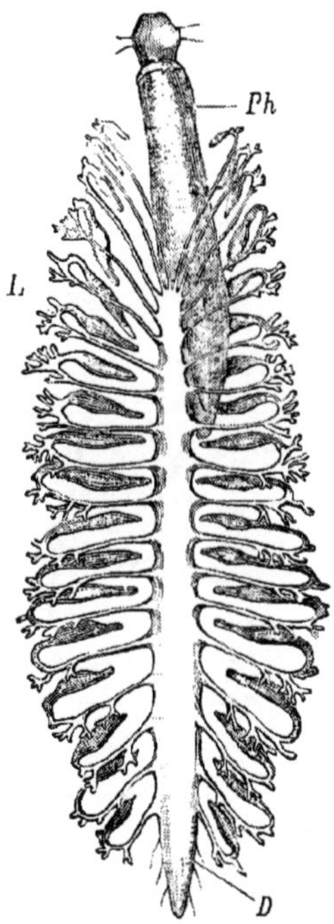

FIG. 368.—Alimentary canal of *Aphrodite aculeata* (after M. Edwards). *Ph* pharynx; *D* intestine; *L* intestinal (hepatic) caeca.

Sense organs. The tentacles, palps, and tentacular cirri must be regarded as tactile organs; and the ciliated pits which are often present on the prestomium and abut upon a special lobe of the brain (*Capitellidae*, *Opheliidae*, *Arenicola*, etc.) as olfactory. In addition to these we sometimes find **otocysts** in the prestomium (*Arenicola* and *Polyophthalmus*), or in some other segment of the body (*Fabricia*, *Myxicola*, *Terebella*), or even segmentally repeated in the dorsal regions of a certain number of segments (*Aricia*). In some species of *Arenicola* they retain the opening to the exterior.*

In the *Capitellidae* groups of sensory cells bearing sense-hairs are

* Ehlers, *Z. f. w. Z.*, Sup. Bd. 53, 1892.

found on the lateral portions of the segments, and are called the *lateral line* organs.

The pigmented bodies, called eyes, and occurring in pairs on some of the segments of *Polyophthalmus*, are probably photogenic organs.

Eyes.* Paired eyes are often present on the prestomium, especially in the free-living forms; but they are also found in other places, *e.g.*, the anal segment (*Myxicola*, *Fabricia*), the gills of some Sabellids (*Branchiomma*, *Dasychone*). The cephalic eyes attain a large size and complex development in the pelagic *Alciopidae*,†

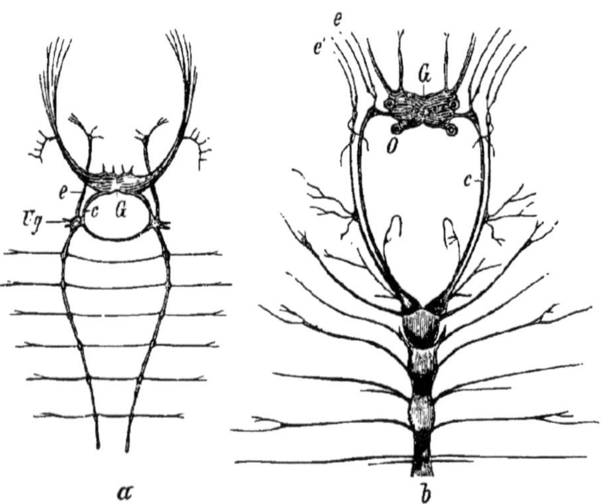

Fig. 309.—Brain and anterior portion of the ganglionic chain of *a*, *Serpula*; *b*, *Nereis* (after Quatrefages). *O* eyes; *G* cerebral ganglion (brain); *c* oesophageal (or pharyngeal) commissures; *Ug* suboesophageal ganglion; *e*, *e'* nerves to the tentacular cirri and peristomium.

where they have a large lens and a complex retina (Fig. 389). In most forms the prestomial eyes are simple cups of the epidermal layer, which may be open or filled by a substance continuous with the outer cuticle and constituting a lens, as in the eye of some *Gastropoda* (*vide* Fig. 290, *B*).

The **vascular system** is constructed on the usual type: there are dorsal and ventral contractile vessels giving off vessels to the skin and internal organs. In some forms there is a continuous blood sinus round the intestine instead of the usual capillary plexus

* E. Andrews, "Eyes of Polychaetous Annelids," *Journ. Morph.* 7, 1892, p. 169.

† Greef. "*Üb. d. Auge d. Alciopiden*," *etc.*, Marburg, 1876.

(Terebellids, *Cryptocephala*, etc.). The blood is usually red, owing to the presence of haemoglobin in solution; but in some forms (*Chlorhaemidae*, and some *Sabelliformia*) it is green, owing to the presence of a green pigment called *chlorocruorin*. In *Aphrodite* the blood fluid is yellow, and in *Magelona* it is pink, owing to the presence of a pigment called *haemerythrin*. The vascular system is said to be absent in the *Glyceridae*, *Capitellidae*, *Polycirrus*, and *Tomopteris*.

The **respiratory function** is discharged by the gills (see above), and probably by the whole surface of the skin.

The **perivisceral cavity (coelom)** is well developed, and is frequently divided by transverse segmental septa into chambers, but in some of the tubicolous forms the septa are deficient, many of them being absent. Muscular bands pass from the median ventral line, where they are inserted on each side of the nerve cord obliquely dorsally and outwards, where they spread out in a fan-like manner to be inserted into the dorso-lateral body-wall. They divide the body-cavity more or less completely into three regions, two lateral, and a median containing the alimentary canal (Fig. 364). The coelomic fluid contains amoeboid corpuscles, and is usually colourless; but in the *Glyceridae*, *Capitellidae*, *Polycirrus*, the coelomic corpuscles are coloured with haemoglobin.

Median dorsal and ventral mesenteries passing from the body-wall to the gut-wall, and dividing the body-cavity into a right and left half, are present in some forms (*Capitellidae*, *Sabella*, etc.).

Excretory organs. In many Polychaets the nephridia occur in all the segments except a few of the anterior and posterior; but in some they are reduced in number, *e.g.*, to six pairs in *Arenicola*, one pair in *Sternaspis*, and one or two pairs in *Chlorhaemidae*. In the *Capitellidae* there may be more than one pair in a segment, and a single nephridium may have more than one internal opening; moreover, adjacent nephridia may be connected together.

The nephridia of Polychaets are either long convoluted tubes (*e.g.*, *Nereis*, Fig. 370) with a small internal funnel, or short and wide tubes (*e.g.*, *Arenicola*, Fig. 371) with a wide funnel. The nephridia of the former kind cannot serve as escape-ducts for the ova, and it becomes a question, in animals which possess them alone, how the eggs pass outwards.* In some tube-dwellers one or more

* It is possible that the dorsal ciliated organs of *Nereis*, which occur in pairs in most of the segments as ciliated patches of coelomic epithelium, may really have undetected external apertures (Goodrich, *Q.J.M.S.*, 34, 1893.)

pairs of the anterior nephridia become much enlarged as renal organs, while the nephridia of the posterior part of the body become much reduced, and serve as genital ducts.

Reproductive organs. The Polychaets are mostly dioecious, but a few are hermaphrodite (some Sabelliformia, *e.g.*, *Amphiglena*,

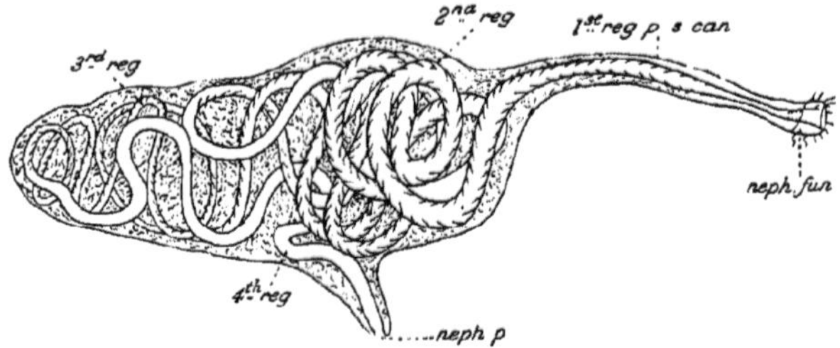

FIG. 370.—Nephridium of *Nereis* (after Goodrich, diagrammatic). The nephridial tube opens internally at *neph. fun.*, externally at *neph. p.*, and is divided into four regions; *p. s. can.* postseptal canal.

Salmacina, *Protula*, *Spirorbis*, and some *Hesionidae*). The gonads, which are differentiations of the coelomic epithelium, are not so narrowly localized as in *Oligochaeta*, but are more diffused over the body—on the walls of blood vessels, on the intestinal wall, or even on the body-wall.* The products, when ripe, are dehisced into the

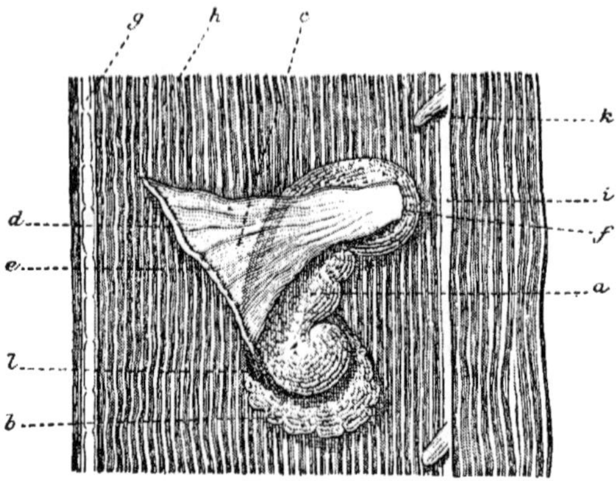

FIG. 371.—Nephridium of *Arenicola marina* (after Vogt and Yung). *a* vesicle; *b* glandular appendage; *c* funnel; *d* lower, *e* upper lip of funnel; *f* opening of funnel into the vesicle; *g* nerve cord; *h* longitudinal muscles; *i* line of insertion of setae; *k* setigerous sacs; *l* aggregation of reproductive cells.

* Cosmovici, *Arch. Zool. Expér.*, 8, 1879–80.

body-cavity, where they mature; they eventually escape either by rupture of the body-wall, or through the nephridia. There are no special generative ducts. Only a few Polychaets are viviparous (*e.g.*, *Syllis vivipara* Kr., *Marphysa sanguinea* Mont., etc.); all the rest are oviparous; many lay their eggs in a jelly (*Aricia*,

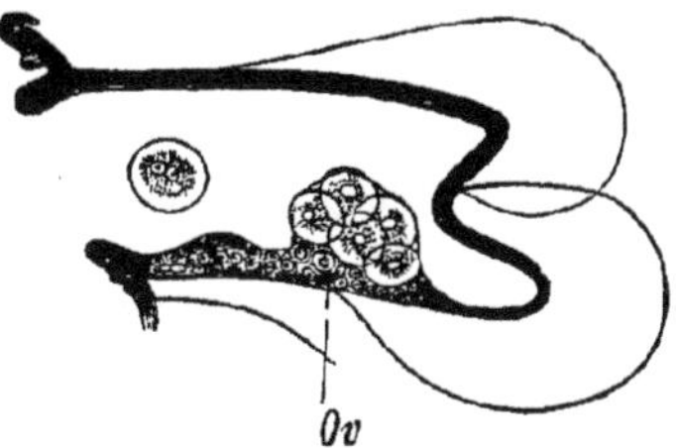

FIG. 372.—A parapodium of *Tomopteris* with a mass of ova *Ov*, and one free ovum (after Gegenbaur).

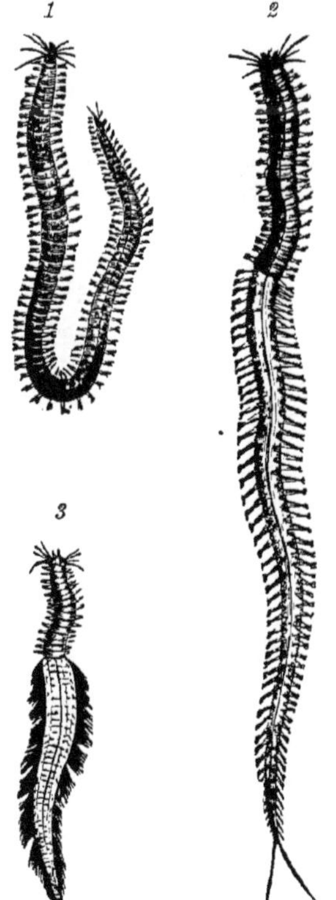

FIG. 373.—Different forms of *Nereis dumerilii* (after Claparède, from Perrier). *1*, young form; *2*, heteronereid female; *3* heteronereid male.

Ophelia, *Phyllodoce*), others attach them to their own body, *e.g.*, to the back beneath the elytra (*Polynoe cirrata*), to the dorsal or ventral cirri (*Exogone*), in a ventral brood-sac (*Autolytus*), in the operculum (*Spirorbis*), or in various tubicolous forms to the tube.

Polymorphism. The forms formerly placed in the genus *Heteronereis* have been shown by Malmgren to be merely the sexually mature individuals of certain species of *Nereis*. The genus has consequently been given up, but the name has been retained to denote the sexual phase in the life-history of these Nereids.

The changes which the worms undergo in passing from the immature condition to the mature Heteronereid condition chiefly affect the posterior part of the body in which the generative organs are contained; in this part (Fig. 373) the parapodia become larger and acquire flattened foliaceous outgrowths, while the setae are thrown off and replaced by new setae of a flattened form and a fan-like arrangement. Moreover, the eyes become enlarged, the dorsal cirri altered, and the animal passes from a creeping to a free-swimming existence. Further, there is a sexual dimorphism, inasmuch as the sexually mature male is more altered from the immature form than the female, having fewer unaltered anterior segments.

In *Nereis dumerilii* Claparède has shown that there is a still more remarkable phenomenon of the same kind. In this species it appears that there are four distinct sexual forms, differing from one another in size, form, mode of life, etc.; they are (1) a sexual, dioecious *Nereis* form, distinguished by its small size; (2) a sexual hermaphrodite *Nereis* form; (3) a dioecious *Heteronereis* which is small, and swims on the surface; (4) a dioecious *Heteronereis* which is larger, and creeps on the bottom. It is not known how these forms are related to each other.

In the *Syllidae* phenomena of a very similar kind have been observed, and there is a Heterosyllid sexual condition in which the posterior sexual segments acquire a dorsal bundle of specially long setae; but in this family the phenomenon is often accompanied by asexual reproduction, which we will now proceed to consider.

Asexual reproduction. The power of asexual reproduction is closely associated with the power of reproducing lost parts, and with the power of indefinite growth, *i.e.* of growth after the adult condition has been attained. In many Polychaetes the number of segments continues to increase throughout the life of the animal by the addition of new segments between the penultimate and anal segments; and in most, if not in all of them, the power of reproducing lost parts is very extensive. For instance, in many of them it has been observed that if the body be cut into two parts, the anterior part will produce a new hind end, and the posterior part a new head.

These two phenomena—the reproduction of lost parts and the protracted formation of new segments—are both instances of the phenomenon of budding, and it is not surprising to find reproductive gemmation normally occurring in the group.

The simplest cases are those in which a zone of fission is formed between two segments, which becomes differentiated into an anal region for the part of the worm in front, and a cephalic region for the part of the worm behind. This results in the division of the worm into two (*Salmacina dysteri*, *Filigrana implexa*). Sexual reproduction does not appear to take place in a worm undergoing fission.

In some genera of the *Syllidae* the process is more complicated.* The Heterosyllid condition is assumed (Fig. 375, *I*), and the worm divides into an anterior non-sexual portion, and into a posterior sexual portion (Fig. 375, *II*), which after separation acquires a new head, and becomes an independent male or female worm. The anterior non-sexual portion develops after separation new anal

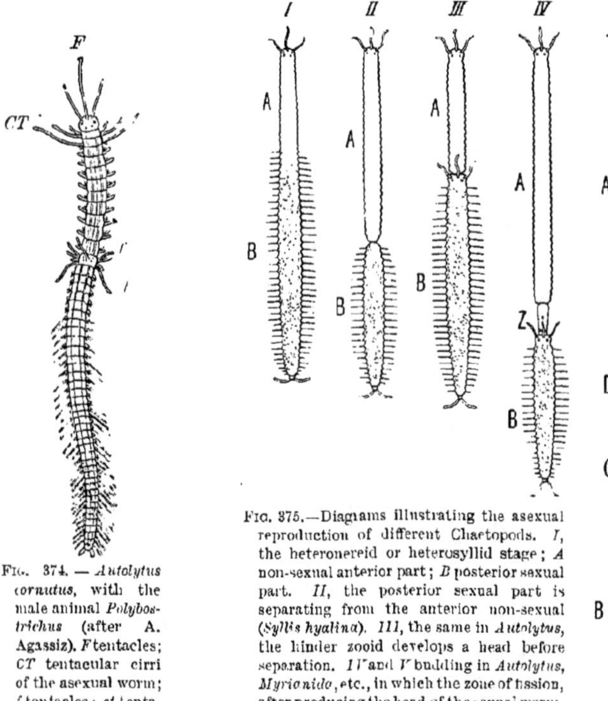

FIG. 374. — *Autolytus cornutus*, with the male animal *Polybostrichus* (after A. Agassiz). *F* tentacles; *CT* tentacular cirri of the asexual worm; *f* tentacles; *ct* tentacular cirri of the male *Polybostrichus*.

FIG. 375.—Diagrams illustrating the asexual reproduction of different Chaetopods. *I*, the heteronereid or heterosyllid stage; *A* non-sexual anterior part; *B* posterior sexual part. *II*, the posterior sexual part is separating from the anterior non-sexual (*Syllis hyalina*). *III*, the same in *Autolytus*, the hinder zooid develops a head before separation. *IV* and *V* budding in *Autolytus*, *Myrianida*, etc., in which the zone of fission, after producing the head of the sexual worm, persists (*Z*), and gives rise to a chain of sexual zooids (*C*, *D*). Altered from Benham.

segments, which eventually acquire generative organs and Heterosyllid setae, and in their turn separate from the anterior portion to form an independent sexual worm, male or female; in this case the two sexes are similar. In some cases the sexual form may never

* **Schizogamy** is the name given to that method of reproduction in which a sexual worm is produced by fission (fissiparous) or by gemmation (gemmiparous) from a sexless worm; in other words, schizogamy is synonymous with metagenesis.

develop a head, but as a rule it develops a head, and is then known, according to the characters of the head, as the *Tetraglena*, *Chaetosyllis*, *Syllis amica*, or *Ioda*.

In *Autolytus* (Fig. 375, *III*) the process is still more complicated,

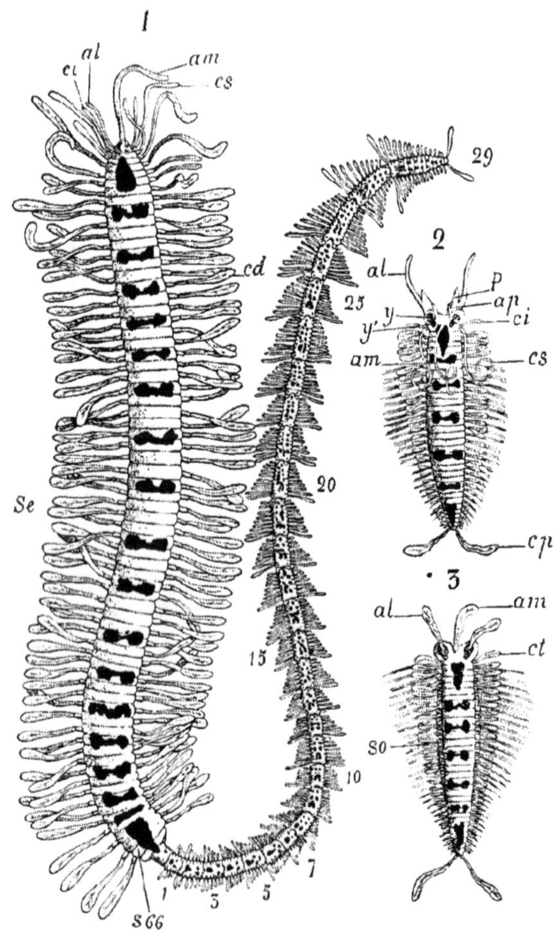

Fig. 376.—*Myrianida fasciata.* *1*, sexless budding form with a chain of 29 budded sexual forms. *2*, male form produced by budding, and called *Polybostrichus.* *3*, a female form, called *Sacconereis.* *am* unpaired antenna; *al* lateral antennae; *ap* posterior lateral antennae; *ci*, *cs*, *ct* tentacular cirri; *cp* terminal cirri; *cd* dorsal cirri; *Se* sexless zooid; *so* ovigerous sac; *y*, *y'* eyes (after Malaquin, from Perrier).

for the formation of the new segments begins before the two parts of the worm separate, so that the posterior sexual worm acquires its head while still attached to its sexless nurse. In other words a zone of fission is formed, and this zone may produce not only the head of the hinder worm, but a number of segments which become marked out into a series of zooids in which genital organs eventually appear. In this way a chain of sexual zooids, attached to the asexual "scolex" in front, is formed; the zooids are all of the same sex, and the posterior of them is the oldest and most developed (Fig. 375, *IV*, *V*). The male zooids differ very remarkably from the female, and were formerly relegated to a distinct genus called *Polybostrichus* (Fig. 374), the female being known as *Sacconereis* and characterized by the possession of a ventral brood-sac. In this case we have a combination of *fission*, by which the first-formed zooid is formed, and *gemmation*, by which the succeeding worms are formed from the segments successively budded off at the zone of fission.

In *Myrianida* a very similar process takes place, and a chain of sexual zooids attached to an anterior sexless zooid is formed; in this case, however, there is no fission, but gemmation only from a zone of fission, for none of the segments of the original worm, except the anal, enter into the constitution of the first formed sexual zooid; the process starts with the formation, in the gemmiparous zone between the penultimate and anal segment, of new segments, which give rise successively to the first formed and all subsequent zooids of the chain. In this case, as in *Autolytus*, the males and females are never formed on the same stock, and are distinguished as *Polybostrichus* and *Sacconereis* forms (Fig. 376). Moreover, in each zooid there is a region in which new segments are formed, so that the posterior and oldest zooid of the chain is also the largest.

Syllis ramosa—a form obtained by the "Challenger" in the canal-system of an Hexactinellid sponge—possesses the peculiar property of forming lateral buds, which may themselves produce buds before separation: in this manner a branched colony arises, some of the individuals of which develop genital organs, and probably become free.

Almost all Polychaets are marine, but a few fresh-water forms are known, *e.g.*, a *Nereis* and *Lumbriconereis* from Trinidad (Kennel), *Manayunkia* in N. America (Leidy). A few are parasitic, *e.g.*, *Oligognathus bonelliae* (one of the *Eunicidae*) in the body-cavity of *Bonellia*; *Labrorostratus parasiticus*, another Eunicid, in the body-cavity of *Odontosyllis ctenostomatus*; larvae of the *Alciopinae* in the *Cydippidae*. Many are commensals, living in the tubes of other worms, and in the ambulacral grooves of starfishes; this is especially common with the Polynoids. A few are pelagic, *e.g.*, *Alciopidae*, some *Phyllodocidae*, *Tomopteris*, *Typhloscolex*, and some sexual Nereids and Syllids (Heteronereids and Heterosyllids). A few are borers, *e.g.*, *Polydora ciliata* in chalk, limestone, shells, etc., *Sabella saxicava*, etc.

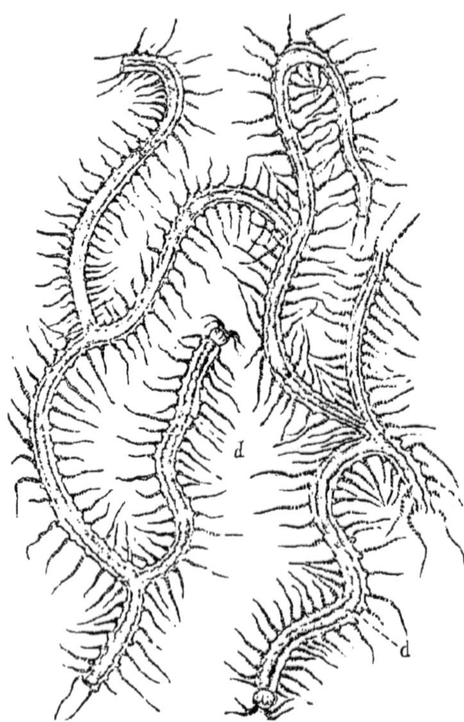

FIG. 377.—Part of a stock of *Syllis ramosa* (somewhat diagrammatic, after McIntosh, from Korschelt and Heider). *d* intestine which branches through the entire stock. The stock is injured in some places.

The majority are littoral, but they are found at all depths to 3000 fathoms. They either live freely, creeping on the bottom, or inhabit tubes. The tube is formed of mucus which is secreted by the ventral gland-shield, or other glands. The mucus hardens to form the tube, which may be strengthened by the incorporation of mud, sand-grains, or pieces of shell, which are actively collected for the purpose by the animal itself. In the Serpulids the tube is fortified by the presence of calcareous matter secreted by the animal.

Some worms are phosphorescent, *e.g.*, *Chaetopterus*, many *Polynoids*, *Syllids*, *Terebellids*, etc., and probably *Tomopteris* by special organs on the parapodia, and *Polyophthalmus* by the so-called segmental eyes.

The *Polychaeta* are found fossil from the Silurian onwards.

Development. There are two main types of embryonic development met with in this group. In one of them the ovum is small and has little food-yolk, the gastrula arises by invagination, and the blastopore remains open as the mouth (*Eupomatus uncinatus*, *Pomatoceros triqueter*); in the other the ovum is larger and has more food-yolk, the gastrula arises by epibole, and the blastopore closes (*Psygmobranchus protensus*, *Nereis cultrifera*, etc.). In all cases hatching takes place at an early stage, and there is a prolonged period of larval life. The larva has the trochosphere form, which has already been described in the case of *Polygordius* (p. 453), or a form which can be readily reduced to that of the trochosphere.

The trochosphere of the marine Chaetopods presents two distinct types: (1) larvae in which there is an extremely large blastocoele, *i.e.*, space between the ectoderm and endoderm, derived from the

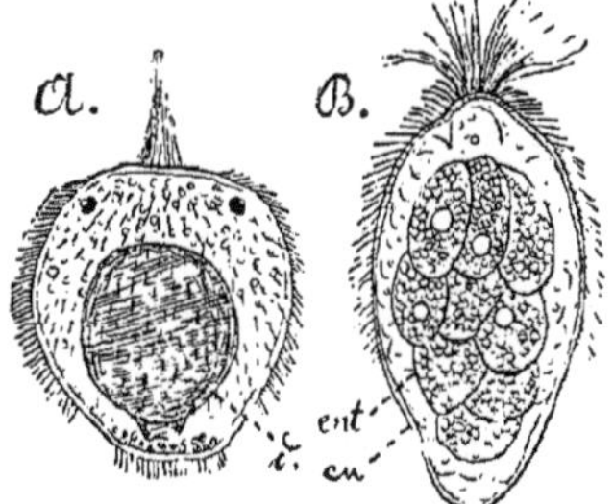

FIG. 378.—Atrochal larvae. *A*, of *Lumbriconereis* (after Claparède and Metschnikoff); *B*, of *Sternaspis scutata* (after Vejdovsky). *cu* cuticle; *d* intestine; *ent* endoderm.

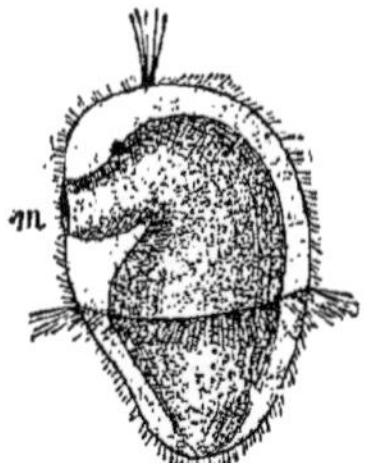

FIG. 379.—Mesotrochal larva of *Chaetopterus pergamentaceus* (after Wilson). *m* mouth.

segmentation cavity of the embryo; (2) larvae in which there is no blastocoele, the ectoderm and endoderm being in contact except where separated by the larval mesoderm.

Larvae of the first type, in all cases in which their embryonic development is known, proceed from embryos with an invaginate gastrula, *i.e.*, from embryos which develop on the type first described. The best known instances of this larval type are afforded by *Eupomatus*, *Pomatoceros*, and *Polygordius* (Fig. 454).

In like manner larvae of the second type, in all cases in which the embryonic development is known, proceed from embryos with an epibolic gastrula. As instances of this larval type, the larvae of *Psygmobranchus*, *Nereis*, etc., described by Salensky, may be mentioned. The larva of *Lopadorhynchus* (Kleinenberg) is of this type, but its embryonic development is unfortunately not known.

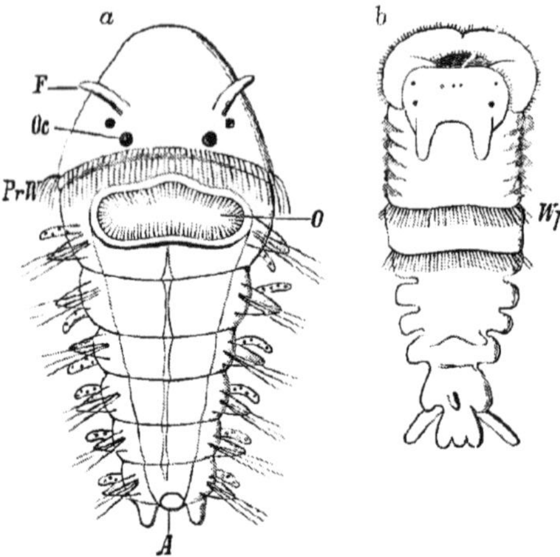

FIG. 380.—Larvae of *Polychaeta* (after Busch.). *a*, larva of *Nereis*. *F* tentacle; *Oc* eyes; *PrW* preoral ring; *O* mouth; *A* anus. *b*, mesotrochal larva of *Chaetopterus*; *Wp* circles of cilia.

In the larvae cilia are rarely distributed over the whole surface of the body (*Atrocha*,* Fig. 378). They are usually confined to special rings. Sometimes, as in Lovén's larva, there is one row placed in front of the mouth at some distance from the anterior end of the body (*e.g.*, larva of *Polynoe*, etc.). Sometimes there are two rows,

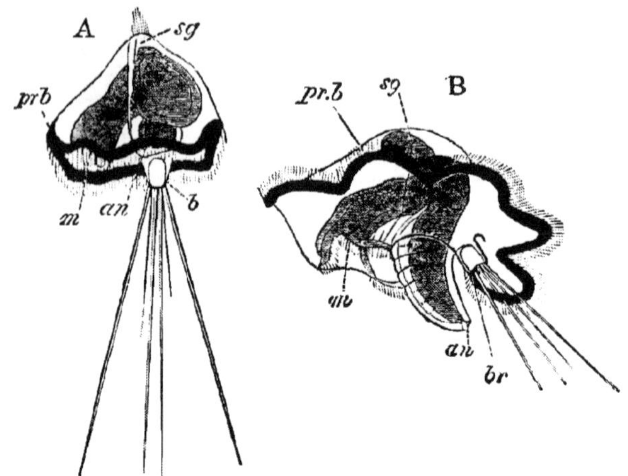

FIG. 381.—Lateral view of *Mitraria* larvae (after Metschnikoff, from Balfour). *an* anus; *b* and *br* the processes carrying the provisional setae; *m* mouth; *pr.b* preoral ring; *sg* apical plate.

* *Cf.* E. Claparède and E. Metschnikoff, "Beiträge zur Entwickelungsgesch. der Chaetopoden," *Z. f. w. Z.*, 19, 1869.

one at each end of the body, constituting a preoral and perianal ring (*Telotrocha*, *e.g.*, *Spio*, *Nephthys* larva). In addition to these two rings of cilia, incomplete rings may also be present on the ventral surface (*Gastrotrocha*), or both ventrally and dorsally (*Amphitrocha*). In other cases one or more rows of cilia surround the middle of the body (*Mesotrocha*), while the terminal rings (preoral and perianal) are absent, *e.g.*, *Chaetopterus* larvae (Fig. 379). Many larvae are provided with long provisional setae, which are later replaced by permanent structures, *e.g.*, the *Mitraria* larva (Fig. 381), and the larva of *Nerine*, etc. In some cases the larvae have true mesoderm segments, each of which is provided with a ring of cilia (*Polytrocha*, Fig. 383).

FIG. 382.—Annelid larva with provisional setae (after Agassiz, from Balfour).

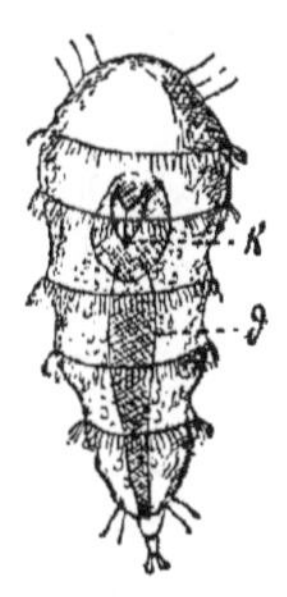

FIG. 383.—Polytrochal larva of *Ophryotrocha puerilis* (after Clap. and Metsch.). *d* intestine; *k* jaws.

In spite of their great diversity of form the Chaetopod larvae can in their later development also be reduced to the type of the larva of Lovén (larva of *Polygordius*, Fig. 454).

Branch A.* PHANEROCEPHALA.

Prestomium free and exposed, generally with eyes, tentacles, and palps. The body segments are more or less alike.

Sub-order 1. NEREIDIFORMIA.

With well-developed tentacles and palps; the peristomium generally possesses cirri. The parapodia are well-marked locomotor organs, supported by acicula; they carry dorsal and ventral cirri. The setae are usually jointed; uncini are never present. There is an eversible buccal region and a muscular pharynx, which is usually

* The classification here adopted is that of Dr. W. B. Benham in his admirable account of the Polychaeta in the *Cambridge Natural History*, Worms, Rotifers, and Polyzoa, Macmillan & Co., London, 1896.

armed with chitinous jaws (Fig. 367). The septa and nephridia are regularly repeated throughout the body. For the most part predaceous; a few form tubes.

Fam. 1. **Syllidae.** Small worms with elongated, flattened body; head usually with three tentacles, palps, and two to four tentacular cirri. The palps are often fused with each other, or with the prestomium. Parapodia with a single setigerous branch (neuropodium), often with a dorsal and ventral cirrus; the dorsal cirrus always appears with sexual maturity. Following the pharynx, which has one or more teeth, is a special gizzard with thick muscular walls. The oesophagus receives in many genera a pair of T-shaped diverticula. Sexual and asexual forms are sometimes found in the same species. Many carry their eggs about till the young are hatched. *Syllis* Sav., tentacles and cirri moniliform, three tentacles; *Odontosyllis* Clap.; *Trypanosyllis* Clap.; *Sphaerosyllis* Clap.; *Grubea* Clap.; *Syllides* Oersted; *Paedophylax* Clp.: *Anoplosyllis* Clp.; *Eusyllis* Mlg.; *Pionosyllis* Mlg.; *Pterosyllis* Clp.; *Exogone* Oerst. *Autolytus* Gr., palps small, and fused with prestomium, no ventral cirri; *A. prolifer* O. F. M., asexual form, the male has been described as *Polybostrichus Mülleri* Kef., the female as *Sacconereis helgolandica* Müll. (Fig. 384); *Proceraea* Ehlers. *Myrianida* M. Edw.; *M. fasciata* M. Edw., with foliaceous cirri (Fig. 376). The position of *Nerilla* O. Schm. (*Dujardinia* Clap.) is uncertain.

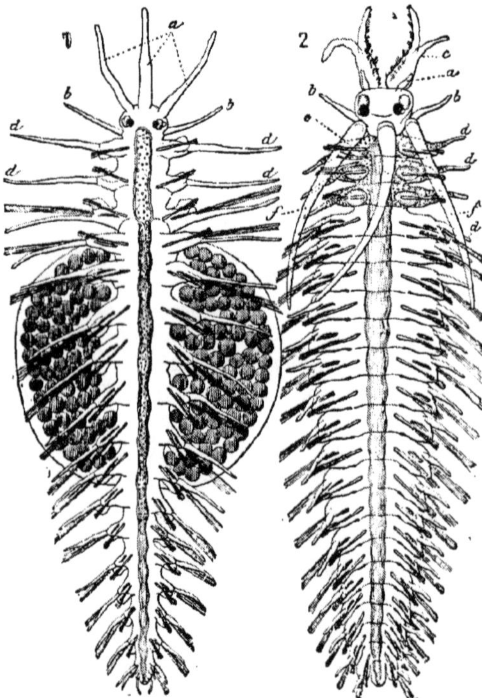

Fig. 384.—The two sexes of *Autolytus prolifer.* *1*, female carrying its sac of eggs (*Sacconereis helgolandica* Müller). *2*, male (*Polybostrichus Mülleri* Kef.). *a* antennae; *b* first pair of dorsal cirri; *c* bifurcated palps of the male; *d* normal dorsal cirri; *e* median antenna of male; *f* tentacular cirri. (From Perrier.)

Fam. 2. **Sphaerodoridae.** Dorsal and ventral cirri spherical. *Sphaerodorum* Oerst. (*Pollicita* Johnst.) *Ephesia* Rathke.

Fam. 3. **Hesionidae.** Body

short, with only a few segments; prestomium with 4 eyes, 2 or 3 tentacles, and generally with jointed palps; peristomium and 2 or 3 following segments achaetous, with long tentacular cirri. Parapodium usually uniramous, or, if biramous, notopodium small, dorsal cirri long and multi-articulate. Eversible region short, pharynx long, unarmed; two diverticula behind pharynx often containing air (air-bladders*). Anal segment with two cirri, and often with reduced parapodia. *Hesione* Sav. (Fig. 385); *H. sicula*, D. Ch., Med.; *Psamathe* Johnst.: *Tyrrhena* Clp.; *Telamone* Clp.; *Stephania* Clp. (*Ophiodromus* Sars); *Dalhousia* M'Int.; *Salvatoria* M'Int.; *Magalia* Mar. and Bob.: *Oxydromus* Gr.; *Kefersteinia* Qtf.

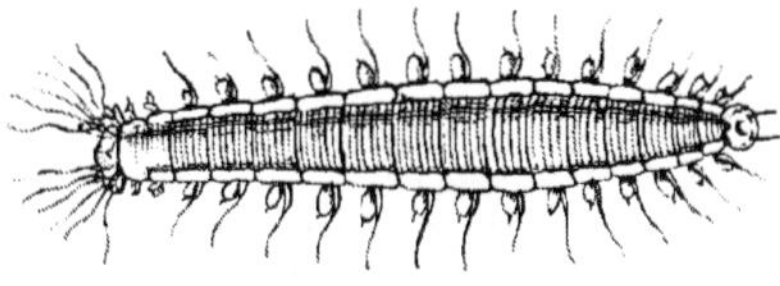

FIG. 385.—*Hesione splendida* Sav. (Règne Animal).

Fam. 4. **Aphroditidae.** With broad scales (*elytra*, *i.e.*, flattened dorsal cirri) on the notopodia; these are usually placed on alternate segments, often only on the anterior part of the body. Prestomium with eyes, with one unpaired and usually two lateral tentacles, to which may be added two palps. Facial tubercle in front of mouth. Pharynx thick-walled, with two upper and two lower jaws. The intestine is provided with a number of paired caeca. Very commonly as epiparasites on other animals, *e.g.*, *Malmgrenia castanea* M'Int, in the mouth of *Spatangus purpureus*; *Hermadion assimile* M'Int., round the peristome of *Echinus esculentus*, *Halosydna Bairdi* between the mantle and foot of *Fissurella cratitia*; *Acholoe astericola* Clp. in the ambulacra of *Astropecten*, *Polynoe cirrata* in those of *Spatangus spinosissimus*, *Hermadion echini* in those of *Echinus sphaera*; *Evarne pentactes* Giard on the body of *Cucumaria pentactes*; or as commensals with other Annelids, *e.g.*, *Lagisca extenuata* in the tube of *Serpula vermicularis*, *Harmothoe macleodi* in the tube of *Terebella conchilega*, *Polynoe scolopendrina* in the tubes of Terebellids, *Harmothoe Malmgreni*, and *Nychia cirrosa* in the tubes of *Chaetopterus insignis* (? *variopedatus*).

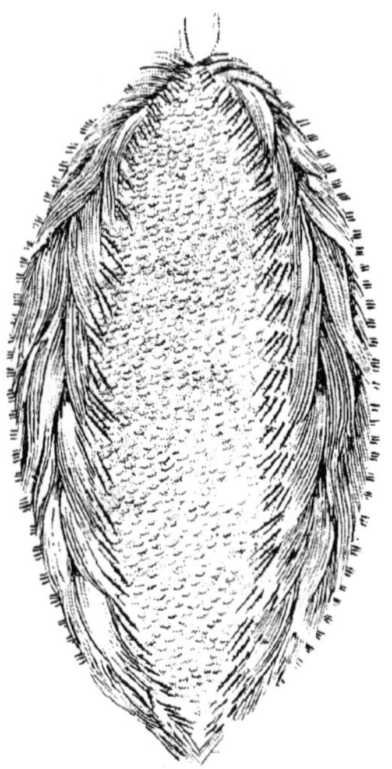

FIG. 386.—*Aphrodite aculeata* Baster (Règne Animal).

* H. Eisig, *Mit. a. d. Zool. Stat. Neapel*, 2, p. 255.

Sub-fam. 1. **Hermioninae.** Body short, oval; notopodial setae directed upwards and backwards to protect the elytra. Prestomium with median tentacle and two long palps, without lateral tentacles. Peristomium setigerous with long cirri. Elytra on segments 2, 4, 5, 7, 9, etc., up to 23, then on every third segment. Jaws not hardened. *Aphrodite* L. (Fig. 386), the back and elytra are covered by a feltwork of chitinous threads arising from the notopodium. *A. aculeata* L., the sea-mouse (*Hystrix marina* Redi), the strong notopodial setae are iridescent, Atlantic and Med.; *Hermione* Blainv. (*Laetmoniee* Kinb.), dorsal feltwork absent; *Pontogenia* Clp.

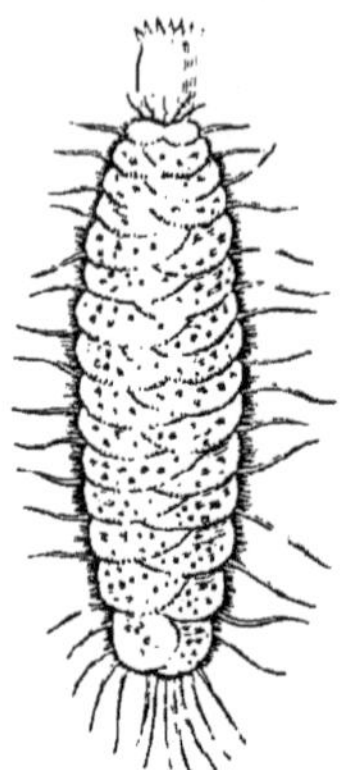

Fig. 387.—*Polynoe setosissima* Sav.(Règne Animal).

Sub-fam. 2. **Polynoinae.** Elytra as in *Hermioninae*; two lateral frontal tentacles, with or without facial tubercle; peristomium with long dorsal and ventral cirri, and the ventral cirri of the next segment elongated. Large teeth on the pharynx. *Polynoe* Sav. (Fig. 387), with three tentacles, posterior region often without elytra; *P. scolopendrina* Sav.; *P.* (*Harmothoe*) *areolata* Gr., in the tubes of *Chaetopterus* and *Terebella nebulosa*; *Iphione* Kinb., two tentacles; *Lepidonotus* Leach; *Halosydna* Sars; *Lepidasthenia* Mgr.; *Lepidametria* Webs.; *Nychia* Mgr.; *Dasylepis* Mgr.; *Harmothoe* Kbg.; *Leucia* Mgr.; *Lagisca* Mgr.; *Hermadion* Kbg.; *Enipo* Mgr.; *Melaenis* Mgr.; *Bylgia* Théel: *Acholoe* Clp.; *Malmgrenia* McInt.; *Eupolynoe* McInt.; *Polyeunoa* McInt.; *Allmaniella* McInt.

Sub-fam. 3. **Acoetinae.** Long vermiform body. The elytriferous segments alternate regularly with segments bearing dorsal cirri throughout the body. With two pedunculated eyes, without facial tubercle. Three tentacles. *Acoeta* M.-Edw.; *Eupompe* Kinb.; *Polyodontes* Renier; *Panthalis* Kinb.

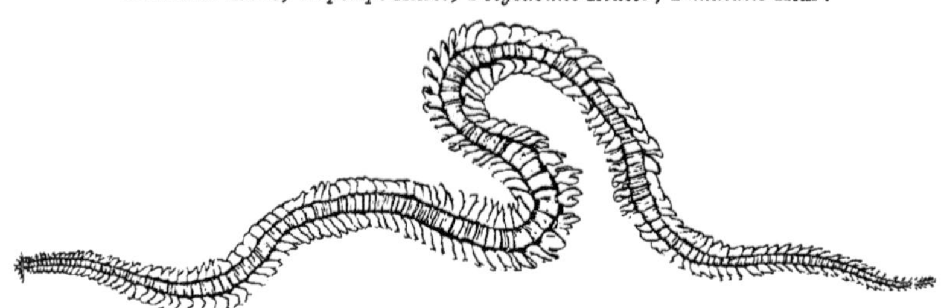

Fig. 388.—*Phyllodoca Paretti* Blv. (Règne Animal).

Sub-fam. 4. **Sigalioninae.** Long vermiform body. Elytra anteriorly on alternate segments, posteriorly on all the segments with or without gills. Without facial tubercle. *Sigalion* Aud. and Edw.; *Sthenelais* Kinb.; *Psammolyce* Kinb.; *Thalenessa* Baird; *Leanira* Kinb.; *Pholoe* Johnst.

Sub-fam. 5. **Polylepinae.** Elytra on all segments, no dorsal cirri. *Lepidopleurus* Clap.; *Pelogenia* Schm.

Fam. 5. **Palmyridae.** No elytra; strong setae disposed in a fan-like manner arise from all the parapodia dorsal to the dorsal cirrus. *Chrysopetalum* Ehl.; *Palmyra* Sav.

Fam. 6. **Phyllodocidae.** Body elongated, with numerous segments; prestomium rounded with 4 or 5 tentacles; peristomium with 4 tentacular cirri; the cirri of the parapodia are foliaceous.

Sub-fam. 1. **Phyllodocinae.** With small eyes. *Phyllodoce* Sav. (Fig. 388), with 4 tentacles; *Eulalia* Sav., with 5 tentacles; *Eteone* Sav.; *Anaitis* Mgr.; *Genethyllis* Mgr.; *Mystides* Théel; *Notophyllum* Oerst.; *Lacydonia* Mar. and Bob.

Sub-fam. 2. **Lopadorhynchinae.** Small transparent pelagic forms with small eyes; intermediate between *Phyllodocinae* and *Alciopinae*. *Lopadorhynchus* Gr.; *Hydrophanes* Clp.; *Pelagobia* Grf.

Sub-fam. 3. **Alciopinae.*** Transparent, pelagic worms with one pair of large hemispherical projecting eyes (Fig. 389). There are long peristomial cirri and 5 prestomial tentacles. Found in most oceans, but rare in the North Sea. The larvae are in part parasitic in *Cydippidae*. *Alciopa* Aud. and Edw.; *Alciopina* Clp. and Panc.; *Asterope* Clp.; *Vanadis* Clp. (Fig. 389); *Halodora* Grf.; *Greefia* (*Nauphanta*) M'Int.; *Callizona* Grf.; *Rhynchonerella* A. Costa.

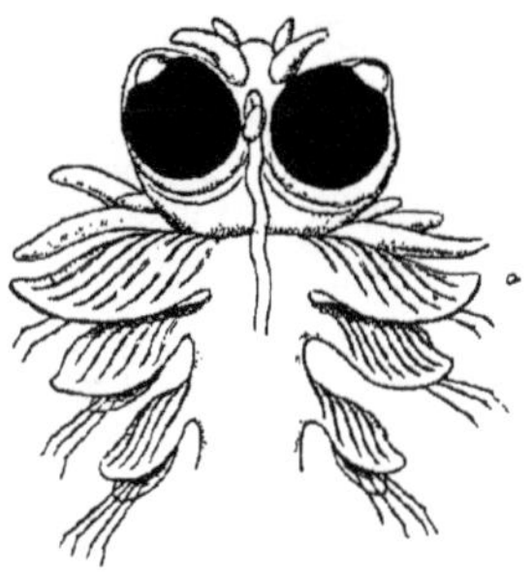

FIG. 389.—Head and anterior end of *Vanadis melanophthalmus* Greeff (after Greeff). Showing the large eyes.

Fam. 7. **Tomopteridae.**† Transparent, pelagic forms, with two eyes, bifid prestomium (the lobes of the prestomium may be tentacles; their cavity is continuous with the body-cavity), and four tentacular cirri, of which the posterior are much the longest and contain one seta, while the anterior contain two setae. The mouth is without proboscis and jaws. The parapodia are large, biramous, and without setae; each branch contains a yellow rosette-shaped organ, probably photogenic. *Tomopteris* Eschsch. (Fig. 390).

Fam. 8. **Nereidae** (*Lycoridae*). Body elongated, with many segments. Prestomium with two tentacles, two palps and four eyes; peristomium with four pairs of tentacular cirri. Parapodia uni- or biramous with dorsal and ventral cirri and composite setae. Proboscis usually with spines, and always with two jaws. *Nereis* L.; *N. diversicolor* Mull., burrows in mud and sand, fleshy red colour; *N. cultrifera* Gr., greenish grey, with small rectangular light spots along the dorsal surface; 6 inches. *N. dumerilii* Aud. and Edw. (Figs. 391, 392), reddish violet with darker transverse lines in each segment; peristomial cirri very long. *N. pelagica* L., bronze, large, widest in middle of body, back arched, on rocky ground. *N.* (*Nereilepas*) *fucata* Sav., lives on whelk shells. *N.* (*Alitta*) *virens* Sars—the creeper—may reach length of 18 inches, burrows in clay, etc., parapodia with large foliaceous lobes. *Lycastis* Sav.; *Dendronereis* Peters.

Fam. 9. **Nephthydidae.** Body elongated, quadrangular in section with flat dorsal and ventral surfaces. Prestomium inconspicuous, with 2 or 4 small tentacles. Peristomium with reduced parapodia bearing setae and two tentacular cirri. Parapodia biramous; the two branches are widely separated, and each is fringed with a membrane. The notopodium has a curved gill on its under

* R. Greeff, *Nova Acta Acad. Car. Leop.*, 29, 1876, p. 35.
† R. Greeff, *Zeit. f. wiss. Zool.*, 42, 1885, p. 432.

side, and a small cirrus. The neuropodium has a ventral cirrus. Pharynx large, with papillae and two small jaws. *Nephthys* Cuv.; *N. hombergii* Aud. and Edw., burrows in the sand near the shore to 20 fathoms; *Portellia* Qtg.

Fam. 10. **Amphinomidae.** Body vermiform or oval and flattened, with a small number of segments. The mouth is shifted backwards and surrounded by several similarly-built segments. The prestomium is indistinct, or represented dorsally by a sense-organ—the caruncle—which extends over several segments. Usually three tentacles, two palps, and one or two pairs of eyes. Proboscis well-developed, without teeth. Parapodia biramous, with dorsal and ventral cirri, and dorsal branchiae. Most of the species are tropical. *Euphrosyne foliosa* Aud. and Edw., and *Eurythoe borealis* Oerst., are found in the British area.

Sub-fam. 1. **Amphinominae.** With caruncle: with 2 branchiae on each segment. *Amphinoma* Brug. (*Pleione* Sav.); *Hermodice* Kinb.; *Eurythoe* Kinb.; *Notopygos* Gr.; *Chloeia* Sav.

Sub-fam. 2. **Euphrosyninae.** With caruncle; with numerous branchiae on each segment. *Euphrosyne* Sav.

Sub-fam. 3. **Hipponinae.** Without caruncle. *Hipponoa* Aud. and Edw.; *Spinther* Johnst.; *Aristenia* Sav.

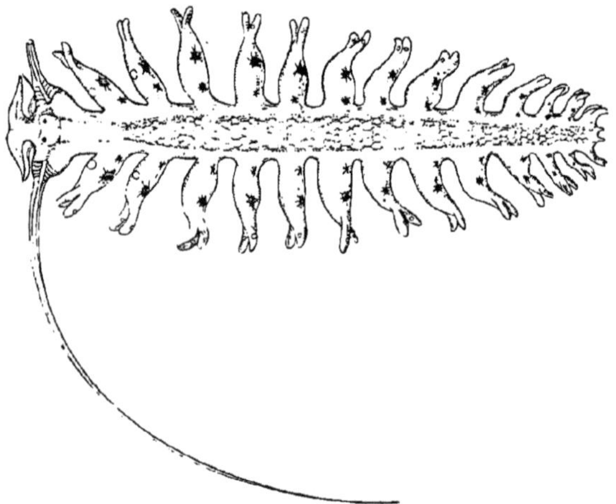

FIG. 390.—*Tomopteris rolasi* Greeff (after Greeff).

Fam. 11. **Staurocephalidae.** Prestomium with two dorsal jointed tentacles, and two inferior lateral tentacles (? palps). The two first segments without appendages. Parapodia biramous, with two kinds of setae. Without branchiae. *Staurocephalus* Gr. (*Anisoceras* Gr.; *Prionognathus* Kef.); *Paractius* Lev.

Fam. 12. **Lysaretidae.** Prestomium with three tentacles and four eyes. The two first segments without appendages. Parapodia uniramous with one kind of seta. Dorsal cirri foliaceous, branchial. *Halla* Costa; *Lysarete* Kinb.; *Danymene* Kinb.

Fam. 13. **Lumbriconereidae.** Without branchiae, without or with reduced cirri, and without tentacles. The two first segments without appendages.

Arabella Gr.; *Lumbriconereis* Gr. prestomium conical without tentacles and palps; *Drilonereis* Clp.; *Notocirrus* Schm.; *Laranda* Kinb.; *Ophryotrocha* Clp. and Meczn., with segmental ciliated rings.

Fam. 14. **Eunicidae.** Elongated body, with numerous segments. Prestomium distinct and projecting, with 3 or 5 tentacles, or with 5 tentacles and 2 palps, or without appendages; usually with eyes. The first (peristomium) or the two first segments without parapodia, but they may have cirri. In *Eunice* and *Diopatra* the second segment has a dorsal cirrus—the nuchal cirrus. Parapodia uniramous, rarely biramous, with ventral cirri and branchiae (the so-called dorsal cirrus contains an aciculum, and is the reduced notopodium). Usually 4 cirri below the anus. An upper jaw formed of several pieces, and a lower jaw of 2 lamellae are placed in a pocket attached below the oesophagus (Fig. 393). In most genera there are parchment-like tubes. *Diopatra*

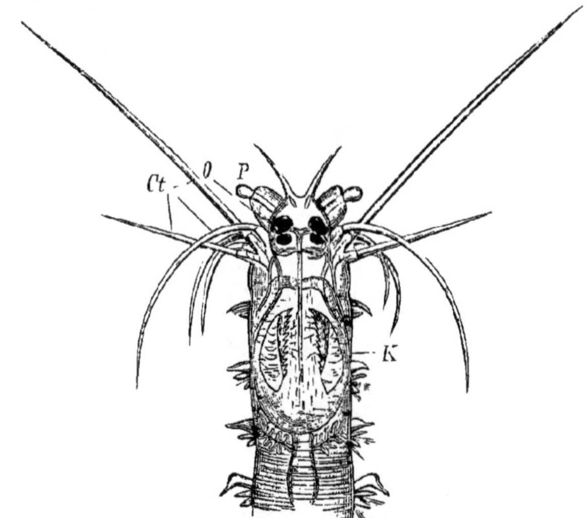

FIG. 391.—Head and anterior body segments of *Nereis dumerilii* (after Claparède). *O* eyes; *P* palps; *Ct* tentacular cirri; *K* pharyngeal jaws.

Aud. and Edw.; *Eunice* Cuv., 5 tentacles, 2 palps, and 2 nuchal cirri; *Marphysa* Qtf., like *Eunice*, but without nuchal cirri; *Hyalinoecia* Malmg., 5 tentacles, 2 palps, 2 frontal palps, no nuchal cirri; *Onuphis* Aud. and Edw.: *Nicidion* Kinb.; *Macduffia* McInt.; *Lysidice* Sav., 3 tentacles, 2 palps, no gills; *Nematonereis* Schmarda, one tentacle; *Palola* Gray, 3 tentacles. A species of this worm lives in fissures in the coral-reefs of certain parts of the Pacific (Samoa, Fiji). Twice in every year—once in October and once in November—these worms swarm out to the surface of the sea in enormous numbers to spawn, and are caught by the natives for food. In each of these months the worms appear on two successive days, viz., at dawn on the day on which the moon is in her last quarter and at dawn on the day before (*vide* S. J. Whitmee, "On the habits of *Palola viridis*," *Proc. Zool. Soc.*,

1875, p. 496, and A. Collin, "Bemerk. üb. d. essbaren Palolowurm," in Appendix to Krämer's work on the Coral-reefs of Samoa, Kiel and Leipzig, 1897, Lipsius and Fischer).

Fam. 15. **Glyceridae.** Body elongated, with numerous segments. Prestomium is long, conical, annulated, and carries at its apex four small tentacles, at its base two palps. The pharynx can be protruded to a great length, and is provided with four, or several, strong teeth. Parapodia of first two segments

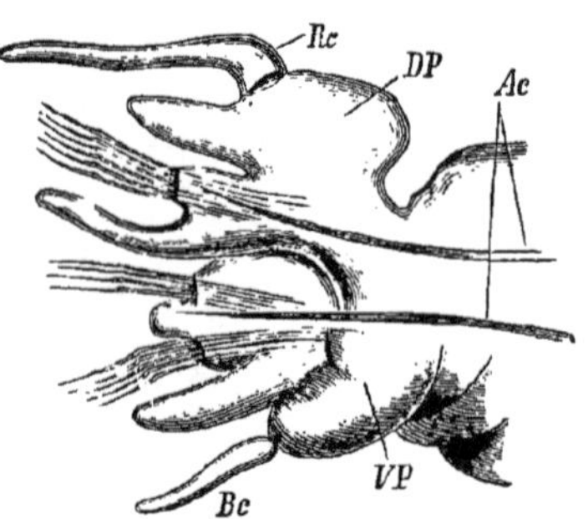

Fig. 392. — Parapodium of *Nereis* (after Quatrefages). *DP* notopodium; *VP* neuropodium, with bundles of setae; *Ac* aciculum; *Rc* dorsal cirrus; *Bc* ventral cirrus.

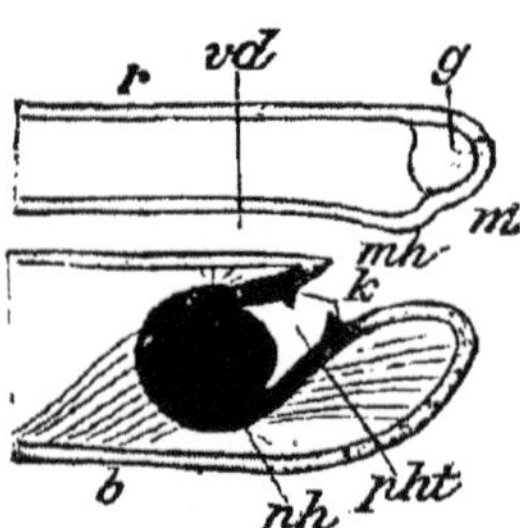

Fig. 393.—Pharyngeal apparatus of *Eunice* in a retracted condition. *g* brain; *vd* oesophagus; *ph* pharynx; *pht* pharyngeal pocket; *k* jaws; *mh* buccal cavity; *m* mouth; *r* dorsal, *b* ventral surface. (After Lang.)

incomplete, without tentacular cirri, uni- or biramous. Two anal cirri. Special retractile gills are present. The corpuscles of the coelomic fluid are coloured red with haemoglobin, and there are said to be no blood-vessels. *Glycera* Sav., parapodia all alike; *G. gigantea* Qfg., and *G. capitata* Oerst., under stones and in sand, English Channel; *Hemipodus* Qfg.; *Goniada* Aud. and Edw., anterior and posterior parapodia not alike.

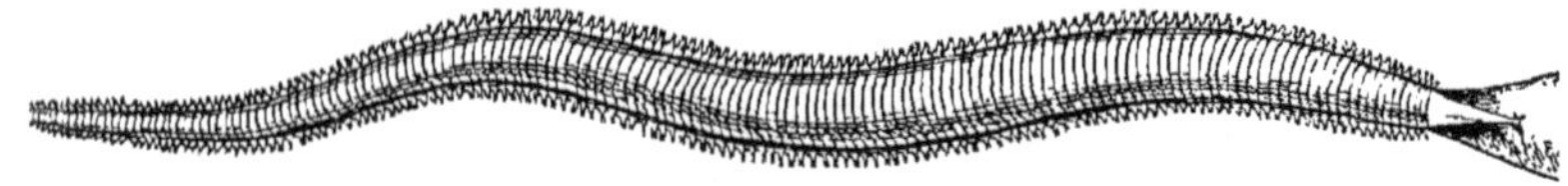

Fig. 394.—*Glycera Meckelii* Aud. and Edw. (Règne Animal), with everted pharynx.

Fam. 16. **Ariciidae.** Numerous segments. Prestomium without tentacles, or with small tentacles, or with tentacular cirri. Peristomium with setigerous tubercles. Pharynx short, unarmed, slightly or non-retractile. Parapodia short, biramous. In all but the first few segments the parapodia are somewhat dorsally placed. Burrow in sand. *Aricia* Sav.; *Theodisca* F. Müll.; *Scoloplos* Blainv.; *Sc. armiger* Müll., is common on our coast, length one inch, red gills; *Aricidea* Webs.

Fam. 17. **Typhloscolecidae.*** Pelagic forms of uncertain affinities. *Typhloscolex* Busch; *Sagitella* Wag.; *Travisiopsis* Lev.

* Reibisch, *Phyllodociden u. Typhloscoleciden d. Plankton Expedition*, 1895.

Sub-order 2. SPIONIFORMIA.

Without tentacles or palps; the peristomium usually carries a pair of long tentacular cirri, and extends forwards at the sides of the prestomium. The parapodia project but slightly; the dorsal cirri may attain a considerable size, and act as gills throughout the greater part of the body. The setae are unjointed; uncini are present only in *Chaetopterus*.

The body may present two regions more or less distinctly marked externally, but without corresponding internal differences.

The buccal region may be eversible, but there are no jaws. Septa and nephridia are regularly developed. The worms burrow or are tubiculous.

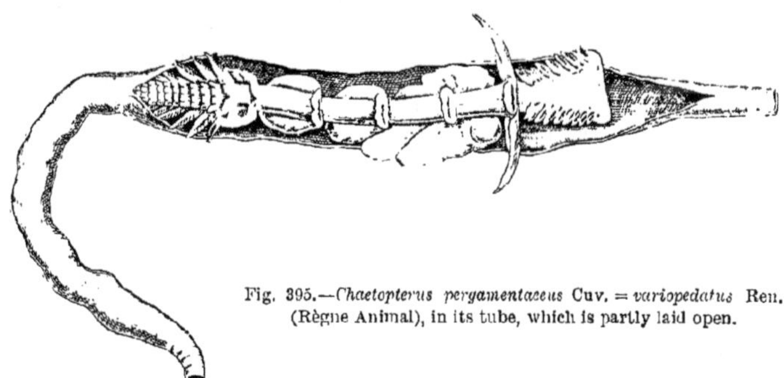

Fig. 395.—*Chaetopterus pergamentaceus* Cuv. = *variopedatus* Ren. (Règne Animal), in its tube, which is partly laid open.

Fam. 1. **Spionidae.** Prestomium small, without tentacles or palps, though sometimes with tentaculiform projections, usually with small eyes. Peristomium extends forward on each side of the prestomium, and bears two long tentacular cirri. Parapodia generally biramous, with simple setae. The notopodial cirri are long, finger-shaped, and ciliated, and function as gills. There are no ventral cirri. The buccal region is eversible. They burrow in mud and sand. The females lay their eggs in the tubes. *Polydora* Bosc. (*Leucodore* Johnst.). Prestomium with two tentaculiform projections. The 5th setigerous segment is enlarged, and bears specially strong setae arranged in a comb-like manner. *P. ciliata* Johnst., in soft mud tubes in U-shaped galleries in stones and shells; length ½-inch; 40 segments; anus surrounded by incomplete funnel; world-wide; *P. coeca* Oerst. *Spio* Fabr., *S. seticornis* Fabr., Greenland and Scandinavia; *Prionospio* Mlg.; *Scolecolepis* Blainv.; *Nerine* Johnst.; *N. vulgaris* Johnst., under stones and on seaweed, 3 or 4 inches long, N. Atlantic.

Fam. 2. **Chaetopteridae.*** Body divided into two or three unequal regions. Prestomium small, often with eye-spots. Peristomium prolonged round the mouth in a funnel-like manner, and bearing two or four tentacular cirri. The fourth segment behind the peristomium with specially strong setae as in *Polydora*. Neuropodia bifid in the posterior region, and sometimes in the middle

* Joyeux-Laffuie, Arch. Zool. Exp., (2), 8, 1890, p. 244.

region. Notopodia of the middle region in the form of fins, often lobed. They inhabit U-shaped parchment-like tubes, the two ends of which are placed at some distance from each other, and project above the sand. The skin of the worm is soft and delicate. The larvae are mesotrochal. *Telepsavus* G. Costa, two long tentacular cirri; body divided into two regions; *Spiochaetopterus* Sars; *Phyllochaetopterus* Gr., two pairs of tentacular cirri; body divided into three regions. *Chaetopterus* Cuv. (Fig. 395), two tentacular cirri. Body divided into three regions. Phosphorescent. *Ch. variopedatus* Ren., Channel Islands, English coast, etc., anterior region with 9 pairs* of setigerous conical parapodia; middle region of 5 segments, of which the first carries two large wing-like processes directed forwards (notopodia), the second carries a dorsal and ventral sucker (modified parapodia), the third, fourth, and fifth a membranous fin-like notopodium—the fans, and a ventral bilobed neuropodium; the posterior region consists of from 25 to 35 less modified, similar segments, with notopodium and bilobed neuropodium.

Fam. 3. **Magelonidae.**† Body divisible into two regions by differences in the setae. Prestomium large and flat, two long peristomial cirri. Large eversible buccal region, blood of madder-pink colour when oxygenated, and colourless when deoxidized (*haemerythrin*, found also in *Sipunculus*); the colouring matter being contained in globules which float in the colourless plasma; isolated nuclei are also found in the plasma. Live buried in sand. Sole genus, *Magelona* F. Mull.

Fam. 4. The **Ammocharidae** may be placed here. They live in sandy tubes, and the mouth is surrounded, except ventrally, by a membrane, which is frayed out marginally into filaments, and is probably an appendage of the peristomium. The anterior segments longer than the hinder ones. Sole genus *Owenia* D. Ch.

Sub-order 3. TEREBELLIFORMIA.

The prestomium is a more or less prominent lobe (upper lip) with or without tentacles, but without palps. The peristomium may carry cirri or tentacular filaments. The parapodia are feebly developed; there are no ventral cirri; the dorsal cirri may exist and function as gills on some of the segments. The setae are unjointed, and uncini are usually present. The buccal region is not eversible, and there are no jaws. The septa are usually incomplete, with the exception of one strongly-developed diaphragm anteriorly. The nephridia are dimorphic, those of the prediaphragmatic segments being of large size and excretory, while the posterior nephridia are mere funnels which serve as generative ducts. They are burrowers or tube-formers, and in the majority the tube-forming glands are grouped on the ventral surface of the anterior segments, where they form *gland-shields*.

* The number is variable; specimens have been found with 12, or with 9 on one side and 8 on the other.

† W. B. Benham, *Q. J. M. S.*, 39, 1896, p. 1.

Fam. 1. **Cirratulidae.** Cylindrical body pointed at both ends. Prostomium elongated, conical, without tentacles, or with two tentacles (tentacular cirri, Grube). Parapodia small, with simple setae. On some of the segments the dorsal cirri are long and filamentous, and function as gills. A single pair of anterior nephridia; the septa and genital ducts are repeated through the hinder part of the body. Inhabit burrows. *Cirratulus* Lam.; *Audouinia* Qfg.; *Chaetozone* Malmg.; *Dodecaceraea* (*Dodecaceria*) Oerst.; *Hekaterobranchus* F. Buchanan.

Fam. 2. **Terebellidae.** Body vermiform and thicker anteriorly. The thinner posterior portion is sometimes distinctly marked off as an appendage devoid of setae. The prostomium, which is indistinctly separated from the peristomium, forms an upper lip and carries numerous filiform tentacles arranged in a transverse row; it may have four eye spots. There are pectinate or branched, rarely filamentous, gills on one to three of the anterior segments; they contain a prolongation of the coelom. Dorsal prominences (notopodia) with simple setae, and ventral transverse ridges (neuropodia) with hooked setae. One to three pairs of large nephridia. From three to twelve pairs of small nephridia (gen. ducts) behind the diaphragm. Tubicolous, tube formed of foreign bodies.

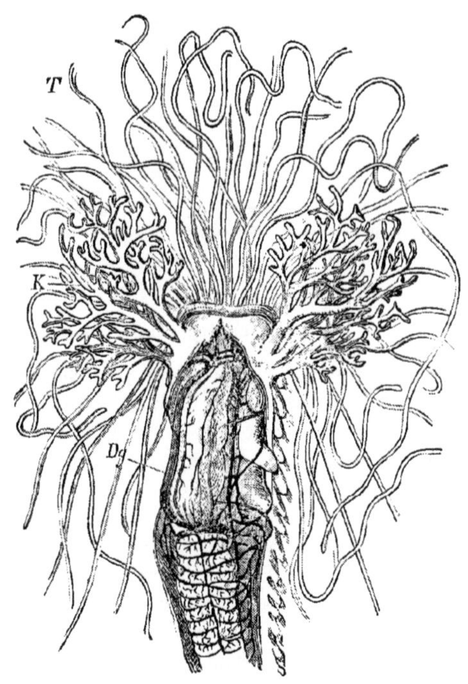

Fig. 396.—*Terebella nebulosa* opened from the dorsal side (after M. Edwards). *T* tentacles; *K* gills; *Dg* dorsal vessel or heart.

Sub-fam. 1. **Amphitritinae.** Almost always with gills; cephalic lobe short, with numerous tentacles. *Amphitrite* O. F. Müll., simple setae in anterior region only, 3 pairs of gills of nearly the same size, no eyes; *A. johnstoni* Mlmg. *Terebella* L. (Fig. 396), distinguished by the small size of the posterior of the three pairs of branchiae; *T.* (*Lanice*) *conchilega* Pall., English coast; *Nicolea* Mlmg., two pairs of equal gills; *Heteroterebella* Qtf.; *Thelepus* Leuck. (incl. *Heterophenacia* Qtf., *Phenacia* Qtf., *Neottis* Mlmg., etc.), filiform gills in two segments; *Euthelepus* McInt.; *Leaena* Mlmg.; *Lanassa* Mlmg.; *Lolmia* Mlmg.; *Pista* Mlmg.; *Nicolea* Mlmg.

Sub-fam. 2. **Polycirrinae.** Without branchiae. The prestomium forms a large upper lip, and gives origin to many long tentacles. *Polycirrus* Gr. (including *Leucariste* Mlmg.; *Ereutho* Mlmg.); *P. aurantiacus*, bright red, with orange tentacles.

Malmgren distinguishes 3 other sub-families, **Artamaceae, Trichobranchidae, Canephoridae** (*Terebellides*, Sars.).

Fam. 3. **Ampharetidae.** Body formed of two regions, of which the anterior is broad and bears simple setae and hooks, while the posterior part is narrow and has only hooks. Numerous filiform tentacles spring from the prestomium. The peristomium forms a kind of lower lip. Four filiform branchiae are placed on each side of the anterior segments, and in front of them are some strong setae. Inhabit tubes in mud. *Amphareta* Mlmg.; *Amphicteis* Gr. (*Lysippe, Sosane* Mlmg.); *Sabellides* Gr.; *Melinna* Mlmg.; *Melinnopsis*, McInt.; *Phyllocomus* Gr.; *Grubianella* McInt.; *Samythopsis* McInt.; *Eusamytha* McInt.

Fam. 4. **Amphictenidae.** Distinguished from *Terebellidae* mainly by the presence of tentacles disposed in two bundles on the buccal ring, of two pairs of tentacular cirri and of pectinated branchiae on the second and third segments. The anterior end bears two rows of golden setae, which act as an operculum to the tube. The tubes are straight or slightly curved, and composed of small grains of sand. *Pectinaria* Lam.; *Amphictene* Sav.; *Petta* Mlmg.

Sub-order 4. CAPITELLIFORMIA.

Blood-red worms, with a conical prestomium. Without prestomial processes, but with a pair of large retractile ciliated organs. The segment next the prestomium is without setae or cirri. The parapodia do not project; the setae are unjointed, and are hair-like in the anterior segments, and hooked posteriorly. This external division of the body does not correspond with internal differences. There are no cirri, though special, sometimes retractile, gills are frequently present. The buccal region is eversible; there is no armed pharynx. A pair of prominences with sensory hairs, placed one on each side, in most segments. The nephridia are small, and sometimes more than one pair in a segment; special genital funnels exist in more or fewer of the anterior segments of the hind body. There are no blood-vessels; the coelomic corpuscles are red. The larvae are telotrochous, and are ciliated on the ventral surface. They are burrowers.

Fam. **Capitellidae,*** with characters of sub-order. *Capitella* Blainv.; *C. capitata* v. Ben., North Sea, etc.; *Notomastus* Sars; *Dasybranchus* Grube; *Mastobranchus* Eisig; *Capitomastus* Eisig.

* H. Eisig, "Die Capitelliden," *Fauna and Flora d. G. v. Neapel*, xvi., 1887.

Sub-order 5. SCOLECIFORMIA.

With a prestomium which rarely (*Chlorhaemidae*) carries sensory processes; the peristomium is without cirri (except perhaps in the *Chlorhaemidae*). The parapodia are reduced, and may be absent; dorsal cirri (acting as gills) are rarely present; ventral cirri are absent. Setae unjointed; uncini are not present. Buccal region eversible—without armed pharynx. The septa are not developed regularly; the nephridia, which are all alike, are considerably reduced in number, it may be to a single pair (*Sternaspidae* and some *Chlorhaemidae*). Mostly burrowers.

This is the most unsatisfactory of Benham's sub-orders. Its merit is that it comprises a number of families which cannot be placed with any of the other sub-orders.

Fam. 1. **Opheliidae.** Short worms; prestomium conical, without appendages, but with two ciliated pits. Parapodia reduced, with simple setae in two groups. Dorsal cirri acting as gills, or reduced, usually present. They live in the sand. *Ophelia* Sav.; *Ammotrypane* Rathke; *Travisia* Johnst.; *Armandia* Fil.; *Polyophthalmus* Qtf., without gills, eye-like pigment spots on the sides of some of the segments; circular muscles absent; oesophageal commissures and ventral cords in contact with the ectoderm; septa absent, except three in front.

Fam. 2. **Maldanidae*** **(Clymenidae).** Body cylindrical, divided into two or three regions. The segments towards the middle of the body may be longer than the rest. Prestomium small, fused with peristomium; often with eye spots.

* St. Joseph, *Ann. Sci. Nat.* (7), 17, 1894, p. 130.

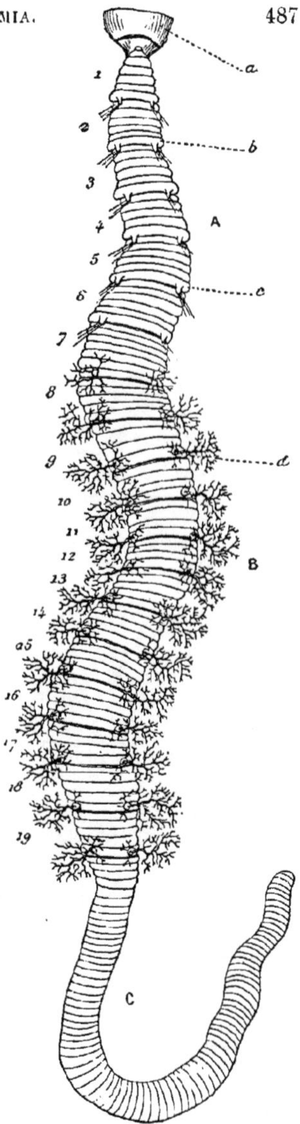

Fig. 397.—*Arenicola piscatorum*, dorsal view. *A* anterior region, comprising six segments; *B* middle or branchial region, with thirteen segments; *C* tail region of variable length *a* eversible buccal region; *b*, *c* notopodia with bundles of setae; *d* gills. (From Vogt and Yung.)

A horny plate may be developed on upper side of head. Anus usually surrounded by a papillated funnel. Tentacles and branchiae absent. Notopodium small with simple or pinnate setae, absent posteriorly; neuropodium as transverse ridge with hooked setae, absent in anterior region. Inhabit long, sandy tubes. *Mitraria* is a clymenid larva. *Clymene* Sav., head with plate; *C. lumbricoides* Qtf.; *Praxilla* Mlmg.; *Leiocephalus* Qtf.; *Maldane* Gr., head without plate; *Nicomache* Mlmg.; *Axiothea* Mlmg.; *Leiochone* Gr.; *Petaloproctus* Qtf.

Fam. 3. **Arenicolidae (Telethusidae).** Prestomium small, without tentacles, fused with peristomium. Peristomium without setae; eversible buccal region covered with papillae, without jaws. Notopodia small, with a bundle of simple setae; neuropodia with hooks in rows. Branched gills on middle segments. Posterior segments without parapodia and gills. Burrow in sand. *Arenicola* Leach; *A. marina** L. (*A. piscatorum* Lam.). North Sea and Mediterranean. Lob- or lug-worm. Length, 5–6 inches (Naples specimens smaller), colour, brownish green. The burrows are U-shaped, one opening being marked by a heap of vermicular, sandy ejecta. Body divided into three regions: an anterior thicker than the rest, consisting of the peristomium and 6 chaetiferous segments; a middle, not so thick, of 13 chaetiferous and branchiferous; a posterior much thinner and devoid of setae and branchiae, with segments less clearly marked and variable in number. In the anterior and middle region of the body there are 3 coelomic septa, the 1st behind the peristomium, the 2nd between the 2nd and 3rd setigerous segments, and the 3rd between the 3rd and 4th setigerous segments. In the caudal regions the coelomic septa are segmental. There are 6 pairs of nephridia in segments 4–9.

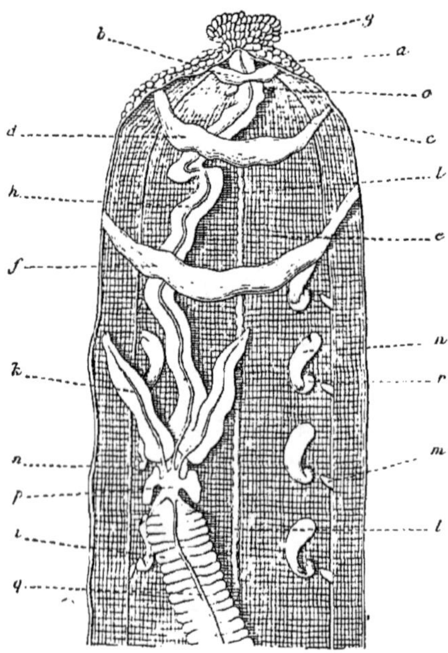

Fig. 398.—Anterior end of *Arenicola* opened from the dorsal side. *a* anterior small chamber of the coelom; *b*, *d*, and *f* first, second, and third septa; *c* and *e* second and third chambers; *g* papillae of proboscis; *h* oesophagus; *i* gut; *k* gland; *l* nerve cord; *m* setigerous sacs; *n* nephridia; *o* anterior glands; *p* heart; *q* dorsal vessel; *r* vesicles of nephridia. (From Vogt and Yung.)

* Vogt and Yung, *Praktische Vergl. Anat.*, Bd. 1, 1888, p. 487.

Fam. 4. **Scalibregmidae.*** *Scalibregma* Rathke; *Eumenia* Oerst.; *Sclerocheilus* Gr.; *Lipobranchius* Cunn. and Ram.

Fam. 5. **Chlorhaemidae**† **(Pherusidae).** Elongated cylindrical body with green blood (due to chlorocruorin). Prestomium with a pair of long processes (? palps) and several green tentacles arranged in a transverse row and acting as gills. Peristomium achaetous. The head can be withdrawn into the body. The setae of the anterior segments are long and directed forwards. Skin papillated. Only one or two coelomic septa, and two or four nephridia. *Flabelligera* Sars (*Siphonostoma* Voigt; *Chlorhaema* Duj.); *F. affinis* Sars, transparent body-wall, skin with long papillae, which traverse a thick jelly-like envelope secreted by it, under stones at low tide; *Trophonia* M.-Edw. (*Pherusa* Blv.); *Brada* Stimp.; *Buskiella* McInt.; *Stylarioides* D. Ch.

Fam. 6. **Sternaspidae.**‡ Body short, anterior region thickened and carrying on each side 3 rows of setae; on the ventral surface of the hind end there is a bilobed horny plate, round the edges of which are placed a number of bundles of long setae. Mouth overhung by a small knob-like prestomium; anus on a papilla dorsal to the horny shield. Two bundles of filamentous branchiae just anterior to, and on each side of the anus on the dorsal surface. About 30 segments; segments 2, 3, 4 have a row of strong setae on each side, segments 5, 6, 7 are without setae, segments 8–15 have setae which do not pierce the cuticle: the remaining segments are marked by the bundles of long setae round the horny plate. Coelomic septa absent; alimentary canal coiled; one pair of nephridia attached to the skin between segments 6 and 7, and opening internally. Projecting externally between segments 8 and 9 on the ventral surface are two processes which contain the terminal parts of the generative ducts. The generative organ is a sac which communicates with the visceral coelom by a narrow canal; the walls of the sac are continued into the two generative ducts. *Sternaspis* Otto, *S. scutata* Ranzani, Mediterranean.

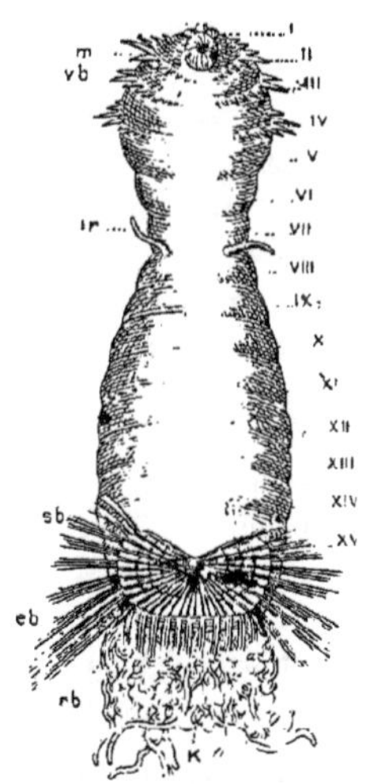

FIG. 390.—*Sternaspis scutata*, ventral view × 2 (after Vejdovsky). The segments in front of the shield are numbered I–XV. *K* branchial filaments; *lr* processes containing the generative ducts; *m* mouth; *rb*, *sb*, *eb*, setae of horny shield; *vb* rows of setae on segments *II*, *III*, *IV*.

* Levinsen, *Syst. geog. Overs. over de Nord. Annul* (*Vidensk. Meddelser*) for 1883, Copenhagne, 1884, p. 130.

† E. Grube, "Mitth. über d. Fam. Chlorhaemiden," *Sitz. ber. d. Schlesischen Gesellschaft*, Breslau, 1876.

‡ Vejdovsky, "*Dk. Akad. Wien*, 43, 1882. Rietsch, *Ann. Sci. Nat.* (6), 13, 1882. Goodrich, *Q. J. M. S.*, 40, p. 233.

Branch B. CRYPTOCEPHALA.

Prestomium more or less hidden by the peristomium which grows forward. The tentacles are reduced, but the palps are greatly developed and subdivided, forming the crown of gills. The body is distinguishable into two regions—the thorax and abdomen characterized by the form and arrangement of the setae, and by certain internal differences.

Sub-order 1. SABELLIFORMIA.

Prestomium entirely hidden by the forward extension of the peristomium; tentacles very small, being often represented by small knobs of sense-cells; palps large and vascular, acting as respiratory and sensory organs. Peristomium without cirri or setae; it is usually raised into a projecting collar used in fashioning the lip of the tube. The parapodia are small; cirri are absent except in the *Serpulidae.* The setae are of two kinds—hair-like and uncini; by their arrangement the body is divided into a thorax, typically of nine segments, and into an abdomen; in the thorax the hair setae are dorsal, in the abdomen ventral. The buccal region is not eversible; there is no pharynx. Septa absent in thorax, present in abdomen. There are two large nephridia in the thorax, opening by a median dorsal pore just above the brain; while the nephridia of the abdomen are small funnels which act as generative ducts. Tubicolous; gland-shields are present on the thoracic segments. Some genera are hermaphrodite.

Fam. 1. **Sabellidae*** (including *Eriographidae* and *Amphicorinidae*). Tube formed of mucin of variable consistence and more or less transparent, covered or not with mud, sand, or pieces of shells. Without thoracic membrane and operculum. Branchial filaments usually arise from a semicircular base. The ventral gland-shields are continued on to the abdomen. A median ciliated groove, usually situated on the ventral surface (it sometimes bends to one side on reaching the thorax, and may extend forwards on the dorsal surface to the head), starts from the anus and conducts the faecal matters to the opening of the tube. The species are highly variable: the individuals of the same species often vary in the following features; colour, number of branchiae, form of collar, number of thoracic segments, number and position of eyes and otocysts.

Spirographis Viv., the basal lamella of the gill filaments describes several turns in spiral. The branchiae of the two sides are of unequal size; *Sp. spallanzanii* Viv., Naples; *Bispira* Kr. Like preceding, except the two branchiae are equal. *Sabella* L., gill-filaments arise from a semicircular base, gills equal, two filamentous dorsally-placed lip-processes (found in other Sabellids: they are not tentacles); *S. pavonina* Sav.; *Potamilla* Mgr.; *Hypsicomus* Gr.; *Potamis* Ehlers; *Branchiomma* Köll., with a compound eye close to the tip.

* St. Joseph, *Ann. Sci. Nat.* (7), 17, 1894, p. 248.

of each gill-filament; *Protulides* Webst.; *Dasychone* Sars, gill-filaments with dorsal appendices, *D. bombyx* Dal., with compound eyes at the base of the dorsal appendices of the gills; *Laonome* Mgr.; *Notaulax* Tauber; *Sabellastarte* Kr.; *Eurato* St. Jos.; *Haplobranchus* A. G. Bourne, estuarine; *Manayunkia* Leidy, fresh-water; *Caobangia* Giard, fresh-water, anus ventral and anterior; *Fabricia* Blv.; *Othonia* Johnst.; *Amphicora* Ehr., with two eyes on the last segment; *Oria* Qtf.; *Jasmineira* Lang; *Myxicola* Koch (*Eriographis* Gr.; *Arippasa* Johnst.), gill-filaments connected by a membrane which reaches nearly to their tips; *Chone* Kröyer; *Euchone* Mgr.; *Dialychone* Clp.

Fam. 2. **Serpulidae.*** Tube formed of mucine indurated with calcareous matter. Almost always with thoracic membrane and operculum. Without a ciliated groove, but the ventral surface (and sometimes the dorsal) is in part ciliated. The thoracic membrane is formed by the fusion of the dorsal and ventral cirri of the thoracic segments. The operculum consist of a terminal dilatation of one of the branchial filaments. *Serpula* L.; sub-genera *Hydroides* Gunn. (*Eupomatus* Phil.; *Eucarphus* Mörch.; *Polyphragma* Qtf.), and *Crucigera* Benedict.; *Protis* Ehl.; *Chitinopoma* Lev.; *Filograna* Oken, two or several opercula; *Salmacina* Clp., no operculum; *Spirorbis* Daudin (Fig. 400), operculum grooved on one side, serving as brood pouch; *Pileolaria* Clp.; *Janua* St. Joseph; *Omphalopoma* Mörch; *Circeis, Omphalopomopsis, Janita, Leodora, Mera, Hyalopomatopsis*, all St. Jos.; *Vermilia* Lmk., etc.; *Pomatoceros* Phil.; *Spirobranchus* Blv.; *Pomatostegus* Schm.; *Placostegus* Phil.; *Protula* Risso; (*Psygmobranchus* Phil.), no operculum; *Apomatus* Phil.

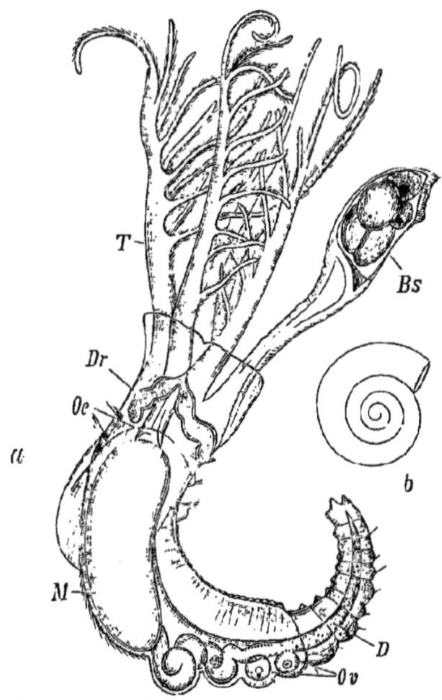

FIG. 400.—*Spirorbis laevis* (after Claparède). *a*, the animal removed from its tube, strongly magnified; *b*, its tube. *T* tentacles; *Bs* brood pouch, with operculum; *Dr* glands; *Ov* ova; *Oe* oesophagus; *M* stomach; *D* intestine.

Sub-order 2.

HERMELLIFORMIA.

The peristomium is enormously developed, and forms a bilobed hood capable of closing over the mouth; the truncated free end of each lobe carries three semicircles of peculiar setae, which act as a protection when the worm is withdrawn into its tube. The prestomium is small, but possesses a pair of well-

* St. Joseph, *Ann. Sci. Nat.* (7), 17, p. 259.

developed tentacles; the palps, which are subdivided as in the *Sabelliformia*, have become fused with the ventral edges of the peristomium, and appear as a series of ridges on each side, carrying numerous filaments. The thorax consists of five segments, the notopodia of three of which are well developed, and bear strong setae; dorsal cirri are present along the greater part of the body, and act as gills. The worms form tubes of sand.

Fam. **Hermellidae.** *Sabellaria* Lam. (*Hermella* Sav.).

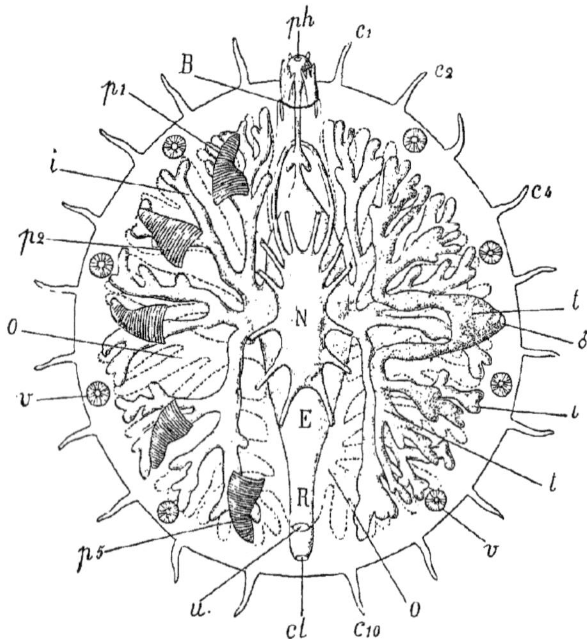

FIG. 401.—Diagrammatic figure representing the organisation of *Myzostoma* (after v. Graff, from Perrier). *B* mouth; *c1, c10* marginal cirri; *cl* cloaca; *E* stomach; *i* branches of the intestine; *N* ventral nervous system; *o* coelom filled with ova; *p1, p5* parapodia; *ph* pharynx; *R* rectum; *t* testis; *u* female generative opening; *v* suckers; ♂ male opening.

MYZOSTOMIDA.*

The *Myzostomida*, though differing in many respects from the *Polychaeta*, may most conveniently be treated in an appendix to that group. There are two genera, *Myzostoma* Leuckart, and *Stelechopus* (parasitic on *Hyocrinus*) von Graff. They are all parasitic on Crinoids, external for the most part

* v. Graff., "Myzostomida," *Challenger Reports*, vol. x., 1884, and Supplement, vol. 20, 1887. Wheeler, *Naples Mitth.*, 12, 1896, p. 227.

(*M. pulvinar* v. Gr., lives in the intestine). They infest the disc and arms, and may be free or enclosed in globular cysts.

They possess a soft and ciliated skin; five pairs of ventrally placed parapodia (Fig. 401), each of which bears a hooked seta and an aciculum; four pairs of laterally placed suckers; and ten or more pairs of cirri placed on the margin of the body.

The alimentary canal consists of a protrusible pharynx, a stomach which gives off the radially arranged caeca, and a rectum. The mouth is anterior and ventral, and the anus posterior and ventral (except in a few species in which one or both openings may be dorsal).

The nervous system consists of a nerve ring surrounding the pharynx, slightly enlarged dorsally, and ventrally continued into a large ganglionic mass (Fig. 401), which represents the ventral cords.

The animals are hermaphrodite and protandrous, the testes maturing before the ovaries. The testes are branched glands on each side, and lead into two ducts which open laterally, one on each side, just external to the third parapodium.

The ovaries are two small patches of coelomic epithelium, and the eggs are dehisced into the coelom, which they fill. From the coelom a tube—the oviduct—passes dorsal to the rectum, and opens either into the latter, or by a median pore just dorsal to the anus.

There are two nephridia, opening with ciliated funnels into the coelom and into the rectum behind.

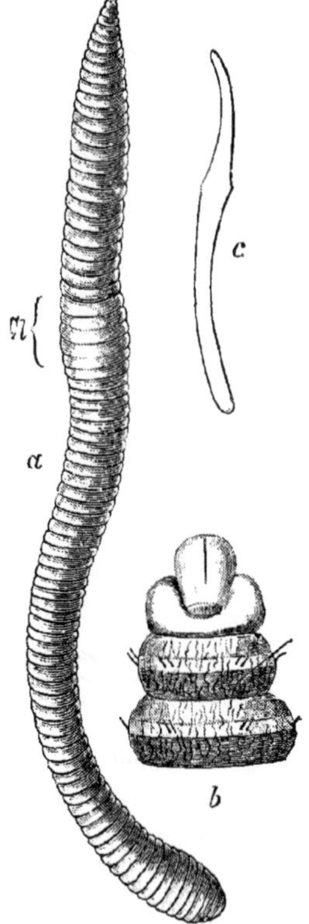

FIG. 402.—*Lumbricus rubellus* (after G. Eisen). *a*, the whole worm; *Cl* clitellum. *b*, the anterior end from the ventral side. *c*, isolated seta

Order 2. OLIGOCHAETA.*

Hermaphrodite Chaetopoda without pharyngeal armature and parapodia. Without tentacles and cirri, rarely with branchiae. Gonads limited in number, and restricted in position. The development is direct.

The *Oligochaeta* comprise certain small, and for the most part fresh-water worms, and all the worms known as earthworms,

* E. Claparède, "*Recherches Anatomiques sur les Oligochaetes,*" Geneva, 1862. F. Vejdovsky, *System u. Morphologie der Oligochaeten,* Prag, 1884. W. B. Benham, "An attempt to classify Earthworms," *Q. J. M. S.*, vol. 31, 1890. F. E. Beddard, "*A Monograph of the Oligochaeta,*" Oxford, 1895, which see for literature.

which are sometimes of considerable size, even reaching a length of from four to six feet (*Microchaeta rappi; Megascolides australis*). The setae are, as a rule, few in number, and are not set in parapodia. The head consists of a prestomium, which is sometimes much elongated (*Nais lacustris*), and is generally marked off from the oral segment (peristomium), or first segment of the body, by a groove. The oral segment is always without setae, and is never reinforced by the fusion with it of posterior body segments. There are no cephalic appendages either on the pre- or peristomium. The segmentation of the body is always well marked externally by circular grooves,* and the coelomic septa are always present. The segments are all alike, and are not divided into regions, though some of the

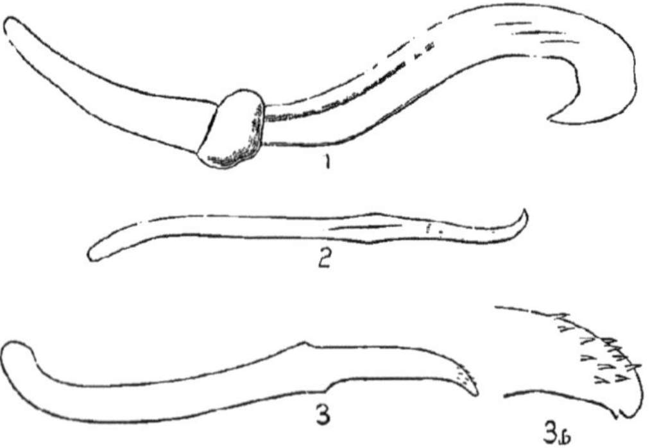

FIG. 408.—Setae of *Oligochaeta* (after Vejdovsky, Stolc, and Michaelsen, from Beddard). *1*, *Onychochaeta windlei*, posterior seta. *2*, ornamented seta of Geoscolecid. *3*, *Trichochaeta hesperidum*; *3b*, end of same more highly magnified.

segments behind the peristomium are occasionally devoid of setae. The gonads are localized thickenings of the coelomic epithelium, and special generative ducts are present to convey their products to the exterior. There is always present a glandular development of the ectoderm, generally having a marked annular form and called the *clitellum* or girdle: it secretes the cocoon in which the eggs are laid.

Considering the great constancy presented by the Oligochaeta in the arrangement and structure of many important organs, *e.g.*

* In some species of *Lumbricus* a curious modification of the metamerism, called spiral metamerism, has been observed as a variation. The grooves between the segments in such cases have a spiral course (*Vide* Bates on "*Materials for the Study of Variation*," p. 157, London, 1894).

the body-form, the ventral nervous system, etc., the variations in structure in the excretory and generative organs are most remarkable.

The **setae** are small chitinous rods projecting from the body-wall. Their lower ends are embedded in small ectodermal pits, the cells of which secrete them. They are absent in the *Discodrilidae* and in the genus *Anachaeta*, otherwise they are present to the number of eight or a larger number on most of the segments. When there are eight they are usually arranged in four groups of two on the ventral side, but the number and arrangement presents great variety, and in some forms (*Perichaetidae*) there is a continuous circle of them round each segment (Fig. 411). The usual form of seta is that of an *ʃ*, but they vary somewhat, as is shown by the accompanying

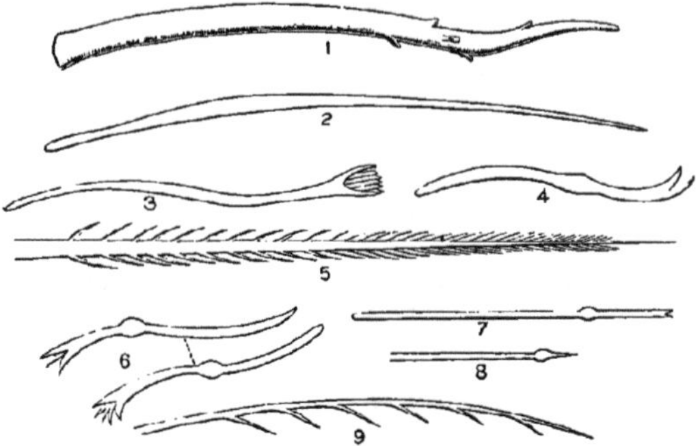

FIG. 404.—*1, 2*, Penial setae of *Acanthodrilus georgianus*. *3*, *Spirosperma ferox*. *4*, *Ilyodrilus coccineus*. *5*, *Lophochaeta ignota*. *6*, *Tubifex rivulorum*. *7, 8*, *Nais*. *9*, *Bohemilla comata*. (From Beddard.)

diagrams. The setae are retractile and protractile by special muscles attached to the walls of the setigerous sacs (Fig. 405). They are organs which assist in locomotion.

Branchiae are found in a few forms, *e.g. Chaetobranchus semperi*, *Branchiura Sowerbyi*, and *Hesperodrilus branchiatus*.

The **body-wall** consists of a cuticle, a layer of ectoderm, an outer circular and an inner longitudinal muscular layer. The ectoderm consists of cells, many of which are glandular. It also contains tactile sense organs in the form of papillae, and organs which resemble taste-buds; and in the aquatic forms many of the ectoderm cells carry sensory hairs. The gland-cells are especially well-developed in the clitellum.

The muscles are constructed upon the nematode type (*i.e.*, the contractile substance partially surrounds the unmodified nucleated protoplasm (Fig. 406, *A*, *B*, *C*). In the earthworms, however, the longitudinal muscular layer presents an appearance, which it is difficult to interpret. In transverse section they appear to consist of a number of lamellae placed like the barbs of a feather upon a central axis, the whole structure being placed in a connective substance with nuclei (Fig. 406, *D*). The explanation appears to be that we have to do in such cases with a folded layer of longitudinal muscles; each lamella consisting of a muscle-cell, from which the granular unmodified protoplasm and nucleus has disappeared (*vide* Hesse, *Z. f. w. Z.*, 58, 1894, p. 394).

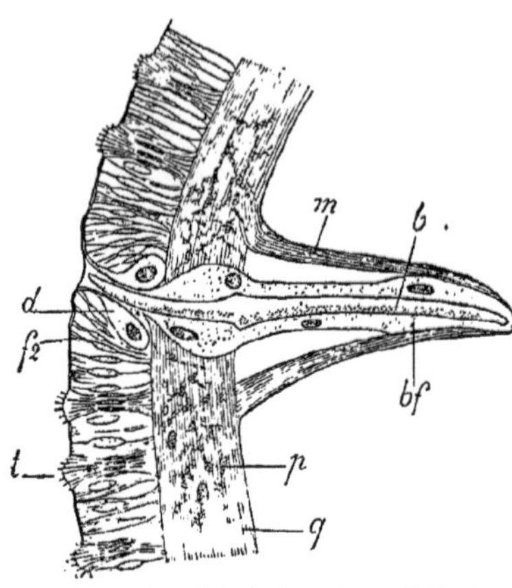

Fig. 405.—Section of the body-wall of *Allolobophora (Dendrobaena) rubida* in the neighbourhood of a seta. *d* glandular cells on each side of the seta; *f*² epidermal cells; *t* tactile organs; *g* circular muscles with pigment *p*; *b* seta; *bf* setigerous sac; *m* its muscles. (From Perrier, after Vejdovsky.)

The **central nervous system** is on the usual Annelidan type, consisting of two ventral cords closely approximated and enclosed in a common sheath. They are swollen in each segment where the main nerves are given off, and in front they separate from one another and embrace the anterior end of the alimentary canal, to be continuous with one another dorsally. At this dorsal point there is generally a bilobed swelling which constitutes the cerebral or suprapharyngeal ganglia, from which the nerves to the prestomium are given off. The first ventral swelling is where the circumpharyngeal commissures join to form the ventral cord, and constitutes the sub-pharyngeal ganglion. It gives off a large number of nerves to the peristomium. In some forms a few large, longitudinal fibres resembling

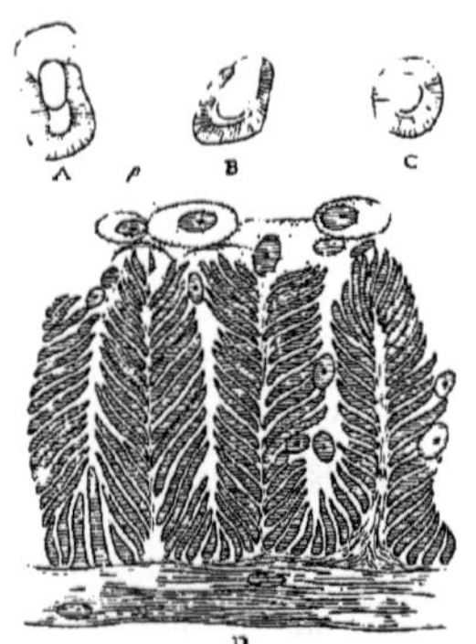

Fig. 406.—*A*-*C*, sections through the circular muscle-cells of *Allolobophora chlorotica*; *D*, section through the longitudinal muscular layer of a young earthworm; *p* coelomic epithelium. (After Hesse.)

tubes, run along the dorsal side of the conjoined nerve cords in the neurilemma. They are called the giant fibres, and are probably large medullated nerve fibres. A number of visceral nerves are given off by the circumpharyngeal commissures to the pharynx and constitute the visceral system.

Eye-spots are present in a few of the aquatic forms, but they are absent in the terrestrial forms; in spite of this fact the earthworms appear to be able to distinguish light from darkness. There appears to be no auditory sense.

The **alimentary canal** is often divided into several regions, the relations of which are most complicated in the *Terricolae*. In *Lumbricus* the buccal cavity leads into a muscular pharynx, which is probably used for sucking. This is followed by a long oesophagus extending into about the fourteenth segment, and furnished with a thick layer of glandular cells, and several dilated glandular appendages, the *calcareous glands*. The oesophagus is dilated behind into a crop which leads into a muscular gizzard. The gizzard (double in *Moniligastridae* and *Benhamia*) has muscular walls and a chitinous lining, and it often contains small stones; it leads into the straight intestine, the dorsal wall of which is pushed inwards so as to form a longitudinal fold, the *typhlosole* (comparable to a spiral valve). The intestine is dilated in each segment and opens at the hind end by a terminal anus. In the *Limicolae* the alimentary canal is simpler by the absence of a gizzard; a pharynx and oesophagus however are always present.

The **vascular system** is a closed system of tubes containing a red fluid, in which colourless corpuscles float. The red (yellow in thin layers) colour is due to haemoglobin in solution in the plasma. Although in solution, the haemoglobin does not appear to diffuse through into the tissues or body spaces, the fluids of which are colourless. The principal blood vessels are (1) a longitudinal dorsal vessel, contractile from behind forwards; (2) a ventral subintestinal vessel also longitudinal, but noncontractile, and sometimes (3) a subneural longitudinal vessel beneath the nerve-cord. These longitudinal vessels give off lateral branches all along their course, which themselves branch in the skin, nephridia, and other organs. In some of the anterior segments there is a single pair of dilated lateral vessels, which pass round the gut and open into the subintestinal vessel: these are the so-called *hearts*, and are contractile from above downwards. The dorsal vessel is sometimes double, either wholly or in part. In some forms there appears to be a perienteric blood-sinus in the intestinal wall, from which the dorsal vessel arises.

In the lower Oligochaeta the walls of even the large vessels consist merely of a thin, apparently structureless, membrane, with here and there a nucleus. In the contractile vessels the contractility would appear to reside in this membrane. In the higher forms the large vessels possess muscular fibres in their walls, and an epitheloid lining.

The **coelom** is spacious, and is always divided by transverse segmental septa into chambers, which communicate usually above the nerve-cord. It contains a colourless fluid with amoeboid cells, and communicates with the exterior through three sets of apertures—the nephridia, the generative ducts, and the dorsal pores. The dorsal pores are present, one in each segment, in many *Terricolae*, but in the *Limicolae* are confined to one, in the prestomium. They are usually placed intersegmentally over the septa. Their function is unknown, but in some worms fluid has been seen to exude from them.

The lining of the coelom consists mainly of a flat epithelium, but on some parts of the alimentary canal it has the form of large, yellow cells—the so-called yellow or **chloragogen** cells—which are loaded with pigment grains. These cells are probably excretory in function, and allow their contents to escape into the body-cavity, and to pass thence to the exterior through the nephridia. The coelom is subdivided, as stated above, by the transverse septa, and in certain forms (*Libyodrilus*, *Branchiura*) there are obliquely placed sheets shutting off a lateral, parapodial-like section of it, as in some Polychaets. Further, portions of the coelom are sometimes shut off from the rest in connection with the generative organs (see below). There is no longitudinal dorso-ventral mesentery (except in *Criodrilus*), but traces of the ventral mesentery are sometimes left.

Sometimes certain of the larger blood vessels are enclosed in perihaemal spaces. Whether or not these are coelomic is unknown.

Nephridia are absent from a certain number of the anterior segments; and in the *Limicolae* the generative segments are without them in the breeding season. They are of two kinds: (1) the *meganephric*, or ordinary kind, such as is found in the *Limicolae* and in many of the *Terricolae*, and (2) the *plectonephric* or *diffuse* variety, found in some members of the families *Perichaetidae*, *Cryptodrilidae*, and *Acanthodrilidae*. The two kinds may co-exist in the same animal, and even in the same segment (*Megascolides*).

In *Lumbricus* we find the *meganephric* condition. The nephridia, of which there is a single pair in each segment, consist of a long winding tube, in which the following regions can be distinguished (Fig. 407): (1) the preseptal portion consisting of the coelomic funnel and a short tube (*a*), leading back to the septum which it perforates, to become continuous with (2) the rest of the nephridium, which lies in the segment behind. This postseptal part presents the following regions: (*b*) the *narrow tube* of considerable length and thin wall; parts of it are ciliated: (*c*) the *middle tube* which is short and confined to the length of one loop only; the middle tube is ciliated: (*d*) the *wide tube*, which is long, non-ciliated, and opens into a short, dilated, muscular-walled portion—the *muscular duct* which opens to the exterior through the ventral body-wall. The whole nephridium, except the muscular duct and the funnel, consists of the so-called perforated or drain-pipe cells.

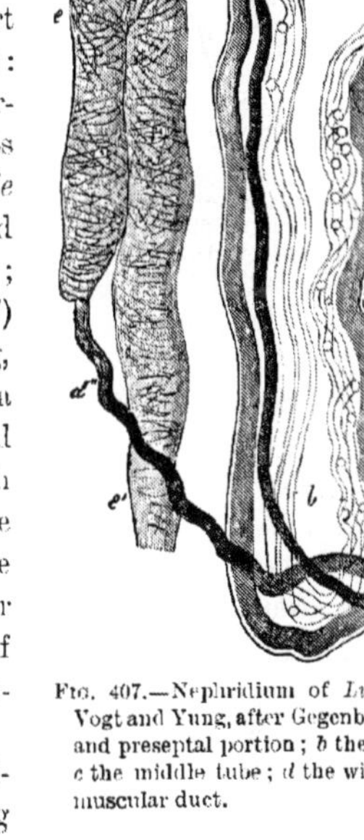

FIG. 407.—Nephridium of *Lumbricus* (from Vogt and Yung, after Gegenbaur). *a* funnel and preseptal portion; *b* the narrow tube; *c* the middle tube; *d* the wide tube; *e* the muscular duct.

In the *Limicolae* the nephridium is generally simpler, consisting of the funnel (preseptal) and a long coiled duct, passing through a protoplasmic tube or mass in which cell limits are not always discernible (Fig. 409), and leading to the wider terminal part.

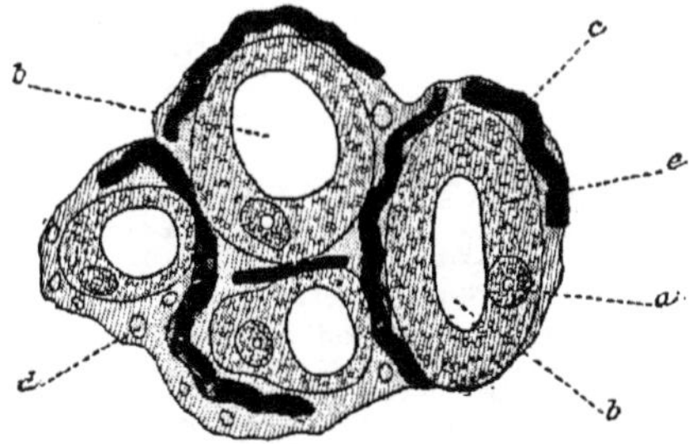

FIG. 408.—Section through a loop of the nephridium of *Lumbricus* (after Claparède, from Vogt and Yung), showing the perforated cellular character; *a* nuclei; *b* lumen; *c* blood vessel; *d* connective tissue.

As a rule there is only one pair of nephridia in each

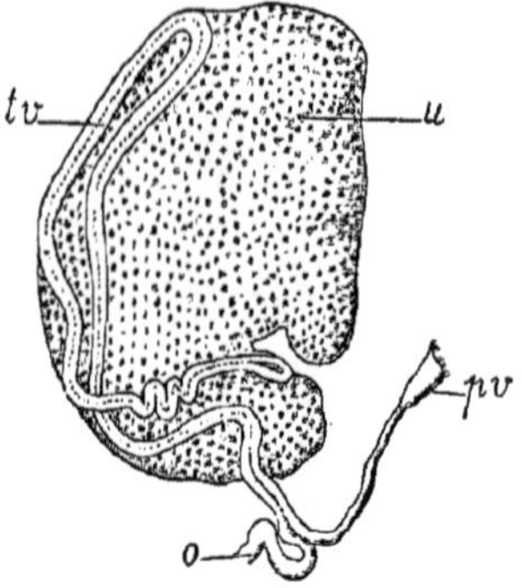

FIG. 409.—Nephridium of *Pontodrilus* (after E. Perrier). *pv* nephrostome; *o* dilated part (vesicle) near the external opening; *tv* the tube of the nephridium passing through the nucleated protoplasmic mass *u*.

segment, but in *Brachydrilus* there are two pairs, and in *Trinephrus* three. The nephridia always extend over two segments, the internal opening being in the segment anterior to that in which the main part of the nephridium lies.

The external opening is usually near the setae of the ventral row, but sometimes it is near the setae of the dorsal row, and even dorsal to these; and nephridia in successive segments may open in different places. This difference in position of the opening may be accompanied by a difference in structure, *e.g.* in *Acanthodrilus novae-zealandiae* the nephridia which open in front of the ventral setae have a caecal appendage, while those which open in front of the dorsal setae are without this appendage.

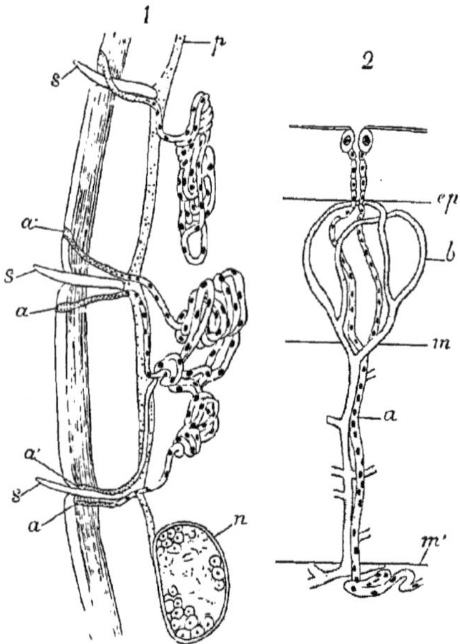

FIG. 410.—The diffuse or plectonephric nephridia of *Oligochaeta*. *1*, portion of a transverse section at the level of the setae of *Octochaetus multiporus*. *S* setae; *a*, *a'* nephridial pores; *n* nerve cord; *p* peritoneum. *2*, section across a nephridial tube of *Perichaeta aspergillum* passing through the wall of the body. *ep* epidermis; *m* layer of circular, *m'* of longitudinal muscles (the lines mark the inner limits of these layers); *a* nephridial tube with its vesicle in the layer of circular muscles; *b* blood vessel. (From Perrier, after Beddard.)

The *plectonephric* or *diffuse* nephridia are networks of fine tubes lying on the body-wall and septa in each segment (Figs. 410, 411). They have several openings to the exterior, and they sometimes open into the coelom (*Perichaeta*, and in *Octochaetus* in the case of the nephridia which open into the rectum), and sometimes they do not (*Octochaetus*, *Megascolides*). The coelomic openings when present are numerous, but whether they have any numerical relation to the external openings is not clear from the published accounts. Nor is it clear whether we are to look upon the plectonephric condition as being due to the presence of numerous small nephridia in a segment which have become connected together, or whether they are derived from the modification of a single pair of nephridia. However this may be, it appears certain that the plectonephric condition is preceded in the embryo by a stage in which there is only one pair of ordinary nephridia in each segment. It is supposed that the adult condition in such forms has arisen by the multiplication of the original simple tubes. The diffuse nephridia may be the only kind of nephridia present; or they may coexist in the same segment with meganephridia (*Perichaeta armata*, etc.); or meganephridia may be present in one part of the body, and plectonephric nephridia in another part, the two conditions gradually passing into one another; *e.g.* in *Megascolides australis* there are plectonephridia without internal funnels in the anterior segments, an intermediate condition in the middle segments, while in the posterior segments the nephridia are meganephric and possess coelomic funnels. It appears that in some cases the nephridial network is continuous from segment to segment.

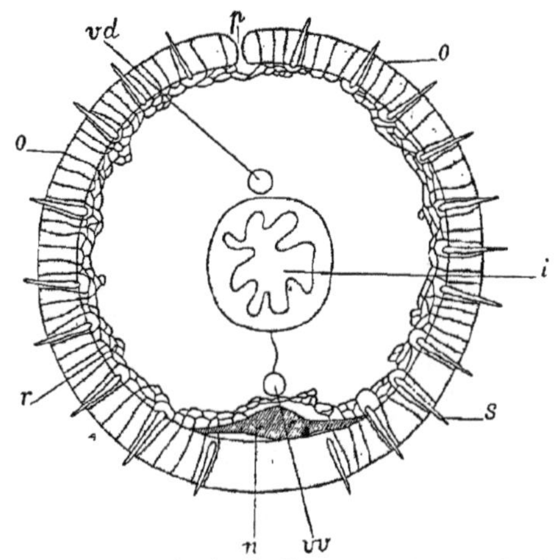

Fig. 411.—Diagram showing the disposition of the nephridia of a segment of *Perichaeta aspergillum* (from Perrier, after Beddard). *p* dorsal pore; *o* external openings of nephridia; *i* intestine; *s* setae; *vv* ventral blood vessel; *n* nerve cord; *r* nephridial network; *vd* dorsal blood vessel.

In some *Oligochaeta* the external opening of a certain number of nephridia is into the alimentary canal. This is the case in the *Enchytraeidae*, *Octochaetus multiporus*, species of *Rhinodrilus*, *Acantho-*

drilus, *Megascolides*, etc., in which the salivary glands are supposed to be modified nephridia, and in *Octochaetus multiporus* there are nephridia in the posterior segments, which open both to the exterior and into the rectal part of the gut, as well as internally into the coelom (recalling the respiratory trees of *Gephyrea*), see p. 530.

In the *Eudrilidae*, which have paired meganephridia, the ducts of the nephridia branch in the body-wall, and form a network which has several openings to the exterior.

The nephridia are developed from certain structures in the embryo called *pronephridia*. These are small cellular aggregations attached to the septa, one pair in each segment, except in front. The first pair constitutes the head-kidney: they open on the head, sometimes on the dorsal surface by the head pore (*Lumbricus*), and occupy two or three segments. They may persist or disappear. In the hinder segments they are at first straight rows of cells ending anteriorly on the septum in a flame-cell with a vacuole containing a flame-shaped flagellum (*Rhynchelmis*). This large terminal cell eventually divides up, loses its flagellum, and forms the ciliated coelomic opening.

As stated above, the plectonephric condition is preceded by a stage with one pair of pronephridia in the segment.

In *Megascolides australis* the pronephridia in the anterior part of the body give off loops, which become disconnected, so as to form several separate nephridia, of which one retains the coelomic opening, and forms the large nephridium of the segment.

Generative organs. The *Oligochaeta* are all hermaphrodite. The position of the gonads varies in the *Limicolae;* but it is fairly constant in the *Terricolae*, there being in this group usually (Fig. 412) two pairs of testes placed in segments 10 and 11 respectively, and one pair of ovaries in segment 13 (except in the *Moniligastridae*).

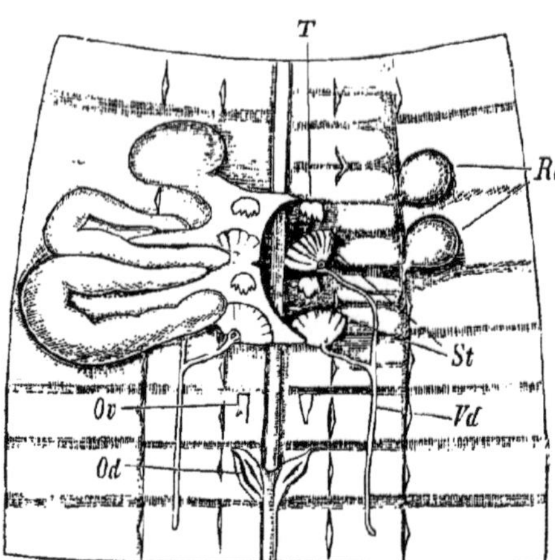

FIG. 412.—Generative organs of *Lumbricus* (after Hering). *T* testes in *X* and *XI*; *St* the two sperm-rosettes (funnels of vas deferens); *Vd* vas deferens; *Ov* ovary; *Od* oviduct; *Rc* receptacula seminis. One half of the sperm-reservoir and the sperm sacs of one side have been removed.

When there is only one pair of testes, it is in 10 in *Typhaeus*, in 11 in *Diachaeta* and *Eudriloides*, and in 12 in *Pontoscolex* and *Geoscolex*. In *Brachydrilus* the ovaries are said to be in 12, but this is doubtful.

The reproductive glands are definitely localised thickenings of the coelomic epithelium, and are placed upon the posterior faces of certain septa. The part of the body-cavity in which the testes lie may, in some *Terricolae*, be cut off from the rest by a development of the septa which forms the median part of the vesicula seminalis or sperm reservoir (Figs. 412, 413).

The *vesiculae seminales* or *sperm sacs* are developments of the septa of the testes segments. They are paired structures in *Terricolae*, and unpaired in most *Limicolae*. They communicate with the coelom, but in many *Terricolae* the part of the coelom into which they open is, as stated above, cut off from the rest by a membrane which

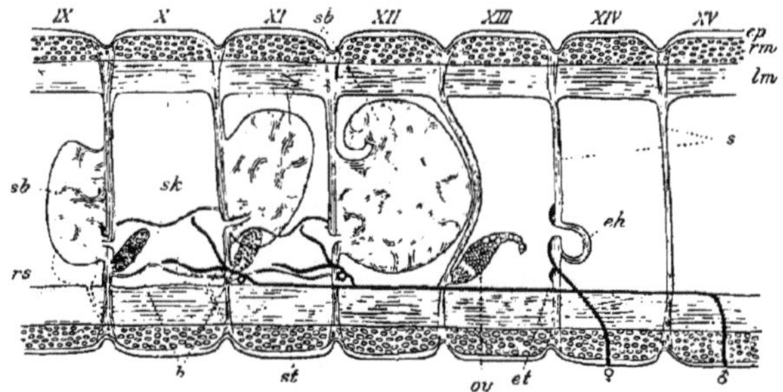

FIG. 413.—Diagram of a longitudinal section through the generative segments of *Lumbricus herculeus* to show the relation of the generative ducts and sperm sacs, etc. The segments are numbered in roman numerals. *eh* ovisac; *ep* ectoderm; *et* internal opening of oviduct; *h* testes; *lm* longitudinal muscles; *ov* ovary; *rm* circular muscles; *rs* spermathecae; *sb* sperm sacs; *sk* sperm reservoir; *st* sperm rosette (internal opening of vas deferens) in the sperm reservoir. (After Hesse.)

encloses a space containing the testes and the sperm rosette. This is the *sperm reservoir* (Fig. 413). A sperm reservoir is not developed in *Allolobophora*, *Acanthodrilus*, and some other forms (Fig. 414). Sperm sacs are absent in the *Enchytraeidae* (except *Mesenchytraeus*).

The vesiculae seminales or sperm sacs do not necessarily correspond in number with the testes; they vary in number from one pair to four (Fig. 414). They may attain a great size, extending back through a great number of segments (60 in a species of *Geoscolex*). In the *Limicolae* the sperm sac is, as stated above, generally unpaired.

A similar sac is formed in connection with the ovary in some forms. This ovisac in the *Limicolae* is unpaired, and may be very large (*Rhynchelmis*). In the *Terricolae* the ovisacs, when present, are usually small and paired (Fig. 413, etc.). They are diverticula of the septum at the hind end of the ovarian segment, and in some *Limicolae* the sperm-sac projects into them.

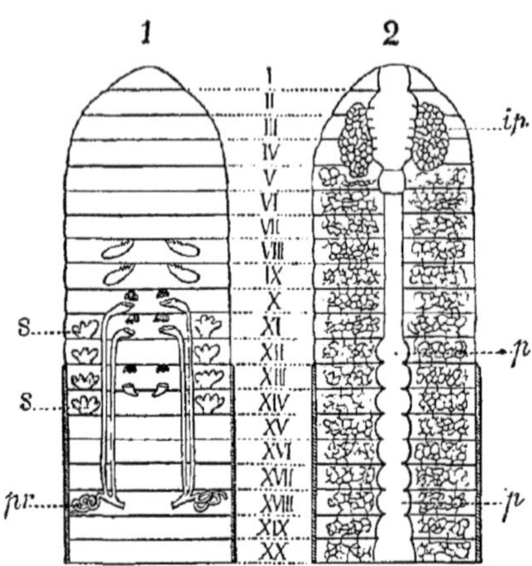

Fig. 414.—Diagram of the organisation of *Megascolides* (from Perrier, after Benham). *1*, generative apparatus; *2*, renal organs. Segments numbered in roman numerals. In *VIII* and *IX* are the spermathecae; in *X* and *XI* the testes; in *XI* to *XIV* the four sperm-sacs *S*; in *XIII* the ovaries and funnels of the oviducts; in *XVIII* the male generative opening and spermiducal glands *pr*. *ip* the pepto-nephridia; *p* dilatations of the intestine. The nephridial network is shown in *2*.

The generative ducts correspond in number to the generative glands. They usually open into the general body-cavity of the segment containing the gonads, but in some *Terricolae* (*Lumbricus*, etc.) the vasa deferentia open into the median part of the sperm-sacs (sperm-reservoir). In the *Terricolae* there are two oviducts and two pairs of vasa deferentia. The oviducts, as in the *Limicolae*, are short, and open externally on the segment behind that containing the ovaries, usually on the 14th; and internally into the coelomic chamber of the ovarian segment, usually the 13th. The two vasa deferentia of the same side join behind to form a single duct, except in *Phreoryctes*, where they open on two consecutive segments, and in *Pelodrilus*, in which they open on the same segment. The opening of the vas deferens is variable in position*; in *Lumbricus* it opens on segment 15. The internal openings of the vasa deferentia—or *sperm rosettes* as they are called—are in the coelom of the segments of the testis (generally 10 and 11), and if there is a sperm reservoir they open into it. There is in many forms attached to and opening into the vas deferens near its external opening a glandular appendage called the *prostate*, or *spermiducal gland* (Fig. 414). In some genera the spermiducal gland opens to the exterior independently of the sperm ducts.

* From segment 11 in *Moniligaster* to 22 in *Diachaeta*.

In the *Limicolae* the *atrium*, which is a dilatation on the vas deferens close to its opening, seems to correspond to the spermiducal gland.

In the *Limicolae* there is usually one pair of ovaries (except *Phreoryctes*) and one pair of testes (except *Lumbriculidae*), and their position varies somewhat in the different families. They are placed further forward than in the *Terricolae* (for position see account of families), and their ducts are short, opening on the segment following that containing the gland. The nephridia of the segments of the ovaries and testes disappear as soon as these organs begin to develop.

The ova of the *Limicolae* are much larger, and have more yolk than the ova of the *Terricolae*.

According to Beddard a trace of a second pair of ovaries can be made out in the young of some earthworms (*Lumbricus* and *Octochaetus*) in the 12th segment. In this case the number of ovaries will equal the number of testes, and the series of gonads becomes a continuous one.

The spermatozoa, which are filiform, develop in the sperm sacs, into which the cells of the testis (progametes) are dehisced. When ripe they are drawn into the vasa deferentia by the cilia of the sperm rosettes, and passed down these tubes, which their presence renders of an opaque white colour and so visible to the naked eye in earthworms, to the exterior. Here they are taken up, in the manner described below, by the *spermathecae* (receptacula seminis) of another worm, where they are stored until they are required for use. The spermathecae are small vascular sacs, which open externally near one end of the segment which contains them (usually the anterior); the number of them varies considerably. In the *Terricolae* there are usually several pairs (4 or 5). Generally there is only one pair in a segment. In the *Limicolae* they are generally contained in the testicular segment. In *Enchytraeus*, and possibly other genera, the spermathecae open into the gut as well as to the exterior.

The *capsulogenous glands*, or as they are now called from their supposed function, the *albumen glands*, are glandular developments in the ventral body-wall in the generative segments of some earthworms. They probably secrete the albuminous matter found in the cocoon.

Spermatophores have been observed in some forms, both aquatic and terrestrial (*e.g. Tubifex*, some *Lumbricidae*). They have a somewhat elongated shape, and often have a certain complexity of structure. They consist essentially of spermatozoa cemented together into a packet. They are found on the segments which during copulation are placed opposite to the male genital pores. The *tubercula pubertatis* are papillae on the ventral side of the clitellar segments.

The *clitellum* or girdle is a glandular thickening of the integument of one or more segments of the body. In the *Limicolae* it is on the generative segments themselves. In the *Terricolae* it varies in position, but is never in front of the 12th segment, except in the *Moniligastridae*.

The male and female openings are on the clitellum in *Typhaeus, Megascolides, Eudrilus, Pontoscolex, Criodrilus, Moniligaster*, etc. (in *Moniligaster* the male openings are between 10 and 11, the female openings between 11 and 12); in *Hormogaster, Diachaeta, Rhinodrilus, Brachydrilus*, etc., the male openings are on it, the female being in front of it; the female openings are on it, and the male behind it in *Cryptodrilus, Perichaeta, Deinodrilus*, etc.; finally in the *Lumbricidae* both orifices are in front of it.

As a rule the clitellum is saddle-shaped, the glandular development being confined to the dorsal and lateral regions, but in some forms it is a complete girdle. Its secretion serves to make the cocoon, and in some earthworms at any rate to bind the worms together in copulation. Long modified setae are found in some forms in the neighbourhood of the male pores, and protruding through them; and in a few species bundles of similar setae are found near the spermathecae. They are termed *genital setae*, those near the male pore being male penial setae, and those near the spermatheca pore female copulatory setae. When these modified setae are present, the ordinary setae of the segment are absent. The cocoons in which the eggs are enclosed may be spheroidal; more usually they are ellipsoidal, the two poles being prolonged into short projections. They are formed by the secretion of the clitellum which the worm slips over its head. As the cocoon passes over the openings of the generative organs, it receives the ova, spermatozoa, and albumen which form its ultimate contents. Being elastic it shrinks when free of the worm and forms a compact case, traces of its origin being generally retained in the two projections from its poles. The cocoons are attached to aquatic plants (*Limicolae*), or deposited in damp earth (*Terricolae*). It is common, though not universal, to find that only one embryo in each cocoon attains full development. There is no free larval stage: the young leave the egg as minute fully formed worms.

Earthworms copulate. In the common earthworm of this country the process may be observed as follows: two worms in neighbouring burrows apply the anterior ends of their bodies together (the posterior ends remaining in the burrows) in such a way that the openings of the spermathecae are opposite the clitellum of the other worm. The bodies become adherent, probably by suction of the ventral surface of the clitella, with the assistance of a secretion of the clitellum, and sperm is poured out of the openings of the vasa deferentia. This is passed back along the ventral body of the worm in grooves which can be faintly discerned in sexually mature worms, leading back from the male genital opening. These grooves end at the clitellum, and the transmission of the sperm backwards is probably effected by the coordinated contraction of their sides. However that may be, the sperm is transported backwards until it comes to the ventral surface of the clitellar region; here it is opposite the openings of the spermathecae of the adherent worm. These appear to take it up by a suctorial action, and so to become filled with the sperm of the other worm. The use of spermatophores in such cases is unknown.*

Asexual reproduction is found in some of the aquatic forms, and the earthworms seem to have considerable power of reproducing lost parts.

The asexual reproduction consists of a process of fission combined with gemmation. It takes place at certain periods of the year under certain climatic conditions, and is exhibited by worms which are immature in the fact that the generative organs are not developed. In the *Naididae* (*Nais*, *Chaetogaster*, etc.) a zone of fission is formed, *i.e.* one of the somites enlarges and becomes divided into two parts: the anterior part forms the anal segment of the anterior worm, and the posterior gives rise to the head and anterior segments of the posterior portion. A chain of zooids may be formed in this way, the individuals of which eventually separate. In the earthworms the anterior end, including the pharynx and cerebral ganglia, can be reproduced after excision, and the same remark applies to the hind end.†

The *Oligochaeta* are distributed all over the world in fresh-water and on the land. Some are amphibious, and a few species are marine; but the majority are terrestrial, and are earthworms in the ordinary sense of the word.

Earthworms are found in the surface soil of the earth, in which they form burrows, coming to the surface to get rid of their faeces (worm casts), and to procure food, and to copulate. Their food consists of decaying vegetable substances, which they find in the form of leaves and twigs on the surface and in the dark vegetable surface soil, large quantities of which they swallow. The part which earthworms play in modifying the surface soil is well known, and has been fully treated by Darwin in his celebrated work on the subject. Briefly their action may be summed up as follows: they

* Andrews, *American Naturalist*, 1895. † Bülow, *Z. f. w. Z.*, 1883.

are continually bringing up to the surface some of the deeper soil, and depositing it in the form of casts; they thus assist in turning over the soil, and in mixing together the different parts of it, and the soil of the deeper layers is thus exposed to the atmosphere. Phosphorescent earthworms have been observed, but they are not common.

Sub-order 1. **Limicolae.**

Oligochaeta with a clitellum commencing not later than the tenth or eleventh segment, and consisting only of a single layer of cells; the sperm-ducts only occupy two segments, the external pore being on the segment following that into which the funnel opens; the male pore is situated in front of the female pore; eggs generally large, always provided with abundant yolk: egg-sacs large; spermiducal glands, when present, possess a muscular layer interposed between the inner epithelium and the glandular layer; sexual maturity at a fixed period. For the most part fresh-water forms.

Fam. 1. **Phreoryctidae.** Aquatic or terrestrial, of slender form, often very long. Setae in four rows of single setae or of couples, sigmoid. Testes in 10 and 11. Ovaries in 12 and 13, or in 13 only. Sperm-ducts, two pairs opening separately, without spermiducal glands. Spermathecae in front of testes, without diverticula. No genital setae. *Phreoryctes* Hoffmeister, Eur., N. Am., N. Z., clitellum 11–14, 4 ovaries: *Pelodrilus* Beddard, New Zealand, clitellum 11–13.

Fam. 2. **Lumbriculidae.** Aquatic worms of moderate size. Setae paired and *f*-shaped, sometimes with the free extremity bifid. The dorsal blood vessel or the transverse vessels with blind contractile appendages (except *Stylodrilus* and (?) *Alluroides*). Testes in 9 and 10; ovaries in 11. Two pairs of sperm-ducts (exc. *Alluroides*) uniting to open by a single spermiducal gland on each side, which lies in front of the oviducal pores usually in 10. *Lumbriculus* Gr., male pore in 8; *Rhynchelmis* Hoffmeister: *Trichodrilus* Clap.; *Phreatothrix* Vejd.; *Claparedilla* Vejd.; *Stylodrilus* Clap.; *Sutroa* Eisen: *Alluroides* Bed., this genus presents certain characteristics of Terricolae. *Eclipidrilus* Eisen.

Fam. 3. **Tubificidae.** Aquatic, fresh-water or marine, of small size and slender build. Setae of three kinds—capilliform, pectinate, and uncinate—of which the first two (when present) are only found in the dorsal bundles. Dorsal and ventral blood vessels connected in every segment. Testes in 10, ovaries in 11, sperm-ducts always ending in spermiducal gland opening on 11. Oviducts open between 11 and 12. Spermathecae one pair in 10. *Tubifex* Lam., dorsal bundles with capilliform, pectinate, and uncinate setae; *Clitellio* Sav.; *Cl. arenarius* Sav., coasts of Europe: *Limnodrilus* Clap., with uncinate setae only, Eur. and California; *Hesperodrilus* Bedd., Falkland Isl. and S. Am.: *Heterochaeta* Clap., marine, coasts Engl. and Belg.; *Peloscolex* Leidy, N. Am.: *Psammoryctes* Vejd., setae capillif., unc., palmate, and pectinate, Eur.; *Hemitubifex* Eis.; *Telmatodrilus* Eis.; *Spirosperma* Eis.; *Ilyodrilus* Eis.; *Bothrioneuron* Stolc; *Lophochaeta* Stolc; *Branchiura* Bedd., with median dorsal and ventral gills, Victoria regia tank in Regent's Park, London;

Vermiculus Goodrich, marine, Weymouth; *Embolocephalus* Randolph, Lakes of Geneva and Zurich; *Phreodrilus* Bedd., subterranean water, N. Zeal.

Fam. 4. **Naididae.** Small aquatic worms. Setae usually in four groups upon each segment—sigmoid, bifurcate, hastiform, and capilliform. In most genera the dorsal setae are absent from a variable number of the anterior segments. Sexual reproduction at fixed intervals, between which asexual reproduction by fission occurs. Sexual organs (only known in a few types) far forward, commencing even in the 5th segment. Testes in 5, ovaries in 6. There is sometimes only one nephridium in a segment, and in *Uncinais littoralis* they are said to be absent altogether. *Nais* O. F. Müller (including *Slavina* Vejd.; *Stylaria* Lamk., etc.), the first 5 segments lack dorsal setae; *N. lacustris* L.; *Pristina* Ehrbg. (incl. *Naidium* Schm.); *Ripistes* Duj.; *Uncinais* Lev.; *Bohemilla* Vejd.; *Dero* Oken, gills as processes surrounding the anus; *D. Mülleri*, Gt. Brit.; *Chaetobranchus* Bourne, gills as hollow processes of body-wall enclosing the dorsal setae, Madras; *Amphichaeta* Tauber, several segments without setae; *Chaetogaster* v. Baer, ventral setae only present, uncinate, some segments without setae, ventral ganglia more numerous than apparent segments, as also in *Vetrovermis* Imhof., which is probably allied here.

Fam. 5. **Enchytraeidae.** Small worms; setae (absent in *Anachaeta*) short, straight or curved, not bifid. A single pair of calciferous glands sometimes present. Dorsal vessel present only anteriorly, sometimes with cardiac body. Testes in 11, male pores in 12; a reduced spermiducal gland present; oviducts represented by pores. Spermathecae one pair, in 5, generally opening into gut.* In most there is a single pore on the prestomium. Dorsal pores occasionally present. Fresh-water, marine, damp earth, littoral. In the presence of calciferous glands and in the distance between the spermathecae and the male pores this family approaches the *Terricolae*. The presence of dorsal pores in some genera (*Fridericia*) points in same direction. *Distichopus* Verrill; *Chirodrilus* Verrill; *Mesenchytraeus* Eisen; *Stercutus* Michaels., in manure, lumen of al. canal sometimes obliterated; *Pachydrilus* Clap.; *Marionia* Michaels.; *Buchholzia* Michaels.; *Enchytraeus* Henle, setae hooked at free extremity; *Fridericia* Michaels.; *Henlea* Michaels.; *Anachaeta* Vejd. (*Achaeta* Vejd.), setae absent, represented by large cells; *Bryodrilus* Ude; *Parenchytraeus* Hesse.

The **Discodrilidae**† are allied here; they are small parasitic forms, without setae, with a small number of segments; cephalic lobe divided into two; a sucker at the hind end. Pharynx with two jaws, one above the other, formerly mistaken for Leeches. *Branchiobdella* Odier (*Astacobdella* Villot), parasitic on gills of the Crayfish; *Bdellodrilus* Moore; *Myzobdella* Leidy.

Sub-order 2. **Terricolae.**

Earthworms. *Oligochaeta* with a clitellum which never commences before the twelfth segment (except in *Moniligastridae*), and always consists of two layers of cells. The sperm-ducts traverse two or more segments on their way to the exterior. Ova small and with little yolk; egg-sacs small. Spermiducal glands, when present, have not a muscular layer interposed between the inner epithelium and

* Found also in *Rhynchelmis* and *Sutroa*.

† J. P. Moore, *Journ. Morph.*, 10, 1895, p. 497.

the glandular layer. Sexual maturity seems to be more or less continuous. Oviducal pores always on segment 14, and ovaries in segment 13.

Fam. 1. **Moniligastridae.** Large or small earthworms with paired setae, 8 per segment; clitellum 10–13, inconspicuous, one cell thick. One pair of male pores in *Moniligaster*, between 10 and 11, testes in 10; two pairs in *Desmogaster*, on 11 and 12, testes in 10 and 11. Spermiducal gland showing the same structure as in the *Lumbriculidae*, with sometimes a protrusible penis. Ovaries in 13 in *Desmogaster* and some species of *Moniligaster*, in 11 in other species of *M*. Egg-sacs large; eggs smallish, larger than those of other Earthworms, smaller than those of aquatic forms. India, Ceylon, Sumatra, Borneo, Burmah, Bahamas. *Moniligaster* Per.; *Desmogaster* Rosa. This family of earthworms is placed by some zoologists amongst the *Limicolae*, because of the position and structure of the clitellum, the length of the vas deferens, and the position of the ovaries in some species. Terricolous.

Fam. 2. **Perichaetidae.** With numerous setae in each segment arranged in a continuous circle, or with dorsal and ventral gaps; male pore nearly always on 18. Gizzard always present: intestine frequently with two (sometimes more) caeca and a rudimentary typhlosole. Nephridia diffused or paired. Spermiducal glands generally lobate; penial setae present or absent; spermathecae with one or two diverticula.

Perichaeta Schmarda, very active, setae usually in a continuous row, nephridia diffuse, mainly tropical Oriental, also in Australia, Europe and America; *Megascolex* Templeton, setae interrupted dorsally and ventrally, nephridia diffuse and paired, Australian and Oriental; *Pleionogaster* Michaels., Philippines; *Perionyx* Perr., tropics Old World; *Diporochaeta* Bedd., N. Z. and Australia.

Fam. 3. **Cryptodrilidae.** Terrestrial, sometimes aquatic, with 8 setae per segment. Clitellum variable in extent, occupying some or all of segments 12 to 23, usually complete anteriorly. Spermathecae one to five pairs, placed anteriorly, nearly always with one or two diverticula. Male pores on segment 17 or 18, opening independently of or into the distal non-glandular part of the spermiducal gland, which is either tubular or lobate. Penial setae generally present. Nephridia paired or diffuse. The large number of new forms which have been described make it a difficult matter to separate the genera. In most parts of the world, but mainly tropical, specially Australian. This family interdigitates with the *Acanthodrilidae* and with the *Perichaetidae*, from which the only difference is the character of 8 setae per segment, which, indeed, is shared by the anterior (though not the posterior part of the body) of *Megascolex*. *Microscolex* Rosa, mainly S. America: *Pontodrilus* Perr.; *Typhaeus* Bedd.; *Dichogaster* Bedd.; *Deodrilus* Bedd.; *Millsonia* Bedd.; *Fletcherodrilus* Michaels; *Trinephrus* Bedd., 3 pairs of nephridia in each segment; *Digaster* Perr.; *Megascolides* McCoy; *M. australis* Spencer, the giant earthworm of Gippsland, 1·23 m.; *Cryptodrilus* Fletcher; the last five genera are mainly confined to Australia; *Microdrilus* Bedd., Java, very active; *Gordiodrilus* Bedd., mainly trop. Africa; *Ocnerodrilus* Eisen, America and Africa; *Nannodrilus* Bedd., West Africa, aquatic.

Fam. 4. **Acanthodrilidae.** Large or small, usually terrestrial, rarely aquatic. Setae 8, 12, or numerous. Male pore on 18; pores of spermiducal glands, which are tubular structures generally with penial setae, on 17 and 19.

Spermathecae generally 2 pairs in 8 and 9, with diverticulum or diverticula. This family is closely connected with the last. *Acanthodrilus* Perr., paired nephridia, entirely from the Southern hemisphere; *Diplocardia* Garman; *Octochaetus* Bedd., nephridia diffuse, N. Z.; *Kerria* Bedd., American; *Deinodrilus* Bedd., with pericardium, N. Z.; *Plagiochaeta* Benh., N. Z.; *Trigaster* Benh., St. Thomas; *Benhamia* Michaels, diffuse nephridia, mainly tropical Africa.

Fam. 5. **Eudrilidae.** Size variable, with paired nephridia; in some genera nephridial ducts as networks in body-wall, with numerous external openings; setae in couples, sigmoid; spermiducal glands always present; sperm-ducts open into these glands at some distance from external orifice. Spermathecae (if present) unpaired, and opening outwards near the female genital pores, generally replaced by, or, if present, enclosed in coelomic sacs, which are frequently connected with the other parts of the female reproductive system. In some genera the male and female generative openings are unpaired. With the exception of *Eudrilus* the family is restricted to Africa.

Sub-fam. 1. **Pareudrilini.** Integumental sense organs rarely present. Sperm-ducts without dilatation; ducts of nephridia branching in body-walls; calciferous glands modified or absent. *Eudriloides* Michaels, ; *Reithrodrilus* Mich.; *Megachaeta* Mich.; *Metadrilus* Mich.; *Notykus* Mich.; *Pareudrilus* Bedd.; *Unyoria* Mich.; *Platydrilus* Mich.; *Nemertodrilus* Mich.; *Libyodrilus* Bedd.; *Stuhlmannia* Mich.

Sub-fam. 2. **Eudrilini.** Integumental sense organs present. Calciferous glands, paired and unpaired, of the usual structure. No integumental nephridial network; sperm-ducts with dilatation immediately following funnel. *Eudrilus* Perr.; *Eminoscolex* Mich.; *Polytoreutus* Mich.; *Preussia* Mich.; *Paradrilus* Mich.; *Hyperiodrilus* Bedd.; *Heliodrilus* Bedd.; *Alvania* Bedd.; *Teleudrilus* Rosa.

Fam. 6. **Geoscolicidae.** Size various, sometimes aquatic. Setae 8 in a segment, paired or irregular in arrangement. Clitellum saddle-shaped except in *Diachaeta* and *Ilyogenia*, usually rather extensive, often furnished with modified setae. Nephridia always paired, rarely more than one pair in a segment. Gizzard anterior. Male pores generally within the clitellum, with or without spermiducal glands. Spermathecae without diverticula.

Sub-fam. 1. **Geoscolicinae.** Spermathecae one to four pairs in the neighbourhood of the gizzard. All tropical American except *Ilyogenia* from Natal. Prestomium absent in *Pontoscolex*, *Diachaeta*, and *Onychochaeta* (as in *Deodrilus*). *Rhinodrilus* Perr.; *Geoscolex* Leuck.; *Sparganophilus* Benham; *Sp. tamesis* Benham, England; *Trichochaeta* Bedd.; *Onychochaeta* Bedd.; *Ilyogenia* Bedd.; *Tykonus* Mich.; *Anteus* Perr.; *Pontoscolex* Schm.; *Urobenus* Benh.; *Diachaeta* Benh.

Sub-fam. 2. **Microchaetinae.** Spermathecae, if present, usually near female apertures, nearly always minute, and generally numerous in each segment. Copulatory papillae furnished with special glands, and usually with modified setae. Old world. *Criodrilus* Hoffmeister; *Microchaeta* Perr; *Callidrilus* Mich.; *Kynotus* Mich.; *Glyphidrilus* Horst; *Annadrilus* Horst; *Hormogaster* Rosa, Italy; *Alma* Gr. (*Siphonogaster* Levinsen), with long paired penial processes; *Bilimba* Rosa; *Brachydrilus* Benh., ovaries said to be in 12.

Fam. **Lumbricidae.** Terrestrial (rarely aquatic), of moderate size. Setae 8 in a segment. Male pores usually on 15, rarely on 12 or 13. Clitellum saddle-shaped, begins some way behind the segment of the male pores. Dorsal pores present. Gizzard single, at end of oesophagus. Nephridia paired and similar. No spermiducal glands or penial setae. Tubercula pubertatis nearly always present. Cosmopolitan, but most characteristic of Palaearctic and Nearctic regions. *Allurus* Eisen, male pores on 13; *Tetragonus* Eisen, male pores on 12; *Allolobophora* Eisen, prestomium incompletely divides buccal lobe, male pores on 15, no median seminal reservoirs; *A. terrestris* Sav., spermathecae in 9 and 10 resp.; *Lumbricus* Eisen, prestomium completely divides buccal segment, male pores on 15, two median seminal reservoirs occupying segments 10 and 11 into which open three pairs of sperm-sacs, spermathecae two pairs, in 9 and 10 resp.; *L. rubellus* Hoffm., clitell. 27–32, tubercula pubertatis 28–31, 1st dorsal pore 7/8. Male pores almost invisible; *L. herculeus* Sav., the common earthworm of this country, number of segments 180, clitellum 32–37, tubercula pubertatis 33–36, 1st dorsal pore 7/8, male pores conspicuous.

Class III. HIRUDINEA.*

Elongated, vermiform Annelida with anterior and posterior suckers, without setae (except Acanthobdella) or parapodia, and with median genital openings. Coelom usually broken up into numerous intercommunicating spaces. Hermaphrodite.

The *Hirudinea* have, owing to certain superficial resemblances, often been associated with the *Trematoda*, with which group they have however nothing in common. They are in reality segmented, coelomate animals with distinct affinities to the Oligochaetous *Chaetopoda*. These affinities are specially obvious in the genus *Acanthobdella*, in which there are setae in the anterior segments embedded in pits in the body-wall, and a spacious body-cavity divided by transverse septa into chambers. In the typical Leeches the coelom is difficult to trace completely in the adult, but it is undoubtedly present in the embryo in the usual Annelidan form of a series of paired cavities in the mesoblastic somites; and these cavities give rise in the adult to the sinus system, and indirectly to the tubes of the botryoidal tissue; likewise, though in a modified and veiled form, to the gonads and nephridia. A vascular system is present, and by many anatomists is regarded as being continuous with the sinus system; but this continuity is doubtful, and in the

* S. Apathy, "Süss-wasser Hirudineen." Zool. Jahrb., 3, 1888, p. 725. Id., "Analyse der äusseren Körperform der Hirudineen," *Naples Mittheilungen*, 8, 1888, p. 153. A. Oka, "Beiträge z. Anat. d. Clepsine," *Z. f. w. Z.*, 58, 1894, p. 79. Fr. Leydig, "Circulations—u. Respirationssystem v. Nephelis u. Clepsine," *Ber. d. K. Zool. Anst. zu Würzburg*, 1849. Id., "Zur Anat. v. Piscicola geometrica, etc.," *Z. f. w. Z.*, 1. E. Ray Lankester, "On the connective and vasifactive tissue of the medicinal leech," *Q. J. M. S.*, 20, 1880, p. 307. A. G. Bourne, "Contributions to the Anatomy of the Hirudinea," *Q. J. M. S.*, 24, 1884, p. 419. C. O. Whitman, "The Leeches of Japan," *Q. J. M. S.*, 26, 1886, p. 317. A. Moquin-Tandon, *Monographie des Hirudinées*, Paris, 1846.

present state of our knowledge cannot be admitted. Externally the body is marked by a number of transverse rings—the *annuli*—which are short, and may be more or less indistinct. These rings do not correspond with the internal segments, which may be indicated by incomplete partitions; but they constitute much shorter portions of the body, three (*Branchellion, Clepsine*), six (*Ichthyobdella, Calliobdella, Pontobdella*), five (*Hirudo*), twelve (*Piscicola*) of them corresponding to one internal segment. Towards the two ends of the body the segments are reduced in the number of rings, and in other respects.

There is always a posterior sucker, into the constitution of which a number of true segments (six or seven) enter, and generally an anterior sucker at the anterior end of the body; both are ventral. The anus opens dorsal to the posterior sucker, and the mouth is on the ventral surface in the anterior sucker. The animal moves in a looping manner by means of its suckers. There are no parapodia, or setae (except in *Acanthobdella*), or head appendages, but in one genus (*Branchellion*) there are leaf-like branchial processes on the middle part of the body.

On the annuli in transverse rows are arranged the so-called segmental sense-organs or sensory papillae. The arrangement of these differs on the different annuli of a segment, but is the same for corresponding annuli of different segments. On the head some of these organs are modified into eyes and into the cup-shaped organs. The

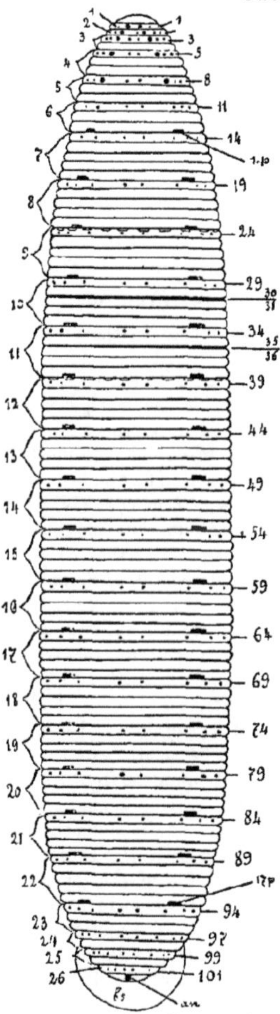

Fig. 415.—*Hirudo medicinalis* showing the segments according to Whitman, and illustrating the enumeration adopted in the text (after Whitman, from Lang). The 5 pairs of eyes are marked by the larger, the segmental sense-organs by the smaller black dots; the openings of the nephridia by oblong marks (*1p–17p*). The rings are counted on the right, the segments on the left. *30/31* marks the position of the male, *35/36* of the female genital opening. *an* anus; *bs* posterior sucker which is supposed to consist of 7 segments fused together.

skin is also often highly coloured, owing to the presence of pigment, not only in the cutis, but to a certain extent also in the epidermis.

The *clitellum* consists of a glandular development of the skin of the 9th, 10th, and 11th segments (?10th, 11th, and 12th). Amongst the *Gnathobdellidae* it is most developed in *Nephelis*; in *Hirudo* it is inconspicuous. The genital openings are on the 10th and 11th segments* (Whitman, Oka) the male opening being in front of the female (Fig. 415).

According to the most recent statements the *Hirudinea* possess, without exception, 33 segments, some of which are however always reduced. A true segment is determined by the presence of a ganglion, so that 33 ventral ganglia can be made out. It must, however, be admitted that these ganglia are not easy to see. The number is arrived at in this way: the suboesophageal ganglion consists of 5 ganglia (Whitman, Oka); then follow 21 single free ganglia, and then one composite one consisting of 7 fused ganglia.

Apathy also makes 33 segments, but according to him the suboesophageal ganglion consists of 6, and the last ganglion also of 6. His enumeration is as follows. The body is divisible into six regions, viz. the region of (1) the head, (2) the clitellum and preclitellum, (3) the stomach, (4) the intestine, (5) the anus, and (6) the posterior sucker; of these the anal region contains three segments, the rest six each. The head region consists of six more or less reduced and shortened segments, four or five of which may enter into the formation of the sucker. Segment 1 is always eyeless, and segment 6 possesses eyes only in the ten-eyed *Gnathobdellidae*. The mouth is on the ventral surface, and its anterior limit is at least two segments from the front end of the body. The preclitellar segments are 7, 8, 9, and the clitellar segments are 10, 11, 12. Of the 12 segments of the middle-body, *i.e.*, of the stomach and intestine, the first and last are usually reduced, but the 2nd to the 11th are complete, and each contains the number of rings characteristic of the genus. The anal region contains three segments much reduced; the anus is in the dorsal middle line at most two segments in front of the sucker. The post-anal segments are all incorporated in the sucker—some of them being invaginated into it; they are six in number, as is shown by the constitution of the last ganglion.

The **body-wall** consists of a single layer of epidermis cells with a cuticle; a cutis containing pigment cells and blood vessels, both of which intrude into the epidermis; an outer circular and an inner longitudinal layer of muscles. The muscular fibres are tubular, *i.e.*, they consist of an outer cortex of striated, presumably differentiated, contractile substance, and an inner core of granular matter, in which lies the nucleus. Unicellular glands, opening on the surface and secreting a mucous fluid, are found in great numbers in the epidermis

* Apathy makes the genital openings in segments 11 and 12, and the clitellum to include 10, 11, and 12. Bourne agrees with the position in the text, if we allow, as Oka does, that the suboesophageal ganglion consists of 5 fused ganglia.

or sunk in the cutis, or even amongst the muscles. The clitellar glands are specially enlarged modifications of the ordinary skin gland, and are found deeply situated in the clitellar region; they secrete a clear viscid substance which quickly hardens outside the body and is used to form the cocoons when the eggs are laid. Within the muscles there is a connective tissue, in which the various organs are embedded. These organs we may now proceed to consider.

The central **nervous system** has the typical Annelidan character of two ventral cords closely approximated, but diverging from one another in front to pass round the oesophagus. There is a bilobed swelling dorsal to the alimentary canal constituting the supra-oesophageal ganglion, and supplying the cephalic sense-organs; and a series of swellings on the ventral cords, each of which gives off two pairs of nerves. These ventral ganglia are segmental in their arrangement, but are partly fused anteriorly and posteriorly. Thus in the suboesophageal ganglion and circum-oesophageal commissures five (or six) ganglia can be made out (Fig. 416); these belong to the first five (or six) segments of the head region. The last ganglion, which supplies the sucker, consists of seven (or six) ganglia fused, and between these two composite ganglia there are twenty-one separate ventral ganglia* (Fig. 424). There is an unpaired median longitudinal nerve passing from ganglion to ganglion between the two halves of the ventral cord and a system of *visceral nerves* which was discovered by Brandt. The latter consists of an intestinal nerve, which arises from the brain and runs close to and above the ganglionic chain; it sends branches to supply the caeca of the stomach. Three ganglia, which

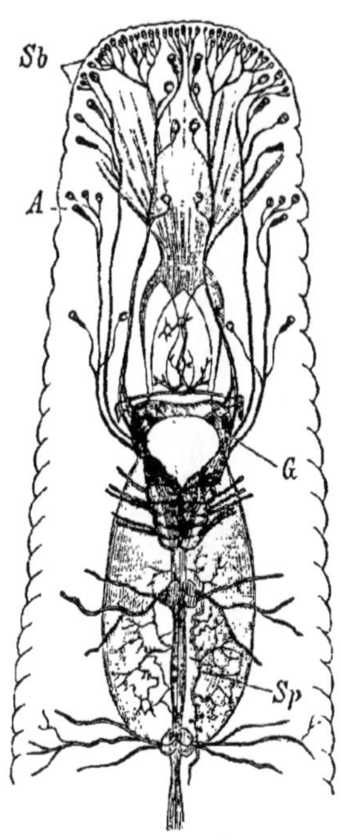

FIG. 416.—Anterior end of *Hirudo* (after Leydig). *G* cerebral ganglion, with suboesophageal ganglionic mass; *Sp* visceral nerves; *A* eyes; *Sb* sense-organs.

* According to Bourne's enumeration there are 23 ventral ganglia, but he counts the suboesophageal and posterior ganglion as single ganglia; if the former be allowed at 5 and the latter at 7, it makes the total number 33.

in the common leech lie in front of the brain, and send their nerve-plexuses to the jaws and pharynx, are considered by Leydig to be enlargements of cerebral nerves, and very likely control the movements which occur in swallowing.

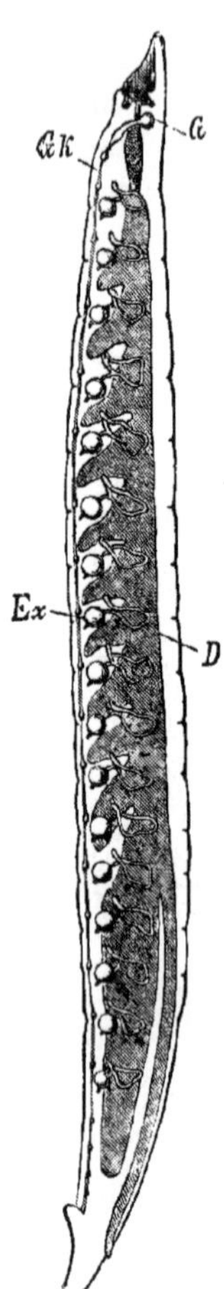

FIG. 418.—Diagram of a longitudinal section through the Medicinal Leech (after Leuckart). *D* stomach with 11 lateral caeca; *G* cerebral ganglion; *Gk* ventral nerve-cord; *Ex* nephridia. The intestinal caeca are not shown.

The **alimentary canal.** The mouth leads into a muscular pharynx provided with salivary glands. In the *Rhynchobdellidae* the anterior part of the pharynx is protractile and highly muscular; it forms the so-called proboscis. In the *Gnathobdellidae* the anterior part of the pharynx projects as three prominent ridges, over which lie the three serrated chitinous plates or jaws (Fig. 417). Occasionally the jaws are reduced in number (*Geobdella*), or they may be vestigial (*Trochetia*), or entirely absent (*Leptostoma edentulum*). The pharynx (Fig. 418) is followed by a tubular oesophagus (inconspicuous in *Hirudo*), which leads into a thin-walled stomach (sometimes called the crop or proventriculus) provided with paired, segmentally arranged lateral caeca. These vary in number in different forms, and the last pair are much the largest, and extend backwards on each side of the intestine. The blood swallowed by the animal is stored up in these caeca, consequently they are best developed in forms which feed most rarely; in *Aulastoma*, which does not suck blood at all, but feeds on small worms, etc., the last pair alone are present. From the stomach a short intestine, the anterior part of which is often provided with two dorsal caeca (*Hirudo*), leads back to the anus. The

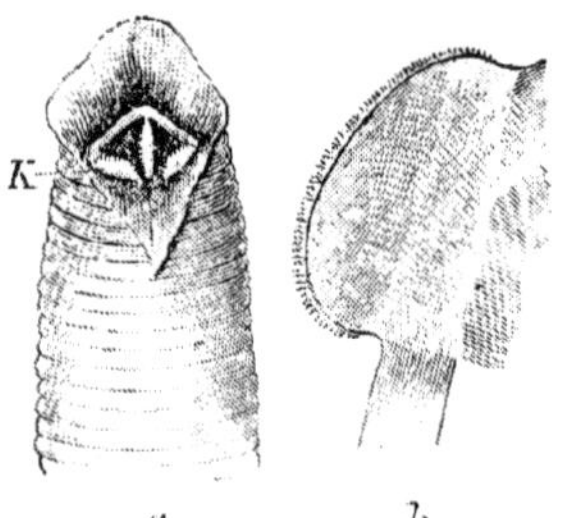

FIG. 417.—*a*, cephalic region of the Medicinal Leech; the three jaws are shown. *b*, one of the jaws isolated with the finely serrated free edge.

anterior part of the intestine is sometimes called the stomach, and the hinder part may be distinguished as the rectum.

The **perivisceral** or **body-cavity** is a coelom, as is shown by its development and its relations to the renal organs; but in *Acanthobdella* alone is it a spacious cavity divided by septa into chambers, as in the Oligochaetes. In all other leeches it is much broken up, and though retaining its relations with some of the organs, *e.g.* ventral nerve cords, principal blood vessels, etc., it is largely without a perivisceral character, and has the form of longitudinal canals or sinuses connected by a complicated system of intercommunicating spaces. This system of canals and spaces is commonly called in leech morphology the *sinus system*; the arrangement of it varies somewhat in the different genera.

Owing to the fact that the sinus system contains a fluid closely resembling the fluid in the blood-vessels, it has often been held to be a part of the vascular system. Leydig was the first to recognize the distinction between the two, and in this he was followed by subsequent workers. But although it is now recognized by the majority of workers that the two systems are distinct, it is still generally held that they are continuous through their finer branches. This view is, as we shall see, based on insufficient evidence, and having regard to the statements of Oka and Burger, it seems safe to assert as a fact that the two systems are separate, as has been shown to be the case for all other groups in which the matter has been closely investigated.

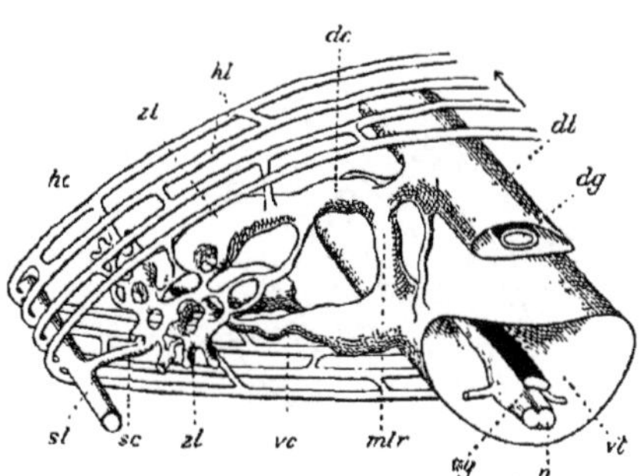

Fig. 419.—Diagram illustrating the arrangement of the coelom or sinus system in the middle part of the body of *Clepsine complanata* (after Oka). *dc*, *mlr*, *vc*, *sc*, *hc* transverse sinuses connecting the main longitudinal sinuses; *dg* dorsal blood vessel; *dl* dorsal sinus; *vg* ventral blood vessel; *vt* ventral sinus; *n* nerve cord; *zl* intermediate sinus as a network of canals in this part of the body; *sl* lateral sinus; *hl* hypodermal sinuses.

In *Clepsine* (Fig. 419) there are five longitudinal canals: (1) a median sinus, which in the region of the gut diverticula, *i.e.*, in the middle part of the body, is divided into a dorsal sinus enclosing the dorsal blood vessel, and a ventral having relation to the ventral blood vessel, nerve cord, and some of the reproductive organs (ovaries and vas deferens); (2 and 3) a pair of lateral sinuses placed at the two margins of the body; and (4 and 5) a pair of intermediate sinuses, which, in the region of the nephridia, are broken up into a network of canals. The median sinus is divided up by a few incomplete, but where they occur segmental, septa. These five sinuses communicate with one another by numerous transverse sinuses, and there is a system of hypodermal sinuses communicating with the others. The fluid in the sinuses contains two kinds of corpuscles, the smaller are exactly similar to the corpuscles of the blood, the larger are apparently detached from the epithelium lining the sinuses, and are not found in the blood. That these sinuses are coelomic is shown by the fact that the nephridia open into them, and that they do not communicate with the vascular system. The latter fact, though long unrecognized on account of the difficulty of distinguishing between blood-vessels and the finer ramifications of the sinus system, can hardly be disputed if Oka is correct in his statements that the two can easily be distinguished in stained preparations, and that the corpuscles of the sinus fluid differ as above explained from those of the blood. Moreover, developmentally the two systems are absolutely distinct.

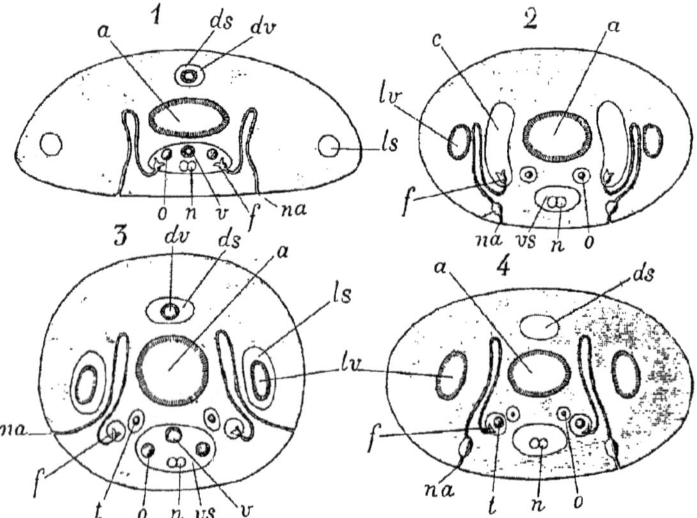

FIG. 420.—Diagrams of transverse sections through various leeches to show the relation of the sinus system (coelom) to the other organs. *1*, *Clepsine*. *2*, *Nephelis*. *3*, *Pontobdella*. *4*, *Hirudo*. (After Bourne, from Perrier). *a* alimentary canal; *ds* dorsal sinus; *dv* dorsal vessel; *f* nephridial funnel; *c*, *ls* lateral sinus; *lv* lateral vessel; *n* nerve cord; *na* nephridiopore; *o* ovary; *t* testis; *vs* ventral sinus; *v* ventral vessel.

In other leeches the coelom is very similar (Fig. 420). In *Pontobdella* it is even more complete than in *Clepsine*. In *Hirudo* the lateral sinuses are absent, and in *Nephelis* the dorsal.

Vascular system. The blood is colourless in the *Rhynchobdellidae*, and coloured red with haemoglobin in the *Gnathobdellidae*. In all cases it has approximately the same colour as the coelomic fluid.* Since Leydig's work the main blood vessels have been distinguished

* In the young *Nephelis* the blood is red and the coelomic fluid yellow (Bürger).

from the longitudinal canals of the sinus system by having thicker, often more muscular walls. The main longitudinal trunks vary considerably in different genera (Fig. 420), but however many there are they all communicate with one another and form a continuous system, and are often contained in one or more of the canals of the coelomic system.

In *Pontobdella* there are four main trunks, a dorsal and ventral and two lateral, all contained in sinuses; in *Clepsine* there is a dorsal and a ventral, both contained in sinuses; in *Hirudo* and *Nephelis* there are no dorsal and ventral vessels, but only two lateral, which are not contained in sinuses.

In *Clepsine* the dorsal vessel is dilated into fifteen small chambers, each of which possesses a cellular projection of its wall, which is called a *valve* and probably buds off blood corpuscles. Moreover, the hinder part of the dorsal vessel in this animal spreads out into a sinus, which completely surrounds the intestine and opens into the ventral vessel.

The **botryoidal tissue** is found in the *Gnathobdellidae*. It is a brown pigmented tissue which embraces the alimentary canal, the blood vessels, the main lacunae of the coelom, etc., "in a word, it is a pigmented tissue, which in the absence of a proper perivisceral cavity fills up all the interstices situated between the organs it invests" (Leydig quoted by Lankester). It consists of swollen cells joined together in anastomosing rows, and containing tubular cavities continuous along the cells of a row. These botryoidal vessels contain a red fluid, and constitute a tubular network lying between the alimentary canal and body-wall, and extending into the muscles of the latter. In places they are swollen into the botryoidal sinuses, which are vesicles lined by botryoidal cells. These are conspicuously developed in *Nephelis* and *Trochetia* (forms without the dorsal sinus), and some of them contain and receive the opening of the funnels of the nephridia. They are comparable to the so-called perinephrostomial sinuses, or parts of the sinus system which contain the nephridial funnels in other *Gnathobdellidae*.

In *Hirudo* a similar, but less conspicuous structure, formed of botryoidal tissue and containing the funnel of the nephridium, is present on the dorsal side of the testis.

The botryoidal vessels undoubtedly communicate with the dorsal and ventral sinuses, or with extensions of these. From their development (see below) there can be no doubt of their coelomic nature, and they clearly correspond to the network of tubes which connect together the main sinuses of the *Rhynchobdellidae*, in which group botryoidal tissue is absent. As in this group they contain a fluid similar in colour and character to the blood (red in the *Gnathobdellidae*), and this similarity has led to the view that the sinus system and vascular system are continuous in both divisions of the *Hirudinea*. As we have already seen (p. 517), no observations have been brought in support of this for the

Rhynchobdellidae; on the contrary, Oka has shown that in *Clepsine* no such continuity exists; but for the *Gnathobdellidae* we have the statements of Bourne and Lankester that they have seen botryoidal tubes and small blood vessels in direct continuity, and the statement of Bürger, in direct opposition to the above, that in the young *Nephelis* the two systems are separate, and contain differently coloured fluids.

In weighing these statements, we must not forget (1) that it is extremely unlikely that animals so closely similar in other respects as the *Gnathobdellidae* and *Rhynchobdellidae* should differ in such an important point of structure; (2) that the sinus network and the small blood vessels are intermingled in the greatest complexity of arrangement, and that the walls of both have a very similar structure, and that they contain a very similar fluid, and therefore might very easily be mistaken for one another, and (3) that a continuity between the vascular system and the undoubted coelom of the sinuses would be a unique phenomenon in the structure of the animal kingdom. The bearing of these considerations is obvious, their weight indubitable; and it is clear that until more detailed and elaborate observations in support of the view of continuity are forthcoming, we are bound to hold, provisionally at any rate, that in the Leeches, as in other animals, the blood system and coelom are separate from one another. This position is still further strengthened when we remember that developmentally the two systems are separate, and if continuity exists it is only established comparatively late in life, after the larval and embryonic period.

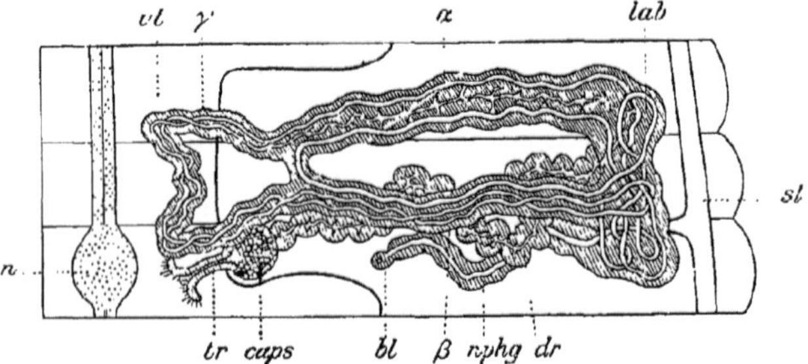

FIG. 421.—Diagrammatic representation of a nephridium of *Clepsine* (after Oka). *α* the anterior division; *β* the posterior division; *γ* the median division; *bl* the invagination of skin; *caps* capsule; *dr* glandular cells of the nephridium; *lab* labyrinthic coil of nephridial tube; *n* nerve cord; *nphg* duct of nephridium; *sl* lateral sinus; *tr* funnel; *vl* ventral sinus.

The **nephridia** vary in number in the different genera, and even in species of the same genus. They are absent from the anterior and posterior part (Fig. 424), and are best developed in the middle region of the body. They consist (Fig. 421) of a long convoluted canal, which opens at one end to the exterior on the ventral surface of the body, and at the other internally into some part of the sinus system (see above, p. 518). The part of the organ next the external opening (sometimes as in *Hirudo* dilated into a vesicle) is probably derived from an invagination of the skin, and is intercellular; the remainder is intracellular, and is described as consisting of a row of tubular cells fitted end to end like drain-pipes. On account of the long convoluted character of the nephridium, different parts of the cell-rows are in contact, and where this happens the cells of the different rows are not distinguishable save by their nuclei. At some parts of its course the main tube of the nephridium gives off branching ductules in the protoplasm of the nephridial cells. Just before the internal opening is reached, the nephridium often becomes dilated into the capsule (Oka), which consists of a protoplasmic wall with embedded nuclei and an internal mass of cells, which are partially, sometimes completely, fused with one another (Fig. 421). On reaching the capsule the intracellular tube of the nephridium branches out into a number of branching and anastomosing fine ductules, which open into a space between the wall and central mass; this space leads into the funnel, and is most developed near the point, where that structure is connected to the capsule. The funnel itself and its stalk consist of protoplasm without cell-limits, but containing a few nuclei; it is ciliated, and the cilia are directed towards the capsule.

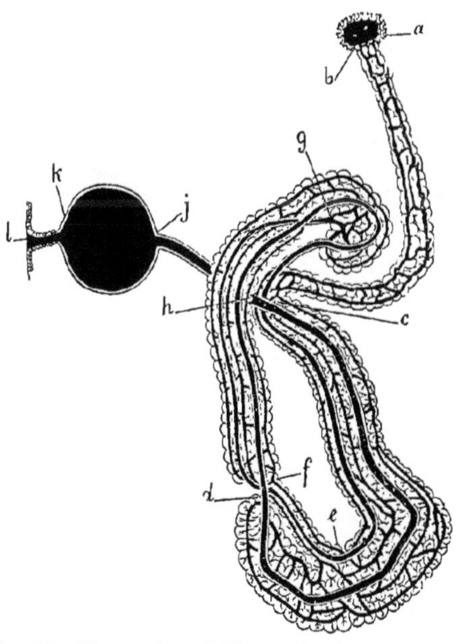

FIG. 422.—Diagram of a nephridium of Hirudo (from Perrier, after Bourne). *a* capsule and funnel; *b–c* testis lobe; *c–d* main lobe; *e–g* recurrent lobe; *f–g* apical lobe; *h–j* the duct; *j–k* the vesicle; *l* the external opening.

In *Pontobdella* (Fig. 423) and *Branchellion* the nephridial system consists of a network of hollow cells extending from segment to segment, and across the middle line. In each of the segments in which the nephridial network is found there are connected to the network two internal openings (of the usual structure, a syncytial ciliated funnel and neck, and a capsule), and two ventral external openings.

In *Clepsine* (Fig. 423 *C*) and *Hirudo* (Fig. 423 *B*) the nephridial tube is much twisted, and gives off branching ductules just before reaching the capsule (Fig. 422).

In *Hirudo* the internal opening is a large structure, and is sessile upon the capsule (Fig. 422); it is divided up into several ciliated spoon-shaped openings, and has the usual syncytial character.

In *Hirudo medicinalis* there are 17 pairs of nephridia; the first one is in the segment of the 2nd free ganglion, the last one in that of the 18th ganglion. The first six are without internal funnels; the remainder, the first nine of which are in the segments of the testes, and the last two in the segments behind the testes, possess internal openings at the end of the testis-lobe. The testis-lobe is a process of the nephridium which projects towards the middle line, and in the testis segments abuts upon the testis.

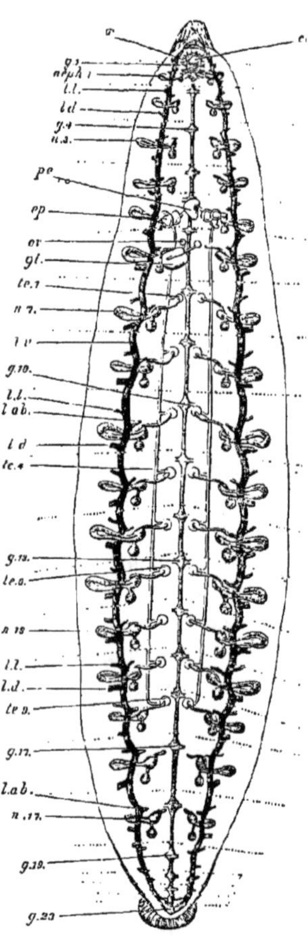

FIG. 424.—Diagram illustrating the arrangement of the renal, generative, vascular, and nervous organs of *Hirudo medicinalis* (from Perrier, after Bourne). *cer.g.* passage of the oesophagus through the nerve-collar; *ep* epididymis; *g. 1-23* the ventral ganglia; *gl.* glandular dilation of oviduct: *l.v.* lateral blood vessel; *l.l.*, *l.ab.*, *l.d.* branches of the same; *neph.*, or *n. 1-17* the 17 pairs of nephridia; *oe* circum-oesophageal commissure of the left side; *ov* ovary; *pe* penis; *te 1-9* the 9 pairs of testes.

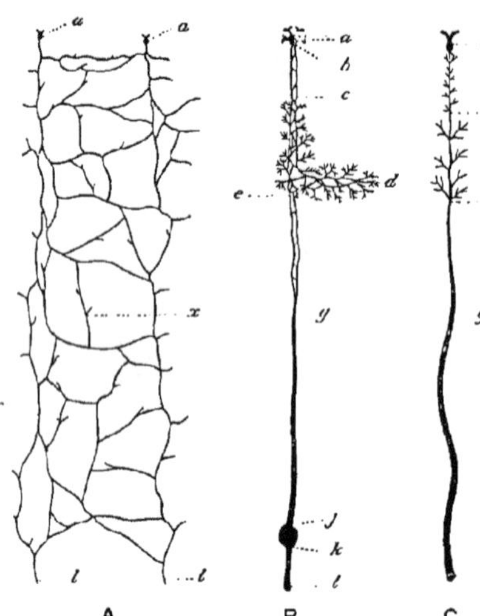

FIG. 423.—Diagrams of the nephridia. *A*, of *Pontobdella* (two contiguous nephridia are shown); *B*, of *Hirudo*; *C*, of *Clepsine*. *a*, *a* capsules and funnels; *b–c* canal in testis lobe; *c–e* canal in main lobe; *d* in *Hirudo*, canals in caecal end of the "main lobe"; *e–g* in *Hirudo* ducts in the apical lobe; *g* the apex; *e–l*, in *Clepsine*, and *g–l*, in *Hirudo*, unbranched canal passing to the external aperture; *j-k*, in *Hirudo*, the vesicle; *x*, *x* caecal ductules in *Pontobdella*. (After Bourne.)

Generative organs. The *Hirudinea* are hermaphrodite. As in many marine *Planaria*, the openings of the male and female organs are placed one behind the other, the male being anterior, in the middle ventral line of the anterior region of the body. In *Hirudo medicinalis* the male opening, which in the *Gnathobdellidae* is provided with a protrusible penis, is in the segment of the 6th distinct postoral ganglion (11th segment counting the suboesophageal ganglion as consisting of 6 fused ganglia), and the female opening in the segment behind (Fig. 424).

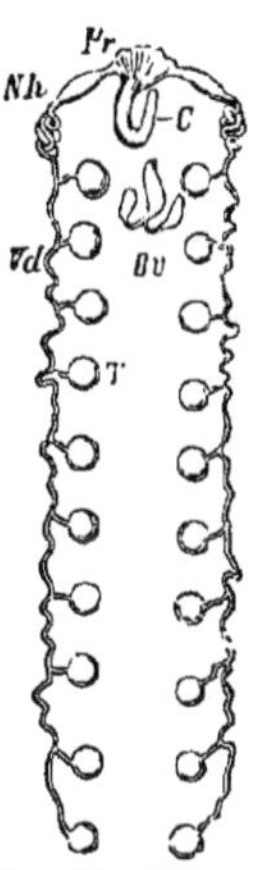

FIG. 425.—Generative organs of the Medicinal Leech. *T* testis; *Vd* vas deferens; *Nh* epididymis; *Pr* prostate; *C* penis; *Ov* ovaries with vagina and female genital opening.

The reproductive glands are hollow structures, and are continuous with their ducts.

There are usually several pairs of testes arranged segmentally, but in *Nephelis* they are arranged irregularly. In *Hirudo* (Fig. 424) there are nine or ten pairs of testicular vesicles, which are connected by short ducts with a sinuous vas deferens on either side. Each vas deferens is coiled in front to form a kind of epididymis (Fig. 425), and is then prolonged into a muscular portion, the ductus ejaculatorius, which unites with that of the other side to form an unpaired eversible organ called the penis, with the internal end of which is connected a well-developed gland—the prostate. The female apparatus (Fig. 426) consists either of two long tubular ovaries extending over several segments (*Clepsine*, *Nephelis*), or of two short saccular structures. The ovaries may either join together at the genital opening (*Clepsine*), or their walls may be continued as the two oviducts, which soon join. The single oviduct so formed becomes convoluted, and passes through the albumin gland, after which it becomes dilated and constitutes the vagina which opens to the exterior (Fig. 426).

The reproductive cells are produced by the cells lining the reproductive glands, which must be regarded as special parts of the coelom (sinus system) which have become shut off from the rest in continuity with their ducts. This separation of the generative part of the coelom from the rest takes place as we have seen in some *Oligochaeta*.

The spermatozoa are united together in packets—the spermatophores. There does not appear to be any true copulation, but the spermatophores are deposited on the body of another leech, whence the spermatozoa appear to make their way through the skin and tissues to the ovarian tubes, where it is probable that fertilization occurs.* This process of cutaneous injection of spermatozoa is found in other groups, amongst which may be mentioned *Turbellaria* and *Peripatus*.

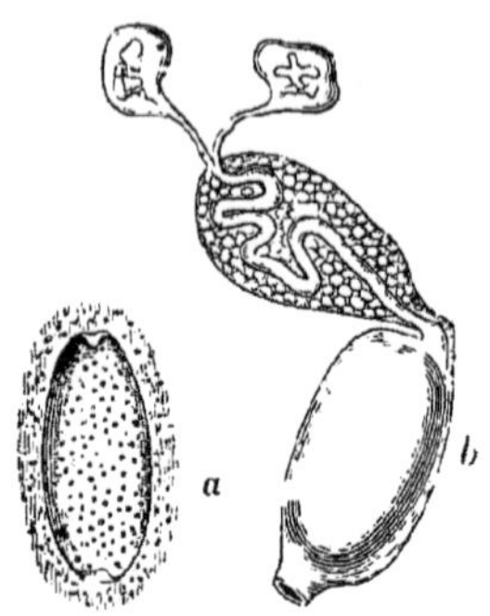

FIG. 426.—*a*, cocoon; *b*, female organs of *Hirudo medicinalis* (after Leuckart).

The eggs are usually laid in cocoons (Fig. 426 *a*) either on stones or plants (*Nephelis*, *Clepsine*), or in damp earth, the animal, if aquatic, leaving the water for the purpose. The cocoon is formed by the clitellar glands. When the eggs are about to be laid the leech attaches itself firmly by its ventral sucker, and, twisting itself about, envelops the anterior part of the body with a viscid mass, which covers especially the genital rings like a girdle and gradually hardens to form a firm membrane. A number of small eggs and a considerable quantity of albuminous matter in the case of the *Gnathobdellidae*, then pass out, and the animal withdraws its anterior end from the barrel-shaped membrane, which now contains eggs, etc., and which, after the animal has left it, becomes in consequence of the narrowing of the external openings a tolerably completely closed cocoon.

The number of eggs contained in a cocoon varies, but is never large. The eggs of the *Gnathobdellidae* are small, and the young are hatched early and float as larvae in the albumen which they swallow. In the *Rhynchobdellidae* the eggs are larger and contain more yolk; they are hatched at a later stage. *Clepsine* (see p. 526) attaches its cocoon to stones and broods over it till the young are hatched, which takes place at an early stage; the young then attach themselves to the ventral side of the mother, and are carried about by her, living on albumen secreted by her.

The **development** is characterized by the early specialization of the so-called pole-cells (neuro-nephroblasts, mesoblasts), and by the formation from them of cords of cells with a definite destination. This characteristic is not confined to the embryo, but is found also in the adult, where the rows of cells of which the nephridia and botryoidal tissue are both formed, not to speak of the egg-strings,

* *Vide* Whitman, "Spermatophores as a means of hypodermic impregnation," *Journ. Morph.*, 4, 1891.

are striking features. The segmentation is unequal. The mouth is formed early, and through it, after the formation of the pharynx and rudiments of the alimentary canal, the albumen contained in the cocoon is taken into the intestine of the growing embryo by means of swallowing movements of the pharynx.

In the *Gnathobdellidae* it appears that a considerable part of the larva is cast off in the attainment of the adult stage, *e.g.* the ectoderm, muscles, pharynx, and provisional kidneys. The development* resembles in many respects that of the Oligochaetes. This particularly applies to the origin and fate of the mesoblastic bands. They arise from two pole-mesoblasts and become segmented into *somites*, each of which soon acquires a cavity—the coelom. The somites spread ventrally, and unite with one another across the middle line. In this ventral part the walls between successive somites break down, and a continuous tube is formed—the ventral sinus—while in the lateral parts they give rise to the septa, which in an incomplete manner persists throughout life. In this way arise the ventral sinus and the lateral sinuses, while partial obliteration of the connection between these (intermediate sinus of *Clepsine*) and diverticula of them give rise to the complicated sinuses of the adult.

The nephridia arise from a single large cell of the somite-wall, which proliferates and produces the characteristic cord, which becomes the nephridium—the strings of botryoidal tissue appear to arise in a very similar manner from large cells of the somite wall which produce strings of cells by proliferation. In both cases the cell-cords become hollow, the lumen communicating with the coelom. The botryoidal sinus in *Nephelis*, which contains the nephridial funnel, is directly derived from the lateral part of the primitive somites, the segmentation of which is retained. If this account of the origin of the botryoidal cords is correct, it would appear that they resemble the nephridia closely in their development, structure, and relations to the coelom, and we must regard them, like the nephridia, as special organs of the coelom.

The ovaries arise as thickened patches of the coelomic epithelium in parts of the coelom which become constricted off from the rest; the ducts arise by prolongation of the sac so formed to the skin of the ventral surface. The testes arise on each side from a continuous ridge of cells, which separates from the peritoneum and becomes hollowed out into the testes and vas deferens.

The terminal parts of the nephridia and generative ducts are developed from ectodermal ingrowths.

Asexual reproduction is unknown in the group. The leeches live for the most part in water, or temporarily in damp earth. A few of the *Gnathobdellidae* are true land forms, and have lost the power of swimming. Comparatively few are marine, and they belong to the *Ichthyobdellidae*. They move partly by "looping," with the help of their suckers, and partly by swimming with active undulations of the body. Many of them are parasitic on the skin or gills of aquatic animals; most, however, are only occasionally parasites of the outer skin of warm-blooded animals. They do not feed exclusively on any special genus of animals, and their diet is not always the same in the different periods of their existence. The caeca of the stomach

* O. Bürger, *Zoologische Jahrbucher*, 4, 1891, and *Z. f. w. Z.*, 58, 1894.

enable them to store up a large quantity of food, and many of them are able to go without food for a long period.

Fam. 1. **Rhynchobdellidae.** With colourless blood, with a protrusible proboscis, without jaws. Each typical segment consists of 3, 6, or 12 rings. The 6-ringed forms are all marine except *Haementaria.* Organs concerned in the formation of blood corpuscles (so-called valves) are often present in the dorsal vessel. Without botryoidal tissues.

Sub-fam. 1. **Ichthyobdellidae.** Marine and fresh-water leeches, parasitic for the most part on fishes. With cylindrical body. The proboscis is not longer than the preclitellum. *Ichthyobdella* Blainv., on marine fish; *Piscicola* Blainv., on fresh-water fish; *Pontobdella* Leach, marine, on sharks and rays; *Branchellion* Sav., with foliaceous lateral appendages; *B. torpedinis* Sav.; *Calliobdella* v. Ben. and Hesse; *Ozobranchus* Qfg., with branchiae, in the mouth of tortoises, crocodiles, pelicans.

Sub-fam. 2. **Clepsinidae.** Fresh-water leeches, generally parasitic on snails, but also on fish. With dorsoventrally compressed, never cylindrical body. The proboscis is longer than the preclitellum. *Clepsine** Sav., common fresh-water form. They brood over their eggs, which are attached to some foreign body. In two species *bioculata* and *heteroclita* the eggs are attached to the ventral surface of the body. After hatching the young attach themselves to the ventral surface of the mother. *Batrachobdella* Viguier, on Batrachians, Algeria; *Haementaria* de Fil.; *H. mexicana* de Fil., *H. officinalis* de Fil, both in the lagunes of Mexico, the latter used for medicinal purposes; *H. Ghilanii* de Fil., in the Amazon.

Fam. 2. **Gnathobdellidae.** Fresh-water and land leeches with jaws, without protrusible proboscis, with red blood; the complete segments have five rings. With botryoidal tissue.

Sub-fam. 1. **Nephelidae.** With 19 complete segments; the sexual openings are separated from one another by two rings. Eyes are never present on the 6th segment, 4 pairs only. *Nephelis* Sav. (*Herpobdella* Blainv.), without hard jaws, feeds on snails and planarians; *Liostomum* Wagler; *L. lumbricoides*, lives in the earth after the fashion of earthworms, which it resembles, Brazil. *Macrobdella* Phil.; *Trochetia* Dutroch., jaws vestigial, without teeth, 4 pairs of eyes; it is a water-leech, but leaves the water to feed on earthworms, continent of Europe, has been introduced into England; *Archaeobdella* O. Grimm, *A. esmonti* O. Grimm, from the mud at the bottom of the Caspian.

Sub-fam. 2. **Hirudinidae.** With 16 complete segments; sexual openings separated by 5 rings; 5 pairs of eyes. *Hirudo* Ray and Linnaeus, with 102 rings; eyes on the three anterior rings, and on the fifth and eighth; male opening between rings 30 and 31, female opening between rings 35 and 36. The three jaws are finely serrated, and can be moved like a circular saw in a manner well-adapted to make a wound, which readily heals, in the external skin of man; aquatic, but deposits its cocoons in damp earth. *H. medicinalis* Ray, with the variety distinguished as *officinalis*, possesses 80-90 fine teeth on the free edge of the jaws, and attains a length of about 6 inches. They were formerly common in Germany, and are still frequently

* The genus *Glossiphonia* Johnson has priority over *Clepsine*, but whether it is advisable to give up a name so well-known as *Clepsine* is a doubtful question.

to be found in Hungary and France. They are cultivated in special ponds, and take three years to attain sexual maturity. In the young stage they live on the blood of insects, then on that of frogs, and only when they have attained sexual maturity is a diet of warm blood necessary to them. *Aulastoma* Moq. Tand., often called the horse-leech, feeds on worms and molluscs, last pair of gut-caeca alone present; *A. gulo* M. T. *Haemopis* Sav.; *H. vorax* M. T., the horse-leech, indigenous in Europe and N. Africa, attaches itself to the interior of the pharynx of horses, cattle, and men; jaws with 30 denticles. *Haemadipsa* Tennent, the land-leech, in forests or damp districts in or near the tropics; *Geobdella* Whitman, Australian land-leech; *Leptostoma* Whitman, jaws vestigial; *Limnatis* M. T., 4 pairs of eyes, jaws without denticles, fresh-water; *L. nilotica* M. T.

Fam. 3. **Acanthobdellidae.*** Fish parasites from Siberia, on caudal and anal fins of Salmo salvelinus, with two double rows of setae on each side on the first 5 segments of the body (Fig. 427). The setae are embedded in setigerous sacs which are provided with retractor muscles. Body composed of 20 segments; male opening on segment 7, female on segment 8. Anus dorsal to sucker (Kowalevsky). 4 or 5 rings to one segment. Body-cavity spacious and incompletely divided by 20 segmental transverse septa. The nephridia are in the dissepiments, and open between the segments; internal openings not observed. The visceral peritoneum consists of chloragogen cells. The vascular system consists of a dorsal and ventral vessel. The nervous system consists of 20 ventral ganglia, of which the first and last are composite. Testes as two continuous tubes in segments 6–15. Ovaries as two tubes in segments 8–13. The genital ducts are presumably continuous with the glands. *Acanthobdella* Grube.

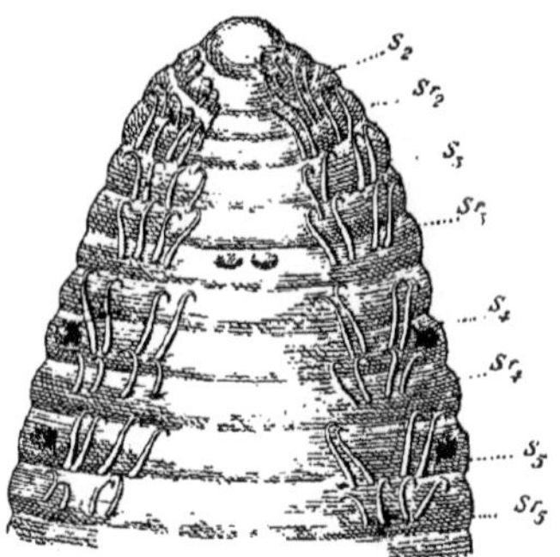

FIG. 427.—Ventral view of anterior end of *Acanthobdella*, showing the setae *S*, and reserve setae *Sr* (after Kowalevsky).

Class IV. ECHIUROIDEA† (GEPHYREA ARMATA).

Annelida with variable traces of segmentation in the adult; with a well-marked preoral lobe and a pair of ventral hooked setae.

The *Echiuroidea* were formerly united with the *Sipunculoidea* in the class *Gephyrea*. In the present work, however, it has been

* A. Kowalevsky, *Bull. Acad. Imp. des Sciences St. Petersbourg*, June and Nov., 1896, T. v.

† R. Greef, "Die Echiuren," *Nov. Act. Leop. Car.*, 41, 1879. J. Spengel, "Die Organisation des Echiurus Pallasii," *Z. f. w. Z.*, 34, 1880. D. C. Danielssen and J. Koren, "Gephyrea," *Norwegian North Atlantic Expedition*, 1881. B. Hatschek, "Ueb. Entwick. von Echiurus," *Arb. Zool. Inst. Wien*, 3, 1881. E. Selenka, "Report on the Gephyrea," *Challenger Reports*, vol. 13, 1885. M. Rietsch, "Étude sur les Gephyriens armés," *Recueil z. Suisse*, 3, 1886. A. E. Shipley, "Gephyrea" in *Cambridge Natural History*, 2, 1896.

deemed advisable, for reasons set forth on another page (p. 533), to break up this class, and to establish the *Sipunculoidea* as an independent phylum of the animal kingdom, with affinities to *Phoronis* and perhaps to the *Annelida*, and the *Echiuroidea* as an aberrant class of the phylum *Annelida*.

The *Echiuroidea* have a somewhat cylindrical body, which is prolonged anteriorly into a long, highly contractile preoral proboscis (Figs. 428, 429). They are exclusively marine, and for the most part live in holes and fissures in rocks and between stones; *Echiurus*, however, frequents sand or mud, in which it forms tubes with two openings. Many members of the class appear to have the habit of frequently changing their residence. The proboscis has a ciliated groove on its ventral surface, which leads behind into the **mouth** situated at its base. The animals appear to subsist on organisms and organic particles, which are brought to the mouth by the currents of water driven along the groove of the proboscis by the cilia. The **anus** is posterior and terminal.

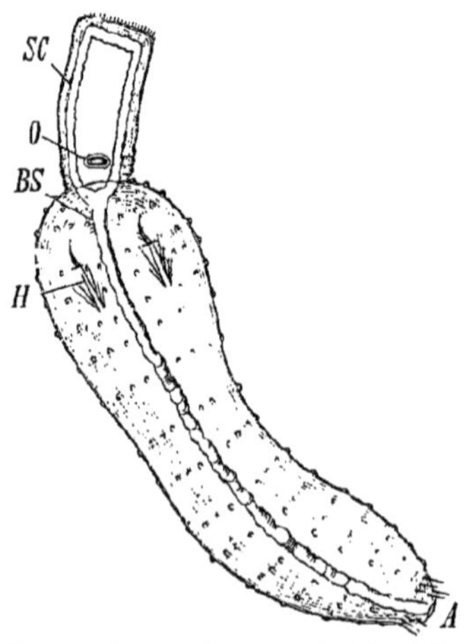

FIG. 428.—Young *Echiurus*, ventral view (after Hatschek). *O* mouth at the base of the proboscis; *SC* circumoesophageal commissures; *BS* ventral cord; *A* anus; *H* ventral hooks.

There are almost always two hooked setae on the ventral surface not far from the front end (Fig. 428 *H*); they are embedded in pits of the skin, by the lining of which they are secreted in the typical annelidan manner. In *Echiurus* there is in addition a single or double row of setae round the hind end; these are the anal setae. A short distance behind the hooked setae are the openings of the anterior nephridia; these vary in number from four pairs to a single one.

The skin is covered with small papillae, which are often arranged in rings. The body-wall is highly muscular, and consists of the usual layers.

The **central nervous system** consists, as in the *Sipunculoidea*, of a *single* ventral cord, lying entirely within the body-wall. In front this cord divides into two, which pass round the oesophagus and extend quite to the anterior end of the proboscis, where they join one another (Fig. 428). There are no ganglionic swellings, but the whole system has a uniform coating of nerve-cells. The ventral portion, moreover, contains a fine canal, which extends into the commissures, but is absent from the hind end of the cord, and from the supra-oesophageal portion. The central organ gives off nerves to the adjacent parts of the body.

There are no special sense-organs.

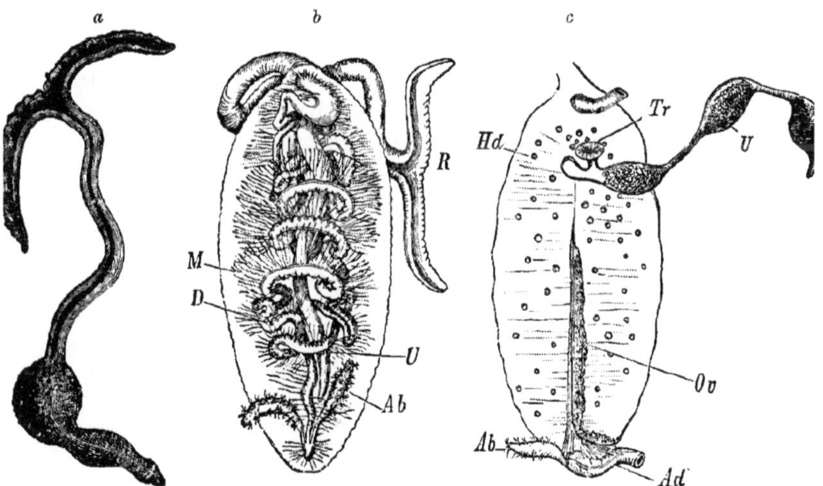

FIG. 429.—Female of *Bonellia viridis*; *a*, the whole animal; *b*, anatomy; *c*, integument and generative organs after removal of the intestine (after L. Duthiers). *Ab* anal vesicles; *Ad* rectum; *D* alimentary canal; *Hd* cutaneous glands; *M* mesentery; *Ov* ovary; *R* proboscis (preoral lobe); *Tr* ciliated funnel of the single brown tube or anterior nephridium (uterus); *U* uterus (anterior nephridium).

The **alimentary canal**, in which various divisions can be made out, consists of a long, thin-walled, much convoluted tube, along the ventral wall of part of which is a ciliated groove. There is also a *siphon*, or accessory intestine, like that found in Echinids and some Chaetopods; this is a tube given off ventrally from the anterior part of the long and coiled intestine; it lies ventral to the ciliated groove, and opens posteriorly into the hind part of the intestine. There is no special mesentery, but strands of tissue run from all parts of the body-wall across the body-cavity, to be inserted into the walls of the alimentary canal.

The rectum resembles the skin in its lining epithelium, having many unicellular glands. It receives the openings of the two **anal vesicles** (Fig. 429, *Ab*) — one on each side; these are elongated tubular, sometimes branched, contractile structures which open into the body-cavity by many ciliated openings. They are supposed to be a pair of modified nephridia, the openings of which, as in some Oligochaetes (p. 502), have got shifted into the rectum.

The **perivisceral cavity** is spacious and entirely coelomic, and contains a corpusculated fluid, the corpuscles of which are said, in some forms, to contain haemoglobin.

The **vascular system** consists of a dorsal vessel, lying on the anterior part of the alimentary canal and continued along the proboscis, and of a ventral supra-neural vessel. These two vessels communicate in front and behind, and doubtless give off branches to the organs, though such have not been described.

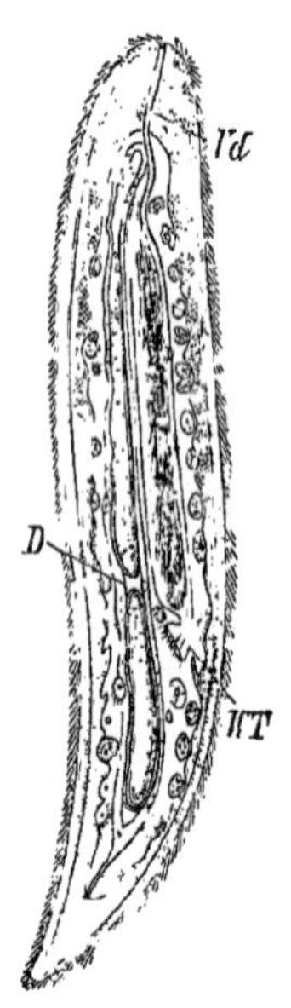

Fig. 430.—Planarian-like male of *Bonellia* (after Spengel). *D* intestine; *WT* ciliated funnel of the vas deferens (*Vd*), which is filled with sperm.

The **renal organs** are constituted by the so-called brown tubes or anterior nephridia, and the anal vesicles or posterior nephridia. The anterior nephridia closely resemble in form and relations the brown tubes of *Sipunculus*; they are attached to the ventral body-wall at their external openings, and project back into the body-cavity as long blind tubes; the internal opening, as in *Sipunculus*, is close to the external. They function also as generative ducts, and the generative cells are collected in them before extrusion, for which reason they are sometimes called uteri. They vary in number from one to four pairs. In *Echiurus* there are two pairs, but in *Thalassema* the number varies in the different species from one to four pairs. In *Bonellia* there is only one brown tube (Fig. 429), sometimes on the right side and sometimes on the left. The posterior nephridia or anal vesicles have already been described.

The **sexes** are separate, and the generative cells are derived from that portion of the coelomic epithelium which overlies the ventral (supra-neural) vessel (Fig. 429 *b*). They are dehisced into the body-cavity, where they mature and escape by the anterior nephridia.

The males and females are not distinguishable externally except in *Bonellia* and *Hamingia*, in which there is a very pronounced sexual dimorphism. In these genera the female has the ordinary form of the species; but the male (Fig. 430) is a minute (in *Bonellia* 1–5 mm. long) planarian-like organism which lives in considerable numbers as a parasite, principally in the uterus and pharynx of the female. It possesses an alimentary canal which is without any opening to the exterior; the skin is ciliated, and contains muscular fibres; there is a body-cavity, the lining cells of which give rise to spermatozoa; there is also a single anterior

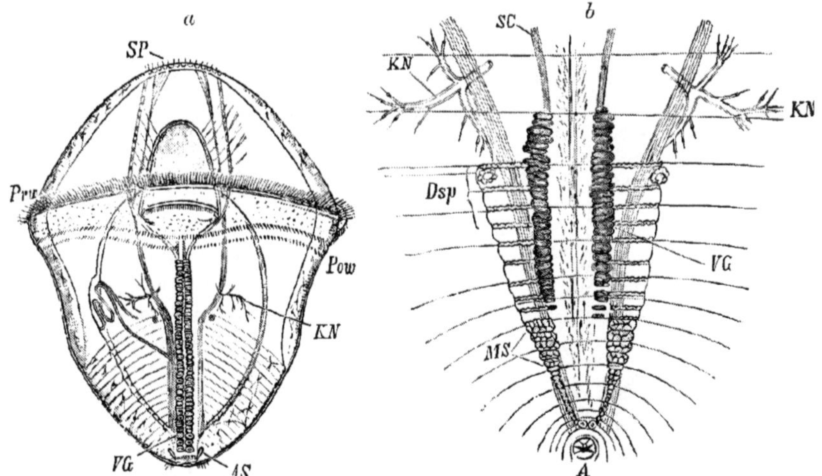

FIG. 431.—*a*, Larva of *Echiurus* from the ventral side (after Hatschek). *SP* apical plate; *Prw* preoral; *Pow* postoral ring of cilia; *KN* head-kidney; *VG* ventral nerve cord connected with the apical plate by the long oesophageal commissures; *AS* anal vesicles. *b*, ventral region of the *Echiurus* larva with segmented mesodermal bands; *SC* oesophageal commissures; *Dsp* dissepiments of the anterior body segments; *MS* mesodermal bands; *A* anus (after Hatschek).

nephridium which opens to the exterior at the front end of the body, and internally into the body-cavity at its hind end. Small vestiges of the anal vesicles, which open directly on the surface of the body, can be made out, and in both the genera the hooked setae are present.

Development. In the *Echiuroidea* a typical trochosphere larva (Fig. 431) has been observed in *Echiurus* and *Thalassema*, with a strong preoral circle of cilia (*Prw*), in addition to which there is also a delicate postoral circle (*Poc*). In *Echiurus* early in larval life an excretory organ, the head-kidney or pronephros (*KN*) is developed, one on either side; and behind it a pair of mesoblastic bands make their appearance, and become divided up in subsequent development into the rudiments of 15 pairs of mesoblastic somites (Fig. 431, *b*). In the terminal segment, which is surrounded by a circle of cilia, there appear in the somatic mesoderm the rudiments of the anal vesicles (Fig. 431, *AS*). The rudiments of both the cerebral ganglion and of the ventral cord are derived from growths of the ectoderm—the former from the apical plate, and the latter from paired thickenings of the ventral ectoderm. The two are connected by the oesophageal ring, which is also provided with ganglion cells. In later stages the segments disappear, and the ciliary apparatus degenerates and vanishes; after which two strong hooked setae appear at the side of the nerve cord, not far from the mouth, and two circles of shorter setae are formed at the hind end of the body (Fig. 432). The preoral lobe of the larva becomes the proboscis of the young *Echiurus*. In *Thalassema* the development is very similar, and the mesoblastic bands are segmented. In *Bonellia*, on the other hand, no traces of segmentation have as yet been observed, and the larva is not so obviously built on the trochosphere type.

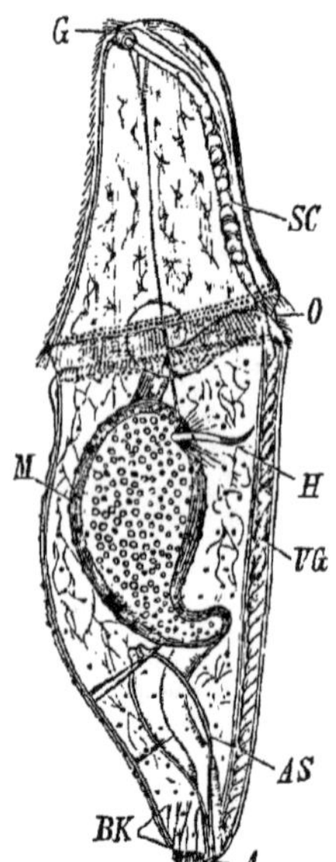

FIG. 432.—Older *Echiurus* larva from the side (after Hatschek). The head-kidney is atrophied. *A* anus; *AS* anal vesicles; *BK* circles of setae; *G* cerebral ganglion developed from the apical plate; *H* ventral hooks; *M* stomach; *O* mouth; *SC* oesophageal commissures; *VG* ventral nerve cord.

In view of the development of *Echiurus* and *Thalassema*, it is difficult to resist the conclusion that the *Echiuroidea* are *Annelida*, a conclusion which is further emphasized by the presence in all genera of typical Annelidan setae embedded in and secreted by pits of the skin. It is curious that in *Bonellia* the trochosphere stage is so much modified, and especially that no traces of segmented mesoblastic bands should be seen. With regard to the latter point, however, it must not be forgotten that a renewed investigation may yet bring to light some traces of larval segments. We may therefore regard the *Echiuroidea* as Annelids in which the segmentation is feeble, showing faintly in the young, but except in the repetition of the nephridia (*e.g.* *Thalassema* with three pairs of anterior and one pair of posterior nephridia), being absent from the adult.

The resemblances between the *Sipunculoidea* and *Echiuroidea* have been referred to above. They formerly led to the union of the two groups in the class *Gephyrea*, and they consist mainly in the resemblance between the brown tubes of the former and the anterior nephridia of the latter, in the spacious character of the coelom in the two groups, and in the singleness of the ventral nerve cord. As to these we may remark that nephridia tend to resemble each other in groups much more remote from each other than those under discussion, that a spacious coelom is found in many widely divergent forms, and that the singleness of the ventral cord, though an important feature, is not of itself sufficient to outweigh the following most important differences: (1) the difference in the position of the anus; (2) the absence of a preoral lobe in the *Sipunculoidea*; (3) the absence of anal vesicles in the same group; (4) the absence of any traces of Annelidan setae; and (5) the total absence of any trace of segmentation in the larva and in the adult.

Echiurus Cuv. (Fig. 428). Proboscis not bifurcated; anal setae in one or two rows; two pairs of anterior nephridia; males and females alike. *E. pallasii* Guérin, N. Sea, English Channel, etc.

Thalassema Gaert. Proboscis not bifurcated; no anal setae; one to four pairs of anterior nephridia; sexes alike. *Th. neptuni* Gaert., English Channel, etc.

Bonellia Rolando (Fig. 429). The animals are of a green colour, owing to the presence of a pigment distinct from chlorophyll, and called bonellein. Proboscis bifurcated; no anal setae; a single anterior nephridium; sexual dimorphism pronounced. *B. viridis* Rolando, North Sea, etc.

Hamingia Dan. and Kor. Proboscis not bifurcated; no ventral hooks and no anal setae; one or two anterior nephridia; sexual dimorphism pronounced; males provided with ventral hooks. *H. arctica* D. and K.

Saccosoma Dan. and Kor. Proboscis absent; no hooks or setae; one anterior nephridium; male unknown. *S. vitreum* D. and K.

Epithetosoma Dan. and Kor., may be placed here, though it differs considerably from other *Echiuroidea*. The proboscis is long and whip-like, and contains a cavity continuous with the body-cavity. Posterior to this tubular proboscis, on either side of the anterior extremity of the trunk, is a fissure, the bottom of which is pierced with apertures leading into the body-cavity. There are no hooks or setae, and no anal vesicles. The mouth is at the base of the proboscis, and the anus is posterior; the alimentary canal is straight. The nerve-cord contains a tube. Vascular system not observed.

CHAPTER XII.

SIPUNCULOIDEA* (GEPHYREA ACHAETA).

Unsegmented vermiform animals with a spacious coelom, antero-dorsal anus, and one pair of nephridia. The anterior part of the body is invaginable.

As stated before on p. 451, we have thought it necessary, in the present state of our knowledge, to break up the old group *Gephyrea* into (1) the *Echiuroidea* (*Gephyrea armata*) which remains with the *Annelida*, (2) the *Sipunculoidea* (*Gephyrea Achaeta*), (3), the *Priapuloidea*, and (4) *Phoronidea* (*Gephyrea tubicola*). The three last of these we advance provisionally to phyletic rank, not being able to subordinate them to any other group. The *Sipunculoidea* and *Phoronidea* are undoubtedly *Coelomata;* but of the *Priapuloidea* we cannot say this, and the fact that we deal with them in the part of the work devoted to coelomate animals must be regarded as a concession to the older view of their relationship to the *Sipunculoidea*, rather than as an expression of opinion on our part that they have a coelom.

The *Sipunculoidea* are elongated, vermiform animals and live in sand and ooze in the sea; they also bore in coral rock. The anterior part of the body is of a different appearance to the posterior or main portion; it is called the *introvert*, because it can be invaginated into the larger posterior portion by means of special retractor muscles. The **mouth** is placed at the anterior end of the introvert and is in relation with a row of ciliated, hollow tentacles, which may be arranged in a circle or in the form of a double horseshoe, the concavity of which is dorsal.

* Keferstein, "Beiträge z. Anat. u. syst. Kenntniss d. Sipunculiden," *Z. f. w. Z.*, xv., 1865. E. Selenka, "Die Sipunculiden," *Semper's Reisen*, 1883. D. C. Danielssen and J. Koren, "Gephyrea," *The Norwegian N. Atlantic Expedition*, 1876–78, 1881. B. Hatschek, "Üb. Entwick. v. Sipunculus nudus," *Arb. Zool. Inst. Wien*, 5, 1884. A. E. Shipley, "On Phymosoma varians," *Q. J. M. S.*, 31.

There is no trace of a preoral lobe. The **anus** is approximated to the mouth, being placed on the dorsal surface about one-third of the length of the body from the anterior end. The body is devoid of setae, but the introvert is often covered by rows of horny hooks or by small, scale-like papillae which are directed backwards and overlap one another. There is no trace of segmentation, but the skin is often thrown into ridges, both transverse and longitudinal. There are no special organs of sense. The body-wall is highly muscular, and consists of an external cuticle, a single-layered epidermis, a cutis, a layer of circular muscles, a thin layer of oblique muscles, and a thick layer of longitudinal muscles which are often arranged in bundles; lastly there is a layer of flat coelomic epithelial cells bearing isolated cilia. The cutis contains numerous unicellular or multicellular glandular organs which open to the exterior by pores; these structures have often a marked nervous supply and have been mistaken for sense-organs.

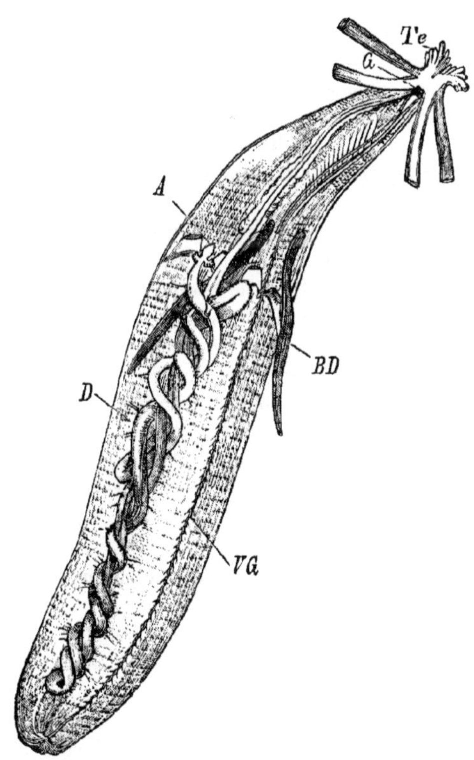

FIG. 433.—*Sipunculus nudus*, laid open from the side (after W. Keferstein). *Te* tentacles; *G* cerebral ganglion; *VG* ventral nerve cord; *D* intestine; *A* anus; *BD* brown tubes (nephridia).

The **alimentary canal** is a thin-walled tube in which three regions may be distinguished—the anterior portion (oesophagus) straight and uncoiled, the middle portion (intestine) coiled and bent on itself, and the terminal portion (rectum) short and straight, leading to the anus. The intestine passes back to the hind end of the body and then turns forwards, the two limbs being coiled spirally round one another. Delicate muscular bands pass from the body-wall to the intestine, and there is often a muscle—the spindle muscle—which extends along the axis of the intestinal spiral from the hind end of the body to the rectum. There are no special glands, but a caecal diverticulum, which varies much in size in different individuals and opens into the rectum, is often present. Two tufts of tissue are also attached to the rectum; they were formerly supposed to be vestiges of the anal vesicles of the *Echiuroidea*. The alimentary canal—particularly the intestinal portion—is generally full of sand, excepting along a ciliated groove, which extends along its whole length.

The retractor muscles are usually four in number; they are inserted into the alimentary canal just behind the mouth, and arise from the body-wall a little behind the anus (Fig. 433).

The **nervous system** consists of a single ventral cord without ganglionic swellings, and showing no sign of being formed of two halves; it divides below the oesophagus into two cords which embrace the oesophagus and unite in a single supra-oesophageal ganglion on the dorsal side just behind the attachment of the retractor muscles. The cord and ganglion give off numerous nerves to adjacent parts. In the dorsal middle line, outside the tentacular circlet, there is an ectodermal pit which reaches as far as the supra-oesophageal ganglion; it is called the cerebral organ, and is lined by ciliated cells. In *Physcosoma* (*Phymosoma*) the inner end of this pit is bilobed and lined with cells containing a black pigment; it is embedded in the substance of the supra-oesophageal ganglion.

The **body-cavity** is entirely coelomic and very spacious; it contains a richly corpusculated fluid, which often has a pink colour. The corpuscles are of two kinds—biconcave discs, which contain the red pigment called haemerythrin, and amoeboid cells. In addition there are found floating in the coelomic fluid the reproductive cells, and in *Sipunculus* and *Phascolosoma* some peculiar ciliated structures called "*urns*"; these are apparently budded off from the coelomic epithelium overlying the dorsal blood-vessel.

The **vascular system** is closed, and contains a corpusculated fluid; it consists of a contractile vessel lying on the dorsal side of the oesophagus, and of an annular vessel surrounding the mouth. The dorsal vessel ends blindly behind, and in front opens into the circumoral vessel, which sends prolongations into the tentacles; it gives off no branches. There is often present in addition a vessel on the ventral side of the oesophagus, which also opens into the circumoral vessel. Inasmuch as this system gives no branches to the viscera, it cannot be regarded as a nutritive or respiratory organ; it is probably mainly concerned with expanding the tentacles. A vascular system is absent in *Onchnesoma* and in *Tylosoma*. There are no special organs of **respiration,** unless the tentacles are such.

The **nephridia** are two in number; they are often called the *brown tubes*, and open externally on the ventral surface in the anterior region. They project backwards in the body-cavity from their attachment to the skin, and they possess an internal opening into the perivisceral coelom at their anterior end close to the internal opening. They have glandular walls, and serve to transmit the reproductive cells to the exterior.

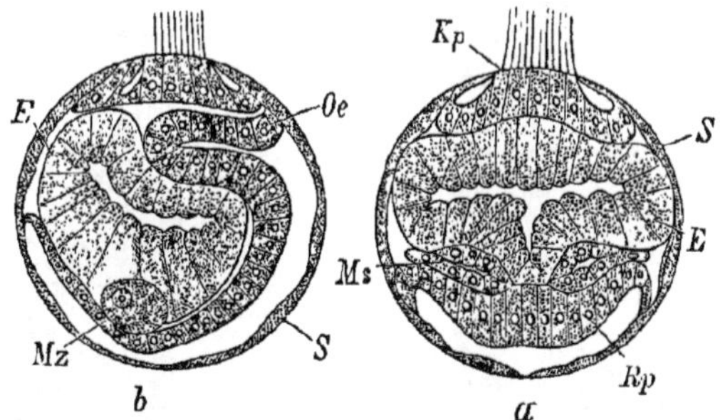

FIG. 434.—Stage of *Sipunculus* embryos, in which the cephalic and ventral plates are beginning to fuse with one another. The vitelline membrane and the cilia of the serosa which perforate it are omitted (after Hatschek). *a*, In transverse section; *Kp* cephalic, *Rp* ventral plate; *E* endoderm; *Ms* mesoderm; *S* serosa. *b*, In median longitudinal section; *Oe* oesophagus; *Mz* pole cells of the mesoderm.

The sexes are generally separate, and the **reproductive glands** are simply proliferations of the coelomic epithelium. In *Sipunculus* these proliferations form an annular thickening of the peritoneal lining at the point of attachment of the retractor muscle to the body-wall. The reproductive cells are dehisced from the generative ridge before they are ripe, and mature in the body-cavity.

Fertilization is probably external. In *Sipunculus*, the development of which was observed by Hatschek on larvae taken in the Pantano, near the lighthouse at the north end of the Straits of Messina, there is an invaginate gastrula. The blastopore closes; the mouth and anus being new formations. There are two primary mesoblast cells which give rise to two mesoblastic bands, which do not undergo segmentation. Invaginations of the ectoderm of the animal pole and ventral surface of the embryo give rise to cephalic and ventral plates respectively (Fig. 434), while the remainder of the ectoderm cells grow round these, and form an external envelope for the embryo of the nature of a serous membrane (serosa). Cilia project from the latter through the pores of the vitelline membrane, and are employed

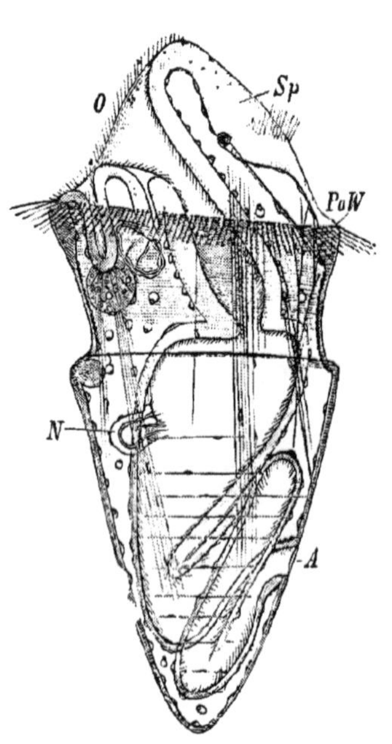

FIG. 435.—Larva of *Sipunculus* (after Hatschek). *O* mouth ; *Sp* apical plate ; *A* anus; *PoW* postoral circle of cilia ; *N* kidney.

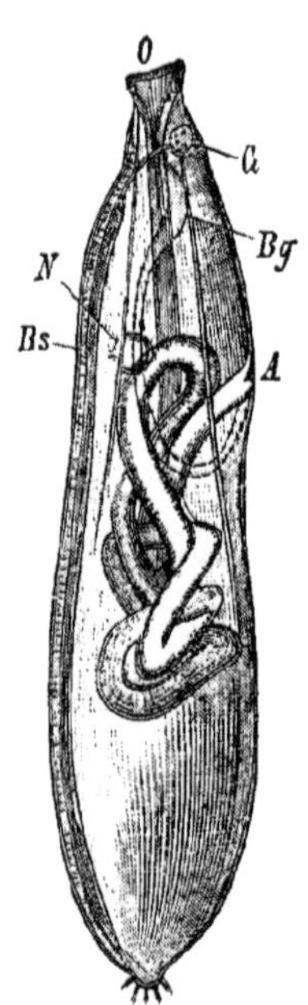

FIG. 436.—Quite young *Sipunculus* still without tentacles (after Hatschek). *O* mouth ; *A* anus; *Bs* ventral nerve cord; *N* nephridium (brown tube); *G* cerebral ganglion ; *Bg* blood vessel.

by the embryo in swimming. The cephalic and ventral plates soon grow together. The mesodermal bands soon split into somatic and splanchnic layers, between which is the coelom, and give rise to the rudiments of the two nephridia ; while the oesophagus arises as an invagination of the ectoderm of the antero-ventral region, and a postoral circlet of cilia is formed behind it (Fig. 435). The serous membrane is cast off with the vitelline membrane, and the larva then contains all the essential organs of the young *Sipunculus*, except the ventral cord and the blood vessels. At a later stage, during the growth of the larva, the ventral nerve cord is developed from the ectoderm, the circle of cilia disappears, the first tentacles sprout out at the edge of the mouth, and the metamorphosis of the free-swimming larva into the creeping young *Sipunculus* (Fig. 436) is completed. The anus is never posterior, but is dorsal at its first appearance. There is a small preoral lobe with an apical plate in the larva, which disappears in the adult, the mouth becoming terminal. It is important to notice that the anus is on the same side of the body as the preoral lobe, and is therefore truly dorsal. Whether it marks the morphological hind end of the body, as it does in *Phoronis*, or whether it is, as described in the text, on the anterior part of the true dorsal surface, cannot be certainly determined, though from its position at its first appearance it is reasonable to conclude that it is at least dorsal in position.

Fam. **Sipunculidae.** With the characters of the order.

(*a*) Longitudinal muscles of the body-wall divided into bands. *Physcosoma* Sel. (*Phymosoma* Sel.), with papillae; *Sipunculus* L., without papillae; *S. nudus* L., English Channel, North Sea.

(*b*) Longitudinal muscles continuous. *Phascolosoma* Leuckart, two nephridia, numerous tentacles; *P. vulgare* Blainv., English Channel and N. Sea; *P. elongatum* Kef.; *P. papillosum* Thompson, E. Channel; *Dendrostoma* Grube, two nephridia, 4–6 tentacles; *Phascolion* Théel, right nephridium only; *P. strombi* Mont., E. Channel.

(*c*) A distinct shield at the anus and at the hind end, or a calcareous ring round the anus. *Aspidosiphon* Grube, with both shields; *Clocosiphon* Grube, a calcareous ring in front of the anus; *Golfingia* Lankester, a preanal horny ring, and a posterior horny spike.

(*d*) Two foliaceous tentacles. *Petalostoma* Keferstein; *P. minutum* Kef., E. Channel.

(*e*) No tentacles or vascular system; one retractor muscle and nephridium. *Onchnesoma* Kor. and Dan.; *Tylosoma* Kor. and Dan.

CHAPTER XIII.

PRIAPULOIDEA.*

Unsegmented vermiform animals with an anterior terminal mouth and a posterior terminal anus. The central nervous system is not separated from the ectoderm, and the renal and reproductive organs are entirely separate from the body-cavity.

These are vermiform animals which live in sand and mud in the sea. They have a short introvert, and the mouth, which is terminal and anterior, is surrounded by chitinous teeth. The anus is terminal and posterior, and in *Priapulus* there is a hollow caudal appendage attached ventral to the anus, and covered with a number of hollow papillae. There are no tentacles. The body-wall is muscular, and the skin is transversely ridged.

The **alimentary canal** is straight, or but slightly looped, and is divided into a muscular pharynx lined with chitinous teeth, an intestine, and a rectum.

The central **nervous system** consists of a circumpharyngeal ring and a ventral cord, and is throughout in continuity with the surface ectoderm.

There are no special sense-organs.

There is a spacious **body-cavity**, which is probably a haemocoele, inasmuch as it has no connection with the renal or generative organs. There is no canalicular **vascular system**, and if the body-cavity be not a haemocoele it must be admitted that the vascular system is absent.

The **renal** and **reproductive** organs are paired, and have common ducts, which open posteriorly, one on either side of the anus. These ducts are beset with two kinds of caecal diverticula—the one generative and the other renal. The generative caeca produce the

* W. Apel, "Zur Anat. d. Priapulus caudatus u. d. Halicryptus spinulosus," *Z. f. w. Z.*, 42, 1885. E. Ehlers, "Ueb. d. Gattung Priapulus," *Z. f. w. Z.*, 11, 1861. H. Schauinsland, "Die Excret. u. Geschlechtsorgane d. Priapuliden," *Zool. Anz.*, 9, 1886.

generative cells from their lining epithelium, and the renal caeca are branched and ciliated, and terminate in pear-shaped cells, which contain a long, flame-shaped cilium projecting into the lumen of the canal. It therefore appears that in the *Priapuloidea* the generative glands are continuous with their ducts, and not connected in any way with the body-cavity; while the renal organs resemble those of the *Platyhelminthes.*

The sexes are separate, and the development is unknown.

Fam. **Priapulidae.** With the characters of the group. *Priapulus* Lam., *Halicryptus* v. Sieb., both found in the North Sea.

CHAPTER XIV.

PHORONIDEA.*

Tubicolous, hermaphrodite Coelomata, with a tentaculated, horseshoe-shaped lophophore, a dorsal anus, and one pair of nephridia.

It is convenient in the present state of our knowledge to give *Phoronis* the rank of an independent phylum of the animal kingdom. It has been relegated by various naturalists to the *Gephyrea*, the *Polyzoa*, the *Brachiopoda*, and even to the *Enteropneusta*, but to not one of these groups are its affinities close enough to justify a phyletic association.

Phoronis is a coelomate animal with a vascular system, and an ectodermal nervous system. In the adult state it is sedentary, and inhabits leathery tubes, to which particles of foreign matter, such as sand grains and sponge spicules, are often found adhering (Fig. 437). The animal can protrude the anterior part of its body from the opening of the tube, to which it appears to be but loosely attached.

FIG. 437.—Three tubes of *Phoronis psammophila* (after Cori), natural size.

A number of individuals are commonly associated together, their tubes being twisted round one another; but their bodies are not connected in any way. We are ignorant of the cause of this association of individuals; it is not known that they possess the power of budding at any stage of their existence. The tube is secreted by the animal, very possibly

* W. H. Caldwell, "On the Structure, Development, and Affinities of Phoronis," *Proc. Roy. Soc.*, 1882. *Id.*, "Blastopore, Mesoderm, and Metameric Segmentation," *Q. J. M. S.*, 25, 1885. W. C. McIntosh, "Phoronis Buskii," *Challenger Reports*, 28, 1888. W. B. Benham, "The Anatomy of Phoronis australis," *Q. J. M. S.*, 30, 1889. E. Metschnikoff, "Ueb. d. Metamorphose einiger Seethiere," *Z. f. w. Z.*, 21, 1871. *Id.*, "Vergl. Emb. Studien," *Ibid.*, 37, 1882.

by cutaneous glands in the anterior region.

At the anterior or oral end of the animal there is a horseshoe-shaped lophophore entirely attached to the oral surface (Fig. 438), no part of it projecting as in fresh-water Polyzoa (see Fig. 441). The concavity of the horseshoe is dorsal. In some forms, *e.g.* *Ph. australis*, the two ends of the horseshoe are curved inwards into a spiral. The lophophore carries two rows of hollow ciliated tentacles, between which is a groove leading into the median **mouth.** Overhanging the mouth on its dorsal side, *i.e.* between it and the inner tentacles of the lophophore, is a laterally extended flap of the body-wall—the **epistome** (Fig. 438, *ep.*). The **anus** (*an*) is dorsal and outside the lophophore, at the summit of a median longitudinal ridge (*Rr*), on each side of which there is a lateral ridge (*nr*), each bearing at its oral end a pore, the aperture of a nephridium (*neo*). The row of tentacles on the inner, *i.e.* concave side of the lophophore, is incomplete in the middle line.

The body is elongated aborally, and may attain a length of 6 inches. Development shows us that the dorsal surface* is the area between the mouth and

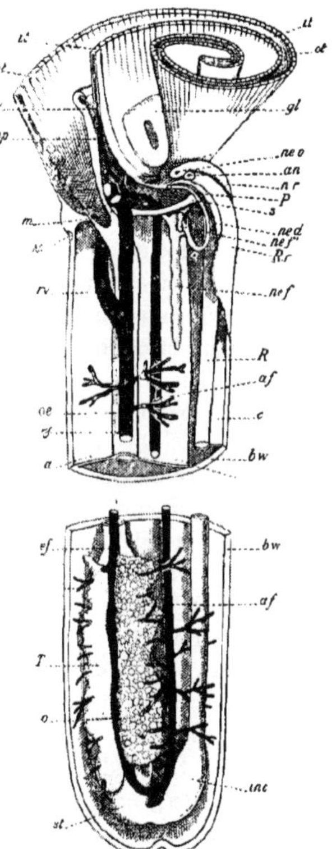

FIG. 438.—Diagram showing the anatomy of *Phoronis*; the left side of the body-wall and the left half of the lophophore is supposed to be cut away (after Benham). *a* oesophageal mesentery; *af* afferent blood vessel; *an* anus; *b* right lateral mesentery; *bw* body-wall; *c* rectal mesentery; *ef* efferent blood vessel; *ep* epistome; *gl* glandular ridge and pit; *int* intestine; *it* inner series of tentacles; *m* mouth; *N* nerve band; *n.r.* right nephridial ridge; *ne.d* left nephridial duct; *ne.f* large funnel of the left nephridium; *ne.f′* small funnel of the left nephridium; *ne.o* right nephridiopore; *o* ovary; *oe* oesophagus; *ot* outer tentacles; *R* rectum; *R.r* rectal ridge; *s* transverse septum; *st* stomach; *tv* tentacular vessel; *T* testis.

* Caldwell states that the epistome is derived from the preoral lobe of the larva, but this is inconsistent with the statement he also makes that the preoral lobe is swallowed and digested at the metamorphosis.

the anus, the whole of the aboral extension of the body being ventral. The tentacles possess within the epidermis a ring of skeletal tissue of mesoblastic origin, and contain a blood vessel and an extension of the oral coelom. The body-wall consists of an ectodermal epithelium, a basement membrane, the usual two layers of muscle, and internally the coelomic epithelium.

The **nervous system** lies in the skin immediately within the epidermis, and outside the basement membrane. There is a special concentration of it round the mouth in the form of a circumoral ring (*N*), which is enlarged on the dorsal side between the mouth and the anus into what may be called a dorsal or supra-oesophageal ganglion. This ring follows the curve of the lophophore near the base of the outer tentacles, and gives off nerves to the tentacles and kidneys. There are also two tubes lying in the skin, and extending aboralwards from the nerve ring along the insertion of the lateral mesenteries. These are probably large nerve-fibres comparable to the giant-fibres of earthworms, etc. Some punctated nervous tissue lies internally to the tube of the left side.

There are no organs of special sense.

On the dorsal side of the base of the inner series of tentacles are two ciliated pits, which have been interpreted by some observers as glandular structures, by others as sense-organs (*gl*).

The alimentary canal is a **U**-shaped structure occupying the aboral extension of the body. It presents four regions—the oesophagus; the stomach (*st*), at the base of the proximal limb of the **U**; the intestine, or second stomach (*int*); and the rectum, which leads to the anus. The walls of the stomach are glandular.

The **perivisceral cavity** is entirely coelomic; it is divided at the level of the lophophore by a transverse septum (*s*) into a small oral section, which is continued into the epistome and tentacles, and a larger posterior (aboral) part in relation with the alimentary and other viscera. This septum is perforated by the oesophagus, but not by the rectum, the anus being placed aborally to it.

The posterior (aboral) section is further subdivided by longitudinal mesenteries, which run from the gut-wall to the body-wall. One of these is a *median ventral mesentery* (*a* and *c*) attaching the outside of both descending and ascending limbs of the alimentary canal to the body-wall. Besides this there are two *lateral mesenteries* passing from the oesophagus and stomach to the body-wall (Fig. 438, *b*). The body-cavity is thus divided into three chambers—a rectal chamber lying between the right and left lateral mesenteries and containing the rectum, and a right and left chamber between the oesophageal and stomach part of the median mesentery and the right and left lateral mesenteries. As the lateral mesenteries end before the aboral apex of the body is reached, all three chambers are in communication. In addition to the mesenteries, bridles of connective tissue pass across the body-cavity from the body- to the gut-wall. The coelom contains a corpusculated fluid, in which the reproductive cells are often found floating.

The **vascular system** consists of two main longitudinal trunks, one lying in the rectal chamber between the two limbs of the alimentary canal (*af*), and the other on the left side of the oesophagus (*ef*). These vessels are contractile, and are continuous aborally; they both give branches to the gut-wall and gonads. Anteriorly they pass through the septum, and are both connected with the single vessel found in each tentacle. The blood consists of a colourless plasma, containing nucleated red (haemoglobin) corpuscles in suspension.

The **renal organs** consist of two nephridia, which also act as generative ducts and open externally on each side of the anus. Each nephridium, after passing aboralwards through the septum, opens into the body-cavity by two openings. One of these—the smaller (*ne.f'*)—is into the lateral chamber of the body-cavity, the other—and larger (*ne.f*)—into the rectal chamber.

Phoronis is hermaphrodite, and the **gonads** lie on the left side of the stomach in the aboral region of the body-cavity—the ovaries (*o*) on one side of the oesophageal blood vessel, and the testes (*T*) on the other. They consist of special developments of the coelomic epithelium overlying certain caecal branches of the oesophageal vessel. The reproductive cells are dehisced into the coelom, and pass out by the nephridia.

Development. The eggs are small, and are probably fertilized externally. They undergo the first part of their development entangled in the arms of the mother. They are then hatched out as free-swimming larvae, which after a certain period of free life undergo a remarkable metamorphosis, and acquire the adult form and habit. The larva is called *Actinotrocha*.

There is an invaginate gastrula, and the blastopore, which assumes a slit-like form, closes up behind, but remains open in front as the mouth. The anus is formed as a pit at the hind end of the closed-up portion. The mesoderm arises partly as cells budded off from the endoderm on each side, and partly as a pair of diverticula from the anal pit.

The fully-formed larva has a preoral lobe covered with cilia, a ventral mouth, and a posterior anus, and, like so many larvae, has a gelatinous, transparent appearance. The preoral lobe possesses an apical thickening of ectoderm—the larval ganglion—and in some species eye-spots.

Behind the mouth there is a circlet of ciliated larval tentacles (Fig. 439, *Lt*), and behind these again the rudiments of the adult tentacles may be discerned. On the ventral surface behind the tentacles an invagination of skin is formed which projects into the body of the larva. There is a circlet of long cilia round the anus, and the cilia on the margin of the preoral lobe are longer than the rest, constituting a velar or preoral ring. It is this ring, combined with the preoral lobe and apical plate, which has led naturalists to regard *Actinotrocha* as a modified trochosphere larva. But this type of larva is found in so many widely divergent groups (*e.g. Mollusca, Annelida, Echinodermata, Enteropneusta*), that too much attention must not be paid to it as a sign of phyletic affinity. In this connection it should be noted that *Actinotrocha* differs from the trochosphere of Molluscs and Annelids in possessing a coelom in the form of two pairs of mesoblastic sacs which not only

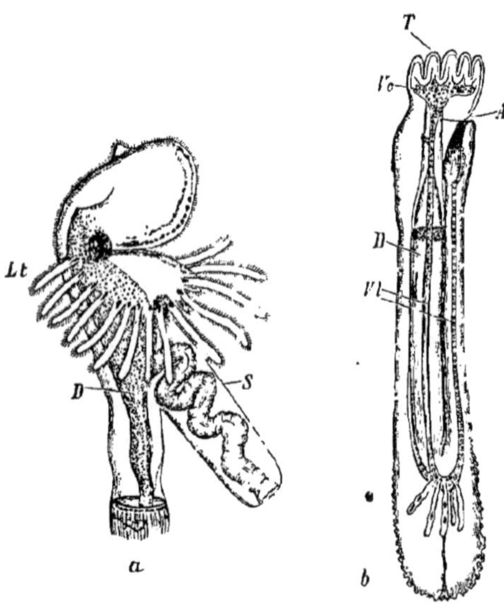

FIG. 439.—*a*, Metamorphosing *Actinotrocha* with ventral invagination evaginated at *S* (after Schneider). *a*, anus with circle of cilia; *D* intestine; *Lt* larval tentacles. The preoral lobe is the part of the larva above *Lt* (in the position of the drawing). It overhangs the mouth, which is on the right hand side of it in the drawing. *b*, Young *Phoronis* (after Metschnikoff). *D* intestine; *A* anus; *T* adult tentacles; *Vc* circular blood vessel; *Vl* longitudinal blood vessels.

constitute a body-cavity for the larva, but also give rise to the body-cavity of the adult *Phoronis*.

After a certain period of free life the larva sinks to the bottom, and undergoes a sudden change of form and character which constitutes the metamorphosis. The ventral invagination becomes evaginated, and forms a tubular projection standing out from the ventral surface almost at right angles to the long axis of the larva (Fig. 439, *a*). Into this projection, which will constitute the body of the adult, and which soon becomes larger than the larval body, the alimentary canal passes and forms the **U**-shaped loop characteristic of the adult. At the same time the hind part of the larval body, or, as we may now call it, the anal papilla, becomes less and less conspicuous, and eventually reduced to the anal projection of the adult. While these changes are going on, the preoral lobe with its ganglion and sense-organs passes into the stomach by the oesophagus and is digested. The larval tentacles likewise pass into the stomach. Meanwhile the adult tentacles have become fully formed, attachment effected by the aboral apex of the new body and the adult form acquired.

Phoronis is exclusively a marine animal, and about 6 species are known.

Phoronis S. Wright; *Ph. hippocrepia* S. Wright, embedded in coral or limestone, Devonshire Coast, Firth of Forth; *Ph. kowalevskii* Caldwell, tubes coated with sand, Mediterranean.

It has been suggested by Caldwell and others that *Phoronis* has affinities to the *Polyzoa* and to the *Brachiopoda*. According to this view in all these animals the dorsal surface of the body is reduced to the space between the mouth and the anus, and the preoral lobe of the larva almost or entirely disappears in the adult. There is much to be said for this view, which is fully discussed under the two groups concerned (pp. 571 and 584).

A. T. Masterman has quite recently (*Q. J. M. S.*, 40, 1897, p. 281) endeavoured to show that *Actinotrocha* has Enteropneust features. The principal data upon which he bases his view are partly new and partly old. The old are the presence of two glandular pockets opening into the anterior end of the stomach, which he identifies as a double notochord; on what substantial grounds it is difficult to see. The new data are (1) the presence of three sections of the body-cavity: (*a*) one, unpaired, situated in the preoral lobe and opening to the exterior by two pores placed on each side of the preoral ganglion, (*b*) a second chamber—which he calls collar body-cavity—divided by a dorsal mesentery but not by a ventral, and opening to the exterior ventrally by a pair of nephridial tubes; and (*c*) a posterior chamber divided by a ventral mesentery and possessing paired nephridia or the rudiments of them. (2) The presence of an ectodermal pit in front of the preoral ganglion, which he calls the neuropore; and (3) the presence of an ectodermal pit on the ventral side of the preoral lobe, the opening of which shifts into the stomodaeum, and which he calls the subneural gland.

Of these new facts the second and third derive any importance they may have in the discussion from the names which Masterman has given them. The statements under (1) are more important, but they are supported by evidence which is very insufficient when one remembers the size of the edifice which he erects upon them; *e.g.*, the external openings of his proboscis pores are not shown in section, nor is either of the openings of the collar nephridium. Further, one sees no reason for speaking of a collar region of the body; and lastly, the development of these most important spaces has not been examined. All we know on this subject is based on Caldwell's observations that there are two pairs of coelomic cavities in the embryo. The method of development of these spaces in *Actinotrocha* is a highly important one from Masterman's point of view, and should, in our opinion, have been examined by him. For, in the *Balanoglossus* larva, the collar coelom and the trunk coelom are paired, whereas the spaces in *Actinotrocha* with which he compares them are not paired, but, however they may arise in the embryo, are continuous in one case dorsally and in the other ventrally to the alimentary canal.

CHAPTER XV.

PHYLUM POLYZOA.*

Small animals usually united together in colonies. With ciliated tentacular crown, U-*shaped alimentary canal, and simple ganglion. With coelom, but without vascular system. Asexual reproduction by budding always found.*

The *Polyzoa* (J. V. Thompson) or *Bryozoa* (moss-like animals) as they are sometimes called (Ehrenberg), are with rare exceptions (*Loxosoma*) colonial in habit. The colonies, which may present a superficial resemblance to colonies of Hydroids, may have a foliaceous or dendritic appearance, or they may form crusts on the surface of foreign objects. The erect forms may have a calcareous ectocyst, and be rigid, in which case owing to their brittle character they are not found between tide-marks; or they are flexible. Flexibility may coexist with a calcified ectocyst, but in this case there are horny joints at intervals. In such jointed colonies the part between any two joints is an *internode.*

The individual zooids of the colony (Fig. 440) are small polyp-like organisms, and usually possess a horny, or coriaceous, frequently calcareous, rarely gelatinous exoskeleton, which is really the cuticle of the zooids, and is secreted by the ectoderm. This cuticular layer is called the **ectocyst,** and the case formed by it is the **cell.**

The soft part of the body-wall, which consists of ectoderm and mesoderm, lies close beneath the ectocyst, and is called the **endocyst.** The ectocyst and endocyst together constitute the **zooecium**; while

* T. Hincks, "*A History of the British Marine Polyzoa,*" London, 1880, 2 vols. G. J. Allman, *Monograph of Fresh-water Polyzoa,* Ray Society, 1856. G. Busk, "Report on the Polyzoa," *Challenger Reports,* vols. 10 and 17, 1884 and 1886. K. Kraepelin, "Die Deutschen Süsswasser-Bryozoen," *Abh. Naturwiss. Ver. Hamburg,* Bd. x., 1887, and Bd. xii., 1892. E. C. Jelly, *Synonymic Catalogue of the Recent Marine Polyzoa,* London, 1889. S. F. Harmer, "On the Nature of the Excretory Processes of Marine Polyzoa," *Quart. J. Mic. Sci.,* vol. 33; and "On the Occurrence of Embryonic Fission in Cyclostomatous Polyzoa," *Ibid.,* vols. 34 and 39, 1893 and 1896. S. F. Harmer, "*Polyzoa,*" in the *Cambridge Natural History,* London, 1896.

the alimentary canal and tentacles are called the **polypide** (for origin of this absurd nomenclature see below, p. 555). The cell at, or near, its anterior end has an opening called the *orifice*, through which the anterior part of the zooid can be protruded and retracted. This anterior invaginable part of the body, which carries a crown of tentacles, is also covered by cuticle, which is continuous with the stiffer ectocyst at the orifice, but distinguished from the latter by its softness and flexibility. At the orifice the body-wall (with its soft cuticle) is invaginated inwards, and passes thence on to the anterior and extrusible part of the body. In many *Polyzoa* this reduplication of the body-wall is present even when the zooid is protruded (Fig. 440). The greater part of the anterior region of the body, with its crown of tentacles, can however be protruded from the cell and retracted into it again by special muscles—the *parieto-vaginal* and the *retractor* muscles—which traverse the body-cavity (Fig. 440). That part of the body-wall which encloses the space in which the tentacles lie when completely retracted is called the *tentacle-sheath*.

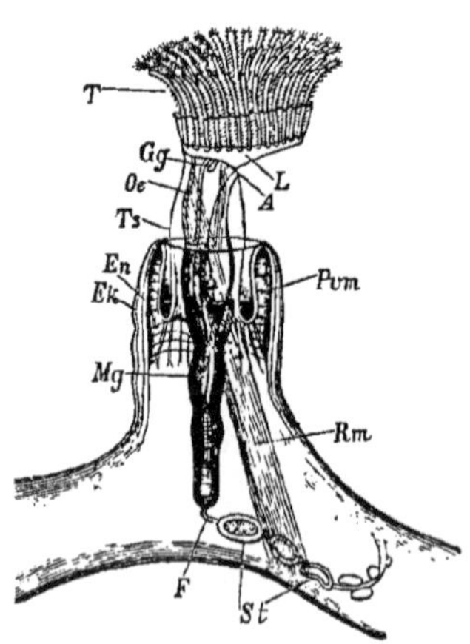

FIG. 440.—*Plumatella repens* (after Allman). *T* tentacles; *L* lophophore; *Oe* oesophagus; *A* anus; *Mg* stomach; *F* funiculus; *St* statoblasts; *Ts* tentacle sheath; *Ek* ectocyst; *En* endocyst; *Gg* ganglion; *Pvm* parieto-vaginal muscles; *Rm* retractor muscle.

In the *Cheilostomata* there is a movable chitinous lid, the *operculum*, which can shut down over the orifice. Sometimes the ectocyst is raised into a ridge—the peristome—round the orifice: the tube thus formed is the *secondary orifice*. **Zoarium** is the term applied to the whole colony. The term **ooecium** or **ovicell** is applied to the receptacles in which the ova undergo their development. The ooecia are of very different kinds: in the *Cheilostomata* they are merely pouches of the *zooecia* into which the eggs pass. In some—if not all—*Cyclostomata* (*Crisia*) they are special zooecia in which the polypide is rudimentary or degenerates. This is probably the case with all ovicells in the *Cyclostomata*, but it has not been proved in all cases. In *Phylactolaemata*, in which the body-cavities of the zooecia are in wide communication, a rudimentary polypide bud is formed near the ovum. This passes round the egg and invests it, but remains rudimentary. In the *Ctenostomata* the eggs develop in the tentacular sheath or in the sea.

The body-wall consists of cuticle, ectoderm, and a delicate layer of cells which line the body-cavity. The peritoneal cells are ciliated in the *Phylactolaemata*, in which group there is also a thin muscular layer in the body-wall.

The mouth is placed at the anterior end of the body in the midst of the circlet of ciliated tentacles, and the disc bearing it and the tentacles is called the **lophophore** (Fig. 440). The lophophore is either circular (*Gymnolaemata*), or is drawn out into two lobes (horseshoe-shaped lophophore, Fig. 441, *Phylactolaemata*), and the tentacles are set along its edge. The latter are simply hollow processes of the body-wall; they are provided with longitudinal muscles, and their cavity communicates with a circular canal which surrounds the body at their base. This canal communicates with the body-cavity in the *Phylactolaemata*, but is separate from it in the *Gymnolaemata*. The tentacles serve both for procuring food (setting up by means of their cilia whirlpools in the water) and for respiration.

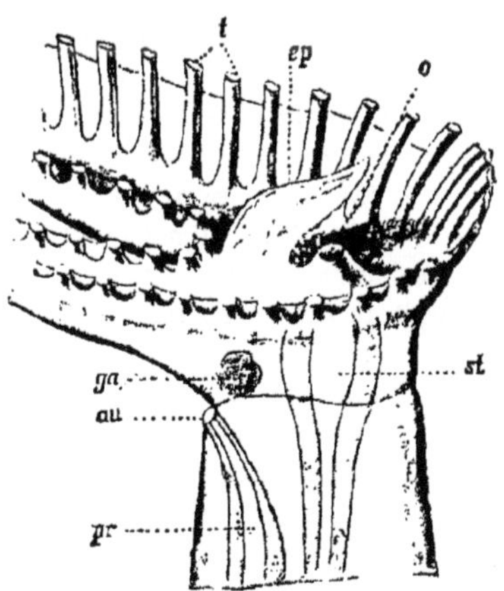

FIG. 441.—Anterior part of the body of *Lophopus* (from Lang, after Allman) from the right side. *t* tentacles cut off near the base; *o* mouth; *ep* epistome; *st* pharynx; *ga* ganglion; *au* anus; *pr* rectum.

The **digestive organs** lie freely in the body-cavity, and are attached to the body-wall by a mesodermal reticulum which traverses the body-cavity. The term **funiculus** is applied to one or two special strands (Fig. 440, *F*) of this tissue. The mouth is placed in the centre of the circular or horseshoe-shaped lophophore, and a movable epiglottis-like process, called the **epistome**, is in the fresh-water forms placed on the dorsal side of it, and projects over it.

The alimentary canal is bent on itself, and consists of (1) an elongated ciliated oesophagus, often dilated at its upper end into a muscular pharynx; (2) a spacious stomach with a blind backward prolongation, the hind end of which is attached to the body-wall by the funiculus; and (3) a narrow intestine which is bent nearly parallel with the pharynx, and is directed upwards. The intestine opens by the anus, which is placed either within the tentacular circlet (*Entoprocta*), or near but outside it on the body-wall (*Ectoprocta*, Fig. 440). The anal side of the body is called the **dorsal** side. In the *Phylactolaemata* the epistome and concavity of the horseshoe lophophore are dorsal.

The **nervous system** consists of a ganglion placed on the oesophagus between the mouth and the anus (Fig. 441, *ga*). It lies in the space (ring-canal of lophophore) already described (p. 551), which surrounds the oesophagus and is developed as a part of the body-cavity. In the *Phylactolaemata* the ganglion is placed in the concavity of the lophophore, and is attached to the oesophagus by a delicate circumoesophageal ring: it contains a cavity which arises as an ectodermal pit in the bud, and it sends off numerous nerves to the tentacles and oesophagus. According to Fr. Müller there is in *Serialaria* (*Zoobotryon*) a so-called colonial nervous system which connects the individual zooids of a colony, and enables them to coordinate their activities. Claparède describes the same in some forms. Special organs of sense have not been recognized.

Heart and **vascular system** are absent. The **body-cavity** is absent in the *Entoprocta*. In the *Ectoprocta* it is spacious and filled with fluid, and its lining gives rise to the generative cells; it is therefore to be regarded as coelomic, which view of it is justified by the *intertentacular organ* of the *Ectoprocta*. This is a fine ciliated tube leading outwards from the body-cavity, and opening externally between two of the tentacles on the anal side of the lophophore. It is not found in all specimens, and, according to Prouho, only in individuals containing an ovary. It functions as an oviduct, and probably also as an outlet for the sperm.

The body-cavity is divided, as stated above, into two parts, (1) the general body-cavity and (2) the *ring-canal* of the lophophore which is prolonged into the tentacles. The ring-canal is completely shut off from the general body-cavity in the *Gymnolaemata*, but in the *Phylactolaemata* the two are in communication on the dorsal (anal) side of the lophophore except in the middle line, where the ring-canal is cut off from the general body-cavity by the prolongation of the latter into the epistome. For the epistome of the *Phylactolaemata* is hollow and contains a prolongation from the general body-cavity which is continued upwards into it as a tube which passes on the anal side of the ganglion, between the ganglion and the median dorsal part of the ring-canal. In the *Phylactolaemata* this median dorsal part of the ring-canal communicates with the exterior by a pore which together with some columnar ciliated epithelium of the adjacent parts of the ring-canal constitutes the so-called **nephridium.*** In some *Gymnolaemata* the **intertentacular organ** opens externally in a corresponding position, and internally into the body-cavity. Into what part of the body-cavity the intertentacular organ opens is not clear from the accounts available, but presumably into the general body-cavity, for it is said to form an escape-duct for the ova.

The zooids are usually hermaphrodite. The **testes** are developed either on the upper part of the funiculus, or near the point of attachment of the latter to the body-wall. The **ovaries** are placed on the body-wall near the anterior end of the zooecium. Both kinds of generative cells are dehisced into the body-cavity. There do not appear to be any special **renal organs,** unless the intertentacular organ be such, and the fine ciliated canal of the *Entoprocta.*

Many forms of *Polyzoa* present examples of a well-marked polymorphism. In *Zoobotryon* and its allies the joints of the stalk represent a special form of individual; they have a considerable size and a simplified organization, and serve as the ramified substratum on which the nutritive zooids are placed. In addition there are here and there joints of the roots, which, under the form of tendril and stolon-like processes, serve to attach the colony. The stalk of *Kinetoskias* is a modified stalk-zooid. In addition to these stem- and root-forming zooids, there are the peculiar **avicularia** (Fig. 442) and **vibracula** (Fig. 443). These are only found in the *Cheilostomata.* They are both modifications of an ordinary zooecium, with the operculum as the mandible in the one case, and as the lashing filament in the other. A fully-developed avicularium resembles a bird's head; the upper beak is hooked, and the lower one, or mandible, is spiked. There is a strong adductor muscle arising in the "head" of the avicularium, and inserted into the

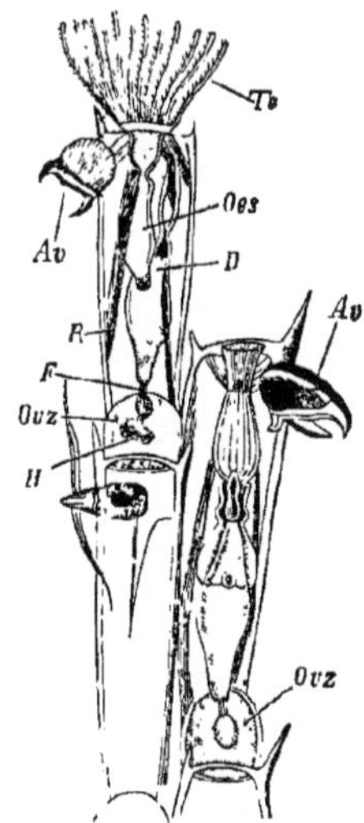

FIG. 442.—*Bugula avicularia* (after v. Nordmann). *Te* tentacular crown; *R* retractor muscle; *D* alimentary canal; *F* funiculus; *Av* avicularia; *Oes* oesophagus; *Ovz* ovicells.

* Verworn, *Z. f. w. Z.*, 46. Braem, *Bibliotheca Zoologica*, Heft 6. Cori, *Z. f. w. Z.*, 55. Oka, *Journal Coll. Sci. Japan*, 8.

middle of the mandible; the contraction of this muscle causes the lower jaw to close upon the upper, and there are two small muscles which bring about the opening.

There are all stages between such an avicularium as that just described, and a small zooecium with a movable operculum. From *Flustra foliacea*, *Cellaria*, etc., in which the avicularium occupies the place of an ordinary zooecium (vicarious avicularia), we pass through stages in which the avicularium, though still a recognizable zooecium, has lost the place of a zooecium in the colony, and is attached to some part of an ordinary zooecium (adventitious) until we reach the highly specialized form just described.

Vibracula (Fig. 443) are merely avicularia with the beak much elongated, as a long whip-like seta, which has the power of sweeping through the water. The function of avicularia is obscure. They have been observed to seize small organisms, *e.g.* worms, and to hold them until they are dead; the decomposing organic remains may possibly be swept into the mouth by the currents caused by the cilia of the tentacles.

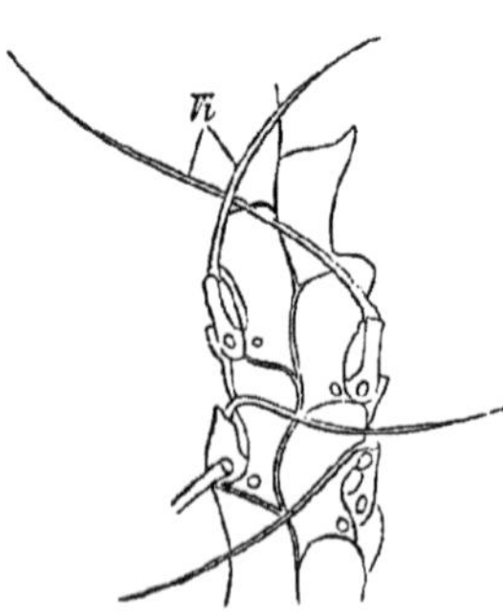

FIG. 443.—*Scrupocellaria ferox* (after Busk). *Vi* vibracula.

Finally there are the ovicells of some *Cyclostomata* (see above, p. 550), which are zooecia with a vestigial polypide.*

The form of the cells, and the manner in which they are connected together, are very different in the different groups, and give rise to a great variety in the form of the colonies. The zooecia are usually completely cut off from one another, though the soft parts are continuous through pores in the ectocysts of adjacent zooecia (*rosette plates, communication plates*). In the *Phylactolaemata* the body-cavities of neighbouring zooecia are in open communication. These two conditions result from the manner of budding: in the first case the bud, which arises as an outgrowth of the body-wall, becomes completely cut off by an ingrowth of body-wall; while in the second no such partition is formed, so that the body-cavity of the whole colony is a single, continuous space.

The **brown bodies**, which are present in most colonies of the *Polyzoa*,

* In some Cyclostomes (*Lichenopora*, etc.) the polypide is fully formed and functional in the young ovicells.

consist of brown pigment masses contained in the zooecia, and are derived from the breaking down of the tentacles, alimentary canal, and nervous system of the polypides. A new set of these organs may be formed from the persistent body-wall of such partially degenerated polypides, as an internal bud. The brown body finally breaks up, and its pigments may pass into the new stomach and so out by the anus. It is this degeneration of the tentacles, digestive organs, etc., and the subsequent acquisition of a new set by growth from the body-wall, which gave rise to the idea that the zooid of a Polyzoan colony really consists of two individuals—one the polypide, and the other the zooecium or house of the polypide, which has the power of budding new polypides (tentacles, alimentary canal, etc.).

The funicular tissue contains numerous transparent cells with processes, by means of which they are suspended in the funicular network. They may be faintly yellow in colour, and are regarded as excretory * in function. As there is no organ for the ejection of such excreted matters to the exterior, it is supposed that the disruption of the internal organs resulting in the formation of the brown bodies, and the subsequent ejection of the latter through the alimentary canal of the new polypide, are the means by which excretory products are removed from the colony.†

Reproduction is partly sexual and partly asexual; in the latter case it may be effected by the so-called *statoblasts* or by budding. The generative cells are products of the coelomic lining (*vide* p. 553), and are dehisced into the coelom.

The name **statoblast** (Fig. 444) was given by Allman to certain peculiar reproductive bodies which were formerly regarded as hard-shelled winter eggs, but by him were recognized to be multinuclear and of the nature of internal buds. These statoblasts are found only in the *Phylactolaemata*: they arise from masses of cells which appear, mainly towards the end of summer, on the funiculus (Fig. 440). They usually possess a lens-like, biconvex form, and are enclosed by

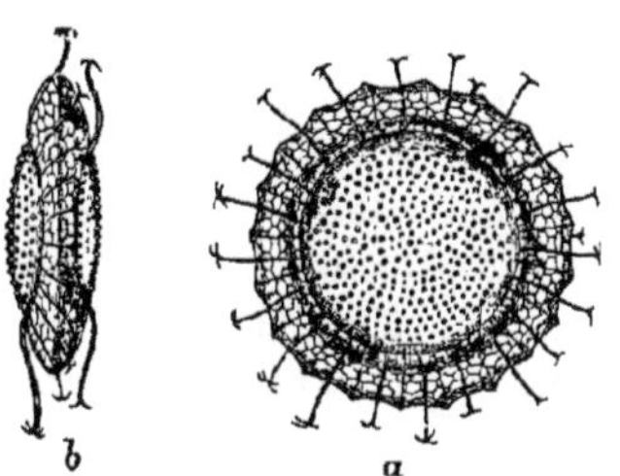

Fig. 444.—Statoblasts of *Cristatella mucedo* (from Allman). *a*, from the surface *b*, from the side.

* S. F. Harmer, "On the Nature of the Excretory Processes of Marine Polyzoa," *Q. J. M. S.*, 33, p. 123.

† *Cf.* Tunicates, in which there is no apparatus for voiding the excretory concretions which are stored in various parts of the body.

two watchglass-shaped, hard chitinous shells, the edges of which are usually enclosed by a flat ring formed of cells containing air (float), and sometimes provided with a crown of projecting spines (Fig. 444). The *Phylactolaemata* generally die down in the winter, and the statoblasts, which germinate in spring and give rise to new colonies, serve the purpose of perpetuating the species. In the fresh-water Ctenostomes, *Paludicella* and *Victorella*, the colonies also die down in winter, and the species is continued by means of certain external buds, which are arrested in development and last through the winter: they are called **hibernacula**. In some cases (*Cristatella*, *Lophopus*) parts separated off from the colony are able to develop into new colonies, and in *Lophopus* the colonies divide spontaneously.

The *Polyzoa* are for the most part marine organisms. They are found between tide-marks, and on the floor of the ocean to a depth of 3000 fathoms. *Cheilostomata* only are recorded from depths of

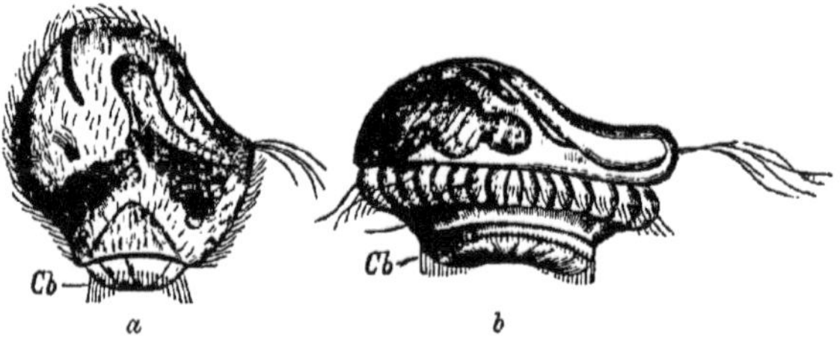

FIG. 445.—*a*, Larva of *Scrupocellaria reptans*. *b*, Larva of *Schizoporella* (after Barrois). *Cb* ciliated disc.

over 3000 fathoms; *Cyclostomata* are found in depths of 1400 fathoms and less. Ctenostomes are mostly found in depths of less than 40 fathoms.

One order—the *Phylactolaemata*—is exclusively fresh-water, and a few fresh-water forms are known in the *Gymnolaemata*. A few forms are phosphorescent. Some forms (*Terebripora*) excavate passages in the shells of Mollusca; and *Hypophorella* is found in passages in the tubes of *Lanice* and *Chaetopterus*.

The eggs of the *Ectoprocta* are sometimes laid and develop freely in the sea (*Hypophorella expansa*, *Alcyonidium albidum*, *Membranipora pilosa*, *Farrella repens*). But more often the early stages of development take place under the protection of the parent; either in the tentacle-sheath (some *Ctenostomata*), or, in many *Cheilostomata*, in a pouch of the zooecium; or in special zooecia (*ooecia*), of which the polypide is rudimentary or degenerate (*Cyclostomata*); or in a rudimentary polypide bud (*Phylactolaemata*, see p. 551). It is not certain where fertilization is effected, but it is generally described as taking place in the body-cavity, and the eggs or larvae escape either by the intertentacular canal, or by the rupture of the body-wall of the parent, or by the degeneration of its polypide. In *Alcyonidium duplex* the sexual zooecia possess two polypides; the first of these produces spermatozoa and then undergoes disruption into a brown body, whereupon a second polypide is formed which produces ova.

In the oviparous forms, to whatever sub-order they belong, in which the early development takes place in the sea, a free-swimming larva with a functional alimentary canal is rapidly produced. This larva has, generally speaking, always the same form, and is called *Cyphonautes*. On the other hand, in all forms in which the early stages are passed through under the protection of the parent, the larva is without a functional alimentary canal, and the other larval organs present varying degrees of development in different cases.

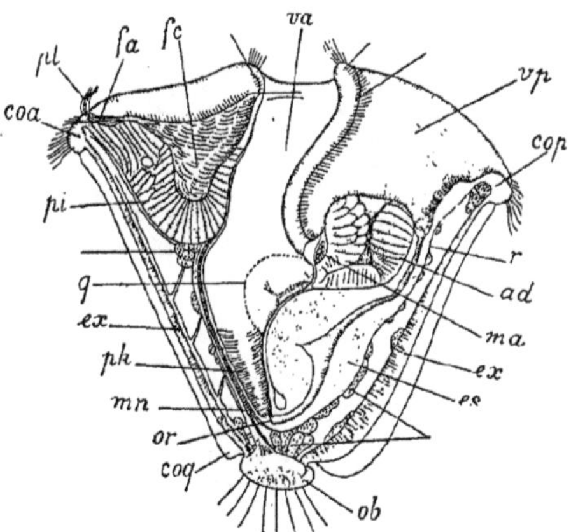

FIG. 446.—Optical section of a *Cyphonautes* (larva of *Membranipora* or of *Alcyonidium albidum*). (From Perrier, after Prouho.) *ad* adhesive organ; *coa* anterior, *cop* posterior part of the ciliary ring (velum); *coq* chitinous valves of the shell; *ex* ectoderm; *es* stomach; *fa* anterior pit; *fc* cavity of pyriform organ; *ma* adductor muscle of the shell-valves; *mn* musculo-nervous tract; *ob* aboral disc; *or* entry into oesophagus; *ph* pharynx; *pi* pyriform organ; *pl* vibratile tuft of pyriform organ; *q* horns of the adhesive organ; *r* rectum; *va* anterior, *vp* posterior chamber of vestibule.

Cyphonautes (Fig. 446) has the form of a laterally compressed bell with a circle of cilia round the base. The base of the bell is the ventral surface; it has two openings both enclosed by a special lobe of the ciliated ring. The larger and posterior of these leads into a depression called the vestibule; the smaller and anterior into an ectodermal depression called the pyriform organ (*pi*). The vestibule is divided into two parts, the anterior chamber (*va*), at the bottom of which is the mouth, and a posterior chamber (*vp*), into

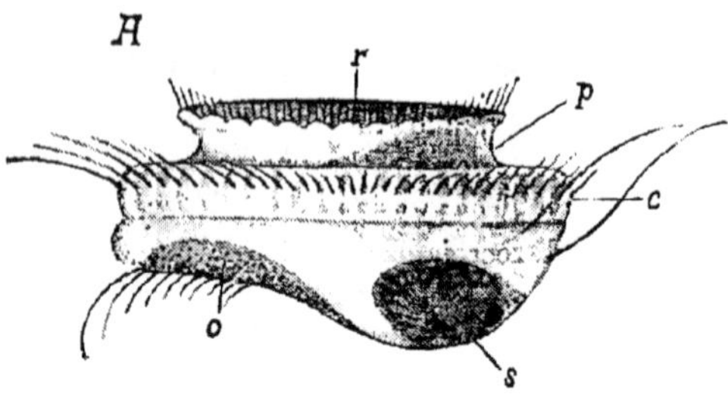

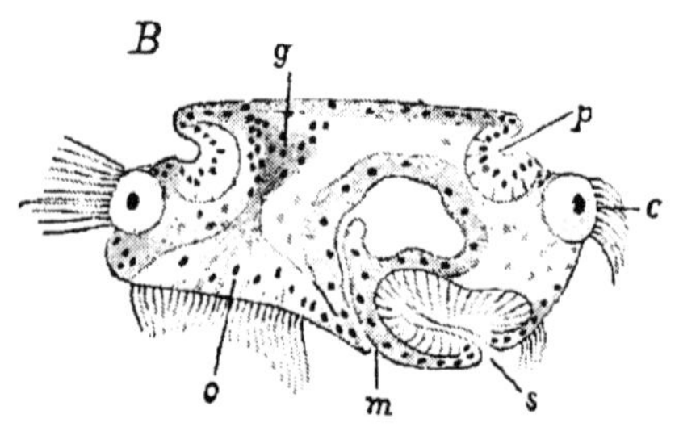

FIG. 447.—*A*, larva of *Alcyonidium mytili* (after Barrois). *B*, diagram of longitudinal section through the larva of *Alcyonidium* (after Harmer), both from Korschelt and Heider. *c* cells of the velum; *g* ganglion (? cerebral) in connection with the ciliated disc and pyriform organ *o*; *m* mouth; *p* mantle-groove; *r* ciliated disc; *s* adhesive organ (sucker). In these figures the oral surface is directed downwards, whereas in Figs. 445 and 446 it is upwards.

which the rectum (*r*) opens. The alimentary canal consists of the pharynx (*ph*), oesophagus, stomach (*es*) and rectum (*r*). Just in front of the anus on the floor of the vestibule there is an invagination of thickened ectoderm; this is the adhesive organ (*ad*), by which the larva attaches itself. There are two shell-plates placed right and left on the aboral surface. They

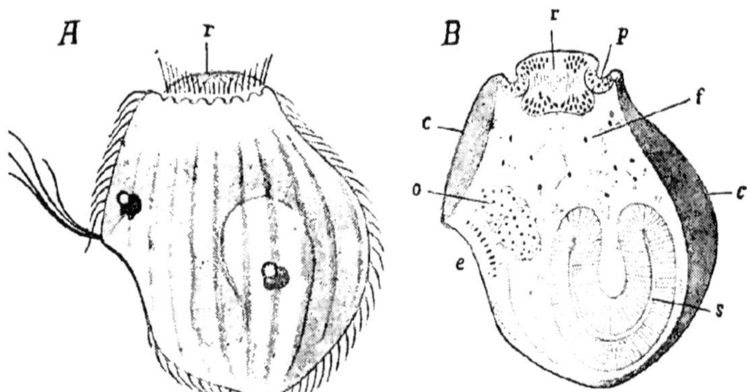

FIG. 448.—*A*, larva of *Bugula plumosa* (after Barrois). *B*, diagram of median section through a larva of *Bugula* (after Vigelius), both from Korschelt and Heider. *c* cells of velum; *e* groove of pyriform organ; *f* internal tissue of the larva; *o* pyriform organ; *p* mantle-groove; *r* ciliated disc; *s* adhesive organ or sucker.

meet in front and behind, but gape ventrally. At the aboral apex there is an opening between the two valves, through which a ciliated disc (*ob*) projects.

In the larva of *Alcyonidium mytili* (Fig. 447) we find all the parts present in *Cyphonautes*, except the bivalve shell. But it must be noted that the alimentary canal is partly aborted, there being no rectum or anus. Also the aboral structure (*r*) which corresponds to the ciliated disc is larger, and surrounded by a groove called the mantle-groove (*p*).

The larva of *Flustrella* is intermediate between the above forms; it has the shell plates but not the complete alimentary canal.

Finally, in some *Cheilostomata* the specialisation of the larva is carried still further in the great development of the ciliary circle, and in the absence of an alimentary canal. In these larvae (Fig. 448) there is a pyriform organ (*o*) with its tuft of cilia; an adhesive organ (*s*), an aboral, ciliated disc (*r*) surrounded by a mantle-groove (*p*), but there is no trace of alimentary canal, mouth, or anus.

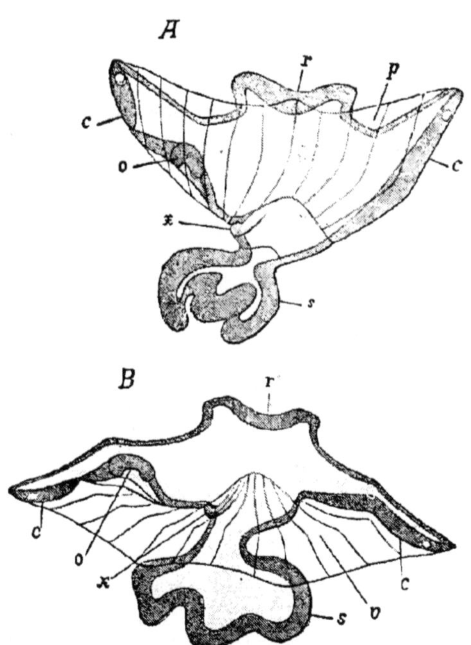

FIG. 449.—Metamorphosis of the larva of *Schizoporella* (?) *unicornis* (after Barrois, from Korschelt and Heider). *A*, first stage of the metamorphosis, showing the evagination of the adhesive organ *s*. *B*, next or so-called umbrella stage; the velum has spread out and become bent towards the sucker, giving the umbrella shape. *c* cells of velum; *o* pyriform organ; *p* mantle-groove; *r* ciliated disc; *s* evaginated sucker; *v* vestibule or concavity of umbrella—the edge of the umbrella eventually applies itself to the sucker, and the inner walls of the umbrella fuse and disappear; *x* unknown organ.

After a certain duration of free life, the object of which is to distribute the species, the larva comes to rest and attaches itself to some foreign object. The attachment is effected by the ventral adhesive organ, which is evaginated and applied to the substratum (Fig. 449). At the same time the larval organs are in process of disappearance. The alimentary canal, if present in the larva, totally disappears, and is in no way concerned in the formation of the digestive tube of the adult. The pyriform organ and the velum share in the general fate, and the larva becomes reduced to a layer of epithelium surrounding a central mass of cells and broken-down larval organs.

The attached organism is now called the primary zooecium, and the degenerated larval organs in its interior constitute the first brown

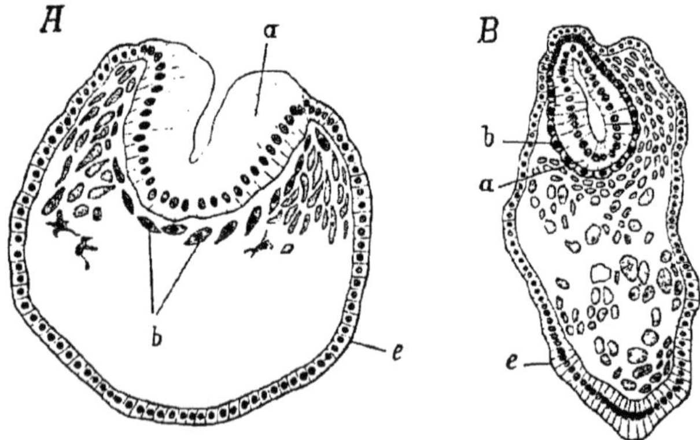

FIG. 450.—Two stages of the primary zooecium, showing the development of the first polypide of *Bugula calathus* (after Vigelius, from Korschelt and Heider). *A*, invagination of ectoderm for the formation of the vesicular rudiment of the first polypide; *b* cells of the mesoderm of the future polypide; *e* ectoderm. *B*, somewhat older stage, showing the two-layered vesicular rudiment of the first polypide; *a* ectoderm lining the vesicle; *b* mesoderm. In *A* the internal mass resulting from the degenerated larval organs is omitted.

body. The next stage consists in the invagination inward of the aboral ciliated disc* (Fig. 450 *A*, *a*); the pit so formed becomes converted into a vesicle (Fig. 450 *B*), round the walls of which some of the internal cells arrange themselves. These latter give rise to the mesoderm of the future polypide, while from the lining cells of the vesicle, which are ectodermal in origin, are developed

* According to Prouho the aboral disc, after invagination, shares the fate of the other larval organs, and the polypide rudiment is derived from an invagination of ectoderm of the primary zooecium, or it may be a new formation possibly from the internal tissue.

the lining of the alimentary canal, the ganglion, and the covering of the tentacles and the tentacle-sheath (Fig. 451).

The muscles, funiculus, and somatic mesoderm of the animal are derived from other cells of the internal mass, while the body-cavity comes from the space in the same mass.

In the *Phylactolaemata* the embryo develops in the brood pouch into a vesicle, the outer wall of which is ectoderm, the inner mesoderm, and the cavity coelom. The larva which issues, either by the degeneration of the polypide or the opening out of the zooecium, is covered with cilia by means of which it swims. It is the primary zooecium: during its free-swimming life it develops a complete polypide, or it may be two polypides (*Plumatella*), or even a greater number (*Cristatella*), before fixation takes place; thus differing from the primary zooecium of the *Gymnolaemata*, which does not develop the polypide until after fixation. As in other forms, an invagination occurs, from the walls of which the polypide is developed, *i.e.* the lining of the alimentary canal,* the covering of the tentacles and tentacle-sheath, and the nerve ganglion are alike derived from so-called ectoderm.

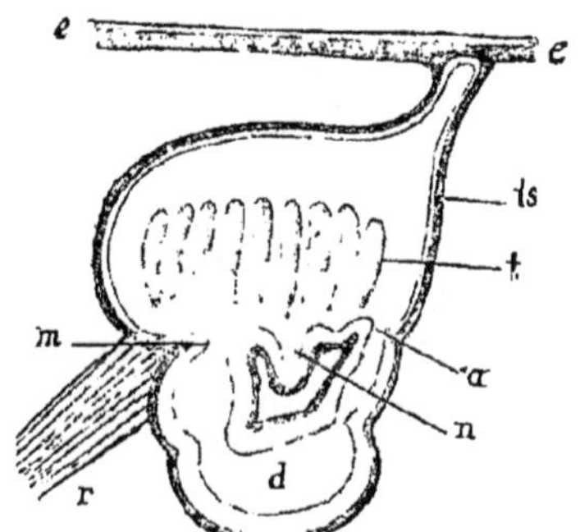

FIG. 451.—Diagram showing the developing first polypide in connection with the wall of the primary zooecium. *a* anus; *d* rudiment of stomach; *ee* wall of primary zooecium (endocyst); *m* mouth; *n* rudiment of ganglion; *r* retractor muscle; *t* tentacles; *ts* tentacle-sheath. (From Korschelt and Heider.)

From the foregoing account a certain number of conclusions may be drawn. (1) The free-swimming larva of the *Ectoprocta* is a trochosphere with a preoral ciliated ring, and an apical sense-organ—the aboral disc.† (2) The larva attaches itself by its oral face, and becomes the primary zooecium. (3) The larval organs disappear in the primary zooecium. (4) The formation of the adult organs (except the body-wall) is either to be looked upon as a case of

* In a recent work on the development of *Plumatella* (Leuckart and Chun's *Bibliotheca Zoologica*, heft 23, 1897) Braem states that endoderm tissue can be made out in the embryo.

† It is not quite certain that the aboral disc is the apical plate, but if Harmer's observation that it is connected with nerve tissue (Fig. 447) is correct, and further, if Prouho is right in describing its degeneration in the primary zooecium (Note, p. 560), this conclusion may reasonably be drawn.

budding from the primary zooecium, or as a process of metamorphosis in which the larval organs undergo degeneration and new formation in consequence of the extensive changes compressed into the metamorphosis. Until this point is settled it is impossible to relate with certainty the adult organs and surfaces to those of the larva, because of the extensive degenerative changes and new formations which take place at the metamorphosis.

If the formation of the first polypide is a case of budding, we ought, in determining the homologous surfaces of the primary zooecium and first polypide, to attach no more importance to the fact that apparently the oro-anal surface of the first polypide is derived from the aboral surface of the larva, because it is formed from it, than we should to the fact that in a bud developed from the dorsal surface of the adult, the oro-anal surface of the budded form is derived from the dorsal (aboral) surface of the parent. Consequently the argument of Korschelt and Heider, which is based upon this consideration, and from which they deduce most far-reaching consequences, that the oro-anal line of *Cyphonautes* and other Ectoproct larvae is really dorsal — has no value. Moreover, on the same hypothesis, the fact that in the formation of the first polypide the ectoderm of the primary zooecium apparently gives rise to the alimentary canal—has no more importance than a similar phenomenon would have in ordinary gemmation, in which the value of the layers is not the same as in embryonic development.

It must be noted here that in the embryonic origin (not described in this work) of the larvae of the *Ectoprocta* without alimentary canal, the endoderm is, in some cases, not formed at all; the internal tissue being, to judge by its fate, entirely mesodermal, and any cavity in it entirely coelomic (*vide* especially the larva of *Phylactolaemata*). The endoderm appears only in the first bud, and is in that derived from ectoderm.

Finally, to sum up the matter, if the larva is a trochosphere and the ciliated disc an apical plate, and if we regard the formation of the first polypide as a case of budding from the primary zooecium, which is the metamorphosed larva; and if moreover there is in the *Polyzoa*, as there is in other animals, an homology between the mouth and anus and ventral surface of an organism produced by budding and the corresponding structures in the parent organism; then it follows that the oro-anal line of the *Ectoprocta* is ventral, and the ganglion a suboesophageal ganglion, and to be compared to the suboesophageal ganglion of other types: that is to say, it is not dorsal as might be surmised from the terminology applied to the parts in the adult (see p. 552). On the other hand, it may be argued with equal force that the oro-anal surface of the adult is equivalent to the dorsal surface of the larva, for the larva attaches itself by a sucker on the ventral surface between the mouth and anus. In this case we should have to regard the formation of the first polypide as a case of larval *metamorphosis*, comparable to that of a holometabolous insect, in which the changes are so extensive that many of the larval organs degenerate completely, and are re-formed in the imago. If this view is correct the reduction of the dorsal surface and the approximation of the mouth and anus in the adult as compared with the larva, would be comparable to the phenomena which take place in *Actinotrocha* at the metamorphosis into the adult *Phoronis*. On the whole we are inclined to the view that the first polypide is formed by budding from the primary zooecium, which is the metamorphosed larva, and that consequently the ganglion of the Polyzoa is really a ventral, suboesophageal ganglion, and that the surface between the mouth and anus of the adult is the ventral surface, and comparable to the surface on which the mouth and anus of the larva are placed. In coming to this conclusion we attach great importance to the fact that the formation of the first polypide from the primary zooecium is closely similar to the budding which takes place in the formation of the adult colonies.

Before leaving the subject of the development of the *Ectoprocta*, attention must be called to a most important process of embryonic fission discovered by Harmer in the *Cyclostomata*. In this sub-order the ovicell is a zooecium with a rudimentary polypide (*Crisia*), in the funicular tissue of which the ovum arises. Fertilisation has not been observed, but the alimentary canal of the polypide grows round the egg and forms a multinucleated follicle for it. The ovum segments and becomes a multinucleated mass, in which cell limits cannot be seen. This buds off small pieces, which become ciliated, escape from the zooecium in a manner not clearly ascertained, and swim away as ciliated larvae.

Asexual reproduction by budding is of course a most important process in the *Ectoprocta*. As shown above, it begins even in the larva, and is a marked feature of the metamorphosis; and the formation of the colonies is entirely due to it. Indeed, we may say that the *Ectoprocta* resemble the *Siphonophora* and other Coelenterates in possessing that power of multiplication and of differentiation of parts (*e.g.*, avicularia, etc.) which is so preeminently a characteristic of the vegetable kingdom.

The actual process of budding appears to be similar to the process by which the first polypide arises from the primary zooecium, that is to say the budding polypide is derived from a two-layered vesicle attached to the inner side of the body-wall of the parent zooecium. The inner wall of this vesicle is formed from the ectoderm, and gives rise to the alimentary canal, ganglion, and covering of the tentacles and tentacle-sheath of the new form, while the outer wall constitutes the mesodermal structures of the new polypide. The endoderm does not take part in the gemmation of the Ectoprocta.

Class I. ECTOPROCTA.

Anus outside the tentacular circlet.

This group includes by far the greater number of the Polyzoa, and their structure has been specially referred to in the precedent description of the class. The anus always opens outside the ring of tentacles, which are either arranged in a circle (*Gymnolaemata*) or on a two-armed horseshoe-shaped lophophore The coelom is well developed, and the reproductive cells are developed from its lining.

Order 1. Gymnolaemata.

Ectoprocta with circular lophophore, without epistome.

The *Gymnolaemata* are, with a few exceptions, marine forms. *Alcyonidium* and *Bowerbankia* are found in estuaries, and *Victorella* and *Paludicella* are only known in fresh or brackish water. Statoblasts are never found; but in the fresh-water forms the whole colony dies down in the winter with the exception of certain external buds, which are called *hibernacula* and give rise in the spring to new colonies.

The cuticle is sometimes horny, and sometimes encrusted with calcareous matters, over which a layer of cells may even extend. The *zooecia* present a great variety of form.

Sub-order 1. CHEILOSTOMATA.

The orifice of the zooecium can be closed by an operculum. Avicularia, vibracula, and ovicells are often present.

Tribe A. **CELLULARINA**. With corneous or corneo-calcareous infundibuliform cells, the inferior part of which below the aperture is tubular or obconic.

Fam. 1. **Aeteidae.** Zooecia tubular, with a lateral membranous area: orifice terminal. Tentacle-sheath terminating above in a circle of setae, which are everted during the expansion of the polypide. *Aetea* Lamouroux.

Fam. 2. **Eucrateidae.** Zooecia uniserial, or in two series placed back to back, with a terminal or subterminal and usually oblique aperture. Avicularia and vibracula absent. Zoaria as slender, branching, phytoid tufts. *Eucratea* Lamx.; *Gemellaria* Sav.; *Scruparia* Hincks; *Huxleya* Dyster; *Brettia* Dyster; *Pasythea* Lamx.

Fam. 3. **Chlidoniidae.** Zooecium composed of upright, free, segmented stems springing from a stolonate network. Zooecia bicamerate; unarmed. *Chlidonia* Sav.

Fam. 4. **Catenariidae.** Zooecium radicate, segmented; internodes except at a bifurcation formed of a single zooecium. *Catenicella* Blainv.; *Catenaria* Sav.

Fam. 5. **Cellulariidae.** Zooecia in two or more series, closely united and ranged in the same plane; avic. and vibrac., or avic. only, almost always present and sessile. Zoarium erect, dichotomously branched. *Cellularia* Pall.; *Menipea* Lamx. (*Emma* Gray); *Scrupocellaria* v. Ben.; *Caberea* Lamx.; *Canda* Lamx.; *Nellia* Busk.

Fam. 6. **Bicellariidae.** Zooecia rather loosely united in two or more series, or disjunct; obconic or boat-shaped; the aperture usually occupying a large proportion of the front. Avic. when present, capitate, pedunculate, and jointed. Zoarium not articulated, erect and phytoid, or composed of a number of cells connected by tubular processes. *Bicellaria* Blainv.; *Bugula* Oken.; *Beania* Johnston (*Diachoris* Busk); *Kinetoskias* Kor. and Dan.; *Ichthyaria* Busk.

Fam. 7. **Farciminariidae.** *Farciminaria* Busk.

Fam. 8. **Notamiidae.** Zooecia in pairs, each pair arising by tubular prolongations from the pair next but one below it; at each bifurcation a new series of cells intercalated into the branches. *Notamia* Fleming.

Tribe B. **FLUSTRINA.** With quadrate cells, the front surface of which is flat and equals the area of the primitive aperture.

Fam. 9. **Cellariidae.** Zooecia usually rhomboidal or hexangular, disposed in series round an imaginary axis, so as to form cylindrical shoots. Zoarium erect, calcareous, dichotomously branched. *Cellaria* Lamx. (*Salicornaria* Cuv.); *Melicerita* M.-Edw.

Fam. 10. **Flustridae.** Zoarium corneous and flexible, expanded, foliaceous, erect. Zooecia contiguous, multiserial. Avicularia usually of a very simple type. *Flustra* L., *F. foliacea* L.; *Carbasea* Gray.

Fam. 11. **Membraniporidae.** Zoarium calcareous or membrano-calcareous. Zooecia forming an irregular continuous expansion, or in linear series, with raised margins, and more or less membranous in front. *Membranipora* Blainv.; *Electra* Lamx.; *Megapora* Hincks; *Amphiblestrum* Gray; *Biflustra* d'Orb.; *Foveolaria* Busk.

Fam. 12. **Microporidae.** Zooecia with the front wall wholly calcareous; margins elevated. *Micropora* Gray; *Steganoporella* Smith; *Setosella* Hincks; *Vincularia* Defrance; *Calcschara* MacGillivray.

Tribe C. **ESCHARINA.** With calcareous cells, the aperture of which about equals the operculum in size, no membranous area being left.

Fam. 13. **Bifaxariidae.** *Bifaxaria* Busk; *Calymmophora* Busk.

Fam. 14. **Tubucellariidae.** *Tubucellaria* d'Orb.; *Siphonicytara* Busk.

Fam. 15. **Onchoporidae.** *Onchopora* Busk; *Onchoporella* Busk.

Fam. 16. **Reteporidae.** Zoarium calcareous, erect, fixed; foliaceous and fenestrate, unilaminar; or reticulately or freely ramose in one plane. Zooecia secund. *Retepora* Imperato; *Reteporella* Busk; *Turritigera* Busk.

Fam. 17. **Cribrilinidae.** Zoarium adnate, forming an indefinite crust, or erect. Zooecia with the front wall more or less fissured or traversed by radiating furrows. *Cribrilina* Gray; *Membraniporella* Smitt.

Fam. 18. **Microporellidae.** Zoarium incrusting. Zooecia with a semicircular aperture, the lower margin entire, and a semilunate or circular pore below it. *Microporella* H.; *Diporula* H.; *Chorizopora* H.; *Flustramorpha* Gray.

Fam. 19. **Porinidae.** Zoarium encrusting, or erect and ramified. Zooecia with a raised tubular or subtubular orifice, and frequently a special pore on the front wall. *Porina* d'Orb. (*Tessaradoma* Norman); *Anarthropora* Smitt; *Lagenipora* Hincks; *Celleporella* Gray.

Fam. 20. **Myriozoidae.** Zoarium encrusting, or rising into foliaceous expansions, or dendroid. Zooecia calcareous, destitute of membranous area and of raised margins; orifice with a sinus on the lower lip. *Schizoporella* H.: *Mastigophora* H.; *Rhynchozoon* H.; *Hippothoa* Lamx.; *Myriozoum* Donati: (*Gephyrophora* B.; *Haswellia* B.; *Gemellipora* Smitt).

Fam. 21. **Escharidae.** Zoarium calcareous, encrusting, or erect and lamellate, or ramose. Zooecia without a membranous area or raised margins. Without special pores and orifice without a true sinus. *Lepralia* Johnston; *Umbonula* H.; *Porella* Gray; *Escharoides* Smitt; *Smittia* H.; *Phylactella* H.; *Mucronella* H.; *Palmicellaria* Alder; *Eschara* Pallas; *Aspidostoma* H.

Fam. 22. **Adeonidae.** *Adeona* Lamx.; *Adeonella* B.; *Reptadeonella* B.

Fam. 23. **Celleporidae.** Zooecia calcareous, more or less vertical to the plane of the colony, irregularly heaped together, with a terminal orifice. *Cellepora* Fabricius.

Fam. 24. **Selenariidae.** *Cupularia* and *Lunularia* Lamx.; *Selenaria* Busk.

Sub-order 2. CYCLOSTOMATA.

Zooecia tubular, with a plain terminal orifice without operculum. Without movable appendages. The ovicells are modified zooecia. Many fossil species.

Tribe A. **ARTICULATA.** With erect branches divided at intervals by chitinous joints.

Fam. 1. **Crisiidae.** Zoarium dendroid, calcareous, composed of segments united by corneous joints. *Crisia* Lamx.

Tribe B. **INARTICULATA.** With unjointed zoarium.

Fam. 2. **Tubuliporidae.** Zoarium entirely adherent, or more or less free and erect, multiform, often linear, or flabellate or lobate, sometimes cylindrical. Zooecia tubular, disposed in contiguous series, or in single lines. Ooecium formed by the inflation of the branch. *Alecto* Lamx.; *Tubulipora* Lamx.; *Stomatopora* Bronn; *Idmonea* Lamx.; *Entalophora* Lamx.; *Diastopora* Lamx.; *Pustulopora* Blainv.

Fam. 3. **Horneridae.** Zooecia opening on one side only of a ramose zoarium, never adnate and repent. *Hornera* Lamx.

Fam. 4. **Lichenoporidae.** Zoarium discoid, simple or composite, adnate, or partially free and stipitate. Zooecia tubular, erect or suberect, disposed in more or less distinct series, which radiate from a free central area; the intermediate surface cancellated or porous. *Lichenopora* Defrance; *Domopora* d'Orb.

Fam. 5. **Frondiporidae.** *Frondipora* Imperato; *Fasciculipora* d'Orb.; *Supercytis* d'Orb.

Sub-order 3. CTENOSTOMATA.

When the tentacle-sheath is retracted, the orifice of the zooecium is closed by a folded membrane as by an operculum. Zoarium never calcareous. Ovicells and appendages absent.

Tribe 1. **HALCYONELLEA.** Zoarium fleshy or membranous. Zooecia developed by budding from other zooecia, and not from the internodes of a stolon.

Fam. 1. **Alcyonidiidae.** Zooecia more or less closely united, immersed in an expanded and adherent gelatinous crust, or forming an erect cylindrical or compressed zoarium; orifice closed by the mere invagination of the tentacular sheath; not protected by external labia. *Alcyonidium* Lamx.

Fam. 2. **Flustrellidae.** Zooecia immersed in a gelatinous crust; orifice bilabiate. Larvae with a bivalve shell. *Flustrella* Gray; *F. hispida* Fabr., on Fuci between tide-marks.

Fam. 3. **Arachnidiidae.** Zooecia usually more or less distant, membranous. *Arachnidium* Hincks.

Tribe 2. **STOLONIFERA.** Zoarium horny or membranous. Zooecia developed by budding from the internodes of a distinct stolon or stem.

A. *All the tentacles erect.*

Fam. 4. **Vesiculariidae.** Zooecia contracted below, not closely united to the stem at the base, deciduous, destitute of a membranous area. Zoarium repent or erect. All with gizzard except *Farrella*. *Vesicularia* J. V. Thompson; *Amathia* Lamx.; *Zoobotryon* Ehr.; *Bowerbankia* Farre; *Avenella* Dalyell; *Farrella* Ehrenb.

Fam. 5. **Buskiidae.** Zooecia contracted below, not continuous with the creeping stolon, with an aperture on the ventral surface. *Buskia* Alder.

Fam. 6. **Cylindroeciidae.** Zooecia not contracted below, closely united to the stem at the base, not deciduous; destitute of a membranous area. *Cylindroecium* Hincks; *Anguinella* v. Ben.

Fam. 7. **Triticellidae.** Zooecia horny, with an aperture and membranous area on the ventral aspect; borne on a rigid peduncle, to which they are attached by a movable joint, deciduous. *Triticella* Dalyell; *Hippuraria* Busk.

B. *Two of the tentacles everted.*

Fam. 8. **Valkeriidae.** Zooecia contracted below, deciduous, destitute of a membranous area. *Valkeria* Fleming.

Fam. 9. **Mimosellidae.** Zooecia contracted below, movable, deciduous, with an aperture on the ventral side. *Mimosella* Hincks.

Fam. 10. **Victorellidae.** Zooecia originating in an enlargement of the creeping tubular stem, with which they are continuous at the base; above free and cylindrical; not deciduous. *Victorella* S. Kent, fresh-water form, with hibernacula. *Paludicella* Gerv. is allied here.

Order 2. Phylactolaemata.

Fresh-water Polyzoa with horseshoe-shaped lophophore and epistome.

The *Phylactolaemata* are mainly distinguished by the bilateral arrangement of the numerous tentacles on the two-armed lophophore. There is always present above the mouth a movable tongue-shaped process—the *epistome*, whence the name given by Allman to this sub-order. The zooids are usually of considerable size, and are all alike (*i.e.* there is no polymorphism). The body-cavities of the zooids of a colony are in wide communication, and the colonies are ramified or massive, often transparent, and of a horny, or leathery, or gelatinous consistency. Statoblasts are present. The cerebral ganglion contains a cavity which is developed as a pit in the ectoderm of the bud.

Fam. 1. **Cristatellidae.** Free-moving colonies, on the upper surface of which the zooids are arranged in concentric ellipses. *Cristatella* Cuvier.

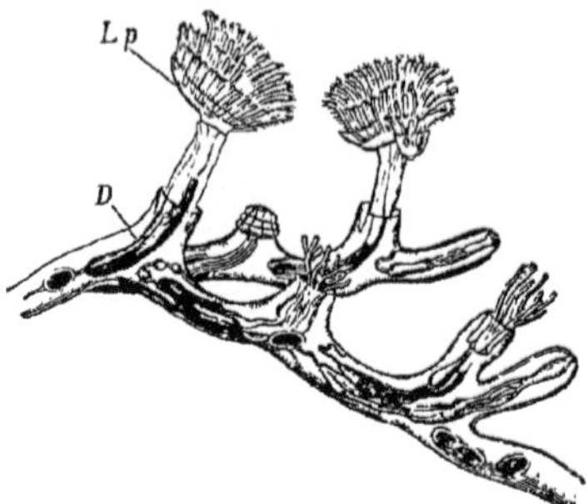

Fig. 452.—*Plumatella repens* slightly magnified (after Allman). *Lp* lophophore; *D* alimentary canal.

Fam. 2. **Plumatellidae.** Attached, massive or ramified colonies of fleshy or coriaceous consistency. *Pectinatella* Leidy; *Lophopus* Dumortier, discovered by Trembley in 1741, and is the first Polyzoon of which we have any notice; *L. crystallinus* Pall.; *Plumatella* Lamk., including *Alcyonella* Lamk., *A. fungosa* Pall.; *P. repens* L. (Fig. 452); *Fredericella* Gervais, differs from the other members of the family in having a nearly circular lophophore; the statoblasts are without a ring of air-cells.

Class II. ENTOPROCTA.

Anus within the tentacular circlet.

The *Entoprocta* are more simply organised than the *Ectoprocta.* The tentacular circlet surrounds a kind of vestibule into which the mouth and anus, the generative and excretory organs open. The alimentary canal is a simple U-shaped tube, differentiated into oesophagus, stomach, and rectum (Fig. 453). There is a ganglion in the floor of the vestibule between the mouth and anus. There is no body-cavity, but only a slight amount of connective tissue between the body-wall and gut-wall. There is a pair of fine, ciliated canals, which are described by some observers as intracellular, by others as intercellular, which lie on the ventral face of the stomach and open into the vestibule on the oral side of the ganglion.

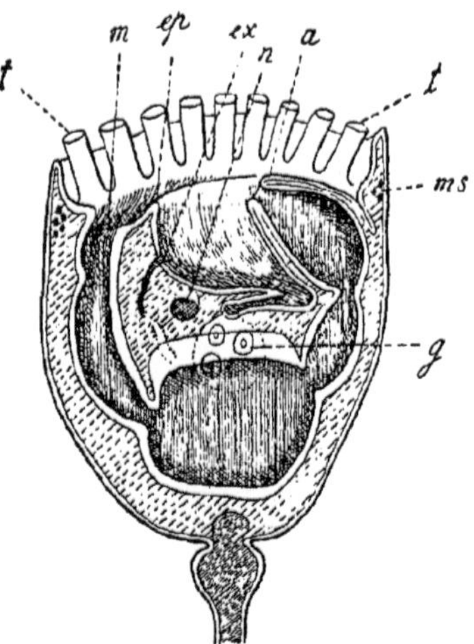

Fig. 453.—Diagrammatic longitudinal section through *Pedicellina* (after Ehlers, from Korschelt and Heider). *a* anus; *ep* epistome; *ex* excretory canal: *g* gonad; *m* mouth; *ms* fibres of the circular muscle; *n* ganglion; *t* tentacles.

The generative organs are continuous with their ducts, and open on the ventral surface between the mouth and anus; in *Pedicellina* they open into a kind of brood-pouch in the ventral wall of the vestibule on the anal side of the ganglion. They are paired glands, and the ducts of the testes open into a vesicula seminalis. *Pedicellina* is hermaphrodite, but the organs of the two sexes do not appear usually to mature at the same time. *Loxosoma* is dioecious.

The early stages of development take place in the brood-pouch. The ovum undergoes total cleavage, and an invaginate gastrula with two pole cells is formed. The blastopore remains open in the

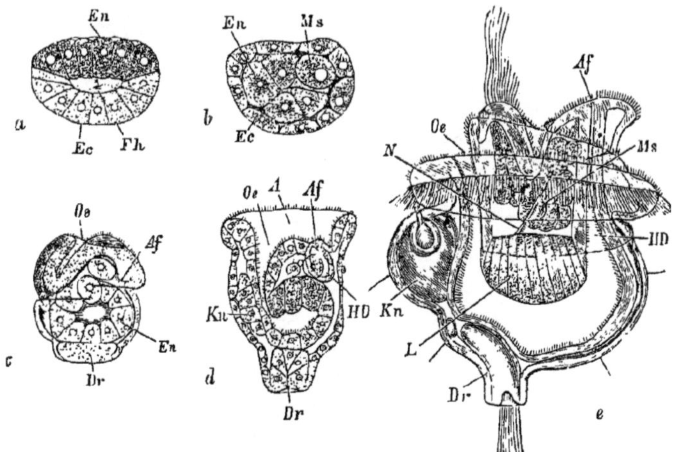

Fig. 454.—Development of *Pedicellina echinata* (after Hatschek). *a*, blastosphere. *b* and *c*, later stages in optical section. *d*, young larva in median optical section. *e*, free-swimming larva extended. *A* vestibule; *Af* rudiment of the rectum; *Dr* apical thickening; *Ec* ectoderm; *En* endoderm; *Fh* segmentation cavity; *Hd* rectum; *Kn* dorsal organ; *L* liver cells; *Ms* mesoderm; *N* excretory canal; *Oe* oesophagus.

position of the anus, and the pole cells give rise to two short mesoblastic bands. The embryo leaves the brood-pouch as a trochosphere larva (Fig. 454), with a well-marked velar ring of cilia having the same relation to the mouth and anus that the tentacles have in the adult, an apical thickening of ectoderm (*Dr*), which carries a tuft of cilia and consists of large glandular cells, and a pit of thickened ectoderm on the anterior side of the dorsal surface (*kn*) called the dorsal organ. The apical thickening is variously interpreted as a sense organ and an organ for adhering; the dorsal organ is provided with two pigmented eye-spots in

Loxosoma, and is interpreted by Harmer as the rudiment of the supraoesophageal ganglion; it is connected by two nervous tracts round the oesophagus with the persistent ganglion, which probably develops from the floor of the vestibule between the mouth and anus as a thickening of ectoderm, and on the same view may be interpreted as the suboesophageal ganglion.

As was first made known by the independent researches of Barrois and Harmer, the larva fixes itself by the oral surface and then

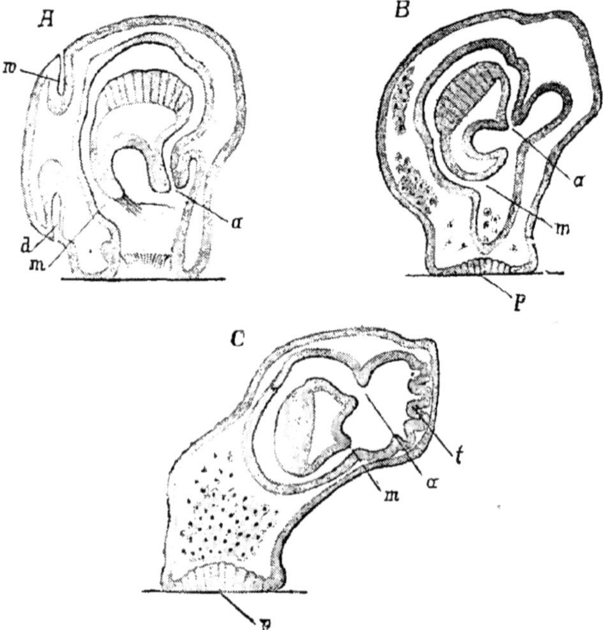

Fig. 455.—Three stages in the metamorphosis of *Pedicellina* (after Harmer, from Korschelt and Heider). *A*, just fixed larva. *B*, beginning of the rotation of the alimentary canal, and disappearance of larval organs. *C*, formation of new opening of vestibule and of tentacles. *a* anus; *d* dorsal organ; *m* mouth; *p* gland of attaching surface; *t* rudiments of tentacles; *w* ciliated disc.

proceeds to undergo the following metamorphosis (Fig. 455). The opening of the vestibule closes, the dorsal organ and apical plate disappear, and the alimentary canal apparently undergoes a process of rotation of such a kind that the atrium comes to lie on the surface opposite to that by which the animal is attached. Meanwhile a stalk is formed by the elongation of the *new* aboral region, and the tentacles and a new opening for the vestibule are established. During the whole of this process the alimentary tube, with its mouth and anus, persist.

Here then is a metamorphosis very similar in its results to the process which takes place in the *Ectoprocta*, viz., an entire shifting of the oral surface round to the opposite side of the larva; but differing from that, in the fact (1) that the attachment is effected by the edge of the ventral* surface, and not by an adhesive organ between the mouth and anus, and (2) that the larval organs do not undergo disruption, but allow us to perceive that the alteration is due to the relative growth of the parts rather than to the formation of a new mouth and anus on the aboral side of the larva.

As to the actual relations between this larva and the *Cyphonautes*, they appear to resemble one another very nearly, but on a critical examination certain points of difference become obvious. These are (1) the absence in the Entoproct larva of an adhesive organ; (2) the presence of a rudiment of the cerebral ganglion (dorsal organ), and (3) the absence of the pyriform organ in the Entoproct larva. The pyriform organ, which has been supposed to have something to do with the cerebral ganglion,† cannot be homologous with the dorsal organ of *Entoprocta*, because it is within the velar circlet. So that, on the whole, we are inclined to think that the resemblances between these larvae are superficial ones, and cannot be used as important arguments in favour of the view that the *Polyzoa* constitute a monophyletic group. On the other hand, it is held by many zoologists that the *Ectoprocta* and *Entoprocta* should not be associated in the same group, and that there is really no close relationship between them. The principal argument used in favour of this contention is that the *Entoprocta* are without a coelom, and should be removed altogether from the *Coelomata*. This view of the matter is further borne out by the relations of the gonads and the presence of the two ciliated excretory canals. We, however, while admitting that there is much to be said for it, do not regard it as proven, and cannot give it expression in this work.

The comparison which Caldwell has instituted between *Phoronis* and the *Polyzoa* is based upon the supposed resemblance between *Phoronis* and the

* This attachment must have been confined to the anterior side of the ventral edge, otherwise it is impossible to conceive how the process of relative shifting of the mouth and anus above described in its abbreviated embryonic form could have happened in the ancestral history.

† The cerebral ganglion on the view here adopted of the *Ectoprocta* is not, of course, present in the adult, for as pointed out above, the ganglion of the *Ectoprocta*, though sometimes called dorsal, must be really suboesophageal.

Phylactolaemata. In both there is a horseshoe-shaped lophophore, an epistome, a division of the body-cavity by a septum into a small part at the oral end, and the main part around the alimentary canal, and the presence of a pair of very similar nephridia. Of these resemblances the first two are not important, while the third and fourth concern doubtful points of anatomy. Against this comparison is to be set the development. In *Phoronis* the line between the mouth and anus is dorsal, and would contain, if it still existed, the preoral lobe of the larva. In the *Ectoprocta*, on the other hand, the line between the mouth and anus is ventral, though this would possibly be disputed by some zoologists (see p. 563); while in the *Entoprocta* the oro-anal surface is unquestionably ventral, the preoral lobe and its sense organ belonging demonstrably to the surface between the anus and the attachment of the adult (Fig. 455).

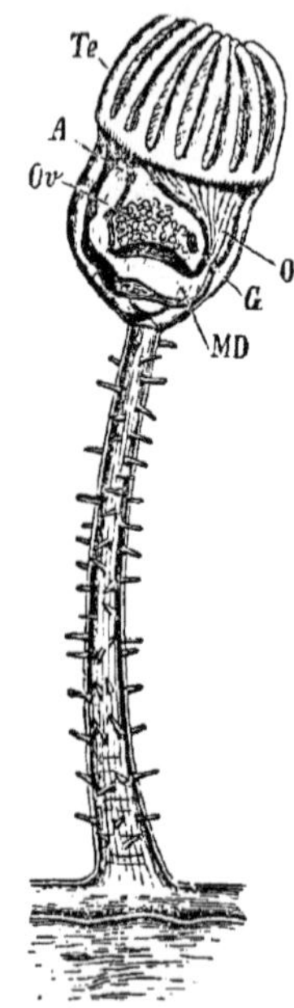

FIG. 456. — *Pedicellina echinata.* *Te* tentacular crown; *O* mouth; *Md* alimentary canal; *A* anus; *Ov* ovary; *G* ganglion. (From Claus.)

The *Entoprocta* are, with the exception of *Urnatella*, marine animals. *Pedicellina* is stalked and colonial. *Loxosoma* is solitary; the buds, which are formed, separating from their parent. *Loxosoma* is generally a commensal.

Fam. 1. **Loxosomatidae.** Solitary forms budding from the body; the buds separate from the parent. *Loxosoma* Kef., *L. phascolosomatum* at the end of the body of *Phascolosoma*, *L.* (*Cyclatella*) *annelidicola*, on the *Maldanidae*, etc.

Fam. 2. **Pedicellinidae.** Stalked, colonial forms, attached to a creeping stolon; budding at the growing point of the stolon. *Pedicellina* M. Sars.

Fam. 3. **Urnatellidae.** Stalked colonial, fresh-water forms, N. America. *Urnatella* Leidy.

CHAPTER XVI.

PHYLUM BRACHIOPODA.*

Fixed, solitary, apparently unsegmented Coelomata, with a tentaculated buccal groove often prolonged into arms, and a bivalve shell.

The *Brachiopoda* constitute an isolated phylum of the animal kingdom. They were formerly placed with the *Mollusca*, with which they have clearly no affinities. Later they were associated with the *Polyzoa* under the heading *Molluscoidea;* but this grouping must also be regarded as unsatisfactory, for they differ from the *Polyzoa* in several important points, amongst which may be mentioned the presence of a vascular system, of paired nephridia, of setae embedded in pits of the skin, of a shell composed of two pieces, and in not forming colonies.

On the whole, the increased knowledge of their anatomy and development which has been acquired of late years, though still far from satisfactory, points to the view that we must assign to the group the position of an independent phylum of the animal kingdom with affinities, by the form of their central nervous system, and by their setae, by the presence of a well-developed perivisceral coelom and a canalicular haemocoel, and by the traces of an imperfect segmentation, to the *Annelida;* though at the same time it must be pointed out that by the presence of longitudinal dorsal

* R. Owen, "On the Anatomy of the Brachiopoda," *Trans. Zool. Soc. London* 1835. A. Hancock, "On the Organisation of the Brachiopoda," *Phil. Trans.* 148, London, 1858. T. Davidson, "A Monograph of the British Fossil Brachiopoda," I.-VI., Palaeontographical Society, London, 1851-84. Id., "A Mon. of recent Brachiopoda," Pts. 1, 2, *Trans. Lin. Soc. London*, 1885-8. T. H. Huxley, "Contributions to the Anatomy of the Brachiopoda," *Proc. Roy. Soc.*, 7, 1854 (2), 14, 1854. H. Lacaze-Duthiers, "Histoire Nat. d. Brachiopodes de la Méditerranée," *Ann. Sci. Nat.* (4), 15, 1861. J. F. van Bemmelen, "Unters. üb. d. Anat. u. hist. Bau d. Brachiopoda Testicardinia," *Jen. Zeits.*, 16, 1883. A. E. Shipley, "On the Structure and Devel. of Argiope," *Mit. Zool. Stat. Neapel*, 4, 1883. A. E. Shipley and F. R. C. Reed, "Brachiopoda Recent and Extinct," in *Cambridge Natural History*, vol. 3, 1895.

and ventral mesenteries, and by the development of the coelom as enteric outgrowths, they recall the *Chaetognatha*. The possible affinities of the group with *Phoronis*, in favour of which there is much to be said, are discussed at the end of the chapter.

The mouth is placed in a buccal groove, one side of which (the dorsal) is provided with a lip, and the other with a row of tentacles. The groove is either continued as a circle on the dorsal lobe of the mantle or is drawn out on each side into the characteristic arms which project into the mantle-cavity. The anus is absent, or present usually on the right side of the front surface of the body, in a position which is described as ventral to the mouth (in *Crania* it is median, slightly dorsal and posterior). The two valves of the shell are commonly described as being dorsal and ventral, though it must be borne in mind that there are embryological reasons for regarding the surface between the mouth and anus as dorsal, and the whole of the surface on which the shell-valves lie as ventral. There is a circumoesophageal nerve-collar, not separated from the ectoderm and bearing supra-oesophageal and suboesophageal ganglionic swellings. There are no organs of special sense in the adult, though the larva possesses eye spots. The edges of the mantle-lobes, which are folds of the body-wall, contain setae embedded in ectodermal pits. There is a heart on the so-called dorsal side of the stomach giving off vessels. The perivisceral cavity is coelomic, and extends into the mantle-folds; it communicates with the exterior by one pair (in *Rhynchonella* two pairs) of nephridia; the generative cells are developed on its walls and dehisced into it, and carried outwards by the nephridia. Asexual reproduction and parthenogenesis are unknown in the group.

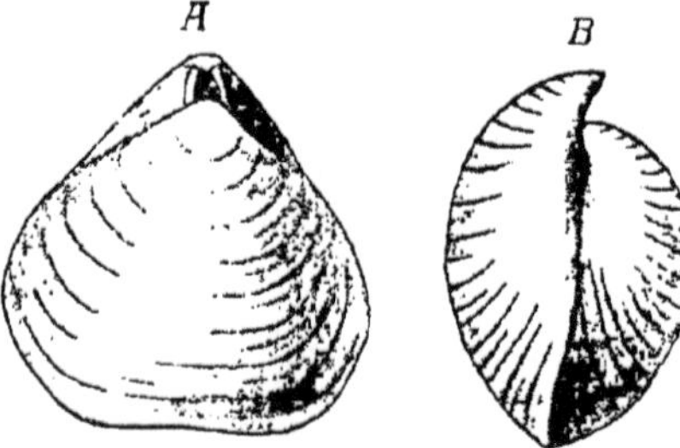

FIG. 457.—*Rhynchonella psittacea*. *A*, from above; *B* from the left side (after Lang).

It is commonly said of the *Brachiopoda* that they are pre-eminently an ancient group, and some importance is attached to this fact. It is true that they are an ancient group, representatives of it being known in the Cambrian, and the number of fossil forms is vastly greater than that of the living; but no special distinctive importance can be attached to these facts, for similar statements may be made of almost all the great groups, representatives of which are known from the earliest formations.

The body is enclosed in a bivalve shell, of which one valve is called dorsal, and the other ventral (Fig. 457). Both valves lie upon corresponding folds of the integument called the *mantle lobes*, and are often connected posteriorly by a kind of hinge, beyond which the usually more arched ventral valve projects like a beak (Fig. 458, *St*). The ventral valve is usually larger than the dorsal, and is either

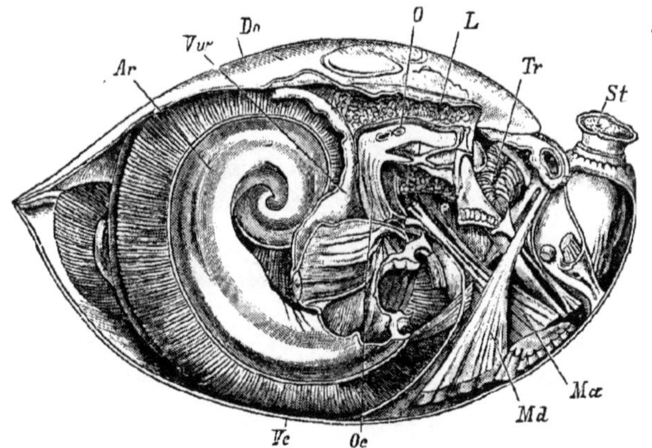

FIG. 458.—Anatomy of *Waldheimia australis*, side view of partly dissected specimen (after Hancock). *Ar* arms (the proximal loop of the left arm has been cut away); *Do* dorsal valve; *L* liver; *Ma* adductor, *Md* divaricator muscles; *O* the liver ducts of the left side cut short; *Oe* oesophagus; *St* peduncle; *Tr* folded margins of the funnel of the right nephridium (oviduct); *Ve* ventral valve; *Vw* anterior wall of body.

directly fused with foreign bodies, or the animal is attached by a peduncle projecting through the opening in the beak. The peduncle may however pass out between the two valves (*Lingula*). In *Crania*, which is attached by its ventral valve, the peduncle is absent. It is doubtful which surface of the body the stalk belongs to; according to the view stated below, which is admittedly highly speculative, it is a projection of the ventral surface of the body. The valves of the shell are cuticular structures secreted by the skin and impregnated with calcareous salts; they are not opened by a ligament, but by special groups of muscles (Fig. 458, *Md*); they are also closed by muscles, which are placed near the hinge and pass transversely from the dorsal to the ventral surface through the body-cavity (*Ma*).

In many forms the shell is perforated by minute pores, which however are closed externally by the outer uncalcified layer of the shell. They transmit tubular prolongations of the mantle. The body is bilaterally symmetrical, excepting for the anus which in some forms is on the right side, and is enclosed by the shell, of which it only occupies the posterior part; it possesses the two large reduplications of the integument, the two mantle lobes already mentioned, which are applied to the inner surface of the shell. The edges of the mantle are thickened and carry setae, which are regularly arranged and secreted by the lining of pits of the skin. The mantle may also produce within its own substance calcareous spicules, or a continuous calcareous framework.

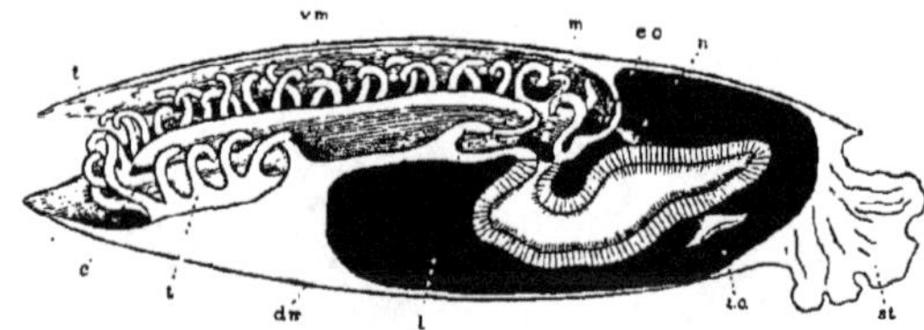

Fig. 459.—Diagram of a view of the left half of *Cistella* (*Argiope*) *neapolitana*, showing the alimentary canal, the left nephridium, and the buccal groove. The animal is supposed to have been removed from its shell, and to be cut in two by a median longitudinal incision. *e.o* external, *i.o* internal opening of nephridium *n*; *d.m* dorsal, *v.m* ventral lobe of the mantle; *l* and *c* lip; *t* tentacles; *m* mouth; *st* stalk. (Modified from Shipley.)

The mouth is placed in a transversely directed groove, bounded dorsally by a lip, and ventrally by a row of tentacles. In the simplest cases (*Argiope*, *Cistella*) this groove has a somewhat circular course (Figs. 459, 460 *a*), and lies along the dorsal part of the anterior wall of the body which is spread out on the dorsal valve. The mouth is in the posterior part of it, and the anterior part of it is bent backwards. In other forms the buccal groove does not form a complete circle, but is incomplete anteriorly, and the two ends so formed are coiled into a spiral, which is variously disposed in different forms. Moreover, in these forms it is not united to the dorsal wall, but projects into the mantle-cavity.

In *Waldheimia* (Figs. 460, 462) the buccal groove with its dorsal lip and ventral row of tentacles passes outwards (*j*) on each side of the mouth, and forwards towards the opening of the shell; it then bends on itself and passes backwards (*t*, *h*) on the ventral side of the proximal part, extending almost as far as the mouth. This may be called the proximal loop of the buccal groove; its two limbs are united by a membrane. The groove now turns ventralwards, and again passes forwards into a vertically placed spiral, which is coiled on to the dorsal valve and at the apex of which it terminates. The spiral part is united with the spiral part of the opposite side by a membrane. The course of the buccal groove in *Waldheimia* is shown in the diagram (Fig. 460). The proximal loop of the apparatus is supported in *Waldheimia* and *Terebratula* by a calcareous process of the dorsal valve (Fig. 461). In *Rhynchonella* there is no calcareous support for the arms, but the buccal groove is prolonged into a long, horizontally placed spiral on each side, the apex of the spiral being directed to the dorsal side (Figs. 460 *c*, 463). In *Lingula* the arms are very much as they are in *Rhynchonella*, the spiral being horizontally placed with its apex towards the dorsal valve.

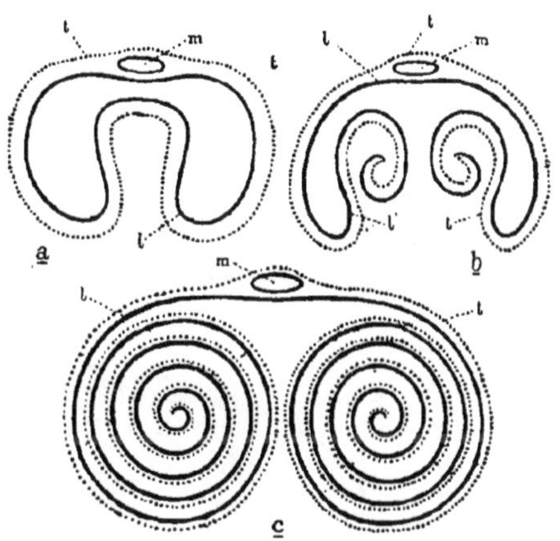

FIG. 460.—Diagrams showing course of buccal groove in *a*, *Argiope*; *b*, *Waldheimia*; and *c*, *Rhynchonella*. *m* mouth; *l* lip; *t* tentacles.

FIG. 461.—Dorsal valve of the shell of *Waldheimia australis*, with the brachial skeleton (after Hancock).

In *Lingula* and *Rhynchonella* it appears that the animal has the power of protruding its arms from the shell. The epithelium of the buccal groove and the tentacles is ciliated, and the whole apparatus is to be regarded as a food-procuring organ.

The mouth leads into an oesophagus which, passing dorsalwards, opens into the stomach. The stomach receives the ducts of a paired gland called the liver (Fig. 458, *o*), and passes into the intestine which is directed ventralwards. The intestine is short, and ends blindly in the *Testicardines*; while in the *Ecardines* it is long and coiled, and opens by an anus into the mantle-cavity. The anus opens on the right side in *Lingula* and *Discina*; in *Lingula* it is placed between the margins of the mantle. In *Crania** the rectum is in the middle line and opens at the hind end in the middle line into a space between the two valves where the hinge would be. Sometimes the end of the intestine is continued as a cord (*Thecidium*).

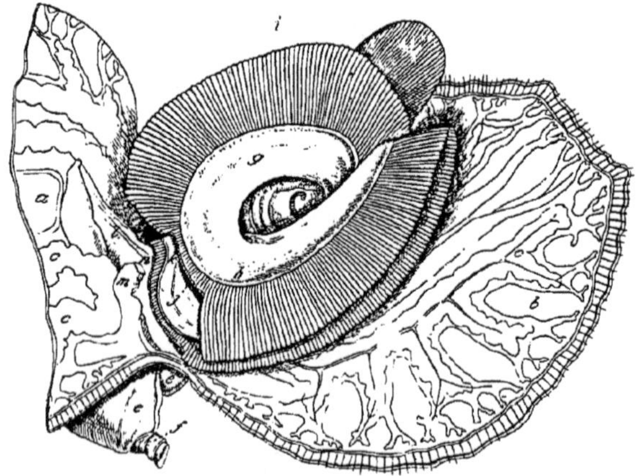

FIG. 462.—*Waldheimia australis* removed from its shell, and with the mantle lobes turned back so as to expose the arms and anterior wall of the body. *a* ventral, *b* dorsal lobe of mantle; *c, c* gonads; *f* peduncle; *g* spiral part of the left arm; *h* distal limb of the proximal loop; *i* tentacles; *j* lip, a few tentacles cut away to show it; *k* position of mouth, which is concealed by the lip; *l* occlusor muscles seen through the anterior wall of body; *m* left nephridium opening at *n*. (After Hancock.)

The alimentary canal is supported by a median dorsal and ventral mesentery, which partially divides the body-cavity into a right and left half; and by two incomplete transverse septa which pass, as does the dorsal mesentery, from the body-wall to the gut-wall. Of the two latter the anterior is called the *gastroparietal*, and the posterior the *ilioparietal* band, from their relations to the stomach and intestine respectively.

The **nervous system** consists of a circumoesophageal ring, upon

* Joubin, *Arch. Zool. Exp.* (2) 4.

which are developed a variable number of ganglia. The most general arrangement appears to be one small supra-oesophageal and one larger suboesophageal swelling. The suboesophageal gives off nerves to the dorsal mantle-lobe, the arms, and adductor muscles, and to two small ganglia which supply the ventral mantle-lobe and the peduncular muscles. The ganglia and commissure are in immediate contact with the ectoderm. Special organs of sense have not been described.

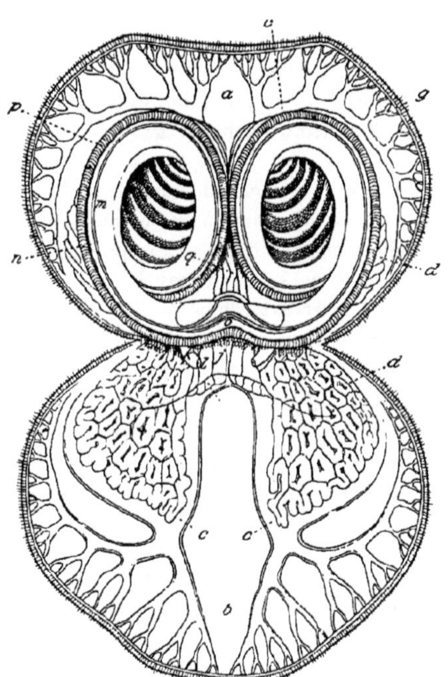

Fig. 463.—View of mantle-chamber of *Rhynchonella psittacea*, the ventral mantle lobe turned back (after Hancock). *a* dorsal, *b* ventral mantle lobe; *c*, *c* mantle sinuses; *d*, *d* genital glands; *g* circumpallial vessel; *j* occlusor muscles; *k* ventral nephridia opening at *l*; *m* arms; *n* lip of brachial groove; *o* position of mouth concealed by lip; *p* tentacles contracted; *q* oesophagus.

The **vascular system** has a canalicular character. There is a heart placed on the dorsal side of the stomach in the dorsal mesentery. It gives off a vessel which passes forwards along the oesophagus in the dorsal mesentery; this divides into two, which are continued into the arms as the tentacular vessels, and are connected with a vessel which surrounds the oesophagus. In addition to the above there are two pairs of lateral vessels running from the heart to the generative organs.

The blood is colourless.

Our knowledge of the vascular system and its ramifications is very small, as may be gathered from the statement made by some observers that the blood vessels communicate with the coelom. This is a statement which has been made of other groups, and has always been disproved with the progress of knowledge; and in the present instance, having regard to the difficulties of the observations required, and to the fact that the vascular system is essentially an organ separate from the coelom and from other organs derived from the primitive enteron, it cannot be accepted (it should not have been made at the present day by any well-informed Zoologist) without a good deal more evidence than has yet been adduced in its favour.

The **body-cavity** is well developed, and is a coelom. As already stated it is divided, incompletely it is true, into two lateral halves by a longitudinal mesentery, and into three successive chambers by two imperfect transverse septa, the gastroparietal and ilioparietal bands. The latter structures are, however, merely bands, there being three (or two) gastroparietal and two ilioparietal. They are important, because they may turn out to be remains of transverse septa dividing up the coelom in the embryo (see account of development). The body-cavity contains a corpusculated fluid, and possesses a ciliated epithelial lining. It is prolonged into the mantle lobes as the *pallial sinuses*, which may be much branched (*Lingula*). With regard to other extensions of the body-cavity we know very little. For instance, there are two channels running along each arm, the one small and in relation with the row of tentacles, into each of which it sends a prolongation, and the other larger and in relation with the lip (?epistome). It is disputed whether these canals communicate with each other or with the body-cavity. It is said, however (see above), that the tentacular channel is a blood vessel, and is a continuation of the vessel which leaves the anterior end of the heart.

As in coelomate animals the renal organs are **nephridia**. There is usually one pair, but in *Rhynchonella* there are two. They open externally on each side of the mouth, near the base of the arms, and internally by funnel-shaped apertures into the body-cavity. When there is one pair, the internal opening is on the ilioparietal band, into the posterior chamber of the body-cavity; when there are two pairs the openings of the second pair is supported by the gastroparietal bands, and is into the middle chamber of the body-cavity. The external openings of the posterior pair, *i.e.* the pair usually present, are a little on the ventral side of the mouth, while those of the anterior pair are a little nearer the dorsal valve. The nephridia function also as generative ducts, and were called *oviducts* by Hancock.

Generative organs. In all probability most *Brachiopoda* are dioecious, but possibly a few of them may be hermaphrodite. The generative organs are developments of the coelomic epithelium. They are thick yellow bands or ridges which project into the body-cavity, and extend into the lacunae of the mantle where they may be considerably ramified. The ripe sexual cells are dehisced into the coelom, whence they pass to the exterior through the nephridia. In some genera (*Thecidium*, *Cistella*, *Argiope*) the eggs undergo the

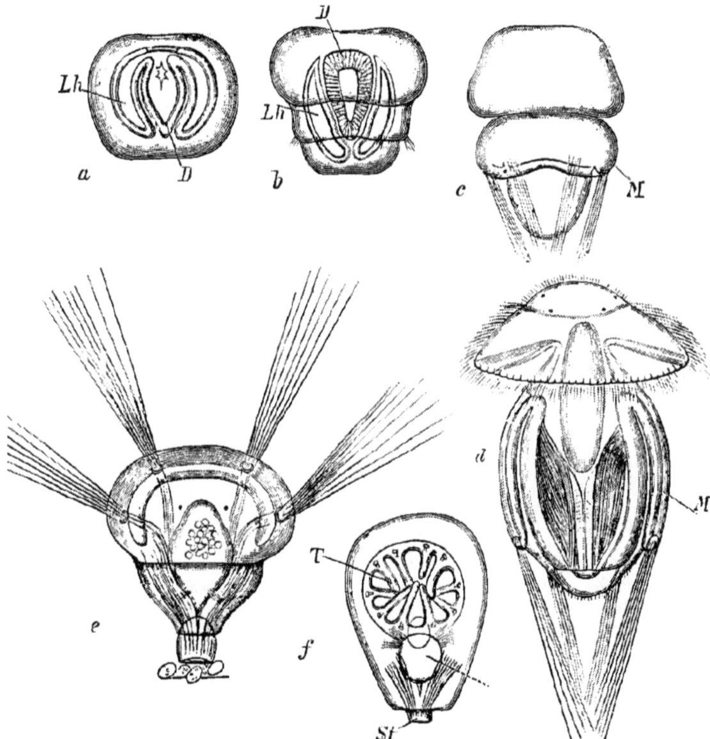

FIG. 464.—Development of *Argiope* (after Kowalewsky). *a*, embryo in which the enteron *D* still communicates in front with the coelomic sacs *Lh*. *b*, embryo with three segments, and commencing mantle fold with setae. *c*, older embryo showing the same features; *M* mantle fold. *d*, free larva with eye spots, cephalic umbrella, ciliated ring, and large mantle lobes. *e*, attached larva with mantle lobes bent anteriorly. *f*, later stage with tentacles *T*, and commencing peduncle *St*.

first stages of their development in brood-pouches placed near the openings of the nephridia. In *Thecidium* the brood-pouch is median and ventral. In *Argiope* and *Cistella* it is paired.

Our knowledge of the **development** is very incomplete, and what we do know relates almost entirely to the *Testicardines*. The egg is small and the segmentation is complete, the gastrula is formed usually by invagination, and the blastopore closes at the anterior end of the ventral surface. The archenteron gives off two lateral diverticula, which are gradually constricted off from it as the coelomic sacs (Fig. 464, *a*, *b*).

The coelomic sacs extend posteriorly behind the enteron, but the last remnant of the communication between the two is at the front end of the body. The embryo now becomes constricted by an annular furrow into two parts, the anterior of which soon divides in a similar manner, so that three segments are formed (Fig. 464, *b*). It does not appear, however, that the coelomic sacs participate in this segmentation. The enteron is confined to the two anterior segments, and does not extend into the caudal segment.

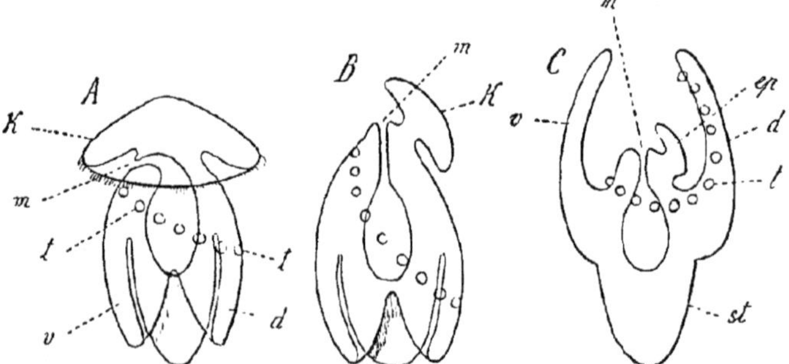

FIG. 465.—Three diagrams illustrating the hypothetical metamorphosis of a Brachiopod in relation to that of *Actinotrocha* (after Korschelt and Heider). *A*, free-swimming larva showing the mouth-opening and commencing tentacles of the lophophore, neither of which are present at this stage. *B*, a transitional stage showing reduction of supposed preoral lobe. *C*, young Brachiopod after the turning forward of the mantle lobes. The preoral lobe (cephalic umbrella) has shrunk to the epistome. *d* dorsal mantle lobe; *ep* epistome; *k* preoral lobe (cephalic umbrella) with its ring of cilia (trochosphere ring); *m* mouth; *st* stalk; *t* rudimentary tentacles; *v* ventral mantle lobe.

On the first segment (head) there is developed an umbrella-like disc, the edge of which becomes ciliated, and four eye spots. On the second segment (thorax) there is an annular fold (Fig. 464, *c*), which soon grows out into a dorsal and ventral lobe—the mantle-lobes. On the ventral part of this are developed four bundles of provisional setae.

The mantle-lobes project back over the caudal region (Fig. 464, *d*). The embryo now leaves the brood-pouch and enters upon its free-swimming life.

The larval stage is of short duration, and during it the enteron remains an entirely closed sac. Attachment is effected by the hind (caudal) end (Fig. 464, *e*, *f*), and as soon as it has occurred the mantle-lobes turn forward and envelop the first segment.

The larva now assumes the adult condition; the shell valves are developed on the mantle-folds, the provisional setae are replaced by the permanent ones, and a mouth is formed at the front end. The exact relation of the mouth to the umbrella-like head segment is obscure. It has been suggested that the mouth is formed upon the ventral side of it, and that the head segment (Fig. 465) is really a preoral lobe, like that of *Phoronis*—which soon disappears or is reduced to the epistome (lip of the buccal groove). The tentacles of the arms appear on the inner side of the dorsal mantle lobe (Fig. 465, *C*).

Nothing is known of the embryonic development of the *Ecardines*. Their larvae differ from those of the *Testicardines* in possessing a shell while free-swimming. A larva of *Discina radiata* with a large process overhanging the mouth has been described by F. Müller (*Müller's Arch.*, 1860), and we owe an account of the larva of *Lingula* to Brooks (Chesapeake Zool. Lab. Results, 1878).

A consideration of the above facts does not lead us very far in deciding the question of the affinities of the *Brachiopoda*. If we adopt the suggestion made above, that the mouth is formed on the ventral surface—which, it must be remembered, is opposed to Kowalevsky's statement that it is at the anterior end—we may regard the larva as a trochosphere with a preoral ring of cilia and preoral sense-organs.

FIG. 466.—Larva of *Lingula* (after Brooks). *T* tentacles; *O* mouth; *D* alimentary canal; *Af* anus; *L* liver; *St* rudiment of peduncle.

The presence of setae secreted in cutaneous sacs, the indications of segmentation, and the relations of the coelom in the adult, further suggest Annelidan affinities. Finally, the form of the central nervous system is not opposed to this view of the relationship of the group. If we adopt this view we should regard the *Brachiopoda* as Annelids with three segments, marked in the embryo by the annular constrictions of the integument, and in the adult by the imperfect septa represented by the gastro- and ilioparietal bands, and by the two pairs of nephridia in *Rhynchonella*, which have relations to these bands.

But it must be remembered that according to present accounts there are no traces of mesoblastic segments in the embryo, the coelomic sacs being continuous and not divided up by septa. This, however, may be due to imperfect observation, a possibility which becomes the more likely when we remember that the larvae are extremely minute, the observations very difficult to make, and that there is no account of the formation of the structures which are supposed to be transverse septa in the adult, or of the nephridia.

But this Annelid view of the group is not the only one which has been put forward. It has been suggested that the arms, with their tentacles, are comparable to the lophophore, and the lip to the epistome of *Phoronis* (Fig. 465). There is much to be said for the affinities with *Phoronis* which this view suggests, but it must be remembered that the view is one which itself rests upon a basis of hypothesis (see p. 583), for in neither group has the development of the lip been satisfactorily followed.

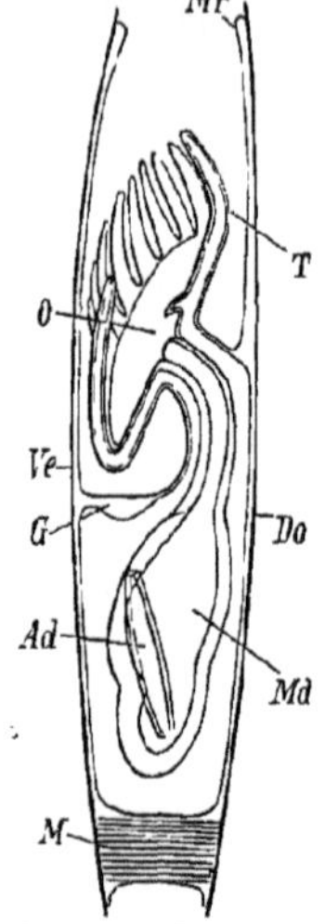

FIG. 467.—Longitudinal section of an older larva of *Lingula* (after Brooks). *Do* dorsal; *Ve* ventral valve of shell; *Mr* thickened edge of mantle; *T* tentacles; *O* mouth; *Md* stomach; *Ad* intestine; *M* posterior muscle; *G* ganglion.

The most important objection to it is that the flexure of the intestine, and the position of the anus nearer the ventral valve, is different to the arrangement in *Phoronis*, in which the intestine has the opposite flexure, and the anus is on the epistome side (dorsal) of the mouth. It may fairly be urged, however, that, in our present state of ignorance of the development of the Brachiopods with an anus, and consequently of the real position of that organ, too much weight must not be attached to this objection. The possible presence of three mesoblastic segments, and the relations of the nephridia and coelom, are not opposed to it, nor is the mode of attachment by the caudal region; for it is argued the caudal region might fairly be regarded as equivalent to the evaginated foot of *Phoronis*. We think the suggestion that the Brachiopods are allied to Phoronis an attractive, but at the same time a highly speculative one. For the principal arguments in its favour, viz., that in both the preoral lobe shrivels up or disappears, leaving at most the epistome as an indication of its presence, and that in both the dorsal surface between the mouth and anus is extraordinarily shortened, are open to the serious reply, that although these statements may fairly be made of *Phoronis*, they can only be put forward in a hypothetical form with regard to Brachiopods.

The *Brachiopoda* are found in all seas at different depths. The larger number live at moderate depths, down to 500 or 600 fathoms. Beyond this depth they are rare, though species of *Discina* and *Terebratula* have been taken at a depth of over 2000 fathoms. *Lingula* and *Glottidia* are found between tide-marks. For the most part they live on rocky ground, and are found in great numbers together. *Lingula* burrows in the sand.

The number of living Brachiopods is small, as compared with the much larger number found in the earlier geological formations, certain species of which have great importance as characteristic fossils. The oldest known fossils are Brachiopods, and certain genera which first appeared in the Cambrian have persisted to the present day.

Order 1. Ecardines (Inarticulata).*

Shell without hinge and brachial skeleton. Alimentary canal with anus either ventral and to the right of the middle line, or posterior and median (*Crania*).

Fam. 1. **Lingulidae.** Shell thin, horny, almost equivalve, a long peduncle passing between the two valves. Live in tubes in sand. *Lingula* Brug.; *L. anatina* Lam.; *Glottidia* Dall.

Fam. 2. **Discinidae.** Shell fixed by a peduncle passing through a hole in the ventral valve. *Discina* Lam.; *D. atlantica* King, 1330 fathoms, off West of Ireland; *Discinisca* Dall.

Fam. 3. **Craniidae.** Shell orbicular, calcareous; ventral valve adherent; without peduncle. *Crania* Retz.; *C. anomala* Müller, Loch Fyne.

Order 2. Testicardines (Articulata).*

The shell is calcareous, with hinge and brachial skeleton. The intestine ends blindly.

The exclusively fossil family **Productidae**, the edge of the shells of which have no hinge, form the transition between the two orders.

Fam. 1. **Terebratulidae.** Shell usually biconvex; punctate, with complete hinge; beak of ventral valve perforated for short peduncle; calcareous loop bent back on itself. *Waldheimia* King (Figs. 461, 462), *W. cranium* Müll., North British seas; *Terebratula* Brug., *T. capsula* Jeff., English Channel; *Terebratulina* d'Orb., *T. caput serpentis* L. Oban; *Terebratella* d'Orb., *T. Spitzbergensis* Dav., N. British seas; *Argiope* Deslongchamps, *A. decollata* Chemnitz, English Channel; *Cistella* Gray (Fig. 459), *C. cistellula* S. Wood, N. British seas, English Channel; *Megerlia* King; *Kraussina* Dav.

Fam. 2. **Thecidiidae.** Shell plano-convex, usually fixed by beak of ventral valve. *Thecidium* Defr.

Fam. 3. **Rhynchonellidae.** Shell biconvex, hinge usually curved; beak of ventral valve incurved with foramen; calcareous brachial supports as a pair of short curved crura; arms coiled in spiral. *Rhynchonella* Fisch. (Fig. 463), *R. psittacea* Gmelin, N. British seas; *Atretia* Jeff., *A. gnomon* J., west of Donegal Bay in 1400 fathoms.

* The localities mentioned in this list refer only to the places in the British area, where the species have been obtained. They are taken from Shipley's article in the *Camb. Nat. Hist.*

CHAPTER XVII.

CHAETOGNATHA.*

Hermaphrodite Coelomata, in which the body is divided into three regions or segments. The head carries two groups of sickle-shaped setae, and the central nervous system consists of cerebral and sub-oesophageal ganglia connected by circumoesophageal commissures.

The *Chaetognatha* are a very small group of transparent pelagic organisms. Though small in numbers and presenting but little variety of organisation, the group is one which, both in its adult structure and embryonic development, is of the greatest interest to naturalists. It is impossible to relate it to any of the great phyla of the animal kingdoms, and, like so many of the groups which we have recently considered, it must have assigned to it the dignity of phyletic rank.

In certain features of structure, *e.g.*, in the form of the nervous system, and in the relations and character of the coelom, it resembles the *Annelida*, but it differs from them in the small number and in the character of the segments of which the body is composed, in the absence of distinct nephridia and a vascular system, and in the structure of the embryo. For though the egg is small, and the whole development takes place in the sea, there is no larval stage which in any way recalls the trochosphere, and the coelom is developed from archenteric diverticula, a feature which is found in no Annelid.

As in the *Brachiopoda* the coelom consists of three pairs of chambers separated by transverse septa, and as in that group the division of the coelom is effected subsequently to its establishment as a single enteric diverticulum on each side.

* A. Krohn, "*Anat.-physiol. Beobachtungen üb. die Sagitta bipunctata,*" Hamburg, 1844. R Wilms, "*De Sagitta mare germanicum circa insulam Helgoland incolente,*" Berolini, 1846. A. Kowalewski, "Embryologische Studien an Würmern u. Arthropoden," *Mém. de l'Acad. Petersbourg*, 16. O. Hertwig, "*Die Chaetognathen, eine Monographie,*" Jena, 1880. B. Grassi, "I. Chetognathi," *Flora und Fauna d. Golfes von Neapel*, 1883. S. Strodtmann, "Die Systematik der Chaetognathen etc.," *Arch. Naturg.*, 58, 1892.

The body is elongated and divided into three regions—the head, the trunk, and the tail (Fig. 468). The mouth is on the ventral surface of the head, and has the form of an elongated slit (Fig. 469). The anus is on the ventral surface, at the junction of the trunk and the tail.

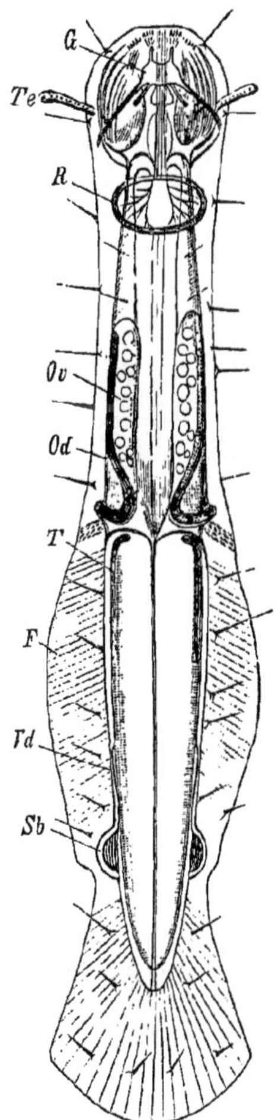

Fig. 468.—*Spadella cephaloptera* (after O. Hertwig), dorsal view. *F* lateral fin (in caudal region only, in this genus); *G* cerebral ganglion; *Te* tentacles; *Ov* ovary; *Od* oviduct; *T* testis; *Vd* vas deferens; *Sb* vesicula seminalis; *R* olfactory organ.

The head is slightly swollen, and is partly covered by a prepuce-like fold of the skin (Fig. 469), which arises from the dorsal side of the head and forms a kind of hood more developed laterally than ventrally. It carries two laterally placed groups of sickle-shaped setae and a number of short spines. Moreover, there are two eyes in close connection with the cerebral ganglion. Close behind the head on the dorsal surface there is an annular modification of the ectoderm (Fig. 468 *R*), the cells of which bear cilia; this is supposed to be an olfactory organ. The body possesses one or two pairs of flat cutaneous fin-like folds—the lateral fins—and the tail is surrounded by a caudal fin. The fins consist of a fold of the ectoderm containing a gelatinous substance, on each surface of which lie some chitinous rod-like rays beneath the ectoderm.

The skin consists of an epithelium, the cells of which may be arranged in one or in more than one layer, of a basement membrane, and of a layer of longitudinally disposed muscular fibres; the latter are disposed in the trunk and tail in four bundles—two dorsal and two ventral (Fig. 470), and the fibres are cross-striped; these muscles, both in their arrangement and appearance in transverse section, are very like those of the Nematodes. In the head the muscles are broken up into bundles which work upon the jaws.

The alimentary canal is a straight tube, and is divided into oesophagus and intestine. There are no special glands opening into it.

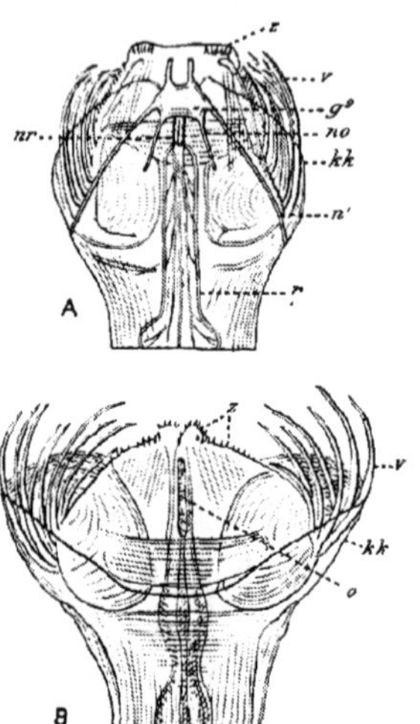

FIG. 469.—Head of *Sagitta bipunctata*. *A* dorsal, *B* ventral view. g^2 cerebral ganglion; *kk* cephalic hood; n^1 circum-oesophageal commissures; *no* optic nerve; *nr* olfactory nerve; *o* mouth; *r* olfactory organ; *v* hooks; *z* spines. (After Hertwig.)

The **nervous system** consists of a cerebral ganglion in the head, and a large elongated ventral ganglion—the suboesophageal ganglion—placed in about the middle of the body length. These two ganglia are connected by long circum-oesophageal commissures, and the whole of this part of the nervous system lies in the ectoderm (Fig. 470). There is in addition a pair of ganglia connected with the cerebral, and with each other below the oesophagus; these are placed in the mesoderm and supply the muscles of the head.

The sense-organs consist of a pair of eyes on the dorsal surface of the head, innervated from the cerebral ganglion; the annular tract of ciliated ectoderm on the dorsal surface already mentioned—it is supposed to be olfactory in function, and varies in shape in different species; and tactile papillae consisting of ectoderm cells, and distributed all over the surface.

There is no vascular system.

The **body-cavity** is well developed, and, as is shown by its development and its relation to the male generative organs, is a coelom. It is lined by a layer of flat epithelium, and is divided into two lateral halves by a longitudinal septum, in which the alimentary canal is placed (Fig. 470); this constitutes the dorsal and ventral mesentery, and in the tail the septum.

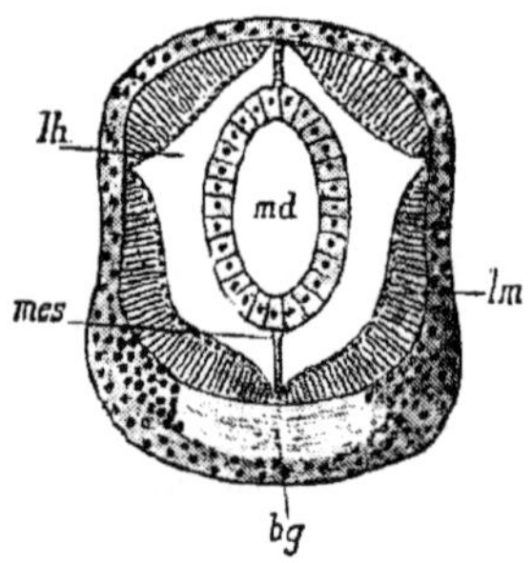

FIG. 470.—Transverse section through the trunk of *Sagitta* (from Lang, after Hertwig). *lh* coelom; *mes* mesentery; *md* intestine; *lm* longitudinal muscles; *bg* suboesophageal ganglion.

It is further divided by two transverse septa into three successive chambers: one in the head, another in the trunk, and the third in the tail (Fig. 468). These chambers are indications of segmentation, though the exact mode of their origin from the original continuous coelom has not been made out. Of these three coelomic chambers or segments, the posterior alone has, so far as is known, a communication with the exterior through the male generative ducts.

Generative organs. The animals are hermaphrodite. The female organs are contained in the middle segment of the body, and the oviduct opens on each side of the anus at the junction of the middle and caudal regions. The ovaries are placed one on each side (Fig. 468); they are solid and are attached to the wall of the body, projecting into the body-cavity. The oviducts are narrow

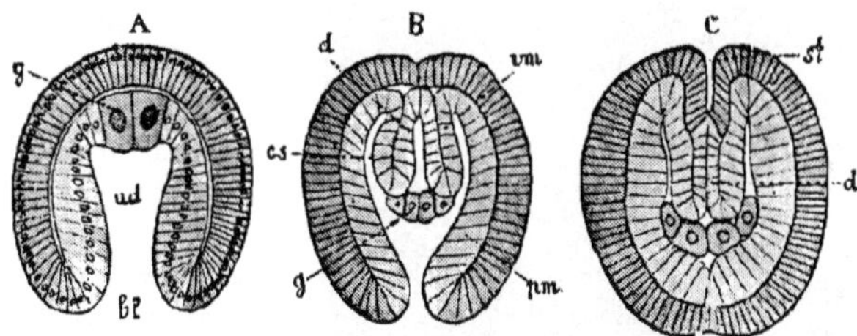

FIG. 471.—Three embryos of *Sagitta* (from Lang, after Hertwig). *bl* blastopore; *ud* archenteron; *g* primitive generative cells; *vm* splanchnic, *pm* somatic (parietal) layer of mesoderm; *d* rudiment of intestine; *cs* coelomic sacs; *st* stomodaeum (ectodermal ingrowth which forms the anterior part of the alimentary canal).

tubes extending along the outer sides of the ovaries and lined by an epithelium. They end blindly in front, and open to the exterior behind; they sometimes contain spermatozoa. The ova are not dehisced into the coelom, and it is not known how they get into the oviducts, which appear to be much too narrow to contain them.

The testes are thickenings of the parietal coelomic epithelium in the anterior part of the caudal chambers of the body-cavity. Cells break off from them and fall into the body-cavity, where they develop into spermatozoa. The male generative ducts (Fig. 468) are paired tubes opening on the side of the body in the caudal region, and internally by a ciliated opening into the caudal region of the coelom. The vesicula seminalis is a dilatation on the course of the short vas deferens.

No separate renal organs are known.

Development. The ova are transparent structures which float on the surface of the sea, except in *Spadella cephaloptera*, which attaches them to seaweeds. The mode of oviposition and the place of fertilization is not known. The whole development takes place in the sea, and there is no larval stage. The segmentation is complete, and leads to the formation of a hollow blastosphere, and an invaginate gastrula. The blastopore closes at the posterior end of the embryo, and the mouth is a new formation at the anterior end. The archenteron gives off two pouches from its posterior end (Fig. 471), and thus becomes trilobed. The middle lobe acquires an opening to the exterior at the front end, while posteriorly it opens into the part of the archenteron which opens into the lateral pouches (Fig. 471). This hind opening becomes closed, so that the middle lobe forms a tube opening in front by the mouth, and behind ending blindly, and thus gives rise to the enteron. The anus is formed later. The portion of the archenteron thus separated off is the coelom; it consists of two lateral sacs opening into an unpaired portion (Fig. 471) behind. From the anterior end of the two lateral sacs a portion is cut off, which apparently becomes the cephalic sections of the adult coelom (Fig. 472). The further changes of the hinder part of the coelom into the adult condition have not been followed.

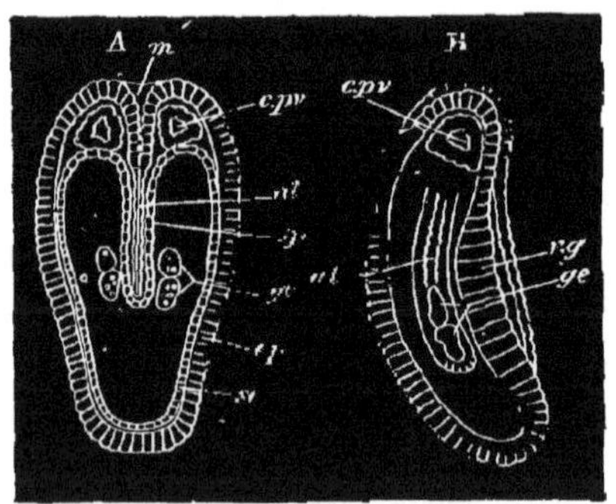

FIG. 472.—*A*, dorsal; *B*, lateral views of an advanced embryo of *Sagitta* (from Balfour, after Bütschli). *m* mouth; *al* alimentary canal; *vg* ventral ganglion; *ep* ectoderm; *cpv* cephalic section of the body-cavity; *so* somatic; *sp* splanchnic layer of mesoderm; *ge* generative organs.

One of the most remarkable features of the development of *Sagitta* is the early appearance of the generative cells, if indeed they be such. These appear in the gastrula stage as a prominence of six cells projecting from the hypoblast at the anterior end of the archenteron. This mass, on the folding of the archenteric wall, is placed at the hind end of the median lobe (Fig. 471), and when the division of the archenteron takes place it remains in the coelomic portion (Fig. 472), and divides into an anterior and posterior part on each side. The former is said to constitute the ovary, and the latter the testis.

As stated above the *Chaetognatha* are essentially pelagic organisms. They swim by flexion of the body, and have been taken at considerable depths (600 fathoms). One species is littoral (*Spadella cephaloptera*).

The *Chaetognatha*, in possessing three divisions of the coelom, resemble the *Brachiopoda*, on the assumption that the gastro- and älioparietal bands of the latter group are the remains of transverse coelomic septa (p. 580). In the same feature they also resemble *Balanoglossus*. The resemblance to the latter is, however, diminished by the fact that the division of the coelom into three chambers is not an original one, but occurs subsequently to the separation from the archenteron. There is this further difference, viz., in *Balanoglossus* it is the anterior portion of the embryonic coelom which is unpaired and median, while in *Chaetognatha* it is the posterior part. With regard to the latter point it is possible that further investigations of the embryos of the *Chaetognatha* with modern methods may disclose some errors in the old accounts, and with regard to the first, it must not be forgotten that even in the same group the coelomic sacs may differ as to when they divide up transversely; *e.g.*, in the *Echinodermata*—in which group it is probable that the typical arrangement of coelomic sacs is one unpaired and two paired—the sacs may in some forms arise separately from the archenteron, and in others arise by the division of one archenteric diverticulum on each side.

Three genera of *Chaetognatha* are known.

Sagitta Slabber, with two pairs of lateral fins.

Krohnia Langerhans, with one lateral fin on each side extending on to the caudal region.

Spadella Langerhans, with one pair of lateral fins on the tail, and a thickening of epidermis on each side of the body from the head to the fin.

END OF VOL. I.

INDEX.

Lightning Source UK Ltd.
Milton Keynes UK
UKHW022133160921
390713UK00002B/294